Factoring Formulas

$ab + ac = a(b + c)$ common factor

$a^2 - b^2 = (a + b)(a - b)$ difference of two squares

$a^2 + 2ab + b^2 = (a + b)^2$

$a^2 - 2ab + b^2 = (a - b)^2$ perfect square trinomials

$a^3 + b^3 = (a + b)(a^2 - ab + b^2)$ sum of two cubes

$a^3 - b^3 = (a - b)(a^2 + ab + b^2)$ difference of two cubes

Laws of Exponents

$a^m a^n = a^{m+n}$

$\dfrac{a^m}{a^n} = a^{m-n}, \quad a \neq 0$

$(a^m)^n = a^{mn}$

$(ab)^n = a^n b^n$

$\left(\dfrac{a}{b}\right)^n = \dfrac{a^n}{b^n}, \quad b \neq 0$

$a^0 = 1, \quad a \neq 0$

$a^{-n} = \dfrac{1}{a^n}, \quad a \neq 0$

Laws of Radicals

$a^{1/n} = \sqrt[n]{a}$

$a^{m/n} = \sqrt[n]{a^m} = (\sqrt[n]{a})^m$

$\sqrt[n]{a^n} = a$

$\sqrt[n]{ab} = \sqrt[n]{a}\,\sqrt[n]{b}$

$\sqrt[n]{\dfrac{a}{b}} = \dfrac{\sqrt[n]{a}}{\sqrt[n]{b}}, \quad b \neq 0$

$\sqrt[m]{\sqrt[n]{a}} = \sqrt[mn]{a}$

Properties of Logarithms

$\log_b xy = \log_b x + \log_b y$

$\log_b \dfrac{x}{y} = \log_b x - \log_b y$

$\log_b x^r = r \log_b x$

$\log_b 1 = 0 \quad \log_b b = 1$

$\log_b x = \dfrac{\log_a x}{\log_a b}$ (change of base)

Quadratic Formula

The solutions of the quadratic equation

$ax^2 + bx + c = 0, \quad a \neq 0$

are given by

$x = \dfrac{-b \pm \sqrt{b^2 - 4ac}}{2a}$

Basic Technical Mathematics

Books in the Kuhfittig series in technical mathematics

Basic Technical Mathematics
Basic Technical Mathematics with Calculus
Technical Calculus with Analytic Geometry

The cover illustration was created at Digital Image Corporation in Grand Rapids, Michigan. Digital Image uses Z80 microprocessors and starts with a mathematical equation. Creating the basic image takes about half an hour; the computer can then produce several thousand color combinations per second.

Basic Technical Mathematics

Peter K. F. Kuhfittig
Milwaukee School of Engineering

*Brooks/Cole Publishing Company
Monterey, California*

To Dagmar

Brooks/Cole Publishing Company
A Division of Wadsworth, Inc.

© 1984 by Wadsworth, Inc., Belmont, California 94002. All rights reserved. No part of this book may be reproduced, stored in a retrieval system, or transcribed, in any form or by any means—electronic, mechanical, photocopying, recording, or otherwise—without the prior written permission of the publisher, Brooks/Cole Publishing Company, Monterey, California 93940, a division of Wadsworth, Inc.

Printed in the United States of America

10 9 8 7 6 5 4 3 2 1

Library of Congress Cataloging in Publication Data

Kuhfittig, Peter K. F.
   Basic technical mathematics.

   Includes index.
   1. Engineering mathematics. I. Title.
TA330.K84 1984   620′.0042   83-20975
ISBN 0-534-03074-2

Sponsoring Editor: *Craig Barth*
Production Editors: *Richard Mason and David Hoyt*
Manuscript Editor: *Paul Monsour*
Permissions Editor: *Carline Haga*
Interior and Cover Design: *Jamie Sue Brooks*
Cover Photo: © *Digital Image Corporation, Grand Rapids, Michigan*
Art Coordinator: *Rebecca Tait*
Interior Illustration: *ANCO/Boston*
Typesetting: *Bi-Comp, Inc., York, Pennsylvania*
Cover Printing: *Phoenix Color Corp., Long Island City, New York*
Printing and Binding: *R.R. Donnelley & Sons Co.*

Preface

Purpose of This Book

This book is designed for students in technical and engineering technology programs. It provides a thorough coverage of the main precalculus topics from beginning algebra through trigonometry, and it includes an introduction to analytic geometry.

Main Features

Just as the need for persons with a technical background is on the rise, so is the need for textbooks that help students prepare for technical and scientific careers. Traditional approaches to algebra and trigonometry are not enough: The student has to see how the mathematics is actually applied in technology. This book not only fills the needs of today's technology student; the presentation is exceptionally student-oriented.

1. The approach is concrete and intuitive.
2. Drill exercises make use of notations commonly encountered in technical areas.
3. Most sections contain exercises that illustrate how mathematics is applied to technical problems.
4. Whenever feasible, sections end with a technical application. (In some cases, chapters have a separate section devoted entirely to applications instead.)
5. The most important concepts are boxed and labeled for easy reference; other concepts are identified by marginal labels.

6. Color ink is not used in a merely decorative way. Its main function is to help explain difficult steps.
7. Important procedures are summarized; step-by-step procedures are provided whenever appropriate.
8. Calculator operations are discussed throughout the book. They are usually identified by the marginal label *Calculator Comment*.
9. Examples are worked out in great detail. Marginal notes are used to help explain the steps.
10. A large number of drill exercises is included to help reinforce basic concepts.
11. Common pitfalls are pointed out in special segments called *Common Errors*.

The use of realistic notations, even in drill exercises, is particularly important, since students tend to have great difficulty in transferring skills to technical problems that use a different notation. No prior knowledge of the different technical areas is assumed, however.

Most books on algebra and trigonometry encourage occasional use of calculators; some even go so far as to specify exercises to be solved with the aid of a calculator. Since this book is written for students in technology, it is assumed that they will be using scientific calculators throughout. Calculator procedures are explained whenever appropriate, and many problems are specifically designed for calculator use.

Some other features are:

1. Throughout the book, the use of graphs has been given particular attention.
2. Both metric and English units are employed, with SI notation generally used for the former.
3. Every chapter begins with a set of clearly stated objectives and ends with a set of review exercises.
4. Cumulative Review Exercises are given at the end of every third chapter.
5. A background in algebra and geometry is assumed for most of the material in this book. However, the discussion begins at a sufficiently low level to help overcome deficiencies in some areas.

Flexibility

This book is planned to give the teacher considerable flexibility in the order of presenting topics. In this way, certain topics can be studied at the most appropriate times. For example, although scientific notation is introduced in Chapter 1 and basic trigonometry in Chapter 4, these topics can be studied later in the course. On the other hand, trigonometric identities (Chapter 16) can be taken up after Chapter 10. Since determinants up to the third order are discussed in Chapter 3, higher-order determinants and matrices, which are first discussed in Chapter 15, can be introduced earlier or omitted altogether. The relationships between the chapters are summarized in the following chart:

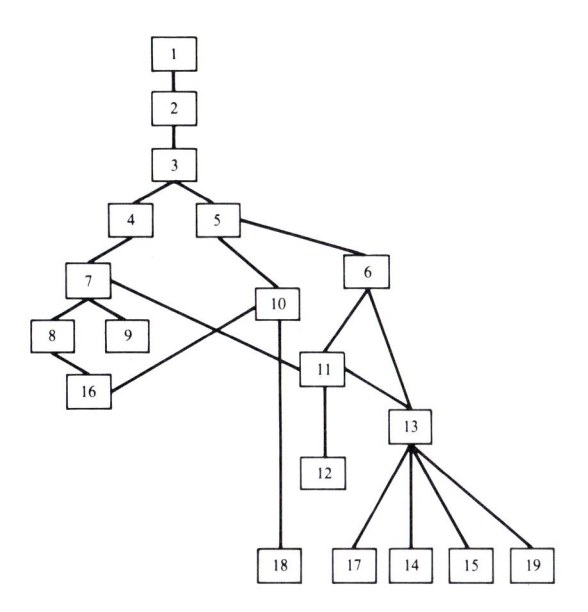

Coverage and Scope

Chapter 1 is a basic introduction to algebra. The topics covered range from signed numbers to operations with polynomials. Zero and negative exponents are introduced in case the teacher wishes to cover scientific notation early in the course (Section 1.11), but these topics may be postponed until Chapter 10 is covered. Because of the large number of applied problems in the book, rounding off numbers and significant figures are also discussed in Chapter 1.

Chapter 2 begins with the solution of linear equations, which is followed by a section on ratio and proportion and a section on variation. These topics are a natural continuation of the discussion of equations and formulas and provide a concrete basis for the definition of a function in Section 2.6.

Chapter 3 deals with systems of linear equations and their applications. The methods of solution discussed are graphing, addition or subtraction, and substitution. Determinants up to the third order are also introduced. (Higher-order determinants are covered in Chapter 15.)

Chapter 4 is an introduction to right-triangle trigonometry. The topics covered are basic definitions, values of trigonometric functions, and right-triangle applications.

Chapter 5 on factoring and fractions is organized to allow a gradual mastery of factoring: Basic special products, introduced in Section 5.1, are followed by the corresponding factoring cases in Section 5.2. Section 5.3 introduces more special products, and Section 5.4 the corresponding factoring cases. Factoring by grouping is then discussed in Section 5.5. The rest of the chapter is devoted to operations with fractions, complex fractions, and fractional equations.

Chapter 6 covers quadratic equations and their applications. The methods of solution discussed are factoring, completing the square, and the use of the quadratic formula.

Chapter 7 expands the discussion of trigonometric functions begun in Chapter 4. The topics covered are the functions of angles in any quadrant, functions of special angles, radian measure, and applications of radian measure, including linear and angular velocity. The emphasis is on the use of calculators rather than of tables.

Chapter 8 covers the graphs of trigonometric functions, as well as graphing by addition of coordinates. Considerable space is given to applications of sinusoidal functions.

Vectors and applications of vectors are treated in Chapter 9. The sections on the sine and cosine laws contain additional vector applications.

Chapter 10 comprises a detailed treatment of exponents and radicals: zero, negative, and fractional exponents, and fundamental operations with radicals.

Chapter 11 covers complex numbers: rectangular, polar, and exponential forms of complex numbers and powers and roots by De Moivre's theorem. The chapter ends with a brief discussion of phasors.

Chapter 12 contains a thorough treatment of logarithmic and exponential functions with special emphasis on applications. Also included are discussions of natural logarithms, properties of logarithms, exponential and logarithmic equations, and graphing on logarithmic paper. As in the case of trigonometric functions, the emphasis is on the use of calculators rather than of tables. (Computations with logarithms are introduced only briefly.)

Chapter 13 begins with a brief discussion of conic sections to provide a basis for solving systems of two quadratic equations by means of graphs. Algebraic methods are discussed next, followed by equations in quadratic form and fractional equations.

Chapter 14 contains a detailed treatment of higher-order equations. The method of approximating irrational roots by linear interpolation, which has become highly practical with the rise of calculators, is also discussed in detail.

Higher-order determinants and a detailed treatment of matrices can be found in Chapter 15.

Chapter 16 is devoted to trigonometric identities, trigonometric equations, and inverse trigonometric functions.

Chapter 17 on inequalities covers both graphical and algebraic methods of solution. The chapter ends with an optional section on linear programming.

Geometric series, the binomial theorem, and binomial series are treated in Chapter 18.

Chapter 19 is an introduction to analytic geometry: the distance formula, slope, lines, conics, the translation of axes, and polar coordinates.

Supplements

The answers to the odd-numbered exercises are given in the answer section; the answers to the even-numbered exercises are published in a separate Instructor's Manual. A Student Solutions Manual, with approximately every fourth problem worked in detail, is also available.

Acknowledgments

I am indebted to the students at the Milwaukee School of Engineering for having provided me with the stimulus to write this book. I am especially grateful to my colleagues at the Milwaukee School of Engineering for checking all the answers to the exercises and providing numerous valuable suggestions and constructive criticisms. In particular, I wish to thank Professors George L. Edenharder, Stanley J. Guberud, Richard B. Hernday, Dorothy J. Johnson, Ralph J. Jondle, Janet G. Klein, Robert P. Schilleman, and Andrew B. Schmirler.

Special thanks goes to Ms. Sally Canapa for typing the entire manuscript with great competence and care. I should also like to thank the staff of Brooks/Cole Publishing Company and the following reviewers for their cooperation and help in the preparation of the manuscript: Dale E. Boye, Schoolcraft College; William Brower, New Jersey Institute of Technology; Cheryl Cleaves, State Technical Institute, Memphis, Tennessee; Douglas Cook, Michael J. Owens Technical College; A. V. Galanti, Pennsylvania State University; Harold B. Hackett, Alfred State College, State University of New York; Marjorie Hobbs, State Technical Institute, Memphis, Tennessee; Jeffrey Jablonski, Michael J. Owens Technical College; Larry R. Lance, Columbus Technical Institute; Gordon O. Schlafmann, Western Wisconsin Technical Institute; Merriline Smith, California State Polytechnic University; and Roman Voronka, New Jersey Institute of Technology. I am especially indebted to Mr. Craig Barth, mathematics editor, who devoted so much of his time and energy to this project.

Peter K. F. Kuhfittig

Contents

Chapter 1   **Fundamental Operations**   1

- **1.1** Numbers   1
- **1.2** The Four Fundamental Operations   3
- **1.3** Operations with Zero   7
- **1.4** Polynomials   9
- **1.5** Addition and Subtraction of Polynomials   11
- **1.6** Symbols of Grouping   15
- **1.7** Exponents and Radicals   18
- **1.8** Multiplication of Polynomials   25
- **1.9** Division of Polynomials   30
- **1.10** Approximation and Significant Figures   34
- **1.11** Scientific Notation   38
- Review Exercises   43

Chapter 2   **Equations, Ratio and Proportion, Variation, and Functions**   45

- **2.1** Equations   45
- **2.2** Transposition and Formulas   49
- **2.3** Applications of Equations   54
- **2.4** Ratio and Proportion   60
- **2.5** Variation   65
- **2.6** Functions and Graphs   69

2.7   Notation for Functions   75
      Review Exercises   77

Chapter 3   Systems of Linear Equations and Introduction to Determinants   79

3.1   Simultaneous Linear Equations   79
3.2   Algebraic Solutions   84
3.3   Introduction to Determinants   91
3.4   Applications of Systems of Linear Equations   96
3.5   Systems of Linear Equations with More Than Two Unknowns   101
3.6   More on Determinants   103
      Review Exercises   114

Cumulative Review Exercises/Chapters 1–3   118

Chapter 4   Introduction to Trigonometry   120

4.1   What Is Trigonometry?   120
4.2   Angles   121
4.3   Trigonometric Functions   123
4.4   Values of Trigonometric Functions   130
4.5   Applications of Right Triangles   138
      Review Exercises   146

Chapter 5   Factoring and Fractions   149

5.1   Special Products   150
5.2   Common Factors and Factors of a Binomial   151
5.3   More Special Products   157
5.4   Factoring Trinomials   161
5.5   Factoring by Grouping   166
5.6   Equivalent Fractions   169
5.7   Multiplication and Division of Fractions   175
5.8   Addition and Subtraction of Fractions   179
5.9   Complex Fractions   185
5.10 Fractional Equations   190
      Review Exercises   195

Chapter 6   Quadratic Equations   198

6.1   Solution by Factoring and Pure Quadratic Equations   198
6.2   Solution by Completing the Square   203

6.3 The Quadratic Formula 208
6.4 Applications of Quadratic Equations 214
Review Exercises 219

Cumulative Review Exercises/Chapters 4–6 220

Chapter 7 More on Trigonometric Functions 222

7.1 Algebraic Signs of Trigonometric Functions 222
7.2 Special Angles 229
7.3 Trigonometric Functions of Arbitrary Angles 235
7.4 Radians 239
7.5 Applications of Radian Measure 243
Review Exercises 248

Chapter 8 Graphs of Trigonometric Functions 250

8.1 Graphs of Sine and Cosine Functions 250
8.2 Phase Shifts 257
8.3 Applications of Sinusoidal Functions 260
8.4 Graphs of Other Trigonometric Functions 265
8.5 Addition of Coordinates 270
Review Exercises 272

Chapter 9 Vectors and Oblique Triangles 274

9.1 Vectors 274
9.2 Vector Components 277
9.3 Basic Applications of Vectors 280
9.4 The Law of Sines 284
9.5 The Law of Cosines 291
Review Exercises 296

Cumulative Review Exercises/Chapters 7–9 298

Chapter 10 Exponents and Radicals 299

10.1 Review of Exponents 299
10.2 Zero and Negative Integral Exponents 302
10.3 Fractional Exponents 308
10.4 The Simplest Radical Form and Addition and Subtraction of Radicals 313
10.5 Multiplication of Radicals 322

	10.6	Division of Radicals 324	
		Review Exercises 328	

Chapter 11 Complex Numbers — 331

	11.1	Basic Concepts and Definitions 331
	11.2	The Fundamental Operations 337
	11.3	Polar and Exponential Forms 341
	11.4	Products and Quotients of Complex Numbers 346
	11.5	Powers and Roots of Complex Numbers 348
	11.6	Phasors (Optional) 353
		Review Exercises 355

Chapter 12 Logarithmic and Exponential Functions — 358

	12.1	The Definition of Logarithm 358
	12.2	Graphs of Exponential and Logarithmic Functions 361
	12.3	Properties of Logarithms 366
	12.4	Common Logarithms 370
	12.5	Computations with Logarithms (Optional) 374
	12.6	Natural Logarithms, Change of Base, and Powers of e 376
	12.7	Special Exponential and Logarithmic Equations 380
	12.8	Graphs on Logarithmic Paper (Optional) 385
		Review Exercises 391

Cumulative Review Exercises/Chapters 10–12 394

Chapter 13 More on Equations and Systems of Equations — 396

	13.1	Solving Systems of Equations Graphically 396
	13.2	Solving Systems of Equations Algebraically 403
	13.3	Equations in Quadratic Form 407
	13.4	Radical Equations 411
		Review Exercises 416

Chapter 14 Higher-Order Equations in One Variable — 418

	14.1	Introduction 418
	14.2	The Remainder and Factor Theorems 419
	14.3	Synthetic Division 423
	14.4	The Fundamental Theorem of Algebra 428
	14.5	Rational Roots 432

14.6 Irrational Roots   439
Review Exercises   445

Chapter 15 Determinants and Matrices 447

15.1 Introduction   447
15.2 Review of Determinants   447
15.3 Properties of Determinants   449
15.4 Cramer's Rule   456
15.5 Matrix Algebra   461
15.6 Inverse of a Matrix and Row Operations   470
15.7 Solving Systems of Equations by Matrix Methods   477
Review Exercises   483

Cumulative Review Exercises/Chapters 13–15   486

Chapter 16 Additional Topics in Trigonometry 488

16.1 Fundamental Identities   488
16.2 Proving Identities   493
16.3 The Sum and Difference Formulas   499
16.4 Double-Angle Formulas   506
16.5 Half-Angle Formulas   511
16.6 Trigonometric Equations   515
16.7 Inverse Trigonometric Relations   520
16.8 Inverse Trigonometric Functions   523
Review Exercises   530

Chapter 17 Inequalities 534

17.1 Basic Properties of Inequalities   534
17.2 More Inequalities   539
17.3 Absolute Inequalities   544
17.4 Graphing Inequalities   547
17.5 Linear Programming (Optional)   552
Review Exercises   559

Chapter 18 Geometric Progressions and the Binomial Theorem 561

18.1 Geometric Progressions   561
18.2 Geometric Series   566

	18.3	Binomial Theorem and Series 568	
		Review Exercises 573	
		Cumulative Review Exercises/Chapters 16–18 575	
Chapter 19		**Introduction to Analytic Geometry**	**576**
	19.1	Introduction 577	
	19.2	The Distance Formula 577	
	19.3	The Slope 578	
	19.4	The Straight Line 582	
	19.5	The Conics 586	
	19.6	The Circle 587	
	19.7	The Parabola 590	
	19.8	The Ellipse 595	
	19.9	The Hyperbola 601	
	19.10	Translation of Axes 606	
	19.11	Polar Coordinates 610	
	19.12	Curves in Polar Coordinates 614	
		Review Exercises 620	
Appendix A		**Units of Measurement and Conversions**	**623**
Appendix B		**Tables**	**628**
		Table 1 Trigonometric Functions 628	
		Table 2 Common Logarithms 634	
		Table 3 Natural Logarithms 636	
Appendix C		**Answers to Odd-Numbered Exercises**	**637**
		Index 681	

Basic Technical Mathematics

CHAPTER 1

Fundamental Operations

Objectives Upon completion of this chapter, you should be able to:

1. State whether a given number is rational or irrational.
2. Find the absolute value of a given number.
3. Carry out the four fundamental operations with signed numbers.
4. Identify a monomial, binomial, trinomial, multinomial, and polynomial.
5. Add and subtract polynomials.
6. Remove and insert symbols of grouping.
7. Perform basic operations with exponents and radicals.
8. Multiply and divide polynomials.
9. Round off certain approximate values to the proper number of significant figures.
10. Multiply and divide numbers written in scientific notation.

1.1 Numbers

While the nature of mathematics has changed over the centuries, it is generally agreed that mathematics has always been based on the concepts of number and form. We will begin our study with a brief review of the number concept.

Rational number

First recall that a number is called **rational** if it can be written as a ratio of two integers. For example, the numbers

$$\frac{2}{3}, \quad \frac{7}{11}, \quad \frac{19}{6}, \quad 3 \text{ or } \frac{3}{1}, \quad 10 \text{ or } \frac{10}{1}$$

Irrational number

are rational. A number that cannot be written as a ratio of two integers is called **irrational.** Examples of irrational numbers are

$$\sqrt{2}, \quad \sqrt{5}, \quad \sqrt[3]{4}, \quad \pi$$

(Thus $\frac{22}{7}$ is not equal to π; it is only an approximation.)

Rational numbers have a unique repeating decimal expansion. For example,

$$\frac{1}{11} = 0.090909\ldots, \text{ or } 0.09\overline{09}$$

the three dots indicating an omission of the terms that complete the expression. (This equality can be checked by direct division.) Note that the cycle 09 repeats indefinitely. Similarly,

$$\frac{1}{5} = 0.200\overline{0}$$

In this case the zero repeats indefinitely.

Irrational numbers do not have repeating decimal expansions. For example, in

$$\sqrt{2} = 1.4142136\ldots$$

there are no cycles of repeating integers.

Although we cannot pursue this idea any further at this point, it should be noted that irrational numbers fall into two categories: algebraic and transcendental. The former is a root of an algebraic equation, which will be defined later. (Thus rational numbers are by definition algebraic.) The number π and most trigonometric ratios, on the other hand, are transcendental.

Negative number

Another important type of number that we need to consider is a **negative number,** which is most easily defined geometrically. Let us assume that the set of all positive numbers can be placed on a line so that each positive number corresponds to a unique point. To accomplish this, we select an arbitrary point, called the **origin,** to be associated with the number 0. The positive integers are then placed at equal intervals to the right of the origin (Figure 1.1). The other numbers are assigned to appropriate points between the integers. The number $\frac{3}{2}$, for example, is located halfway between 1 and 2.

Origin

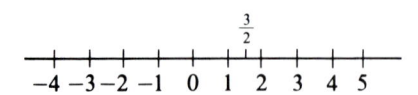

Figure 1.1

Negative numbers are now defined as follows: A positive number a has a "mirror image" on the left side of the origin. This mirror image is denoted by $-a$. Also, $a + (-a) = 0$.

> **Negative number:** For every number a there exists a number $-a$ such that $a + (-a) = 0$. (The number 0 itself is neither positive nor negative.)

Greater than and less than

Recall that the symbol $>$ means **greater than** and the symbol $<$ means **less than.** These relations carry over to negative numbers. For example, since -5 is to the left of -4, we have $-5 < -4$. For the same reason, $-2 < 3$, while $1 > -3$. Also, the symbol $\geq$ means **greater than or equal to** and $\leq$ means **less than or equal to.**

Real numbers

The numbers considered so far are collectively known as the set of **real numbers,** which are distinguished from **complex numbers.** Although not part of the real number system, complex numbers will play an important role in our later work.

The composition of the real number system is shown in Figure 1.2.

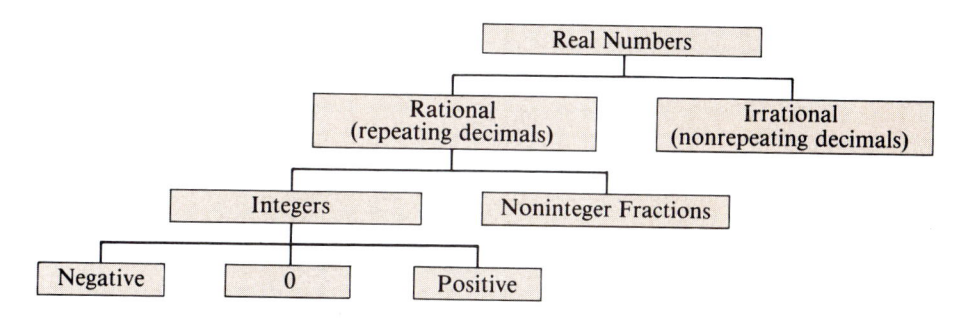

Figure 1.2

1.2 The Four Fundamental Operations

In performing the fundamental operations of addition, subtraction, multiplication, and division, we follow certain rules that we know and use intuitively. These rules are called the commutative, associative, and distributive laws.

For example,

$$2 + 3 = 3 + 2 \quad \text{and} \quad 4 \times 3 = 3 \times 4$$

In other words, addition and multiplication are **commutative**—they are independent of order.

Nor does the sum or product of any two numbers depend on the order of grouping. For example,

$$2 + (4 + 7) = (2 + 4) + 7$$

and

$$(3 \times 6) \times 5 = 3 \times (6 \times 5)$$

These properties are called the **associative laws** of addition and multiplication, respectively.

Multiplication and addition are also connected by the **distributive law,** which states that the product of one number and the sum of two or more other numbers is equal to the sum of the products of the first number and each of the numbers in the sum. For example,

$$2 \times (7 + 9) = (2 \times 7) + (2 \times 9)$$

These properties, together with the identity and inverse properties, constitute the properties of real numbers. They are summarized in the accompanying table.

The properties of real numbers

Addition	Multiplication
Commutative law	
$a + b = b + a$	$ab = ba$
Associative law	
$(a + b) + c = a + (b + c)$	$(ab)c = a(bc)$
Distributive law	
$a(b + c) = ab + ac$	
$(b + c)a = ba + ca$	
Identity	
There exists a unique real number 0 such that $a + 0 = a$ for every a.	There exists a unique real number 1 such that $a \cdot 1 = a$ for every a.
Inverse	
For every real number a there exists a unique real number $-a$ such that $a + (-a) = 0$.	For every nonzero real number a there exists a unique real number $1/a$ such that $a \cdot 1/a = 1$.
Multiplication by 0	
	$a \cdot 0 = 0$ for every real number a.

Although first defined for positive numbers, *the fundamental operations are also valid for negative numbers.* In fact, the rules for signed numbers were designed to preserve the rules for positive numbers.

To state these rules, we need the concept of **absolute value,** which is defined next.

1.2 THE FOUR FUNDAMENTAL OPERATIONS

Absolute value: The absolute value of a positive number is the number itself. The absolute value of a negative number is the corresponding positive number. The absolute value of 0 is 0.

For example, the absolute value of 3 is 3, and the absolute value of -5 is 5. The absolute value is denoted by the symbol $|\ \ |$. Thus

$$|3| = 3 \quad \text{and} \quad |-5| = 5$$

Similarly,

$$|4| = 4, \quad |-8| = 8, \quad \text{and} \quad |-20| = 20$$

Stated informally, the absolute value of a number is the number without the minus sign. A formal definition of absolute value is the following:

$$|x| = \begin{cases} x, & \text{if } x \geq 0 \\ -x, & \text{if } x < 0 \end{cases}$$

For example, we know that $|-2| = 2$. By the formal definition, since $-2 < 0$, we have

$$|-2| = -(-2) = 2$$

The concept of absolute value is used in the rules for operating with signed numbers, which are summarized next.

Rules for operating with signed numbers

Rule 1. To add two real numbers with like signs, find the sum of their absolute values and affix their common sign to the result.
Rule 2. To add two real numbers with unlike signs, find the difference of their absolute values and affix the sign of the number with the larger absolute value.
Rule 3. To subtract two real numbers, change the sign of the subtrahend (the number to be subtracted) and proceed as in addition.
Rule 4. The product (or quotient) of two real numbers with like signs is the product (or quotient) of their absolute values. The product (or quotient) of two real numbers with unlike signs is the negative of the product (or quotient) of their absolute values.

The following example illustrates the rules for operating with signed numbers:

Example 1 Perform the indicated operations:

    a. $-3 - 7$ **b.** $-4 - (-9)$ **c.** $(-4)(-5)(-8)$

Solution. **a.** By the rule for subtraction, we change the sign of the subtrahend, 7, and add:

$$-3 - 7 = -3 + (-7) = -10$$

b. Again, changing the sign of the subtrahend, we get

$$-4 - (-9) = -4 + (+9) = -4 + 9 = 5$$

c. -4 and -5 have like signs, so that $(-4)(-5) = 20$. We now have

$$(-4)(-5)(-8) = (20)(-8) = -160$$

since 20 and -8 have unlike signs.

Example 2 Carry out the indicated operations:

a. $-2 - (-3) + 5$ **b.** $-10 - (-4) - 6 - 8 - (-3)$

Solution.   **a.** By the rule for subtraction

$$-2 - (-3) + 5 = -2 + (3) + 5 = 6$$

b. $-10 - (-4) - 6 - 8 - (-3) = (-10) + (4) + (-6) + (-8) + (3)$
$= [(-10) + (-6) + (-8)] + (4 + 3)$
$= (-24) + 7 = -17$

Note the order in which the numbers are combined:

1. Add all negative numbers and all positive numbers.
2. Add the resulting sums.

Order of operations   If the order of operations is not indicated by any specific grouping, perform all multiplications and divisions first; then perform addition and subtraction.

Example 3 Carry out the indicated operations:

$$\frac{(6)(-8)}{(-2)4} - 5 - (-8)$$

Solution. By the rules for multiplication and division,

$$\frac{(6)(-8)}{(-2)4} = \frac{-48}{-8} = \mathbf{6}$$

Since multiplication and division are always performed first, we get

$$6 - 5 - (-8) = 6 + (-5) + (8) = 9$$

Example 4　Evaluate

$$\frac{8}{-4} - |-4|$$

Solution. Since $|-4| = 4$, we get

$$\frac{8}{-4} - |-4| = -2 - 4 = -2 + (-4) = -6$$

Example 5
The formula relating Celsius and Fahrenheit temperatures is $C = \frac{5}{9}(F - 32)$. What is $-13°F$ in degrees Celsius?

Solution. If we substitute -13 for F in the above formula, we get

$$C = \frac{5}{9}(-13 - 32) = \frac{5}{9}(-45) = -25°C$$

1.3　Operations with Zero

For any number a, the following operations with zero hold:

$$a + 0 = a$$
$$a - 0 = a$$
$$a \cdot 0 = 0$$
$$\frac{0}{a} = 0$$

Only division by zero ($\frac{a}{0}$) is undefined. Let's consider the reason. We know that $12 \div 4 = 3$ because $4 \times 3 = 12$. In general,

$$\frac{a}{b} = c \quad \text{if, and only if,} \quad a = bc$$

As a consequence,

$$\frac{a}{0} = x \quad \text{is equivalent to} \quad a = 0 \cdot x = 0$$

So if $a \neq 0$, we conclude that $a = 0$, which is a contradiction. On the other hand, if $a = 0$, then no contradiction results; however, we are then forced to draw another implausible conclusion:

$$\frac{0}{0} = x \text{ for any number } x$$

8 CHAPTER 1 FUNDAMENTAL OPERATIONS

For this reason $\frac{0}{0}$ is called *indeterminate*, since no specific value can be determined.

To summarize, ***never divide by zero.***

Exercises / Section 1.3

In Exercises 1–10, find the absolute value of:

1. 3
2. -5
3. -7
4. 10
5. $-\pi$
6. $-\sqrt{3}$
7. $-(-\sqrt{3})$
8. $\sqrt[3]{-8}$
9. $\sqrt[3]{-27}$
10. $-\sqrt[3]{27}$

In Exercises 11–20, state whether the given numbers are rational or irrational.

11. $\frac{2}{3}$
12. $\frac{1}{4}$
13. $7\frac{1}{2}$
14. $4\frac{1}{3}$
15. $\sqrt{10}$
16. $\sqrt{7}$
17. $\sqrt{9}$
18. $\sqrt{25}$
19. $\sqrt[3]{9}$
20. $\sqrt[3]{25}$

In Exercises 21–58, carry out the indicated operations.

21. $|-2| - 5$
22. $5 - |-4|$
23. $|1 - 9|$
24. $|-1 - (-3) + (-20)|$
25. $2 + |3 - 12|$
26. $-4 - 7 - 10$
27. $(-5) + (-8) + (-1) + (-3)$
28. $-5 - 8 - 1 - 3$
29. $3 + 9 - 20 - (-3)$
30. $-2 - 6 + 0 - 3$
31. $3 - 4 + (-4) - 6 + 0$
32. $-20 - 1 + 15 - (-6) - (-9)$
33. $15 + (-7) + (-12) - (-20)$
34. $18 - (-30) + 7$
35. $3.2 - (-4.5) + 0.7$
36. $1.74 + (-3.6) - (-2.13)$
37. $-\frac{5}{7} - \left(-\frac{3}{14}\right)$
38. $\frac{1}{15} + \left(-\frac{1}{6}\right)$
39. $-3(7 + 19)$
40. $-4(-25 + 20)$
41. $8(-5 - 10 + 3)$
42. $(6 - 7 + 0) \cdot 6$
43. $(8)(-3)(-4)$
44. $(-1)(-6)(-10)$
45. $(1.3)(-0.7)(-1)$
46. $(7.3)(0)(-8.96)(3.2)$
47. $(-3.22)(-8.7)$
48. $\dfrac{-3}{-9}$
49. $\dfrac{(-6)(15)}{(-3)(-18)}$
50. $\dfrac{(-14)(-1)}{(-2)(7)}$
51. $\dfrac{(3)(-9)}{(-3)(0)}$
52. $\dfrac{\left(\frac{1}{2}\right)\left(\frac{1}{3}\right)}{\left(-\frac{2}{3}\right)\left(-\frac{1}{4}\right)}$
53. $\dfrac{\left(-\frac{2}{7}\right)\left(-\frac{7}{15}\right)}{\left(\frac{15}{16}\right)(-4)}$
54. $\dfrac{\left(-\frac{1}{2}\right)\left(-\frac{1}{10}\right)}{(0)\left(-\frac{1}{2}\right)}$
55. $\dfrac{(3)(0)(-6)}{(-21)(27)}$
56. $\dfrac{(-2)(0)(-6)}{(-8)(7)(0)}$

57. $\dfrac{(2)(-10) - (3)(-5)}{3 - 8}$ **58.** $\dfrac{6 + 8 - 3(-7)}{-2 - 2(-5)}$

59. The highest point in the United States is Mt. McKinley at 20,320 ft, and the lowest point is Death Valley at -282 ft. What is the difference in altitude between these two points?

60. A technician in a liquid oxygen manufacturing plant checks the temperature of incoming air hourly during the first cooling stage. During his eight-hour shift he obtains the following readings (in degrees Celsius): $-2.2, -1.3, 0.5, -1.5, 0.6, 1.3, 2.1,$ and -0.3. Determine the average temperature, which is the algebraic sum of all the readings divided by the number of readings.

61. The overnight low temperatures (in degrees Fahrenheit) in Walhalla, North Dakota, during the first week in January were $-10, -12, -1, 2, 4, -8,$ and -3. Determine the average low temperature during that week.

62. The current in a certain circuit changes from -4 to 3 amperes. What is the change in the current?

63. For a short time interval, the velocity v (in feet per second) of a certain particle moving along a line is $v = 2.0 - 1.0t - 1.0t^2$, where t is measured in seconds. Determine the velocity a second ago (at $t = -1$).

1.4 Polynomials

In algebra, the letters of the alphabet are customarily used to represent numbers. The use of letters requires some additional definitions and conventions. The basic terms will be discussed in this section.

Our first notion is that of "power." Consider the following notations and definitions:

$$a \cdot a = a^2 \quad \text{is read} \quad \text{``}a \text{ squared''}$$
$$a \cdot a \cdot a = a^3 \quad \text{is read} \quad \text{``}a \text{ cubed''}$$
$$a \cdot a \cdot a \cdot a = a^4 \quad \text{is read} \quad \text{``}a \text{ to the fourth power''}$$

or simply "a to the fourth," and so on.

> **The nth power:** $a^n = a \cdot a \cdot a \cdot \ldots \cdot a$ (n factors) is read "a to the nth power"; a is called the **base** and n the **exponent.**

Numerical coefficient A number multiplied by a given factor is called a **numerical coefficient** or simply a **coefficient** of that factor. For example, 2 is the coefficient of $2xy$, 3 is the coefficient of $3abc$, and -5 is the coefficient of $-5x^2y^3$.

Expression If two or more numbers or literal symbols are combined by means of the four algebraic operations, the combination is called an **expression.** (Single numbers and literal symbols are also expressions.) Examples of expressions are

$$5, \quad x, \quad 2x, \quad 3x^4y^3, \quad \dfrac{\sqrt{x}}{ab}, \quad 5\sqrt{a} + b$$

Multinomials

Monomial
Multinomial

An expression that does not involve addition or subtraction is called a **monomial.** The sum of two or more monomials is called a **multinomial.** Each monomial in a multinomial, together with the sign that precedes it, is called a **term.** A multinomial consisting of exactly two terms is called a **binomial,** and one with exactly three terms a **trinomial.**

Term, Binomial
Trinomial

An important special case of a multinomial is a **polynomial,** which is defined below.

> A **polynomial** is a multinomial containing only terms with positive integral powers (that is, the exponents are positive integers).

For example, $x^2 + x$ and $2x^3 - 3x^2 + 7$ are polynomials.

By these definitions, binomials and trinomials are also multinomials. However, a monomial is not a multinomial, since it contains only one term.

Example 1

a. $3x/y$ is a monomial.
b. $a/b + c^2 + 3$ is a trinomial and hence a multinomial.

Example 2 Examples of polynomials are

$$3x^2 + 2x^2y^3 - y^4 \quad \text{and} \quad 2a^2x^2 - 4a^2y^2z^3 + 5$$

since all the exponents are integers.

Example 3

a. $2\sqrt{x} + x/y$ is a binomial but not a polynomial (because of the $\sqrt{x}$ and the y in the denominator).
b. $3\sqrt{a} + a\sqrt{b} + 3$ is a trinomial but not a polynomial (because of the $\sqrt{a}$ and $\sqrt{b}$).

Historical note. Much of the material discussed in this chapter, together with equations of first and second degree, had already been developed in the early Middle Ages. Some of these developments took place after the expansion of the Arab state, when Baghdad became a new center of learning about the middle of the eighth century. Of the many scholars in Baghdad, the best-remembered is al-Khowarizmi (who died before 850 A.D.). His books, particularly the *Al-jabr w'al muqabala,* were to have a profound influence on the development of algebra. In fact, our word *algebra* comes from the name of this book, and the word *algorithm* comes from the author's name.

Our symbols for the digits 0 and 1 through 9 also date from this period. Most of our other algebraic notations, however, were introduced only gradually over many centuries.

1.5 Addition and Subtraction of Polynomials

In this section we will study two of the fundamental operations involving polynomials, addition and subtraction. To state the rules for addition and subtraction, we need the definition of *like terms*.

> **Like terms**: Two or more terms are said to be **like terms** if the literal (nonnumerical) parts are the same.

For example,

$$5x, \quad 10x, \quad -3x, \quad \text{and} \quad -7x$$

are like terms, but

$$5x \quad \text{and} \quad 10y$$

are not like terms;

$$7xy^2, \quad -2xy^2, \quad \text{and} \quad 3xy^2$$

are also like terms, but

$$7xy^2 \quad \text{and} \quad -2xy$$

are not because the y-factors have different exponents.

Addition and subtraction of polynomials are carried out in accordance with the following rule.

> **Addition and subtraction of polynomials**: When adding (subtracting) polynomials, add (subtract) the coefficients of like terms.

For example, the sum $2x + 5x = 7x$ is obtained by adding the coefficients. This addition can be justified by means of the distributive law: $(b + c)a = ba + ca$.

Using the distributive law in reverse, we have

$$2x + 5x = (2 + 5)x = 7x$$

Another way to see this is by noting that $2x = x + x$, since multiplication is actually repeated addition. Thus,

$$2x + 5x = (x + x) + (x + x + x + x + x) = 7x$$

On the other hand,

$$2x + 5y$$

cannot be combined, since neither x nor y is common to both terms. This can also be seen from the fact that

$$2x + 5y = x + x + y + y + y + y + y$$

which cannot be written as a single term.

> When adding algebraic expressions, only like terms can be combined to form a new term.

Example 1 Add the polynomials

$$-3x^2y + 4xy^2 \quad \text{and} \quad -7x^2y - 2xy^2$$

Solution. Since the first terms and the last terms in the expressions are like terms, the polynomials may be arranged in columns. Now add the coefficients in each column:

$$\begin{array}{ll} -3x^2y + 4xy^2 & (-3) + (-7) = -10 \\ \underline{-7x^2y - 2xy^2} & 4 + (-2) = 2 \\ -10x^2y + 2xy^2 & \end{array}$$

Note that only the like terms are added. The two terms in the sum cannot be combined, since x^2y and xy^2 are not the same.

Example 2 Add the polynomials

$$10x - 8y + 5z \quad \text{and} \quad -9x - 2y + w$$

Solution. Arranging the expressions so that the like terms are in the same column, we get

$$\begin{array}{ll} 10x - 8y + 5z \\ \underline{-9x - 2y + w} \\ 1x - 10y + 5z + w & \text{adding coefficients} \end{array}$$

Note that $5z$ and w are unlike terms and do not appear in the same column. Also, since $1 \cdot x = x$, the sum should be written

$$x - 10y + 5z + w$$

When adding polynomials, we use the fact that $x = 1x$ and $-x = -1x$, as shown in the next example.

1.5 ADDITION AND SUBTRACTION OF POLYNOMIALS

Example 3 Add $-x$ and $6x$.

Solution. Since $-x = -1x$, we get

$$-x + 6x = -1x + 6x = (-1 + 6)x = 5x$$

As we saw in Section 1.2

$$a - b = a + (-b)$$

Subtraction So whenever we subtract one term from another, we simply change the sign of the subtrahend and add algebraically.

Example 4 Subtract $-5x - y + z$ from $-6x + 2y + 2z$.

Solution. If we arrange the work in columns and obey the rules regarding the sign changes, we obtain

$$\begin{array}{r} -6x + 2y + 2z \\ \underline{5x + y - z} \\ -1x + 3y + 1z \end{array} \quad \begin{array}{l} -(-5x - y + z) \\ \\ \text{adding coefficients} \end{array}$$

or $-x + 3y + z$.

Example 5 The voltage drops (in volts) across two resistors are given by $V_1 = 2.1R - 0.03$ and $V_2 = 1.5R + 0.19$, respectively. Find $V_1 + V_2$, which is the voltage drop across the two resistors connected in series. (The subscripts indicate that V_1 and V_2 are different quantities.)

Solution.
$$V_1 + V_2 = (2.1R - 0.03) + (1.5R + 0.19)$$
$$= 3.6R + 0.16 \text{ (in volts)}$$

Exercises / Section 1.5

In Exercises 1–8, simplify the given algebraic expressions.

1. $2x - 5y - 8x$
2. $3x - 4y + 2x$
3. $-7x - 3y + 5x - y$
4. $x - 5y - 2y + x$
5. $2a^2 - 3a - 8a^2 + a$
6. $-3a^2b + 10ab^2 - 2a^2b + 20ab^2$
7. $9xy^2 - 9x^2y - 8xy^2 + 6x^2y$
8. $-19xyz + 5xy - xyz$

In Exercises 9–20, find the sum of each of the given polynomials.

9. $\begin{array}{r} 2x - 5y - 3z \\ -4x + y - 2z \\ x - y + 6z \end{array}$

10. $\begin{array}{r} 5ab - 9b^2 \\ -4ab - 9b^2 \\ ab + 18b^2 \end{array}$

11. $x^2 - y^2 + 2z^2$
    $3x^2 + z^2$
    $ - 5y^2 - 6z^2$

12. $9a - 2b - 9x^2$
    $a - 5b$
    $6a - 8x^2$

13. $3a - 5b - c$, $-7a + 2b + 2c$, $5a + 3b - c$
14. $5a^2 - 10b^2 + 2c$, $-6a^2 - 9b^2 + c$, $-a^2 + 20b^2 - 4c$
15. $2x^2 - 3xy + y^2$, $8x^2 + 5xy + 2y^2$, $-10x^2 - 3xy - y^2$
16. $2a^2 - 5x$, $a^2 - 5x + 2y$, $-x + y$, $3a^2 + 3y$
17. $3x - 2y + w$, $-5x + z - 3w$, $-4x - 3y + 2w$
18. $5a^3 - 3c^2 + x$, $a^3 - 6d + 2x$, $-6a^3 + 5c^2 - 4x$
19. $a - b$, $b + 2c$, $2a - 2b - 3c$, $-5a - 6c$
20. $2x^2 + y$, $-3x^2 - y + z$, $-x^2 - 6z$, $y + z$

In Exercises 21–28, subtract the second expression from the first.

21. $5a - 7b + c$, $-6a - 6b - 6c$
22. $x^2 - y^2 + a^2$, $-x^2 - y^2 - 2a^2$
23. $6a + 5b - 3c$, $5a - b - 4c$
24. $w + z + 2a$, $-w + z + 2a$
25. $-3a^2 + 5c^2 - d^2$, $4a^2 - 5c^2 - e$
26. $2x^2$, $-3x^2 + 5b - d$
27. $-3a + 3b^2 + c$, $-3b^2$
28. $7x^2 - 2xy + 6y^2$, $-2x^2 - 6y^2$

In Exercises 29–32, subtract the third expression from the sum of the first two.

29. $a - 10b - 11c$, $a + b - c$, $2a - 8b - 8c$
30. $-x^2 + 2xy + 2y^2$, $-2x^2 + 4xy - 5y^2$, $-4x^2 + 5xy - 4y^2$
31. $20a^2 + b^2 + c^2$, $a^2 + 10b^2 - 2c^2$, $-20a^2 - b^2 - c^2$
32. $-5ab + 6b^2 - 3ac$, $-4ab + b^2 + 2ac$, $-ab - ac$

33. Write the polynomial representing the perimeter of the foundation wall in Figure 1.3.

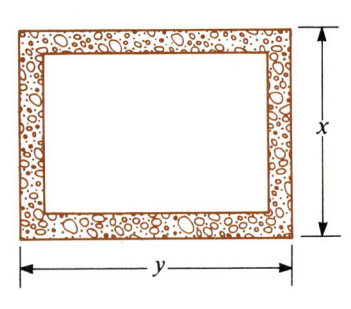

Figure 1.3

34. A shipment contains x gears costing $3 each and y gears costing $4 each. What does the polynomial $3x + 4y$ represent?

35. A box contains x bolts weighing 2 oz each and y bolts weighing 4 oz each. What does the polynomial $2x + 4y$ represent?

36. The capacitance in microfarads (μF) of two capacitors is $C_1 = 0.013C + 3$ and $C_2 = 0.198C - 2$, respectively. Find $C_1 + C_2$, the combined capacitance if the two capacitors are connected in parallel. (C_1 and C_2 denote different quantities.)

37. In determining the currents I_1 and I_2 in a certain two-loop circuit, the following expressions have to be added: $-I_1 + 3I_2$ and $6I_1 - 7I_2$. Perform the indicated operation.

1.6 Symbols of Grouping

In algebra, terms are frequently grouped to indicate that they are to be considered as a single quantity or number. For example, the parentheses in the expression $3(2 + 6)$ indicate that the addition operation is to be carried out before multiplication. Thus, $3(2 + 6) = 3 \times 8 = 24$. Without parentheses we would have

$$3 \cdot 2 + 6 = 6 + 6 = 12 \neq 24$$

since multiplication always precedes addition.

Symbols of grouping

The most common symbols of grouping are parentheses ( ), brackets [ ], and braces { }; these symbols sometimes occur in combination. When simplifying certain algebraic expressions, symbols of grouping must often be removed. This is usually accomplished by means of the **distributive law.** Consider, for example, the expression

$$3[a + 2(a + b)]$$

If we first use the distributive law on the terms inside the parentheses, we get

$$3[a + 2a + 2b] = 3[3a + 2b] = 9a + 6b$$

If the expression to be simplified contains several sets of grouping symbols, we use the distributive law to remove the **innermost** set of grouping symbols and then combine similar terms, if possible. We then repeat the procedure for the resulting expression.

> **Removing symbols of grouping**: When removing symbols of grouping, work from the inside out.

Example 1

$$-2[3(x - 2y) - y] \qquad \text{original expression}$$
$$= -2[3x - 6y - y] \qquad \text{removing inner parentheses}$$
$$= -2[3x - 7y] \qquad \text{simplifying}$$
$$= -6x + 14y \qquad \text{removing brackets}$$

An important special case involves a set of parentheses preceded by a negative sign, as in the expression

$$a - (a + b)$$

Since the quantity $a + b$ is subtracted from a, we can apply the rule from the previous section and conclude that the signs of both terms have to be changed. Thus

$$a - (a + b) = a - a - b = -b$$

A simple alternative is to write

$$-(a + b) = -1(a + b) \qquad (1.1)$$

and then apply the distributive law to get $-a - b$. If the set of parentheses is preceded by a plus sign (or no sign), then no sign change is required and the parentheses may simply be dropped: $+ (a + b) = a + b$.

Example 2 Remove parentheses and simplify

$$-\{x^2 - 2[x + (-x + 3)]\}$$

Solution. Since several sets of grouping symbols have to be removed, it is necessary to work from the inside out.

$-\{x^2 + 2[x - (-x + 3)]\}$	original expression
$= -\{x^2 + 2[x + x - 3]\}$	removing inner parentheses
$= -\{x^2 + 2[2x - 3]\}$	simplifying
$= -\{x^2 + 4x - 6\}$	distributive law
$= -x^2 - 4x + 6$	removing braces

Inserting parentheses

It is occasionally necessary to reverse the above procedure and *insert* parentheses. Consider the next example.

Example 3 Enclose the last two terms in parentheses preceded by a minus sign

$$2x^2y - 3 - ab + c$$

Solution. We know from the procedure for removing parentheses that the signs of the last two terms have to be changed. We therefore get

$$2x^2y - 3 - (ab - c)$$

which can be checked by removing the parentheses.

1.6 SYMBOLS OF GROUPING

Common errors
1. Leaving out parentheses in $a(b + c)$ and incorrectly writing $ab + c$. The two expressions are not the same.
2. Changing the sign of only the first term when removing parentheses. Thus, $-(-a - b) = a + b$, not $a - b$.

Example 4 If two forces $F_1 = 4f_1 - 2f_2$ and $F_2 = -f_1 + 6f_2$ are acting on the same object in opposite directions, then the net force F_n is given by $F_n = F_1 - F_2$. Find the expression for F_n.

Solution.
$$F_n = F_1 - F_2 = (4f_1 - 2f_2) - (-f_1 + 6f_2)$$
$$= 4f_1 - 2f_2 + f_1 - 6f_2$$
$$= 5f_1 - 8f_2$$

Exercises / Section 1.6

In Exercises 1–16, remove symbols of grouping and simplify.

1. $3(-4 + x)$
2. $2(a - 4)$
3. $2(x + 2y) - 3(x - 2y)$
4. $3(-a - b) + 4(a + 2b)$
5. $-(x - y) - (-2x + 3y)$
6. $x^2 - y^2 - (x^2 - y^2)$
7. $(x^2 - xy) - (3x^2 + 3xy)$
8. $-(a + b) - (2a - 2b) - 3a$
9. $2[a - (2a - b)] - (a + b)$
10. $-[x^2 - x(2x + y)] + (x^2 - xy)$
11. $-(x - xy) - [3 - (x + y) - xy]$
12. $2[x - 3(x - y) + y]$
13. $-\{x - [(y - x) - y]\}$
14. $-\{2a - 3[-b - (-2b + 4a)]\}$
15. $4\{2a - 2[2a - 3(2a - 3) + 4a] + 3\}$
16. $-3\{x + 2(y - 2z) + 3[x - 4(y + 2z)]\}$

In Exercises 17–30, enclose the last two terms in parentheses preceded by (a) a plus sign and (b) a minus sign.

17. $a + b - c + 2d$
18. $2 + 3x + 2a - 5$
19. $x - 2y - 3x^2 - 2y^2$
20. $5 - 3x + 2y$
21. $7 + 2a - 5b$
22. $3x - 7y - 3 + c$
23. $6w + 5z - 3x + 5y$
24. $5r^2 - 7s^2 + 2a - 1$
25. $7x^2y + 2xy^2 - 6xz + 5yz$
26. $2pq^2 - 3q^2p - 9pq + 3$
27. $-3x^2 + 2y^2 - 7z^2 - 8w^2$
28. $-9 - x^2y - xy^2$
29. $-5x - 7a - 7b$
30. $-6y + 2a + 5b$

31. Simplify the following expression, which comes from a problem in electrical circuits: $6R_1 + 3R_2 - (4R_1 - R_2)$.
32. Simplify the following expression, which arose in a problem involving forces: $-(F_1 - 2F_2) - (3F_1 + 5F_2)$.
33. Write the following expression for the current in a circuit with a minus sign in front: $-R_1I_1 + R_2I_2 - R_3I_3$.
34. Repeat Exercise 33 for the expression $-3R_1I_1 - 2R_2I_2 + 5R_3I_3$.

1.7 Exponents and Radicals

The purpose of this section is to introduce the basic operations with exponents and radicals. A more detailed study will be undertaken in Chapter 10.

1. Positive Integral Exponents

Recall the definition of exponent from Section 1.4:

$$a^n = a \cdot a \cdot a \cdot \ldots \cdot a \quad (n \text{ factors})$$

This definition leads at once to the laws of exponents for multiplication and division. For example,

$$a^2 \cdot a^4 = (a \cdot a)(a \cdot a \cdot a \cdot a) = a^{2+4} = a^6$$

Similarly,

$$\frac{a^6}{a^3} = \frac{\cancel{a} \cdot \cancel{a} \cdot \cancel{a} \cdot a \cdot a \cdot a}{\cancel{a} \cdot \cancel{a} \cdot \cancel{a}} = a^{6-3} = a^3$$

Finally,

$$(a^2)^4 = a^2 \cdot a^2 \cdot a^2 \cdot a^2 = a^{4 \cdot 2} = a^8$$

The basic laws of exponents are summarized next.

Laws of exponents: If m and n are positive integers,

$$a^m a^n = a^{m+n} \tag{1.2}$$

$$\frac{a^m}{a^n} = a^{m-n} \quad (m > n, \; a \neq 0) \tag{1.3}$$

$$(a^n)^m = a^{nm} \tag{1.4}$$

$$(ab)^n = a^n b^n \tag{1.5}$$

$$\left(\frac{a}{b}\right)^n = \frac{a^n}{b^n} \quad (b \neq 0) \tag{1.6}$$

These rules can be used to simplify certain algebraic expressions. For example, the expression $2x^3 x^4$ can be written in simpler form by adding exponents: $2x^3 x^4 = 2x^{3+4} = 2x^7$. Consider another example.

Example 1

1. $a^2 a^8 = a^{2+8} = a^{10}$ adding exponents

2. $\dfrac{2x^6}{3x^4} = \dfrac{2}{3} x^{6-4} = \dfrac{2}{3} x^2$ subtracting exponents

1.7 EXPONENTS AND RADICALS   **19**

3. $(a^3)^5 = a^{3\cdot 5} = a^{15}$ multiplying exponents
4. $(-2x)^3 = (-2)^3 x^3$ since $(ab)^n = a^n b^n$
$ = -8x^3$ $(-2)^3 = -8$
5. $\left(\dfrac{3}{x}\right)^3 = \dfrac{3^3}{x^3} = \dfrac{27}{x^3}$ since $\left(\dfrac{a}{b}\right)^n = \dfrac{a^n}{b^n}$

Example 2   Simplify the following expressions:

    1. $\left(\dfrac{2x}{y}\right)^5$ 2. $\dfrac{-3(x^3 y^2)^3}{6(xy^2)^2}$

Solution.   1. $\left(\dfrac{2x}{y}\right)^5 = \dfrac{(2x)^5}{y^5}$ since $\left(\dfrac{a}{b}\right)^n = \dfrac{a^n}{b^n}$

$\phantom{\left(\dfrac{2x}{y}\right)^5} = \dfrac{32 x^5}{y^5}$ since $(ab)^n = a^n b^n$

2. $\dfrac{-3(x^3 y^2)^3}{6(xy^2)^2} = \dfrac{-3(x^3)^3 (y^2)^3}{6 x^2 (y^2)^2}$ since $(ab)^n = a^n b^n$

$\phantom{\dfrac{-3(x^3 y^2)^3}{6(xy^2)^2}} = \dfrac{-3 x^9 y^6}{6 x^2 y^4}$ multiplying exponents

$\phantom{\dfrac{-3(x^3 y^2)^3}{6(xy^2)^2}} = \dfrac{-x^7 y^2}{2}$ subtracting exponents

$\phantom{\dfrac{-3(x^3 y^2)^3}{6(xy^2)^2}} = -\dfrac{x^7 y^2}{2}$ $\dfrac{-a}{b} = \dfrac{a}{-b} = -\dfrac{a}{b}$

Common error   Failing to distinguish between $(-ax)^2$ and $-(ax)^2$. Note that

$$(-ax)^2 = (-ax)(-ax) = a^2 x^2$$

but

$$-(ax)^2 = -1(ax)^2 = -a^2 x^2$$

So $(-ax)^2 \neq -(ax)^2$.

Example 3 $(-x^2)^7 = -x^{14}$

but

$$(-x^2)^6 = x^{12}$$

In other words, an even power of a negative number is positive and an odd power of a negative number is negative.

Example 4

1. $(-3a)^2 = 9a^2$, but $(-3a)^3 = -27a^3$
2. $(-ab^2)^4 = a^4b^8$, but $(-ab^2)^5 = -a^5b^{10}$
3. $(-a^2b^3)^4 = a^8b^{12}$, but $-(a^2b^3)^4 = -1(a^2b^3)^4 = -a^8b^{12}$

2. Radicals

Radical

Square root

The other important concept to be discussed briefly here is that of **radical** or **root**. Consider the problem of what number multiplied by itself equals 16. This number, namely 4, is called the **square root** of 16, and is written

$$\sqrt{16} = 4$$

n^{th} root

Similarly, $\sqrt[3]{8}$, read "cube root of 8," is the number whose cube is equal to 8. The **n^{th} root** of a is denoted by $\sqrt[n]{a}$. If $n = 2$, we omit the index and write $\sqrt{a}$.

Remark. The square root of 16 could quite reasonably be given as 4 or -4, and there are times when both roots are important. However, whenever the symbol $\sqrt{\ }$ is employed, by convention only the positive root is intended. Thus

$$\sqrt{16} = 4, \quad \sqrt{16} \neq -4$$

Principal square root

The positive square root is known as the **principal square root.** Similar conventions apply to the nth root. Thus $\sqrt[4]{16} = 2$, not -2, but $\sqrt[5]{-32} = -2$, since $(-2)^5 = -32$, while $2^5 = 32$.

As in the case of exponents, there exist several laws of operations with radicals, some of which will be introduced in this section. However, the justification for these laws must be postponed until Chapter 10, which contains a more elaborate discussion of exponents and radicals.

Laws of radicals

$$\sqrt[n]{ab} = \sqrt[n]{a} \cdot \sqrt[n]{b} \tag{1.7}$$

$$\sqrt[n]{\frac{a}{b}} = \frac{\sqrt[n]{a}}{\sqrt[n]{b}} \quad (b \neq 0) \tag{1.8}$$

$$(\sqrt[n]{x})^n = \sqrt[n]{x^n} = x \tag{1.9}$$

At this point we use these laws only to simplify certain algebraic expressions. For example, since $50 = 25 \cdot 2$ and $\sqrt{25} = 5$, we have by rule (1.7)

$$\sqrt{50} = \sqrt{25 \cdot 2} = \sqrt{25}\sqrt{2} = 5\sqrt{2}$$

1.7 EXPONENTS AND RADICALS

Example 5 Simplify the following radicals:

   a. $\sqrt{48}$ **b.** $\sqrt{125}$ **c.** $\dfrac{1}{\sqrt{40}}$

Solution.   **a.** Since $48 = 16 \cdot 3$ and $\sqrt{16} = 4$, we get
$$\sqrt{48} = \sqrt{16 \cdot 3} = \sqrt{16}\,\sqrt{3} = 4\sqrt{3}$$
   b. $\sqrt{125} = \sqrt{25 \cdot 5} = \sqrt{25}\,\sqrt{5} = 5\sqrt{5}$

   c. $\dfrac{1}{\sqrt{40}} = \dfrac{1}{\sqrt{4 \cdot 10}} = \dfrac{1}{2\sqrt{10}}$

Rationalizing the denominator

We can use law (1.9) to perform an operation called **rationalizing the denominator**: A fraction of the form

$$\dfrac{1}{\sqrt{x}}$$

can be written without a radical in the denominator. To eliminate the radical, we **multiply numerator and denominator by** $\sqrt{x}$. (Multiplying both numerator and denominator of a fraction by the same number does not affect the value of the fraction.) Observe that

$$\dfrac{1}{\sqrt{x}} = \dfrac{1 \cdot \sqrt{x}}{\sqrt{x} \cdot \sqrt{x}} = \dfrac{\sqrt{x}}{(\sqrt{x})^2} = \dfrac{\sqrt{x}}{x}$$

since $(\sqrt[n]{x})^n = x$. For example,

$$\dfrac{1}{\sqrt{2}} = \dfrac{1}{\sqrt{2}} \cdot \dfrac{\sqrt{2}}{\sqrt{2}} = \dfrac{\sqrt{2}}{(\sqrt{2})^2} = \dfrac{\sqrt{2}}{2}$$

Example 6 Rationalize the denominator in each of the following fractions:

   a. $\dfrac{1}{\sqrt{10}}$ **b.** $\dfrac{1}{2\sqrt{x}}$ **c.** $\dfrac{3}{2\sqrt{5}}$ **d.** $\dfrac{2}{\sqrt{2}}$

Solution.   **a.** $\dfrac{1}{\sqrt{10}} = \dfrac{1 \cdot \sqrt{10}}{\sqrt{10}\,\sqrt{10}} = \dfrac{\sqrt{10}}{(\sqrt{10})^2} = \dfrac{\sqrt{10}}{10}$

   b. $\dfrac{1}{2\sqrt{x}} = \dfrac{1 \cdot \sqrt{x}}{2\sqrt{x}\,\sqrt{x}} = \dfrac{\sqrt{x}}{2(\sqrt{x})^2} = \dfrac{\sqrt{x}}{2x}$

   c. $\dfrac{3}{2\sqrt{5}} = \dfrac{3\sqrt{5}}{2\sqrt{5}\,\sqrt{5}} = \dfrac{3\sqrt{5}}{2(\sqrt{5})^2} = \dfrac{3\sqrt{5}}{2(5)} = \dfrac{3\sqrt{5}}{10}$

   d. $\dfrac{2}{\sqrt{2}} = \dfrac{2\sqrt{2}}{\sqrt{2}\,\sqrt{2}} = \dfrac{2\sqrt{2}}{(\sqrt{2})^2} = \dfrac{2\sqrt{2}}{2} = \sqrt{2}$

22 CHAPTER 1 FUNDAMENTAL OPERATIONS

Caution. When writing $\sqrt{a}$, we assume that a is positive or zero, since we cannot find the square root of a negative number. For example, there is no (real) number equal to $\sqrt{-4}$. Thus $\sqrt{x-3}$ is valid only for $x \geq 3$, and $\sqrt{1-x}$ is valid only for $x \leq 1$.

Complex number   Numbers of the form $a + b\sqrt{-1}$ are called *complex numbers* and will be studied in detail in Chapter 11. The square root of a negative number such as $\sqrt{-4}$ is called an *imaginary number*.

Example 7   The frequency of a series resonance circuit is given by $f = 1/(2\pi\sqrt{LC})$. Write the expression for f without a radical in the denominator.

Solution. $f = \dfrac{1}{2\pi\sqrt{LC}} \cdot \dfrac{\sqrt{LC}}{\sqrt{LC}} = \dfrac{\sqrt{LC}}{2\pi(\sqrt{LC})^2} = \dfrac{\sqrt{LC}}{2\pi LC}$

3. Zero and Negative Exponents.*

If the laws of exponents are applied to a^4/a^4, we get

$$\frac{a^4}{a^4} = a^{4-4} = a^0$$

On the other hand,

$$\frac{a^4}{a^4} = 1$$

since any number divided by itself is 1. These results suggest the definition

$$a^0 = 1, \quad a \neq 0 \tag{1.10}$$

Similarly,

$$\frac{a^2}{a^5} = a^{2-5} = a^{-3} \quad \text{and} \quad \frac{a^2}{a^5} = \frac{1}{a^3}$$

This suggests that $a^{-3} = 1/a^3$. In general,

$$a^{-n} = \frac{1}{a^n}, \quad a \neq 0 \tag{1.11}$$

---

* Zero and negative exponents are needed at this time only for the discussion of scientific notation in Section 1.11. Consideration of these topics may be postponed until Chapter 10.

The rules for multiplication and division hold for negative exponents. Consider the following example:

Example 8

a. $x^0 x^4 = x^{0+4} = x^4$ adding exponents

b. $\dfrac{a^5}{a^0} = a^{5-0} = a^5$ subtracting exponents

c. $a^5 a^{-3} = a^{5+(-3)} = a^2$ adding exponents

d. $\dfrac{a^5}{a^{-3}} = a^{5-(-3)} = a^8$ subtracting exponents

e. $\dfrac{b^{-4} b^3}{b^{-5}} = b^{-4+3-(-5)} = b^4$

Other operations with negative exponents will be discussed in Chapter 10.

Notation. So far we have used letters such as x, y, and z as variables. In technology, however, it is customary to use letters that suggest the meaning of physical quantities directly. For example, the formula for the power delivered to a resistor is given by $P = RI^2$. Here P stands for the power in watts, I for the current in amperes, and R for the resistance in ohms. The use of these letters makes the formula easier to comprehend and to remember. *From now on many of the exercises will contain notations that are frequently encountered in technical fields.*

Exercises / Section 1.7

In Exercises 1–48, simplify each of the given expressions.

1. $(-2)^3$
2. $\left(-\dfrac{1}{4}\right)^4$
3. $\sqrt[4]{16}$
4. $\sqrt[5]{32}$
5. $x^2 x^6$
6. $2x(x^3)$
7. $-2(xy)^3$
8. $(-x)^4$
9. $\dfrac{x^{10}}{x^5}$
10. $\dfrac{4y^5}{2y^2}$
11. $\dfrac{-4z^5}{8z}$
12. $\dfrac{-3x^{12}}{-2x^{12}}$
13. $\dfrac{-(2x)^3}{(-2x)^3}$
14. $\dfrac{(-3x)^2}{-(3x)^2}$
15. $\dfrac{(-x)^3 y^6}{-2x^3 y^6}$
16. $\dfrac{-4x^6 y^6}{-2x^6 y^5}$
17. $\dfrac{(-2x)^2 y^3}{(-2x)^3 y^2}$
18. $\dfrac{-2(x^2 y^3)^3}{3(xy^2)^2}$
19. $\dfrac{x^4}{x^5}$
20. $\dfrac{2x^6}{-x^{10}}$
21. $\dfrac{-10 x^6 y^3}{-5 x^8 y}$

22. $\dfrac{-(xy)(-2x^2y^2)}{2x^3y^3}$
23. $\dfrac{(-5ab)(-10ab)}{25a^5b^4}$
24. $\dfrac{(-2x)(-b)^3}{3x^2b}$

25. $\sqrt{8}$
26. $\sqrt{24}$
27. $\sqrt{54}$

28. $\sqrt{18}$
29. $\sqrt{75}$
30. $\sqrt{72}$

31. $\sqrt[3]{16}$
32. $\sqrt[4]{32}$
33. $\sqrt[5]{64}$

34. $\sqrt[3]{32}$
35. $\sqrt{32}$
36. $\sqrt{128}$

37. $\dfrac{(-\omega^2)\alpha^3}{-5\omega^2(-\alpha)^4}$
38. $\dfrac{-14L_1{}^2(-L_2)^3}{7L_1L_2{}^2}$
39. $\dfrac{(4A_1)^2(-A_2)^4}{(-2A_1A_2)^2}$

40. $\dfrac{(-I_1)^5}{(-2I_1)^3 I_2}$
41. $\dfrac{(\sqrt{3}v_1)^2(-v_2)^2}{3v_1v_2{}^2}$
42. $\dfrac{(-2xy^2)(-2x^2y^3)^3}{(x^3y)^4}$

43. $\dfrac{(-a^3b^4)^2(-2a^4b^2c)^3}{-(a^2bc)^3}$
44. $\dfrac{(-xyz)^2(-xyz)^3}{-3x^2(yz)^2}$
45. $\dfrac{(\sqrt{2}xy^2)^2(\sqrt{3}xy)}{\sqrt{3}x^2y^2}$

46. $\dfrac{(2x^3y^6)^4(x^2y)^2}{(-x^2y^2)^3}$
47. $\dfrac{(v_1v_2)^2(-2v_1v_2)}{(-v_1v_2)^2}$
48. $\dfrac{(-i_1i_2)^2(-i_1i_2)}{3i_1{}^2i_2{}^3}$

In Exercises 49–60, simplify the given radicals by rationalizing denominators.

49. $\dfrac{1}{\sqrt{5}}$
50. $\dfrac{2}{\sqrt{7}}$
51. $\dfrac{3}{\sqrt{3}}$

52. $\dfrac{1}{2\sqrt{3}}$
53. $\dfrac{a}{\sqrt{b}}$
54. $\dfrac{a}{b\sqrt{c}}$

55. $\dfrac{2}{3\sqrt{5}}$
56. $\dfrac{1}{\sqrt{8}}$
57. $\dfrac{2}{\sqrt{32}}$

58. $\dfrac{4}{\sqrt{20}}$
59. $\dfrac{3}{\sqrt{12}}$
60. $\dfrac{2}{\sqrt{50}}$

61. A box in the shape of a cube has a volume of 729 cubic centimeters. Find the length of one side.

62. The frequency, or number of vibrations per unit time, of a mass hanging on a spring (Figure 1.4) is given by

$$f = \dfrac{1}{2\pi}\sqrt{\dfrac{k}{m}}$$

where m is the mass and k is a constant. Simplify this expression by rationalizing the denominator.

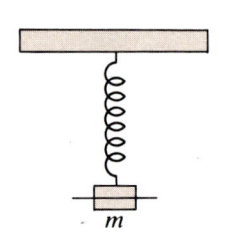

Figure 1.4

63. In determining the velocity of a water wave, the expression $\sqrt{4\pi/ad}$ has to be simplified. Rationalize the denominator and simplify.

64. Simplify the following expression from a problem in mechanics:

$$\frac{(3m^2v)^3}{(4m^2v)^2}$$

65. The area of a triangle with sides of length a, b, and c is

$$A = \sqrt{s(s-a)(s-b)(s-c)}$$

where $a + b + c = 2s$. Find the area of a triangle with sides of length 7, 9, and 12 centimeters.

In Exercises 66–82, simplify the expression (see Example 8).

66. $a^0 a^5$

67. $\dfrac{a^6}{a^0}$

68. $a^0 x^2$

69. $a^{-2} a^6$

70. $x^{-4} x^7$

71. $a^{-4} a^5$

72. $\dfrac{x^{-2}}{x^{-8}}$

73. $\dfrac{x^2}{x^{-4}}$

74. $\dfrac{3x^4}{x^{-1}}$

75. $\dfrac{4a^0 x^{-2}}{x^{-5}}$

76. $\dfrac{5x^0 y^{-1}}{y^{-2}}$

77. $\dfrac{6a^{-2} y^0}{a^{-4}}$

78. $\dfrac{2x^2 y^3}{x^{-2} y^{-2}}$

79. $\dfrac{7p^{-1} q^{-1}}{p^{-5} q^{-3}}$

80. $\dfrac{3R_1^{-2} R_2}{R_1^{-2} R_2^{-1}}$

81. $\dfrac{9V_1^{-4} V_2}{V_1^{-4} V_2^{-2}}$

82. $\dfrac{L^0 L^3 M^{-2}}{L^{-1} M^{-4}}$

1.8 Multiplication of Polynomials

Earlier we considered the addition and subtraction of polynomials. We will complete our study of the fundamental operations by taking up multiplication in this section and division in the next.

The basic rules of multiplication and division of single terms can be applied to polynomials by means of the distributive law $a(b + c) = ab + ac$. In particular, to multiply a polynomial by a monomial, we multiply *each* term of the polynomial by the monomial. For example,

Multiplication by a monomial

$$4(2x + y) = 4(2x) + 4(y) = 8x + 4y$$

Consider the following example.

Example 1 Multiply $-2x(x^2 - 5)$.

Solution. Since $a(b + c) = ab + ac$, we have

$$-2x(x^2 - 5) = (-2x)(x^2) + (-2x)(-5)$$
$$= -2x^3 + 10x$$

(Always place either a plus or a minus sign between terms.)

Remark. It is probably simpler to skip the intermediate step and to multiply directly, since direct multiplication avoids possible confusion on the signs. In other words, it is easy to see that $(-2x)(x^2) = -2x^3$ and $(-2x)(-5) = 10x$. So it follows in just one step that

$$-2x(x^2 - 5) = -2x^3 + 10x$$

Polynomials can be multiplied by the rule given next.

> **Multiplication of polynomials:** To multiply two polynomials, multiply each term of one polynomial by each term of the other and add the results.

As an example, to multiply

$$(x - 2y)(2x + y)$$

the expression $x - 2y$ may be treated as a single number. In other words, in the distributive law $a(b + c) = ab + ac$, let

$$a = x - 2y, \quad b = 2x, \quad \text{and} \quad c = y$$

Then

$$(x - 2y)(2x + y) = (x - 2y)(2x) + (x - 2y)(y)$$

Now apply the distributive law again to obtain

$$(2x^2 - 4xy) + (xy - 2y^2)$$

so that

$$(x - 2y)(2x + y) = 2x^2 - 4xy + xy - 2y^2$$
$$= 2x^2 - 3xy - 2y^2$$

To make this procedure more systematic, let us arrange our work so that it resembles ordinary multiplication in arithmetic. First, let us write the polynomials in two rows:

$$\begin{array}{r} x - 2y \\ \underline{2x + y} \end{array}$$

Now, working from left to right, multiply $2x$ (in the second row) by both terms in the first row and write the result in the first row below the line:

$$\begin{array}{r} x - 2y \\ \underline{2x + y} \\ 2x^2 - 4xy \end{array}$$

Next, multiply $+y$ (in the second row) by both terms in the first row and write the result in the next row under the line, taking care to place like terms in the same column to facilitate the addition step:

$$\begin{array}{r} x - 2y \\ 2x + y \\ \hline 2x^2 - 4xy \\ xy - 2y^2 \\ \hline 2x^2 - 3xy - 2y^2 \end{array} \qquad xy = 1xy$$

after drawing another line and adding.

Example 2 Multiply $(2x - 5y)(-x + 2y)$.

Solution. First we write the polynomials in two rows:

$$\begin{array}{c} 2x - 5y \\ -x + 2y \end{array}$$

Working from left to right, starting with $-x$, the multiplication is carried out as follows:

$$\begin{array}{r} 2x - 5y \\ -x + 2y \\ \hline -2x^2 + 5xy \\ 4xy - 10y^2 \\ \hline -2x^2 + 9xy - 10y^2 \end{array}$$

Example 3 Multiply the following polynomials:

$$\begin{array}{r} 3x^2 - 2xy + y^2 \\ 2x - 3y \\ \hline 6x^3 - 4x^2y + 2xy^2 \\ - 9x^2y + 6xy^2 - 3y^3 \\ \hline 6x^3 - 13x^2y + 8xy^2 - 3y^3 \end{array}$$

The characteristic pattern (in particular, the indented row of intermediate products) is always obtained if the terms in the polynomials appear in *descending* powers of x and in *ascending* powers of y. For example,

$$2x^3 - 3x^2y + 2xy^2 - 5y^3$$

has x^3 in the first term, x^2 in the second, x in the third, and no x in the fourth. The powers of y follow the opposite order. If the polynomials are not in this form, then the pattern becomes disrupted.

Example 4

$$\begin{array}{rrrrr} 2x^2 & + & xy & + & z \\ & & x & - & y \\ \hline 2x^3 & + & x^2y & + & xz \\ & - & 2x^2y & & & - xy^2 - yz \\ \hline 2x^3 & - & x^2y & + & xz - xy^2 - yz \end{array}$$

Note that unlike terms are placed in separate columns.

Example 5 Multiply $(3x^2 - xy + 2y^2)(-2x^2 + 4xy - y^2)$.

Solution. After writing the polynomials in two rows, we work again from left to right, starting with $-2x^2$. Note that the intermediate products take up three rows.

$$\begin{array}{rrrrrr} 3x^2 & - & xy & + & 2y^2 & \\ -2x^2 & + & 4xy & - & y^2 & \\ \hline -6x^4 & + & 2x^3y & - & 4x^2y^2 & \\ & & 12x^3y & - & 4x^2y^2 & + 8xy^3 \\ & & & - & 3x^2y^2 & + xy^3 - 2y^4 \\ \hline -6x^4 & + & 14x^3y & - & 11x^2y^2 & + 9xy^3 - 2y^4 \end{array}$$

Example 6 A circuit has a variable resistance $R = 2.5t + 1.6$ (in ohms), where t is measured in seconds. If the current in the circuit is $I = 0.01t - 0.02$ (in amperes), find the voltage $V = IR$ across the resistor.

Solution. From $V = IR$, we get

$$V = (0.01t - 0.02)(2.5t + 1.6)$$
$$= (0.01)(2.5)t^2 - (0.02)(2.5)t + (0.01)(1.6)t - (0.02)(1.6)$$
$$= 0.025t^2 - 0.05t + 0.016t - 0.032$$
$$= 0.025t^2 - 0.034t - 0.032 \text{ (in volts)}$$

Exercises / Section 1.8

In Exercises 1–60, multiply the given expressions.

1. $(-3a^2b^3)(-4ab)$
2. $(-x^2y^2)(-xy)(3xy^3)$
3. $(2ab)(-3a^3b^4)(-4x)$
4. $(abc)(abc)(-abc)$

5. $xy(3x + 2xy)$
6. $2y(-3x^2 + 10xy)$
7. $-5x^2y^2(6xy^2 - 4x^2y)$
8. $-10ab^2(2ab + 5a^3b)$
9. $ab(2a^2b + 3ab - 4ab^2)$
10. $-5xyz(3x^2y^2z^2 - 15x^3y^3z^3)$
11. $-2a^2x(6a^2b - 5a^3x + 2ab)$
12. $axy(axy - 2a^2x^2y^2 - 3ax)$
13. $2xy(3xy - 5x^2y^2) - 5xy(-2xy + 10x^2y^2)$
    (Combine like terms after multiplying.)
14. $ab(-ab + ab^2) - ab(2ab - 5ab^2)$
15. $a^2b(2a^3b - 3ab^2) - ab^2(-5a^4 + 2a^2b)$
16. $-2xy(5x^2y - 2xy^2) - x^2y^2(x + 2y)$
17. $(x - 1)(x - 2)$
18. $(x + 2)(x - 5)$
19. $(2x - 1)(x - 3)$
20. $(3x - 7)(2x + 2)$
21. $(2x + 2)(5x - 7)$
22. $(x - 5)(2x + 2)$
23. $(x - y)(x + y)$
24. $(x - y)^2$
25. $(2x - 1)(2x + 1)$
26. $(2x - y)(2x + y)$
27. $(3x - 2y)(3x + 2y)$
28. $(4x + y)(4x - y)$
29. $(3x - 1)^2$
30. $(3x + y)^2$
31. $(2x - 2y)^2$
32. $(5x + 1)^2$
33. $(2x + y)^2$
34. $(x + 2b)(x - 2b)$
35. $(x^2 - y^2)(x^2 + y^2)$
36. $(2x^2 + y^2)(2x^2 - y^2)$
37. $(x - 2b)(5x - b)$
38. $(2x - z)(3x + 2z)$
39. $(x - 1)(x^2 - 2x - 3)$
40. $(2x + 2)(x^2 + 5x - 4)$
41. $(a - 3b)(a^2 - ab + b^2)$
42. $(c - 2d)(c^2 - 3cd + 5d^2)$
43. $(5x - 3y)^2$
44. $(3a^2 + 4b^2)^2$
45. $(5x - 2y - 3z)(x - y + z)$
46. $(3a^2 - 2ab - 5b^2)(a^2 - ab + b^2)$
47. $(4x^4 - 2x^2 + 5)(2x^4 - x^2 - 1)$
48. $(2a^4 - 3a^2 - 1)(a^4 - 6a^2 + 3)$
49. $(2x^3y - 4x^2y^2 - xy^3)(x^2y + 2xy^2 - 2y^3)$
50. $(2ab + 3ac - 4bc)(3ab - ac - 2bc)$
51. $(x^4 - x^3 + x^2 - 1)(x + 1)$
52. $(x^3 - x^2y + 2xy^2 + 2y^3)(x^2 + xy - y^2)$
53. $(x + y + z)^2$
54. $(a - b - c)^2$
55. $(2a - 3b + 2c)^2$
56. $(-a - 4c - 6d)^2$
57. $3(l - 2w)(l + 3w)$
58. $(I_1 + I_2)^2 - 2(I_1 + I_2)^2$
    (Combine like terms after multiplying.)
59. $3(R_1 + R_2)(2R_1 + 4R_2)$
60. $(2i_1 + 3i_2 + i_3)^2$

61. Find an expression for the area inside the concrete form shown in Figure 1.5.
62. Find an expression for the volume illustrated in Figure 1.6.
63. The voltage (in volts) across a resistor is given by $V = IR$, where I is measured in amperes and R in ohms. If $I = 0.1t - 2.0$ and $R = 0.4t + 9.0$ (t measured in seconds), find V.
64. The power P (in watts) delivered to a resistor is given by $P = RI^2$, where I is in amperes and R in ohms. If $R = 10$ ohms and $I = 2.0t^2 - 4.0t + 1.0$, find P.

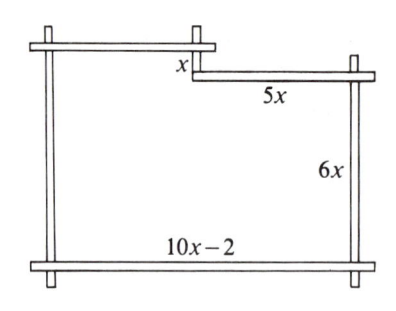

Figure 1.5

Figure 1.6

65. The energy E radiated by a blackbody can be written $E = k(T^2 + T_0^2)(T - T_0)(T + T_0)$, where T is the temperature of the body, T_0 the temperature of the surrounding medium, and k a constant. Simplify the expression for E.

66. Simplify the following expression for the deflection of a beam: $d = (L - ax/2)(L - ax^2)$.

67. In computing the center of mass of a plate of uniform density, the expression $\frac{1}{2}(y_2 - y_1)(y_2 + y_1)$ has to be simplified. Carry out the simplification.

1.9 Division of Polynomials

Division of polynomials can be carried out by a scheme analogous to that for multiplication: We set up the procedure so that it resembles long division in arithmetic. But first we need to discuss the simpler case of division by a monomial. Consider, for example,

$$(4x^3 - 8x^2) \div 2x, \quad \text{or} \quad \frac{4x^3 - 8x^2}{2x}$$

By the distributive law, this problem is equivalent to

$$\frac{1}{2x}(4x^3 - 8x^2) = \frac{4x^3}{2x} - \frac{8x^2}{2x} = 2x^2 - 4x$$

Division by a monomial — In summary, the quotient of a polynomial and a monomial is found by dividing each term of the polynomial by the monomial and adding the results. The division should be carried out in one step to avoid sign errors.

Example 1

$$\frac{15a^3b^4 - 20a^6b^7}{-5a^2b} = -3ab^3 + 4a^4b^6$$

To carry out the division

$$(x^3 - 4x^2 + x + 2) \div (x - 1)$$

we set up the division problem as follows:

$$x - 1 \overline{)x^3 - 4x^2 + x + 2}$$

(Both polynomials should be written in descending powers of x.) Next, we divide x in the divisor into the first term, x^3, of the dividend and place the resulting quotient, x^2, on top. Now we multiply x^2 by the entire divisor and place the product in the row below:

$$\begin{array}{r} x^2 \\ x-1\overline{\smash{\big)}x^3 - 4x^2 + x + 2} \\ x^3 - x^2 \end{array}$$

As in regular long division, we now subtract. (Remember to change the signs of the terms in the subtrahend.) After subtracting we bring down the next term:

$$\begin{array}{r} x^2 \\ x-1\overline{\smash{\big)}x^3 - 4x^2 + x + 2} \\ \underline{x^3 - x^2} \\ -3x^2 + x \end{array}$$

The procedure is now repeated. We divide x into the first term of the last row to obtain $-3x$, the next term in the quotient.

$$\begin{array}{r} x^2 - 3x \\ x-1\overline{\smash{\big)}x^3 - 4x^2 + x + 2} \\ \underline{x^3 - x^2} \\ -3x^2 + x \\ \underline{-3x^2 + 3x} \end{array}$$

Note that $-3x$ has been multiplied by the entire divisor and the result placed in a new row. As before, we now subtract and bring down the next term:

$$\begin{array}{r} x^2 - 3x \\ x-1\overline{\smash{\big)}x^3 - 4x^2 + x + 2} \\ \underline{x^3 - x^2} \\ -3x^2 + x \\ \underline{-3x^2 + 3x} \\ -2x + 2 \end{array}$$

Finally, we divide x into $-2x$ to obtain -2, the last term of the quotient.

Division procedure

$$\begin{array}{r} x^2 - 3x - 2 \\ x-1\overline{\smash{\big)}x^3 - 4x^2 + x + 2} \\ \underline{x^3 - x^2} \\ -3x^2 + x \\ \underline{-3x^2 + 3x} \\ -2x + 2 \\ \underline{-2x + 2} \\ 0 \end{array}$$

Since the remainder is 0, we conclude that

$$(x^3 - 4x^2 + x + 2) \div (x - 1) = x^2 - 3x - 2$$

Example 2 Carry out the following division:

$$(2x^3 + 2y^3 + 2xy^2 - 6x^2y) \div (2x - 2y)$$

Solution. Arranging the terms in descending powers of x, we get

$$
\begin{array}{r}
x^2 - 2xy - y^2 \\
2x - 2y \overline{\smash{)}2x^3 - 6x^2y + 2xy^2 + 2y^3} \\
\underline{2x^3 - 2x^2y } \\
-4x^2y + 2xy^2 \\
\underline{-4x^2y + 4xy^2 } \\
-2xy^2 + 2y^3 \\
\underline{-2xy^2 + 2y^3} \\
0
\end{array}
$$

If one of the powers in the dividend is missing, we need to leave a space, as shown in the next example. (A missing term merely indicates a coefficient of 0.)

Example 3 Divide: $(4x^3 - 5x + 2) \div (2x - 1)$.

Solution. Note that the x^2 term is missing.

$$
\begin{array}{r}
2x^2 + x - 2 \\
2x - 1 \overline{\smash{)}4x^3 - 5x + 2} \\
\underline{4x^3 - 2x^2 } \\
2x^2 - 5x \\
\underline{2x^2 - x } \\
-4x + 2 \\
\underline{-4x + 2} \\
0
\end{array}
$$

Recall from arithmetic that whenever the remainder is not 0, we divide the remainder by the divisor and add the resulting fraction to the quotient.

Example 4

$$\begin{array}{r}2x^2 - xy - y^2\\x^2 - 3xy - 2y^2\overline{)2x^4 - 7x^3y - 2x^2y^2 + 5xy^3 + y^4}\\\underline{2x^4 - 6x^3y - 4x^2y^2}\\-x^3y + 2x^2y^2 + 5xy^3\\\underline{-x^3y + 3x^2y^2 + 2xy^3}\\-x^2y^2 + 3xy^3 + y^4\\\underline{-x^2y^2 + 3xy^3 + 2y^4}\\-y^4\end{array}$$

The resulting expression is

$$2x^2 - xy - y^2 + \frac{-y^4}{x^2 - 3xy - 2y^2}$$

or

$$2x^2 - xy - y^2 - \frac{y^4}{x^2 - 3xy - 2y^2}$$

Exercises / Section 1.9

In Exercises 1–12, find each of the quotients.

1. $\dfrac{a^4}{a^2}$

2. $\dfrac{3a^6b^5}{ab}$

3. $\dfrac{8x^4y^2}{-4x^2y}$

4. $\dfrac{-10a^6x^5y^4}{-5ax^2}$

5. $(-8x^6 - 4x^4) \div 2x^2$

6. $(24a^2b^2 - 12a^3b^2) \div a^2b^2$

7. $(-12abc + 9a^2bc^2) \div (-3ab)$

8. $(-36ab - 27a^2b^2 + 18ab^2) \div (9ab)$

9. $\dfrac{-17x^2y^3 + 34x^3y^5 - 51x^5y^5}{-17x^2y^3}$

10. $\dfrac{25x^2y^2z^2 - 15x^3y^3z^3}{-5x^2z}$

11. $\dfrac{12x^7y^6a^5 + 6x^6y^4a^3 - 9x^5y^3a^7}{3x^2y^3a^3}$

12. $\dfrac{5a^2y^2z - 10a^3yz^2 + 12a^4y^3z}{a^2yz}$

In Exercises 13–40, divide the first expression by the second.

13. $6x^2 - x - 2$, $\quad 2x + 1$
14. $3x^2 + 10x - 8$, $\quad x + 4$
15. $2x^2 - 5x - 3$, $\quad x - 3$
16. $8x^2 + 2x - 6$, $\quad 2x + 2$
17. $2x^2 - 8$, $\quad x - 2$
18. $3a^2 - 27$, $\quad a - 3$
19. $4x^2 - 4$, $\quad x + 1$
20. $2x^2 - 18$, $\quad 2x + 6$
21. $3x^2 - 5x - 12$, $\quad x - 3$
22. $2x^2 + xy - 6y^2$, $\quad x + 2y$
23. $12x^2 + 18xy - 12y^2$, $\quad 4x - 2y$
24. $6x^2 - 11xy + 4y^2$, $\quad 2x - y$
25. $x^3 - x^2y - xy^2 + y^3$, $\quad x + y$
26. $a^3 - a^2b - 3ab^2 - 9b^3$, $\quad a - 3b$

27. $3a^3 + 5a^2b - 17ab^2 + 5b^3$, $3a - b$
28. $2x^3 + 3x^2y - xy^2 + 2y^3$, $x + 2y$
29. $4x^3 + xy^2 + y^3$, $2x + y$
30. $9x^3 - 13xy^2 - 6y^3$, $3x + 2y$
31. $x^3 + y^3$, $x + y$
32. $x^3 - y^3$, $x - y$
33. $2x^4 + 2x + 2x^2 - 5x^3 - 1$, $x^2 - x - 1$
34. $4x + 1 + 2x^4 - 5x^3$, $2x^2 - 3x - 1$
35. $6x^3 + 7xy^2 - 11x^2y - 6y^3$, $3x^2 + 2y^2 - xy$
36. $x^3y + x^4 + 14xy^3 - 8x^2y^2 - 8y^4$, $x^2 + 3xy - 4y^2$
37. $y^3 - 3xy^2 + x^3$, $x + 2y$
38. $x^3 - 7x^2y - 2y^3$, $x - 3y$
39. $x^4 - 3x^3 - 3x^2 - 6x + 3$, $x - 4$
40. $4x^3 - 3x^2y + 4xy^2 - y^3$, $x + 2y$

1.10 Approximation and Significant Figures

This section briefly discusses rounding off numbers and significant figures. To appreciate the problem, suppose that the measured length of a beam has been rounded off to 32.6 ft. We conclude that the correct length lies between 32.55 ft and 32.65 ft. Thus all three figures contain significant information, and we say that 32.6 has three **significant figures.** If the length was rounded off to 30 ft, then we would know only that the length is closer to 30 ft than it is to 20 ft or 40 ft. So the final zero serves only to place the decimal point and is therefore not significant. Put another way, we can "believe" the 6 in 32.6, but we cannot "believe" the 0 in 30.

In general, the digits 1 through 9 are always significant. The digit 0 may or may not be significant: If its only function is to place the decimal point, then it is not significant. For example, 0.0032 has two significant figures, since the zeros are only placeholders. On the other hand, 1,031.5 has five significant figures, since the digit 0 is not just a placeholder in this case.

A length of 920 ft appears to have only two significant figures, since the true length is only known to lie between 910 ft and 930 ft, but can we always be sure? For example, if the length was really rounded off to the nearest foot, then the zero would indeed be significant; in other words, we could then "believe" the zero. However, unless this information is provided, we must presume that 920 has only two significant figures.

On the other hand, final zeros are significant if they occur to the right of the decimal point. For example, 1.3200 has five significant figures, since the zeros do more than just place the decimal point. The rules for determining significant figures are listed next.

Significant figures

1. The digits 1 through 9 are always significant.
2. Zeros are not always significant:
   a. Zeros are significant if they lie between two significant figures.
   b. Unless otherwise specified, final zeros are significant only if they lie to the right of the decimal point.
   c. Initial (beginning) zeros are not significant.

1.10 APPROXIMATION AND SIGNIFICANT FIGURES

Remark: Our discussion of significant figures applies only to approximate numbers (obtained by measurement), not to exact numbers. For example, since 1 ft = 12 in. by definition, there are exactly 12 inches in a foot.

If you carefully follow the above rules, you should not have any trouble determining the number of significant figures. Study the following example:

Example 1

a. 12.4 has three significant figures. (Nonzero digits are always significant.)
b. 102 has three significant figures, and 102.1 has four significant figures, since the zero lies between two significant figures in each case.
c. 302.00 has five significant figures, since the final zeros lie to the right of the decimal point.
d. 30,200 has only three significant figures. (The final zeros do not lie to the right of the decimal point and are therefore not significant.)
e. 0.013 has two significant figures, since the initial zeros are not significant.
f. 0.0130 has three significant figures. (Note that the final zero is significant since it lies to the right of the decimal point.)
g. 1.0020 has five significant figures.

Accuracy

Significant figures are often discussed in terms of accuracy. The **accuracy** of a given number refers to the number of significant figures. Numbers must frequently be rounded off to a certain level of accuracy, that is, to a certain number of significant figures. In such cases zeros may be needed as placeholders.

Example 2

a. 9,043 rounded to three significant figures is 9,040.
b. 9,046 rounded to three significant figures is 9,050.
c. 999 rounded to one significant figure is 1,000.
d. 31,512 rounded to two significant figures is 32,000.
e. 1.35 rounded to two significant figures is 1.4.
f. 1.256 rounded to three significant figures is 1.26.
g. 1.012 rounded to two significant figures is 1.0.
h. 21.97 rounded to three significant figures is 22.0.

CALCULATOR COMMENT

Since you will probably be using a scientific calculator for most of the applied problems in this book, the use of calculators will be discussed whenever appropriate. You must realize, however, that a calculator often produces answers that contain more digits than the input data warrant. For example, if the dimensions of a room are 10.1 ft by 15.9 ft, then the calculated area of

$$(10.1 \text{ ft}) \times (15.9 \text{ ft}) = 160.59 \text{ ft}^2$$

is more accurate than is warranted by the given information and therefore should be rounded off. Since both dimensions are given to three significant figures, the answer should also be rounded off to three significant figures, namely, 161 ft². Any more accuracy cannot be justified. For example, the measurement of 160.6 ft² may seem acceptable, since both dimensions are expressed to the nearest tenth of a foot. However, 10.1 is only known to lie between 10.05 and 10.15, and 15.9 between 15.85 and 15.95. So the area must lie between

$$(10.05) \times (15.85) = 159.3 \text{ ft}^2$$

and

$$(10.15) \times (15.95) = 161.9 \text{ ft}^2$$

So 160.9 ft² is not correct.

The rule for multiplication can be stated in terms of accuracy:

> When approximate numbers are multiplied or divided, the result is no more accurate than the least accurate number in the input data.

Example 3 The dimensions of a box are 3.2 ft × 0.98 ft × 1.98 ft. Find the volume.

Solution. Using a calculator, we have the following sequence:

3.2 $\boxed{\times}$ 0.98 $\boxed{\times}$ 1.98 $\boxed{=}$

Display: 6.20928

Since the least accurate number is 3.2 (two significant figures), we round off 6.20928 to two significant figures and obtain 6.2 ft³.

Precision

The rule for addition and subtraction can be stated in terms of precision: The **precision** of a number refers to the position of the last significant figure with respect to the decimal point. Thus 10.4 in. is more precise than 104 in.

Example 4 0.0014 ft is more precise than 1.4 ft. However, the accuracy is the same (two significant figures).

The rule for addition and subtraction is given next.

> When approximate numbers are added or subtracted, the result is only as precise as the least precise number.

1.10 APPROXIMATION AND SIGNIFICANT FIGURES

Example 5 Add the following weights:

10.132 lb, 1.95 lb, 3.7 lb, and 0.13 lb

Solution. The sequence is

10.132 $\boxed{+}$ 1.95 $\boxed{+}$ 3.7 $\boxed{+}$ 0.13 $\boxed{=}$

Display: 15.912

Since 3.7 lb has only one decimal place, the answer is rounded off to 15.9 lb.

Exercises / Section 1.10

In Exercises 1–12, round off to the number of significant figures indicated.

1. 3.1416   (four)
2. 0.039   (one)
3. 1.039   (three)
4. 7.173   (three)
5. 0.000155 (two)
6. 1.000155 (six)
7. 985,732   (three)
8. 1,325   (three)
9. 876,512   (two)
10. 13,705   (four)
11. 0.0201   (two)
12. 1.396   (three)

In Exercises 13–36, carry out the indicated operations using a calculator. Round off the answers according to the rules given in this section.

13. $1.35 + 7.94 - 1.987 + 0.128$
14. $0.31 + 4.9 - 1.782$
15. $8.965 + 2.1314$
16. $35.86 + 4.037 - 10.1176$
17. $10.98 - 1.71 + 11.3$
18. $9.99 - 2.651 + 0.3$
19. $(2.1)(5.6)(1.11)$
20. $(0.28)(0.165)(1.1)$
21. $(15.3)(14.6)(0.12)$
22. $\dfrac{100.7}{3.55}$
23. $\dfrac{87.51}{2.71}$
24. $\dfrac{(8.34)(0.73)}{19.3}$
25. $\dfrac{(68.1)(0.3)}{(0.5)}$
26. $\dfrac{(100.3)(500.4)}{(10.11)(3.4456)}$
27. $(606.6)(0.26)$
28. $(98.34)(0.92)$
29. $(22.9)(3.80)$
30. $(0.110)(0.091)$
31. $(2.600)(0.250)$
32. $(3.20)(0.70)$

33. The respective resistances of three resistors connected in series are 15.3 ohms, 19.21 ohms, and 19.0 ohms. Find the resistance of the combination, which is the sum of the individual resistances.

34. Two forces of 9.732 lb and 6.60 lb are acting on an object in opposite directions. Determine the net force, which is (9.732 lb) − (6.60 lb).

35. The power (in watts) developed in a circuit is given by $P = RI^2$. If $I = 0.132$ amperes and $R = 9.6$ ohms, find P.

36. A circular metal plate has a diameter of 10.6 centimeters. Find the area ($A = \pi r^2$).

1.11 Scientific Notation

Many scientific and technical applications involve very large or very small numbers. A convenient way to write such numbers is by means of scientific notation, which is defined below.

> **Scientific notation**: A number in **scientific notation** has the form
>
> $$n \times 10^k, \quad 1 \leq n < 10 \quad \quad (1.12)$$
>
> where k is a positive or negative integer.

Before examining scientific notation in detail, let us consider the following powers of 10:

$$10 = 10^1 \quad \quad \frac{1}{10} = 10^{-1}$$

$$100 = 10^2 \quad \quad \frac{1}{100} = 10^{-2}$$

$$1{,}000 = 10^3 \quad \quad \frac{1}{1{,}000} = 10^{-3}$$

$$10{,}000 = 10^4 \quad \quad \frac{1}{10{,}000} = 10^{-4}$$

To write the number 30,000 in scientific notation, observe that $30{,}000 = (3)(10{,}000) = 3 \times 10^4$. Similarly, $31{,}000 = (3.1)(10{,}000) = 3.1 \times 10^4$.

Now consider a small number such as 0.0054. Written as an ordinary fraction, we get

$$0.0054 = \frac{54}{10{,}000} = \frac{54}{10^4} = \frac{5.4}{10^3} = 5.4 \times 10^{-3}$$

> To write a number in **scientific notation**
>
> $$n \times 10^k, \quad 1 \leq n < 10$$
>
> move the decimal point so that only one nonzero digit remains to the left of the decimal point. The number of places that the decimal point is moved is equal to the exponent k. If the decimal point is moved to the left, k is positive; if the decimal point is moved to the right, k is negative.

Example 1 $246{,}000 = 2.46 \times 10^5$ (decimal point moved five places to the left)
$0.00000129 = 1.29 \times 10^{-6}$ (decimal point moved six places to the right)

To change a number from scientific to regular decimal notation, the above procedure is reversed.

Example 2 $9.3 \times 10^4 = 93{,}000$ (decimal point moved four places)
$2.6 \times 10^{-3} = 0.0026$ (decimal point moved three places)

Scientific notation is particularly useful for indicating significant figures, especially when ambiguous final zeros are involved.

> When a number is written in scientific notation $n \times 10^k$, n indicates the number of significant figures.

For example, **8.30×10^9** has three significant figures.

Example 3   Since

$$3{,}140{,}000 = 3.14 \times 10^6,$$

$n = 3.14$ and indicates three significant figures.

Scientific notation deals with final zeros in a simple way. Consider the next example.

Example 4   The number 3,140,000 in Example 3 is written

     a. 3.140×10^6 to indicate four significant figures.
     b. 3.1400×10^6 to indicate five significant figures.

Multiplication and division can be carried out directly with numbers written in scientific notation, but the final result must be expressed in the proper form.

Example 5 Carry out the following multiplication and express the answer in scientific notation:

$$(5.26 \times 10^{10}) \times (9.35 \times 10^{-4})$$

Solution.
$$\begin{aligned}(5.26 \times 10^{10}) \times (9.35 \times 10^{-4}) &= (5.26)(9.35) \times (10^{10} \times 10^{-4}) \\ &= (5.26)(9.35) \times 10^{6} \\ &= 49.181 \times 10^{6} \\ &= 49.2 \times 10^{6}\end{aligned}$$

to three significant figures. Since the first number has to be between 1 and 10, we write

$$\mathbf{49.2 \times 10^{6} = (4.92 \times 10) \times 10^{6}} = 4.92 \times 10^{7}$$

Example 6 Carry out the following division:

$$\frac{4.360 \times 10^{3}}{8.290 \times 10^{12}}$$

Solution. $\dfrac{4.360 \times 10^{3}}{8.290 \times 10^{12}} = \dfrac{4.360}{8.290} \times 10^{-9} = 0.5259 \times 10^{-9}$

to four significant figures. Since the first number has to be between 1 and 10, we write

$$\begin{aligned}\mathbf{0.5259 \times 10^{-9}} &= \frac{5.259}{10} \times 10^{-9} \\ &= \mathbf{(5.259 \times 10^{-1}) \times 10^{-9}} \\ &= 5.259 \times 10^{-10}\end{aligned}$$

Example 7 Express the given number in scientific notation before multiplying:

$$(29{,}600)(0.00000000816)$$

Solution.
$$\begin{aligned}(29{,}600)(0.00000000816) &= (2.96 \times 10^{4})(8.16 \times 10^{-9}) \\ &= 24.1536 \times 10^{-5} \\ &= 2.41536 \times 10^{-4} \\ &= 2.42 \times 10^{-4}\end{aligned}$$

to three significant figures.

The next example illustrates the addition of numbers expressed in scientific notation.

Example 8 Add

$$2.78 \times 10^{-5} \quad \text{and} \quad 3.12 \times 10^{-5}$$

Solution.

Numbers	Algebraic comparison	
2.78×10^{-5}	$2.78x$	
$+\ 3.12 \times 10^{-5}$	$+\ 3.12x$	
5.90×10^{-5}	$5.90x$	adding coefficients

Note that to add numbers expressed in scientific notation, the powers of 10 must be the same.

CALCULATOR COMMENT

Scientific calculators are programmed to express numbers in scientific notation. In fact, large and small numbers resulting from a calculation are automatically displayed in this form. Moreover, very large and very small numbers must first be converted to scientific notation before they can be entered.

To calculate numbers expressed in scientific notation, use the $\boxed{\text{EE}}$, $\boxed{\text{EEX}}$, $\boxed{\text{EXP}}$, or similar key to enter the exponent. For example, to enter 9.316×10^{-20}, use the following sequence:

9.316 $\boxed{\text{EE}}$ 20 $\boxed{+/-}$

Display: 9.316 -20

Example 9 Use a calculator to multiply

$$(2.75 \times 10^{8})(4.372 \times 10^{-25})$$

Solution. The sequence is

2.75 $\boxed{\text{EE}}$ 8 $\boxed{\times}$ 4.372 $\boxed{\text{EE}}$ 25 $\boxed{+/-}$ $\boxed{=}$

Display: 1.2023 -16

The result to three significant figures is 1.20×10^{-16}.

Example 10

Light travels at the rate of 6.70×10^{8} mi/hr. The average distance from the earth to the sun is 9.29×10^{7} miles. How long does it take for the sun's light to reach the earth?

Solution. The time is found by dividing the distance to the sun by the given rate:

$$t = \frac{9.29 \times 10^{7} \text{ mi}}{6.70 \times 10^{8} \text{ mi/hr}}$$

The sequence is

9.29 $\boxed{EE}$ 7 $\boxed{\div}$ 6.70 $\boxed{EE}$ 8 $\boxed{=}$

Display: 0.1386567 (hours)

So the time is about 0.139 hr = 8.34 min

Exercises / Section 1.11

In Exercises 1–8, express each number in scientific notation.

1. 26,000
2. 9,230,000
3. 379,200,000
4. 43,000,000
5. 0.00013
6. 0.00000126
7. 0.00008927
8. 0.0000000347

In Exercises 9–16, express each number in decimal notation.

9. 1.2×10^6
10. 3.98×10^{-4}
11. 6.273×10^{-3}
12. 7.8×10^8
13. 2.7×10^{-7}
14. 5.17×10^5
15. 9.56×10^5
16. 3.743×10^{-7}

In the statements of Exercises 17–23, change the given number from scientific to decimal notation or from decimal to scientific notation. (For the units employed, see the table on the inside cover.)

17. The distance from the earth to the moon is 240,000 mi.

18. The mean distance from the earth to the sun is 93 million miles.

19. The gravitational constant G is $G = 6.670 \times 10^{-11}$ N · m²/kg².

20. The capacitance of a certain capacitor is 4.1×10^{-6} F (farad).

21. 1 kilowatt · hour = 3.6×10^6 J (joules).

22. A typical steel wire will break if pulled with a force of 60,000 lb/in².

23. A pressure of 1 atmosphere = 1.013×10^5 newtons per square meter.

In Exercises 24–37, express the given numbers in scientific notation and carry out the indicated operations with a calculator. Round off the answers to the appropriate number of significant digits. (See Examples 7 and 9.)

24. (34,000)(9,630,000)
25. (823,000)(4,760,000)
26. (2,760,000)(0.00052)
27. (0.0024)(0.00096)(0.00082)
28. (31,600)(0.0000126)(0.0009116)
29. $\dfrac{0.00127}{54,300}$
30. $\dfrac{723,000}{0.003864}$
31. $\dfrac{(2,340)(52,300)}{0.001763}$
32. $\dfrac{(0.00148)(72,800)}{0.0000264}$
33. $\dfrac{(0.00026)(0.00092)}{2,008}$
34. (0.0000007132)(0.000002516)
35. (0.000003651)(0.00000081)

36. $\dfrac{856{,}000{,}000}{36{,}000{,}000{,}000}$

37. $\dfrac{792{,}500{,}000}{0.00000039}$

In Exercises 38–40, perform the indicated operations. (See Example 8.)

38. $(3.785 \times 10^{-5}) + (1.720 \times 10^{-5})$
   $(3.785 \times 10^{-5}) - (1.720 \times 10^{-5})$

39. $(8.13 \times 10^{-6}) + (0.13 \times 10^{-6})$
   $(8.13 \times 10^{-6}) - (0.13 \times 10^{-6})$

40. $(1.35 \times 10^{7}) + (2.78 \times 10^{7})$
   $(1.35 \times 10^{7}) - (2.78 \times 10^{7})$

Review Exercises / Chapter 1

In Exercises 1–8, carry out the indicated operations.

1. $\dfrac{12}{|-4|} - 3$

2. $|-3| - |-5|$

3. $(-5) + (-10) - 1$

4. $-5 - 10 - 1$

5. $(-4) + (-3) + (2) - 5$

6. $-4 - 3 + 2 - 5$

7. $\dfrac{(-28)(-3)}{(-5)(-14)}$

8. $\dfrac{(-7)(9)(-3)}{(2)(0)(-6)}$

In Exercises 9–14, remove symbols of grouping and simplify.

9. $(2x - y) + (5x - 7y)$

10. $(3x - 2y) - (6x + 5y)$

11. $(2x^2 - 4xy) - (-3x^2 + 4xy)$

12. $(-a + 2b) + (5a - b) - (7a + 2b)$

13. $-[a - (a + b)]$

14. $-\{x - [y - (2x + y)]\}$

In Exercises 15–16, enclose the last three terms in parentheses preceded by (a) a plus sign and (b) a minus sign.

15. $2x + a - x^2 + y - 4w$

16. $a - 3b - 2a^2 - b^2 + c^2$

In Exercises 17–24, simplify each of the given expressions.

17. $\dfrac{(-5s^2 t)(3s^3 t^3)}{-25s^{10} t^4}$

18. $\dfrac{(-6R_1^2 R_2 R_3)(2R_1^3 R_2^2 R_3)}{12 R_1^5 R_2^4 R_3}$

19. $\sqrt{50}$

20. $\sqrt{48}$

21. $\sqrt[3]{81}$

22. $\dfrac{2}{\sqrt{A}}$

23. $\dfrac{1}{\sqrt{5}}$

24. $\dfrac{5}{\sqrt{50}}$

In Exercises 25–36, carry out the indicated operations.

25. $x(x^2 - 2y)$

26. $\dfrac{6v^3 - 12v^2}{3v^2}$

27. $(2x - 6y)(-x - 2y)$

28. $(3a^2 + 2a + 1)(a - 2)$

29. $(4s^2 - 2st - t^2)(s + 3t)$

30. $(x - 2y)(3x - a)$

31. $(x^2 + 3xy - 10y^2) \div (x + 5y)$

32. $(x^2 - xy - 12y^2) \div (x - 4y)$

33. $\dfrac{5a^2 + 2ab - b^2}{a + b}$

34. $\dfrac{4ab^2 + 2a^3 - 5a^2b - b^3}{a - b}$

35. $(a^3 - 8b^3) \div (a - 2b)$

36. $(2x + y)(4x^2 - 2xy + y^2)$

37. Find an expression for the area in Figure 1.7.

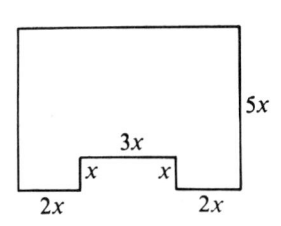

Figure 1.7

38. The voltage across a resistor is the product of the resistance and current. If the current i (in amperes) and the resistance R (in ohms) vary with time t (in seconds) according to the relations $i = 2 - 3t^2$ and $R = 5 + 3t$, find an expression for the voltage.

39. Simplify the following expression for the deflection of a beam: $d = (L - x)(2L - 2x^2)$.

40. The area A of a circular ring with inner radius r_1 and outer radius r_2 is $A = \pi(r_2 - r_1)(r_2 + r_1)$. Simplify this expression.

In Exercises 41–48, carry out the indicated operations with a calculator. Round off the answers to the appropriate number of significant figures.

41. $0.135 + 2.13 - 0.72$

42. $10.19 + 3.471 - 2.98$

43. $(1.23)(0.72)(3.980)(0.11)$

44. $(10.10)(20.3)(0.123)$

45. $\dfrac{(7.21)(1.32)}{1,001}$

46. $(3,200)(5.3)$

47. $(2.70)(0.37)$

48. $(4,900)(0.57)$

In Exercises 49–54, change the given numbers from scientific to decimal notation or from decimal notation to scientific notation.

49. The mass of the earth is 5.98×10^{24} kilograms.

50. The diameter of the sun is 864,000 mi.

51. The rest mass of a lithium nucleus is 0.00000000000000000000001164 gram.

52. The diameter of an oxygen molecule (O_2) is 3×10^{-8} centimeter.

53. The charge of a single electron is 1.60×10^{-19} C (coulomb).

54. The speed of light is approximately 186,000 mi/sec.

55. Multiply $(0.000000792)(0.000000031)$.

56. Multiply $(0.00000085)(0.000000760)$.

CHAPTER 2

Equations, Ratio and Proportion, Variation, and Functions

Objectives Upon completion of this chapter, you should be able to:

1. Solve first-order equations by performing the same algebraic operations on both sides.
2. Solve first-order equations by transposing terms.
3. Solve a given formula for one variable in terms of the remaining variables.
4. Solve certain stated problems by expressing the written statement as an equation.
5. Form ratios of given quantities.
6. Set up proportions and solve stated problems that involve proportions.
7. Use ratios and proportions to convert units.
8. Express direct, inverse, and joint variations in algebraic form.
9. Write a given statement as a function.
10. Use functional notation.
11. Graph basic functions by plotting points.

2.1 Equations

Most of this chapter is devoted to the study of various types of equations. While important in mathematics, equations play an equally prominent role in science and technology, as we shall see later in this chapter.

An **equation** is an equality between two expressions. Examples of equations are

1. $\dfrac{1}{2} = \dfrac{2}{4}$

2. $x = 1$
3. $2y + 3y = 5y$
4. $3x + 1 = 4$

Identity
Conditional equation

Note that equation 3 is valid for all numbers y and for that reason is called an **identity.** Equation 4, on the other hand, is only valid for $x = 1$ and is therefore called a **conditional equation.** The word *equation* is commonly restricted to refer to conditional equations. Let us now consider some of the basic terms.

> The literal symbol in a conditional equation is called the **variable** (which may represent some unknown quantity). The value of the unknown for which the equality holds is called the **solution** or **root** and is said to **satisfy** the equation. Finding the solution is referred to as **solving** the equation.

The most direct way to solve an equation is to make use of the fact that both sides of the equation are numbers. Consequently, the left side of the equation must be equal to the right side. This equality can be preserved at all times if we perform certain algebraic operations on both sides. More precisely:

> The equality can be preserved if we add the same number to both sides of the equation, subtract the same number from both sides, multiply both sides by the same number, or divide both sides by the same nonzero constant.

To illustrate these operations, consider the equation

$x - 3 = 5$

Even if we cannot see offhand what the solution is, we still know that $x - 3$ (the left side) must be equal to 5 (the right side). So adding 3 to both sides will not destroy this equality. Thus

$x - 3 + 3 = 5 + 3$

or

$x = 8$

Similarly, to solve the equation

$2x = 7$

we divide both sides by 2 to get

$$2x = 7 \quad \text{or} \quad x = \frac{7}{2} = \frac{7}{2}$$

The equation

$$4x + 5 = -13 \qquad (2.1)$$

requires more than one step. First subtract 5 from both sides (or add -5):

$$4x + 5 - 5 = -13 - 5$$
$$4x = -18$$

Now divide both sides of the equation by **4** to yield the solution:

$$x = -\frac{18}{4} = -\frac{9}{2}$$

The solution found can always be checked by substituting the root into the original equation. Letting $x = -(9/2)$ in the left side of equation (2.1), we get

$$4\left(-\frac{9}{2}\right) + 5 = -18 + 5 = -13$$

which is indeed equal to the right side.

Summary   In summary, we solve an equation by writing all terms containing the variable on one side of the equation and all constants on the opposite side. After combining the terms, we obtain the form $ax = b$. The resulting solution is $x = b/a$.

Example 1   Solve the equation $3x - 3 = x - 6$.

Solution. In this equation the unknown appears on both sides of the equal sign. One of these terms should be eliminated so that the terms containing the unknown are on one side only:

$$\begin{aligned}
3x - 3 &= x - 6 &&\text{given equation} \\
3x - x - 3 &= x - x - 6 &&\text{subtracting } x \text{ from both sides} \\
2x - 3 &= -6 &&\text{combining terms} \\
2x - 3 + 3 &= -6 + 3 &&\text{adding 3 to both sides} \\
2x &= -3 &&\text{combining terms} \\
x &= -\frac{3}{2} = -\frac{3}{2} &&\text{dividing by 2}
\end{aligned}$$

To check the solution, we substitute $x = -(3/2)$ into the given equation.

Left Side	Right Side
$3\left(-\dfrac{3}{2}\right) - 3$	$-\dfrac{3}{2} - 6$
$= -\dfrac{9}{2} - \dfrac{6}{2}$	$= -\dfrac{3}{2} - \dfrac{12}{2}$
$= -\dfrac{15}{2}$	$= -\dfrac{15}{2}$

The solution checks.

Example 2 Solve the equation $2x + 5 = 5x - 7$.

Solution. In this equation, it is helpful to eliminate the $2x$ from the left side to avoid negative coefficients:

$2x + 5 = 5x - 7$	given equation
$2x - 2x + 5 = 5x - 2x - 7$	subtracting $2x$ from both sides
$5 = 3x - 7$	combining terms
$5 + 7 = 3x - 7 + 7$	adding 7 to both sides
$12 = 3x$	combining terms
$4 = x$	dividing by 3
$x = 4$	

Check:

Left Side	Right Side
$2(4) + 5 = 13$	$5(4) - 7 = 13$

The solution checks.

Exercises / Section 2.1

Solve the following equations for x and check.

1. $x - 6 = 1$
2. $2x = 8$
3. $2x + 3 = -9$
4. $3x - 2 = 8$
5. $2x + 5 = x + 3$
6. $2x - 7 = 3x - 5$
7. $2x - 2 = 4x + 6$
8. $7x - 3 = 1 - 4x$
9. $x + 2 = 5x - 3$
10. $5x - 1 = -x - 2$
11. $1 - 2x = 3x - 4$
12. $2 - 3x = 7 - 5x$
13. $1 - 2x = 6 - 3x$
14. $2x - 1 = 2 - 6x$
15. $2x - 3 = 4 + 6x$

2.2 Transposition and Formulas

The procedure for solving equations studied in the last section can be shortened considerably. To see how, consider the equation

$$5x + 5 = 1 - x$$

Adding x to both sides, we get

$$5x + 5 + x = 1$$

Comparing this equation with the original, it is apparent that moving $-x$ to the other side of the equal sign and changing the sign has the same effect as adding x to both sides. The resulting equation is $6x + 5 = 1$.

Next, subtract 5 from both sides of

$$6x + 5 = 1$$

to obtain

$$6x = 1 - 5$$

Again, moving 5 to the other side of the equal sign and changing the sign has the same effect as subtraction. Finally, from $6x = -4$, we get $x = -\frac{2}{3}$.

The procedure just described is called **transposition**. Note that the division step is still carried out last.

> **Transposition**: To **transpose** a term, move the term to the other side of the equal sign and change the sign.

Example 1 Solve the equation $3 - x = 2x + 5$.

Solution.

$3 - x = 2x + 5$	given equation
$3 = 2x + x + 5$	transposing $-x$
$3 = 3x + 5$	simplifying
$3 = 3x + 5$	
$3 - 5 = 3x$	transposing 5
$-2 = 3x$	simplifying
$x = -\dfrac{2}{3}$	dividing by 3

Check:

Left Side   Right Side

$3 - \left(-\dfrac{2}{3}\right) = \dfrac{11}{3}$ $2\left(-\dfrac{2}{3}\right) + 5 = \dfrac{11}{3}$

50 CHAPTER 2 EQUATIONS, RATIO AND PROPORTION, VARIATION, AND FUNCTIONS

If more than one term must be transposed, all transpositions can be carried out in one step, as shown in the next example.

Example 2 Solve the equation $x - 2(x - 3) = 5x - 12$.

Solution. The first step is to remove the parentheses:

$x - 2(x - 3) = 5x - 12$ original equation
$x - 2x + 6 = 5x - 12$ removing parentheses
$-x + 6 = 5x - 12$ simplifying
$\mathbf{-x} + 6 = 5x \mathbf{- 12}$
$6 + 12 = 5x \mathbf{+ x}$ transposing $-x$ and -12
$18 = 6x$ simplifying
$3 = x$ dividing by 6
$x = 3$

Check:

$3 - 2(\mathbf{3} - 3) = 3, \quad 5(\mathbf{3}) - 12 = 3$

The equations we have considered so far are called **first-degree equations,** since the variable x is raised to the first power. The procedure for solving first-degree equations to this point is summarized next.

Procedure for solving first-degree equations

1. Remove symbols of grouping.
2. Use transposition to collect all terms containing x on one side of the equation and all constants on the other side. The resulting equation has the form $ax = b$.
3. Divide both sides by the coefficient of x to obtain the solution $x = b/a$.
4. Check the solution by substituting the value of x into the original equation.

Sometimes equations have to be solved in terms of letters, as shown in the next example.

Example 3 Solve the following equation for x, given that a is a constant:

$2a(x - 3) = ax + 5$

2.2 TRANSPOSITION AND FORMULAS

Solution. $\quad 2a(x - 3) = ax + 5 \quad$ given equation
Step 1. $\quad 2ax - 6a = ax + 5 \quad$ removing parentheses
Step 2. $\quad 2ax - ax = 6a + 5 \quad$ transposing
$\quad\quad\quad\quad\quad ax = 6a + 5 \quad$ combining terms
Step 3. $\quad\quad\quad x = \dfrac{6a + 5}{a} \quad$ dividing by a

Step 4. Check:

Left Side

$$2a\left(\dfrac{6a + 5}{a} - 3\right)$$
$$= 2(6a + 5) - 6a$$
$$= 12a + 10 - 6a$$
$$= 6a + 10$$

Right Side

$$a\left(\dfrac{6a + 5}{a}\right) + 5$$
$$= 6a + 5 + 5$$
$$= 6a + 10$$

Clearing fractions

If an equation contains fractions, it is best to simplify the equation first. This can be done by *multiplying both sides of the equation by the lowest common denominator*. This procedure is called **clearing fractions.**

Example 4 Solve the equation

$$\dfrac{1}{5}x + 1 = \dfrac{1}{15}x - \dfrac{1}{5}$$

Solution. We multiply both sides of the equation by 15, which is the lowest common denominator. Remember to multiply all the terms!

$$\mathbf{15} \cdot \dfrac{1}{5}x + \mathbf{15} \cdot 1 = \mathbf{15} \cdot \dfrac{1}{15}x - \mathbf{15} \cdot \dfrac{1}{5} \quad \text{multiplying by 15}$$
$$3x + 15 = x - 3$$
$$3x - x = -3 - 15 \quad\quad \text{transposition}$$
$$2x = -18$$
$$x = -9$$

The unknown can also occur in the denominator. Such equations are important enough to warrant a separate section in Chapter 5. However, the simplest cases can be adequately handled with our present techniques.

Example 5 Solve the equation
$$\frac{1}{x} + 2 = 1 - \frac{1}{2x}$$

Solution. All the fractions can be cleared by multiplying both sides by $2x$.

$\frac{1}{x} + 2 = 1 - \frac{1}{2x}$ original equation

$2x\left(\frac{1}{x}\right) + 2x(2) = 2x(1) - 2x\left(\frac{1}{2x}\right)$ clearing fractions

$2 + 4x = 2x - 1$

$4x - 2x = -1 - 2$ transposition

$2x = -3$

$x = -\frac{3}{2}$

Some equations have no solutions at all. For example, the equation

$x + 1 = x + 2$

leads to

$x - x = 2 - 1$ or $0 = 1$

which is impossible. In other words, this equation is not satisfied by any value of x.

Formula   A particularly useful type of equation is a formula. A **formula** is a relationship between variable quantities, often expressing a geometric or physical law. For example, one form of Newton's second law is $F = ma$, where F is the force applied to a body, m the mass of the body, and a the resulting acceleration. (For a discussion of the units, see Appendix A.)

To see why a given formula may have to be "solved" for a given letter, consider the formula $C = \frac{5}{9}(F - 32)$, which is used for converting degrees Fahrenheit to degrees Celsius. If this formula is solved for F, we obtain the formula for converting degrees Celsius to degrees Fahrenheit. This is shown in the next example.

Example 6 The formula for converting degrees Fahrenheit to degrees Celsius is given by

$$C = \frac{5}{9}(F - 32)$$

Solve the formula for F in terms of C.

Solution.

$$C = \frac{5}{9}(F - 32) \quad \text{given formula}$$

$$\frac{9}{5}C = F - 32 \quad \text{multiplying by } \frac{9}{5}$$

$$\frac{9}{5}C + 32 = F \quad \text{transposing}$$

$$F = \frac{9}{5}C + 32$$

This is the formula for converting degrees Celsius to degrees Fahrenheit.

The formulas (and equations) in this section can be solved for the indicated letter by executing the following steps:

Summary
1. Remove symbols of grouping.
2. Clear fractions if necessary.
3. Write the terms containing the letter to be solved on one side of the equation and collect the remaining terms on the other side.
4. Divide both sides by the coefficient of the letter to be solved.

Example 7 Solve the formula

$$B = S_0[a + d(D - D_0)]$$

for D.

Solution. Using the four steps listed above, we get the following solution:

$$S_0[a + d(D - D_0)] = B \quad \text{given formula}$$

Step 1. $\quad S_0[a + dD - dD_0] = B \quad$ } eliminating symbols of grouping

$$S_0a + S_0dD - S_0dD_0 = B$$

Step 2. (not necessary in this formula)

Step 3. $\quad S_0dD = B - S_0a + S_0dD_0 \quad \text{transposing}$

Step 4. $\quad D = \dfrac{B - S_0a + S_0dD_0}{S_0d} \quad \text{dividing by } S_0d$

Exercises / Section 2.2

In Exercises 1–40, solve each equation for x.

1. $x - 5 = 4$
2. $x + 5 = -5$
3. $3x = 12$
4. $5x = 16$
5. $2x - 3 = 5$
6. $5x + 2 = 12$
7. $4x - 5 = 7$
8. $3x - 1 = 5$
9. $3 - x = 7$

10. $1 - 5x = 5$
11. $2x - 3 = 5x + 6$
12. $1 - x = -5 + 6x$
13. $-2x - 3 = 5x + 1$
14. $2(x + 5) = -3(x + 2)$
15. $3(2 - x) = 5(2x + 1)$
16. $7(3 + x) = 8(4 - x)$
17. $6(2 - 2x) = 3(2 - 4x)$
18. $6(2 - 3x) = -(x + 2)$
19. $-(x - 1) = -(x + 3)$
20. $2(x + 3) = 2(x + 1)$
21. $\frac{1}{2}x + 1 = x + 3$
22. $\frac{1}{3}x - \frac{1}{6} = \frac{1}{6}x - \frac{1}{3}$
23. $\frac{1}{4}x - \frac{1}{12} = \frac{1}{6}x - \frac{1}{3}$
24. $\frac{1}{2}x - \frac{1}{3}x = \frac{1}{6}x - \frac{1}{6}$
25. $\frac{1}{5}x - 7 = x - 3$
26. $\frac{1}{3}x - \frac{1}{5} = \frac{1}{7}$
27. $\frac{1}{4}(x - 2) = \frac{1}{8}(2x + 1)$
28. $\frac{1}{3}(2x - 7) = \frac{1}{4}(1 - 3x)$
29. $ax = b$
30. $x - 3 = a$
31. $2ax - 1 = 3ax + 2$
32. $5ax + 2 = 8ax - 7$
33. $3ax + b = ax - b$
34. $b - ax = 3b - 5ax$
35. $bx + a = a - 2bx$
36. $2(ax + b) = 3(ax - 5b)$
37. $a(2x - 3) = a - 5ax$
38. $b(ax + 1) = b(3 - 2ax)$
39. $2(bx + 3) = 4(bx - 1)$
40. $b(1 - x) = b(2 - 2x)$

In Exercises 41–52, solve each formula for the letter indicated.

41. $V = IR$, I
42. $A = lw$, w
43. $A = \pi r^2$, π
44. $V = \pi r^2 h$, h
45. $A = \frac{1}{2}h(b_1 + b_2)$, b_2
46. $v = v_0 + at$, a
47. $A = p + prt$, r
48. $F = \frac{9}{5}C + 32$, C
49. $\frac{P_1 V_1}{T_1} = \frac{P_2 V_2}{T_2}$, P_2
50. $E = (m - m_0)c^2$, m_0
51. $V = \frac{4}{3}\pi r^3$, π
52. $S = \frac{1}{2}at^2 + v_0 t$, a

In Exercises 53–60, solve each equation for x.

53. $\frac{2}{x} + 1 = \frac{1}{x}$
54. $\frac{1}{x} - 1 = 1 - \frac{2}{x}$
55. $\frac{1}{2x} - \frac{1}{2} = \frac{1}{2x}$
56. $\frac{1}{2x} - \frac{1}{4} = \frac{1}{2} - \frac{1}{4x}$
57. $\frac{1}{5x} - \frac{1}{x} = \frac{1}{15}$
58. $\frac{1}{6x} - \frac{1}{8x} = \frac{1}{24}$
59. $\frac{1}{2x} - \frac{1}{4} = \frac{1}{x}$
60. $\frac{1}{3x} - \frac{1}{x} = \frac{1}{6}$

2.3 Applications of Equations

Our discussion of formulas in the previous section demonstrated the importance of first-degree equations in science and technology. Equations are also used to solve "word problems." These problems are solved by translating the given statements from ordinary language into algebraic language. For example, if we know that the length of a rectangle is one unit longer than the width, we write $L = W + 1$.

The translation from verbal statements to mathematical equations can be difficult. Although no general rule can cover all cases, here are some guidelines.

> **Guidelines for solving word problems**
> 1. Read the problem carefully; make sure you understand the situation described. Drawing a figure may help.
> 2. Identify the known and unknown quantities. Assign a letter to one of the unknown quantities and express the others, if any, in terms of this quantity.
> 3. Look for information that tells you which quantity or quantities are equal. Use this information to write the equation.
> 4. Solve the equation.
> 5. Check the result in the original problem.

Example 1 Two less than twice a given number is 10. What is the number?

Solution.

Step 1. Is the statement of the problem clear? We are looking for a number. If this number is doubled, and if 2 is subtracted from the result, we get 10.

Step 2. The number we are looking for is unknown. Let's denote this number by x. (The known quantities are 2 and 10.)

Step 3. What quantities are equal? Note first that twice the unknown number is $2x$ and 2 less than this is $2x - 2$. So the statement "Two less than twice a given number is 10" translates to

$$2x - 2 = 10$$

Step 4. We solve the equation:

$$2x - 2 = 10$$
$$2x = 10 + 2 \quad \text{transposing}$$
$$x = 6 \quad \text{dividing by 2}$$

Step 5. Check: The number we found is 6. Since $2 \times 6 = 12$ and $12 - 2 = 10$, the number 6 satisfies the conditions in the given problem.

Example 2 Three resistors are connected in series. The resistance (in ohms) of the second is $2\frac{1}{2}$ times that of the first, and the resistance of the third is 1.2 ohms more than that of the first. The total resistance is 19.2 ohms. Find the resistance of the first resistor.

Solution.
Step 1. Since the resistors are connected in series, we make use of the following information: The total resistance R_T of two or more resistors connected in series is equal to the sum of the individual resistances. (See Figure 2.1.)

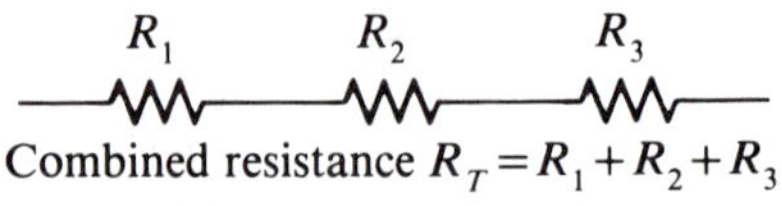

$$R_1 \quad R_2 \quad R_3$$
$$-\!\!\!\!\bigwedge\!\!\!\bigwedge\!\!\!\!- -\!\!\!\!\bigwedge\!\!\!\bigwedge\!\!\!\!- -\!\!\!\!\bigwedge\!\!\!\bigwedge\!\!\!\!-$$
Combined resistance $R_T = R_1 + R_2 + R_3$

Figure 2.1

Step 2. We are looking for the resistance of the first resistor. Let's denote this unknown quantity by R. (The known quantities are 1.2 ohms and 19.2 ohms.) Since R is the resistance of the first resistor, we have

$2.5R$ = resistance of second resistor

and

$R + 1.2$ = resistance of third resistor

Step 3. From Step 1 and Figure 2.1:

sum of the three resistances = total resistance

or

$R + 2.5R + (R + 1.2) = 19.2$ (in ohms)

Step 4.

$$4.5R + 1.2 = 19.2$$
$$4.5R = 19.2 - 1.2$$
$$4.5R = 18$$
$$R = 4.0 \text{ ohms}$$

Step 5. To check the solution, we should always return to the given problem rather than the equation, since the equation itself may be wrong. If 4.0 ohms is the resistance of the first resistor, then $(4.0)(2.5) = 10$ ohms and $4.0 + 1.2 = 5.2$ ohms are the respective resistances of the other two. The total is 19.2 ohms, as required.

SI units   From now on we are going to use the following SI (International System of Units) notations: Ω for ohm, V for volt, A for ampere, F for farad, and C for coulomb. (See the table on the inside front cover. Also, metric units are discussed in Appendix A.)

Example 3 A young woman can row at the rate of 4 mi/hr in still water. Rowing downstream in a river, she can travel 3 times as far in 1 hr as she can rowing upstream. What is the rate of flow of the river?

Solution. The basic relationship in this kind of problem is distance = rate × time, or

$$d = rt$$

Let x = the rate of flow of the river (in miles per hour). The rate downstream is $4 + x$ (in miles per hour) and the rate upstream $4 - x$ (in miles per hour). In 1 hr the respective distances are $(4 + x) \frac{\text{mi}}{\text{hr}} \cdot 1 \text{ hr} = (4 + x)$ mi and $(4 - x) \frac{\text{mi}}{\text{hr}} \cdot 1 \text{ hr} = (4 - x)$ miles. From the given information:

$$\text{distance (downstream)} = 3 \times \text{distance (upstream)}$$
$$4 + x = 3(4 - x)$$
$$4 + x = 12 - 3x$$
$$x + 3x = 12 - 4$$
$$4x = 8$$
$$x = 2$$

So the rate of flow of the river is 2 mi/hr.

Word problems in algebra have often been called artificial. No doubt many of them are. For example, probably no one could row at the same rate for a whole hour regardless of physical condition. Yet this problem does not differ substantially from the problem of detecting the speed of the earth through the "ether" (or the apparent rate of flow of the ether past the earth) if the speed of light is substituted for the speed of the boat. This was the famous Michelson-Morley experiment. Its outcome was negative: The speed of light proved to be the same in the direction of the earth's motion as in the lateral direction, thus opening the door to the discovery of the special theory of relativity by Albert Einstein in 1905.

On a more modest scale, skill in solving algebra problems is often valuable in setting up more interesting problems in calculus. Of course, many problems in algebra are of interest in their own right. Consider the following examples.

Example 4 A storage tank can be filled in 18 hr and drained in 6 hr. If the tank is initially full and both the drain and the intake valve are open, how long will it take to drain the tank?

58 CHAPTER 2 EQUATIONS, RATIO AND PROPORTION, VARIATION, AND FUNCTIONS

Solution. The best procedure in this kind of problem is to examine the situation after one time unit. So if

$$x = \text{time taken to drain tank}$$

then $1/x$ is the fractional part drained after 1 hr. Similarly, $\frac{1}{6}$ of the tank is drained after 1 hr and $\frac{1}{18}$ is filled. Thus

$$\frac{1}{x} = \frac{1}{6} - \frac{1}{18}$$

$$\frac{1}{x} = \frac{3}{18} - \frac{1}{18} = \frac{2}{18} = \frac{1}{9}$$

$$x = 9 \text{ hr}$$

Alternatively, we can clear fractions by multiplying by $18x$. Then

$$18x \left(\frac{1}{x}\right) = 18x \left(\frac{1}{6} - \frac{1}{18}\right)$$

$$18 = 3x - x$$

$$18 = 2x$$

$$x = 9$$

Example 5 One brine solution contains 15% salt by volume, and another solution 25%. How many liters of each must be mixed to produce 40 L of brine containing 18% salt?

Solution. Let

$$x = \text{number of liters of 15\% solution}$$

Then

$$40 - x = \text{number of liters of 25\% solution}$$

The best way to obtain an equation is to work directly with the quantities involved—in this case, the amount of salt. For example, in the 18% solution the fractional part by volume consisting of salt is $(0.18)(40) = 7.20$ L. (See Figure 2.2.)

x L   + $(40-x)$ L   =   40 L

Amount of salt: Amount of salt: Amount of salt:
0.15x L   0.25(40−x) L   0.18(40) L

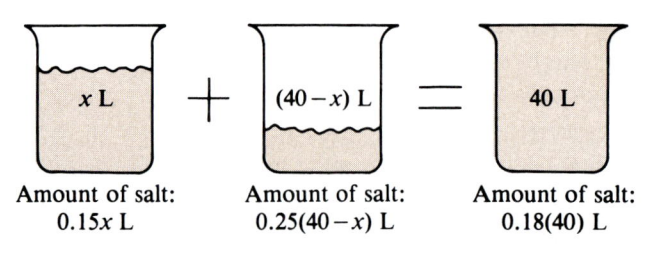

Figure 2.2

amount of salt (before mixing) = amount of salt (after mixing)
$$0.15x + 0.25(40 - x) = 0.18(40)$$
$$0.15x + 10 - 0.25x = 7.20 \quad \text{removing parentheses}$$
$$15x + 1000 - 25x = 720 \quad \text{multiplying by 100}$$
$$-10x = -280$$
$$x = 28 \text{ L}$$

Hence 28 L of the 15% solution must be mixed with 12 L of the 25% solution.

Exercises / Section 2.3

1. The sum of two numbers is 30. If one is 3 less than the other, what are the numbers?
2. One machine part costs twice as much as the other. If the total price is $9.63, what is the cost of each?
3. In 3 years Jane will be twice as old as she was 5 years ago. How old is she now?
4. The perimeter of a rectangle is 21.0 ft. If the length is 2.5 ft more than the width, what are the dimensions?
5. One alloy contains 10% brass and another 15% brass (by weight). How many pounds of each must be combined to form 100 lb of an alloy containing 12% brass?
6. One brine solution contains 10% salt by volume, and another contains 18%. How many gallons of each must be mixed to produce 16 gal containing 15% salt?
7. How many gallons of a 20% salt solution must be added to 5 gal of a 10% salt solution to produce a 16% salt solution?
8. How much water must be added to 20 gal of a 60% alcohol solution to produce a 45% alcohol solution?
9. A 1-L bottle contains a 20% alcohol solution. How much of the solution must be drained off and replaced by an 80% solution to produce a 50% alcohol solution?
10. A chemist ordered a 5% sulfuric acid solution from the supply room but received an 8% solution by mistake. How much must be drawn off from a 1-L bottle and replaced by distilled water to produce the right concentration?
11. The River Queen, a ship for sightseers, averages 15 mi/hr in still water. Traveling downstream for 2 hr, she can go twice as far as traveling upstream for 2 hr. How fast is the river flowing?
12. John can row at the rate of 5 mi/hr in still water. Rowing downstream in a certain river, he can travel 4 times as fast as upstream. Find the rate at which the river is flowing.
13. A river flows at the rate of 3 mi/hr. A motorboat traveling downstream for 2 hr requires 6 hr to get back to the starting point. Find the speed of the boat.
14. A car traveling at 40 mi/hr leaves a certain intersection 45 min before a second car traveling at 52 mi/hr. How long will it take for the second car to overtake the first?
15. Paul usually takes his bicycle to school, which is only 5 mi from his home. One morning he is in a hurry and asks his mother to drive him. If the car goes 3 times as fast as the bike and the trip takes 20 min less, what is the speed of the bike?
16. A man earns $80 plus board every day that he works. On idle days he is charged $25 for board. In one 30-day period he received $1,560. How many days did he work?
17. A piece of wire 38 in. long is bent into the shape of a rectangle that is twice as long as it is wide. Find its dimensions.

18. One inlet valve can fill a tank in 10 hr, and a second valve can fill the same tank in 16 hr. How long will it take to fill the tank if both valves are open?

19. An inlet valve can fill a tank in 20.0 hr, while it takes 25.0 hr to drain the tank. If the tank is initially empty and both the valve and the drain are open, how long will it take to fill the tank?

20. A chemical tank can be filled in 12.0 min and drained in 15.0 min. If both the drain and the valve are open and if the tank is initially one-fourth full, how long does it take to fill the rest of the tank?

21. If the drain in Exercise 19 is left open for 8.00 hr and then closed while the valve is open, how long will it take to fill the tank?

22. Jane has taken three out of the first five tests in her physics class, and her test average is 91%. What will the average grade on her last two tests have to be for her to receive an A (93%)?

23. Two cars traveling toward each other at 36 mi/hr and 44 mi/hr, respectively, are 200 mi apart. How much of the 200-mi distance has each car covered when they meet?

24. How much water must be evaporated from a 1-gal can containing a 20% salt solution to produce a 25% salt solution?

25. The sum of two currents is 4.3 A. If one of the currents is 1.9 A more than the other, find each of the currents.

26. The sum of two resistors in series is 97.2 Ω. If the resistance of one is 3 times that of the other, find each of the resistances.

27. Four years from now a copying machine will be twice as old as it was two years ago. How old is the copying machine?

2.4 Ratio and Proportion

In this section we are going to discuss two important concepts—**ratio** and **proportion.**

> A **ratio** is a quotient of two quantities.

So if the quantities are denoted by a and b, then the ratio is a/b.

Ratios occur frequently in everyday life. For example, if you peddle your bicycle at the rate of 15 ft/sec, then you change your distance by 15 ft every second. So the velocity

$$v = \frac{15 \text{ ft}}{1 \text{ sec}} = 15 \text{ ft/sec}$$

is a ratio. Similarly, 30 mi/gal is a ratio. Since the number π is defined to be the circumference of a circle divided by its diameter, π is also a ratio.

2.4 RATIO AND PROPORTION

Example 1 If Joan travels 100 miles in 2 hr, then her average velocity, which is the ratio of distance to time, is given by

$$\frac{100 \text{ mi}}{2 \text{ hr}} = 50 \text{ mi/hr}$$

As shown in Example 1, if the ratio involves dissimilar quantities, then proper units must be assigned to the ratio. If the ratio involves measurements of the same kind, then it should be expressed as a *dimensionless number*. For example, the definition of π involves two lengths measured in the same unit of measure. Consider another example.

Example 2 The *specific gravity* of a substance is the ratio of the density of the substance to the density of water (1.94 slugs/ft^3). Given that the density of copper is 17.25 slugs/ft^3, what is its specific gravity?

Solution. Since the ratio involves units of the same kind, we get the following dimensionless number:

$$\frac{17.25}{1.94} = 8.9$$

(Other examples of standard ratios will be mentioned in the exercises.)

Many problems in science and technology lead to two equal ratios. This condition is called a **proportion.**

> The equality of two ratios is called a **proportion.**

It follows from the definition that a proportion has the form

$$\frac{a}{b} = \frac{c}{d}$$

Proportions can be used to convert units. For example, to convert $2\frac{1}{2}$ ft to inches, we set up the proportion as follows: Let x be the number of inches. Then, since 12 in. = 1 ft, we have

$$\frac{x}{2.5 \text{ ft}} = \frac{12 \text{ in.}}{1 \text{ ft}}$$

$$x = \frac{(2.5 \text{ ft})(12 \text{ in.})}{1 \text{ ft}} = 30 \text{ in.}$$

(Note that units can be canceled as if they were numbers.)

Since proportions are a convenient way to carry out certain conversions, some basic conversion units are summarized below.*

Basic conversion units

12 in. = 1 ft	1 km = 1,000 m
3 ft = 1 yd	1 cm = 10 mm
1 in. = 2.54 cm	1 kg = 1,000 g
39.37 in. = 1 m	1 lb = 454 g
3.28 ft = 1 m	1 kg = 2.2 lb
1 m = 100 cm	

Example 3 Convert 5.00 ft to meters.

Solution. Let x = number of meters. Since 3.28 ft = 1 m,

$$\frac{x}{1\text{ m}} = \frac{5.00\text{ ft}}{3.28\text{ ft}} = \frac{5.00}{3.28} = 1.52$$

$$x = 1.52 \times 1\text{ m} = 1.52\text{ m}$$

Example 4 Convert 1,000.0 g to pounds.

Solution. Let x = the number of pounds. Then

$$\frac{x}{1\text{ lb}} = \frac{1{,}000.0\text{ g}}{454\text{ g}}$$

$$x = 2.20\text{ lb}$$

Example 5 It is known that 88 ft/sec = 60 mi/hr. Convert 48 mi/hr to feet per second.

Solution. Let x = the number of feet per second. Then

$$\frac{x}{48\text{ mi/hr}} = \frac{88\text{ ft/sec}}{60\text{ mi/hr}}$$

* Other conversion techniques are discussed in Appendix A.

2.4 RATIO AND PROPORTION

Multiplying both sides by 48 mi/hr, we get

$$x = \left(\frac{88 \text{ ft/sec}}{60 \text{ mi/hr}}\right) 48 \text{ mi/hr}$$

$$x = 70 \text{ ft/sec}$$

Many ordinary situations lead to proportions. Here is one example.

Example 6 A recipe for cocoa fudge calls for $\frac{2}{3}$ cup of cocoa and $1\frac{1}{2}$ cups of milk, in addition to other ingredients. If you had $3\frac{1}{2}$ cups of cocoa to use up, how much milk would you need?

Solution. Let x be the amount of milk. Then

$$\frac{x}{3\frac{1}{2} \text{ cups}} = \frac{1\frac{1}{2} \text{ cups}}{\frac{2}{3} \text{ cups}}$$

$$\frac{x}{\frac{7}{2} \text{ cups}} = \frac{3/2}{2/3}$$

$$x = \frac{3}{2} \cdot \frac{3}{2} \cdot \frac{7}{2} \text{ cups} = \frac{63}{8} \text{ cups}$$

$$x = 7\frac{7}{8} \text{ cups of milk}$$

Proportions play an important role in science and technology. Consider the following example:

Example 7 The resistance in a wire is proportional to its length. If the resistance in a wire 10.2 ft long is 0.560 Ω, determine the resistance in a wire 16.3 ft long if the wire is made of the same material.

Solution. Let R denote the resistance in the wire. Then

$$\frac{R}{16.3 \text{ ft}} = \frac{0.560 \text{ }\Omega}{10.2 \text{ ft}}$$

and

$$R = \frac{(16.3)(0.560)}{10.2} \Omega = 0.895 \text{ }\Omega$$

64 CHAPTER 2 EQUATIONS, RATIO AND PROPORTION, VARIATION, AND FUNCTIONS

Exercises / Section 2.4

1. The *Mach number* is the ratio of the velocity of an object to the velocity of sound. Given that the velocity of sound is 330 m/sec, what is the Mach number of a rocket whose velocity is 750 m/sec?

2. The *density* of a substance is the ratio of its mass to its volume. Find the density of gold, given that 2.60 cm^3 of gold has a mass of 50.18 g.

3. *Pressure* is defined as the ratio of force to area. The force against a horizontal plate with an area of 5.00 ft^2 submerged 10.0 ft below the surface of the water is 3,120 lb. Find the pressure on the plate.

4. The *compression ratio* of a car engine is the ratio of the cylinder volume to the compressed volume. If the cylinder volume is 50.0 in.^{3} when the piston is at the bottom of its stroke and 5.75 in.^{3} when it is at the top, what is the compression ratio?

5. The *intelligence quotient* (IQ) for children is defined as the ratio of mental age to chronological age multiplied by 100. If the scholastic ability of a boy aged 6 years and 9 months corresponds to that of a child aged $8\frac{1}{2}$ years, what is his IQ?

6. Given that 1 in. = 2.54 cm, convert 3.72 in. to centimeters.

7. Convert 10.0 cm to inches.

8. Given that 1 oz = 28.35 g, convert 2.25 oz to grams.

9. Convert 50.0 g to ounces. (See Exercise 8.)

10. Medieval records of the city of Goslar, a former seat of the Holy Roman Empire situated in northern Germany, show that 1.2 silver marks had the buying power of $1,500 (in 1950 dollars). Determine the buying power of 20 silver marks, a typical yearly income of a medieval count.

11. A plumber charges $24 for 30 min of labor. How much would the plumber charge for a job taking 1 hr and 46 min?

12. The owner of a small car gets 29.4 miles per gallon of gasoline. How far can he travel on a full tank of 15.0 gal?

13. Mr. Veldboom figures that a 4-day vacation will cost his family $460. How much would a 9-day vacation cost?

14. Mr. and Mrs. Benot can wash 2 windows in 9 min. Working at that rate, how long will it take them to wash all 16 windows in their house?

15. The resistance (in ohms) in a wire is proportional to its length. If 12 ft of wire have a resistance of 0.10 Ω, what is the resistance in a wire 20 ft long?

16. A restaurant manager knows that she needs 3 lb of rice to make rice pudding for 21 people. How much rice does she need to serve 50 people?

17. Given that 60 mi/hr = 88 ft/sec, what is 25.0 mi/hr in feet per second?

18. Suppose you were pedaling your bicycle at the rate of 2.2 revolutions per second to attain a speed of 10 mi/hr. How fast would you have to pedal to attain a speed of 16 mi/hr?

19. A Wheatstone bridge (Figure 2.3) is a convenient instrument for measuring resistance. It was invented by the English scientist Charles Wheatstone in 1843. In the figure both R_1 and R_2 have a known constant resistance, R is an adjustable resistor, and X is the unknown resistance. R is adjusted so that the current from a to b is zero, as measured by galvanometer G. The relationship between the resistances is given by the

proportion

$$\frac{R_1}{R_2} = \frac{R}{X}$$

If $R_1 = 10 \, \Omega$, $R_2 = 100 \, \Omega$, and R was found to be $7 \, \Omega$, determine the resistance X.

20. A transformer consists of two coils. The current in one coil induces a current in the other coil. (See Figure 2.4.) The following proportion gives the relationship between the voltages in each coil and the number of windings:

$$\frac{V_1}{V_2} = \frac{N_1}{N_2}$$

If $V_1 = 80$ V, $N_1 = 500$ (first coil), and $N_2 = 1,200$ (second coil), show that $V_2 = 192$ V. Because of the increase in voltage, this is called a step-up transformer.

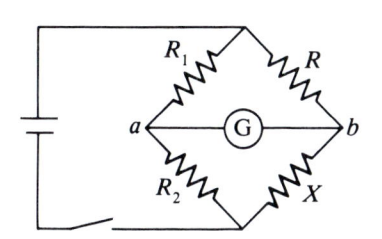

Figure 2.3

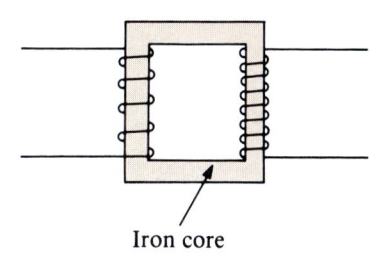

Figure 2.4

2.5 Variation

Many situations in science involve variable quantities. In this section we shall be concerned with a special group of relationships called **variations.** The first type, called **direct variation,** is really a proportion.

> **Direct variation**: If two variables x and y are related so that $y = kx$, then y is said to vary directly as x or to be directly proportional to x. The constant k is called the **constant of proportionality.**

If we are told that y varies directly as x, we write $y = kx$, where k is the constant of proportionality. If we are given a pair of values for x and y, then k can be evaluated, thereby yielding a specific equation. For example, given that

$$y = kx, \quad x = 2, \quad \text{and} \quad y = 4$$

we get

$$4 = k \cdot 2, \quad \text{or} \quad k = 2$$

So the relationship is $y = 2x$. Consider another example.

Example 1 The force against a horizontal plate varies directly as the depth d. If the force is 1,560 lb at a depth of 2.5 ft, what is the force at a depth of 7.4 ft?

Solution. By definition, $F = kd$. The first task in a problem on variation is to compute the constant k. From the given information, $F = 1{,}560$ lb and $d = 2.5$ ft. We now get

$$1{,}560 = 2.5k$$

so that $k = 624$. The resulting equation is

$$F = 624d$$

Finally, if $d = 7.4$, $F = (624)(7.4) = 4{,}600$ lb to two significant figures.

The **constant of proportionality** occurs in all variations. If the variable quantities have units, then so does k. In Example 1, if F is in pounds and d in feet, then k is expressed in pounds per foot.

The variable x in a direct variation can take on various forms. For example, "y is directly proportional to x^2" is written $y = kx^2$, "y varies directly as x^3" is written $y = kx^3$, and "y varies directly as $\sqrt{x}$" is written $y = k\sqrt{x}$. Consider another example.

Example 2 A freely falling body is subject to a retarding force due to air resistance. This force depends on the size and shape of the object. For some objects this force is directly proportional to the square of the velocity v, or

$$f = kv^2$$

If y is directly proportional to $1/x$, we write

$$y = k\left(\frac{1}{x}\right) \quad \text{or} \quad y = \frac{k}{x}$$

This relationship is called an **inverse variation.**

> **Inverse variation:** If two variables x and y are related so that $y = k/x$, then we say that y varies inversely as x or that y is inversely proportional to x.

As in the case of direct variations, the variables can take on various forms. For example, if y varies inversely as the cube root of x, we write

$$y = \frac{k}{\sqrt[3]{x}}$$

Consider another example.

Example 3 The gravitational force between two objects is inversely proportional to the square of the distance between them, or

$$F = \frac{k}{x^2}$$

If the force is 3 lb at a distance of 100 mi, what is the force at a distance of 150 mi?

Solution. The first step is to calculate the constant k from the given information. Thus

$$3 = \frac{k}{(100)^2} \quad \text{or} \quad k = 30{,}000$$

and

$$F = \frac{30{,}000}{x^2}$$

If $x = 150$, then

$$F = \frac{30{,}000}{(150)^2} = 1\frac{1}{3} \text{ lb}$$

The last type of variation we will discuss here is called **joint variation.**

> **Joint variation:** If $z = kxy$, then we say that z varies jointly as x and y or that z is directly proportional to the product of x and y.

Example 4 The rate of heat W developed in a circuit varies jointly as the resistance R and the square of the current i. If an element is drawing 2 A and develops heat at the rate of 8 cal/sec when $R = 10 \, \Omega$, what is the heat developed when R is decreased to $5 \, \Omega$ and the current is doubled?

Solution. By definition

$$W = kRi^2$$

Then

$$8 = k(10)(2)^2$$

or $k = \frac{1}{5}$. Thus

$$W = \frac{1}{5} Ri^2$$

From the given information

$$W = \frac{1}{5}(5)(4)^2 = 16 \text{ cal/sec}$$

The variations considered may occur in combination, as shown in the next example.

Example 5 The gravitational force F between two masses m_1 and m_2 varies jointly as the masses and inversely as the square of the distance r between them. Thus

$$F = k\frac{m_1 m_2}{r^2}$$

Here k is called the *gravitational constant*.

Exercises / Section 2.5

In Exercises 1–12, express each statement in the form of an equation.

1. y is directly proportional to x. $y = kx$
2. y varies directly as x^2. $y = kx^2$
3. w varies directly as i^2. $w = ki^2$
4. y is directly proportional to x^3. $y = kx^3$
5. y varies inversely as x^2. $y = \frac{k}{x^2}$
6. y is inversely proportional to x^3. $y = \frac{k}{x^3}$
7. y varies jointly as x and w. $y = kxw$
8. y is directly proportional to the product of w and z. $y = kwz$
9. A varies directly as a and inversely as b. $A = k\frac{a}{b}$
10. Z varies jointly as s and t and inversely as r. $Z = k\frac{st}{r}$
11. F is directly proportional to the product of m and M and inversely proportional to r^2. $F = k\frac{mM}{r^2}$
12. E varies jointly as m and v^2. $E = kmv^2$

In Exercises 13–16, find the equation, including the constant of proportionality.

13. y varies directly as x; $y = 8$ when $x = 2$. $y = kx$
14. s is directly proportional to t^2; $s = 64$ when $t = 2$. $s = kt^2$ $64 = k(2)^2$ $64 = k4$ $k = 16$
15. P varies inversely as V; $P = 1$ when $V = 2$. $P = \frac{k}{V}$
16. F varies directly as s; $F = 15$ when $s = 1.5$. $F = ks$
17. The force exerted by a spring is directly proportional to the spring's extension. If a force of 2.5 lb is required to stretch a spring 1.8 in., what is the force required to stretch it 5.0 in.?
18. The acceleration of a particle varies directly as the force applied (Newton's second law). If for some object an acceleration of 30.0 m/sec² results from a force of 10.0 N, what is the force required to yield an acceleration of 1.5 m/sec²?

19. The strength S of a rectangular beam varies jointly as the width w and the square of the depth d. Find an expression for S.

20. The voltage V across a resistor varies directly as the current I. If $V = 2.4\ \Omega$ when $I = 1.9$ A, find V when $I = 5.2$ A.

21. Within a small range the demand for a product varies inversely as the price. If a store can sell 90 units of a certain commodity per week at $1.50 apiece, how many units per week can it expect to sell if the price is reduced to $1.35?

22. For two gears in mesh, the speed of each is inversely proportional to the number of teeth. If a gear having 30 teeth and turning at 100 rpm (revolutions per minute) is in mesh with a gear having 50 teeth, what is the speed of the second gear?

23. The repulsive force f between two charged particles having like charges q_1 and q_2, respectively, varies jointly as q_1 and q_2 and inversely as the square of the distance r between them. Find an expression for the repulsive force.

24. The kinetic energy (ergs) of a particle is directly proportional to the square of the velocity (in centimeters per second). If the kinetic energy of a particle moving at 100 cm/sec is 15,000 ergs, what is the kinetic energy if the velocity is increased to 120 cm/sec?

25. Boyle's law states that the pressure of an ideal gas varies inversely as the volume, provided that the temperature remains constant. If a pressure of 3 atmospheres corresponds to a volume of 10 m^3, what is the pressure if the volume is decreased to 7.5 m^3?

26. The volume V of a sphere is directly proportional to the cube of the radius r. Determine the relationship if $V = 36\pi$ when $r = 3$.

27. The simple interest earned during a certain time period varies jointly as the principal and the interest rate. If $500 earned $120 at 8% interest during a certain time period, find the interest earned on $666.67 at 6%.

28. One of Kepler's three laws states that the square of the time required for a planet to make one revolution about the sun is directly proportional to the cube of its average distance from the sun. Given that Mars is approximately $1\frac{1}{2}$ times as far from the sun as the earth, find the number of days required to make one revolution.

29. The resistance R in a wire varies directly as the length L and inversely as the square of the diameter D. If a wire 50.0 ft long with a diameter of 0.0140 in. has a resistance of 5.00 Ω, find the relationship.

30. The resistance in a wire varies directly as the length and inversely as the square of the diameter. If a wire 100 ft long with a diameter of 0.028 in. has a resistance of 10 Ω, find the resistance in a wire 150 ft long with a diameter of 0.016 in. if it is made of the same material.

2.6 Functions and Graphs

In this section we will briefly look at functions and graphs. Both concepts play a major role in science and technology and are encountered time and again in more advanced mathematics courses.

We shall define a function as a special type of relationship between two variables. Fortunately, the basic idea is already familiar from our study of variation. First note that in the equation

$$y = 3x$$

(y varies directly as x) we get, for every x, a unique value for y. This leads to the following definition.

> **Definition of function**: If for every value of the variable x there corresponds one, and only one, value of the variable y, then we call y a **function** of x.

The variable x is called the **independent variable,** and y is called the **dependent variable.** The letters x and y are representative and can stand for various physical quantities, as shown in the first two examples.

Example 1 The formula for the area of a circle

$$A = \pi r^2$$

defines a function, since for every $r > 0$ there exists a unique value for A. The independent variable is r and the dependent variable is A.

Example 2 The power P developed in a 10-Ω resistor as a function of the current I is $P = 10I^2$. The formula is a function because every value of I yields a unique value of P. The independent variable is I, and the dependent variable is P.

Functions do not necessarily have to be in the form of equations. Tables, charts, and graphs may also describe functions.

Rectangular coordinate system

Our main concern in this section is the graphical representation of functions on the **rectangular** or **Cartesian coordinate system.** This coordinate system is named after the French mathematician and philosopher René Descartes. Descartes was the founder of *analytic geometry,* the study of geometry by means of coordinate systems.

Recall from Chapter 1 that the set of real numbers can be made to correspond to the points on a line. To locate points in a plane, we construct a reference system consisting of two number lines that intersect at right angles at their respective origins. (See Figure 2.5.) The horizontal line is called the **x-axis** and the vertical line the **y-axis;** together these lines are known as the

Coordinate axes

coordinate axes. The positive numbers are on the right side of the origin on the x-axis and above the origin on the y-axis. Points can be described in the plane by giving their perpendicular distances and directions from the coordinate axes. The distance from the y-axis, with the proper sign, is called the

Abscissa
Ordinate

x-coordinate or **abscissa,** and the distance from the x-axis the **y-coordinate** or **ordinate.**

To designate a point, it is customary to place the two coordinates in parentheses with the x-coordinate given first. The two coordinates are separated by a comma. For example, the point P in Figure 2.5 is denoted by $(1, 4)$, Q by $(-2, 3)$, and R by $(-4, -2)$.

Quadrant

The coordinate axes divide the plane into four parts, called **quadrants,** numbered I, II, III, and IV (see Figure 2.5). Note that both coordinates are

2.6 FUNCTIONS AND GRAPHS

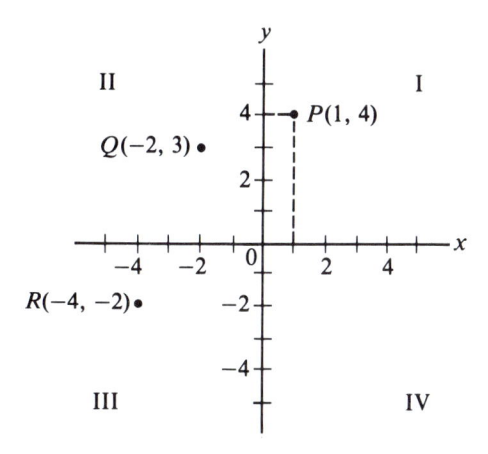

Figure 2.5

positive in the first quadrant and negative in the third quadrant. In the second quadrant the x-coordinates are negative and the y-coordinates positive; in the fourth quadrant the x-coordinates are positive and the y-coordinates negative.

By using the rectangular coordinate system, functions can be graphed in a simple and natural way. Consider, for example, the function

$$y = 2x + 1$$

Since for any x there exists a unique value for y, we can find any number of such pairs and plot them. A few such pairs are given in the following table:

Plotting points

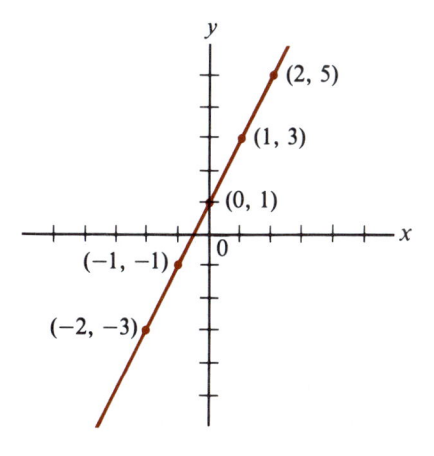

x:	-2	-1	0	1	2
y:	-3	-1	1	3	5

Figure 2.6

CHAPTER 2 EQUATIONS, RATIO AND PROPORTION, VARIATION, AND FUNCTIONS

Once these points are plotted, they can be connected by a smooth curve, which turns out to be a straight line (Figure 2.6).

Example 3 Graph the function $y = x^2 - 1$.

Solution. First we need to construct a table of values:

x:	-3	-2	-1	0	1	2	3
y:	8	3	0	-1	0	3	8

These points are now plotted and connected by a smooth curve; see Figure 2.7.

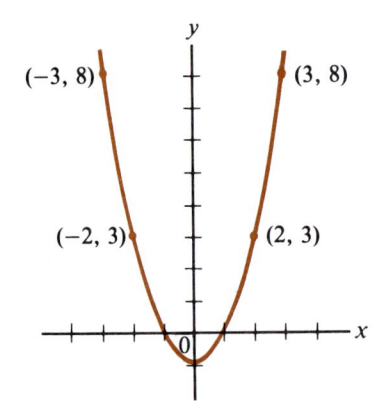

Figure 2.7

Example 4 Graph the function $y = \sqrt{1 - x}$.

Solution. Recall that the symbol $\sqrt{}$ stands for the principal, or positive, square root. Hence for every x there is only one y, so that the equation is indeed a function. However, if $x > 1$, then the values of $1 - x$ become

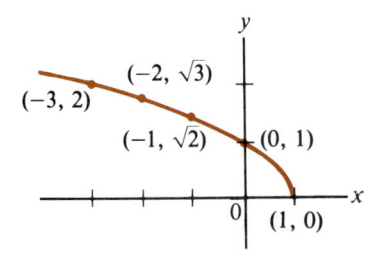

Figure 2.8

negative, so that the resulting y-values are imaginary. (See Section 1.7.) We conclude that $x \leq 1$.

x:	1	0	-1	-2	-3
y:	0	1	$\sqrt{2}$	$\sqrt{3}$	2

The graph is shown in Figure 2.8.

Example 4 shows that a function may not be defined for all values of x in the equation. The set of all x-values for which the y-values exist is called the **domain** of the function. The domain of the function $y = \sqrt{1 - x}$ in Example 4 is $x \leq 1$.

Domain

Example 5 State the domain of the function

$$y = \frac{1}{x - 2}$$

Solution. To avoid division by 0, x cannot be equal to 2. So the domain is the set of all x such that $x \neq 2$.

Representing a function in graphical form can be very useful in technology, since a graph gives a revealing picture of the behavior of the function. Consider the next example.

Example 6 The output P (in watts) of a certain battery is given by $P = 4I - I^2$, where I is measured in amperes. Sketch the graph. (For physical reasons, the domain is $0 \leq I \leq 4$.)

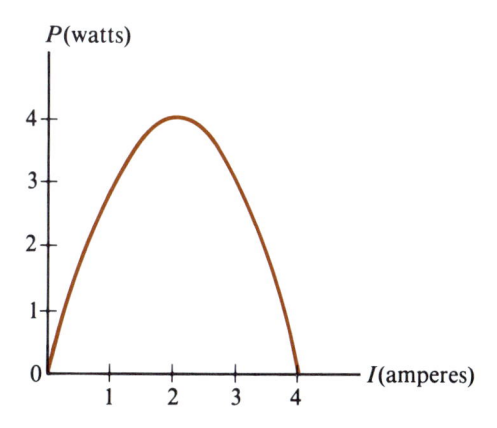

Figure 2.9

Solution. The graph (Figure 2.9) is plotted from the following table of values:

I (amperes):	0	1	2	3	4
P (watts):	0	3	4	3	0

Note that the graph shows the output to be at a maximum when $I = 2$ A.

Range

If y is a function of x, then the set of admissible y-values is called the **range** of the function. For example, the range of the function $y = x^2$ is $y \geq 0$.

While functions play an important role in our later work, *not every equation in x and y represents a function of x*. Consider, for example, the equation $y^2 = x$. If $x = 4$, then $y^2 = 4$; thus, $y = 2$ or $y = -2$ (usually written $y = \pm 2$). In fact, for every $x > 0$, we obtain two y-values. Some of these values are listed in the following table:

x:	0	1	2	3	4
y:	0	± 1	$\pm \sqrt{2}$	$\pm \sqrt{3}$	± 2

The graph is shown in Figure 2.10.

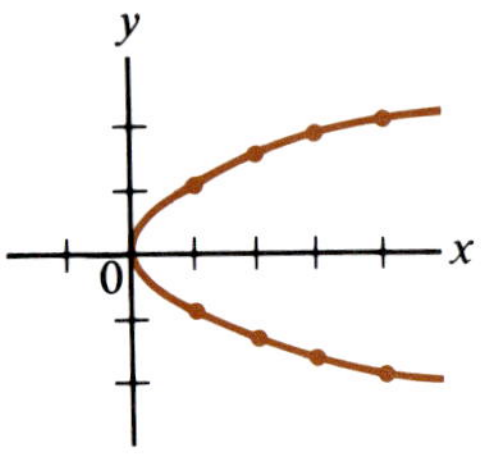

Figure 2.10

Exercises / Section 2.6

In Exercises 1–4, write each of the given statements as a function. Identify the independent and dependent variables.

1. The volume V of a sphere as a function of its radius r.
2. The perimeter P of a square as a function of its side s.
3. The volume V of a cylinder of radius 2 as a function of its height h.
4. The area A of a trapezoid with bases 2 and 4 as a function of its altitude h.

In Exercises 5–10, state the domain and range of each function.

5. $y = 2x$
6. $y = 1 - 3x$
7. $y = \sqrt{x + 2}$
8. $y = \sqrt{2 - x}$
9. $y = \dfrac{1}{x}$
10. $y = \dfrac{1}{x - 2}$

In Exercises 11–32, graph each of the functions.

11. $y = x - 1$
12. $y = 3x + 2$
13. $y = 3x$
14. $y = 1 - 2x$
15. $y = x^2 + 1$
16. $y = 4x^2 - 1$
17. $y = 1 - x^2$
18. $y = x(x - 1)$
19. $y = x^3$
20. $y = x^3 - 1$
21. $y = \sqrt{x}$
22. $y = \sqrt{x - 1}$
23. $y = \sqrt{4 - x}$
24. $y = \sqrt{x + 3}$
25. $y = \dfrac{1}{x}$
26. $y = \dfrac{1}{x - 1}$

27. $y = \dfrac{1}{x^2}$ **28.** $y = \dfrac{1}{x^2 + 4}$ **29.** $y = 1 - \sqrt{x}$ **30.** $y = \dfrac{1}{x^2 - 1}$

31. $y = \dfrac{x}{x^2 + 1}$ **32.** $y = \dfrac{x^2}{x^2 + 1}$

33. The power P (in watts) drawn by a certain variable resistor is given by

$$P = \frac{2R}{(R + 1)^2} \quad (R \geq 0)$$

where R is measured in ohms. Sketch P as a function of R and estimate the setting on R for which the power drawn is a maximum.

34. For a projectile shot directly upward with a velocity of 10 m/sec, the distance s (in meters) from the ground as a function of t (in seconds) is $s = 10t - 5t^2$ ($t \geq 0$). Sketch the curve.

2.7 Notation for Functions

Since functions are encountered so frequently in mathematics and technology, we need a convenient notation for expressing them.

Suppose y is a function of some variable x. For convenience this statement is written $y = f(x)$.

> **Notation for functions:** If y is a function of x, we write $y = f(x)$, which is read "y equals f of x."

Thus $y = x^2$ can be written $f(x) = x^2$, and $y = \sqrt{x + 1}$ can be written $f(x) = \sqrt{x + 1}$.

One of the main advantages of the notation $y = f(x)$ is that a function value can be specified in a natural way. For example, if $f(x) = x^2$, then $f(2) = 2^2 = 4$.

Example 1 Refer to the function $f(x) = \sqrt{1 - x}$ in Example 4, Section 2.6. The values in the table can now be expressed as follows:

$(1, 0)$: $\quad f(1) = \sqrt{1 - 1} = 0$
$(0, 1)$: $\quad f(0) = \sqrt{1 - 0} = 1$
$(-1, \sqrt{2})$: $\quad f(-1) = \sqrt{1 + 1} = \sqrt{2}$
$(-2, \sqrt{3})$: $\quad f(-2) = \sqrt{1 + 2} = \sqrt{3}$
$(-3, 2)$: $\quad f(-3) = \sqrt{1 + 3} = 2$

If more than one function occurs in a particular discussion, we need to use different letters. For example, $y = G(x)$, $w = h(y)$, and $z = f(r)$ all represent functions.

Example 2 If $H(z) = 2 - 3z^2$, find $H(2)$ and $H(-1)$.

Solution.

$$H(\mathbf{2}) = 2 - 3(\mathbf{2})^2 = -10$$
$$H(\mathbf{-1}) = 2 - 3(\mathbf{-1})^2 = -1$$

Sometimes functions are evaluated in terms of letters, as shown in the next example.

Example 3 If $f(x) = \sqrt{1 - x^2}$, find $f(t)$ and $f(a^2)$.

Solution. If we replace x by t, we get

$$f(t) = \sqrt{1 - t^2}$$

Replacing x by a^2, we have

$$f(a^2) = \sqrt{1 - (a^2)^2} = \sqrt{1 - a^4}$$

Exercises / Section 2.7

1. If $f(x) = x^3 - x + 1$, find $f(1), f(-1), f(0), f(4)$.
2. If $f(x) = \sqrt{x - 3}$, find $f(3), f(4), f(7), f(10)$.
3. If $f(x) = \sqrt{2 - x}$, find $f(2), f(-7)$.
4. If $g(x) = \sqrt{x} - 3$, find $g(0), g(4)$.
5. If $h(x) = \dfrac{1}{x + 2}$, find $h(0), h(-1)$.
6. If $f(x) = \dfrac{1}{\sqrt{x}}$, find $f(1), f(9)$.
7. If $i(t) = 2t + 1$, find $i(0), i(1)$.
8. If $q(t) = \sqrt{1 + t}$, find $q(0), q(3)$.
9. If $H(z) = z^3 - 1$, find $H(1), H(-3)$.
10. If $G(w) = \sqrt{2w + 3}$, find $G(0), G(7)$.
11. If $f(x) = x^2 + 2$, find $f(a), f(a^2)$.
12. If $g(x) = 1 - 2x^2$, find $f(t), f(t^3)$.
13. The resistance in ohms in a certain wire as a function of temperature in degrees Celsius is given by

$$R(T) = 2.00 + 0.132T + 0.00100T^2$$

Find $R(10.0°)$ and $R(100.0°)$.

14. The relation between the tensile strength (in pounds) of a piece of material and the temperature (in degrees Fahrenheit) is given by

$$S(T) = 515.0 - 0.07500\sqrt{T}$$

Find $S(112.1°)$ and $S(248.3°)$.

Review Exercises / Chapter 2

In Exercises 1–12, solve each equation for x.

1. $2x - 3 = x + 6$
2. $4x - 5 = 2x - 7$
3. $2(x + 2) = 5x + 1$
4. $x - 3 = 3(x + 1)$
5. $\frac{1}{2}x - \frac{1}{4} = \frac{1}{4}x + \frac{1}{2}$
6. $\frac{x}{3} + \frac{1}{6} = \frac{2x}{3} + \frac{1}{3}$
7. $\frac{1}{x} = \frac{2}{x} + \frac{1}{2}$
8. $\frac{1}{2x} = \frac{1}{2} + \frac{1}{4x}$
9. $5ax + b = 4ax + 2b$
10. $2(ax + 1) = 4(ax - 3)$
11. $3(ax + b) = 5ax - 3$
12. $4(ax - 2b) = 2(ax - 2b)$

In Exercises 13–16, solve each of the given formulas for the letter indicated.

13. $V = IR$, R
14. $A = an + bn$, a
15. $Q = \frac{3a}{b} + C$, b
16. $A = p + prt$, t

17. The sum of two electric currents is 3.2 A. If one current is 1.8 A more than the other, find the values of the two currents.

18. Two cars 190 mi apart are traveling toward each other at speeds of 45 mi/hr and 50 mi/hr, respectively. When will they meet?

19. How many pounds of an alloy containing 20% brass must be combined with 20 lb of an alloy containing 14% brass to produce an alloy containing 18% brass?

20. Jane has $1.75 in change. She has twice as many dimes as quarters and two more nickels than dimes. How many of each does she have?

21. Otto requires 50 min to wash the family car, and his brother Otis only 40 min. How long would it take for them to wash the car together?

22. If the side of a square is increased by 4 ft, the area of the square is increased by 48 ft². How long is the side of the square?

23. The *efficiency* of an engine is defined as the ratio of output to input; this ratio is commonly expressed as a percentage. If the output of a certain engine is 9,000 W and the input 15,000 W, find its efficiency.

24. The *relative error* is defined as the ratio of the error of measurement to the correct value. If the side of a square is 10.0 ft with a possible error of 0.2 ft, find the resulting relative error in the area.

25. We know from geometry that two triangles are similar if the corresponding parts are proportional. Suppose a man 6 ft tall casts a shadow 4 ft long. At the same time a tree casts a shadow 20 ft long. How tall is the tree?

26. A flagpole casts a shadow 8 ft long. At the same time a yardstick casts a shadow 6 in. long. Determine the height of the flagpole. (Refer to Exercise 25.)

78   CHAPTER 2   EQUATIONS, RATIO AND PROPORTION, VARIATION, AND FUNCTIONS

In Exercises 27–32, express each statement in the form of an equation.

27. y varies inversely as r and jointly as s and t.
28. y is directly proportional to x^2; $y = 9$ when $x = 3$.
29. The weight w of a body above the earth's surface varies inversely as the square of the distance d of the body from the center of the earth.
30. The capacitance C of two parallel plates varies inversely as the distance d between the plates.
31. The resistance R in a wire varies directly as its length L and inversely as the square of its diameter D.
32. The period P of a pendulum varies directly as the square root of its length L.
33. Graph the function $y = 2\sqrt{x - 4}$.
34. Graph the function $y = 1/(x^2 + 1)$.
35. State the domain of the function in Exercise 33.
36. If $g(z) = \sqrt{1 - z}$, find $g(0)$, $g(1)$.

CHAPTER 3

Systems of Linear Equations and Introduction to Determinants

Objectives Upon completion of this chapter, you should be able to:
1. Solve systems of two linear equations graphically.
2. Solve systems of two linear equations algebraically by means of:
   a. Addition or subtraction.
   b. Substitution.
   c. Determinants.
3. Solve word problems leading to systems of equations.
4. Solve systems of three or four equations algebraically.
5. Expand third-order determinants by minors.
6. Solve systems of three equations by Cramer's rule.

3.1 Simultaneous Linear Equations

In Chapter 2 we studied the solution of first-order equations. However, many applications in technology require the solution of systems containing two or more equations. In this section we will study the geometric basis of systems containing two equations, as well as how to solve such equations by drawing graphs.

It is revealed through analytic geometry that the general equation of a straight line is

$$ax + by = c \tag{3.1}$$

Linear equation

(A detailed discussion of the line is given in Chapter 19.) Since equation (3.1) represents a straight line, it is called a **linear equation** or a **linear equation in two variables.** For example, the equation $3x - 2y = 6$ fits form (3.1) and therefore represents a line. Suppose we graph this line from the following table of values:

x:	0	1	2
y:	-3	$-\frac{3}{2}$	0

The graph is shown in Figure 3.1. Note that the coordinates of the points $(0, -3)$, $(1, -\frac{3}{2})$, and $(2, 0)$ in the table satisfy the equation. Moreover, the coordinates of *every* point on the line satisfy the equation.

Now consider another line, $x + y = 2$, shown in Figure 3.2. As before, the coordinates of *every* point on this line satisfy the equation $x + y = 2$.

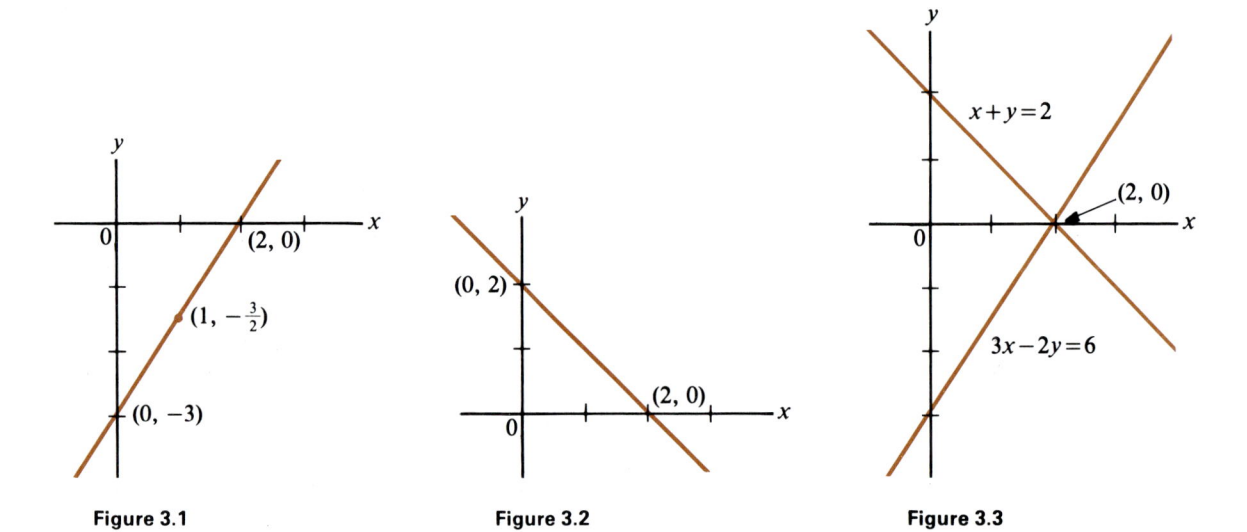

Figure 3.1

Figure 3.2

Figure 3.3

According to Figure 3.3, which shows both lines, the two lines intersect at $(2, 0)$. So the coordinates of the point $(2, 0)$ satisfy *both* equations. Consequently, $x = 2$ and $y = 0$ is called the **common solution** of the system

$$3x - 2y = 6$$
$$x + y = 2$$

In general, two linear equations in two unknowns are referred to as a system of two **simultaneous linear equations.**

Simultaneous linear equations

$$a_1 x + b_1 y = c_1$$
$$a_2 x + b_2 y = c_2$$

(3.2)

Common solution

Any pair (x, y) of values that satisfies both equations is called a **common solution** of the system, or simply a **solution.** If a common solution exists, then it follows from the foregoing discussion that the point (x, y) is the intersection of the two lines. If the lines are distinct, then this point, and hence the solution, is necessarily unique. On the other hand, since two lines may be parallel, a system may not have any solution.

> **Graphical solution**: To find the common solution of a system of two linear equations graphically, draw the two lines and determine the point of intersection from the graph.

To be able to sketch the graphs of linear equations rapidly, we need the notion of **intercept,** which is defined next.

> **Definition of intercept**: An **intercept** is a point at which the graph crosses a coordinate axis. A point $(a, 0)$ is called an ***x*-intercept,** and a point $(0, b)$ is called a ***y*-intercept.**

To find the *x*-intercept, we let $y = 0$ and solve the resulting equation for *x*. To find the *y*-intercept, we let $x = 0$ and solve the resulting equation for *y*.

For example, in $3x - 2y = 6$, if **y = 0**, then **x = 2**. So the point **(2, 0)** is the *x*-intercept. (See Figure 3.4.) If **x = 0**, then **y = −3**, so that **(0, −3)** is the *y*-intercept. Since two distinct points determine a straight line, these are the only points needed to draw the graph. A third point, $(1, -\frac{3}{2})$, is included only as a check. (See Figure 3.4.)

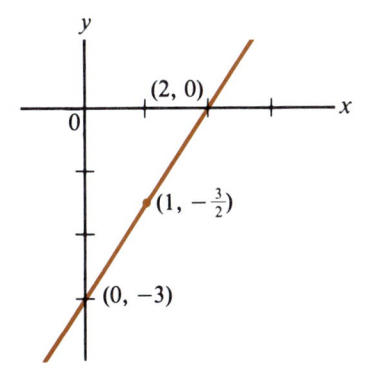

Figure 3.4

Example 1 Determine the common solution of the system

$$2x + y = 5$$
$$x + 3y = 5$$

by drawing the graph of each line and estimating the coordinates of the point of intersection.

Solution. As indicated earlier, the simplest way to draw a line is to find the intercepts. Letting $x = 0$ in the first equation, we find that $y = 5$. Similarly, if we let $y = 0$, then $x = \frac{5}{2}$. (See Figure 3.5.) For the second equation, if $x = 0$, then $y = \frac{5}{3}$; and if $y = 0$, then $x = 5$. (See Figure 3.5.) To get a check point for the first equation, we let $x = 1$, so that $y = 3$. A check point for the second line is $(-1, 2)$; both points are shown in Figure 3.5.

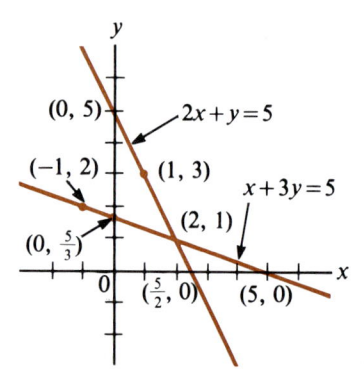

Figure 3.5

We now draw the two lines and observe that they appear to cross at $(2, 1)$, at least as closely as can be determined from the graph. As a check, let us substitute the coordinates of this point into the given equations:

$$2(\mathbf{2}) + \mathbf{1} = 5$$
$$\mathbf{2} + 3(\mathbf{1}) = 5$$

So the common solution is indeed given by

$$x = 2 \quad \text{and} \quad y = 1$$

Example 2 Determine the solution of the system

$$2x - y = -8$$
$$x - 3y = 3$$

graphically to the nearest tenth of a unit.

Solution. To obtain the intercepts for the first equation, let $x = 0$, so that $y = 8$. If $y = 0$, then $x = -4$. The intercepts are $(-4, 0)$ and $(0, 8)$.

Now assign some arbitrary value to x, such as $x = 1$. Then $2(1) - y = -8$, and $y = 10$. So the check point for the first equation is $(1, 10)$.

Similarly, for the second equation we find that the intercepts are $(3, 0)$ and $(0, -1)$. A check point is $(6, 1)$.

Now we draw the two lines and estimate the coordinates of the point of intersection. To the nearest tenth the coordinates appear to be $(-5.4, -2.8)$. (See Figure 3.6.)

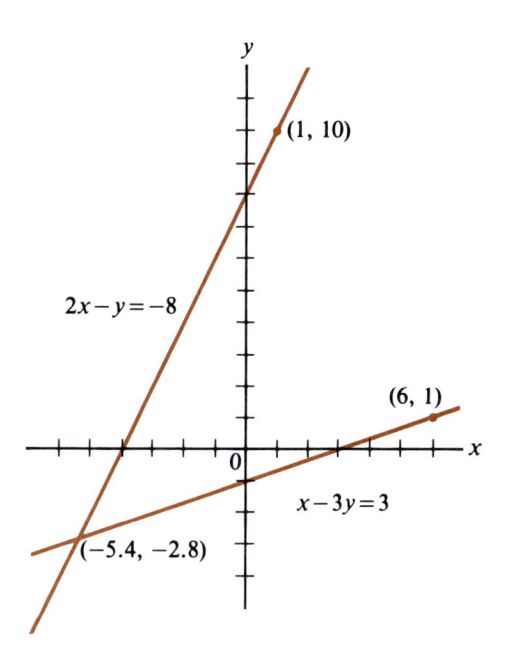

Figure 3.6

We know from geometry that two distinct lines are either parallel or intersecting. If they are parallel, then a common solution cannot exist. Whether two lines really are parallel cannot be determined graphically, since two lines may look parallel and yet intersect at some distant point. Moreover, Example 2 shows that the graphical method for solving simultaneous equations is awkward at best. Fortunately, there are algebraic methods for solving systems directly, which we will study in the next section. However, the graphical approach has provided us with the necessary geometric background.

Exercises / Section 3.1

Solve the following systems of equations graphically. Estimate the answers to the nearest tenth of a unit.

1. $2x - y = 1$
   $x - y = 2$
2. $x - y = -3$
   $x + y = -1$
3. $x - 2y = -4$
   $x + 3y = 11$
4. $2x - y = 2$
   $x + y = 1$
5. $-x - 3y = 4$
   $2x + 2y = 5$
6. $x + 2y = 13$
   $x - 3y = 2$
7. $2x + 3y = 2$
   $3x + 2y = 1$
8. $3x + 3y = 19$
   $-x + 2y = 1$
9. $5x - 2y = 9$
   $4x - 3y = 4$
10. $6x + 2y = 1$
    $3x + 4y = 10$
11. $3x + y = 0$
    $x - 2y = 20$
12. $x - 3y = 10$
    $3x - y = 3$

3.2 Algebraic Solutions

In the last section we considered the graphical solution of two simultaneous linear equations. In this section we shall turn our attention to solving such systems algebraically.

The real difficulty in solving two equations simultaneously is that two different unknowns are involved. Algebraic solutions resolve this difficulty by eliminating one of the unknowns, thereby reducing the problem to solving a single equation with one unknown.

Addition or Subtraction

The first method to be considered is the **method of addition or subtraction.**

Method of addition or subtraction

1. Multiply both sides of the equations (if necessary) by constants so chosen that the coefficients of one of the unknowns are numerically equal.
2. If the coefficients have opposite signs, add the corresponding members of the equations. If the coefficients have like signs, subtract the corresponding members of the equations.
3. Solve the resulting equation in one unknown for the unknown.
4. Substitute the value of the unknown in either of the original equations and solve for the second unknown.
5. Check the solution in the original system.

To see how addition or subtraction can eliminate one of the unknowns, consider the system

$$2x + 3y = 1$$
$$x + 3y = 2$$

Note that the y-coefficients are the same. If the second equation is subtracted from the first, y is eliminated:

$$
\begin{aligned}
2x + 3y &= 1 \\
\underline{x + 3y} &= \underline{2} \\
x \phantom{{}+3y} &= -1 \quad \text{subtracting}
\end{aligned}
$$

We conclude that $x = -1$. From the second equation ($x + 3y = 2$) we get $(-1) + 3y = 2$, so that $y = 1$. The common solution is therefore $x = -1$ and $y = 1$.

Consider another example.

Example 1 Solve the following system:

$$
\begin{aligned}
2x - y &= 1 \\
x - 3y &= -2
\end{aligned}
$$

Solution. We can eliminate x as follows: Multiply both sides of the second equation by 2, thereby making the coefficients the same, and then subtract the second equation from the first. Thus

$$2x - y = 1$$
Step 1. $\quad \underline{2x - 6y = -4} \quad 2(x - 3y) = 2(-2)$
Step 2. $\quad 0 + 5y = 5 \quad$ subtracting

Step 3. Solving the resulting equation, we get $y = 1$

Step 4. To find the corresponding x-value, substitute $y = 1$ into either of the given equations and solve for x. Using the second equation, we get

$$x - 3(1) = -2$$

which yields $x = 1$. The solution is therefore given by $(1, 1)$.

Step 5. As a check, let us substitute these values into the given equations:

$$2(1) - 1 = 1 \quad \text{and} \quad 1 - 3(1) = -2$$

The solution checks.

If the coefficients of one of the two variables are numerically equal but have opposite signs, this variable is eliminated by adding the two equations.

Example 2 Solve the system of equations

$$
\begin{aligned}
3x - 2y &= 5 \\
5x + 2y &= 1
\end{aligned}
$$

Solution. Since the coefficients of y are numerically equal but have opposite signs, we can eliminate y by adding the equations. Thus

$$3x - 2y = 5$$
$$5x + 2y = 1$$
$$8x \phantom{{}-2y} = 6 \qquad \text{adding}$$

and $x = \frac{3}{4}$. Substituting into the first equation, we get

$$3\left(\frac{3}{4}\right) - 2y = 5$$

$$-2y = 5 - \frac{9}{4} \qquad \text{transposing}$$

$$-2y = \frac{11}{4} \qquad \text{simplifying}$$

$$y = -\frac{11}{8} \qquad \text{dividing by } -2$$

As a check, substitute $(\frac{3}{4}, -\frac{11}{8})$ into the second equation:

$$5\left(\frac{3}{4}\right) + 2\left(-\frac{11}{8}\right) = \frac{15}{4} - \frac{11}{4} = 1$$

in agreement with the right side.

In some cases both equations have to be multiplied by a constant before one of the variables can be eliminated.

Example 3 Solve the system

$$3x + 2y = 1$$
$$4x - 3y = 7$$

Solution. In this example, direct addition or subtraction will not eliminate either variable. The simplest approach is to eliminate y by multiplying the first equation by 3 and the second by 2. Thus

$$9x + 6y = 3 \qquad 3(3x + 2y) = 3(1)$$
$$8x - 6y = 14 \qquad 2(4x - 3y) = 2(7)$$
$$17x \phantom{{}- 6y} = 17 \qquad \text{adding}$$

It follows that $x = 1$ and $y = -1$.

Sometimes a system of equations does not have any solution, as we can see in the next example.

Example 4

Solve the system

$$-5x + 2y = 7$$
$$10x - 4y = 5$$

Solution. In this example y appears to be the easier of the two unknowns to eliminate:

$$-10x + 4y = 14 \quad \text{multiplying by 2}$$
$$\underline{10x - 4y = 5}$$
$$0 = 19 \quad \text{adding}$$

There is something wrong, since 0 cannot be equal to 19. It follows that the system has no solution. Geometrically, the equations represent two parallel lines. (See Figure 3.7.)

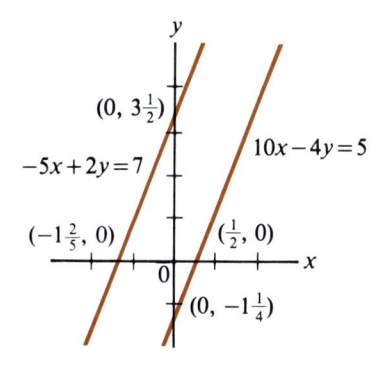

Figure 3.7

Example 5

Compare the system

$$-5x + 2y = -\frac{5}{2}$$
$$10x - 4y = 5$$

to that in Example 4.

Solution.
$$-10x + 4y = -5 \quad \text{multiplying by 2}$$
$$\underline{10x - 4y = 5}$$
$$0 = 0 \quad \text{adding}$$

This time no contradiction results. In fact, we have merely shown that the two equations represent exactly the same line. As a result, the coordinates of any point on this line satisfy the given system. (See Figure 3.8.)

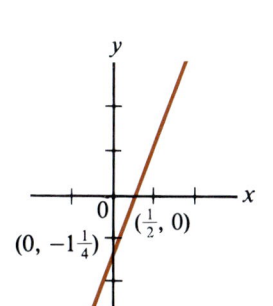

Figure 3.8

The system in Example 4 is said to be **inconsistent,** which means that the system has no solution. Geometrically, an inconsistent system consists of two parallel lines. The system in Example 5 is said to be **dependent;** that is, the lines coincide, and the system has infinitely many solutions.

In both cases the coefficients of the respective unknowns are multiples of each other. These ideas are summarized next.

A **dependent system** has the form

$$ax + by = c$$
$$kax + kby = kc \quad (k \neq 0)$$

An **inconsistent system** has the form

$$ax + by = c$$
$$kax + kby = d, \quad d \neq kc$$

Substitution

At this point we seem to have covered all cases. Indeed, the method of addition or subtraction works with any system of linear equations. However, if one equation is easily solved for one of the unknowns, it may be more convenient to solve by the **method of substitution.**

Method of substitution

1. Solve one of the equations for one of the unknowns in terms of the other.
2. Substitute the expression obtained into the other equation.
3. Solve the resulting equation in one unknown for the unknown.
4. Substitute the value of the unknown in either of the original equations and solve for the second unknown.
5. Check the solution in the original system.

To see how the method of substitution can be used to eliminate one of the unknowns, consider the system

$$2x - y = 2$$
$$6x + 2y = 1$$

Note that the first equation is readily solved for y to yield

Step 1. $y = 2x - 2$

Substituting this expression for y in the second equation results in an equation containing only one unknown:

$$6x + 2y = 1 \quad \text{second equation}$$

Step 2. $\quad 6x + 2(\mathbf{2x - 2}) = 1 \quad$ substituting $2x - 2$ for y

Step 3. Solve for x:

$$6x + 4x - 4 = 1$$
$$10x = 5$$
$$x = \frac{1}{2}$$

Step 4. From the first equation, rewritten as $y = 2x - 2$, we get

$$y = 2\left(\frac{1}{2}\right) - 2 = -1$$

The solution is therefore given by $(\frac{1}{2}, -1)$.

Step 5. Check:

$$2\left(\frac{1}{2}\right) - (\mathbf{-1}) = 2 \quad 6\left(\frac{1}{2}\right) + 2(\mathbf{-1}) = 1$$

This example shows that the method of substitution is most convenient if one of the variables has a coefficient of 1. (Otherwise solving for one of the unknowns may lead to an expression involving fractions.)

Some equations have unknowns occurring in the denominator:

$$\frac{1}{z} + \frac{2}{w} = 1$$
$$\frac{3}{z} - \frac{2}{w} = 7$$

Such systems can be solved by the usual method if we let $x = 1/z$ and $y = 1/w$. Consider the following example:

Example 6 Use the method of substitution to solve the system

$$\frac{3}{s_1} - \frac{2}{s_2} = 1$$
$$\frac{16}{s_1} - \frac{12}{s_2} = 5$$

Solution. This system can be written in the usual form by letting $x = 1/s_1$ and $y = 1/s_2$:

$$3x - 2y = 1$$
$$16x - 12y = 5$$

Suppose we solve the first equation for y in terms of x. Then

$$-2y = 1 - 3x$$

$$y = -\frac{1}{2}(1 - 3x) \tag{3.3}$$

Substituting into the second equation, we get

$$16x - 12\left[-\frac{1}{2}(1 - 3x)\right] = 5 \quad \text{substituting } -\frac{1}{2}(1 - 3x) \text{ for } y$$

$$16x + 6(1 - 3x) = 5$$

$$16x + 6 - 18x = 5$$

$$-2x = -1$$

$$x = \frac{1}{2}$$

Since $y = -\frac{1}{2}(1 - 3x)$, we now get

$$y = -\frac{1}{2}\left(1 - \frac{3}{2}\right) = \frac{1}{4}$$

Finally, since $x = \dfrac{1}{s_1}$, it follows that $s_1 = 1/x = 2$. Similarly, $s_2 = 1/y = 4$.

Remark. The method of substitution is particularly important for solving equations of the second degree, where the variables have the form x^2 and y^2. Since these will be taken up in Chapter 13, you need to become familiar with the method of substitution.

Exercises / Section 3.2

In Exercises 1–10, solve each system of equations by the method of addition or subtraction.

1. $x + y = 4$
   $2x - y = 5$

2. $3x - 2y = 6$
   $3x - 4y = -2$

3. $3x - 2y = 1$
   $4x - 3y = 4$

4. $2x + 7y = 0$
   $3x - 2y = 25$

5. $3x - 2y = 21$
   $4x - 5y = 42$

6. $3x - 2y = 7$
   $4x - 6y = 11$

7. $4x - 3y = -11$
   $12x + 25y = 69$

8. $4x + 3y = 1$
   $5x + 8y = 10$

9. $2x + 2y = 1$
   $5x - 5y = 1$

10. $3x + 5y = 10$
    $5x + 3y = 10$

In Exercises 11–20, solve each system of equations by the method of substitution.

11. $x - 3y = 4$
    $2x - y = 3$

12. $x + 2y = 12$
    $x - 3y = 2$

13. $3x + 4y = 21$
    $-x + 2y = 3$

14. $x + 3y = 1$
    $2x - y = -5$

15. $2x + y = 1$
    $x + 3y = 8$

16. $x + 2y = 13$
    $3x - y = -31$

17. $8x - 10y = -13$
    $x + 2y = 0$

18. $-x - 2y = 3$
    $2x + y = 4$

19. $5x + 2y = 3$
    $6x + 3y = 2$

20. $2x - 3y = 3$
    $5x - 4y = 1$

In Exercises 21–40, solve each system of equations by either method.

21. $6x - 7y = 49$
    $8x - 9y = 63$

22. $3x + 5y = 0$
    $x + 4y = 0$

23. $3x - 2y = 1$
    $6x - 4y = 5$

24. $2x + 4y = 1$
    $6x + 12y = -1$

25. $\dfrac{4}{x} - \dfrac{3}{y} = 1$
    $\dfrac{5}{x} - \dfrac{4}{y} = 1$

26. $\dfrac{5}{x} - \dfrac{25}{y} = 51$
    $\dfrac{10}{x} - \dfrac{55}{y} = 112$

27. $\dfrac{2}{x} - \dfrac{3}{y} = 1$
    $\dfrac{3}{x} - \dfrac{2}{y} = 2$

28. $\dfrac{2}{x} - \dfrac{1}{y} = 2$
    $\dfrac{4}{x} + \dfrac{5}{y} = 6$

29. $\dfrac{2}{p} - \dfrac{1}{q} = 1$
    $\dfrac{4}{p} - \dfrac{3}{q} = 4$

30. $\dfrac{2}{R} - \dfrac{3}{P} = 12$
    $\dfrac{6}{R} - \dfrac{5}{P} = 15$

31. $3F_1 - 2F_2 = -7$
    $3F_1 - 12F_2 = -37$

32. $12s - 8t = 19$
    $4s - 12t = 25$

33. $3A_1 + 2A_2 = 2$
    $5A_1 + 3A_2 = 3$

34. $-2y + 4w = 1$
    $-3y + 5w = 2$

35. $2w - 3z = 5$
    $4w - 6z = 10$

36. $7m_1 + 2m_2 = 20$
    $9m_1 + 3m_2 = -15$

37. $-2v + 5w = 10$
    $4v - 10w = 15$

38. $2v_0 - 6v_1 = 3$
    $-3v_0 + 3v_1 = 1$

39. $3I + 2I_0 = 7$
    $2I + I_0 = 4$

40. $-2r + 4s = -6$
    $r - 2s = 3$

3.3 Introduction to Determinants

So far we have considered three methods for solving systems of equations. In this section we are going to study yet another, the method of **determinants.** This method is not only elegant but also quite easy to use and to apply to larger systems.

Determinants were discovered by Gottfried Leibniz, the codiscoverer of the calculus, and promptly forgotten. They were rediscovered by the Swiss mathematician Gabriel Cramer (1704–1752) in 1750. Some time later the English algebraist Arthur Cayley (1821–1895) undertook a systematic study of determinants and matrices. Cayley's work eventually led to a separate branch of mathematics called *linear algebra*.

To see the value of determinants, consider the general solution of the class of systems

$$a_1 x + b_1 y = c_1$$
$$a_2 x + b_2 y = c_2 \qquad (3.4)$$

Gabriel Cramer

We can eliminate y by multiplying the first equation by b_2 and the second by b_1 to obtain

$$a_1 b_2 x + b_1 b_2 y = c_1 b_2$$
$$a_2 b_1 x + b_2 b_1 y = c_2 b_1$$

Subtracting, we get

$$a_1 b_2 x - a_2 b_1 x = c_1 b_2 - c_2 b_1$$

It now follows from the distributive law that

$$(a_1 b_2 - a_2 b_1)x = c_1 b_2 - c_2 b_1$$

and therefore

$$x = \frac{c_1 b_2 - c_2 b_1}{a_1 b_2 - a_2 b_1} \tag{3.5}$$

By eliminating x, one can show that

$$y = \frac{a_1 c_2 - a_2 c_1}{a_1 b_2 - a_2 b_1} \tag{3.6}$$

We have found the solution of the general system (3.4). Thus anyone who memorizes formulas (3.5) and (3.6) would never again have to solve such a system. Unfortunately, memorizing such formulas is probably more trouble than solving the system unless, of course, some kind of simple pattern can be discovered. Such a pattern is provided by a square array of numbers called a **determinant.** Thus $a_1 b_2 - a_2 b_1$, which appears in each of the denominators, is a quantity denoted by the symbol given next.

Definition of a 2 × 2 determinant

$$\begin{vmatrix} a_1 & b_1 \\ a_2 & b_2 \end{vmatrix} = a_1 b_2 - a_2 b_1 \tag{3.7}$$

The determinant

$$\begin{vmatrix} a_1 & b_1 \\ a_2 & b_2 \end{vmatrix}$$

Elements
Row and column

is called a 2 × 2 (two-by-two) determinant (larger determinants will be taken up later). The entries are called **elements.** The elements a_1 and b_1 form the first **row,** and the elements a_1 and a_2 the first **column.** (The second row and column are then defined in a similar way.) Writing the square array as a number, called the **expansion** of the determinant, can best be remembered by means of the following diagram:

Expansion of a
2 × 2 determinant

$$\begin{vmatrix} a_1 & b_1 \\ a_2 & b_2 \end{vmatrix} = a_1 b_2 - a_2 b_1 \tag{3.8}$$

The arrows indicate which elements are to be multiplied, and the sign at the tip of each arrow indicates which sign is to be affixed to the product.

Example 1 Expand the determinants

a. $\begin{vmatrix} 0 & 2 \\ -3 & 5 \end{vmatrix}$ b. $\begin{vmatrix} 2 & -1 \\ -6 & -4 \end{vmatrix}$

Solution. By expansion (3.8)

a. $\begin{vmatrix} 0 & 2 \\ -3 & 5 \end{vmatrix} = (0)(5) - (-3)(2) = 0 + 6 = 6$

b. $\begin{vmatrix} 2 & -1 \\ -6 & -4 \end{vmatrix} = (2)(-4) - (-6)(-1) = -8 - 6 = -14$

Let us now return to the system of equations (3.4):

$$a_1x + b_1y = c_1$$
$$a_2x + b_2y = c_2 \tag{3.9}$$

Observe that the solution, given in statements (3.5) and (3.6), can be written

$$x = \frac{\begin{vmatrix} c_1 & b_1 \\ c_2 & b_2 \end{vmatrix}}{\begin{vmatrix} a_1 & b_1 \\ a_2 & b_2 \end{vmatrix}}, \quad y = \frac{\begin{vmatrix} a_1 & c_1 \\ a_2 & c_2 \end{vmatrix}}{\begin{vmatrix} a_1 & b_1 \\ a_2 & b_2 \end{vmatrix}}$$

Now notice the pattern: The entries in the two denominators are the coefficients of the unknowns arranged as in the original system (3.9). The entries in the numerator of the x-value are obtained by replacing the coefficients of x by the constants on the right side of the system. Similarly, the entries in the numerator of the y-value are obtained by replacing the coefficients of y by the constants on the right. This method of solution is known as **Cramer's rule** and can be extended to larger systems of equations.

Before continuing with an example, let's summarize the solution of systems of two equations by Cramer's rule.

Cramer's rule: The solution of the system

$$a_1x + b_1y = c_1$$
$$a_2x + b_2y = c_2$$

is given by

$$x = \frac{\begin{vmatrix} c_1 & b_1 \\ c_2 & b_2 \end{vmatrix}}{\begin{vmatrix} a_1 & b_1 \\ a_2 & b_2 \end{vmatrix}}, \quad y = \frac{\begin{vmatrix} a_1 & c_1 \\ a_2 & c_2 \end{vmatrix}}{\begin{vmatrix} a_1 & b_1 \\ a_2 & b_2 \end{vmatrix}}$$

Example 2 Solve the system

$$5x - 2y = 10$$
$$-3x + 7y = 0$$

by Cramer's rule.

Solution. By Cramer's rule, the determinant in both denominators consists of the coefficients of the unknowns arranged as in the given system:

$$\begin{vmatrix} 5 & -2 \\ -3 & 7 \end{vmatrix}$$

The numerator for x can be constructed from this determinant by replacing the first column, the coefficients of x, by the constants on the right:

$$\begin{vmatrix} 10 & -2 \\ 0 & 7 \end{vmatrix}$$

Similarly, to find y we construct the determinant in the numerator by replacing the second column, the coefficients of y, by the constants on the right.

$$\begin{vmatrix} 5 & 10 \\ -3 & 0 \end{vmatrix}$$

The solution is now written as follows:

$$x = \frac{\begin{vmatrix} 10 & -2 \\ 0 & 7 \end{vmatrix}}{\begin{vmatrix} 5 & -2 \\ -3 & 7 \end{vmatrix}} = \frac{(10)(7) - (0)(-2)}{(5)(7) - (-3)(-2)} = \frac{70 - 0}{35 - 6} = \frac{70}{29}$$

$$y = \frac{\begin{vmatrix} 5 & 10 \\ -3 & 0 \end{vmatrix}}{\begin{vmatrix} 5 & -2 \\ -3 & 7 \end{vmatrix}} = \frac{(5)(0) - (-3)(10)}{29} = \frac{30}{29}$$

Returning to Cramer's rule, note that the determinant occurring in each denominator has to be different from 0 to avoid division by 0. If the determinant is different from 0, then the solution of system (3.9) is necessarily unique, since the value of a determinant is unique. If

$$\begin{vmatrix} a_1 & b_1 \\ a_2 & b_2 \end{vmatrix} = 0$$

then the system does not have a unique solution. In that case the system is either dependent or inconsistent.

CALCULATOR COMMENT   If the arithmetic involved in expanding a determinant is awkward, a calculator can make the computation more convenient. Consider the determinant

$$\begin{vmatrix} -3.59 & 5.26 \\ 7.35 & 3.98 \end{vmatrix} = -3.59 \times 3.98 - 7.35 \times 5.26$$

Since most scientific calculators perform multiplication and division before addition and subtraction, the sequence is

3.59 $\boxed{+/-}$ $\boxed{\times}$ 3.98 $\boxed{-}$ 7.35 $\boxed{\times}$ 5.26 $\boxed{=}$

Display: -52.9492

For calculators that do not automatically perform the operations in the above order, an alternate sequence is

3.59 $\boxed{+/-}$ $\boxed{\times}$ 3.98 $\boxed{=}$ $\boxed{\text{STO}}$ 7.35 $\boxed{\times}$ 5.26 $\boxed{=}$ $\boxed{+/-}$ $\boxed{+}$ $\boxed{\text{MR}}$ $\boxed{=}$

Display: -52.9492

Exercises / Section 3.3

In Exercises 1–16, expand each determinant.

1. $\begin{vmatrix} 0 & 0 \\ 2 & 1 \end{vmatrix}$
2. $\begin{vmatrix} 1 & -2 \\ 4 & -8 \end{vmatrix}$
3. $\begin{vmatrix} -2 & 4 \\ 4 & -8 \end{vmatrix}$
4. $\begin{vmatrix} 1 & -2 \\ 0 & 2 \end{vmatrix}$
5. $\begin{vmatrix} 2 & 1 \\ 7 & 4 \end{vmatrix}$
6. $\begin{vmatrix} 1 & 5 \\ 2 & 3 \end{vmatrix}$
7. $\begin{vmatrix} 2 & 1 \\ 4 & -3 \end{vmatrix}$
8. $\begin{vmatrix} -2 & 6 \\ 8 & -4 \end{vmatrix}$
9. $\begin{vmatrix} -2 & -1 \\ 12 & 5 \end{vmatrix}$
10. $\begin{vmatrix} -3 & -5 \\ -6 & -4 \end{vmatrix}$
11. $\begin{vmatrix} -6 & -17 \\ 3 & 20 \end{vmatrix}$
12. $\begin{vmatrix} 12 & -3 \\ -15 & 6 \end{vmatrix}$
13. $\begin{vmatrix} 32 & 21 \\ -17 & 16 \end{vmatrix}$
14. $\begin{vmatrix} 21 & 5 \\ 20 & -8 \end{vmatrix}$
15. $\begin{vmatrix} 18 & -6 \\ 75 & 0 \end{vmatrix}$
16. $\begin{vmatrix} -12 & -11 \\ 5 & -18 \end{vmatrix}$

In Exercises 17–41, solve each system of equations by using Cramer's rule.

17. $3x + 4y = 1$
    $2x + 3y = 4$
18. $-x + 3y = 5$
    $-2x - y = 0$
19. $3x - 2y = 4$
    $-7x + 5y = 1$
20. $-2x - 6y = 4$
    $5x + 10y = 5$
21. $-5x + 6y = 7$
    $4x - 5y = 8$
22. $3x - 4y = 20$
    $5x - 6y = 8$
23. $6x + 3y = -1$
    $5x + 3y = 2$
24. $-3x + 4y = 11$
    $6x - 8y = 8$
25. $\dfrac{2}{x} - \dfrac{3}{y} = 7$
    $\dfrac{1}{x} + \dfrac{5}{y} = 3$

26. $\dfrac{4}{x} - \dfrac{4}{y} = 9$

$\dfrac{3}{x} + \dfrac{5}{y} = 8$

27. $\dfrac{8}{x} - \dfrac{7}{y} = 20$

$\dfrac{3}{x} - \dfrac{5}{y} = 16$

28. $\dfrac{2}{x} + \dfrac{5}{y} = 3$

$\dfrac{5}{x} - \dfrac{7}{y} = 6$

29. $4T_1 - 7T_2 = 10$
$5T_1 - 8T_2 = 20$

30. $2w_1 + w_2 = 5$
$w_1 + 3w_2 = 5$

31. $F_1 + 2F_2 = 5$
$2F_1 + F_2 = 6$

32. $2W_1 + 3W_2 = 12$
$2W_1 + 4W_2 = 15$

33. $3R_1 + 4R_2 = 20$
$4R_1 + 2R_2 = 15$

34. $2I_1 + 8I_2 = 25$
$4I_1 + 7I_2 = 30$

35. $3C_1 - 7C_2 = 3$
$-9C_1 + 21C_2 = 5$

36. $2p - 7q = 10$
$5p - q = 15$

(The remaining problems should be solved with a calculator.)

37. $2.73x - 1.52y = 5.02$
$0.130x + 2.49y = 2.98$

38. $0.980x + 0.730y = 1.21$
$-1.32x - 5.21y = -1.11$

39. $-2.10x + 3.64y = 1.32$
$1.00x + 1.78y = -4.05$

40. $6.52x + 3.98y = -1.25$
$1.35x - 1.44y = 2.73$

41. $-7.63x - 5.02y = 1.31$
$2.84x - 1.54y = 3.87$

3.4 Applications of Systems of Linear Equations

While systems of equations have varied applications in technology, the type of problems already discussed in Section 2.3 can be done more conveniently by using systems of equations. As a typical example, suppose that the sum of two numbers is 35, and one of the numbers is 3 more than the other. Let x be the smaller number; thus $x + 3$ represents the larger. It follows that

$$x + (x + 3) = 35 \tag{3.10}$$

and

$$2x + 3 = 35$$

From this we get $x = 16$ and $x + 3 = 19$.

Now consider another way to solve this problem by using systems of equations. Let x be the first number and y the second. Then

$$x + y = 35 \tag{3.11}$$

From the other piece of information, we have

$$y = x + 3 \tag{3.12}$$

Substituting equation (3.12) into equation (3.11), we get equation (3.10). The first method, then, is the same as solving the system by substitution. But as we have seen, the method of substitution is not always as convenient as the method of addition or subtraction or the method of determinants.

3.4 APPLICATIONS OF SYSTEMS OF LINEAR EQUATIONS

Example 1 One alloy contains 20% copper, and another 25% copper. How many pounds of each must be combined to form 60 lb of an alloy containing 22% copper?

Solution. Let x be the number of pounds of the first alloy and y the number of pounds of the second. Then $x + y = 60$. Recall that it is best to work with the quantities directly: For example, $(0.22)(60) = 13.2$ lb, the weight of copper in the 22% alloy. Thus

$$0.20x + 0.25y = (0.22)(60)$$
$$x + y = 60$$

are the resulting equations, which can be solved by addition or subtraction:

$$20x + 25y = (22)(60) \quad \text{multiplying first equation by 100}$$
$$\underline{20x + 20y = 1200} \quad \text{multiplying second equation by 20}$$
$$5y = 120 \quad \text{subtracting}$$
$$y = 24 \text{ lb}$$
$$x = 36 \text{ lb}$$

Example 2 Two resistors are connected in series. The resistance of the second is 5.6 Ω less than that of the first. The total resistance is 50.2 Ω. Find the resistance of each resistor.

Solution. Recall that the total resistance of two or more resistors in series is equal to the sum of the individual resistances. Letting R_1 and R_2 be the individual resistances, we get

$$R_1 - R_2 = 5.6$$
$$\underline{R_1 + R_2 = 50.2}$$
$$2R_1 = 55.8 \quad \text{adding}$$
$$R_1 = 27.9 \ \Omega$$
$$R_2 = 22.3 \ \Omega$$

Example 3 A portion of $13,580 was invested at 8% interest and the rest at 10%. If the total interest income was $1,253, how much was invested at each rate?

Solution. Let

$$x = \text{amount invested at 8\%}$$
$$y = \text{amount invested at 10\%}$$

Then

$$x + y = \$13,580$$
$$0.08x + 0.10y = \$1,253$$

are the equations to be solved.

$$\begin{aligned}8x + 8y &= 108,640 \quad &\text{multiplying first equation by 8}\\ \underline{8x + 10y} &= \underline{125,300} \quad &\text{multiplying second equation by 100}\\ -2y &= -16,660 \quad &\text{subtracting}\\ y &= \$8,330\\ x &= \$5,250\end{aligned}$$

Of course, the solution of a problem can always be checked against the given information. Thus 8% of $5,250 equals (0.08)($5,250) = $420 and 10% of $8,330 is $833, for a total of $1,253.

Many problems in technology involve the lever. A **lever** is a rigid bar supported at a point called the **fulcrum,** which is usually positioned between the ends of the bar. Neglecting the weight of the lever, a weight w at a distance d from the fulcrum has a **moment** given by weight times distance, or wd. (The **moment** is a measure of the tendency of the weight to rotate about the fulcrum.) If two weights w_1 and w_2 are placed on opposite sides of the fulcrum at distances d_1 and d_2, respectively, then

$$w_1 d_1 = w_2 d_2$$

whenever the weights are balanced on the lever. The distance from the fulcrum is called the **moment arm.**

One of the fundamental principles of levers is the fact that moments are additive; that is, $w_1 d_1 + w_2 d_2$ is equal to the total moment. Consider, for example, the weights on the lever in Figure 3.9. For the weights to be balanced on the lever, the moment on the left must be equal to the moment on the right, or

$$w_1 d_1 + w_2 d_2 = w_3 d_3$$

This formula can be extended to any number of weights.

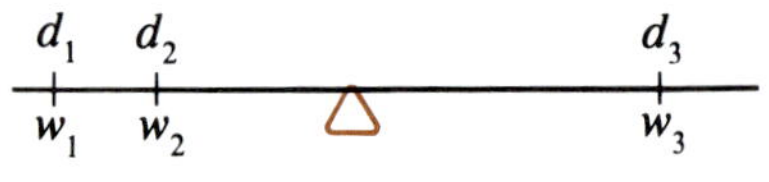

Figure 3.9

Example 4 A weight of 1 lb and a lever are to be used to determine two other weights. Referring to Figure 3.9, the following measurements were taken: given $w_3 = 1$ lb, the lever balances when $d_3 = 31$ in., $d_1 = 5$ in., and $d_2 = 4$ in. Another balance is obtained when $d_3 = 33$ in., $d_1 = 3$ in., and $d_2 = 6$ in.

3.4 APPLICATIONS OF SYSTEMS OF LINEAR EQUATIONS

Solution. From the relationship

$$w_1 d_1 + w_2 d_2 = w_3 d_3$$

and the given measurements, we get the system

$$5w_1 + 4w_2 = 31 \cdot 1$$
$$3w_1 + 6w_2 = 33 \cdot 1$$

By Cramer's rule

$$w_1 = \frac{\begin{vmatrix} 31 & 4 \\ 33 & 6 \end{vmatrix}}{\begin{vmatrix} 5 & 4 \\ 3 & 6 \end{vmatrix}} \quad \text{and} \quad w_2 = \frac{\begin{vmatrix} 5 & 31 \\ 3 & 33 \end{vmatrix}}{\begin{vmatrix} 5 & 4 \\ 3 & 6 \end{vmatrix}}$$

Hence

$$w_1 = \frac{186 - 132}{30 - 12} = \frac{54}{18} = 3 \text{ lb}$$

and

$$w_2 = \frac{165 - 93}{18} = \frac{72}{18} = 4 \text{ lb}$$

Exercises / Section 3.4

1. In Figure 3.10 the moment of weight W is 5. The lever balances when $d_1 = 2$ ft and $d_2 = 1$ ft and when $d_1 = 1$ ft and $d_2 = 3$ ft. Determine the weights w_1 and w_2.

Figure 3.10

Figure 3.11

In Exercises 2–4, refer to Figure 3.11 and find w_1 and w_2 in each case.

2. $w_3 = 2.0$ N; a balance is obtained if $d_1 = 2.0$ m, $d_2 = 2.0$ m, and $d_3 = 3.5$ m and if $d_1 = 2.0$ m, $d_2 = 1.0$ m, and $d_3 = 2.0$ m.

3. $w_3 = 4.000$ lb; a balance is obtained if $d_1 = 3.000$ in., $d_2 = 4.000$ in., and $d_3 = 17.25$ in. and if $d_1 = 5.000$ in., $d_2 = 2.000$ in., and $d_3 = 18.25$ in.

4. $w_3 = \frac{1}{3}$ lb; a balance is obtained if $d_1 = 3$ ft, $d_2 = 4$ ft, and $d_3 = 9$ ft and if $d_1 = 1$ ft, $d_2 = 2$ ft, and $d_3 = 4$ ft.

5. The relationship between the tensile strength S (measured in pounds) of a certain metal rod and the temperature T (in degrees Celsius) has the form $S = a - bT$. Experimenters found that if $T = 50.0°C$, then $S = 565.9$ lb; if $T = 100°C$, then $S = 565.8$ lb. Find the relationship.

6. The relationship between the length of a certain bar (measured in centimeters) and its temperature (in degrees Celsius) is known to be $L = aT + b$. Tests show that if $T = 15°C$, then $L = 50.0$ cm; if $T = 60°C$, then $L = 50.8$ cm. Find the relationship.

7. Two resistors connected in series have a combined resistance of 150 Ω. If the resistance of one resistor is 10 Ω less than that of the other, find the resistance of each.

8. The combined resistance of two resistors in series is 130 Ω. If the resistance of one resistor is 20 Ω less than that of the other, find the resistance of each.

9. A portion of $8,500 is invested at 12% interest and the remainder at 11%. The total interest income is $976.10. Determine the amount invested at each rate.

10. A woman invests a certain amount of money at 10% interest and the rest at 8%. If the first investment is $2,000 more than the second and her total interest income is $740, find how much was invested at each rate.

11. The foreman of a machine shop ordered two sets of machine parts costing $35 per dozen and $50 per dozen. If twice the number in the first set was one dozen more than the number in the second set and if the total bill came to $625, find the number of parts in each set.

12. A machinist has an order for a rectangular metal plate with the following specifications: The length is 1.0 in. less than 3 times the width and the perimeter is 24.4 in. Find the dimensions of the plate.

13. The manager of a shop spends $189 in his budget to buy 80 small castings. Some cost $1.95 apiece, and the rest cost $2.50 apiece. How many of each can he buy?

14. Two separate squares are to be made from a piece of wire 54.0 cm long. If the perimeter of one square is to be 3.0 cm larger than that of the other, how must the wire be cut?

15. The sum of the voltages across two resistors is 55.1 V. It was found that 3 times the first voltage is 9.7 V less than 4 times the second. What are the two voltages?

16. Measurements of the tension, in pounds, of two supporting cables produced the following equations:

$$0.37T_1 - 0.47T_2 = 0$$
$$0.52T_1 + 0.87T_2 = 120.49$$

Find T_1 and T_2.

17. Tickets for an industrial exhibit cost $5.00 for regular admission and $4.00 for senior citizens. On one day 215 tickets were sold for a total intake of $1,050. How many tickets of each type were sold?

18. A technician needs 100 mL (milliliters) of a 16% nitric acid solution (by volume). He has a 20% and a 10% solution (by volume) in stock. How many milliliters of each must he mix to obtain the required solution?

19. How many liters of a 5% solution (by volume) must be added to a 10% solution to obtain 20 L of an 8% solution?

20. One alloy contains 6% brass (by weight) and another 12% brass (by weight). How many pounds of each must be combined to form 50 lb of an alloy containing 10% brass?

21. Two machines have a total of 62 moving parts. If one machine has 2 more than 3 times as many moving parts as the other, how many moving parts does each machine have?

22. John has $6.10 in dimes and quarters. If he has five more dimes than quarters, how many of each does he have?

23. Jane has $3.30 in nickels and dimes. If she has nine more nickels than dimes, determine the number of each.

24. An office building has 20 offices. The smaller offices rent for $300 per month, and the larger offices for $420 per month. If the rental income is $7,440 per month, how many of each type of office are there?

25. One consultant to a firm charges $200 per day, and another consultant charges $250 per day. After 13 days the total charged by the two consultants came to $2,950. Assuming that only one of the two consultants was called in on any one day, how many days did each one work?

3.5 Systems of Linear Equations with More Than Two Unknowns

Our method for solving systems of equations can be readily extended to systems of three or more equations. We shall concentrate on the **method of addition or subtraction** in this section and return to the method of determinants in Section 3.6.

To solve a system of three equations

$$\begin{aligned} a_1x + b_1y + c_1z &= d_1 \\ a_2x + b_2y + c_2z &= d_2 \\ a_3x + b_3y + c_3z &= d_3 \end{aligned} \qquad (3.13)$$

algebraically, we eliminate one of the unknowns between any two of the equations. Then, taking a different pair of equations, we eliminate the same unknown. The resulting system of two equations can then be solved by one of the earlier methods. Consider the following example.

Example 1 Solve the system

(1) $\quad x - 3y - z = -2$
(2) $\quad 4x - y - 2z = 8$
(3) $\quad 3x + 2y + 2z = 1$

Solution. A glance at the different coefficients tells us that z is the easiest of the three unknowns to eliminate.

(4)	$2x - 6y - 2z$	$= -4$	multiplying equation (1) by 2
(5)	$4x - y - 2z$	$= 8$	repeating (2) and (3)
(6)	$3x + 2y + 2z$	$= 1$	
(7)	$-2x - 5y$	$= -12$	subtracting (5) from (4)
(8)	$7x + y$	$= 9$	adding (5) and (6)
(9)	$-2x - 5y$	$= -12$	repeating (7)
(10)	$35x + 5y$	$= 45$	multiplying (8) by 5
(11)	$33x$	$= 33$	adding (9) and (10)
(12)	x	$= 1$	
(13)	$7(1) + y$	$= 9$	substituting in (8)
(14)	y	$= 2$	

(15) $\quad 1 - 3(2) - z = -2 \quad$ substituting in (1)
(16) $\quad\quad\quad\quad\quad\quad z = -3$

The solution is therefore given by $x = 1$, $y = 2$, and $z = -3$. As a check, substitute these values into equations (2) and (3). Then

$$4(1) - 2 - 2(-3) = 8$$

and

$$3(1) + 2(2) + 2(-3) = 1$$

which checks.

Example 2 Solve the system

(1) $\quad 4R_1 - 2R_2 + R_3 = 8$
(2) $\quad 3R_1 - 3R_2 + 4R_3 = 8$
(3) $\quad R_1 + R_2 + 2R_3 = 6$

Solution. We eliminate R_2 as follows:

(4) $\quad 12R_1 - 6R_2 + 3R_3 = 24 \quad$ multiplying (1) by 3
(5) $\quad 6R_1 - 6R_2 + 8R_3 = 16 \quad$ multiplying (2) by 2
(6) $\quad 6R_1 + 6R_2 + 12R_3 = 36 \quad$ multiplying (3) by 6
(7) $\quad 6R_1 \quad\quad\quad - 5R_3 = 8 \quad$ subtracting (5) from (4)
(8) $\quad 12R_1 \quad\quad\quad + 20R_3 = 52 \quad$ adding (5) and (6)
(9) $\quad 24R_1 \quad\quad\quad - 20R_3 = 32 \quad$ multiplying (7) by 4
(10) $\quad 12R_1 \quad\quad\quad + 20R_3 = 52 \quad$ repeating (8)
(11) $\quad 36R_1 \quad\quad\quad\quad\quad = 84 \quad$ adding (9) and (10)

(12) $\quad\quad\quad R_1 = \dfrac{84}{36} = \dfrac{7}{3}$

(13) $\quad 6\left(\dfrac{7}{3}\right) \quad - 5R_3 = 8 \quad$ substituting into (7)

(14) $\quad\quad\quad R_3 = \dfrac{6}{5}$

(15) $\quad \dfrac{7}{3} + R_2 + 2\left(\dfrac{6}{5}\right) = 6 \quad$ substituting into (3)

(16) $\quad\quad\quad R_2 = \dfrac{19}{15}$

As a check, we substitute the values $R_1 = \frac{7}{3}$, $R_2 = \frac{19}{15}$, and $R_3 = \frac{6}{5}$ into equation (1):

$$4\left(\frac{7}{3}\right) - 2\left(\frac{19}{15}\right) + \frac{6}{5} = \frac{140}{15} - \frac{38}{15} + \frac{18}{15} = \frac{120}{15} = 8$$

Equation (2) is checked similarly.

To solve a system with four unknowns, proceed by eliminating one of the unknowns among three different pairs of equations. The resulting system of three equations can then be solved by the methods of this section.

Exercises / Section 3.5

Solve the following systems of equations.

1. $x + y + 2z = 9$
   $x \quad\quad - z = -2$
   $2x - y \quad\quad = 0$

2. $x - y \quad\quad = 2$
   $x + 2y - z = -3$
   $-y + z = 3$

3. $3x \quad\quad + 2z = -1$
   $4x - y - 2z = 7$
   $x + y \quad\quad = 2$

4. $2x - 3y \quad\quad = 8$
   $3x - y + 2z = 8$
   $2x + 3y \quad\quad = -4$

5. $x - 2y - z = 1$
   $2x - y - 2z = -1$
   $3x + y + 2z = 6$

6. $3x - 3y - 2z = 1$
   $-x + y - 6z = 3$
   $x + y + 2z = 3$

7. $2x - y + 3z = 16$
   $3x + 4y + 2z = 7$
   $5x - 6y + 8z = 47$

8. $x - 2y + z = 1$
   $2x + y + 2z = 2$
   $3x + 3y - 3z = 2$

9. $3R_1 - 3R_2 + 7R_3 = 35$
   $6R_1 + 3R_2 - R_3 = -5$
   $9R_1 - 12R_2 + 3R_3 = 38$

10. $2F_1 - 5F_2 + 10F_3 = 21$
    $3F_1 + 5F_2 + 10F_3 = 6$
    $4F_1 - 10F_2 + 40F_3 = 64$

11. $\frac{2}{x} - \frac{1}{y} + \frac{2}{z} = 2$
    $-\frac{4}{x} + \frac{5}{y} - \frac{3}{z} = 1$
    $\frac{3}{x} - \frac{4}{y} + \frac{1}{z} = 3$

12. $\frac{2}{x} + \frac{1}{y} - \frac{1}{z} = 0$
    $\frac{8}{x} - \frac{2}{y} + \frac{1}{z} = 1$
    $\frac{4}{x} - \frac{4}{y} - \frac{1}{z} = -1$

13. $\quad\quad 2y + 3z + w = 1$
    $2x - 2y \quad\quad + w = 1$
    $3x \quad\quad - 2z + 2w = 13$
    $x + 3y - z \quad\quad = 9$

14. $x \quad\quad + 2z - 3w = 12$
    $3x - 2y + z \quad\quad = -3$
    $\quad\quad 3y - 3z - 5w = 12$
    $2x - y \quad\quad - 2w = 3$

15. $2x - y - 2z - 2w = -7$
    $4x - 2y + 3z \quad\quad = 3$
    $6x + 2y + 4z - 3w = 16$
    $8x \quad\quad - 2z - 4w = -6$

3.6 More on Determinants

The method of **determinants** can be extended to systems of more than two unknowns. In this section we shall confine ourselves to systems of three equations. Larger systems will be discussed in Chapter 15.

The easiest way to see the extension of the determinant method is to solve the system

$$a_1x + b_1y + c_1z = d_1$$
$$a_2x + b_2y + c_2z = d_2 \qquad (3.14)$$
$$a_3x + b_3y + c_3z = d_3$$

by the method of addition or subtraction. Although straightforward, the calculation is quite lengthy. The expression for x turns out to be

$$x = \frac{d_1b_2c_3 - d_1b_3c_2 - d_2b_1c_3 + d_3b_1c_2 + d_2b_3c_1 - d_3b_2c_1}{a_1b_2c_3 - a_1b_3c_2 - a_2b_1c_3 + a_3b_1c_2 + a_2b_3c_1 - a_3b_2c_1} \qquad (3.15)$$

A determinant of the third order is defined with Cramer's rule in mind: If Cramer's rule is to carry over, then the denominator of solution (3.15) should be a determinant whose elements are the coefficients of the unknowns. In other words,

$$\begin{vmatrix} a_1 & b_1 & c_1 \\ a_2 & b_2 & c_2 \\ a_3 & b_3 & c_3 \end{vmatrix} = a_1b_2c_3 - a_1b_3c_2 - a_2b_1c_3 + a_3b_1c_2 + a_2b_3c_1 - a_3b_2c_1 \qquad (3.16)$$

The last expression may look like a jumble of symbols without any discernible pattern, but if you look more closely, you will see that *every product consists of exactly one element from each row and one from each column.* (The same holds true for higher-order determinants.) This observation enables us to express a third-order determinant in terms of second-order determinants. For example,

$$a_1 \begin{vmatrix} b_2 & c_2 \\ b_3 & c_3 \end{vmatrix} = a_1(b_2c_3 - b_3c_2) = a_1b_2c_3 - a_1b_3c_2$$

which are the first two terms in expansion (3.16). The determinant

$$\begin{vmatrix} b_2 & c_2 \\ b_3 & c_3 \end{vmatrix}$$

is called the **minor** of the element a_1.

> **Definition of a minor:** The **minor** of a given element is the determinant formed by deleting all the elements in the row and column in which the element lies.

Thus in determinant (3.16) the minor of a_2 is

$$\begin{vmatrix} b_1 & c_1 \\ b_3 & c_3 \end{vmatrix}$$

and the minor of b_2 is

$$\begin{vmatrix} a_1 & c_1 \\ a_3 & c_3 \end{vmatrix}$$

Now observe that, in terms of minors, the determinant (3.16) can be written

$$\begin{vmatrix} a_1 & b_1 & c_1 \\ a_2 & b_2 & c_2 \\ a_3 & b_3 & c_3 \end{vmatrix} = a_1 \begin{vmatrix} b_2 & c_2 \\ b_3 & c_3 \end{vmatrix} - b_1 \begin{vmatrix} a_2 & c_2 \\ a_3 & c_3 \end{vmatrix} + c_1 \begin{vmatrix} a_2 & b_2 \\ a_3 & b_3 \end{vmatrix}$$

In other words, the determinant is expanded by forming the products of the elements in the first row with their corresponding minors and affixing either a plus or minus sign. Moreover, the same expression on the right side of expansion (3.16) can be obtained by using the elements in some other row or even in a column. For example, using the second column, we get the expansion given next.

Typical expansion by minors

$$\begin{vmatrix} a_1 & b_1 & c_1 \\ a_2 & b_2 & c_2 \\ a_3 & b_3 & c_3 \end{vmatrix} = -b_1 \begin{vmatrix} a_2 & c_2 \\ a_3 & c_3 \end{vmatrix} + b_2 \begin{vmatrix} a_1 & c_1 \\ a_3 & c_3 \end{vmatrix} - b_3 \begin{vmatrix} a_1 & c_1 \\ a_2 & c_2 \end{vmatrix}$$

$$= -b_1 a_2 c_3 + b_1 a_3 c_2 + b_2 a_1 c_3 - b_2 a_3 c_1 \\ - b_3 a_1 c_2 + b_3 a_2 c_1 \quad (3.17)$$

Rule for signs

We still need a rule for affixing the sign. It turns out that the sign depends only on the position of the element. Consider the row and column in which the element lies. *If the sum of the number of the row and the number of the column is even, affix a plus sign; if the sum is odd, affix a minus sign.*

Example 1 Expand the determinant

$$\begin{vmatrix} 2 & -3 & 1 \\ -4 & 0 & -7 \\ -3 & -1 & 1 \end{vmatrix}$$

Solution. As already noted, we can expand the determinant along any row or column. Suppose we arbitrarily choose the first column. Then we get

$$\begin{vmatrix} 2 & -3 & 1 \\ -4 & 0 & -7 \\ -3 & -1 & 1 \end{vmatrix}$$

$$= 2 \begin{vmatrix} 0 & -7 \\ -1 & 1 \end{vmatrix} - (-4) \begin{vmatrix} -3 & 1 \\ -1 & 1 \end{vmatrix} + (-3) \begin{vmatrix} -3 & 1 \\ 0 & -7 \end{vmatrix}$$

Take a closer look at how the signs were determined. The first element, **2**, lies in row 1, column 1, and $1 + 1 = 2$, which is even. Hence the element is given a plus sign. The next element, **−4**, lies in row 2, column 1, and $2 + 1 = 3$, which is odd. So the element -4 is given a minus sign to become $-(-4)$. Finally, the third element, **−3**, lies in row 3, column 1, and $3 + 1 = 4$, which is even. So the element is given a plus sign.

No particular row or column offers any obvious advantage with one notable exception: A row or a column containing one or more zeros reduces the number of calculations required. For example, expanding along the second row yields

$$\begin{vmatrix} 2 & -3 & 1 \\ -4 & 0 & -7 \\ -3 & -1 & 1 \end{vmatrix}$$

$$= -(-4)\begin{vmatrix} -3 & 1 \\ -1 & 1 \end{vmatrix} + (0)\begin{vmatrix} 2 & 1 \\ -3 & 1 \end{vmatrix} - (-7)\begin{vmatrix} 2 & -3 \\ -3 & -1 \end{vmatrix}$$

$$= 4(-3 + 1) + 0 + 7(-2 - 9) = -85$$

Example 2 Expand the determinant

$$\begin{vmatrix} -3 & 2 & 1 \\ 4 & -2 & 3 \\ -3 & 1 & 0 \end{vmatrix}$$

Solution. Because of the 0, we expand along the third row, starting with the element -3. This element is situated in row 3, column 1. Since $3 + 1 = 4$, we affix a plus sign.

$$\begin{vmatrix} -3 & 2 & 1 \\ 4 & -2 & 3 \\ -3 & 1 & 0 \end{vmatrix}$$

$$= +(-3)\begin{vmatrix} 2 & 1 \\ -2 & 3 \end{vmatrix} - (1)\begin{vmatrix} -3 & 1 \\ 4 & 3 \end{vmatrix} + (0)\begin{vmatrix} -3 & 2 \\ 4 & -2 \end{vmatrix}$$

$$= -3(6 + 2) - (-9 - 4) + 0 = -11$$

(Note that the signs necessarily alternate, so that only the first one needs to be determined.)

Example 3 Expand the determinant

$$\begin{vmatrix} 7 & -2 & -3 \\ -11 & 0 & 5 \\ 2 & 0 & 2 \end{vmatrix}$$

Solution. If we expand along the second column, then only one minor needs to be evaluated. Note also that the element -2 lies in row 1, column 2, and is therefore given a minus sign.

$$\begin{vmatrix} 7 & -2 & -3 \\ -11 & 0 & 5 \\ 2 & 0 & 2 \end{vmatrix} = -(-2) \begin{vmatrix} -11 & 5 \\ 2 & 2 \end{vmatrix} = 2(-22 - 10) = -64$$

Having defined a third-order determinant, we can now return to Cramer's rule and the solution of equations.

Cramer's rule: The solution of the system

$$a_1 x + b_1 y + c_1 z = d_1$$
$$a_2 x + b_2 y + c_2 z = d_2 \quad (3.18)$$
$$a_3 x + b_3 y + c_3 z = d_3$$

is given by

$$x = \frac{\begin{vmatrix} d_1 & b_1 & c_1 \\ d_2 & b_2 & c_2 \\ d_3 & b_3 & c_3 \end{vmatrix}}{\begin{vmatrix} a_1 & b_1 & c_1 \\ a_2 & b_2 & c_2 \\ a_3 & b_3 & c_3 \end{vmatrix}}, \quad y = \frac{\begin{vmatrix} a_1 & d_1 & c_1 \\ a_2 & d_2 & c_2 \\ a_3 & d_3 & c_3 \end{vmatrix}}{\begin{vmatrix} a_1 & b_1 & c_1 \\ a_2 & b_2 & c_2 \\ a_3 & b_3 & c_3 \end{vmatrix}},$$

$$z = \frac{\begin{vmatrix} a_1 & b_1 & d_1 \\ a_2 & b_2 & d_2 \\ a_3 & b_3 & d_3 \end{vmatrix}}{\begin{vmatrix} a_1 & b_1 & c_1 \\ a_2 & b_2 & c_2 \\ a_3 & b_3 & c_3 \end{vmatrix}} \quad (3.19)$$

As in the case of two equations, the denominators are the same for all the unknowns. The determinants in the numerators are found by replacing the column of coefficients of the unknown to be found by the column of numbers on the right side of the equation. We shall see in Chapter 15 that the rule can be applied to systems of equations of any size.

If the determinant

$$\begin{vmatrix} a_1 & b_1 & c_1 \\ a_2 & b_2 & c_2 \\ a_3 & b_3 & c_3 \end{vmatrix}$$

108 CHAPTER 3 SYSTEMS OF LINEAR EQUATIONS AND INTRODUCTION TO DETERMINANTS

Independent system

in (3.19) is different from zero, then the system has a unique solution, since the values of the determinants are unique. Such a system is called *independent*. If the determinant is zero, then the system is not independent and no unique solution exists. (The system is either inconsistent, having no solution, or dependent, having infinitely many solutions.)

Example 4 Solve the given system by determinants.

$$2x - y + 3z = 2$$
$$x + 2y - z = 1$$
$$3x - 2y = 4$$

Solution. First we evaluate the determinant whose elements consist of the coefficients of the unknowns; this determinant will occur in all the denominators. Expanding along the third row:

$$\begin{vmatrix} 2 & -1 & 3 \\ 1 & 2 & -1 \\ 3 & -2 & 0 \end{vmatrix} = 3 \begin{vmatrix} -1 & 3 \\ 2 & -1 \end{vmatrix} - (-2) \begin{vmatrix} 2 & 3 \\ 1 & -1 \end{vmatrix} + 0 \begin{vmatrix} 2 & -1 \\ 1 & 2 \end{vmatrix}$$

$$= 3(1 - 6) + 2(-2 - 3) + 0 = -25$$

So by Cramer's rule,

$$x = \frac{\begin{vmatrix} 2 & -1 & 3 \\ 1 & 2 & -1 \\ 4 & -2 & 0 \end{vmatrix}}{-25}$$

Expanding along the third column, we get

$$x = -\frac{1}{25} \left[3 \begin{vmatrix} 1 & 2 \\ 4 & -2 \end{vmatrix} - (-1) \begin{vmatrix} 2 & -1 \\ 4 & -2 \end{vmatrix} + 0 \right]$$

$$= -\frac{1}{25} [3(-2 - 8) + 1(-4 + 4)]$$

$$= -\frac{1}{25} (-30)$$

$$= \frac{6}{5}$$

Next,

$$y = -\frac{1}{25} \begin{vmatrix} 2 & 2 & 3 \\ 1 & 1 & -1 \\ 3 & 4 & 0 \end{vmatrix}$$

$$= -\frac{1}{25} \left[3 \begin{vmatrix} 1 & 1 \\ 3 & 4 \end{vmatrix} - (-1) \begin{vmatrix} 2 & 2 \\ 3 & 4 \end{vmatrix} + 0 \right] \qquad \text{expanding along third column}$$

$$= -\frac{1}{25}[3(4-3)+1(8-6)]$$

$$= -\frac{1}{25}(5)$$

$$= -\frac{1}{5}$$

Finally,

$$z = -\frac{1}{25}\begin{vmatrix} 2 & -1 & 2 \\ 1 & 2 & 1 \\ 3 & -2 & 4 \end{vmatrix}$$

$$= -\frac{1}{25}\left[2\begin{vmatrix} 2 & 1 \\ -2 & 4 \end{vmatrix} - (-1)\begin{vmatrix} 1 & 1 \\ 3 & 4 \end{vmatrix} + 2\begin{vmatrix} 1 & 2 \\ 3 & -2 \end{vmatrix}\right] \quad \text{expanding along first row}$$

$$= -\frac{1}{25}[2(8+2)+1(4-3)+2(-2-6)]$$

$$= -\frac{1}{25}(5)$$

$$= -\frac{1}{5}$$

As a check, suppose we substitute $x = \frac{6}{5}$, $y = -\frac{1}{5}$, and $z = -\frac{1}{5}$ in the second equation. Then

$$\frac{6}{5} + 2\left(-\frac{1}{5}\right) - \left(-\frac{1}{5}\right) = \frac{5}{5} = 1$$

The other equations are checked similarly.

The expansion of a 3 × 3 determinant can also be accomplished by rewriting the first two columns to the right of the determinant and then forming products of elements along the resulting full diagonals, as shown in the following scheme:

$$\begin{vmatrix} a_1 & b_1 & c_1 \\ a_2 & b_2 & c_2 \\ a_3 & b_3 & c_3 \end{vmatrix} \begin{matrix} a_1 & b_1 \\ a_2 & b_2 \\ a_3 & b_3 \end{matrix} \quad (3.20)$$

The leftmost arrow pointing upward yields the product $-a_3b_2c_1$, and the leftmost arrow pointing downward yields $+a_1b_2c_3$, and so forth. Comparing

Example 5

Expand the determinant in Example 2 by using the expansion scheme (3.20).

Solution. Rewriting the first two columns, we get

$$\begin{vmatrix} -3 & 2 & 1 \\ 4 & -2 & 3 \\ -3 & 1 & 0 \end{vmatrix} \begin{matrix} -3 & 2 \\ 4 & -2 \\ -3 & 1 \end{matrix} = \begin{matrix} +(-3)(-2)(0) + (2)(3)(-3) + (1)(4)(1) \\ -(-3)(-2)(1) - (1)(3)(-3) - (0)(4)(2) \end{matrix}$$

$$= 0 - 18 + 4 - 6 + 9 - 0$$
$$= -11$$

You will have to judge whether this method is more convenient than expansion by minors.

An important application of systems of equations is the analysis of basic electrical circuits by means of Kirchhoff's laws. Since a discussion of setting up systems of equations using Kirchhoff's laws is too lengthy to consider here, we will work with the given system of equations corresponding to a particular circuit in the next example and the exercises.

Example 6

Consider the circuit in Figure 3.12. If the directions of the currents are as indicated in the diagram, then from Kirchhoff's laws the system of equations is given by

$$I_1 + I_2 - I_3 = 0$$
$$10I_1 - 3I_2 = 4$$
$$ 3I_2 + 5I_3 = 2$$

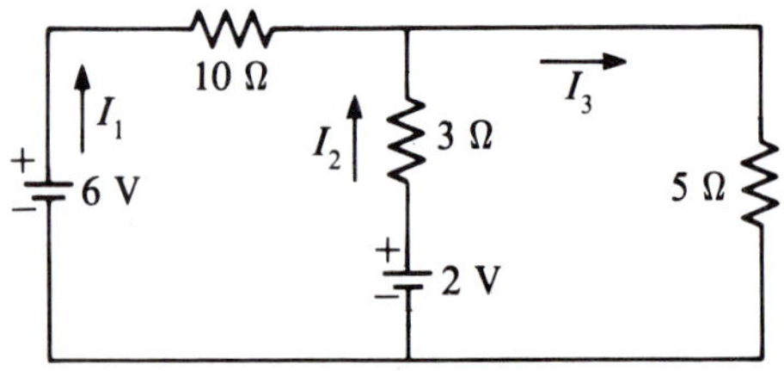

Figure 3.12

By Cramer's rule the solution is found to be $I_1 = \frac{38}{95}$ A, $I_2 = 0$ A, and $I_3 = \frac{38}{95}$ A. (See Exercise 36 in the following exercise set.)

Since the calculated values of I_1 and I_3 are positive, the directions of these currents agree with the directions originally assigned. (If a calculated current is negative, its direction is actually opposite to the direction originally assigned.)

Exercises / Section 3.6

In Exercises 1–12, evaluate each determinant.

1. $\begin{vmatrix} 1 & 0 & 0 \\ 1 & -2 & 3 \\ 1 & -4 & 4 \end{vmatrix}$

2. $\begin{vmatrix} 0 & -1 & 0 \\ 3 & 0 & -6 \\ -7 & 3 & -4 \end{vmatrix}$

3. $\begin{vmatrix} 2 & 1 & 3 \\ 0 & 5 & 0 \\ 3 & 2 & -1 \end{vmatrix}$

4. $\begin{vmatrix} -2 & 3 & -5 \\ 3 & -1 & 4 \\ 0 & -2 & 0 \end{vmatrix}$

5. $\begin{vmatrix} 2 & -1 & 3 \\ 3 & 0 & -5 \\ 10 & 5 & -10 \end{vmatrix}$

6. $\begin{vmatrix} -5 & 12 & 3 \\ 7 & -2 & 1 \\ -3 & 0 & 2 \end{vmatrix}$

7. $\begin{vmatrix} 2 & 3 & 8 \\ -1 & 3 & -2 \\ 5 & -6 & -12 \end{vmatrix}$

8. $\begin{vmatrix} -15 & 20 & 10 \\ 3 & 1 & 7 \\ 9 & -3 & 15 \end{vmatrix}$

9. $\begin{vmatrix} -3 & 2 & 4 \\ 10 & 18 & -20 \\ -11 & -15 & 8 \end{vmatrix}$

10. $\begin{vmatrix} -5 & 0 & 20 \\ 30 & 3 & 40 \\ 5 & -40 & 10 \end{vmatrix}$

11. $\begin{vmatrix} -3 & -4 & -7 \\ 3 & 0 & -6 \\ 10 & 15 & 18 \end{vmatrix}$

12. $\begin{vmatrix} -6 & 13 & -3 \\ -2 & 17 & 6 \\ 0 & 19 & 2 \end{vmatrix}$

In Exercises 13–22, solve each system by Cramer's rule. (Exercises 13–20 are the same as Exercises 1–8 in Section 3.5, respectively.)

13. $x + y + 2z = 9$
$x - z = -2$
$2x - y = 0$

14. $x - y = 2$
$x + 2y - z = -3$
$-y + z = 3$

15. $3x + 2z = -1$
$4x - y - 2z = 7$
$x + y = 2$

16. $2x - 3y = 8$
$3x - y + 2z = 8$
$2x + 3y = -4$

17. $x - 2y - z = 1$
$2x - y - 2z = -1$
$3x + y + 2z = 6$

18. $3x - 3y - 2z = 1$
$-x + y - 6z = 3$
$x + y + 2z = 3$

19. $2x - y + 3z = 16$
$3x + 4y + 2z = 7$
$5x - 6y + 8z = 47$

20. $x - 2y + z = 1$
$2x + y + 2z = 2$
$3x + 3y - 3z = 2$

21. $2x - 3y + z = 1$
$x - 2y - 3z = 1$
$x - 4y + 2z = 2$

22. $3x - y + 4z = 4$
$-x + 2y - 3z = 6$
$-2x - 3y + z = 10$

Solve Exercises 23 and 24; refer to Figure 3.13.

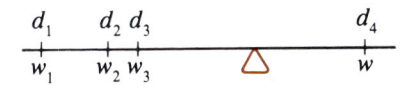

Figure 3.13

23. Find the weights w_1, w_2, and w_3 given the following sets of measurements: $w = 2$ lb and
    (1) $d_1 = 4$ ft, $d_2 = 3$ ft, $d_3 = 2$ ft, $d_4 = 5.5$ ft
    (2) $d_1 = 3$ ft, $d_2 = 2$ ft, $d_3 = 1$ ft, $d_4 = 3.5$ ft
    (3) $d_1 = 5$ ft, $d_2 = 4$ ft, $d_3 = 1$ ft, $d_4 = 5.5$ ft

24. Find the weights w_1, w_2, and w_3 given the following sets of measurements: $w = 11$ N
    (1) $d_1 = 2$ m, $d_2 = 1$ m, $d_3 = 3$ m, $d_4 = 2$ m
    (2) $d_1 = 4$ m, $d_2 = 3$ m, $d_3 = 3$ m, $d_4 = 3\frac{6}{11}$ m
    (3) $d_1 = 6$ m, $d_2 = 2$ m, $d_3 = 6$ m, $d_4 = 4\frac{7}{11}$ m

25. A portion of \$5,950 was invested at 8%, another portion at 10%, and the rest at 12%. The total interest income was \$635. If the sum of the second investment and twice the first investment was \$750 more than the third investment, find the amount invested at each rate.

26. A woman has \$16,750 to invest. She decides to invest the largest portion at 8.25% in a safe investment. She invests the smallest portion at 12% in a high-risk investment and the rest at 10%. In fact, the safe investment is only \$750 less than the other two combined. Determine the amount invested at each rate, given that the total interest income is \$1,605.

27. Three machine parts cost a total of \$40. The first part costs as much as the other two together, while the cost of 6 times the second is \$2 more than the total cost of the other two. Find the cost of each part.

28. The combined ages of three brothers, Richard, Paul, and Craig, total 24 years. Three years ago Richard's age was one sixth of his brothers' combined present ages, while Paul's age was one fourth of what his brothers' combined ages will be two years from now. Find their respective ages.

In Exercises 29–35, find the currents in each of the circuits by solving the system of equations given in each case.

29. $I_1 + I_2 - I_3 = 0$
    $I_1 - 2I_2 = 4$
    $2I_2 + 3I_3 = 2$

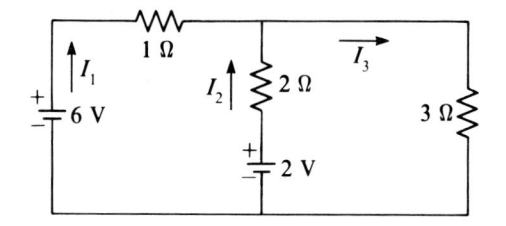

Figure 3.14

30. $I_1 - I_2 + I_3 = 0$
    $2I_1 + 3I_2 = 2$
    $-3I_2 - I_3 = -4$

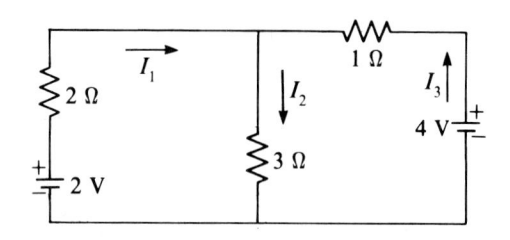

Figure 3.15

3.6 MORE ON DETERMINANTS

31. $I_1 - I_2 + I_3 = 0$
$I_1 + 2I_2 = 10$
$-2I_2 - I_3 = -5$

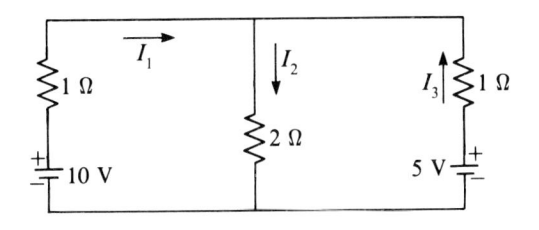

Figure 3.16

32. $-I_1 + I_2 + I_3 = 0$
$-2I_1 - 5I_2 = -20$
$5I_2 - 10I_3 = 10$

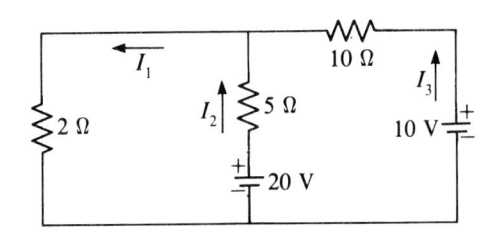

Figure 3.17

33. $-I_1 + I_2 + I_3 = 0$
$-I_1 - 3I_2 = -10$
$3I_2 - 5I_3 = -6$

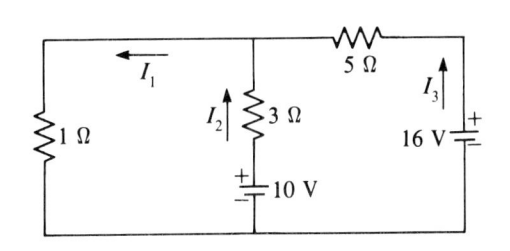

Figure 3.18

34. $I_1 - I_2 - I_3 = 0$
$3I_1 + 4I_2 = -2$
$-4I_2 + 3I_3 = 2$

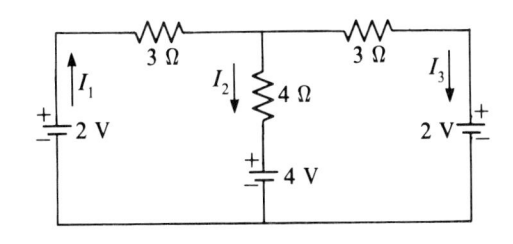

Figure 3.19

35.
$$I_1 + I_2 + I_3 - I_4 = 0$$
$$2I_1 - I_2 = 1$$
$$I_2 - 3I_3 = 3$$
$$3I_3 + I_4 = -2$$

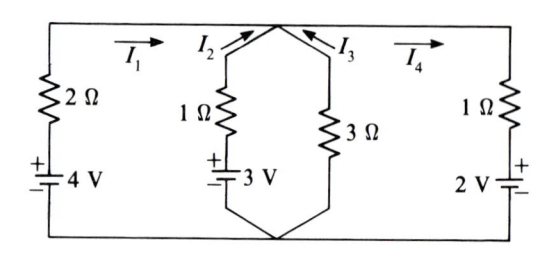

Figure 3.20

36. Solve the system of equations in Example 6.

Review Exercises / Chapter 3

In Exercises 1–8, evaluate each determinant.

1. $\begin{vmatrix} -2 & 1 \\ 5 & 1 \end{vmatrix}$
2. $\begin{vmatrix} 1 & 10 \\ -5 & -3 \end{vmatrix}$
3. $\begin{vmatrix} -8 & 0 \\ 30 & -20 \end{vmatrix}$
4. $\begin{vmatrix} 1 & -7 \\ -10 & 1 \end{vmatrix}$
5. $\begin{vmatrix} 1 & 3 & -2 \\ 0 & 0 & -3 \\ 1 & 4 & 5 \end{vmatrix}$
6. $\begin{vmatrix} 2 & 7 & 1 \\ 3 & -2 & 0 \\ 7 & 8 & -10 \end{vmatrix}$
7. $\begin{vmatrix} 5 & 6 & -1 \\ -4 & 7 & 2 \\ -3 & 0 & 5 \end{vmatrix}$
8. $\begin{vmatrix} 7 & -2 & -11 \\ 10 & 4 & 13 \\ 12 & 6 & 15 \end{vmatrix}$

In Exercises 9–12, solve each system of equations graphically, giving the values to the nearest tenth of a unit.

9. $2x - y = -5$
   $2x + 3y = 3$
10. $3x - 6y = 1$
    $6x - 3y = 5$
11. $2x - y = 3$
    $2x - 5y = -5$
12. $2x - y = 7$
    $4x - y = 11$

In Exercises 13–16, solve each system by the method of substitution.

13. $x - 3y = 1$
    $3x + 2y = 4$
14. $2x - y = 9$
    $x + 2y = 7$
15. $-4x + 2y = 7$
    $x - 3y = 2$
16. $5x - 4y = 5$
    $2x - y = 1$

In Exercises 17–20, solve each system by the method of addition or subtraction.

17. $3x + 2y = 5$
    $x - y = 2$
18. $2x - 4y = 2$
    $3x - 5y = 4$
19. $4x + 3y = 9$
    $5x + 6y = 12$
20. $2x + 6y = 17$
    $-4x + 7y = 23$

REVIEW EXERCISES / CHAPTER 3   **115**

In Exercises 21–24, solve each system by the method of determinants.

21. $5x - 7y = 8$
$4x + 2y = 3$

22. $-x + 3y = 2$
$2x + 4y = 3$

23. $2x - 2y = -7$
$2x + 3y = 6$

24. $x - 3y = 5$
$3x + 2y = 2$

In Exercises 25–36, solve the given systems by any method.

25. $x - 3y = 5$
$2x - y = 5$

26. $\dfrac{1}{x} - \dfrac{3}{y} = 4$
$\dfrac{2}{x} - \dfrac{1}{y} = 5$

27. $-x + 2y = 3$
$2x + 4y = 5$

28. $3x - 4y = 1$
$-2x + 6y = 1$

29. $\dfrac{5}{x} - \dfrac{4}{y} = 9$
$\dfrac{2}{x} - \dfrac{1}{y} = 11$

30. $\dfrac{3}{x} + \dfrac{2}{y} = 3$
$\dfrac{4}{x} + \dfrac{7}{y} = 5$

31. $s_1 + 3s_2 = 14$
$2s_1 - s_2 = 0$

32. $2T_1 - T_2 = 0$
$3T_1 + 2T_2 = 7$

33. $2d_1 + 3d_2 = 13$
$d_1 + 2d_2 = 8$

34. $2C_1 + 2C_2 = 9$
$4C_1 + 3C_2 = 14$

35. $2m + 3n = 8$
$4m + 3n = 7$

36. $r + 2s = 0$
$12r - 6s = -5$

In Exercises 37–40, solve the given systems algebraically.

37. $x - y = 2$
$2x + 3z = 11$
$3x + 2y + 4z = 13$

38. $2w_1 + 2w_2 = 3$
$-w_2 + w_3 = 1$
$4w_1 + 2w_3 = 4$

39. $\dfrac{1}{y} + \dfrac{2}{z} = 7$
$\dfrac{2}{x} + \dfrac{3}{z} = 5$
$\dfrac{3}{x} + \dfrac{4}{y} - \dfrac{1}{z} = -5$

40. $\dfrac{3}{x} - \dfrac{1}{y} = 3$
$-\dfrac{4}{y} + \dfrac{2}{z} = 7$
$\dfrac{1}{y} - \dfrac{1}{z} = -1$

In Exercises 41–44, solve the given systems by using Cramer's rule.

41. $V_1 - V_2 + 4V_3 = -12$
$3V_1 + 2V_2 = 3$
$-2V_1 + 3V_2 - 3V_3 = 17$

42. $a + b + c = 1$
$2a - b + 2c = 2$
$a - b - 4c = 3$

43. $\dfrac{1}{x} - \dfrac{1}{y} - \dfrac{2}{z} = 3$
$\dfrac{2}{x} - \dfrac{4}{z} = 5$
$\dfrac{1}{x} - \dfrac{3}{y} + \dfrac{2}{z} = 2$

44. $T_1 - T_2 = 6$
$2T_1 - 2T_2 + T_3 = 5$
$3T_1 - T_2 + 2T_3 = 1$

45. Determine the value of a that makes the system

$2x + ay = 3$
$4x - 2y = 5$

inconsistent.

46. Show that the system

$x + 2y + 3z = 3$
$4x + 5y + 6z = 1$
$7x + 8y + 9z = 2$

cannot have a unique solution.

47. A portion of $1,014.80 is invested at 10%, another at 8%, and the rest at $12\frac{1}{2}$%. The $12\frac{1}{2}$% investment yields as much interest as the other two investments combined. Find the amount invested at each rate if the total interest income is $103.20.

48. Experimenters found that the velocity (in feet per second) of two moving parts in a machine satisfy the relation

$$\frac{1}{v_1} + \frac{3}{v_2} = 5$$

$$\frac{2}{v_1} + \frac{1}{v_2} = 6$$

Determine v_1 and v_2.

49. Two kinds of milk containing 1% butterfat and 4% butterfat by volume, respectively, are to be mixed to obtain 90 gal of milk containing 2% butterfat. Determine the number of gallons of each required.

50. The perimeter of a triangle is 14 in. The longest side is twice as long as the shortest side and 2 in. less than the sum of the two shorter sides. Find the length of each side.

In Exercises 51–53, find the currents in the given diagrams.

51. $I_1 - I_2 - I_3 = 0$
$2I_1 + I_2 = 3$
$-I_2 + 3I_3 = -1$

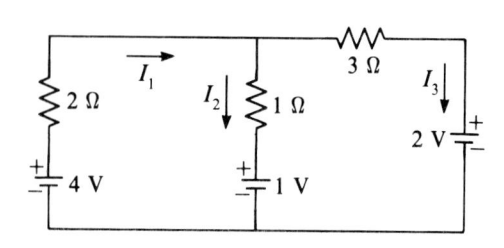

Figure 3.21

52. $I_1 - I_2 + I_3 = 0$
$2I_1 + 4I_2 = -2$
$-4I_2 - 2I_3 = 1$

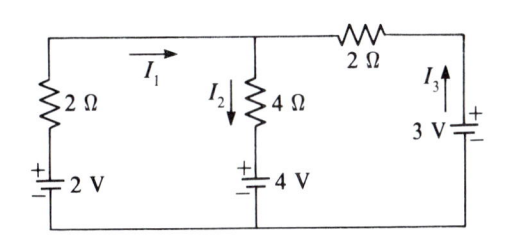

Figure 3.22

53. $I_1 + I_2 + I_3 - I_4 = 0$
$I_1 - 3I_2 = 3$
$3I_2 - 4I_3 = -3$
$4I_3 + I_4 = -1$

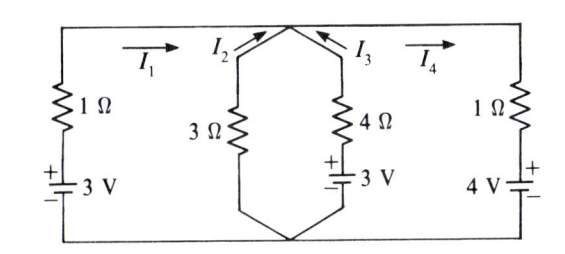

Figure 3.23

54. An experimenter determined that the lever in the figure balances if the weights are positioned as follows:

(1) $d_1 = 3$ in., $d_2 = 2$ in., $d_3 = 1$ in., $d = 2\frac{1}{4}$ in.
(2) $d_1 = 2$ in., $d_2 = 1$ in., $d_3 = 3$ in., $d = 2\frac{1}{2}$ in.
(3) $d_1 = 3$ in., $d_2 = 2$ in., $d_3 = 2$ in., $d = 2\frac{3}{4}$ in.

Find the weights.

Figure 3.24

Cumulative Review Exercises / Chapters 1–3

1. Subtract $5T_a + 2T_b - 3T_c$ from the sum of $-9T_a - 2T_b + 6T_c$ and $7T_a - 6T_b - 7T_c$.
2. Simplify: $-\{L - [(-L - C) - C] - 4L\}$
3. Write $\sqrt{150}$ in its simplest radical form.
4. Simplify by rationalizing the denominator: $\dfrac{L}{2\sqrt{\pi}}$
5. Simplify: $\dfrac{3p^{-2}q^{-1}}{p^{-6}q^{-4}}$
6. Simplify: $\dfrac{28A^2(-AB)^3}{-7A^3B^3}$
7. Perform the following multiplication:
   $(a - 2b) \cdot (a^2 - 3ab - 2b^2)$
8. Perform the following division:
   $(2x^3 - 2x - 12) \div (2x - 4)$
9. Convert the numbers to scientific notation before multiplying: $(0.000000721) \cdot (0.000000089)$
10. Solve for x: $2(2 - 3x) = 5(x - 5)$
11. Solve for x: $\dfrac{1}{4}x - \dfrac{1}{2} = \dfrac{1}{3}x - 2$
12. Solve for t_1: $L = L_1[1 + \beta(t_2 - t_1)]$
13. State the domain of the function $f(x) = \sqrt{4 - x}$.
14. If $f(x) = \sqrt{x^2 + 2}$, find $f(0)$ and $f(\sqrt{2})$.
15. Solve the following system graphically:

    $2x - 3y = 4$
    $x + 2y = 2$

16. Solve the following system algebraically:

    $2r_1 - 3r_2 = 7$
    $3r_1 - 5r_2 = 3$

17. Solve the following system by determinants:

    $3x + 2y - z = 16$
    $2x + 3y + 4z = 7$
    $8x + 5y - 6z = 47$

18. The respective currents in milliamperes (mA) in the two loops of a certain circuit are $1.00I_1 - 2.00I_2$ and $5.00I_1 - 3.00I_2$. Find the sum of the currents.
19. Simplify the following expression from a problem in statics:

    $-[F_1 + 2F_2 - 4(F_1 + F_2)]$

20. The distance to the moon is approximately 239,000 miles. Write the distance in scientific notation, using two significant figures.
21. A rectangular computer chip is 3 times as long as it is wide. If the perimeter is 9.6 mm, find the dimensions.
22. The force exerted by a spring is directly proportional to the extension. If a force of 2.72 lb stretches a spring 1.70 in., determine the force required to stretch it 4.10 in.

CHAPTER 4

Introduction to Trigonometry*

Objectives Upon completion of this chapter, you should be able to:

1. Draw an angle in standard position.
2. Determine the smallest positive angle coterminal with a given angle.
3. Find the values of the six trigonometric functions for an angle in standard position, given one point on the terminal side.
4. Determine the values of trigonometric functions of special angles.
5. Determine the values of trigonometric functions by
   a. Using a table.
   b. Using a calculator.
6. Find an angle, given the value of a trigonometric function of the angle, by
   a. Using a table.
   b. Using a calculator.
7. Solve right triangles.

4.1 What Is Trigonometry?

The literal meaning of **trigonometry** is "triangle measurement." Although the study of triangles is certainly part of the subject, another important aspect is the study of trigonometric functions and their relationships. This part of the subject is called *analytic trigonometry* and will be of great concern to us later. In this chapter we shall concentrate on the basic functions and their applications to right triangles.

Trigonometry originally developed in Egypt and Babylonia, but a systematic study was not undertaken until the second half of the second century B.C. The true originator was Hipparchus of Nicaea (a Greek who lived from

Hipparchus

Note: This chapter may be postponed until the end of Chapter 6.

4.2 Angles

Angle

An **angle** is defined as a geometric figure consisting of two rays with a common endpoint. (A ray is that portion of a line on one side of a fixed point on the line.) Angle measurement is of primary importance in trigonometry, where an angle is better defined in terms of rotation: Start with a single ray, whose original position is called the **initial side,** and rotate this ray about its endpoint to a new position. This new position is called the **terminal side.** The common endpoint is called the **vertex** of the angle (Figure 4.1). The measure, or size, of the angle is the amount of rotation.

Initial side
Terminal side
Vertex

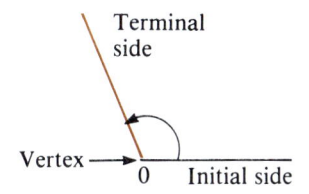

Figure 4.1

The most common units of measurement of angles are the **degree** and the **radian;** in this chapter we will consider only degree measure.

> A **degree** is defined as $\frac{1}{360}$ of a complete rotation.

It follows from this definition that a complete rotation, called a *whole angle,* is 360 degrees. Why the number 360 was chosen is not known. However, Hipparchus, a student of Babylonian astronomy, may have chosen the number since it is close to the number of days in a year.

The Babylonians used a number system with base 60, a system that had a remarkably long life. So it was natural at one time to subdivide an angle of 1 degree into 60 parts, each part called a **minute.** One minute was further subdivided into 60 **seconds.** The symbols °, ′, and ″ are used to denote degrees, minutes, and seconds, respectively. Angles are often denoted by Greek letters, such as θ (theta), α (alpha), β (beta), or γ (gamma).

Minute
Second

Another fruitful idea in trigonometry results from combining the concept of an angle with that of a coordinate system.

CHAPTER 4 INTRODUCTION TO TRIGONOMETRY

> An angle is said to be in **standard position** if its vertex is at the origin and its initial side is on the positive x-axis.

Quadrantal angle

Once in standard position, an angle can be described by specifying one point on the terminal side. For example, in Figure 4.2, the point $(-1, 2)$ uniquely determines an angle (between $0°$ and $360°$). If the terminal side coincides with one of the axes, the angle is called a **quadrantal** angle. For example, the angle θ whose terminal side passes through $(0, -2)$ is quadrantal (Figure 4.3).

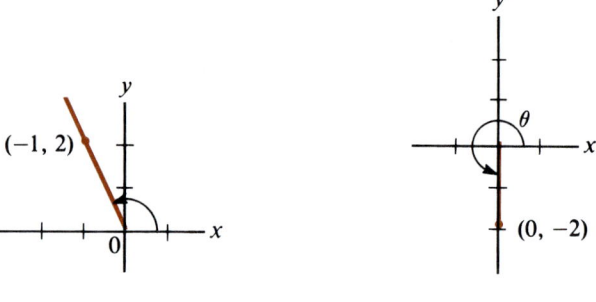

Figure 4.2

Figure 4.3

Positive and negative angles

Since an angle is defined in terms of rotation, there is no limit to its size. For example, the angle $\theta = 380°$ indicates $20°$ more than one complete rotation. (See Figure 4.4.) If the rotation is **counterclockwise,** the angle is considered **positive.** If the rotation is **clockwise,** the angle is considered **negative.** For example, the angle α in Figure 4.5 is negative.

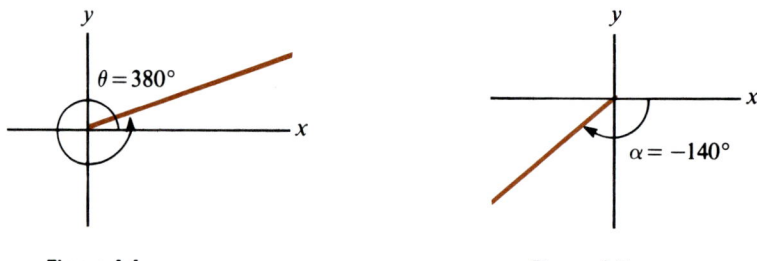

Figure 4.4

Figure 4.5

Coterminal angles

Two different angles can have a common terminal side. For example, the angle $\theta = -45°$ has the same terminal side as $\alpha = 315°$ (Figure 4.6). Such angles are called **coterminal.**

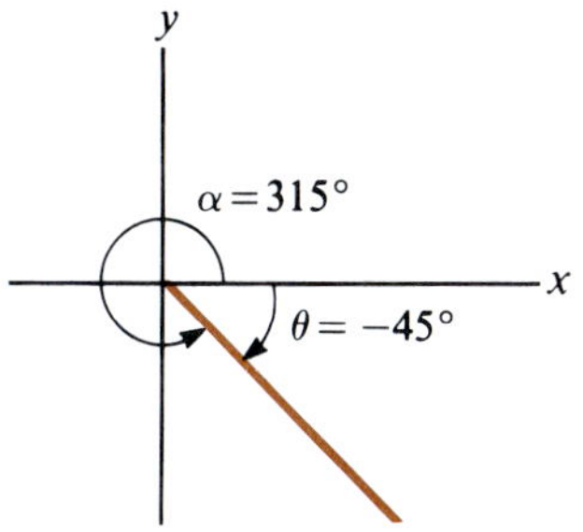

Figure 4.6

Example 1 If $\theta = 120°$, find a negative angle α so that θ and α are coterminal.

Solution. Since $360° - 120° = 240°$, it follows that $\alpha = -240°$. Another possible choice is $-(240° + 360°) = -600°$. In general, $\alpha = -240° - n \cdot 360°$, where $n = 0, 1, 2, \ldots$.

Example 2 Find the smallest positive angle α so that α and $\theta = -45°25'$ are coterminal.

Solution. Write $360°$ as $359°60'$ and subtract:

$$\begin{array}{r} 359°60' \\ -45°25' \\ \hline \alpha = 314°35' \end{array}$$

Exercises / Section 4.2

1. Draw the following angles in standard position:
   a. $45°$
   b. $-45°$
   c. $150°$
   d. $-180°$
   e. $-110°$
   f. $250°$
   g. $450°$
   h. $-400°$
2. Draw the following angles in standard position so that the terminal side passes through the given point.
   a. $(-1, 3)$
   b. $(2, 1)$
   c. $(3, -4)$
   d. $(-5, -6)$

In Exercises 3–16, find the smallest positive angle coterminal with the angle specified.

3. $-155°$
4. $-30°30'$
5. $-72°40'$
6. $-314°23'$
7. $-179°57'$
8. $-90°12'$
9. $-146°5'$
10. $-89°15'$
11. $-181°54'$
12. $-269°16'$
13. $-325°41'$
14. $-359°2'$
15. $-365°$
16. $-400°$

4.3 Trigonometric Functions

The study of trigonometry, as well as the applications of trigonometry, is based on the ratios of the lengths of two sides of a right triangle. These ratios are used in defining the trigonometric functions of an angle.

To understand the idea behind the trigonometric ratios, we need to recall the following property of a 30°–60° right triangle.

> In a 30°–60° right triangle, the side opposite the 30° angle is one-half the hypotenuse.

Now consider the triangle in Figure 4.7. Note that the side opposite the 30° angle is 5 in. long. Consequently, the hypotenuse must be 10 in. long. Similarly, if the side opposite the 30° angle is 7 in. long, then the hypotenuse must be 14 in. long. (See Figure 4.8.) Now consider the ratio of the side opposite the 30° angle to the hypotenuse. In the first case the ratio is

$$\frac{5 \text{ in.}}{10 \text{ in.}} = \frac{1}{2}$$

and in the second case

$$\frac{7 \text{ in.}}{14 \text{ in.}} = \frac{1}{2}$$

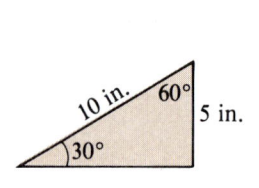

Figure 4.7

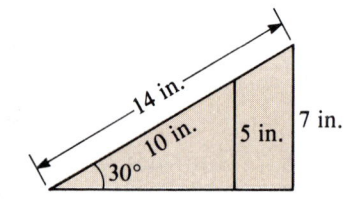

Figure 4.8

The size of the triangle doesn't matter; the ratio of the side opposite the 30° angle to the hypotenuse is always $\frac{1}{2}$.

A similar statement can be made for any angle. If θ is an angle of a right triangle, then the ratio of the side opposite this angle to the hypotenuse is *always fixed*. This ratio has been given a name: It is called the **sine of angle θ**, abbreviated **sin θ**. Referring to the descriptive labels in Figure 4.9, we now write

$$\sin \theta = \frac{\text{opposite side}}{\text{hypotenuse}}$$

For every angle θ, sin θ is a unique number. (Since θ is assumed to be an angle of a right triangle, this definition makes sense only for acute angles—angles between 0° and 90°. Chapter 7 discusses the trigonometric functions of arbitrary angles.)

The sine of θ is only one of several possible ratios. For example, the ratio of the opposite side to the adjacent side (see Figure 4.9) is called the

tangent of θ, abbreviated **tan θ**. The ratio of the adjacent side to the hypotenuse is called the **cosine of θ**, abbreviated **cos θ**.

These ratios turn out to be entirely adequate for the study of right triangles. However, for convenience the reciprocals of the above ratios have also been named. They are called, respectively, the **cosecant of θ (csc θ)**, the **cotangent of θ (cot θ)**, and the **secant of θ (sec θ)**. These ratios should be carefully memorized.

Since the ratios corresponding to each angle are unique, each ratio is a function of the angle. For that reason, we refer to the ratios as the **trigonometric functions**.

Trigonometric functions

Name	Abbreviation	Value
sine of θ	sin θ	$\sin \theta = \dfrac{\text{opposite}}{\text{hypotenuse}}$
cosine of θ	cos θ	$\cos \theta = \dfrac{\text{adjacent}}{\text{hypotenuse}}$
tangent of θ	tan θ	$\tan \theta = \dfrac{\text{opposite}}{\text{adjacent}}$
cosecant of θ	csc θ	$\csc \theta = \dfrac{\text{hypotenuse}}{\text{opposite}}$
secant of θ	sec θ	$\sec \theta = \dfrac{\text{hypotenuse}}{\text{adjacent}}$
cotangent of θ	cot θ	$\cot \theta = \dfrac{\text{adjacent}}{\text{opposite}}$

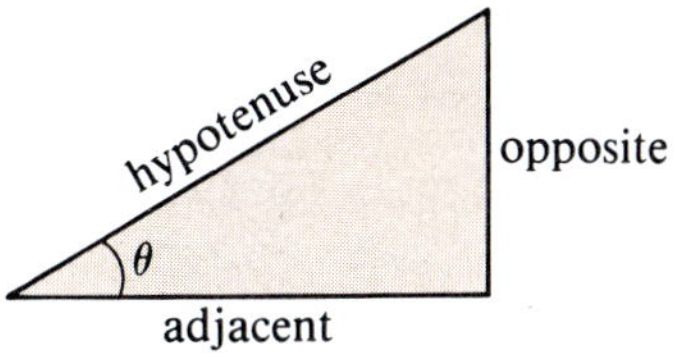

Figure 4.9

In the next section, we will find the values of the functions of an angle. But first we must become acquainted with the trigonometric functions by doing some appropriate exercises.

Let θ be an angle whose terminal side passes through the point $P(x, y)$, which we will place in the first quadrant (Figure 4.10). Dropping a perpendicular line from P to the x-axis, we determine a right triangle. To find the six

126 CHAPTER 4 INTRODUCTION TO TRIGONOMETRY

Pythagorean theorem

ratios, we need to compute the length r of the hypotenuse by using the **Pythagorean theorem:**

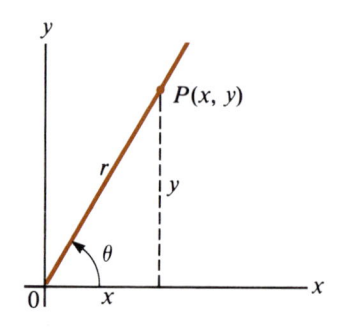

$$r^2 = x^2 + y^2 \quad \text{or} \quad r = \sqrt{x^2 + y^2}$$

Figure 4.10

Radius vector

The distance r from P to the origin is called the **radius vector** and is always taken as positive.

Using the coordinates of P and the radius vector r in Figure 4.10, the definitions of the six trigonometric functions can also be stated in the following form:

$$\sin \theta = \frac{y}{r} \qquad \csc \theta = \frac{r}{y}$$

$$\cos \theta = \frac{x}{r} \qquad \sec \theta = \frac{r}{x}$$

$$\tan \theta = \frac{y}{x} \qquad \cot \theta = \frac{x}{y}$$

These definitions will now be illustrated by several examples.

Example 1 Find the values of the six trigonometric functions of the angle whose terminal side passes through (3, 4).

Solution. Drop a perpendicular from the point (3, 4) to the x-axis (Figure 4.11). Placing the numbers 3 and 4 along the sides, we see that $r = \sqrt{3^2 + 4^2} = \sqrt{25} = 5$. It now follows directly from the definitions that

$$\sin \theta = \frac{4}{5} \qquad \cos \theta = \frac{3}{5} \qquad \tan \theta = \frac{4}{3},$$

$$\csc \theta = \frac{5}{4} \qquad \sec \theta = \frac{5}{3} \qquad \cot \theta = \frac{3}{4}$$

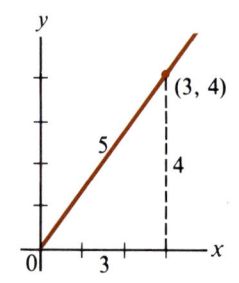

Figure 4.11

Example 2 Find the values of the six trigonometric functions of the angle whose terminal side passes through $(5, \sqrt{3})$.

Solution. Draw Figure 4.12. From the Pythagorean theorem
$$r = \sqrt{5^2 + (\sqrt{3})^2} = \sqrt{25 + 3} = \sqrt{28} = \sqrt{4 \cdot 7} = 2\sqrt{7}$$
Hence
$$\sin \theta = \frac{\sqrt{3}}{2\sqrt{7}} \qquad \cos \theta = \frac{5}{2\sqrt{7}} \qquad \tan \theta = \frac{\sqrt{3}}{5}$$
$$\csc \theta = \frac{2\sqrt{7}}{\sqrt{3}} \qquad \sec \theta = \frac{2\sqrt{7}}{5} \qquad \cot \theta = \frac{5}{\sqrt{3}}$$

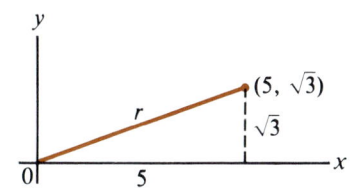

Figure 4.12

Although the values in Example 2 are perfectly satisfactory, recall that it is customary to rationalize denominators. For example,
$$\frac{1}{\sqrt{2}} = \frac{1 \cdot \sqrt{2}}{\sqrt{2} \cdot \sqrt{2}} = \frac{\sqrt{2}}{2}$$

In Example 2
$$\cot \theta = \frac{5}{\sqrt{3}} = \frac{5\sqrt{3}}{(\sqrt{3})^2} = \frac{5\sqrt{3}}{3}$$

and
$$\sin \theta = \frac{\sqrt{3}}{2\sqrt{7}} = \frac{\sqrt{3} \cdot \sqrt{7}}{2\sqrt{7} \cdot \sqrt{7}} = \frac{\sqrt{3 \cdot 7}}{2 \cdot 7} = \frac{\sqrt{21}}{14}$$

and so forth.

If the value of one trigonometric function of an angle is known, the values of the remaining functions can be found, as shown in the next example.

Example 3 Given that $\cos \theta = \frac{3}{7}$, find $\cot \theta$ and $\csc \theta$.

Solution. Since $\cos \theta = $ adjacent/hypotenuse, we need to draw the figure with 3 along the adjacent side and 7 along the hypotenuse (Figure 4.13). Now denote the opposite side by y and use the Pythagorean theorem again. Thus

$$y^2 + 3^2 = 7^2$$
$$y^2 = 49 - 9 = 40$$
$$y = \sqrt{40} = \sqrt{4 \cdot 10} = 2\sqrt{10}$$

Hence

$$\cot \theta = \frac{3}{2\sqrt{10}} = \frac{3\sqrt{10}}{2\sqrt{10} \cdot \sqrt{10}} = \frac{3\sqrt{10}}{2 \cdot 10} = \frac{3\sqrt{10}}{20}$$

and

$$\csc \theta = \frac{7}{2\sqrt{10}} = \frac{7\sqrt{10}}{2\sqrt{10} \cdot \sqrt{10}} = \frac{7\sqrt{10}}{2 \cdot 10} = \frac{7\sqrt{10}}{20}$$

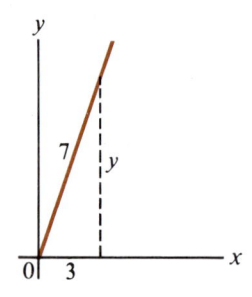

Figure 4.13

Cofunctions

Recall from your study of geometry that two angles A and B are called **complementary angles** if $A + B = 90°$. For example, 30° and 60° are complementary. Suppose α and β are the two acute angles of a right triangle. Since the sum of the angles of a right triangle is 180°, it follows that $\alpha + \beta = 90°$. So the acute angles of a right triangle are complementary.

As we shall see in a moment, if α and β are complementary, then $\sin \alpha = \cos \beta$. For that reason the functions sine and cosine are called **cofunctions** (*cosine* means "complementary sine"). Also, the tangent and cotangent are cofunctions, as are the secant and cosecant. These facts yield the following result.

> Any trigonometric function of an acute angle is equal to the corresponding cofunction of its complement.

To see why this is so, consider angles α and β in Figure 4.14. Note that

$$\sin \alpha = \frac{a}{c}$$

Since a is adjacent to angle β and b is opposite angle β, it follows that

$$\cos \beta = \frac{a}{c}$$

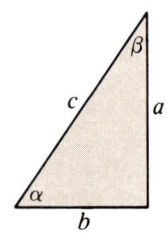

Figure 4.14

Thus sin α = cos β. Using Figure 4.14, we can also show that tan α = cot β and sec α = csc β.

Remark. The word *sine* originated from a mistranslation. A half-chord of a circle was called *jiva* by the ancient Hindus. The Arabs took over this word as *jiba*, which resembles *jaib*, meaning "bay." Confusing the two terms, Robert of Chester, a twelfth-century English mathematician, called the half-chord *sinus*, the Latin word for "bay," in his Latin translation. To see the significance of the half-chord, consider the circle of unit radius in Figure 4.15. Since the radius is 1 unit long, the length of the half-chord is numerically equal to sin θ.

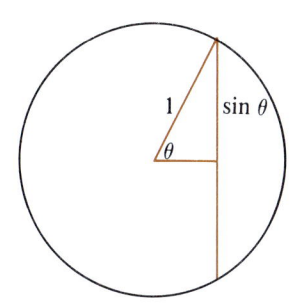

Figure 4.15

Exercises / Section 4.3

In Exercises 1–20, find the values of the six trigonometric functions for angles in standard position and having the given points on their terminal sides.

1. (1, 2)
2. (3, 4)
3. (5, 2)
4. (3, 7)
5. (4, 2)
6. (8, 2)
7. ($\sqrt{2}$, 4)
8. (5, $\sqrt{3}$)
9. (6, $\sqrt{6}$)
10. ($\sqrt{3}$, 3)
11. (7, $\sqrt{10}$)
12. (6, $\sqrt{11}$)
13. ($\sqrt{11}$, 1)
14. (1, $\sqrt{19}$)
15. ($\sqrt{3}$, $\sqrt{5}$)
16. ($\sqrt{5}$, $\sqrt{7}$)
17. (3$\sqrt{2}$, $\sqrt{7}$)
18. (3$\sqrt{5}$, 2)
19. (3$\sqrt{7}$, 2$\sqrt{2}$)
20. (4$\sqrt{2}$, 2$\sqrt{3}$)
21. If sin θ = $\frac{3}{5}$, find tan θ.
22. If tan θ = $\frac{3}{2}$, find cos θ.
23. If sec θ = 2, find sin θ.
    Note: sec θ = 2 = $\frac{2}{1}$.
24. If csc θ = 3, find tan θ.

25. If $\tan \theta = \frac{5}{3}$, find $\csc \theta$.
26. If $\cos \theta = \frac{1}{2}$, find $\cot \theta$.
27. If $\sec \theta = \frac{9}{4}$, find $\cos \theta$.
28. If $\cot \theta = \frac{5}{3}$, find $\sec \theta$.
29. If $\csc \theta = \frac{7}{6}$, find $\sin \theta$.
30. If $\tan \theta = \frac{4}{7}$, find $\cot \theta$.
31. If $\tan \theta = \sqrt{3}/4$, find $\cos \theta$.
32. If $\sin \theta = \sqrt{3}/2$, find $\cot \theta$.
33. If $\csc \theta = \sqrt{3}$, find $\cos \theta$.
34. If $\cot \theta = 2\sqrt{2}/5$, find $\csc \theta$.
35. If $\cos \theta = \sqrt{3}/6$, find $\sin \theta$.
36. If $\sec \theta = 2\sqrt{7}/3$, find $\tan \theta$.
37. If $\sin \theta = 3\sqrt{10}/10$, find $\sec \theta$.
38. If $\tan \theta = 6\sqrt{7}$, find $\cot \theta$.
39. If $\cos \theta = \sqrt{11}/11$, find $\tan \theta$.
40. If $\sec \theta = \sqrt{17}$, find $\tan \theta$.
41. Given that $\tan \theta = 2$, show that $\sin^2\theta + \cos^2\theta = 1$. *Note:* $\sin^2\theta = (\sin \theta)^2$.
42. Given that $\sin \theta = \frac{3}{8}$, show that $1 + \tan^2\theta = \sec^2\theta$.
43. Given that $\cos \theta = \frac{1}{3}$, show that $\cot^2\theta = \csc^2\theta - 1$.

4.4 Values of Trigonometric Functions

So far we have considered only the definitions of the trigonometric functions. We now turn to the problem of finding the values of trigonometric functions of specific angles. We will first consider certain special angles and then discuss arbitrary angles.

1. Special Angles

The values of trigonometric functions of certain special angles—0°, 30°, 45°, 60°, and 90°—can be found by means of diagrams. Consider again the 30°–60° triangle (Figure 4.16). Since the ratios of the sides are the same regardless of the size of the triangle, we can assign a value of 1 unit to the length of the side opposite the 30° angle, thereby making the hypotenuse 2 units long. The length of the remaining side is $x = \sqrt{2^2 - 1} = \sqrt{3}$. For the 60° angle, the adjacent side is 1 unit long and the opposite side $\sqrt{3}$.

Special angles

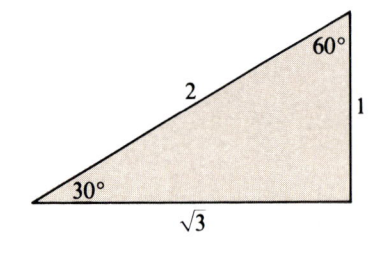

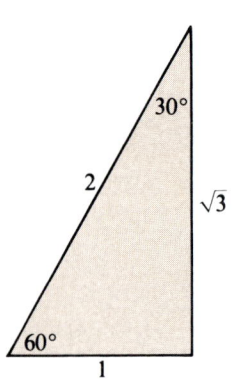

Figure 4.16

4.4 VALUES OF TRIGONOMETRIC FUNCTIONS 131

Given this information, we can find the values of the trigonometric functions of 30° and 60°.

Example 1 From Figure 4.16

$$\sin 30° = \frac{1}{2}, \cos 30° = \frac{\sqrt{3}}{2}, \tan 30° = \frac{1}{\sqrt{3}} = \frac{\sqrt{3}}{3}, \text{ and so forth}$$

$$\sin 60° = \frac{\sqrt{3}}{2}, \cos 60° = \frac{1}{2}, \tan 60° = \sqrt{3}, \text{ and so forth}$$

For $\theta = 45°$, we construct the right triangle in Figure 4.17. Note that the hypotenuse $h = \sqrt{1^2 + 1^2} = \sqrt{2}$.

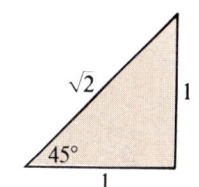

Figure 4.17

Example 2 From Figure 4.17

$$\tan 45° = 1, \sin 45° = \frac{1}{\sqrt{2}} = \frac{\sqrt{2}}{2}, \text{ and so forth}$$

Diagrams can also be used to find the function values of the quadrantal angles 0° and 90°. (We will consider angles outside this range in Chapter 7.) If $\theta = 0°$, we do not get a triangle. However, we can pretend that such a triangle exists by separating the coinciding initial and terminal sides and assigning 0 to the opposite side, as in Figure 4.18. The adjacent side and hypotenuse both have a length of 1 unit.

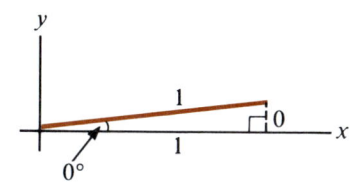

Figure 4.18

Example 3 From Figure 4.18

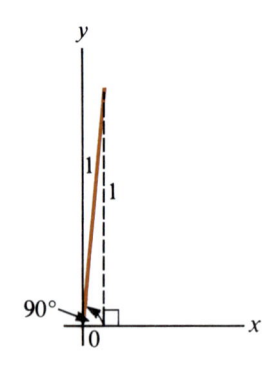

$$\sec 0° = \frac{1}{1} = 1, \sin 0° = \frac{0}{1} = 0, \cot 0° = \frac{1}{0} \text{ (undefined)}$$

For $\theta = 90°$ we do not obtain a triangle, but, as before, we can pretend that such a triangle exists (Figure 4.19). This time the opposite side and hypotenuse have unit length and the adjacent side has a length of 0. Note that the perpendicular is dropped to the x-axis, the same as for all regular triangles.

Figure 4.19

Example 4 From Figure 4.19

$$\cos 90° = \frac{0}{1} = 0, \csc 90° = \frac{1}{1} = 1, \tan 90° = \frac{1}{0} \text{ (undefined)}$$

The following table lists all of the special angles just discussed.

Special angles

θ	0°	30°	45°	60°	90°
$\sin \theta$	0	$\frac{1}{2}$	$\frac{\sqrt{2}}{2}$	$\frac{\sqrt{3}}{2}$	1
$\cos \theta$	1	$\frac{\sqrt{3}}{2}$	$\frac{\sqrt{2}}{2}$	$\frac{1}{2}$	0
$\tan \theta$	0	$\frac{\sqrt{3}}{3}$	1	$\sqrt{3}$	undefined
$\csc \theta$	undefined	2	$\sqrt{2}$	$\frac{2\sqrt{3}}{3}$	1
$\sec \theta$	1	$\frac{2\sqrt{3}}{3}$	$\sqrt{2}$	2	undefined
$\cot \theta$	undefined	$\sqrt{3}$	1	$\frac{\sqrt{3}}{3}$	0

Even though the table lists the values of all the special angles, it would be ludicrous to memorize the whole set. Even if you could remember them, it is much better in the long run to obtain the values directly from a diagram that you have constructed. In addition, more special angles exist having terminal sides in the other quadrants, which makes memorization of all special angles unfeasible anyway.

Common error Mislabeling the diagram for the 30°–60° right triangle, such as placing the 1 or $\sqrt{3}$ along the hypotenuse. With a little care, such errors can be avoided since the hypotenuse is necessarily the longest side and the side opposite the 30° angle the shortest.

2. Arbitrary Acute Angles and Using Tables

Tables

Our final task in this section is to determine the values of trigonometric functions of arbitrary acute angles. Because these values cannot be obtained from a diagram except for the special angles, tables of values have been devised: One such table is given in Appendix B, a portion of which is shown below. Note that angles from 0° to 45° are listed in the left column, while the names of the functions are given on top in the first row. Angles from 45° to 90° are listed in the right column with the names of the functions along the bottom row. The angles themselves are listed in intervals of 10′.

Table 4.1

Degrees	Sin θ	Cos θ	Tan θ	Cot θ	Sec θ	Csc θ	
39°00′	0.6293	0.7771	0.8098	1.235	1.287	1.589	51°00′
10	0.6316	0.7753	0.8146	1.228	1.290	1.583	50
20	0.6338	0.7735	0.8195	1.220	1.293	1.578	40
30	0.6361	0.7716	0.8243	1.213	1.296	1.572	30
40	0.6383	0.7698	0.8292	1.206	1.299	1.567	20
50	0.6406	0.7679	0.8342	1.199	1.302	1.561	10
	Cos θ	Sin θ	Cot θ	Tan θ	Csc θ	Sec θ	Degrees

Example 5 Find sec 39°50′.

Solution. Since the angle is listed in the left column, we need to refer to the secant column listed at the top of the table to get

sec 39°50′ = 1.302

Example 6 Find cos 62°10′.

Solution. Since the angle is listed in the right column, we find the function by looking at the bottom of the table. (See Table 1 in Appendix B.) Thus

cos 62°10′ = 0.4669

It is often necessary to reverse the process, that is, to find an angle whose trigonometric function is known.

Example 7 Find θ, given that $\tan \theta = 1.137$.

Solution. We locate the value by looking down the tangent column until we come to 1.137. Since the column name for this value is at the bottom, we need to read off the angle from the right column. Hence $\theta = 48°40'$.

Interpolation

If the angle is not listed precisely, we obtain the value by "reading between the lines," a process called **interpolation.** Some cases can be seen directly. For example, to find $\cos 19°15'$, we would expect the value to be about midway between

$\cos 19°10' = 0.9446 \quad \text{and} \quad \cos 19°20' = 0.9436$

or

$\cos 19°15' = 0.9441$

Strictly speaking, the desired value does not lie exactly between the other two, but the approximation is totally adequate for most purposes.

If the desired value does not lie midway between two given entries, then we must set up a proportion, as illustrated in the next example.

Example 8 Find $\tan 26°26'$.

Solution. The desired value is between $\tan 26°20' = 0.4950$ and $\tan 26°30' = 0.4986$. Now, since $26'$ is $\frac{6}{10}$ of the way from $20'$ to $30'$, we would expect $\tan 26°26'$ to be $\frac{6}{10}$ of the way from 0.4950 to 0.4986. Ignoring the decimal point and the two digits immediately following it, we see that 36 is the *tabular difference*. The calculation can be done systematically as follows:

$$10 \left[6 \left[\begin{array}{l} \tan 26°20' = 0.4950 \\ \tan 26°26' = \ldots \end{array} \right] x \atop \tan 26°30' = 0.4986 \right] 36$$

From the diagram we set up the proportion

$$\frac{x}{36} = \frac{6}{10}$$

and get $x = 21.6$. Since we cannot obtain accuracy to more than four decimal places, x is rounded off to 22. So 22 is the tabular difference between 0.4950 and the desired value. It follows that

$\tan 26°26' = 0.0022 + 0.4950$

$ = 0.4972$

4.4 VALUES OF TRIGONOMETRIC FUNCTIONS

If the function values decrease as θ increases, special care must be taken when interpolating, as shown in the next example.

Example 9 Find cot 55°3'.

Solution.
$$10 \begin{bmatrix} 3 \begin{bmatrix} \cot 55°\ 0' = 0.7002 \\ \cot 55°\ 3' = \ldots \end{bmatrix} x \\ \cot 55°10' = 0.6959 \end{bmatrix} 43$$

Note that the tabular difference is 102 − 59 = 43. The resulting proportion is

$$\frac{x}{43} = \frac{3}{10}$$

and $x = 12.9 \approx 13$. Since the values of cot θ are decreasing, the number 0.0013 has to be subtracted from 0.7002 to ensure that the desired value falls between 0.7002 and 0.6959. We conclude that

cot 55°3' = 0.6989

Interpolation may also be necessary when determining the angle from the value of a trigonometric function.

Example 10 Find θ such that cos θ = 0.6869.

Solution. We look along the cosine column to find the entry or nearest entry. The given value lies between 0.6884 and 0.6862. This observation suggests the following diagram:

$$10 \begin{bmatrix} x \begin{bmatrix} \cos 46°30' = 0.6884 \\ \cos \theta\ \ \ \ \ = 0.6869 \end{bmatrix} 15 \\ \cos 46°40' = 0.6862 \end{bmatrix} 22$$

For the proportion we get

$$\frac{x}{10} = \frac{15}{22}$$

yielding $x = 6.8 \approx 7$. It follows that

$\theta = 46°37'$

3. Finding Values of Trigonometric Functions with a Calculator

Using a calculator   Scientific calculators are programmed to evaluate the basic trigonometric functions of sine, cosine, and tangent. Pressing the reciprocal key $\boxed{1/x}$ will then yield the values of the cosecant, secant, and cotangent, respectively. Unfortunately, some calculators cannot accept angles expressed in degrees and minutes, although many of them do. For the former type of calculator, the angle must be converted to decimal form.

Example 11 Use a calculator to find:

    a. csc 29°   **b.** cos 43°17′

Solution. We enter 29 and press the sine key, followed by the reciprocal key. The sequence is

29 $\boxed{\text{SIN}}$ $\boxed{1/x}$

Display: 2.0627 (to four decimal places)

    b. 43°17′ = $(43 + \frac{17}{60})°$; the sequence is

43 $\boxed{+}$ 17 $\boxed{\div}$ 60 $\boxed{=}$ $\boxed{\text{COS}}$

Display: 0.7280 (to four decimal places)

To obtain the angle θ from a given trigonometric function, we enter the value, press the inverse key $\boxed{\text{INV}}$ (or similar key), and then press the appropriate function key. Some calculators have the special keys $\boxed{\text{SIN}^{-1}}$, $\boxed{\text{COS}^{-1}}$, and $\boxed{\text{TAN}^{-1}}$ for these operations.

Example 12 Use a calculator to find θ if tan θ = 0.7519.

Solution. The sequence is

0.7519 $\boxed{\text{INV}}$ $\boxed{\text{TAN}}$,   or   0.7519 $\boxed{\text{TAN}^{-1}}$

Display: 36.94 (degrees)

If the calculator does not express this angle in degrees and minutes, we set up the following proportion:

$$0.94 = \frac{94}{100} = \frac{x}{60}$$

which gives x = 56. It follows that θ = 36°56′.

Working with the secant, cosecant, and cotangent functions requires one preliminary step: After entering the function value, press the reciprocal key, as shown in the next example.

4.4 VALUES OF TRIGONOMETRIC FUNCTIONS

Example 13 Find θ, given that $\sec \theta = 1.315$.

Solution. We enter 1.315 and press the reciprocal key. Note that the resulting value is equal to $\cos \theta$, since $\cos \theta = 1/(\sec \theta)$. The sequence is

1.315 $\boxed{1/x}$ $\boxed{\text{INV}}$ $\boxed{\text{COS}}$ or 1.315 $\boxed{1/x}$ $\boxed{\text{COS}^{-1}}$

Display: 40.50 (degrees)

The availability of scientific calculators seems to have made interpolation obsolete. However, since there are many tables besides trigonometric tables, interpolation is still a valuable technique to master. (See, for example, Exercise 81.)

Exercises / Section 4.4

In Exercises 1–20, draw a diagram and determine the exact value of each trigonometric function.

1. tan 30°
2. sec 60°
3. cos 30°
4. cos 45°
5. sin 90°
6. tan 0°
7. csc 30°
8. sec 30°
9. sec 0°
10. cot 90°
11. csc 60°
12. csc 90°
13. tan 45°
14. csc 45°
15. sec 45°
16. tan 60°
17. cos 60°
18. sin 30°
19. sin 45°
20. cot 30°

In Exercises 21–40, use a diagram to find θ.

21. $\tan \theta = 1$
22. $\sin \theta = \dfrac{1}{\sqrt{2}}$
23. $\cos \theta = \dfrac{\sqrt{2}}{2}$
24. $\sin \theta = \dfrac{1}{2}$
25. $\cos \theta = \dfrac{\sqrt{3}}{2}$
26. $\sec \theta = 2$
27. $\csc \theta = \sqrt{2}$
28. $\cot \theta = 1$
29. $\sin \theta = 0$
30. $\sec \theta = 1$
31. $\tan \theta = \dfrac{\sqrt{3}}{3}$
32. $\tan \theta = \sqrt{3}$
33. $\cos \theta = \dfrac{1}{2}$
34. $\csc \theta = 2$
35. $\cot \theta = 0$
36. $\cot \theta = \sqrt{3}$
37. $\sec \theta = \sqrt{2}$
38. $\csc \theta = \dfrac{2\sqrt{3}}{3}$
39. $\sec \theta = \dfrac{2\sqrt{3}}{3}$
40. $\tan \theta$ undefined

In Exercises 41–60, use Table 1 in Appendix B to find the value of the trigonometric function; interpolate if necessary. Check the result by using a calculator.

41. sin 10°40′
42. tan 79°30′
43. cot 37°10′
44. cos 11°20′
45. sec 84°50′
46. csc 15°40′
47. csc 55°15′
48. cos 45°5′
49. sec 17°12′
50. cos 22°31′
51. cot 18°16′
52. sin 85°22′
53. tan 17°35′
54. sec 51°43′
55. cos 5°6′
56. sin 2°28′
57. tan 38°29′
58. sec 46°11′
59. csc 11°55′
60. tan 39°14′

CHAPTER 4 INTRODUCTION TO TRIGONOMETRY

In Exercises 61–80, use Table 1 in Appendix B to find θ; interpolate if necessary. Check the result by using a calculator.

61. $\cos \theta = 0.9377$
62. $\tan \theta = 0.9217$
63. $\sec \theta = 1.595$
64. $\cot \theta = 0.6128$
65. $\sin \theta = 0.9971$
66. $\sin \theta = 0.1984$
67. $\cos \theta = 0.1492$
68. $\cot \theta = 0.1776$
69. $\tan \theta = 0.1066$
70. $\sin \theta = 0.1812$
71. $\csc \theta = 1.619$
72. $\sec \theta = 1.933$
73. $\cos \theta = 0.9476$
74. $\sin \theta = 0.8265$
75. $\cot \theta = 0.9574$
76. $\sec \theta = 2.421$
77. $\csc \theta = 1.399$
78. $\tan \theta = 1.100$
79. $\cos \theta = 0.1310$
80. $\sec \theta = 1.091$

81. In the table below, W (in grams) represents the experimentally determined weight of potassium chloride that will dissolve in 50 g of water at temperature T (in degrees Celsius).

T	0	10	20	30	40	50	60	70	80
W	13.6	15.4	16.8	18.3	19.7	21.1	22.4	24.0	25.2

Use interpolation to estimate W if

a. $T = 15°C$ **b.** $T = 22°C$ **c.** $T = 37°C$ **d.** $T = 66°C$ **e.** $T = 79°C$

82. The *index of refraction* μ of a medium is given by

$$\mu = \frac{\sin \theta_i}{\sin \theta_r}$$

where θ_i is the angle of incidence and θ_r is the angle of refraction. If $\theta_i = 47.0°$ and $\theta_r = 24.1°$, find μ.

83. The instantaneous voltage in a coil rotating in a magnetic field is given by

$$V = V_m \cos \phi$$

where V_m is the maximum voltage and ϕ is the angle that the coil makes with the magnetic field. Given that $V_m = 120$ V, find V at the instant when $\phi = 67.4°$.

84. The largest weight W that can be pulled up an inclined plane making an angle θ with the horizontal by a force F is

$$W = \frac{F}{\mu}(\cos \theta + \mu \sin \theta)$$

where μ is the coefficient of friction. Find the largest weight if $F = 49.5$ lb, $\theta = 15.6°$, and $\mu = 0.276$.

4.5 Applications of Right Triangles

In this section we will study some applications of trigonometric functions to problems involving right triangles.

For our first example, suppose a technician needs to find the height of a condemned building. Because of the condition of the building, it is too dangerous to measure the vertical wall directly. At this point trigonometry comes to the rescue: Our technician measures the distance d (Figure 4.20) and the angle θ determined by the ground and the top of the building. Once d

and θ have been measured, the height x of the building can be calculated by observing that

$$\tan \theta = \frac{x}{d}$$

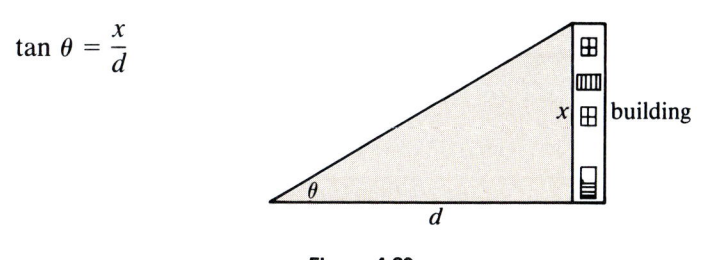

Figure 4.20

so that

$$x = d \tan \theta$$

From this example, we can write the rule for finding the length of an unknown side of a right triangle.

To find the length of an unknown side of a right triangle, use a trigonometric function whose definition involves the ratio of the length of the unknown side and the length of a known side.

We need to illustrate these ideas with some examples. The next three examples refer to Figures 4.21–4.23, respectively. Note that capital letters are used to denote the angles and lowercase letters their respective opposite sides. (Angle C is the right angle.)

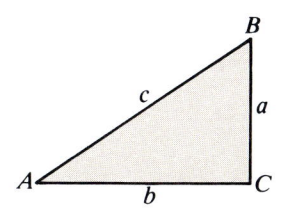

Figure 4.21

Reminder: The terms *opposite* and *adjacent* are relative. For angle A, a is the opposite side and b the adjacent side. For angle B, b is the opposite side and a the adjacent side. Thus $\tan A = a/b$ and $\tan B = b/a$.

Example 1 Given that $B = 36°40'$ and $c = 1.30$, find a.

Solution. We first observe that for angle B, **a is the adjacent side** (and b the opposite side). Since **c is the hypotenuse,** we have from the definition of cosine that

$$\cos 36°40' = \frac{a}{c} = \frac{a}{1.30}$$

Hence

$$a = 1.30 \cos 36°40' = (1.30)(0.8021) = 1.0427$$

Rounding off to three significant figures, we get $a = 1.04$.

Significant figures. Two significant figures in the side measurements corresponds to the nearest degree in the angle measurements; three significant figures corresponds to the nearest multiple of 10' or tenth of a degree; four significant figures corresponds to the nearest minute or hundredth of a degree. The rules are summarized in the following table.

Table 4.2

Significant figures

Accuracy of degree measurement		Significant figures in side measurement
1°	1°	2
10'	0.1°	3
1'	0.01°	4

Example 2 If $A = 59°34'$ and $a = 5.320$, find b (Figure 4.22).

Solution. The two sides involved are a and b, so the appropriate ratio is either $\tan A$ or $\cot A$. Thus

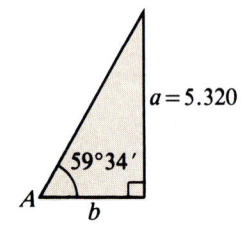

$$\cot 59°34' = \frac{b}{5.320}$$

or

$$b = 5.320 \cot 59°34'$$
$$= 3.125$$

Figure 4.22

to four significant figures.

Example 3 If $a = 3.78$ and $b = 5.289$, find A (Figure 4.23).

Solution. In this problem two sides are known. Angle A can be determined by noting that

$$\tan A = \frac{a}{b} = \frac{3.78}{5.289} = 0.7147$$

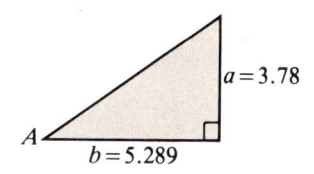

Figure 4.23

From Table 1 of Appendix B or a calculator, we get $A = 35°33'$. Since a has three significant figures, $A = 35°30'$ to the nearest multiple of $10'$.

To make some of the applied problems easier to state, we need to introduce two basic terms. If an object is located above a horizontal plane, then the angle between the horizontal and the line of sight is called the **angle of elevation**. (See Figure 4.24.) If the observer is looking down at an object, then the angle between the horizontal and the line of sight is called the **angle of depression**. (See Figure 4.24.) Note that the angle of elevation in Figure 4.24 is equal to the angle of depression.

Angle of elevation

Angle of depression

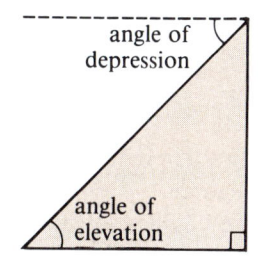

Figure 4.24

Example 4

An observer standing on top of a building is looking down at an intersection. He wants to know the distance from the foot of the building to the middle of the intersection. If the building is 22.0 ft high and the angle of depression is 25°20′, what is the desired distance?

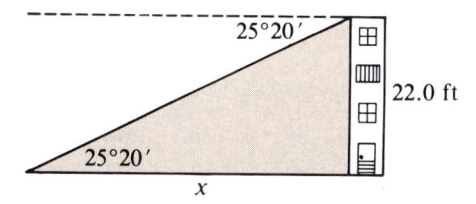

Figure 4.25

142 CHAPTER 4 INTRODUCTION TO TRIGONOMETRY

Solution. In Figure 4.25 let us denote the unknown side by x. From the given angle of depression, note that the angle on the left is also 25°20′. It follows that

$$\frac{x}{22.0} = \cot 25°20'$$

$$x = 22.0 \cot 25°20'$$

$$= (22.0)(2.112)$$

$$= 46.46$$

Since 22.0 has three significant figures, the desired distance is rounded off to 46.5 ft.

Example 5 A 40-ft ladder is leaning against a wall, making an angle of 71° with the ground. How high does the ladder reach up the wall?

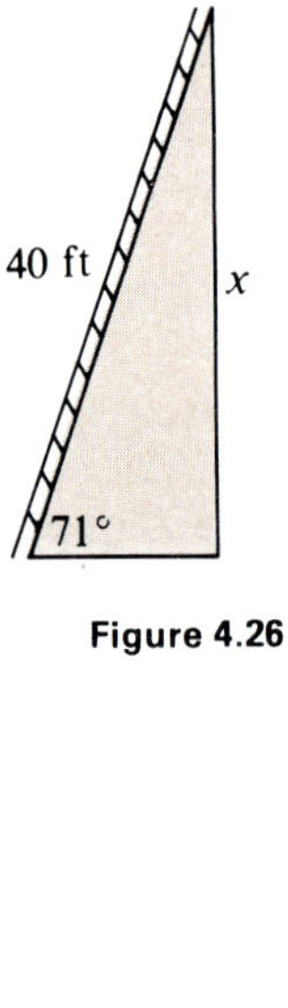

Solution. Draw Figure 4.26 and denote the unknown side by x. It follows that

$$\frac{x}{40} = \sin 71°$$

$$x = 40 \sin 71°$$

$$= (40)(0.9455)$$

$$= 37.82 \text{ ft}$$

Figure 4.26

Assuming the ladder is of standard length (accurate to the nearest foot), we get $x = 38$ ft to two significant figures.

Note that in this problem we could also use the cosecant to find x:

$$\frac{40}{x} = \csc 71°$$

or

$$x = \frac{40}{\csc 71°} = 38 \text{ ft}$$

Some problems in trigonometry involve two triangles, each with an unknown side. The resulting relationships can lead to a system of two equations, as shown in the next example.

Example 6 The angle of depression from the top of a building to a park bench is 19°25′, and that from a window 20.00 ft below the top to the bench is 10°43′. How far is the bench from the foot of the building?

4.5 APPLICATIONS OF RIGHT TRIANGLES 143

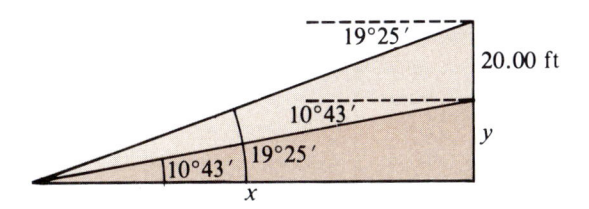

Figure 4.27

Solution. Draw Figure 4.27 and label the angles on the left. (Recall that the angle of depression is equal to the corresponding angle of elevation.) There are two unknown sides, labeled x and y, and two triangles. Consequently, we need to find two relationships that lead to a system of two equations. From the respective triangles, we get

$$\frac{y + 20.00}{x} = \tan 19°25'$$

$$\frac{y}{x} = \tan 10°43'$$

$$\frac{y + 20.00}{x} = 0.3525$$

$$\frac{y}{x} = 0.1893$$

$$y + 20.00 = 0.3525x \quad \text{multiplying both sides by } x$$
$$y = 0.1893x$$
$$20.00 = 0.1632x \quad \text{subtracting}$$
$$x = 122.5 \text{ ft}$$

rounded off to four significant figures. (The other unknown is not needed.)

Exercises / Section 4.5

In Exercises 1–12, refer to Figure 4.28.

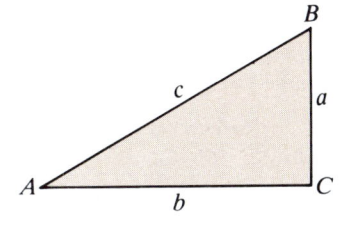

Figure 4.28

1. $A = 37°20'$, $c = 2.76$; find a
2. $B = 65°40'$, $a = 3.96$; find c
3. $B = 9°30'$, $b = 12.3$; find a
4. $A = 15°50'$, $b = 19.2$; find c

In Exercises 5–12, find the remaining parts of each triangle.

5. $A = 30°31'$, $b = 3.520$
6. $B = 50°26'$, $c = 5630$
7. $a = 524$, $c = 941$
8. $b = 22.6$, $c = 38.3$
9. $A = 22°40'$, $c = 2.00$
10. $B = 65°9'$, $b = 0.09410$
11. $A = 5°13'$, $a = 0.008200$
12. $B = 7°55'$, $a = 0.09000$

13. A tree casts a shadow 58.2 ft long. If the angle of elevation of the sun is 48°10′, how tall is the tree?
14. From a vertical cliff 119 m high, the angle of depression of a buoy is 15°50′. Find the distance from the buoy to the base of the cliff.
15. An 8.0 ft ladder is leaning against a building. If the ladder makes an angle of 61° with the ground, how high does the ladder reach on the wall?
16. A guy wire attached to the top of a tower makes an angle of 63°20′ with the ground. If the tower is known to be 50.2 m high, find the length of the wire.
17. Suppose a 19.0-ft telephone pole leans 11.3° from the vertical. What is the length of the shadow if the sun is directly overhead?
18. The Great Pyramid of Cheops has a base of 755 ft and originally had a height of 481 ft. (The top 31 ft have been destroyed.) What angle do the sides make with the ground?
19. If a pendulum 17.9 in. long swings through an arc of 15°20′, how high will the bob rise above its lowest position?
20. The metal plate in Figure 4.29 has the shape of a regular pentagon. Find the perimeter of the plate.

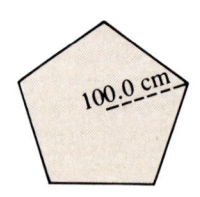

Figure 4.29

21. From a point 220 ft from the base of the Empire State Building and in the same horizontal plane, the angle of elevation of the top of the building is 80.0°. Find the height of the building.
22. Eight screws are equally spaced on the rim of a circular plate. The distance from center to center between two adjacent screws is 2.58 cm. Find the radius of the plate.
23. A platform 15.0 ft high is 211 ft from a building. From the top of the platform, the angle of elevation of the top of the building is 29°30′. Find the height of the building.
24. A woman 5 ft 8 in. tall observes that the angle of elevation of the top of a flagpole is 71°. If she is standing 20 ft from the base, how high is the flagpole?
25. A building is located 151.0 ft from a tall lookout tower. From the top of the building, the angle of depression of the tower base is 25°32′, and the angle of elevation of the top is 61°5′. Find the height of the tower.

4.5 APPLICATIONS OF RIGHT TRIANGLES

26. A war memorial features the figure of a soldier on a pedestal. At a point 19.5 ft from the base of the pedestal, the angles of elevation of the foot and head of the figure are 46°10′ and 57°0′, respectively. Find the height of the figure.

27. In finding the height of the Egyptian pyramid at Chephren, it is observed that the sides make an angle of 53°50′ with the ground. From a point 100 ft from the base of the pyramid and on a line perpendicular to the midpoint of the base, the angle of elevation of the top is measured to be 46°40′. Find the height of the pyramid to the nearest foot.

28. The angle of elevation of a mountaintop is observed to be 4°50′. From a second point 9.20 mi closer to the mountain, the angle of elevation is 6°50′. How high above the plane is the mountaintop?

29. From the top of a lighthouse 29.30 m above the surface of the water, the angles of depression of two ships due east are 12°25′ and 15°32′, respectively. Find the distance between the ships.

30. An observer in a balloon measures the respective angles of depression to two distant buildings as 18°21′ and 25°32′. If the buildings are known to be 1.10 mi apart and in the same vertical plane with the balloon, how high is the balloon?

31. A swimming pool measures 20.1 ft by 45.0 ft. (See Figure 4.30.) If the bottom is flat and inclined 14.0° with the horizontal, what is its area?

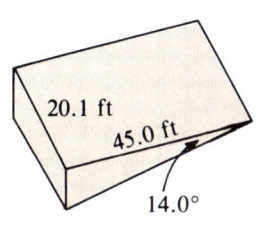

Figure 4.30

32. A dealer in sand and gravel wants to estimate the volume of a conical pile of sand. The circumference of the base measures 20 ft. From one point on the ground, the angle of elevation of the top is 26°; from another point 10 ft further back and in line with the first point and the center of the base, the angle of elevation is 20°. Find the volume of the pile.

33. A castle wall is surrounded by a moat of unknown width. A historian wants to measure the height of the wall by marking off two points 10.0 m apart along a line perpendicular to the wall. The respective angles of elevation of the top were found to be 24°30′ and 16°20′. Find the height of the wall.

34. Archimedes of Syracuse (a Greek who lived between 287? and 212 B.C.) made the best estimate of π in antiquity. He obtained his estimate by using a polygon of 96 sides inscribed in a circle. Show that this method yields the value to two decimal places, that is, 3.14. Assuming the radius of the circle to be 1 unit long, show that for an inscribed polygon of n sides, $\pi \approx n \sin \frac{180°}{n}$.

35. In an alternating current circuit containing a resistance R and an inductive reactance X_L, the *phase angle* ϕ is the angle between the impedance Z and the resistance. Find the phase angle if $R = 10.2 \, \Omega$ and $X_L = 6.21 \, \Omega$. (See Figure 4.31.)

36. A rectangular chip in a computer circuit measures 1.90 mm by 1.08 mm. Find the angle that the diagonal makes with the longer side.

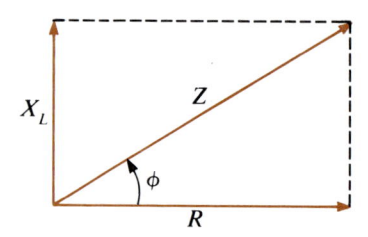

Figure 4.31

Review Exercises / Chapter 4

In Exercises 1–4, find the values of the six trigonometric functions of the angle whose terminal side passes through the given point.

1. $(3, 1)$
2. $(1, \sqrt{7})$
3. $(\sqrt{5}, 2)$
4. $(1, 2\sqrt{2})$
5. If $\sin \theta = \frac{1}{5}$, find $\cos \theta$.
6. If $\sec \theta = \sqrt{7}/2$, find $\csc \theta$.
7. If $\cot \theta = \sqrt{3}/5$, find $\sin \theta$.
8. If $\csc \theta = \sqrt{11}$, find $\tan \theta$.
9. If $\cos \theta = \sqrt{3}/2$, find $\cot \theta$.
10. If $\sin \theta = \sqrt{5}/3$, find $\tan \theta$.
11. If $\tan \theta = \sqrt{2}/6$, find $\sec \theta$.
12. If $\csc \theta = \sqrt{7}/2$, find $\sin \theta$.
13. If $\sin \theta = a/2$, find $\cos \theta$.
14. If $\sec \theta = b$, find $\tan \theta$.
15. Given that $\sin \theta = \frac{1}{3}$, show that $\sin^2\theta + \cos^2\theta = 1$.
16. Given that $\sec \theta = 3$, show that $\tan^2\theta = \sec^2\theta - 1$.

In Exercises 17–24, draw a diagram and determine the exact value of each trigonometric function.

17. $\sin 60°$
18. $\sin 90°$
19. $\cot 45°$
20. $\sec 30°$
21. $\tan 0°$
22. $\csc 45°$
23. $\cos 90°$
24. $\cot 60°$

In Exercises 25–32, use a diagram to find θ.

25. $\sin \theta = \dfrac{\sqrt{3}}{2}$
26. $\cos \theta = 1$
27. $\sec \theta = \sqrt{2}$
28. $\csc \theta = 2$
29. $\cos \theta = \dfrac{\sqrt{3}}{2}$
30. $\sin \theta = \dfrac{1}{\sqrt{2}}$
31. $\sec \theta = \dfrac{2\sqrt{3}}{3}$
32. $\tan \theta = 1$

REVIEW EXERCISES / CHAPTER 4   **147**

In Exercises 33–38, use Table 1 in Appendix B to determine the value of each trigonometric function. Check with a calculator.

33. sin 32°15'
34. tan 62°28'
35. sec 55°19'
36. cos 9°58'
37. csc 8°8'
38. cot 75°31'

In Exercises 39–44, use Table 1 in Appendix B to find θ. Check with a calculator.

39. sin θ = 0.1865
40. cos θ = 0.1914
41. tan θ = 0.1941
42. sec θ = 3.157
43. csc θ = 1.975
44. cot θ = 0.2525

45. A television tower 251 ft high is supported by guy wires extending from the top to points on the ground 94.0 ft away from the base. What angle do the wires make with the ground?

46. A regular pentagon is inscribed in a circle of radius 2.13 cm. Find the length of one side.

47. Find the area of the triangle in Figure 4.32.

48. Find the length of side BC in Figure 4.33.

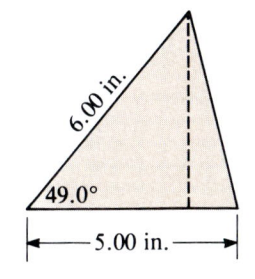

Figure 4.32

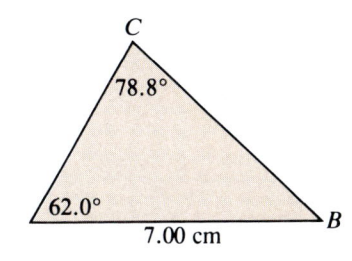

Figure 4.33

49. Find the radius of the circular portion of the machine part pictured in Figure 4.34. (The dotted line is tangent to the circle.)

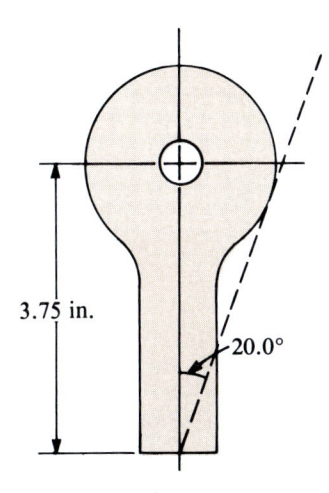

Figure 4.34

50. The angle of depression from the top of a building to a fire hydrant 56.30 ft from the base of the building is 66°39′. How tall is the building?
51. John wants to determine the height of his apartment building, which is located near an observation tower 175 ft high. From the top of the building he measures the angle of elevation of the top of the tower to be 38.0° and the angle of depression of the tower base to be 25.0°. What is the height of his apartment building?
52. A tree is standing on top of a cliff 162.0 ft high. The angle of elevation of the bottom of the tree is 41°38′, and the angle of elevation of the top of the tree is 52°17′. How tall is the tree?
53. Two observers located 2,100 ft apart measure the angle of elevation of a weather balloon located between them and directly over the line joining them. If the angles are 42°10′ and 51°20′, find the height of the balloon above the ground.
54. Find the area of the triangular machine part shown in Figure 4.35.

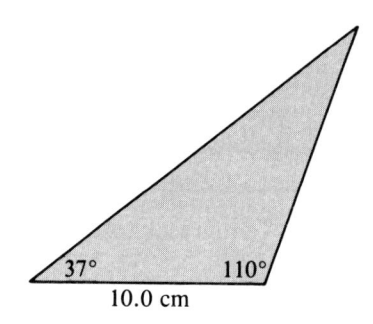

Figure 4.35

CHAPTER **5**

Factoring and Fractions

Objectives Upon completion of this chapter, you should be able to:

1. Write the answers for the following special products directly:
   a. A monomial and a polynomial.
   b. The square of a binomial.
   c. The sum and difference of two terms.
   d. Two binomials.
2. Factor:
   a. Expressions containing a common factor.
   b. A difference of two squares.
   c. A sum or difference of two cubes.
   d. Perfect-square trinomials.
   e. General trinomial forms.
   f. Expressions by grouping terms.
3. Reduce a given fraction to lowest terms.
4. Multiply and divide fractions, expressing the result as a fraction reduced to lowest terms.
5. Find the lowest common denominator of two or more fractions.
6. Add and subtract fractions.
7. Reduce a given complex fraction.
8. Solve fractional equations.

At this point we need to discuss some methods that make certain computations much faster and more accurate. In particular, we need to improve our ability to perform certain multiplications, usually called *special products*. These products occur so often in technology and mathematics that we must be thoroughly familiar with them. Special products lead to the important inverse operation called *factoring*, which we must also study.

5.1 Special Products

Two special products frequently encountered in mathematics and its applications are the *distributive law*

$$a(b + c) = ab + ac$$

and the *product of the sum and difference of two terms*

$$(a - b)(a + b) = a^2 - b^2$$

1. The Distributive Law

$$a(b + c) = ab + ac$$

Example 1
a. $2a(r + s) = 2a(r) + 2a(s) = 2ar + 2as$
b. $xy^2(x^2 - xy) = xy^2(x^2) + xy^2(-xy) = x^3y^2 - x^2y^3$

Example 2
$ab(3a - 2b - 3ab^2) = 3a^2b - 2ab^2 - 3a^2b^3$

2. Product of the Sum and Difference of Two Terms

$$(a - b)(a + b) = a^2 - b^2$$

Note that upon multiplying $(a - b)(a + b)$, the middle terms add to zero. Thus, *the product of the sum and difference of two terms is equal to the difference of their squares.*

Example 3
$(3V - 2R)(3V + 2R) = (3V)^2 - (2R)^2$
$= 9V^2 - 4R^2$

Example 4
$(2x^2 - 5y^3)(2x^2 + 5y^3) = (2x^2)^2 - (5y^3)^2$
$= 4x^4 - 25y^6$

Exercises / Section 5.1

Find the following products in one step.

1. $2(x - 3y)$
2. $4x(x^2 - 2y)$
3. $2a(a^3 - ab^2 - 3ac)$
4. $5x^2y^2(2x^2y^2 - 1)$
5. $V(R_1^2 - R_2^2)$
6. $V^2(3S^3 - 2V^3)$
7. $2P(Q^2 - P^2 + 1)$
8. $3T_1(2T_1 + aT_2 - bT_3)$
9. $2b^2c(a - 2b - 1)$
10. $rs(rs^2 + 3rs + 1)$
11. $(x - 2y)(x + 2y)$
12. $(a + 3b)(a - 3b)$

13. $(6A - B)(6A + B)$
14. $(5P - 3Q)(5P + 3Q)$
15. $(a^2 + b)(a^2 - b)$
16. $(m - n^2)(m + n^2)$
17. $(2x^2 - y^2)(2x^2 + y^2)$
18. $(3M + 2N)(3M - 2N)$
19. $(S - 2T^2)(S + 2T^2)$
20. $(5L^2 + T^3)(5L^2 - T^3)$
21. $(xy + 1)(xy - 1)$
22. $(1 - pq)(1 + pq)$
23. $(3RS^2 - 1)(3RS^2 + 1)$
24. $(5NP - x)(5NP + x)$

5.2 Common Factors and Factors of a Binomial

Factoring plays an important role in algebra. We must be able to factor in order to simplify certain algebraic expressions, to reduce a fraction to lowest terms, and to carry out the four fundamental operations with fractions. In this section we shall see how factoring can be performed on some types of polynomials. All cases considered involve certain special products in reverse: common factors, the difference of two squares, and the sum and difference of two cubes.

Factoring is essentially multiplication in reverse: We start with the product and go backward to recover the factors. For example, we know from the last section that

$$3(a - 2b) = 3a - 6b$$

Therefore *factoring* the binomial $3a - 6b$ means to write $3a - 6b$ as $3(a - 2b)$.

Prime factors

We say that a polynomial has been *factored* if it is written as a product of prime factors. A factor is **prime** or **irreducible** if it contains no other factors than itself and 1.

1. Common Factors

If every term of a polynomial contains the same factor, this **common factor** may be factored out of the polynomial. Consider, for example, the expression $2x + 4y$. Since each term has a factor 2, we may write

$$2x + 4y = 2(x + 2y)$$

by the distributive law in reverse. Similarly, the expression $6x^2 - 9x^3$ has two terms with the common factor $3x^2$. So

$$6x^2 - 9x^3 = 3x^2(2) - 3x^2(3x) = 3x^2(2 - 3x)$$

The general case is just the distributive law stated in reverse.

Common factor:

$$ab + ac = a(b + c)$$

Example 1 a. $2x - 8y = \mathbf{2}(x - 4y) = 2(x - 4y)$ common factor 2
 b. $2x^2 - 8xy = \mathbf{2x}(x - 4y) = 2x(x - 4y)$ common factor $2x$
 c. $3x^2 - 3 = \mathbf{3}(x^2 - 1) = 3(x^2 - 1)$ common factor 3

Common terms can be factored out even if the polynomial contains more than two terms.

Example 2 a. $2x^2y - 4xy^2 + 6x^2y^2 = 2xy(x - 2y + 3xy)$
 b. $9a^3b^3 - 15a^2bc + 3a^2b = 3a^2b(3ab^2 - 5c + 1)$

The expression to be factored out may itself be a polynomial, as in the next example.

Example 3 a. $12a(x^2 + y^2) - 20(x^2 + y^2) = \mathbf{4}[3a(\mathbf{x^2 + y^2}) - 5(\mathbf{x^2 + y^2})]$
 $= \mathbf{4(x^2 + y^2)}(3a - 5)$
 b. $2(x - y) - 6(x - y)z = 2[(x - y) - 3(x - y)z]$
 $= 2(x - y)(1 - 3z)$

Common error Failing to factor out all the common factors. For example,

$$5a^2bc - 15ab^2c - 5a^2b^2c^2 = abc(5a - 15b - 5abc)$$

While this result is not really incorrect, the second factor is not prime. To obtain the necessary prime factors, the 5 must also be factored out to yield

$$5abc(a - 3b - abc)$$

2. Difference of Two Squares

Another factorable expression comes from the special product of the sum and difference of two terms

$$(a + b)(a - b) = a^2 - b^2$$

We see that the product is the difference of the squares of the terms. Consequently, factoring the difference of two squares yields the product of the sum and difference of the terms. For example, $x^2 - 9y^2 = x^2 - (3y)^2$. So x^2 is the square of x and $9y^2$ is the square of $3y$. We therefore get

$$x^2 - (3y)^2 = (x - 3y)(x + 3y)$$

Difference of two squares:
$$a^2 - b^2 = (a + b)(a - b) \tag{5.1}$$

Example 4 Factor $4x^2 - b^2$.

Solution. Since $4x^2 - b^2 = (2x)^2 - b^2$, we see that $4x^2$ is the square of $2x$ and that b^2 is the square of b. It follows that
$$4x^2 - b^2 = (2x + b)(2x - b)$$
or
$$4x^2 - b^2 = (2x - b)(2x + b)$$

If the expressions are more complicated, a preliminary step can be helpful. In particular, a binomial that is a difference of two squares may need to be rewritten to make this fact clear. Some examples follow.

Example 5
  a. $9a^2 - 16b^4 = (\mathbf{3a})^2 - (\mathbf{4b^2})^2 = (3a - 4b^2)(3a + 4b^2)$
  b. $25a^2b^4 - 1 = (\mathbf{5ab^2})^2 - \mathbf{1}^2 = (5ab^2 - 1)(5ab^2 + 1)$

Example 6 Factor $1 - (A - B)^2$.

Solution. Since $1 = 1^2$, the form is a difference of two squares with $a = 1$ and $b = A - B$ in formula (5.1). *In the first step, keep the parentheses around the second term* to avoid possible errors with signs:
$$1^2 - (A - B)^2 = [1 + (A - B)][1 - (A - B)]$$
$$= (1 + A - B)(1 - A + B)$$

3. Sum or Difference of Two Cubes

The factoring cases given next are called, respectively, the **sum** and **difference of two cubes.** They are sometimes listed as special products.

Sum of two cubes:
$$a^3 + b^3 = (a + b)(a^2 - ab + b^2) \tag{5.2}$$

Difference of two cubes:
$$a^3 - b^3 = (a - b)(a^2 + ab + b^2) \tag{5.3}$$

154 CHAPTER 5 FACTORING AND FRACTIONS

To check the sum of two cubes, we multiply the factors on the right.

$$\begin{array}{r} a^2 - ab + b^2 \\ a + b \\ \hline a^3 - a^2b + ab^2 \\ a^2b - ab^2 + b^3 \\ \hline a^3 + b^3 \end{array}$$

Formula (5.3) can be checked similarly.

To remember these forms, note that the binomials have identical signs in each case: The factors of $a^3 + b^3$ are $a + b$ and a trinomial, and those of $a^3 - b^3$ are $a - b$ and a trinomial. Note especially that the trinomials are irreducible.

Example 7 Factor the following expressions:

a. $8x^3 + 1$ **b.** $x^3y^3 - 64b^6$

Solution. All the terms should be written as cubes for two reasons: (1) to check if the binomial really *is* a sum or difference of two cubes and (2) to facilitate the factoring procedure.

a. First note that $8x^3 = (2x)^3$ and $1 = 1^3$. Then in form (5.2), $a = 2x$ and $b = 1$; so $a^2 = (2x)^2$, $ab = (2x)(1)$, and $b^2 = 1^2$. It follows that

$$8x^3 + 1 = \mathbf{(2x)^3 + 1^3}$$
$$= (2x + 1)[(2x)^2 - (2x)(1) + 1^2] \quad \text{sum of two cubes}$$
$$= (2x + 1)(4x^2 - 2x + 1)$$

b. $x^3y^3 - 64b^6 = \mathbf{(xy)^3 - (4b^2)^3}$
$$= (xy - 4b^2)[(xy)^2 + (xy)(4b^2) + (4b^2)^2]$$
$$= (xy - 4b^2)(x^2y^2 + 4xyb^2 + 16b^4)$$

As in the case of the difference of two squares, the terms may themselves be polynomials. Such cases are more complicated, at least in appearance, and extra care must be taken when writing the factors.

Example 8 Factor $z^3 - (x + y)^3$.

Solution. This expression fits form (5.3) with $a = z$ and $b = x + y$; so $a^2 = z^2$, $ab = z(x + y)$, and $b^2 = (x + y)^2$. In the first step, we keep the parentheses around $x + y$. Thus

$$z^3 - (x + y)^3 = [z - (x + y)][z^2 + z(x + y) + (x + y)^2]$$
$$= (z - x - y)(z^2 + zx + zy + x^2 + 2xy + y^2)$$

5.2 COMMON FACTORS AND FACTORS OF A BINOMIAL

Sometimes the factored expressions themselves are factorable, as can be seen in the remaining examples.

Example 9 Factor $x^8 - 16y^{12}$.

Solution.
$$x^8 - 16y^{12} = (x^4)^2 - (4y^6)^2$$
$$= (x^4 - 4y^6)(x^4 + 4y^6) \quad \text{difference of two squares}$$
$$= [(x^2)^2 - (2y^3)^2](x^4 + 4y^6)$$
$$= (x^2 - 2y^3)(x^2 + 2y^3)(x^4 + 4y^6) \quad \text{difference of two squares}$$

Example 10 Factor $2x^2 - 8y^4$.

Solution. At first glance this expression does not appear to fit any of the forms considered. However, if the common factor **2** is factored out, we get
$$2x^2 - 8y^4 = \mathbf{2}(x^2 - 4y^4)$$
$$= 2[x^2 - (2y^2)^2]$$
$$= 2(x - 2y^2)(x + 2y^2)$$

Rule: In all factoring problems, first check for common factors.

Common error Writing
$$a^3 + b^3 \quad \text{as} \quad (a + b)^3$$
Note that
$$a^3 + b^3 = (a + b)(a^2 - ab + b^2)$$
while
$$(a + b)^3 = (a + b)(a + b)(a + b) = (a + b)(a + b)^2$$
$$= (a + b)(a^2 + 2ab + b^2)$$
Thus
$$a^3 + b^3 \neq (a + b)^3$$

Example 11 A cylindrical lead pipe of length L has inner radius r_1 and outer radius r_2. Factor the expression for the volume of the wall of the pipe, which is given by $V = \pi r_2^2 L - \pi r_1^2 L$.

Solution.

$$V = \pi r_2^2 L - \pi r_1^2 L = \pi L(r_2^2 - r_1^2)$$
$$= \pi L(r_2 - r_1)(r_2 + r_1)$$

Exercises / Section 5.2

In Exercises 1–61, factor the expressions completely.

1. $2a + 2b$
2. $3x - 3y$
3. $5x - 5$
4. $ax + ay$
5. $3x^2 - 6x$
6. $3x^3 - 6xy + 9x$
7. $4x^2y^2 - 6x^2y + 12xy^2$
8. $9a^2b^3 - 21a^3b^4 - 12a^2b^2$
9. $14c^3d^2 + 28c^3d^3 + 14c^2d^3$
10. $6R^2C - 15R^3C^2 - 3RC$
11. $11R_1^2R_2 - 22R_1R_2^2 + 11R_1R_2$
12. $mc^2 - m_0c^2$
13. $x^2 - y^2$
14. $4x^2 - y^2$
15. $4x^2 - 4y^2$
16. $x^2 - 16y^2$
17. $64m^2 - 16n^2$
18. $25V^2 - 4C^2$
19. $8f_1^2 - 2f_2^2$
20. $25f^2 - 16g^2$
21. $50x^2 - 32z^2$
22. $D^4 - 1$
23. $\dfrac{1}{4}x^2 - y^2$
24. $a^2 - \dfrac{1}{16}b^2$
25. $\dfrac{1}{9}x^2 - \dfrac{1}{16}y^2$
26. $\dfrac{1}{25}x^2 - \dfrac{1}{49}y^2$
27. $\dfrac{x^2}{a^2} - B^2$
28. $\dfrac{s^2}{t^2} - c^2$
29. $1 - \dfrac{b^2}{a^2}$
30. $4 - \dfrac{x^2}{v^2}$
31. $t_1^4 - t_2^4$
32. $3s^4 - 3$
33. $3x^2 - 27a^2$
34. $a^4 - 16$
35. $64a^4 - 1$
36. $125x^2 - 5$
37. $x^3 - y^3$
38. $z^3 + b^3$
39. $2x^3 + 2a^3$
40. $8x^3 + y^3$
41. $27a^3 - y^3$
42. $64s^3 - 8t^3$
43. $125m^3 - 27a^3$
44. $64D^3 - 1$
45. $a^3b^3 - 1$
46. $1 - x^3y^3$
47. $1 + x^3y^6$
48. $a^3 + x^3y^3$
49. $(a + b)^2 - 1$
50. $(a + b)^3 - 1$
51. $1 - (x + y)^2$
52. $1 - (x + y)^3$

53. $(x - y)^3 + z^3$
54. $(x - y)^2 - z^2$
55. $(v_1 - v_2)^3 - v_3^3$
56. $(l_1 + l_2)^3 + 1$
57. $(F_1 + F_2)^3 - 8$
58. $27 - (m + 2n)^3$
59. $8 + (3r + 2s)^3$
60. $x^4 - 81y^4$
61. $16x^3 - 128a^3$

62. The energy E radiated by a blackbody is $E = kT^4 - kT_0^4$, where T is the temperature of the body, T_0 the temperature of the surrounding medium, and k a constant. Factor the expression for E.

63. The difference E_d in the energy radiated by a filament at temperatures T_1 and T_2, respectively, is given by $E_d = aT_1^4 - aT_2^4$, where a is a constant. Factor this expression.

64. The power delivered to a resistor is given by $P = I^2 R$ (in watts). Factor the expression that represents the difference in the power delivered to a resistor when the current increases from I_1 to I_2.

65. The kinetic energy of a particle of mass m traveling at velocity v is given by $\frac{1}{2}mv^2$. Factor the expression for the change in the kinetic energy of a particle of mass m whose velocity increases from v_1 to v_2.

66. The total resistance in a certain circuit was found to be $4IR_1 - 6IR_2 + 12IR_3$. Factor this expression.

67. The volume of a spherical shell is $V = \frac{4}{3}\pi r_2^3 - \frac{4}{3}\pi r_1^3$, where r_1 is the inner radius and r_2 the outer radius. Factor the expression for V.

68. The area of a rectangle is $x^3 + 27$ square units. One side is $x + 3$ units in length. Find an expression for the length of the other side.

5.3 More Special Products

In Section 5.1 we started our discussion of special products. Now we will look at two more special products, the square of a binomial and the product of two distinct binomials.

1. Square of a Binomial

$$(a + b)^2 = a^2 + 2ab + b^2$$
$$(a - b)^2 = a^2 - 2ab + b^2$$

The square of the sum (or difference) of two terms is the square of the first term, plus (or minus) twice the product of the two terms, plus the square of the second term.

Example 1 To expand $(x + 3y)^2$, we let $a = x$ and $b = 3y$, so that $2ab = 6xy$. Thus

$$(x + 3y)^2 = x^2 + 2(x)(3y) + (3y)^2$$
$$= x^2 + 6xy + 9y^2$$

Example 2
$$(3x^2 - 4y)^2 = (3x^2)^2 - 2(3x^2)(4y) + (4y)^2$$
$$= 9x^4 - 24x^2y + 16y^2$$

by the second form $(a - b)^2$. The first form $(a + b)^2$ could also be used if the binomial is treated as the square of $3x^2 + (-4y)$:

$$[3x^2 + (-4y)]^2 = (3x^2)^2 + 2(3x^2)(-4y) + (-4y)^2$$
$$= 9x^4 - 24x^2y + 16y^2$$

2. Product of Two Binomials

The product of two distinct binomials has the following form:

$$(ax + by)(cx + dy) = acx^2 + (ad + bc)xy + bdy^2$$

The best way to carry out this multiplication directly is with the help of a diagram. First observe that the right side consists of the product of the first terms, plus the product of the last terms, plus a middle term consisting of a sum of two products. These two products are the products of the outer terms ax and dy and the inner terms by and cx. Schematically,

FOIL method

$$(ax + by)(cx + dy) = acx^2 + adxy + bcxy + bdy^2$$

with First, Last, Outer, Inner pairings.

The scheme can be remembered by the acronym **FOIL**, where F stands for "first," O for "outer," I for "inner," and L for "last."

Example 3 Multiply $(2x - y)(x + 3y)$.

Solution. Using the acronym FOIL, we get the following diagram:

$$(2x - y)(x + 3y) = (2x)(x) + [(2x)(3y) + (-y)(x)] + (-y)(3y)$$

Hence

$$(2x - y)(x + 3y) = 2x^2 + (6xy - xy) - 3y^2$$
$$= 2x^2 + 5xy - 3y^2$$

Example 4 Multiply $(3x - 2a)(2x - 6a)$.

Solution.

$$(3x - 2a)(2x - 6a) = \overset{F}{(3x)(2x)} + [\overset{O}{(3x)(-6a)} + \overset{I}{(-2a)(2x)}] + \overset{L}{(-2a)(-6a)}$$

Thus

$$(3x - 2a)(2x - 6a) = 6x^2 + (-18ax - 4ax) + 12a^2$$
$$= 6x^2 - 22ax + 12a^2$$

In practice, *the middle terms should be added mentally.*

Larger polynomials can also be multiplied directly, but this may require an additional step. Of course, the larger the expression, the greater the difficulty in obtaining the product directly. If the factors get too complicated, it is undoubtedly best to return to the method studied in Chapter 1.

Example 5 Multiply $7(x - 5y)(x - 6y)$.

Solution. We multiply the two binomials first and then use the distributive law to multiply by 7:

$$7(x - 5y)(x - 6y) = 7(x^2 - 6xy - 5xy + 30y^2)$$
$$= 7(x^2 - 11xy + 30y^2)$$
$$= 7x^2 - 77xy + 210y^2$$

Example 6 Multiply $(s + t - w)^2$.

Solution. This product can be treated as the square of a binomial if the terms are first grouped as follows:

$$[(s + t) - w]^2 = (s + t)^2 - 2(s + t)(w) + w^2$$
$$= s^2 + 2st + t^2 - 2sw - 2tw + w^2$$

Example 7 Multiply

$$(a + 2b - c)(a + 2b + 5c)$$

Solution. First group the terms as follows:

$$[(a + 2b) - c][(a + 2b) + 5c]$$

Now treat the quantity $a + 2b$ as a single term and multiply by the FOIL scheme:

$$[(a + 2b) - c][(a + 2b) + 5c] = (a + 2b)^2 + [(a + 2b)(5c) + (-c)(a + 2b)] - 5c^2$$

$$= a^2 + 4ab + 4b^2 + (5ac + 10bc - ac - 2bc) - 5c^2$$
$$= a^2 + 4ab + 4b^2 + 4ac + 8bc - 5c^2$$

Common error Forgetting to take twice the product for the middle term or forgetting the middle term altogether. Remember that

$$(a + b)^2 \neq a^2 + ab + b^2$$

and

$$(a + b)^2 \neq a^2 + b^2$$

Instead,

$$(a + b)^2 = a^2 + \mathbf{2ab} + b^2$$

For easy reference, the following box summarizes the different special products.

Special products:

$a(b + c) = ab + ac$	(5.4)
$(a + b)^2 = a^2 + 2ab + b^2$	(5.5)
$(a - b)^2 = a^2 - 2ab + b^2$	(5.6)
$(a + b)(a - b) = a^2 - b^2$	(5.7)
$(ax + b)(cx + d) = acx^2 + (ad + bc)x + bd$	(5.8)
$(ax + by)(cx + dy) = acx^2 + (ad + bc)xy + bdy^2$	(5.9)

Exercises / Section 5.3

In Exercises 1–26, find the special products in *one step*.

1. $(x + 2)^2$ **2.** $(y - 4)^2$ **3.** $(C_1 + C_2)^2$
4. $(v - w)^2$ **5.** $(x + 1)(x - 2)$ **6.** $(x - 4)(x - 1)$

7. $(3x - 2)(5x + 1)$
8. $(8x - 3)(4x - 2)$
9. $(4x - y)(3x - 2y)$
10. $(2a + 5b)(a - 4b)$
11. $(s - 2t)(s - 3t)$
12. $(f_1 - 2f_2)(3f_1 + 5f_2)$
13. $(7i_1 + 2i_2)^2$
14. $(6A - 5B)^2$
15. $(3x - 2y)^2$
16. $(f + 3g)^2$
17. $(2i + 5k)(i + 3k)$
18. $(-6x + 7y)(x - 2y)$
19. $(2x - 10y)(x + y)$
20. $(3x - y)(7x - 2y)$
21. $(2x - 5y)(3x - 8y)$
22. $(v_1 + 2v_2)^2$
23. $(v_1 - 2v_2)(3v_1 + 6v_2)$
24. $(3x + 2y)(3x - 2y)$
25. $(9x - 6y)(9x + 6y)$
26. $(10x - 3y)(10x + 3y)$

In Exercises 27–44, multiply the given polynomials. (More than one step may be required.)

27. $7(x - 2y)(x + 2y)$
28. $3(2x - 5y)(2x + 5y)$
29. $x(3x - y)(3x + y)$
30. $2x(4x - 2y)(4x + 2y)$
31. $2(x - 3y)(x - 2y)$
32. $x(2x + y)(3x + 2y)$
33. $a(4x - y)(3x + 5y)$
34. $a(2C_1 - 3C_2)(C_1 + 4C_2)$
35. $(x + y + 2z)^2$
36. $(2x - 3y + w)^2$
37. $(a + 2b + c)^2$
38. $(t - 3w - 1)^2$
39. $(x + y + 3z)(x + y - 2z)$
40. $(x - 3b + c)(x - 3b + 2c)$
41. $(p + q - 2r)(p + q - 5r)$
42. $(s + 2t + 4w)(s + 2t - w)$
43. $(x + a + b)(2x + a + b)$
44. $(y + 2a - b)(3y + 2a - b)$

5.4 Factoring Trinomials

If two binomials of the first degree are multiplied, the result is a trinomial. For example, $(x - 2)(x + 4) = x^2 + 2x - 8$. In this section we shall develop techniques for the inverse operation—factoring a trinomial into two factors of first degree.

1. Perfect-Square Trinomials

A trinomial is a perfect square if it can be written as the square of a binomial. By reversing the special products (5.5) and (5.6), we can obtain the forms of perfect-square trinomials.

Perfect-square trinomials:
$$a^2 + 2ab + b^2 = (a + b)^2 \qquad (5.10)$$
$$a^2 - 2ab + b^2 = (a - b)^2 \qquad (5.11)$$

Perfect squares

These forms suggest how to recognize a perfect square: *The first and last terms of the trinomial must be perfect squares, and the middle term must be equal to twice the product of a and b.*

$$\underbrace{a^2 + 2ab + b^2}_{\text{perfect squares}}$$

2 times ab

Example 1 Factor:

   a. $x^2 + 4xy + 4y^2$ **b.** $9x^2 - 30xy + 25y^2$

Solution. a. The first term x^2 and the last term $4y^2 = (\mathbf{2y})^2$ are both perfect squares. To check the middle term, note that $2(\mathbf{x})(\mathbf{2y}) = 4xy$. It follows that the given trinomial is indeed a perfect square. Hence

$$x^2 + 4xy + 4y^2 = (x + 2y)^2 \quad \text{by form} \quad (5.10)$$

b. The first term $9x^2 = (\mathbf{3x})^2$ and the last term $25y^2 = (\mathbf{5y})^2$ are both perfect squares; also, $-2(\mathbf{3x})(\mathbf{5y}) = -30xy$, which agrees with the middle term. The form fits (5.11):

$$9x^2 - 30xy + 25y^2 = (3x)^2 - 30xy + (5y)^2$$
$$= (\mathbf{3x})^2 - 2(\mathbf{3x})(\mathbf{5y}) + (\mathbf{5y})^2$$
$$= (3x - 5y)^2$$

Some expressions have a common factor, which must be factored first.

Example 2 Factor $128a^2 - 352ab^2 + 242b^4$.

Solution. It was emphasized earlier that you should always be on the alert for common factors. Only after removing the common factor **2** do the first and last terms of the trinomial become perfect squares. We now have

$$\mathbf{2}(64a^2 - 176ab^2 + 121b^4)$$
$$= 2[(8a)^2 - 2(88ab^2) + (11b^2)^2]$$
$$= 2[(8a)^2 - 2(8a)(11b^2) + (11b^2)^2]$$
$$= 2(8a - 11b^2)^2 \qquad \text{by form} \quad (5.11)$$

2. General Trinomials

Factoring trinomials other than perfect squares is somewhat more difficult. The forms for the general trinomial arise from the special products (5.8) and (5.9).

General trinomials:

$$acx^2 + (ad + bc)x + bd = (ax + b)(cx + d) \qquad (5.12)$$
$$acx^2 + (ad + bc)xy + bdy^2 = (ax + by)(cx + dy) \qquad (5.13)$$

The problem here is that the middle term is really a combination of four different terms. Consequently, it may be necessary to juggle several numbers until the right combination is found. Let's consider the trinomial

$$x^2 + x - 6$$

At this point we know only that the two factors are first-degree binomials (if the expression is factorable). Keeping in mind the acronym FOIL, consider the following diagram:

FOIL method

$$x^2 + x - 6 = (x \quad \overset{F}{}\overset{L}{)(x} \quad)$$

Since $x \cdot x = x^2$, we are satisfied with the F terms. The L terms have several possibilities. Suppose we try -6 and 1. Then the diagram becomes

$$x^2 + x - 6 = (x - 6)(x + 1)$$

Now the L terms check, but the sum of O and I is $-5x$, which does not agree with the middle term x. The combination 6 and -1 also fails.

For the next combination let us try -3 and 2. Then we get

$$x^2 + x - 6 = (x - 3)(x + 2)$$

Although the L terms check, the sum of the inner and outer products is $-x$, which has the wrong sign. This suggests switching signs on the last combination. The result is

$$x^2 + x - 6 = (x + 3)(x - 2)$$

Since we checked all possible combinations in this demonstration, the procedure looks more difficult than it really is. The FOIL scheme tells us that to factor $x^2 + x - 6$, we need two numbers whose *product* is -6 and whose *sum* is 1. These numbers are 3 and -2.

Let's consider some more examples.

Example 3 Factor $x^2 - 6xy + 8y^2$.

Solution. Since the last term is preceded by a plus sign, the two L terms must agree in sign. This common sign must be a minus because the middle term is preceded by a minus sign. So we need two numbers whose *product* is 8 and whose *sum* is −6. The numbers are −2 and −4.

$$x^2 - 6xy + 8y^2 = (x - 2y)(x - 4y)$$

with F (first), L (last), I (inner), O (outer) indicated.

The middle term checks.

Example 4 Factor $x^2 + 5x - 24$.

Solution. This time we need two numbers whose *product* is −24 and whose *sum* is 5. These numbers are 8 and −3, so that

$$x^2 + 5x - 24 = (x - 3)(x + 8)$$

Examples 3 and 4 suggest that there is a simple method for factoring trinomials. Unfortunately, the coefficient of x^2 may be a number other than 1, thereby increasing the number of possible combinations.

Example 5 Factor $4x^2 + 21xy - 18y^2$.

Solution. Since $4x^2$ can be split in more than one way, let us try

$(2x\qquad)(2x\qquad)$

Since $18 = 2 \cdot 9$ and $18 = 3 \cdot 6$, we have the following possibilities:

$(2x + 9y)(2x - 2y)$

$(2x + 6y)(2x - 3y)$

In the first case the resulting middle term is $14xy$, and in the second case $6xy$. Neither is correct. Reversing the signs produces negative middle terms, which makes matters worse. For example,

$$(2x - 9y)(2x + 2y) = 4x^2 - 14xy - 18y^2$$

For our next trial let us start with

$$(4x \quad)(x \quad)$$

Using the same factors of 18, we might try

$$4x^2 + 21xy - 18y^2 = (4x - 3y)(x + 6y)$$

(with F = First, O = Outer, I = Inner, L = Last)

This result is correct.

Example 6 Factor $9x^2 - 219x + 72$.

Solution. First note the common factor **3**:

$$9x^2 - 219x + 72 = \mathbf{3}(3x^2 - 73x + 24)$$

Now we need a combination that produces a very large middle term. As a consequence, even though 24 has many different divisors, only the factors 24 and 1 could produce a sufficiently large middle term. It follows that

$$3(3x^2 - 73x + 24) = 3(3x - 1)(x - 24)$$

Irreducible factor

As noted earlier, not every polynomial is factorable. A nonfactorable polynomial is called **prime** or **irreducible.** For example, $x^2 + x + 7$ is prime.

Example 7 A stone is hurled upward from a height of 32 ft at 56 ft/sec. Its distance s (in feet) above the ground is given by $s = -16t^2 + 56t + 32$, where t is measured in seconds. Factor the expression for s and determine when the stone hits the ground. (Assume that the motion starts at $t = 0$ sec.)

Solution.
$$\begin{aligned} s &= -16t^2 + 56t + 32 \\ &= -8(2t^2 - 7t - 4) \\ &= -8(2t + 1)(t - 4) \end{aligned}$$

Since the stone is at ground level when $s = 0$, we need to find the value of t for which

$$-8(2t + 1)(t - 4) = 0$$

This product is 0 when $t - 4 = 0$ or when $2t + 1 = 0$. Note that $t - 4 = 0$ when $t = 4$. Now, $2t + 1 = 0$ only when $t = -\frac{1}{2}$. Since the motion begins at $t = 0$, the only admissible value is $t = 4$. We conclude that the stone takes 4 sec to reach the ground.

Exercises / Section 5.4

In Exercises 1–50, factor each expression.

1. $x^2 + 4xy + 4y^2$
2. $4x^2 - 4xz + z^2$
3. $9x^2 - 24xy + 16y^2$
4. $16x^2 - 8x + 1$
5. $x^2 - 4x + 3$
6. $x^2 - 5x + 6$
7. $x^2 - x - 12$
8. $x^2 - 2x - 15$
9. $2a^2 - 8ab + 8b^2$
10. $3m^2 + 18mn + 27n^2$
11. $2x^2 + 14x + 12$
12. $ax^2 - 5ax - 14a$
13. $x^2 - 5xy - 6y^2$
14. $a^2 + 3ab - 10b^2$
15. $D^2 + 5D - 14$
16. $T_1^2 + 7T_1T_2 + 12T_2^2$
17. $2x^2 - 3xy + y^2$
18. $3x^2 + xy - 2y^2$
19. $5x^2 - 11xy + 2y^2$
20. $4x^2 - 7xy + 3y^2$
21. $4x^2 + 13xy + 3y^2$
22. $4x^2 - 8xy + 3y^2$
23. $6x^2 - xy - 12y^2$
24. $6p^2 - 13pq + 5q^2$
25. $5w_1^2 - 22w_1w_2 + 8w_2^2$
26. $18x^2 - 21xy - 15y^2$
27. $8L^2 - 8LC - 48C^2$
28. $12R^2 - 18RC - 12C^2$
29. $18f^2 - 48fg + 32g^2$
30. $4\alpha A^2 + 12\alpha AB + 9\alpha B^2$
31. $x^4 - 4x^3 + 4x^2$
32. $\alpha^2 x^2 - 2\alpha\beta x^2 + \beta^2 x^2$
33. $D^2 - D - 8$
34. $y^2 - 3yz + 9z^2$
35. $(a + b)^2 - (a + b) - 6$
36. $(x - y)^2 - (x - y) - 12$
37. $(n + m)^2 - 3(n + m) + 2$
38. $(\beta + \gamma)^2 + 7(\beta + \gamma) + 12$
39. $2(a + b)^2 - 9(a + b) + 4$
40. $3(t_1 - t_2)^2 - 14(t_1 - t_2) + 8$
41. $(f_1 + 2f_2)^3 - (f_1 + 2f_2)^2$
42. $(c - 3d)^2 - (c - 3d)^3$
43. $1 - (x - y)^3$
44. $8 - (x + 2y)^3$
45. $28a^2 - ab - 2b^2$
46. $54x^2 - 3xy - 35y^2$
47. $40x^2 + 7xy - 3y^2$
48. $24x^2 - 38xy + 15y^2$
49. $12\alpha^2 - 23\alpha\beta + 10\beta^2$
50. $42D^2 - 11D - 20$

51. If a stone is hurled upward at a velocity of 24 ft/sec from a height of 72 ft (starting at $t = 0$), then the distance s (in feet) above the ground is given by $s = -16t^2 + 24t + 72$, where t is in seconds. Factor the expression to determine when the stone hits the ground. (See Example 7.)

52. The deflection of a certain beam is $3a^2x^3 - 5ax^2 - 2x$. Factor this expression.

53. The resistance of a variable resistor is given by $R = 18.0t^2 - 48.0t + 32.0$, where t is measured in seconds. Factor the expression for R and determine when the resistance is 0.

5.5 Factoring by Grouping

In this section we are going to discuss **factoring by grouping.** In some cases the terms in a polynomial can be grouped so that the expression can be factored by one of the earlier methods. (This technique is ordinarily used when the polynomial has more than three terms.)

Consider, for example, the expression

$$ax + ay + 3x + 3y$$

Note that the first two terms have a common factor, suggesting that the terms be grouped as follows:

$$(ax + ay) + (3x + 3y)$$

We now see that the last two terms also have a common factor; thus

$$(ax + ay) + (3x + 3y) = a(x + y) + 3(x + y)$$

The resulting expression consists of two terms with the common factor $x + y$. Factoring out $x + y$, we obtain

$$ax + ay + 3x + 3y = a(x + y) + 3(x + y) = (x + y)(a + 3)$$

Example 1 Factor $RV_1 - RV_2 + V_1 - V_2$.

Solution. The common factor R in the first two terms suggests the following grouping:

$$(RV_1 - RV_2) + (V_1 - V_2) = R(V_1 - V_2) + (V_1 - V_2)$$

When removing the common factor $V_1 - V_2$, note that the coefficient of the second term is 1, even though the 1 is not explicitly written. We now get

$$RV_1 - RV_2 + V_1 - V_2 = R(V_1 - V_2) + 1(V_1 - V_2)$$
$$= (V_1 - V_2)(R + 1)$$

The idea in the next example is similar.

Example 2 Factor $a^2 + ab - 2b^2 - 2a + 2b$.

Solution. The first three terms look like a typical trinomial, which is indeed factorable. At the same time, the last two terms have the common factor 2 or -2. Let's group the terms as follows:

$$(a^2 + ab - 2b^2) + (-2a + 2b) \qquad \text{inserting parentheses}$$
$$= (a - b)(a + 2b) + (-2a + 2b) \qquad \text{factoring the trinomial}$$

Now we need to remove -2 from the second expression to obtain a common factor $a - b$:

$$(a - b)(a + 2b) - 2(a - b) \qquad \text{common factor } -2$$
$$= (a - b)(a + 2b - 2) \qquad \text{common factor } a - b$$

Some expressions containing several terms can be grouped in different ways, some of which may not work. Consider the next example.

Example 3 Factor $xz + xy - 2yz - x^2 + 2y^2$.

Solution. If we group the first three terms as in Example 2, we get a nonfactorable combination. A more productive approach is to group the first two terms, which contain a common factor x. Then we get

$$(xz + xy) + (-2yz - x^2 + 2y^2)$$

Unfortunately, the trinomial contains $-2yz$, which is completely different from the other two terms. This observation provides the clue: All the x and y terms should be grouped together. Thus

$$\begin{aligned}
xz - 2yz &- x^2 + xy + 2y^2 & &\text{regrouping} \\
&= (xz - 2yz) - (x^2 - xy - 2y^2) & &\text{inserting parentheses} \\
&= z(x - 2y) - (x - 2y)(x + y) & &\text{factoring} \\
&= (x - 2y)[z - (x + y)] & &\text{common factor } x - 2y \\
&= (x - 2y)(z - x - y)
\end{aligned}$$

In some cases even the methods of Section 5.2 apply if the terms are suitably grouped.

Example 4 Factor $a^2 - 4x^2 + 4xy - y^2$.

Solution. Since the first two terms are a difference of two squares, we try

$$(a^2 - 4x^2) + (4xy - y^2) = (a - 2x)(a + 2x) + y(4x - y)$$

Although a perfectly logical procedure, this attempt leads to a dead end. An alternative is to keep a^2 apart and to group the remaining terms as follows:

$$\begin{aligned}
a^2 &- (4x^2 - 4xy + y^2) & &\text{inserting parentheses} \\
&= a^2 - [(2x)^2 - 2(2xy) + y^2] & &\text{perfect-square trinomial} \\
&= a^2 - (2x - y)^2 & &\text{factoring the trinomial} \\
&= [a - (2x - y)][a + (2x - y)] & &\text{difference of two squares} \\
&= (a - 2x + y)(a + 2x - y)
\end{aligned}$$

Example 5 Suppose the distance s from the origin as a function of time of a particle traveling along a line is $s = 2t(t - 1)^3$. It can be shown by the methods of calculus that the velocity v is given by

$$v = 6t^3 - 12t^2 + 6t + 2(t - 1)^3$$

Write this expression as a product.

Solution.
$$v = 6t^3 - 12t^2 + 6t + 2(t-1)^3$$
$$= 6t(t^2 - 2t + 1) + 2(t-1)^3 \quad \text{common factor } 6t$$
$$= 6t(t-1)^2 + 2(t-1)^3 \quad \text{factoring the trinomial}$$
$$= 2(t-1)^2(3t + t - 1) \quad \text{common factor } 2(t-1)^2$$
$$= 2(t-1)^2(4t - 1)$$

Exercises / Section 5.5

In Exercises 1–30, factor the given expressions by grouping.

1. $ax - ay + bx - by$
2. $xy + x + 3y + 3$
3. $2x^2 + 6xy + x + 3y$
4. $3x - 3y + cx - cy$
5. $4ac - 4bc + a - b$
6. $5bx - 5by + x - y$
7. $5bx - 5by - x + y$
8. $ax + ay - x - y$
9. $2ax + 2ay - 2cx - 2cy$
10. $4ax - 4bx - 4ay + 4by$
11. $3aR - 3ar - 6bR + 6br$
12. $4a_1v_1 + 4a_2v_1 - 4a_1v_2 - 4a_2v_2$
13. $x^2 - y^2 - xz - yz$
14. $x^2 - y^2 - xz + yz$
15. $ax + ay - x^2 + y^2$
16. $zx - zy - x^2 + y^2$
17. $x^2 - 2xz - y^2 + 2yz$
18. $x^2 - 3xy + 2y^2 + 2x - 2y$
19. $x^2 + 3xy - 4y^2 + x - y$
20. $x^2 + 3xy - 4y^2 - x + y$
21. $4x^2 - 4xy + y^2 - z^2$
22. $x^2 - 6xy + 9y^2 - w^2$
23. $x^2 - z^2 + 4y^2 + 4xy$
24. $a^2 - c^2 + 16b^2 - 8ab$
25. $9a^2 - 4b^2 - c^2 - 4bc$
26. $2a - 2b + a^2 + ab - 2b^2$
27. $3x - 3y - x^2 - 3xy + 4y^2$
28. $2x^2 - 3xy - 2y^2 + 6x + 3y$
29. $ax + xy + 2ay - x^2 + 6y^2$
30. $xz + 2xy + 3yz - x^2 + 15y^2$

31. In a problem on cost analysis, the following expression has to be simplified:
$$a^2(A - 1) - [2 + (A - 1)a]^2 + 4 + 2a^2(A - 1)^2$$
Carry out the simplification by factoring.

32. Factor the following expression, which arose in a problem on the strength of materials:
$$2a(x - 2) + 3a(x - 2)^2$$

5.6 Equivalent Fractions

In the earlier chapters we considered the four fundamental operations of addition, subtraction, multiplication, and division on signed numbers and polynomials. We will now learn how to perform these operations on fractions. Just as arithmetic fractions play an important role in everyday life, algebraic fractions play an important role in mathematics and its technical applications.

We would expect operations with algebraic fractions to be quite similar to those with arithmetic fractions. For example, we know that

$$\frac{6}{9} = \frac{2}{3}$$

Similarly, we would expect that

$$\frac{6x}{9x} = \frac{2}{3} \quad (x \neq 0)$$

In other words, algebraic fractions can be reduced by dividing the numerator and denominator by the same factor.

> A fraction is said to be **reduced to lowest terms** if its numerator and denominator have no common factors except 1.

Example 1 Reduce the fraction

$$\frac{14a^3b^2c^4}{21ab^2c^2}$$

to lowest terms.

Solution. Even though this fraction could be reduced directly by inspection, let us rearrange the terms first to illustrate the basic principle. Thus

$$\frac{14a^3b^2c^4}{21ab^2c^2} = \frac{(7ab^2c^2)(2a^2c^2)}{(7ab^2c^2)(3)}$$
$$= \frac{2a^2c^2}{3}$$

obtained by dividing the numerator and denominator by $7ab^2c^2$.

Reducing fractions

The procedure in Example 1 can be applied to fractions containing polynomials. Since fractions are always reduced by dividing the numerator and denominator by the same factor, *both numerator and denominator must first be factored*. This operation is one of the most important applications of the factoring procedure.

(From now on, the numerical values of the variables in a fraction are understood to be restricted so that the denominator is different from zero. This restriction eliminates division by zero.)

5.6 EQUIVALENT FRACTIONS

Example 2 Reduce the fraction

$$\frac{x^2 + xy - 2y^2}{x^2 - 4xy + 3y^2}$$

to lowest terms.

Solution. We first factor both the numerator and denominator to obtain

$$\frac{x^2 + xy - 2y^2}{x^2 - 4xy + 3y^2} = \frac{(x - y)(x + 2y)}{(x - y)(x - 3y)}$$

After dividing both numerator and denominator by the common factor $x - y$, the fraction is reduced to

$$\frac{x + 2y}{x - 3y}$$

Cancellation
The simplest and safest way to reduce this fraction is to cross out the factor $x - y$. This procedure is called **cancellation.** In our problem

$$\frac{x^2 + xy - 2y^2}{x^2 - 4xy + 3y^2} = \frac{\overset{1}{\cancel{(x - y)}}(x + 2y)}{\underset{1}{\cancel{(x - y)}}(x - 3y)} = \frac{x + 2y}{x - 3y}$$

The 1 next to each canceled factor is often omitted:

$$\frac{\cancel{(x - y)}(x + 2y)}{\cancel{(x - y)}(x - 3y)} = \frac{x + 2y}{x - 3y}$$

In this case the line through the canceled factor has to be thought of as a 1.

Example 3 Reduce the fraction

$$\frac{a^2 + ab - 6b^2}{2a^2 + 7ab + 3b^2}$$

by cancellation.

Solution.
$$\frac{a^2 + ab - 6b^2}{2a^2 + 7ab + 3b^2} = \frac{(a + 3b)(a - 2b)}{(2a + b)(a + 3b)} \qquad \text{factoring}$$

$$= \frac{\cancel{(a + 3b)}(a - 2b)}{(2a + b)\cancel{(a + 3b)}} \qquad \text{cancellation}$$

$$= \frac{a - 2b}{2a + b}$$

To avoid possible difficulties with signs, observe that the quantities $a - b$ and $b - a$ are numerically equal but opposite in sign. Indeed, $-(b - a) = -b + a = a - b$, or

$$a - b = -(b - a) \qquad (5.14)$$

Also, recall that

$$\frac{-a}{b} = \frac{a}{-b} = -\frac{a}{b} \qquad (5.15)$$

Example 4 Simplify the fraction

$$\frac{x^3 - y^3}{xy - x^2}$$

Solution. The numerator is a difference of two cubes, and the terms in the denominator have a common factor x. Thus

$$\frac{x^3 - y^3}{xy - x^2} = \frac{(x - y)(x^2 + xy + y^2)}{(y - x)x}$$

$$= \frac{(x - y)(x^2 + xy + y^2)}{-(x - y)x} \qquad \text{since } a - b = -(b - a)$$

$$= -\frac{(x - y)(x^2 + xy + y^2)}{(x - y)x} \qquad \text{since } \frac{a}{-b} = -\frac{a}{b}$$

$$= -\frac{\cancel{(x - y)}(x^2 + xy + y^2)}{\cancel{(x - y)}x} \qquad \text{cancellation}$$

$$= -\frac{x^2 + xy + y^2}{x}$$

Common error Forgetting that

$$\frac{a + b}{a} \neq 1 + b$$

Factors can only be canceled if both numerator and denominator are written as products.

The correct procedure is

$$\frac{a + b}{a} = \frac{a}{a} + \frac{b}{a} = 1 + \frac{b}{a}$$

Example 5 Simplify the fraction

$$\frac{x+2}{(x+3)x+2}$$

Solution. In this example it is particularly tempting to cancel $x+2$. However, the denominator is not a product, since $(x+3)x+2$ is not the same as $(x+3)(x+2)$. The denominator needs to be factored first.

$$\frac{x+2}{(x+3)x+2} = \frac{x+2}{x^2+3x+2} \qquad \text{distributive law}$$

$$= \frac{x+2}{(x+2)(x+1)} \qquad \text{factoring}$$

$$= \frac{\overset{1}{\cancel{x+2}}}{\underset{1}{\cancel{(x+2)}}(x+1)} \qquad \text{canceling}$$

$$= \frac{1}{x+1}$$

Note that since the only factor appearing in the numerator is canceled, it is wise to place a 1 next to the canceled factors. (However, as mentioned earlier, the line through a canceled factor may be thought of as a 1.)

Example 6 Simplify

$$\frac{2x^4 - x^3y - 3x^2y^2}{6x^2y^2 - x^3y - 2x^4}$$

Solution.
$$\frac{2x^4 - x^3y - 3x^2y^2}{6x^2y^2 - x^3y - 2x^4} = \frac{x^2(2x^2 - xy - 3y^2)}{-x^2(2x^2 + xy - 6y^2)}$$

$$= -\frac{\cancel{x^2}\cancel{(2x-3y)}(x+y)}{\cancel{x^2}\cancel{(2x-3y)}(x+2y)}$$

$$= -\frac{x+y}{x+2y}$$

Example 7 If a mass m_1 strikes a stationary mass m_2, then the ratio of the kinetic energy of the system with respect to the floor to the kinetic energy with respect to the center of mass of the system can be expressed as

$$\frac{\tfrac{1}{2}m_1v^2(m_1+m_2) - \tfrac{1}{2}m_1^2v^2}{\tfrac{1}{2}m_1v^2(m_1+m_2)}$$

Reduce this expression.

174 CHAPTER 5 FACTORING AND FRACTIONS

Solution. $\dfrac{\tfrac{1}{2}m_1v^2(m_1+m_2)-\tfrac{1}{2}m_1^2v^2}{\tfrac{1}{2}m_1v^2(m_1+m_2)} = \dfrac{\tfrac{1}{2}m_1^2v^2+\tfrac{1}{2}m_1m_2v^2-\tfrac{1}{2}m_1^2v^2}{\tfrac{1}{2}m_1v^2(m_1+m_2)}$

$= \dfrac{\tfrac{1}{2}\cancel{m_1}m_2\cancel{v^2}}{\tfrac{1}{2}\cancel{m_1}\cancel{v^2}(m_1+m_2)}$

$= \dfrac{m_2}{m_1+m_2}$

Exercises / Section 5.6

In Exercises 1–42, reduce the fractions to lowest terms.

1. $\dfrac{3x^2y^3}{6xy^3}$
2. $\dfrac{-6x^3y^2z}{9x^2y^2z}$
3. $\dfrac{-14a^2b^3x^2}{-28ab^3x^3}$

4. $\dfrac{9a^4bc^3}{-15a^2b^3c}$
5. $\dfrac{x^2+x}{x+1}$
6. $\dfrac{x-3}{2x-6}$

7. $\dfrac{2x^2+8x}{x^3+4x^2}$
8. $\dfrac{ax^2-6ax}{ax+6a}$
9. $\dfrac{x^2-16}{x-4}$

10. $\dfrac{2x^2-8}{x+2}$
11. $\dfrac{x^3+y^3}{x+y}$
12. $\dfrac{x^3-y^3}{x^2+xy+y^2}$

13. $\dfrac{x^2-y^2}{3x+3y}$
14. $\dfrac{s^2-t^2}{vs+vt}$
15. $\dfrac{R^3-8}{R-2}$

16. $\dfrac{v^3-27}{v^2+3v+9}$
17. $\dfrac{x^2-6x+8}{x-2}$
18. $\dfrac{4x^2+11x-3}{x+3}$

19. $\dfrac{i^2+7i+10}{i+2}$
20. $\dfrac{3R^2+13R+4}{R+4}$
21. $\dfrac{6x^2-11x+4}{2x^2+5x-3}$

22. $\dfrac{4x^2-15x-4}{4x^2+13x+3}$
23. $\dfrac{7x^2-40x-12}{2x^2-11x-6}$
24. $\dfrac{3v^2-2v-16}{3v^2-11v+8}$

25. $\dfrac{m_1^2-m_2^2}{m_1^2-3m_1m_2+2m_2^2}$
26. $\dfrac{R^3+r^3}{R^2+4Rr+3r^2}$
27. $\dfrac{(x-3)(x+4)}{(x+4)(3-x)}$

28. $\dfrac{(\beta-2)(2\beta-3)}{(3-2\beta)(2-\beta)}$
29. $\dfrac{2L^2-3LC-2C^2}{3L^2-10LC+8C^2}$
30. $\dfrac{15y^2-7xy-2x^2}{4x^2-4xy-3y^2}$

31. $\dfrac{(x+2)(x^2-7x+12)}{(x-3)(x^2-4x-12)}$
32. $\dfrac{(c-d)(2c^2-3cd-2d^2)}{(c-2d)(3c^2-cd-2d^2)}$
33. $\dfrac{ac+ad-4bc-4bd}{2ac+2ad-bc-bd}$

34. $\dfrac{ac+bc-3ad-3bd}{ac-3ad-bc+3bd}$
35. $\dfrac{x+2}{(2x+5)x+2}$
36. $\dfrac{f+4}{(4f+17)f+4}$

37. $\dfrac{l-3}{(2l-5)l-3}$
38. $\dfrac{s-2}{(3s-5)s-2}$
39. $\dfrac{t+2}{(3t+7)t+2}$

40. $\dfrac{v_0+5}{(3v_0+16)v_0+5}$
41. $\dfrac{2E+1}{3E(2E+1)-2(2E+1)}$
42. $\dfrac{2\pi-3}{2\pi(2\pi-3)+2(2\pi-3)}$

43. The area of a rectangle is $PL - PM - QL + QM$, and the length is $L - M$. Find the width (width = area/length).

44. A common phenomenon in mechanical and electrical systems is simple harmonic motion (which will be discussed in Chapter 8). If a particle is subject to two such motions at right angles, then the resulting motion can be described by

$$y^2 = \frac{4x^2}{A^2}(A^2 - x^2)$$

under certain conditions. Simplify the expression for the ratio of y^2 to $(A - x)/A^2$, which is given by

$$\frac{4x^2(A^2 - x^2)A^2}{A^2(A - x)}$$

45. Suppose a neutron of mass m traveling at velocity v_0 collides with a stationary nucleus of mass M. If the two particles are scattered at right angles to the original motion, then the velocity v of the neutron after the collision satisfies the relation

$$v^2 = \frac{v_0^2 M^2 + v_0^2 m^2}{(M + m)^2}$$

Find the ratio of the kinetic energy of the neutron before the collision to its kinetic energy after the collision; that is, simplify the expression for

$$\frac{E}{E_0} = \frac{mv^2/2}{mv_0^2/2} = \frac{\frac{1}{2}m(v_0^2 M^2 + v_0^2 m^2)}{(M + m)^2(\frac{1}{2}mv_0^2)}$$

46. The voltage drop (in volts) across a certain resistor as a function of time is $V = 16t^2 - 4$ (t in seconds). The variable resistance is given by $R = 4t + 2$ (in ohms). Find a simplified expression for the current $I = V/R$ (in amperes).

47. Repeat Exercise 46 for $V = 2t^2 + 4t - 30$ and $R = t + 5$.

5.7 Multiplication and Division of Fractions

The rules for multiplication and division of fractions carry over from arithmetic:

Multiplication:

$$\frac{a}{b} \cdot \frac{c}{d} = \frac{ac}{bd} \quad (5.16)$$

Division:

$$\frac{a}{b} \div \frac{c}{d} = \frac{a}{b} \cdot \frac{d}{c} = \frac{ad}{bc} \quad (5.17)$$

Multiplication and division of algebraic fractions should be performed by a procedure similar to the reduction of fractions in the previous section. Consider the product

$$\frac{x}{x + y} \cdot \frac{x^2 - y^2}{x^2}$$

176 CHAPTER 5 FACTORING AND FRACTIONS

If the multiplication rule were taken literally, the numerators and denominators would be multiplied together and the resulting fraction simplified. However, multiplying the polynomials is inadvisable, since in the very next step the resulting products must be factored again. Instead, the expressions can be combined as follows:

$$\frac{x(x^2 - y^2)}{(x + y)x^2} = \frac{x(x - y)(x + y)}{(x + y)x^2} = \frac{x - y}{x}$$

It should now be apparent that even the first step is unnecessary. The expressions in the given fractions should be factored and the appropriate factors canceled. Thus

$$\frac{x}{x + y} \cdot \frac{x^2 - y^2}{x^2} = \frac{\cancel{x}}{\cancel{x+y}} \cdot \frac{(x - y)\cancel{(x+y)}}{\cancel{x^2}} = \frac{x - y}{x}$$

Multiplication   *To multiply two or more fractions, factor the expression in the numerator and denominator of each fraction and cancel. Then multiply the resulting fractions by rule (5.16).*

Example 1   Carry out the following multiplication:

$$\frac{x^2 - 3xy}{y - x} \cdot \frac{x^3 - y^3}{x^3 + 2x^2y - 15xy^2}$$

Solution. Following the plan just discussed, let us factor all the polynomials before multiplying. Then we get

$$\frac{x^2 - 3xy}{y - x} \cdot \frac{x^3 - y^3}{x^3 + 2x^2y - 15xy^2}$$

$$= \frac{x(x - 3y)}{y - x} \cdot \frac{(x - y)(x^2 + xy + y^2)}{x(x^2 + 2xy - 15y^2)} \qquad \text{common factors;}$$
$$\text{difference of two cubes}$$

$$= \frac{x(x - 3y)}{-(x - y)} \cdot \frac{(x - y)(x^2 + xy + y^2)}{x(x + 5y)(x - 3y)} \qquad \text{factoring the trinomial;}$$
$$a - b = -(b - a)$$

$$= -\frac{\cancel{x}\cancel{(x - 3y)}}{\cancel{x - y}} \cdot \frac{\cancel{(x - y)}(x^2 + xy + y^2)}{\cancel{x}(x + 5y)\cancel{(x - 3y)}} \qquad \text{cancellation; } \frac{a}{-b} = -\frac{a}{b}$$

$$= -\frac{x^2 + xy + y^2}{x + 5y}$$

Division   **Division**, according to rule (5.17), requires one more step than multiplication: *We invert the divisor and then proceed as in multiplication.* In other words, we multiply by the reciprocal of the divisor.

Example 2   Perform the following division:

$$\frac{x^2 + 2xy + y^2}{x - 2y} \div \frac{x + y}{x^2 - 5xy + 6y^2}$$

5.7 MULTIPLICATION AND DIVISION OF FRACTIONS

Solution.

$$\frac{x^2 + 2xy + y^2}{x - 2y} \div \frac{x + y}{x^2 - 5xy + 6y^2}$$

$$= \frac{x^2 + 2xy + y^2}{x - 2y} \cdot \frac{x^2 - 5xy + 6y^2}{x + y} \qquad \text{inverting the divisor}$$

$$= \frac{(x + y)^2}{\cancel{x - 2y}} \cdot \frac{(x - 3y)\cancel{(x - 2y)}}{\cancel{x + y}} \qquad \text{factoring and canceling}$$

$$= (x + y)(x - 3y)$$

Example 3 Carry out the following division:

$$\frac{a - 4b}{2a - 4b} \div \frac{a^2 - 16b^2}{aL - aC - 2bL + 2bC}$$

Solution. After inverting the divisor, we get

$$\frac{a - 4b}{2a - 4b} \cdot \frac{aL - aC - 2bL + 2bC}{a^2 - 16b^2}$$

$$= \frac{a - 4b}{2a - 4b} \cdot \frac{(aL - aC) + (-2bL + 2bC)}{a^2 - 16b^2} \qquad \text{grouping}$$

$$= \frac{a - 4b}{2(a - 2b)} \cdot \frac{a(L - C) - 2b(L - C)}{(a - 4b)(a + 4b)} \qquad \text{common factors; difference of two squares}$$

$$= \frac{\cancel{a - 4b}}{2\cancel{(a - 2b)}} \cdot \frac{(L - C)\cancel{(a - 2b)}}{\cancel{(a - 4b)}(a + 4b)} \qquad \text{common factor } L - C\text{; cancellation}$$

$$= \frac{L - C}{2(a + 4b)}$$

Example 4 The discovery of deuterium depended on the fact that the motion of the nucleus reduces the mass of an electron in an atom. If m is the mass of an electron, then the reduced mass m' is given by

$$m' = \frac{mM}{M + m}$$

where M is the mass of the nucleus. Determine the ratio m'/m.

Solution. $\dfrac{m'}{m} = \dfrac{\dfrac{mM}{M + m}}{m} = \dfrac{\cancel{m}M}{M + m} \cdot \dfrac{1}{\cancel{m}} = \dfrac{M}{M + m}$

Exercises / Section 5.7

In Exercises 1–34, carry out the indicated operations and simplify.

1. $\dfrac{2x^2y^2}{3z^3w^3} \cdot \dfrac{9z^2w}{4xy}$

2. $\dfrac{2a^3b^2}{5c^2d^3} \cdot \dfrac{5c^3d^3}{a^3b^3}$

3. $\dfrac{6xy}{ab} \cdot \dfrac{8x^2y}{9a^2b} \cdot \dfrac{ab}{2xy}$

4. $\dfrac{5a^4b^6}{x^3y^3} \div \dfrac{15a^2b^5}{x^3y}$

5. $\dfrac{9x^4y^2z^2}{2a^2b^3c^5} \div \dfrac{15x^3yz^2}{4a^2b^4c^6}$

6. $\dfrac{x^2+xy}{a} \cdot \dfrac{b^2}{x^3+x^2y}$

7. $\dfrac{cd^2}{ax-2ay} \cdot \dfrac{x^2-2xy}{c^3d}$

8. $\dfrac{ay^2z^3}{x^2y-x^2z} \cdot \dfrac{xy-xz}{a^2yz^4}$

9. $\dfrac{x^2-y^2}{2x+y} \cdot \dfrac{4x^2+2xy}{x+y}$

10. $\dfrac{x-2y}{x^3-3x^2y} \div \dfrac{x^2-4y^2}{x-3y}$

11. $\dfrac{x^3+y^3}{x^2-y^2} \cdot \dfrac{x+y}{x^2-xy+y^2}$

12. $\dfrac{x^2-16y^2}{x^3-y^3} \cdot \dfrac{x-y}{x-4y}$

13. $\dfrac{a^2-4b^2}{3a-9b} \cdot \dfrac{a-3b}{2b-a}$

14. $\dfrac{a^2-16b^2}{1-2a} \cdot \dfrac{4a^2-1}{a+4b}$

15. $\dfrac{1-9a^2}{x+3y} \div \dfrac{3a-1}{x^3+27y^3}$

16. $\dfrac{x^2-16y^2}{a^2-1} \div \dfrac{x+4y}{a+1}$

17. $\dfrac{2a^2-ab-b^2}{6x^2+x-1} \div \dfrac{a^2-b^2}{8x+4}$

18. $\dfrac{e_1^2+3e_1e_2-10e_2^2}{e_1^2-e_2^2} \cdot \dfrac{e_1-e_2}{2e_1^2+11e_1e_2+5e_2^2}$

19. $\dfrac{3v_0^2-7v_0+4}{4v_1^2-8v_1-5} \div \dfrac{2v_0^2-5v_0+3}{2v_1^2+7v_1-30}$

20. $\dfrac{12s_0^2-13s_0-4}{5s_0^2+14s_0-3} \cdot \dfrac{10s_0^2-27s_0+5}{3s_0^2+2s_0-8}$

21. $\dfrac{6T^2-JT-J^2}{2K^2-9K-35} \div \dfrac{8T^2-2JT-J^2}{10K^2+13K-30}$

22. $\dfrac{2P^2-15PQ+18Q^2}{12R^2-RS-63S^2} \cdot \dfrac{3R^2-13RS+14S^2}{10P^2-33PQ+27Q^2}$

23. $\dfrac{x^2-5xy+6y^2}{x^2-y^2} \div \dfrac{x^3-8y^3}{x-y}$

24. $\dfrac{x^3-27}{x^3+3xy^2} \cdot \dfrac{x^2y+3y^3}{x^2+3x+9}$

25. $\dfrac{x^2-y^2}{x^3+8y^3} \div \dfrac{x+y}{x^2-2xy+4y^2}$

26. $\dfrac{s^2-t^2+2mt-m^2}{s^2+st} \cdot \dfrac{st+t^2}{s-t+m}$

27. $\dfrac{R+r+1}{R^2-Rr-6r^2} \div \dfrac{R^2+2Rr+r^2-1}{R-3r}$

28. $\dfrac{xy-4y+2x-8}{2x+6} \cdot \dfrac{x+3}{2x^2-3x-20}$

29. $\dfrac{2L^2-L-21}{L^2-16} \div \dfrac{LC-5L+3C-15}{2L-8}$

30. $\dfrac{\theta+2\rho}{\theta^2+4\theta\rho+3\rho^2} \cdot \dfrac{2\theta+2\rho}{\theta+3\rho} \cdot \dfrac{\theta^2-9\rho^2}{2\theta+4\rho}$

31. $\dfrac{c^2-4d^2}{c^2-d^2} \cdot \dfrac{c-d}{2c-4d} \cdot \dfrac{4c+4d}{c^2+cd-2d^2}$

32. $\dfrac{2x^2-6xy}{x^2-10xy+24y^2} \cdot \dfrac{x^2-3xy-4y^2}{4x-12y} \cdot \dfrac{2x-12y}{x^2+3xy+2y^2}$

33. $\left(\dfrac{x^2-16}{x^2-25} \cdot \dfrac{x+5}{x-4}\right) \div \dfrac{x+4}{2x-10}$

34. $\left(\dfrac{mv_1^2-mv_2^2}{v_1^2+2v_1v_2} \cdot \dfrac{2v_1v_2^2}{m^2}\right) \div \dfrac{v_1^2+2v_1v_2-3v_2^2}{3mv_1+6mv_2}$

35. Two unequal weights w_1 and w_2 are hanging on a cord passing over a pulley. The tension T on the cord due to the weights, which is the same for both weights, is given by

$$T = \dfrac{2w_1w_2}{w_1+w_2}$$

The acceleration of the system is

$$a = \frac{(w_1 - w_2)g}{w_1 + w_2}$$

Write an expression for $gT(1/a)$ and simplify.

36. In the study of the dispersion of X rays, the expression

$$A = \frac{Ne^2}{\pi m(f_0^2 - f^2)}$$

arises. Multiply A by $m(f_0 + f)$ and simplify.

37. A perfectly flexible cable, suspended from two points at the same height, hangs under its own weight. The tension T_0 at its lowest point is

$$T_0 = \frac{w(s^2 - 4H^2)}{8H}$$

where s is the length of the cable, H the sag, and w the weight per unit length. Find an expression for $T_0[H/(s + 2H)]$.

5.8 Addition and Subtraction of Fractions

In considering the addition and subtraction of algebraic fractions, recall from arithmetic that the fractions $\frac{1}{3}$ and $\frac{1}{6}$ can be added if $\frac{1}{3}$ is changed to $\frac{2}{6}$, so that

$$\frac{2}{6} + \frac{1}{6} = \frac{3}{6} = \frac{1}{2}$$

The number 6 is called the **lowest common denominator (LCD)**.

Since algebraic fractions are added by the same rules, let us first state the definition of the lowest common denominator.

> **Lowest common denominator:** The lowest common denominator (LCD) of two or more fractions is an expression that is divisible by every denominator and does not have any more factors than needed to satisfy this condition.

The LCD can be found by the procedure described next.

> **To construct the lowest common denominator** for a set of algebraic fractions, factor each of the denominators. Then the LCD is the product of the factors of the denominators, each with an exponent equal to the largest of the exponents of any of these factors.

180 CHAPTER 5 FACTORING AND FRACTIONS

To add (or subtract) two or more fractions, we find the LCD for the fractions and change each fraction to an equivalent fraction having the LCD for its denominator. Next, we add (or subtract) the numerators of the fractions, placing the result over the LCD. Finally, we simplify the resulting fraction.

Example 1 Combine

$$\frac{5}{6a^2bc} - \frac{4}{15ab^2c} - \frac{3}{20abc^2}$$

Solution. In terms of prime factors, the different denominators are

$$6a^2bc = 2 \cdot 3 \cdot a^2 \cdot b \cdot c$$
$$15ab^2c = 3 \cdot 5 \cdot a \cdot b^2 \cdot c$$
$$20abc^2 = 2^2 \cdot 5 \cdot a \cdot b \cdot c^2$$

To construct the LCD, observe that the factors are 2, 3, 5, a, b, and c. The largest exponent on the factor 2 is 2, the largest exponent on the factor 3 is 1, and so forth. So the LCD is given by

$$\text{LCD} = 2^2 \times 3 \times 5 \times a^2b^2c^2 = 60a^2b^2c^2$$

Now we write the fractions so that they all have the same denominators. For example, since $60a^2b^2c^2 = (6a^2bc)(10bc)$, we get for the first fraction

$$\frac{5}{6a^2bc} = \frac{5}{6a^2bc} \cdot \frac{10bc}{10bc} = \frac{50bc}{60a^2b^2c^2}$$

The other fractions are adjusted similarly:

$$\frac{5(\mathbf{10bc})}{6a^2bc(\mathbf{10bc})} - \frac{4(\mathbf{4ac})}{15ab^2c(\mathbf{4ac})} - \frac{3(\mathbf{3ab})}{20abc^2(\mathbf{3ab})}$$

$$= \frac{50bc}{60a^2b^2c^2} - \frac{16ac}{60a^2b^2c^2} - \frac{9ab}{60a^2b^2c^2}$$

$$= \frac{50bc - 16ac - 9ab}{60a^2b^2c^2}$$

The procedure for adding or subtracting fractions containing polynomials is similar.

Example 2 Combine

$$\frac{x}{x-y} - \frac{x^2}{x^2-y^2}$$

5.8 ADDITION AND SUBTRACTION OF FRACTIONS

Solution. Factoring the denominator of the second fraction, we get

$$\frac{x}{x-y} - \frac{x^2}{(x-y)(x+y)}$$

The LCD is the product of these factors, each with exponent 1:

$$\text{LCD} = (x-y)(x+y)$$

Since the denominator of the second fraction is the LCD, we need to adjust only the first fraction:

$$\frac{x}{x-y} = \frac{x}{x-y} \cdot \frac{x+y}{x+y}$$

We now have

$$\frac{x}{x-y} - \frac{x^2}{(x-y)(x+y)}$$

$$= \frac{x(x+y)}{(x-y)(x+y)} - \frac{x^2}{(x-y)(x+y)}$$

$$= \frac{x(x+y) - x^2}{(x-y)(x+y)} \qquad \text{combining numerators}$$

$$= \frac{x^2 + xy - x^2}{(x-y)(x+y)} \qquad \text{simplifying the numerator}$$

$$= \frac{xy}{(x-y)(x+y)}$$

Example 3 Combine the following fractions and simplify:

$$\frac{x}{x+3} - \frac{6x+6}{x^2-9} + \frac{x+1}{x-3}$$

Solution. Since the denominator of the middle fraction is

$$x^2 - 9 = (x-3)(x+3)$$

which contains the other denominators, we get

$$\text{LCD} = (x+3)(x-3)$$

CHAPTER 5 FACTORING AND FRACTIONS

Hence

$$\frac{x}{x+3} - \frac{6x+6}{x^2-9} + \frac{x+1}{x-3}$$

$$= \frac{x}{x+3} \cdot \frac{x-3}{x-3} - \frac{6x+6}{(x+3)(x-3)} + \frac{x+1}{x-3} \cdot \frac{x+3}{x+3}$$

$$= \frac{x(x-3)}{(x+3)(x-3)} - \frac{6x+6}{(x+3)(x-3)} + \frac{(x+1)(x+3)}{(x-3)(x+3)}$$

$$= \frac{x(x-3) - (6x+6) + (x+1)(x+3)}{(x+3)(x-3)}$$

$$= \frac{x^2 - 3x - 6x - 6 + x^2 + 4x + 3}{(x+3)(x-3)} \qquad (5.18)$$

$$= \frac{2x^2 - 5x - 3}{(x+3)(x-3)}$$

$$= \frac{(2x+1)(x-3)}{(x+3)(x-3)}$$

$$= \frac{2x+1}{x+3}$$

Another method of combining fractions is to supply each fraction with any factors missing in the denominator until all the denominators are the same.

Example 3 (Alternate) The fractions in Example 3 are

$$\frac{x}{x+3}, \quad \frac{6x+6}{(x-3)(x+3)}, \quad \text{and} \quad \frac{x+1}{x-3}$$

The denominator of the first fraction contains $x+3$ but not $x-3$. Supplying this missing factor, the first fraction becomes

$$\frac{x}{x+3} = \frac{x(x-3)}{(x+3)(x-3)}$$

Similarly, the denominator of the last fraction contains $x-3$ but not $x+3$. Supplying this factor, we get

$$\frac{x+1}{x-3} = \frac{(x+1)(x+3)}{(x-3)(x+3)}$$

Note that the denominator of the middle fraction contains both factors. We now have

$$\frac{x(x-3)}{(x-3)(x+3)} - \frac{6x+6}{(x-3)(x+3)} + \frac{(x+1)(x+3)}{(x-3)(x+3)}$$

The simplification now proceeds as in Example 3.

5.8 ADDITION AND SUBTRACTION OF FRACTIONS

Common errors

(1) Failing to change all the signs when subtracting the numerator. For example, in step (5.18) of Example 3, both signs in the numerator of the middle fraction are changed to become $-6x - 6$.

(2) Forgetting that

$$\frac{1}{x} + \frac{1}{y} \neq \frac{1}{x+y}$$

(3) Forgetting that

$$\frac{1}{x+y} \neq \frac{1}{x} + \frac{1}{y}$$

In case **(2)** the correct procedure is

$$\frac{1}{x} + \frac{1}{y} = \frac{y}{xy} + \frac{x}{xy} = \frac{x+y}{xy}$$

In case **(3)** the fraction

$$\frac{1}{x+y}$$

is already in simplest form and cannot be split up.

Example 4 Two perfectly elastic balls collide with a common velocity v. If their respective masses are m and M, then the velocity of m after the collision is

$$v\left(\frac{M}{M+m} - \frac{m}{M+m}\right) - \frac{2vm}{M+m}$$

Simplify this expression.

Solution.
$$v\left(\frac{M}{M+m} - \frac{m}{M+m}\right) - \frac{2vm}{M+m}$$
$$= \frac{vM}{M+m} - \frac{vm}{M+m} - \frac{2vm}{M+m}$$
$$= \frac{vM - vm - 2vm}{M+m}$$
$$= \frac{vM - 3vm}{M+m}$$

Exercises / Section 5.8

In Exercises 1–35, combine the given fractions and simplify.

1. $\dfrac{1}{2} - \dfrac{1}{18} + \dfrac{5}{9}$

2. $\dfrac{5}{36} - \dfrac{3}{108} + \dfrac{1}{9}$

3. $\dfrac{2x+1}{9} + \dfrac{x}{2} - \dfrac{x}{6}$

4. $\dfrac{1}{x} - \dfrac{3}{2x} - \dfrac{1}{3x}$

5. $\dfrac{3x+3}{4x} - \dfrac{1}{2x} - \dfrac{1}{x}$

6. $\dfrac{x-1}{5y} + \dfrac{x}{10y} - \dfrac{7x}{30y}$

7. $\dfrac{3a}{4b} - \dfrac{4a}{9b} + \dfrac{a-3}{36b}$

8. $\dfrac{x}{x+y} - \dfrac{x-y}{x}$

9. $\dfrac{x+3y}{x-y} - \dfrac{x-3y}{x+y}$

10. $\dfrac{x}{2x+y} - \dfrac{x}{x+y}$

11. $\dfrac{2}{xy} - \dfrac{1}{x} - \dfrac{y^2+2x-2y}{xy(x-y)}$

12. $\dfrac{x}{x-3y} - \dfrac{y}{2x+y}$

13. $\dfrac{x+y}{x+2y} + \dfrac{x}{x-y}$

14. $\dfrac{3}{x-2} + \dfrac{4}{x-3}$

15. $\dfrac{2}{x-3} + \dfrac{1}{x+2} - \dfrac{2x-1}{(x-3)(x+2)}$

16. $\dfrac{x}{(x+y)(x+2y)} + \dfrac{1}{x+y}$

17. $\dfrac{1}{x+3y} + \dfrac{4y}{(x+3y)(x-y)}$

18. $\dfrac{x}{2x-y} - \dfrac{y}{y-3x}$

19. $\dfrac{2x}{3x-y} - \dfrac{2y}{3y-2x}$

20. $\dfrac{1}{x} + \dfrac{2}{y} + \dfrac{1}{x+y}$

21. $\dfrac{1}{x} - \dfrac{1}{y} - \dfrac{y}{x(x+y)}$

22. $\dfrac{y}{x+y} - \dfrac{2y^2}{x(x+y)} + \dfrac{x+2y}{x}$

23. $\dfrac{a+b}{b} - \dfrac{a^2}{b(a+2b)} + \dfrac{a}{a+2b}$

24. $\dfrac{3ab}{a^2-b^2} + \dfrac{a}{b-a} + \dfrac{a-b}{a+b}$

25. $\dfrac{A}{A+B} - \dfrac{B^2}{A(A+B)} + \dfrac{2B}{A}$

26. $\dfrac{3s-t}{s(s+t)} - \dfrac{1}{s+t} + \dfrac{1}{s}$

27. $\dfrac{n}{n-m} - \dfrac{m}{n+m} - \dfrac{2m^2}{n^2-m^2}$

28. $\dfrac{4}{(R+3r)(R+r)} - \dfrac{1}{(R+3r)(R+2r)} - \dfrac{1}{(R+2r)(R+r)}$

29. $\dfrac{3}{(x-y)(x+2y)} - \dfrac{1}{(x+y)(x+2y)} + \dfrac{1}{(y-x)(x+y)}$

30. $\dfrac{2}{x+1} - \dfrac{2x}{x^2-1} + \dfrac{1}{x-1}$

31. $\dfrac{x}{x-2} - \dfrac{2}{x+2} - \dfrac{x^2}{x^2-4}$

32. $\dfrac{T_0}{T_0-2} - \dfrac{2}{T_0+2} + \dfrac{2T_0^2}{4-T_0^2}$

33. $\dfrac{c}{c+2d} - \dfrac{c}{2c+d}$

34. $\dfrac{2}{x^2-y^2} - \dfrac{3}{x^2+xy-2y^2} + \dfrac{1}{x^2+3xy+2y^2}$

35. $\dfrac{1}{2x^2+3xy+y^2} - \dfrac{1}{x^2+4xy+3y^2} + \dfrac{1}{2x^2+7xy+3y^2}$

36. A light rod of length 1 is clamped at both ends and carries a load W at the center. If x is the distance from one end of the rod, then the deflection is

$$y = \dfrac{W}{k}\left(\dfrac{x^3}{12} - \dfrac{x^2}{16}\right), \quad 0 \le x \le \dfrac{1}{2}$$

where k is a constant. Write y as a single fraction.

37. A hydrogen atom dropping from the nth energy level to the $(n - p)$th energy level emits a photon whose frequency is

$$k\left[\frac{1}{(n-p)^2} - \frac{1}{n^2}\right]$$

where k is a constant. Simplify this expression.

38. The distance between two heat sources with respective intensities a and b is L. The total intensity I of heat at a point between the sources is

$$I = \frac{a}{x^2} + \frac{b}{(L-x)^2}$$

where x is the distance from one of the sources. Express I as a single fraction.

39. In determining the force required to rotate a body about a fixed axis, the expression

$$1 - \frac{L^2}{k^2 + L^2}$$

arises. Simplify.

40. A neutron of mass m moving with velocity v_0 collides with a stationary nucleus of mass M. The resulting velocity V of the system is known to be

$$V = v_0 - \frac{Mv_0}{M+m}$$

Combine the two terms on the right side.

5.9 Complex Fractions

If the terms in a fraction are themselves fractions, then the whole expression is called a **complex fraction.** In this section we shall see how to simplify such fractions.

The procedure is best seen by means of an arithmetic example. Consider the complex fraction

$$\frac{\frac{1}{2} - \frac{1}{3}}{\frac{1}{4} - \frac{1}{2}}$$

We combine the fractions in the numerator and denominator separately and divide the resulting simple fractions. Thus

$$\frac{\frac{1}{2} - \frac{1}{3}}{\frac{1}{4} - \frac{1}{2}} = \frac{\frac{3}{6} - \frac{2}{6}}{\frac{1}{4} - \frac{2}{4}} = \frac{\frac{1}{6}}{-\frac{1}{4}} = \left(\frac{1}{6}\right)\left(-\frac{4}{1}\right) = -\frac{2}{3}$$

The procedure for algebraic fractions is similar.

186 CHAPTER 5 FACTORING AND FRACTIONS

Example 1 Simplify

$$\frac{\frac{1}{x} - \frac{1}{y}}{\frac{1}{xy}}$$

Solution. The terms in the numerator may be combined in the usual way:

$$\frac{1}{x} - \frac{1}{y} = \frac{y}{xy} - \frac{x}{xy} = \frac{y-x}{xy} \qquad \text{LCD} = xy$$

We now have

$$\frac{\frac{1}{x} - \frac{1}{y}}{\frac{1}{xy}} = \frac{\frac{y-x}{xy}}{\frac{1}{xy}} = \frac{y-x}{xy} \cdot \frac{xy}{1} = y - x$$

Example 2 Simplify the fraction

$$\frac{\frac{1}{x+y} + \frac{1}{x-y}}{\frac{x}{(x-y)(x+y)}}$$

Solution. For the terms in the numerator we have

$$\frac{1}{x+y} + \frac{1}{x-y} = \frac{x-y}{(x+y)(x-y)} + \frac{x+y}{(x+y)(x-y)}$$

$$= \frac{x-y+x+y}{(x+y)(x-y)}$$

$$= \frac{2x}{(x+y)(x-y)}$$

We now divide numerator by denominator to get

$$\frac{2x}{(x+y)(x-y)} \cdot \frac{(x-y)(x+y)}{x} = 2$$

Example 2 (Alternate) Since the LCD of all the fractions is $(x - y)(x + y)$, a simple alternative to the above method is to multiply the numerator and denominator of the complex fraction by $(x - y)(x + y)$, thereby reducing the complex fraction to a simple fraction. *Remember to multiply all the terms.*

$$\frac{\frac{1}{\cancel{x+y}} \cdot \frac{(x-y)\cancel{(x+y)}}{1} + \frac{1}{\cancel{x-y}} \cdot \frac{\cancel{(x-y)}(x+y)}{1}}{\frac{x}{\cancel{(x-y)(x+y)}} \cdot \frac{\cancel{(x-y)(x+y)}}{1}}$$

$$= \frac{(x-y)+(x+y)}{x} = \frac{2x}{x} = 2$$

The remaining examples further illustrate the alternate technique discussed in the previous example.

Example 3 Simplify the fraction

$$\frac{\dfrac{R}{R+3} + \dfrac{R}{R^2-9}}{\dfrac{1}{R-3} + 1}$$

Solution. Since $R^2 - 9 = (R-3)(R+3)$, it follows that

$$\text{LCD} = (R-3)(R+3)$$

As before, multiplying the numerator and denominator of the complex fraction by the LCD will reduce the fraction directly.

$$\frac{\dfrac{R}{\cancel{R+3}} \cdot \dfrac{(R-3)\cancel{(R+3)}}{1} + \dfrac{R}{\cancel{R^2-9}} \cdot \dfrac{\cancel{(R-3)(R+3)}}{1}}{\dfrac{1}{\cancel{R-3}} \cdot \dfrac{\cancel{(R-3)}(R+3)}{1} + 1 \cdot (R-3)(R+3)}$$

$$= \frac{R(R-3) + R}{(R+3) + (R-3)(R+3)}$$

$$= \frac{R^2 - 3R + R}{R + 3 + R^2 - 9}$$

$$= \frac{R^2 - 2R}{R^2 + R - 6}$$

$$= \frac{R\cancel{(R-2)}}{(R+3)\cancel{(R-2)}}$$

$$= \frac{R}{R+3}$$

Example 4 Just as electrical components can be connected in parallel, blood vessels that branch out and come together again are said to be connected in parallel.

CHAPTER 5 FACTORING AND FRACTIONS

Each such vessel offers a certain resistance to the flow of blood. If r_1, r_2, and r_3 are the respective resistance forces of the blood vessels in parallel, then the combined resistance is given by

$$\frac{1}{\dfrac{1}{r_1} + \dfrac{1}{r_2} + \dfrac{1}{r_3}}$$

Simplify this expression.

Solution.
$$\frac{1}{\dfrac{1}{r_1} + \dfrac{1}{r_2} + \dfrac{1}{r_3}} = \frac{1}{\dfrac{1}{r_1} + \dfrac{1}{r_2} + \dfrac{1}{r_3}} \cdot \frac{r_1 r_2 r_3}{r_1 r_2 r_3}$$

$$= \frac{r_1 r_2 r_3}{r_2 r_3 + r_1 r_3 + r_1 r_2}$$

Exercises / Section 5.9

In Exercises 1–26, simplify the complex fractions.

1. $\dfrac{1 + \dfrac{1}{3}}{2 + \dfrac{2}{3}}$

2. $\dfrac{\dfrac{1}{7} + \dfrac{2}{7}}{1 + \dfrac{2}{7}}$

3. $\dfrac{1 + \dfrac{1}{3}}{2 - \dfrac{1}{6}}$

4. $\dfrac{\dfrac{1}{3} + \dfrac{1}{2}}{1 - \dfrac{5}{6}}$

5. $\dfrac{3 - \dfrac{1}{x}}{9 - \dfrac{1}{x^2}}$

6. $\dfrac{\dfrac{1}{x} - 2}{4 - \dfrac{1}{x^2}}$

7. $\dfrac{1 - \dfrac{16}{x^2}}{1 + \dfrac{4}{x}}$

8. $\dfrac{1 - \dfrac{9}{x^2}}{1 - \dfrac{3}{x}}$

9. $\dfrac{\dfrac{1}{C_1} + \dfrac{1}{C_2}}{\dfrac{1}{C_1 C_2}}$

10. $\dfrac{v_1 - \dfrac{v_2^2}{v_1}}{1 - \dfrac{v_2}{v_1}}$

11. $\dfrac{1 + \dfrac{3}{x} - \dfrac{10}{x^2}}{1 - \dfrac{4}{x} + \dfrac{4}{x^2}}$

12. $\dfrac{2 - \dfrac{11}{y} - \dfrac{6}{y^2}}{2 - \dfrac{5}{y} - \dfrac{3}{y^2}}$

13. $\dfrac{3 - \dfrac{17}{h} + \dfrac{10}{h^2}}{3 + \dfrac{10}{h} - \dfrac{8}{h^2}}$

14. $\dfrac{s - \dfrac{3s}{s+6}}{s + \dfrac{9}{s+6}}$

15. $\dfrac{w - \dfrac{w}{w-5}}{w - \dfrac{6}{w-5}}$

16. $\dfrac{z + \dfrac{4z}{z-3}}{z - \dfrac{4}{z-3}}$

17. $\dfrac{\dfrac{1}{\beta+1} + \dfrac{1}{\beta^2-1}}{\dfrac{\beta}{\beta-1}}$

18. $\dfrac{a - \dfrac{3a+2}{a+2}}{a - \dfrac{2a+1}{a+2}}$

5.9 COMPLEX FRACTIONS   **189**

19. $\dfrac{\dfrac{1}{E-1} + \dfrac{1}{E-2}}{1 + \dfrac{1}{E-2}}$

20. $\dfrac{\dfrac{1}{t+2} - \dfrac{t}{t-3}}{\dfrac{1}{t+2} - 2}$

21. $\dfrac{\dfrac{k}{k+1} - \dfrac{6}{(k+1)^2}}{1 - \dfrac{9}{(k+1)^2}}$

22. $\dfrac{\dfrac{y}{y-3} + \dfrac{1}{y+1}}{\dfrac{y}{y+1} - \dfrac{1}{y+1}}$

23. $\dfrac{\dfrac{x}{x-2} - \dfrac{2}{(x-1)(x-2)}}{\dfrac{x-4}{x-1}}$

24. $\dfrac{\dfrac{b}{a-2b} + 1}{\dfrac{b^2}{a-2b} + a}$

25. $\dfrac{\dfrac{1}{t-3} - 1}{\dfrac{1}{t+1} - 2}$

26. $\dfrac{1 - \dfrac{1}{x-1}}{1 - \dfrac{1}{1 - \dfrac{1}{x}}}$

27. If two resistors R_1 and R_2 (in ohms) are connected in parallel, then the combined resistance (in ohms) is given by

$$\dfrac{1}{\dfrac{1}{R_1} + \dfrac{1}{R_2}}$$

Simplify this expression.

28. When the rate of decrease of the function $f(x) = 1/x$ is studied in calculus, the following expression has to be simplified:

$$\dfrac{\dfrac{1}{x+h} - \dfrac{1}{x}}{h}$$

Carry out this simplification.

29. If p is the distance of an object from a lens whose focal length is f, then the distance of the image from the lens is given by

$$\dfrac{1}{\dfrac{1}{f} - \dfrac{1}{p}}$$

Simplify this expression.

30. Simplify the following expression from a problem on vibrating membranes:

$$\dfrac{1}{1 + \left(\dfrac{y}{x}\right)^2} \left(-\dfrac{y}{x^2}\right)$$

31. The total resistance R of the circuit in Figure 5.1 is given by

$$R = \dfrac{1}{\dfrac{1}{R_1} + \dfrac{1}{R_2 + R_3}}$$

Simplify the expression for R.

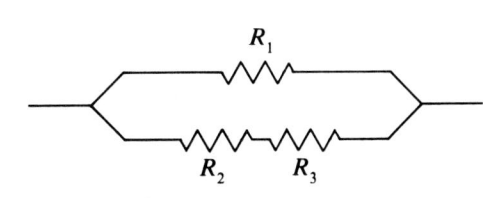

Figure 5.1

32. A light cord wrapped around the rim of a wheel of radius r has a mass m hanging on the end. The acceleration of the mass is given by

$$a = \frac{gm}{m + \dfrac{I}{r^2}}$$

where g is the acceleration due to gravity and I the "moment of inertia" of the wheel. Simplify this expression.

5.10 Fractional Equations

In this section we will continue the study of fractional equations begun in Section 2.2. First let us return to the equation in Example 4, Section 2.2:

$$\frac{1}{5}x + 1 = \frac{1}{15}x - \frac{1}{5}$$

Recall that the simplest way to solve such an equation is to **clear fractions** by multiplying both sides of the equation by the LCD, in this case **15**. Then we get

$$15\left(\frac{1}{5}x + 1\right) = 15\left(\frac{1}{15}x - \frac{1}{5}\right)$$

$$3x + 15 = x - 3$$

$$3x - x = -3 - 15$$

$$2x = -18$$

$$x = -9$$

> **Clearing fractions:** To solve an equation containing fractions, *clear the fractions* by multiplying both sides of the equation by the LCD of all the fractions in the equation.

Example 1 Solve the equation

$$\frac{1}{2(x-1)} + \frac{1}{x-1} = \frac{3}{2}$$

Solution. We multiply both sides of the equation by the LCD, which is $2(x-1)$. Then

$$\frac{2(x-1)}{1}\left(\frac{1}{2(x-1)} + \frac{1}{x-1}\right) = \frac{2(x-1)}{1} \cdot \frac{3}{2}$$

$$\frac{\cancel{2(x-1)}}{1} \cdot \frac{1}{\cancel{2(x-1)}} + \frac{2\cancel{(x-1)}}{1} \cdot \frac{1}{\cancel{x-1}} = \frac{\cancel{2}(x-1)}{1} \cdot \frac{3}{\cancel{2}}$$

$$1 + 2 = 3(x-1)$$

$$3 = 3x - 3$$

$$x = 2$$

Check:

$$\frac{1}{2(2-1)} + \frac{1}{2-1} = \frac{1}{2} + 1 = \frac{3}{2}$$

which is equal to the right side.

Example 2 Solve the equation

$$\frac{1}{x-3} - \frac{4}{x+4} = \frac{x}{x^2+x-12}$$

Solution. The trinomial can be written

$$x^2 + x - 12 = (x+4)(x-3)$$

and turns out to be the LCD. Multiplying both sides of the equation, we get

$$\frac{(x+4)(x-3)}{1}\left(\frac{1}{x-3} - \frac{4}{x+4}\right) = \frac{(x+4)(x-3)}{1} \frac{x}{(x+4)(x-3)}$$

$$(x+4) - 4(x-3) = x$$

$$x + 4 - 4x + 12 = x$$

$$-3x + 16 = x$$

$$-4x = -16$$

$$x = 4$$

Check:

Left Side

$$\frac{1}{4-3} - \frac{4}{4+4} = 1 - \frac{1}{2} = \frac{1}{2}$$

Right Side

$$\frac{4}{4^2+4-12} = \frac{1}{2}$$

The solution checks.

Some fractional equations do not have any solution, as shown in the next two examples.

Example 3 Solve the equation

$$\frac{x-5}{x-6} - 1 = \frac{2}{x-6}$$

Solution. Clearing fractions, we get

$$(x-5) - (x-6) = 2$$

$$x - 5 - x + 6 = 2$$

$$1 = 2$$

which is impossible. We conclude that the equation has no solution.

Example 4   Solve the equation

$$\frac{x+4}{x+3} + 2 = \frac{1}{x+3}$$

Solution. Clearing fractions, we have

$$(x+4) + 2(x+3) = 1$$
$$x + 4 + 2x + 6 = 1$$
$$3x = -9$$
$$x = -3$$

Check: Since substituting into the given equation leads to division by zero, the equation has no solution. The (apparent) root $x = -3$ is called **extraneous**.

Extraneous roots

Remark on extraneous roots. Multiplying both sides of an equation by an expression containing the unknown may give rise to extraneous roots. For example, the equation

$$x - 2 = 0$$

has only one root, $x = 2$. If both sides are multiplied by $x + 1$, we get

$$(x+1)(x-2) = 0$$

which has two roots, $x = 2$ and $x = -1$. The root $x = -1$ does not, of course, satisfy the original equation $x - 2 = 0$. While we normally avoid multiplying both sides of an equation by an expression containing an unknown, such multiplications are necessary with fractional equations. Consequently, extraneous roots can result.

If two resistors R_1 and R_2 (in ohms) are connected in parallel, then the combined resistance R_T (in ohms) is given by

$$\frac{1}{R_T} = \frac{1}{R_1} + \frac{1}{R_2} \tag{5.19}$$

(See Figure 5.2.)

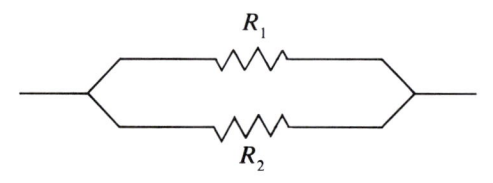

Figure 5.2

Example 5 The combined resistance of two resistors connected in parallel is measured to be 4.0 Ω. If one of the resistors has a resistance of 6.0 Ω, what is the resistance of the other?

Solution. By formula (5.19)

$$\frac{1}{R_1} + \frac{1}{6.0} = \frac{1}{4.0}$$

To clear fractions, we multiply both sides by $12R_1$:

$$12 + 2.0R_1 = 3.0R_1$$
$$12 = 3.0R_1 - 2.0R_1$$
$$R_1 = 12 \text{ Ω}$$

Check: Leaving out final zeros, we get

$$\frac{1}{12} + \frac{1}{6} = \frac{1}{12} + \frac{2}{12} = \frac{3}{12} = \frac{1}{4}$$

which equals the right side.

Exercises / Section 5.10

In Exercises 1–28, solve each equation for x.

1. $\dfrac{4}{x} + \dfrac{2}{3} = 2$

2. $1 - \dfrac{2}{x} = \dfrac{1}{2}$

3. $\dfrac{1}{x-2} = \dfrac{1}{2x}$

4. $\dfrac{x}{x-7} = 2$

5. $\dfrac{x-3}{x+4} = \dfrac{x-4}{x+2}$

6. $\dfrac{x-6}{x+4} = \dfrac{x-4}{x+2}$

7. $\dfrac{x-2}{x} = \dfrac{x-3}{x+1}$

8. $\dfrac{x+6}{x-5} = \dfrac{x}{x-2}$

9. $\dfrac{x-7}{x-4} = \dfrac{x-2}{x}$

10. $\dfrac{1}{x} = \dfrac{1}{x-2}$

11. $\dfrac{x-3}{x+2} = \dfrac{x-4}{x+1}$

12. $\dfrac{2x-3}{x+2} = \dfrac{2x-4}{x-3}$

13. $\dfrac{2x}{x-2} + \dfrac{1}{x+3} = 2$

14. $\dfrac{1}{x+1} + \dfrac{2x}{2-x} + 2 = 0$

15. $\dfrac{x+3}{x-5} - 2 = \dfrac{1}{x-5}$

16. $\dfrac{x-5}{x-6} - 2 = \dfrac{1}{x-6}$

17. $\dfrac{x+1}{x-8} = 4 + \dfrac{9}{x-8}$

18. $\dfrac{x}{x-4} - \dfrac{8}{x^2-4x} = 1$

19. $\dfrac{x}{x-1} - \dfrac{3}{2x^2 - 2x} = 1$

20. $\dfrac{x}{x-2} - \dfrac{x+1}{x^2-4} = 1$

21. $\dfrac{1}{x-1} - \dfrac{x+2}{x^2-1} = \dfrac{1}{x+1}$

22. $\dfrac{x+3}{x^2-4} - \dfrac{1}{x+2} = \dfrac{1}{x-2}$

23. $\dfrac{2}{x-3} - \dfrac{2}{x+3} = \dfrac{x+2}{x^2-9}$

24. $\dfrac{4}{x-2} - \dfrac{2}{x+2} = \dfrac{x+3}{x^2-4}$

25. $\dfrac{2}{x+3} - \dfrac{5}{x-2} = \dfrac{x+1}{x^2+x-6}$

26. $\dfrac{5}{x-1} - \dfrac{6}{x+3} = \dfrac{x+7}{x^2+2x-3}$

27. $\dfrac{4}{x-3} - \dfrac{5}{x+4} = \dfrac{x+13}{x^2+x-12}$

28. $\dfrac{3}{x+4} + \dfrac{x-8}{x^2+6x+8} = \dfrac{1}{x+2}$

In Exercises 29–35, solve each formula for the indicated letter.

29. $\dfrac{1}{c} + \dfrac{1}{d} = 3; \ d$

30. $\dfrac{c_1}{c_1 + c_2} = \dfrac{1}{q}; \ c_1$

31. $\dfrac{1}{R} = \dfrac{R_1 + 3}{3R_1}; \ R_1$

32. $\dfrac{1}{V_1} - \dfrac{2}{V_2} = \dfrac{a}{b}; \ V_2$

33. $\dfrac{p}{q} = \dfrac{f}{q-f}; \ f$

34. $\dfrac{1}{p} + \dfrac{1}{q} = \dfrac{2}{R}; \ p$

35. $\dfrac{1}{R_1} + \dfrac{1}{R_2} = \dfrac{1}{R}; \ R$

36. The combined capacitance C of two capacitors C_1 and C_2 connected in series (Figure 5.3) is given by

$$\dfrac{1}{C} = \dfrac{1}{C_1} + \dfrac{1}{C_2}$$

Solve for C.

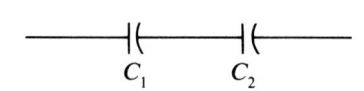

Figure 5.3

37. Two resistors are connected in parallel and have a combined resistance of 5.0 Ω. If one is a 10-Ω resistor, find the resistance of the other.

38. Two blood vessels are connected in parallel (Figure 5.4). If r_1 and r_2 denote the respective resistances of the two vessels to the flow of blood, then the combined resistance r is given by

$$\dfrac{1}{r} = \dfrac{1}{r_1} + \dfrac{1}{r_2}$$

Figure 5.4

If r and r_1 are measured to be 15.0 dynes and 25.0 dynes, respectively, find r_2.

Review Exercises / Chapter 5

In Exercises 1–12, find each product directly (an intermediate step may be needed in some cases).

1. $4s^2t^3(2s - 3st^2 + 5s^2t)$
2. $(F_1 + 4F_2)^2$
3. $(2i_1 - 3i_2)^2$
4. $(x - 3z)(x + 3z)$
5. $(2s - 5t)(2s + 5t)$
6. $(5w_1 - 11w_2)^2$
7. $(2c - 5d)(c + 3d)$
8. $(3a - 7y)(3a + y)$
9. $xy(x - 3y)(x + 3y)$
10. $6(a - 4b)(a - 2b)$
11. $(v + 2w + 1)^2$
12. $(a - 3b + 1)^2$

In Exercises 13–42, factor each expression.

13. $4ax - 4bx$
14. $8R^3C^2 + 12R^2C^3 - 4R^2C^2$
15. $3p^3q^4 + pq$
16. $A^2 - W^2$
17. $4\beta^2 - \gamma^2$
18. $2a^2 - 18b^2$
19. $16L^2 - 64C^2$
20. $b^4 - 1$
21. $27h^3 + g^3$
22. $8F^3 - 1$
23. $(a + 2b)^3 - 1$
24. $1 - (x - y)^3$
25. $a^6 - 1$
26. $x^2 + 2x - 8$
27. $v_0^2 - v_0 - 12$
28. $2a^2 + 7ab - 4b^2$
29. $3v_1^2 + 2v_1v_2 - v_2^2$
30. $P^3S^6 + 1$
31. $C_1^2 - 2C_1C_2 - 15C_2^2$
32. $4x^2 - 9xw + 2w^2$
33. $9y^2 - 3yz + z^2$
34. $2a^2 - 2ac - 24c^2$
35. $x^2 - 2xy + y^2 - 1$
36. $a^2 + 2ad + d^2 - 4$
37. $s^2 - st - 2t^2 + s + t$
38. $(x + b)^2 - (x + b) - 6$
39. $ax + bx - ay - by$
40. $sv - 2sw + 2tv - 4tw$
41. $4x^2 - a^2 + 4xy + y^2$
42. $1 - a^2 - 4b^2 + 4ab$

In Exercises 43–61, carry out the operations and simplify.

43. $\dfrac{x^2 - 9y^2}{x - 3y}$
44. $\dfrac{a^3 - 27c^3}{a^2 + 3ac + 9c^2}$
45. $\dfrac{q^2 + 6qr - 7r^2}{q^2 + 8qr - 9r^2}$
46. $\dfrac{a - 2t}{a - t} \cdot \dfrac{3a - 3t}{a^2 - 4t^2}$
47. $\dfrac{a^2 + 2ad + d^2}{a^3d^3} \div \dfrac{a + d}{2ad^2}$
48. $\dfrac{x^2 - 16w^2}{x^3 - 8w^3} \cdot \dfrac{x - 2w}{x - 4w}$
49. $\left(\dfrac{ax + ay - bx - by}{x - y} \cdot \dfrac{x^2 - y^2}{a - b}\right) \div (x + y)$
50. $\dfrac{4f^2 - 4fg + g^2 - 1}{3f - 6g} \div \dfrac{2f - g - 1}{2f - 4g}$
51. $\dfrac{y}{x + 2y} - \dfrac{2y^2}{x(x + 2y)} + \dfrac{x + y}{x}$

196 CHAPTER 5 FACTORING AND FRACTIONS

52. $\dfrac{a - 3c}{c} - \dfrac{a^2}{c(a - 3c)} + \dfrac{a}{a - 3c}$

53. $\dfrac{y + w}{y - w} - \dfrac{3yw}{y^2 - w^2} - \dfrac{y}{y + w}$

54. $\dfrac{2y}{x + y} - \dfrac{2xy}{x^2 - y^2} + \dfrac{y}{x - y}$

55. $\dfrac{L}{L - 1} - \dfrac{1}{L + 1} - \dfrac{2}{L^2 - 1}$

56. $\dfrac{2}{n^2 - m^2} - \dfrac{3}{n^2 + nm - 2m^2} + \dfrac{2}{n^2 + 3nm + 2m^2}$

57. $\dfrac{1}{a^2 - b^2} + \dfrac{1}{a^2 + 3ab + 2b^2} - \dfrac{1}{a^2 + ab - 2b^2}$

58. $\dfrac{x}{x - 3y} + \dfrac{y}{2x - y}$

59. $\dfrac{v - w}{v - 2w} - \dfrac{v}{v + w}$

60. $\dfrac{2}{\theta + 3} + \dfrac{1}{\theta - 3}$

61. $\dfrac{2}{\omega + 4} + \dfrac{3}{\omega - 2} - \dfrac{4\omega + 4}{(\omega + 4)(\omega - 2)}$

In Exercises 62–68, simplify each fraction.

62. $\dfrac{1 + \dfrac{2}{y} - \dfrac{3}{y^2}}{2 + \dfrac{3}{y} - \dfrac{5}{y^2}}$

63. $\dfrac{i + \dfrac{2i}{i - 4}}{i + \dfrac{4}{i - 4}}$

64. $\dfrac{v - \dfrac{8}{v - 2}}{v + \dfrac{3v - 20}{v - 2}}$

65. $\dfrac{\dfrac{r_2}{r_1 - 2r_2} + 1}{\dfrac{r_1 r_2}{r_1 - 2r_2} + r_1}$

66. $\dfrac{\dfrac{G}{G - 1} - \dfrac{6}{(G - 1)^2}}{1 - \dfrac{4}{(G - 1)^2}}$

67. $\dfrac{\dfrac{V}{T - V} + \dfrac{T}{T + V}}{\dfrac{T^2 + V^2}{T^2 - V^2}}$

68. $\dfrac{\dfrac{2}{A - 2C} - \dfrac{2A}{(A + C)(A - 2C)}}{\dfrac{4}{A + C}}$

In Exercises 69–77, solve each equation for x.

69. $\dfrac{1}{x + 3} = \dfrac{1}{2x}$

70. $\dfrac{2x - 2}{x - 3} = \dfrac{2x + 3}{x - 1}$

71. $\dfrac{x}{x - 4} + \dfrac{1}{x + 4} = 1$

72. $\dfrac{x - 2}{x + 5} - 2 = \dfrac{2}{x + 5}$

73. $\dfrac{3}{2(x + 3)} - \dfrac{x}{x + 3} + 1 = 0$

74. $\dfrac{x}{x - 3} - \dfrac{6}{x^2 - 3x} = 1$

75. $\dfrac{2}{x - 2} - \dfrac{x}{x^2 + x - 6} = \dfrac{3}{x + 3}$

76. $\dfrac{2}{x - 2} + \dfrac{x}{(2 - x)(x + 2)} = \dfrac{3}{x + 2}$

77. $\dfrac{x - 1}{(x - 3)(x + 3)} + \dfrac{1}{3 - x} = \dfrac{2}{x + 3}$

78. Solve for R:

$$\frac{1}{R} = \frac{1}{R_1} + \frac{1}{R_2} + \frac{1}{R_3}$$

79. Solve for a:

$$S = \frac{a - ar^n}{1 - r}$$

80. Solve for m_1:

$$w = \frac{m_1 m_2}{m_1 + m_2}$$

81. The current (in amperes) in a circuit at any time t (in seconds) is $2t^2 - 19t + 45$. Factor this expression and determine when the current is zero.

82. The combined resistance r of two blood vessels in parallel is given by

$$r = \frac{1}{\dfrac{1}{r_1} + \dfrac{1}{r_2}}$$

where r_1 and r_2 are the respective resistances of the individual vessels. Simplify the expression for r.

83. The focal length f of a lens is related to the object distance q and the image distance p by the formula

$$\frac{1}{f} = \frac{1}{p} + \frac{1}{q}$$

If $f = 3.0$ cm and $p = 4.0$ cm, find q.

84. If a body travels along the x-axis so that its position as a function of time is $x(t) = t(t^2 - 4)^4$, the velocity v is given by $v = (t^2 - 4)^4 + 8t^2(t^2 - 4)^3$. Simplify the expression for v.

85. If a mass m_1 strikes a stationary mass m_2, the ratio of the kinetic energy of the system with respect to the floor to the kinetic energy with respect to the center of mass of the system is

$$\frac{T - \dfrac{1}{2}(m_1 + m_2)V^2}{T}$$

where $T = \tfrac{1}{2}m_1 v^2$ and $V = m_1 v/(m_1 + m_2)$. Simplify this expression.

86. If two thin lenses having respective focal lengths f_1 and f_2 are placed in contact, the focal length of the combination is given by

$$\frac{1}{f} = \frac{1}{f_1} + \frac{1}{f_2}$$

Solve this equation for f.

CHAPTER 6

Quadratic Equations

Objectives Upon completion of this chapter, you should be able to:

1. Solve a given quadratic equation by:
   a. Factoring.
   b. Completing the square.
   c. Using the quadratic formula.
2. Solve stated problems leading to quadratic equations.

6.1 Solution by Factoring and Pure Quadratic Equations

All the equations introduced in the earlier chapters were of first degree (involving only the first power of the unknown). Many equations arising in technical problems are of second or higher degree. For example, if an object is tossed upward from the ground with initial velocity v_0, then the distance above the ground as a function of time is given by $s = v_0 t - 16t^2$, where s is measured in feet and t in seconds. This is an example of an equation of second degree. The equation $x^2 - 3x + 4 = 0$ is also of second degree, while $x^3 - 2x^2 + 6x - 1 = 0$ is of third degree. In this chapter we shall study only **quadratic equations,** which are equations of second degree.

> A **quadratic** equation has the form
> $$ax^2 + bx + c = 0, \quad a \neq 0 \tag{6.1}$$

The equation $ax^2 + bx + c = 0$ is called the **standard form** of the quadratic equation.

6.1 SOLUTION BY FACTORING AND PURE QUADRATIC EQUATIONS

In this section we shall confine ourselves to those cases in which the left side of equation (6.1) is factorable. The method of solution depends on the following property of real numbers:

> $ab = 0$ if, and only if, $a = 0$ or $b = 0$.

Consider, for example, the equation

$$x^2 = 8 - 2x$$

For the equation to fit the standard form (6.1), all the terms have to be collected on the left side and written in descending powers of x. Adding $-8 + 2x$ to both sides, we get

Step 1. $\quad x^2 + 2x - 8 = 0 \qquad\qquad (6.2)$

If we now factor the left side, the equation becomes

Step 2. $\quad (x + 4)(x - 2) = 0 \qquad\qquad (6.3)$

Next, we set each factor equal to 0:

Step 3. $\quad \boldsymbol{x + 4 = 0} \quad\quad \boldsymbol{x - 2 = 0}$

Finally, we solve each of the resulting linear equations:

Step 4. $\quad x = -4 \quad\quad x = 2$

The solution is therefore given by two values, $x = -4$ and $x = 2$.

As a check, let us substitute both values into the original equation $x^2 = 8 - 2x$:

$$(-\mathbf{4})^2 = 8 - 2(-\mathbf{4}) \qquad \mathbf{2}^2 = 8 - 2(\mathbf{2})$$
$$16 = 8 + 8 \qquad\qquad 4 = 8 - 4$$
$$16 = 16 \qquad\qquad 4 = 4$$

Since both values check, we see that the equation has two distinct roots.

Finally, note that the roots are unique:

$$(x + 4)(x - 2) = 0$$

if, and only if, $x + 4 = 0$ or $x - 2 = 0$. But $x + 4 = 0$ if, and only if, $x = -4$; and $x - 2 = 0$ if, and only if, $x = 2$.

To illustrate the solution of the equation $x^2 + 2x - 8 = 0$ in (6.2) graphically, consider the function

$$y = x^2 + 2x - 8$$

whose graph appears in Figure 6.1. The solution of $x^2 + 2x - 8 = 0$ consists of the x-intercepts (where $y = 0$).

CHAPTER 6 QUADRATIC EQUATIONS

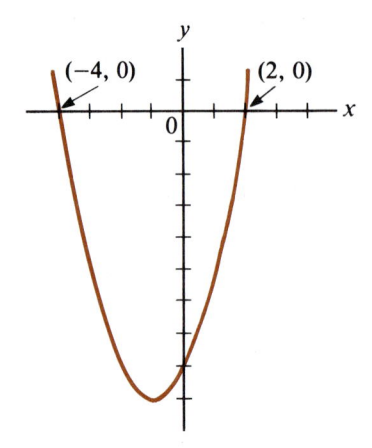

Figure 6.1

Solution by factoring:

1. Write the equation in standard form by collecting all the terms on the left side.
2. Factor the expression on the left side.
3. Set each of the factors equal to zero.
4. Solve the resulting two linear equations.

Example 1 Solve the equation $2x^2 + 15 = 13x$.

Solution.

$2x^2 + 15 = 13x$	given equation
$2x^2 - 13x + 15 = 0$	collecting terms on left side
$(2x - 3)(x - 5) = 0$	factoring left side
$2x - 3 = 0 \quad x - 5 = 0$	setting factors equal to 0
$2x = 3 \quad\quad\quad x = 5$	solving the resulting linear equations
$x = \dfrac{3}{2} \quad\quad\quad x = 5$	

Check:

Left Side

Right Side

$x = \dfrac{3}{2}: \quad 2\left(\dfrac{3}{2}\right)^2 + 15 = \dfrac{39}{2} \quad\quad 13\left(\dfrac{3}{2}\right) = \dfrac{39}{2}$

$x = 5: \quad 2(5)^2 + 15 = 65 \quad\quad\quad 13(5) = 65$

Example 2 Solve the equation $x^2 + 7x = 0$.

Solution. In this equation $c = 0$. As a result, the terms on the left have a common factor x.

$x^2 + 7x = 0$ given equation
$x(x + 7) = 0$ common factor x
$x = 0$ $x + 7 = 0$ setting each factor equal to zero
$x = 0$ $x = -7$

(See Figure 6.2.) The solution can be checked as in Example 1.

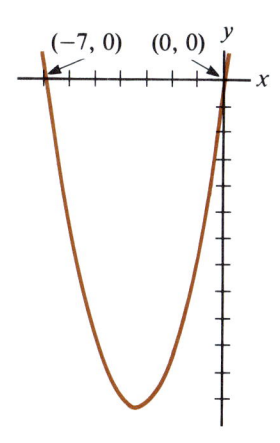

Figure 6.2

Pure quadratic equations

An equation for which $b = 0$ is called a **pure quadratic equation** and may be solved by a different method. For example, to solve the equation

$$x^2 - 4 = 0$$

we first solve for x^2 to obtain

$$x^2 = 4$$

Taking the square root of both sides, we find that $x = \pm\sqrt{4} = \pm 2$, where $\pm$ means *plus or minus*. So the equation has two roots, $x = 2$ and $x = -2$.

Alternatively,

$$x^2 - 4 = (x - 2)(x + 2) = 0$$

leads to the same solution. However, factoring does not work for the equation in the next example.

Example 3 Solve the equation $x^2 - 5 = 0$.

Solution. Adding 5 to both sides of the equation, we get

$$x^2 - 5 + 5 = 0 + 5$$
$$x^2 = 5$$
$$x = \pm\sqrt{5}$$

Remark: The left side of the equation $x^2 + 4x + 4 = 0$ is a perfect-square trinomial leading to identical factors:

$$(x + 2)(x + 2) = 0$$

Double root

Common errors

The solution $x = -2$ and $x = -2$ is called a **repeating root** or a **double root.**

(1) Forgetting the negative root when solving $x^2 - a^2 = 0$. The roots are $x = \pm a$.

(2) Attempting to solve by factoring when the right side is not zero. For example, if

$$(x - 3)(x + 2) = 6$$

we may *not* conclude that $x - 3 = 6$ and $x + 2 = 6$. Instead, we need to write the equation in the form $ax^2 + bx + c = 0$:

$$x^2 - x - 6 = 6$$
$$x^2 - x - 12 = 0$$
$$(x - 4)(x + 3) = 0$$
$$x = 4, -3$$

Historical note. As mentioned in Chapter 1, the first systematic study of second-degree equations was undertaken by al-Khowarizmi in Baghdad. He gave an exhaustive exposition of various cases using ingenious geometric arguments, in the manner of the ancient Greeks. Consequently, al-Khowarizmi's algebra was rhetorical, using words and drawings instead of symbols. Further progress in algebra was slow until algebraic notation was introduced. This far-reaching innovation was due to the French lawyer Franciscus Vieta (1540–1603), who recognized the advantage of using letters to denote both known and unknown quantities.

Other now-familiar symbols were introduced only gradually. In medieval times the letters *p* and *m* were widely used to denote addition and subtraction, and the Latin word *cosa* for the unknown. The signs + and − first appeared in print in a book published in 1489 by Johann Widman, a lecturer in Leipzig. The English mathematician William Oughtred (1574–1660) popularized the symbol × for multiplication, and Johan Rahn of Switzerland first used the sign ÷ for division in 1659. The French philosopher

Franciscus Vieta

René Descartes (1596–1650) introduced the exponential notation, and the Englishman Robert Recorde (1510–1558) used the symbol = for equality because, as he put it, "noe 2. thynges can be moare equalle."

Exercises / Section 6.1

In Exercises 1–12, solve the given pure quadratic equations.

1. $x^2 - 1 = 0$
2. $x^2 - 25 = 0$
3. $x^2 - 36 = 0$
4. $x^2 - 121 = 0$
5. $x^2 - 9 = 0$
6. $x^2 - 16 = 0$
7. $x^2 - 10 = 0$
8. $x^2 - 12 = 0$
9. $2x^2 - 32 = 0$
10. $4x^2 - 1 = 0$
11. $36x^2 - 25 = 0$
12. $49x^2 - 16 = 0$

In Exercises 13–44, solve the given quadratic equations by the method of factoring.

13. $x^2 + x - 2 = 0$
14. $x^2 + 7x + 10 = 0$
15. $x^2 + 2x - 24 = 0$
16. $x^2 + 4x - 21 = 0$
17. $2x^2 - 5x - 3 = 0$
18. $2x^2 - 7x - 15 = 0$
19. $3x^2 + 7x + 2 = 0$
20. $6x^2 + 11x + 4 = 0$
21. $4x^2 + 5x - 6 = 0$
22. $5x^2 + 16x + 12 = 0$
23. $5x^2 + 8x - 21 = 0$
24. $4x^2 + 29x - 24 = 0$
25. $4x^2 + 4x = 15$
26. $6x^2 + 7x = 5$
27. $30x^2 = 7x + 15$
28. $12x^2 = 7x + 10$
29. $8x^2 = 2x + 45$
30. $40x^2 = 67x - 28$
31. $7x^2 + 4x = 11$
32. $14x^2 + 53x + 45 = 0$
33. $18x^2 + 17x = 15$
34. $3x^2 = 10x + 13$
35. $11x^2 = 76x + 7$
36. $4x^2 = 49x - 90$
37. $72x^2 + 13x = 15$
38. $45x^2 + 52x + 15 = 0$
39. $18x^2 = 93x - 110$
40. $21x^2 + 10x = 91$
41. $9x^2 + 24x + 16 = 0$
42. $25x^2 - 10x + 1 = 0$
43. $16x^2 - 8x + 1 = 0$
44. $4x^2 - 20x + 25 = 0$

45. The path of an object tossed at an angle of 45° to the ground is $y = x - 32x^2/v_0^2$, where v_0 is the initial velocity in feet per second and y is the distance in feet above the ground. How far from the starting point ($x = 0$) will the object land?

46. The weekly profit P of a company is $P = x^4 - 30x^3$, $x \geq 1$, where x is the week in the year. During what week is the profit equal to zero? (*Hint:* Factor out x^3.)

47. The load on a beam of length L is such that the deflection is given by $d = 3x^4 - 4Lx^3 + L^2x^2$, where x is the distance from one end. Determine where the deflection is zero. (*Hint:* Factor out x^2.)

48. The formula for the output of a battery is $P = VI - RI^2$. For what values of I is the output equal to zero?

6.2 Solution by Completing the Square

In the last section we restricted our attention to factorable quadratic equations. In this section we shall take up a general method for solving any given quadratic equation. The procedure is referred to as **completing the square.**

Completing the square depends on the fact that any quadratic equation can be written in the form

$$(x + b)^2 = a \tag{6.4}$$

To understand this, recall that the square of a binomial is given by

$$(x + b)^2 = x^2 + 2bx + b^2$$

Perfect-square trinomial Looking at the right side, observe that *a trinomial in x with a coefficient of x^2 equal to 1 is a perfect square if the square of one-half the coefficient of x is equal to the third term.* For example,

$$x^2 + 6x + 9$$

is a perfect square, since $(\frac{1}{2} \cdot 6)^2 = 9$. Similarly,

$$x^2 - \frac{3}{2}x + \frac{9}{16}$$

is a perfect square, since

$$\frac{9}{16} = \left[\frac{1}{2} \cdot \left(-\frac{3}{2}\right)\right]^2$$

so that

$$x^2 - \frac{3}{2}x + \frac{9}{16} = \left(x - \frac{3}{4}\right)^2 \qquad \left(x - \frac{3}{4}\right)^2 = x^2 + 2\left(-\frac{3}{4}x\right) + \left(-\frac{3}{4}\right)^2$$

> The method of completing the square consists of rewriting one side of the equation so that it forms a perfect square.

Consider the next example.

Example 1 Solve the equation $x^2 - 6x + 8 = 0$ by completing the square.

Solution. The first step is to transpose the 8 (or add -8 to both sides) in order to retain only the x^2 and x terms on the left side. Thus

$$x^2 - 6x = -8$$

The critical step is to complete the square on the left side by adding to both sides the square of one-half the coefficient of x, or $[\frac{1}{2} \cdot (-6)]^2 = 9$. We then get

$$x^2 - 6x + 9 = -8 + 9$$

The left side is now a perfect square, so that the equation can be written

$$(x - 3)^2 = 1 \qquad (x - 3)^2 = x^2 + 2(-3x) + (-3)^2$$

The resulting pure quadratic form can be solved by taking the square root of both sides, yielding the two linear equations

$$\sqrt{(x - 3)^2} = \pm\sqrt{1}$$
$$x - 3 = \pm 1$$

Solving, we get

$$x = 3 \pm 1$$

which gives $x = 4$ and $x = 2$.

Check:

$x = 4$: $(4)^2 - 6(4) + 8 = 0$
$x = 2$: $(2)^2 - 6(2) + 8 = 0$

Let us now summarize the procedure for completing the square.

> **Solution by completing the square:**
> 1. Write the equation in the form $ax^2 + bx = -c$.
> 2. Multiply each side by $1/a$.
> 3. Complete the square on the left side by adding the square of one-half the coefficient of x to both sides.
> 4. Write the left side as a square; simplify the right side.
> 5. Take the square root of both sides.
> 6. Solve the resulting two linear equations.

Example 2 Solve the equation $2x^2 + 6x - 3 = 0$ by completing the square.

Solution. Following the procedure for completing the square, we get

$$2x^2 + 6x - 3 = 0 \quad \text{given equation}$$
Step 1. $\quad 2x^2 + 6x = 3 \quad \text{transposing } -3$
Step 2. $\quad x^2 + 3x = \dfrac{3}{2} \quad \text{dividing by 2}$

Note that the square of one-half the coefficient of x is

$$\left(\dfrac{3}{2}\right)^2 = \dfrac{9}{4}$$

This number has to be added to both sides to complete the square. It follows that

Step 3. $\quad x^2 + 3x + \dfrac{9}{4} = \dfrac{3}{2} + \dfrac{9}{4}$

and

Step 4. $\quad \left(x + \dfrac{3}{2}\right)^2 = \dfrac{6}{4} + \dfrac{9}{4} = \dfrac{15}{4} \quad \begin{array}{l}\text{factoring the left side and} \\ \text{simplifying the right side}\end{array}$

Taking the square root of both sides, we get

Step 5. $\quad x + \dfrac{3}{2} = \pm\sqrt{\dfrac{15}{4}} = \pm\dfrac{\sqrt{15}}{2}$

Solving for x,

Step 6. $\quad x = -\dfrac{3}{2} \pm \dfrac{\sqrt{15}}{2}$

or

$$x = \dfrac{-3 \pm \sqrt{15}}{2}$$

The roots can also be written separately as

$$x = \dfrac{-3 + \sqrt{15}}{2} \quad \text{and} \quad x = \dfrac{-3 - \sqrt{15}}{2}$$

(Although highly desirable, checking the solution is difficult at this point, since we do not discuss the multiplication of radical expressions of this complexity until Chapter 10.)

Complex Roots

In Chapter 1 we mentioned that numbers fall into two categories, real and complex. Complex numbers will be studied in detail in Chapter 11. In this chapter we need only to understand the basic concepts and notations.

Consider the pure quadratic equation

$$x^2 + 4 = 0$$

Solving for x, we get $x = \pm\sqrt{-4}$. Since we cannot find a (real) number whose square root is -4, $\sqrt{-4}$ is called a **pure imaginary number**.

For the past two hundred years the letter i (for "imaginary") has been used to denote the imaginary number $\sqrt{-1}$. When imaginary numbers were first introduced in electrical circuit theory, the convention of using i for instantaneous current had already become well established. Consequently, using the letter j to denote $\sqrt{-1}$ became standard in physics and technology, and we shall observe this convention here.

Returning now to $\sqrt{-4}$, observe that

$$\sqrt{-4} = \sqrt{(4)(-1)} = \sqrt{4}\,\sqrt{-1} = 2\sqrt{-1} = 2j,$$

Thus $\sqrt{-4} = 2j$, where $j = \sqrt{-1}$. Imaginary numbers are always written in this form.

> **Basic imaginary unit:**
>
> $\quad j = \sqrt{-1} \quad \text{or} \quad j^2 = -1$
>
> $\quad \sqrt{-a}, \quad a > 0, \quad \text{is written } \sqrt{a}\,j$

6.2 SOLUTION BY COMPLETING THE SQUARE **207**

If a and b are real numbers, then $a + bj$ is called a **complex number**; a is called the **real part** of the complex number, and b is called the **imaginary part**. Thus a pure imaginary number is a complex number whose real part is zero; a real number is a complex number whose imaginary part is zero.

> **Complex number:** A **complex number** has the form $a + bj$, where a and b are real numbers and $j = \sqrt{-1}$.

Some quadratic equations lead to complex roots, as shown in the remaining examples.

Example 3 Solve the equation $x^2 + 4x + 16 = 0$ by completing the square.

Solution.

$x^2 + 4x + 16 = 0$	given equation
$x^2 + 4x = -16$	transposing
$x^2 + 4x + 4 = -16 + 4$	adding $\left(\frac{1}{2} \cdot 4\right)^2$ to both sides
$(x + 2)^2 = -12$	factoring the left side
$x + 2 = \pm\sqrt{-12}$	taking the square root of each side

Recall that $\sqrt{-12} = \sqrt{(-1) \cdot 4 \cdot 3} = 2\sqrt{3}\sqrt{-1} = 2\sqrt{3}j$. It follows that $x = -2 \pm 2\sqrt{3}j$.

Example 4 Solve the equation $2x^2 - 3x + 4 = 0$ by completing the square.

Solution.

$2x^2 - 3x + 4 = 0$	given equation
$x^2 - \frac{3}{2}x + 2 = 0$	dividing by 2
$x^2 - \frac{3}{2}x = -2$	transposing the 2
$x^2 - \frac{3}{2}x + \frac{9}{16} = -2 + \frac{9}{16}$	adding $\left[\frac{1}{2} \cdot \left(-\frac{3}{2}\right)\right]^2$ to both sides
$\left(x - \frac{3}{4}\right)^2 = -\frac{32}{16} + \frac{9}{16} = -\frac{23}{16}$	factoring the left side
$x - \frac{3}{4} = \pm\sqrt{-\frac{23}{16}} = \pm\frac{\sqrt{-23}}{4} = \pm\frac{\sqrt{23}\,j}{4}$	taking the square root of each side
$x = \frac{3}{4} \pm \frac{\sqrt{23}}{4}j$	

Remark. We shall see in the next section that a nonfactorable quadratic equation can be solved directly by a formula. Once you learn this formula, you may feel that completing the square is a waste of time. However, *completing the square is an algebraic technique that arises in contexts other than solving equations*. In fact, solving quadratic equations is merely a convenient way to introduce this technique. It is therefore very important for you to practice solving equations by completing the square in the next exercise set.

Exercises / Section 6.2

Solve each equation by completing the square.

1. $x^2 - 6x + 8 = 0$
2. $x^2 - 6x + 5 = 0$
3. $x^2 + 4x - 12 = 0$
4. $x^2 + 2x - 15 = 0$
5. $x^2 + x - 12 = 0$
6. $x^2 + 3x - 28 = 0$
7. $x^2 + 7x + 10 = 0$
8. $x^2 - 9x + 20 = 0$
9. $x^2 + 5x + 2 = 0$
10. $x^2 - 4x - 6 = 0$
11. $x^2 + 6x + 6 = 0$
12. $x^2 + 3x + 1 = 0$
13. $2x^2 - 6x + 1 = 0$
14. $2x^2 + 5x + 2 = 0$
15. $2x^2 + 3x - 3 = 0$
16. $2x^2 - 3x - 5 = 0$
17. $3x^2 + 2x - 1 = 0$
18. $3x^2 - 2x - 3 = 0$
19. $3x^2 - 4x - 5 = 0$
20. $2x^2 + 5x - 2 = 0$
21. $4x^2 - x - 3 = 0$
22. $5x^2 - 2x - 1 = 0$
23. $6x^2 + x + 2 = 0$
24. $5x^2 + 9x + 1 = 0$
25. $x^2 - 4x + 5 = 0$
26. $3x^2 - 2x + 1 = 0$
27. $4x^2 - 5x + 3 = 0$
28. $2x^2 + 3x + 2 = 0$
29. $7x^2 + 2x - 1 = 0$
30. $8x^2 + 3x + 1 = 0$
31. $6x^2 - 5x - 2 = 0$
32. $7x^2 - 19x - 6 = 0$
33. $6x^2 + 5x - 50 = 0$
34. $8x^2 - 7x + 2 = 0$
35. $5x^2 - x + 1 = 0$
36. $5x^2 + 2x - 3 = 0$
37. $x^2 - bx + 2 = 0$
38. $x^2 - x + c = 0$
39. $ax^2 + 5x - 1 = 0$
40. $x^2 - 3bx + 5 = 0$

6.3 The Quadratic Formula

Completing the square can be used to obtain a general formula for solving any quadratic equation. We start with the standard form:

$$ax^2 + bx + c = 0 \qquad (6.5)$$

$$ax^2 + bx = -c \qquad \text{transposing}$$

$$x^2 + \frac{b}{a}x = -\frac{c}{a} \qquad \text{dividing by } a$$

$$x^2 + \frac{b}{a}x + \left(\frac{b}{2a}\right)^2 = -\frac{c}{a} + \left(\frac{b}{2a}\right)^2 \qquad \text{adding } \left(\frac{1}{2} \cdot \frac{b}{a}\right)^2 \text{ to each side}$$

$$\left(x + \frac{b}{2a}\right)^2 = -\frac{c}{a} + \frac{b^2}{4a^2}$$ factoring the left side and simplifying the right side

$$= -\frac{4ac}{4a^2} + \frac{b^2}{4a^2}$$

$$= \frac{b^2 - 4ac}{4a^2}$$

$$x + \frac{b}{2a} = \pm\sqrt{\frac{b^2 - 4ac}{4a^2}}$$ taking the square root of each side

$$= \pm\frac{\sqrt{b^2 - 4ac}}{2a}$$

$$x = -\frac{b}{2a} \pm \frac{\sqrt{b^2 - 4ac}}{2a}$$

$$x = \frac{-b \pm \sqrt{b^2 - 4ac}}{2a} \tag{6.6}$$

Formula (6.6), known as the **quadratic formula**, should be carefully memorized.

Quadratic formula: The roots of the quadratic equation

$$ax^2 + bx + c = 0, \quad a \neq 0$$

are given by

$$x = \frac{-b \pm \sqrt{b^2 - 4ac}}{2a}$$

By using the quadratic formula, the solutions of a quadratic equation can be written directly, but they usually have to be simplified. The first three examples illustrate the technique.

Example 1 Solve the equation $6x^2 = 2x + 1$ by means of the quadratic formula.

Solution. The equation is first written in the standard form

$$6x^2 - 2x - 1 = 0$$

By equation (6.5), $a = 6$, $b = -2$, and $c = -1$. So by the quadratic formula (6.6), the solution is

$$x = \frac{-(-2) \pm \sqrt{(-2)^2 - 4(6)(-1)}}{2 \cdot 6}$$

$$= \frac{2 \pm \sqrt{28}}{12}$$

To simplify the radical, note that $\sqrt{28} = \sqrt{4 \cdot 7} = 2\sqrt{7}$. Thus

$$x = \frac{2 \pm 2\sqrt{7}}{12}$$

$$= \frac{2(1 \pm \sqrt{7})}{12}$$

$$x = \frac{1 \pm \sqrt{7}}{6}$$

Example 2 Solve the equation $5x^2 + 2x + 4 = 0$ by the quadratic formula.

Solution. From the standard form (6.5), we see that $a = 5$, $b = 2$, and $c = 4$. So by the quadratic formula

$$x = \frac{-2 \pm \sqrt{2^2 - 4(5)(4)}}{2 \cdot 5}$$

$$= \frac{-2 \pm \sqrt{4 - 80}}{10}$$

$$= \frac{-2 \pm \sqrt{-76}}{10}$$

Since $\sqrt{-76} = \sqrt{(-1)(4)(19)} = 2\sqrt{19}j$, we get

$$x = \frac{-2 \pm 2\sqrt{19}j}{10}$$

$$= \frac{-1 \pm \sqrt{19}j}{5}$$

Example 3 Solve the equation $4x^2 - 12x + 9 = 0$.

Solution. Since $a = 4$, $b = -12$, and $c = 9$, we get

$$x = \frac{-(-12) \pm \sqrt{(-12)^2 - 4(4)(9)}}{2 \cdot 4}$$

$$= \frac{12 \pm \sqrt{144 - 144}}{8}$$

$$= \frac{12 \pm 0}{8}$$

$$= \frac{3 \pm 0}{2}$$

Hence $x = \frac{3}{2}, \frac{3}{2}$. (Whenever the radical is 0, we get a double root.)

These examples show that the radical in the quadratic formula
$$x = \frac{-b \pm \sqrt{b^2 - 4ac}}{2a}$$
determines whether the roots are real or complex. The expression $b^2 - 4ac$ under the radical sign is called the **discriminant.** We have seen that a given equation has two distinct real roots if $b^2 - 4ac > 0$ (Example 1) and complex roots if $b^2 - 4ac < 0$ (Example 2). If $b^2 - 4ac = 0$, then the equation has a double root (Example 3).

Example 4

CALCULATOR COMMENT

Use a calculator to solve the equation
$$3.17x^2 - 1.98x - 6.83 = 0$$

Solution. Since $a = 3.17$, $b = -1.98$, and $c = -6.83$, we get
$$x = \frac{1.98 \pm \sqrt{(1.98)^2 - 4(3.17)(-6.83)}}{2(3.17)}$$

The simplest way to carry out this calculation is to find the value of the radical and store it in the memory. Now add 1.98 to the positive value of the radical and divide the sum by $(2 \times 3.17) = 6.34$ to get
$$x = 1.81$$
to two decimal places. Next, transfer the content of the memory to the register, change the sign to minus, and proceed as before. The second root is
$$x = -1.19$$
again to two decimal places.

The sequences are

1.98 $\boxed{x^2}$ $\boxed{-}$ 4 $\boxed{\times}$ 3.17 $\boxed{\times}$ 6.83 $\boxed{+/-}$ $\boxed{=}$
$\boxed{\sqrt{}}$ $\boxed{STO}$ $\boxed{+}$ 1.98 $\boxed{=}$ $\boxed{\div}$ 2 $\boxed{\div}$ 3.17 $\boxed{=}$

Display: 1.8130051

$\boxed{MR}$ $\boxed{+/-}$ $\boxed{+}$ 1.98 $\boxed{=}$ $\boxed{\div}$ 2 $\boxed{\div}$ 3.17 $\boxed{=}$

Display: -1.1883994

The quadratic formula can also be used to solve equations containing two different variables, say x and y. If the equation is to be solved for x in terms of y, then y has to be treated as if it were just another constant. Conversely, to solve for y, x is treated as a constant.

Example 5 Solve the equation
$$x^2 - 3x + xy + 2 - 3y - 2y^2 = 0$$
for x in terms of y.

Solution. We first write the equation in standard form. Noting the common factor x, we get

$$x^2 + (-3 + y)x + (2 - 3y - 2y^2) = 0$$

From this equation we see that

$$a = 1, \quad b = -3 + y, \quad \text{and} \quad c = 2 - 3y - 2y^2$$

By the quadratic formula

$$x = \frac{-(-3 + y) \pm \sqrt{(-3 + y)^2 - 4(2 - 3y - 2y^2)}}{2}$$

$$= \frac{3 - y \pm \sqrt{9 - 6y + y^2 - 8 + 12y + 8y^2}}{2}$$

$$= \frac{3 - y \pm \sqrt{9y^2 + 6y + 1}}{2}$$

The radical can be simplified by noting that

$$9y^2 + 6y + 1 = (3y + 1)^2$$

and

$$\sqrt{(3y + 1)^2} = 3y + 1$$

It follows that

$$x = \frac{3 - y \pm (3y + 1)}{2}$$

Thus

$$x = \frac{3 - y + 3y + 1}{2} \quad \text{and} \quad x = \frac{3 - y - 3y - 1}{2}$$

These fractions simplify to

$$x = 2 + y \quad \text{and} \quad x = 1 - 2y$$

Fractional equations may also lead to quadratic equations, as illustrated in the next example.

Example 6 Solve the equation

$$\frac{1}{x} - \frac{1}{x + 1} = \frac{1}{20}$$

Solution. Recall that the simplest way to solve a fractional equation is to clear the fractions by multiplying both sides by the LCD—in this case

$$20x(x+1). \text{ Then}$$
$$20x(x+1)\left(\frac{1}{x} - \frac{1}{x+1}\right) = 20x(x+1)\frac{1}{20}$$
$$20(x+1) - 20x = x(x+1)$$
$$20x + 20 - 20x = x^2 + x$$
$$20 = x^2 + x$$
$$0 = x^2 + x - 20$$
$$(x+5)(x-4) = 0$$
$$x = 4, -5$$

Exercises / Section 6.3

In Exercises 1–34, solve each equation by the quadratic formula.

1. $x^2 + x - 6 = 0$
2. $x^2 - 3x - 4 = 0$
3. $x^2 - 9x + 20 = 0$
4. $x^2 + 8x + 15 = 0$
5. $2x^2 = 5x - 2$
6. $3x^2 = 13x + 10$
7. $6x^2 - x = 2$
8. $5x^2 - 7x = 6$
9. $2x^2 = 3x + 1$
10. $2x^2 + 2x = 1$
11. $3x^2 + 2x = 2$
12. $x^2 = 3x + 2$
13. $x^2 - 2x + 2 = 0$
14. $x^2 + 3x + 3 = 0$
15. $x^2 + 2x + 4 = 0$
16. $x^2 + 3x + 5 = 0$
17. $3x^2 + 3x + 1 = 0$
18. $3x^2 + 5x + 2 = 0$
19. $4x^2 + 2x + 1 = 0$
20. $4x^2 + 3x - 2 = 0$
21. $4x^2 + 5x = 3$
22. $5x^2 + 3x = 3$
23. $5x^2 + 1 = 0$
24. $3x^2 + 4 = 0$
25. $2x^2 + 3x = 0$
26. $3x^2 - 5x = 0$
27. $2x^2 - 3cx + 1 = 0$
28. $3x^2 + 3x - 2a = 0$
29. $bx^2 + 3x + 1 = 0$
30. $cx^2 + bx - 4 = 0$
31. $4x^2 - 12x + 9 = 0$
32. $9x^2 + 12x + 4 = 0$
33. $4x^2 - 20x + 25 = 0$
34. $9x^2 + 42x + 49 = 0$

In Exercises 35–40, solve the given equations using a calculator. (Find the roots to two decimal places.)

35. $2.00x^2 + 3.12x - 3.19 = 0$
36. $4.12x^2 - 1.30x - 12.1 = 0$
37. $1.79x^2 - 10.0x - 1.91 = 0$
38. $7.179x^2 + 2.862x - 1.998 = 0$
39. $10.103x^2 - 1.701x - 3.28 = 0$
40. $1.738x^2 - 10.162x - 11.773 = 0$

In Exercises 41–48, solve the given equations for x in terms of y.

41. $x^2 - 2x + 1 - y^2 = 0$
42. $x^2 - 2xy + y^2 - 9 = 0$

43. $x^2 - 4xy + 4y^2 - 1 = 0$
44. $x^2 + 4xy + 4y^2 - 9 = 0$
45. $x^2 - 2xy + 3x + y^2 - 3y + 2 = 0$
46. $x^2 - 2xy + 2x + y^2 - 2y - 3 = 0$
47. $x^2 - 3x + 2 - y - y^2 = 0$
48. $x^2 - 3x - xy + 2 + 3y - 2y^2 = 0$

In Exercises 49–56, solve each equation for x.

49. $\dfrac{1}{x} - \dfrac{1}{x+1} = \dfrac{1}{20}$

50. $\dfrac{1}{x} - \dfrac{1}{x+2} = \dfrac{1}{4}$

51. $\dfrac{1}{x+2} + \dfrac{1}{x} = \dfrac{5}{12}$

52. $\dfrac{1}{x} + \dfrac{1}{x+8} = \dfrac{1}{3}$

53. $\dfrac{1}{x} + \dfrac{1}{x-4} = \dfrac{3}{8}$

54. $\dfrac{1}{x} - \dfrac{1}{x+3} = 1$

55. $\dfrac{1}{x} + \dfrac{2}{x-4} = 1$

56. $\dfrac{2}{x} - \dfrac{3}{x+1} = 2$

6.4 Applications of Quadratic Equations

Many physical problems lead quite naturally to quadratic equations. One such case was already mentioned at the beginning of the chapter. Now we will consider a similar example.

Example 1 A rock is hurled upward at the rate of 24 m/sec from a height of 5 m. The distance s (in meters) above the ground as a function of time t (in seconds) is given by $s = -5t^2 + 24t + 5$. (The instant at which the rock is hurled upward corresponds to $t = 0$ sec.) When will the rock strike the ground?

Solution. Since s is the distance above the ground, the problem is to find the value of t for which $s = 0$. Thus

$$-5t^2 + 24t + 5 = 0$$
$$5t^2 - 24t - 5 = 0$$
$$(5t + 1)(t - 5) = 0$$
$$t = -\dfrac{1}{5}, 5$$

Since $t = 0$ corresponds to the instant when the motion begins, the root $t = -\frac{1}{5}$ has no meaning here. We conclude that the rock hits the ground in 5 sec.

Example 2 If the length of a square is increased by 6.0 in., the area becomes 4 times as large. Find the original length of the side.

6.4 APPLICATIONS OF QUADRATIC EQUATIONS

Solution. Let x be the length of the original side. Then $x + 6.0$ in. is the length of the side when increased. Since the new area is equal to 4 times the old area, we get (in square inches, omitting final zeros)

$$(x + 6)^2 = 4x^2$$
$$x^2 + 12x + 36 = 4x^2 \quad \text{expanding the left side}$$
$$-3x^2 + 12x + 36 = 0 \quad \text{subtracting } 4x^2$$
$$x^2 - 4x - 12 = 0 \quad \text{dividing by } -3$$
$$(x - 6)(x + 2) = 0 \quad \text{factoring the left side}$$
$$x = 6, -2$$

Since the root $x = -2$ has no meaning here, the original side is 6.0 in. long (using two significant figures).

Example 3 Two resistors connected in parallel have a combined resistance of 4 Ω, and the resistance of one resistor is 6 Ω more than that of the other. Find the resistance of each.

Solution. Let

R = resistance of first resistor

Then

$R + 6$ = resistance of second resistor

Since the resistors are connected in parallel (Figure 6.3), we have

$$\frac{1}{R} + \frac{1}{R + 6} = \frac{1}{4}$$

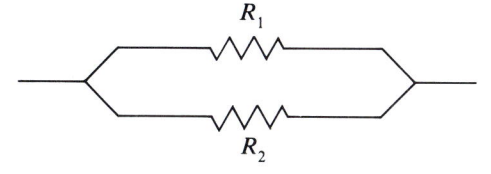

If R_T is the combined resistance, then

$$\frac{1}{R_T} = \frac{1}{R_1} + \frac{1}{R_2}$$

Figure 6.3

This is an example of a fractional equation that reduces to a quadratic equation after clearing fractions. Multiplying both sides of the equation by the LCD = $4R(R + 6)$, we get

$$4(R + 6) + 4R = R(R + 6)$$
$$4R + 24 + 4R = R^2 + 6R$$
$$8R + 24 = R^2 + 6R$$
$$R^2 - 2R - 24 = 0$$
$$(R - 6)(R + 4) = 0$$
$$R = -4, 6$$

Since the negative root has no meaning here, we conclude that $R = 6 \, \Omega$. The resistance of the other resistor is therefore $12 \, \Omega$.

Example 4 A tank can be filled by two inlet valves in 2 hr. One inlet valve requires $7\frac{1}{2}$ hr longer to fill the tank than the other. How long does it take for each valve alone to fill the tank?

Solution. Let x equal the time taken for the faster valve to fill the tank. Then $x + 7\frac{1}{2} = x + \frac{15}{2}$ is the time required for the slower valve to fill the tank.

Now recall from Section 2.3 that an equation can be readily obtained from this information by finding expressions for the fractional part of the tank that can be filled in one time unit, in this case 1 hr. So

$$\frac{1}{x} + \frac{1}{x + \frac{15}{2}} = \frac{1}{2}$$

To clear fractions, note that the LCD equals $2x(x + \frac{15}{2})$. We now get

$$2\left(x + \frac{15}{2}\right) + 2x = x\left(x + \frac{15}{2}\right)$$
$$2x + 15 + 2x = x^2 + \frac{15}{2}x$$

Multiplying both sides by 2, we then have

$$8x + 30 = 2x^2 + 15x$$
$$2x^2 + 7x - 30 = 0$$
$$(2x - 5)(x + 6) = 0$$
$$x = -6, \frac{5}{2}$$

Again, the negative root has no meaning, so we conclude that the times are $2\frac{1}{2}$ hr and 10 hr, respectively.

Example 5 An executive drives to a conference early in the day. Due to heavy morning traffic, her average speed for the first 120 mi is 10 mi/hr less than for the second 120 mi and requires 1 hr more time. Find the two average speeds.

Solution. Recall from Section 2.3 that the basic relationship is distance = rate × time.

If we let x equal the slower rate, then $x + 10$ equals the faster rate. Since the distance is the same in both cases, it follows that $120/x$ is the time required to cover the first 120 mi, and $120/(x + 10)$ the time required to cover the second 120 mi. The difference in the two times is 1 hr. Hence

$$\frac{120}{x} - \frac{120}{x + 10} = 1$$

Clearing fractions,

$$120(x + 10) - 120x = x(x + 10)$$
$$120x + 1{,}200 - 120x = x^2 + 10x$$
$$x^2 + 10x - 1{,}200 = 0$$
$$(x - 30)(x + 40) = 0$$
$$x = 30, -40$$

Taking the positive root again, we conclude that 30 mi/hr is the slower rate and $x + 10 = 40$ mi/hr the faster rate.

Exercises / Section 6.4

1. The current i (in amperes) in a certain circuit at any time t (in seconds) is given by $i = 9.5t^2 - 4.7t$. At what time is the current equal to zero?
2. The sum of two electric currents is 35 A and their product 294 A². Find the two currents.
3. A certain resistance is 2.00 Ω more than another. Their product is 84.0 Ω². Find the two resistances.
4. If an object is hurled vertically downward with velocity v_0, then the distance s that the object falls at any time is $s = v_0 t + \frac{1}{2}gt^2$. Find an expression for t.
5. Recall that the relationship of the focal length f of a lens to the object distance q and the image distance p is

$$\frac{1}{f} = \frac{1}{p} + \frac{1}{q}$$

 If $f = 2.0$ cm and p is 3.0 cm longer than q, find p.
6. A metal shop has an order for a rectangular metal plate of area 84 in.². Find its dimensions if the length exceeds the width by 5.0 in.
7. A parallelogram has an area of 149.0 in.², and the base exceeds the height by 10.00 in. Find the base and height.
8. The difference between a positive integer and its reciprocal is $2\frac{2}{3}$. Find the number.
9. A rectangular casting 0.500 in. thick is to be made from 44.0 in. of forming (Figure 6.4 on page 218). If 42.5 in.³ are poured into the form, find the dimensions of the casting.
10. Suppose the casting in Exercise 9 is 3.00 in. thick and its length is 2.00 in. more than its width. If the volume is 72.0 in.³, find its dimensions.
11. To cover the floor of a new storage area, 100 square tiles of a certain size are needed. If square tiles 2 in. longer on each side are used, only 64 tiles are needed. What is the size of the smaller tile?

218 CHAPTER 6 QUADRATIC EQUATIONS

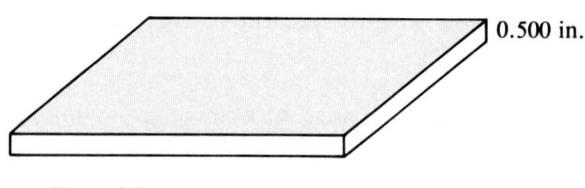

0.500 in.

Figure 6.4

12. A rectangular metal plate is twice as long as it is wide. When heated, each side is increased by 2 mm and the area is multiplied by $\frac{248}{225}$. Find its original dimensions.
13. If both inlets of a tank are open, then the tank can be filled in 2 hr. One of the inlets alone requires $5\frac{1}{3}$ hr more than the other to fill the tank. How long does each one take?
14. Two machines are used to print labels for a large mailing; the job normally takes 2 hr. One day the faster machine breaks down and the slower machine, which takes 3 hr longer than the faster machine to do the job alone, has to be used. How long will it take to complete the job?
15. Two card sorters used simultaneously can sort a set of cards in 24 min. If only one machine is used, then the slower one requires 20 min more than the faster one. How long does each one take?
16. A technician has to order a frame meeting the following specifications: It has the shape of a right triangle with a hypotenuse 26.0 cm long, while the sum of the lengths of the other two sides is 34.0 cm. Find the dimensions.
17. In city traffic, a car travels 15 mi/hr faster than a bicycle. The car can travel 50 mi in 3 hr less time than the bicycle. Find the rate of each.
18. A heavy machine is delivered by truck to a factory 200 mi away. The empty truck makes the return trip 10 mi/hr faster and gets back in 1 hr less time. Find the rate each way.
19. A rectangular enclosure is to be fenced along four sides and divided into two parts by a fence parallel to one of the sides. (See Figure 6.5.) If 170 ft of fence are available and the total area is 1,200 ft², what are the dimensions? (There are two possible solutions.)

x

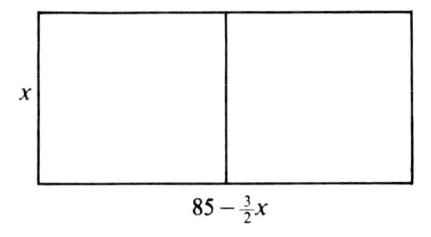

$85 - \frac{3}{2}x$

Figure 6.5

20. A tool shed is adjacent to a building, which serves as the back wall of the tool shed. The total length of the other three walls is 39 ft, and the area of the floor is 180 ft². Find the dimensions.
21. The cost of carpeting an office at $10/ft² was $1,500. If the length exceeds the width by 5 ft, what are the dimensions of the office?
22. Maggie's car gets 5 mi/gal less in the city than on the highway. Driving 300 mi in the city requires 2 gal more gas than driving the same distance on the highway. Determine the gas mileage in the city.

23. An engineer wants to buy $240 worth of stock. One stock costs $10 more per share than another. If she decides to buy the cheaper stock, she can afford four more shares. How many shares of the more expensive stock can she buy?

24. An investor purchases a number of shares of stock for $600. If the investor had paid $2 less per share, the number of shares would have been increased by 10. How many shares of the cheaper stock can he buy?

Review Exercises / Chapter 6

In Exercises 1–8, solve the equations by factoring.

1. $x^2 - 3x - 10 = 0$
2. $2x^2 - 7x - 4 = 0$
3. $5x^2 - 7x - 6 = 0$
4. $2x^2 + 15x - 27 = 0$
5. $6x^2 + 7x + 2 = 0$
6. $4x^2 - 17x + 15 = 0$
7. $6x^2 = 11x - 5$
8. $10x^2 + 3x = 18$

In Exercises 9–16, solve the equations by completing the square.

9. $x^2 - 2x - 8 = 0$
10. $2x^2 + 3x - 5 = 0$
11. $x^2 + 5x + 3 = 0$
12. $2x^2 - 4x + 1 = 0$
13. $2x^2 - 5x + 4 = 0$
14. $3x^2 - 2x + 3 = 0$
15. $4x^2 - x + 3 = 0$
16. $2x^2 + 5x + 2 = 0$

In Exercises 17–24, solve the equations by applying the quadratic formula.

17. $x^2 - 6x + 8 = 0$
18. $6x^2 + 7x - 3 = 0$
19. $3x^2 - 3x + 1 = 0$
20. $x^2 - 8x + 6 = 0$
21. $2x^2 - 4x - 3 = 0$
22. $2x^2 - 4x + 3 = 0$
23. $6x^2 - 8x - 3 = 0$
24. $5x^2 + x + 1 = 0$

In Exercises 25–30, solve the equations using any method.

25. $1.72x^2 + 1.89x - 2.64 = 0$
26. $3.98x^2 + 0.46x - 0.42 = 0$
27. $x^2 - 4x + 4 - y^2 = 0$
28. $x^2 - 2xy + x + y^2 - y - 2 = 0$
29. $\dfrac{1}{x} + \dfrac{1}{x+1} = \dfrac{9}{20}$
30. $\dfrac{1}{x-1} + \dfrac{1}{x+2} = 1$

31. Two currents differ by 2.0 A, while their product is 288 A². Find the two currents.

32. A rectangular enclosure is to be fenced along four sides and divided into three parts by two fences that are parallel to one side. If 80 ft of fence are available and the total area is 200 ft², what are the dimensions of the enclosure?

33. Working together, two men can unload a boxcar in 4 hr. Working alone, one man requires 6 hr more than the other. How long does it take for each man to do the job alone?

34. The cost of tiling a kitchen is $5/ft². If the length of the kitchen exceeds the width by 4 ft and the total cost of tiling comes to $960, what are the dimensions of the kitchen?

35. Early one morning a shop assistant delivers a motor to a garage 70 mi from downtown. Because of lighter traffic, his average speed on the return trip is 15 mi/hr more than on the delivery run, and he returns in 90 min less time. Find his average speed each way.

Cumulative Review Exercises / Chapters 4–6

1. Find the smallest positive angle coterminal with $-237°41'$.
2. Find the values of the six trigonometric functions of the angle in standard position whose terminal side passes through $(\sqrt{13}, \sqrt{3})$.
3. If $\cos \theta = \dfrac{\sqrt{2}}{3}$, find $\csc \theta$.
4. Use a diagram to find $\cot 60°$.
5. Use a calculator to find $\csc 36°19'$.
6. If $\sec \theta = 1.835$, use a calculator to find θ.
7. If $\cot \theta = 2.672$, use a calculator to find θ.
8. Factor the expression $sV_a - sV_b - V_a + V_b$.
9. Reduce the fraction $\dfrac{L^3 - C^3}{2L - 2C}$
10. Perform the following multiplication:

    $$\dfrac{x^2 - y^2}{x^2 + xy - 6y^2} \cdot \dfrac{x - 2y}{2x - 2y}$$

11. Perform the following division:

    $$\dfrac{x - 2y}{ax + 2ay} \div \dfrac{2x^2 - 3xy - 2y^2}{2ax + 4ay - x - 2y}$$

12. Combine the following fractions:

    $$\dfrac{s}{s - t} + \dfrac{1}{s^2 - t^2} - \dfrac{s - t}{s + t}$$

13. Reduce the following complex fraction:

    $$\dfrac{\dfrac{L}{L + 4} + \dfrac{L}{L^2 - 16}}{1 + \dfrac{1}{L - 4}}$$

14. Solve for x:

    $$\dfrac{1}{x - 2} - \dfrac{1}{x} = \dfrac{1}{4}$$

15. Solve for x by the method of factoring:

    $4x^2 + 11x - 3 = 0$

16. Solve for x by completing the square:

    $x^2 - 4x + 3 = 0$

17. Solve for x using the quadratic formula:

    $x^2 - 2x + 2 = 0$

18. Use a calculator to solve the following equation for x:

    $3.12x^2 + 1.71x - 1.30 = 0$

19. The current (in amperes) in a circuit is $i = 0.462 \cos \phi$. Determine the current at the instant when $\phi = 20.4°$.

20. Nine rivets are equally spaced on the rim of a circular metal plate. The distance from center to center between adjacent rivets is 1.06 in. Find the radius of the plate.

21. Three resistors R_1, R_2, and R_3 are connected in parallel. The combined resistance is

    $$\frac{1}{\dfrac{1}{R_1} + \dfrac{1}{R_2} + \dfrac{1}{R_3}}$$

    Simplify this expression.

22. The cost of finishing a metal plate is $4.00 per square inch. If the length of the plate exceeds the width by 2.0 in. and the total cost comes to $73.44, what are the dimensions of the plate?

CHAPTER 7

More on Trigonometric Functions

Objectives Upon completion of this chapter, you should be able to:
1. Determine the sign of the value of a trigonometric function.
2. Find the values of the six trigonometric functions of an arbitrary angle in standard position, given one point on the terminal side.
3. Find the values of the trigonometric functions of a given special angle.
4. Find the values of the trigonometric functions of an arbitrary angle by:
   a. Using a table.
   b. Using a calculator.
5. Find any angle between 0° and 360°, given the value of a trigonometric function by:
   a. Using a table.
   b. Using a calculator.
6. Convert from degree measure to radian measure and from radian measure to degree measure.
7. Find the value of a trigonometric function of a given angle measured in radians.
8. Find the length of an arc of a circle.
9. Find the area of a circular sector.
10. Determine linear velocity from angular velocity.
11. Determine angular velocity from linear velocity.

7.1 Algebraic Signs of Trigonometric Functions

In Chapter 4 we defined the six trigonometric functions for angles with terminal sides in the first quadrant. In this section we shall extend these definitions to apply to angles with terminal sides in any quadrant.

7.1 ALGEBRAIC SIGNS OF TRIGONOMETRIC FUNCTIONS 223

Radius vector

Recall from Chapter 4 that if an angle is in standard position, then the functions can be expressed in terms of the coordinates and radius vector of a point on the terminal side. Consider the angle θ whose terminal side passes through (x, y) (Figure 7.1). The **radius vector** r is the distance from (x, y) to the origin. Using the letters x, y, and r in Figure 7.1, the trigonometric functions have the form given below.

The trigonometric functions

Figure 7.1

$$\sin \theta = \frac{y}{r} \qquad \csc \theta = \frac{r}{y}$$

$$\cos \theta = \frac{x}{r} \qquad \sec \theta = \frac{r}{x} \qquad (7.1)$$

$$\tan \theta = \frac{y}{x} \qquad \cot \theta = \frac{x}{y}$$

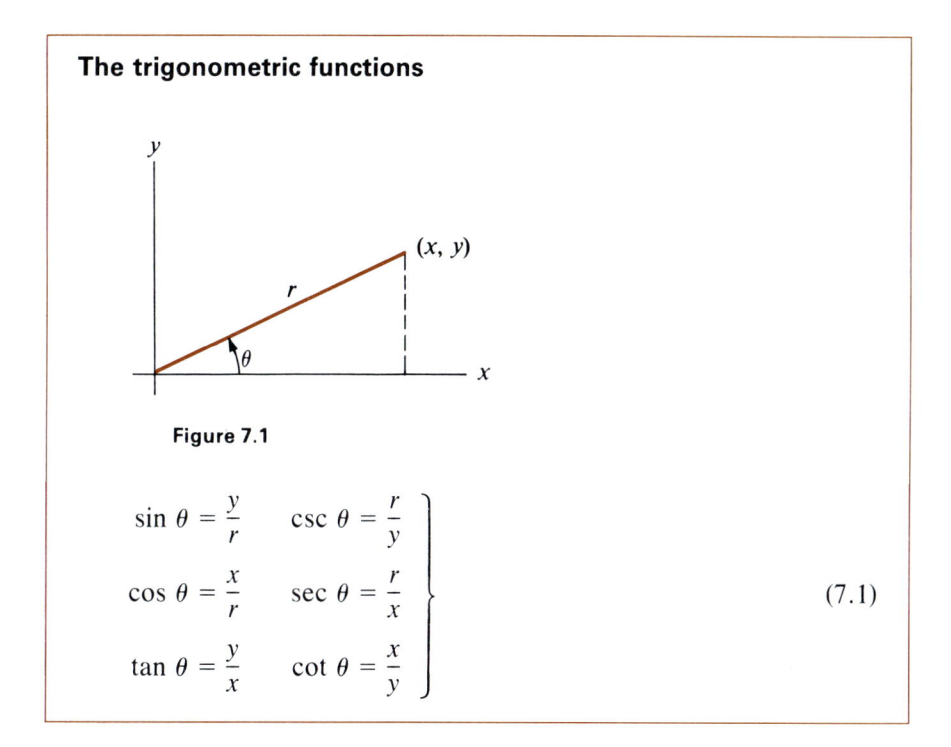

With these definitions, we can find the function values of angles whose terminal sides do not lie in the first quadrant. Consider, for example, the angle θ pictured in Figure 7.2.

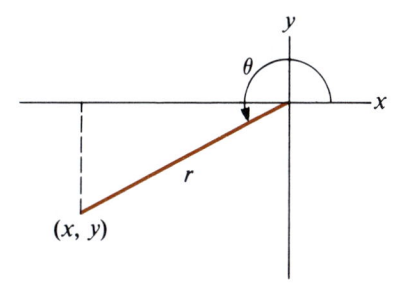

Figure 7.2

CHAPTER 7 MORE ON TRIGONOMETRIC FUNCTIONS

All the definitions in (7.1) carry over automatically. The only difference is that some of the function values are negative, since the coordinates of (x, y) are signed numbers. However, *the radius vector is always positive*. Moreover, by the Pythagorean theorem,

$$r = \sqrt{x^2 + y^2} \tag{7.2}$$

regardless of the position of the point (x, y). Consider the example below.

Example 1 Find the values of the six trigonometric functions for the angle θ in Figure 7.3.

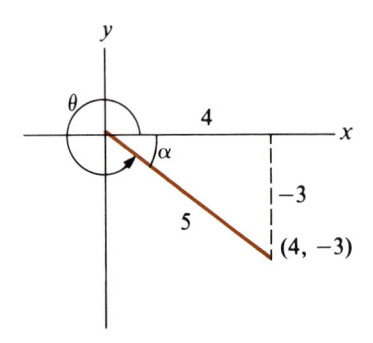

Figure 7.3

Solution. By formula (7.2), $r = \sqrt{x^2 + y^2}$, we have

$$r = \sqrt{4^2 + (-3)^2} = 5$$

Since $x = 4$ and $y = -3$, we obtain directly from the definitions

$$\sin \theta = -\frac{3}{5} \qquad \csc \theta = -\frac{5}{3}$$

$$\cos \theta = \frac{4}{5} \qquad \sec \theta = \frac{5}{4}$$

$$\tan \theta = -\frac{3}{4} \qquad \cot \theta = -\frac{4}{3}$$

Example 1 (Alternate) The descriptive definitions of the trigonometric ratios—using the opposite side, the adjacent side, and the hypotenuse—have proved useful in applied problems and have already become familiar. Can these definitions be preserved? They can indeed by means of the following scheme: Drop a perpendicular from the point $(4, -3)$ to the x-axis and place the numbers 4, -3, and 5 right on the resulting triangle, as shown in Figure 7.4. It now follows that

$$\sin \theta = \frac{\text{opposite}}{\text{hypotenuse}} = \frac{-3}{5} = -\frac{3}{5}$$

$$\cos \theta = \frac{\text{adjacent}}{\text{hypotenuse}} = \frac{4}{5}$$

Reference angle

and so forth. The triangle in Figure 7.4 is called the **reference triangle,** and the acute angle α is called the **reference angle.** To use the descriptive definitions, note that *the opposite side is the side opposite the reference angle and the adjacent side is the side adjacent to the reference angle.*

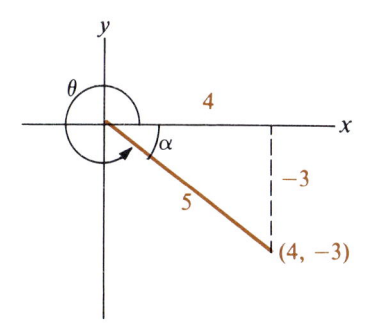

Figure 7.4

Example 2 Find the values of the six trigonometric functions of the angle θ whose terminal side passes through $(-\sqrt{5}, -2)$.

Solution. From $x = -\sqrt{5}$ and $y = -2$, we obtain

$$r = \sqrt{(-\sqrt{5})^2 + (-2)^2} = 3$$

Now drop a perpendicular from $(-\sqrt{5}, -2)$ to the x-axis and label the sides of the resulting triangle as shown in Figure 7.5.

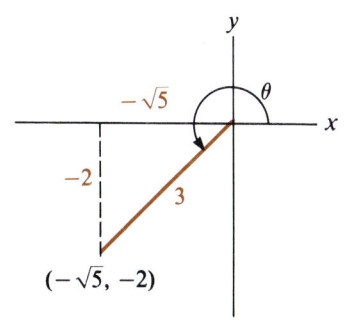

Figure 7.5

Recalling that the terms *opposite* and *adjacent* refer to the reference angle, we get

$$\sin\theta = \frac{\text{opposite}}{\text{hypotenuse}} = \frac{-2}{3} = -\frac{2}{3} \qquad \csc\theta = \frac{3}{-2} = -\frac{3}{2}$$

$$\cos\theta = \frac{\text{adjacent}}{\text{hypotenuse}} = \frac{-\sqrt{5}}{3} = -\frac{\sqrt{5}}{3} \qquad \sec\theta = \frac{3}{-\sqrt{5}} = -\frac{3\sqrt{5}}{5}$$

$$\tan\theta = \frac{\text{opposite}}{\text{adjacent}} = \frac{-2}{-\sqrt{5}} = \frac{2\sqrt{5}}{5} \qquad \cot\theta = \frac{-\sqrt{5}}{-2} = \frac{\sqrt{5}}{2}$$

Alternatively, using the definitions in statement (7.1), we get

$$\sin\theta = \frac{y}{r} = \frac{-2}{3} = -\frac{2}{3} \qquad \csc\theta = \frac{r}{y} = \frac{3}{-2} = -\frac{3}{2}$$

$$\cos\theta = \frac{x}{r} = \frac{-\sqrt{5}}{3} = -\frac{\sqrt{5}}{3} \qquad \sec\theta = \frac{r}{x} = \frac{3}{-\sqrt{5}} = -\frac{3\sqrt{5}}{5}$$

$$\tan\theta = \frac{y}{x} = \frac{-2}{-\sqrt{5}} = \frac{2\sqrt{5}}{5} \qquad \cot\theta = \frac{x}{y} = \frac{-\sqrt{5}}{-2} = \frac{\sqrt{5}}{2}$$

We can see from Examples 1 and 2 that the various trigonometric functions are either positive or negative, depending on which quadrant contains the terminal side of the angle. This sign can be easily determined from the reference angle. (From now on, we shall say that an angle is in the first quadrant if its terminal side lies in the first quadrant, in the second quadrant if its terminal side lies in the second quadrant, and so on.)

The next example illustrates how the signs of trigonometric functions can be determined from the reference angles.

Example 3 Determine the signs of the trigonometric functions for an angle in (a) the second quadrant; (b) the fourth quadrant.

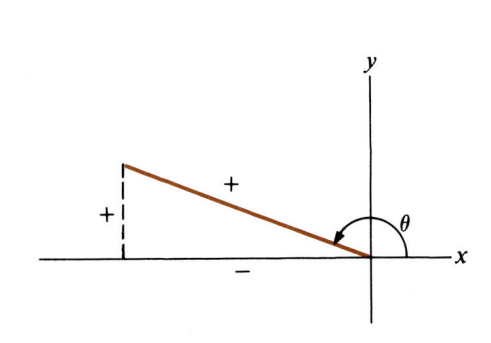

Figure 7.6

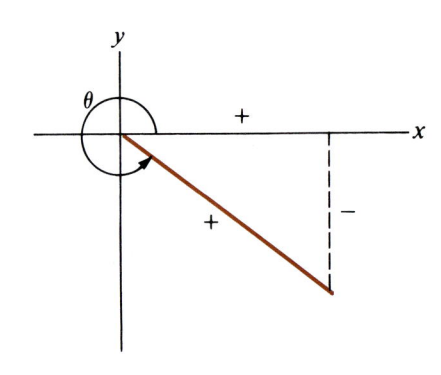

Figure 7.7

7.1 ALGEBRAIC SIGNS OF TRIGONOMETRIC FUNCTIONS

Solution. (a) For a point in the second quadrant, x is negative and y positive. Label the sides of the reference triangle in Figure 7.6 with the proper signs. The signs of the trigonometric functions can then be determined directly from the diagram. They are listed in the following table:

Function:	$\sin \theta$	$\cos \theta$	$\tan \theta$	$\csc \theta$	$\sec \theta$	$\cot \theta$
Sign:	$\dfrac{+}{+} = +$	$\dfrac{-}{+} = -$	$\dfrac{+}{-} = -$	$\dfrac{+}{+} = +$	$\dfrac{+}{-} = -$	$\dfrac{-}{+} = -$

(b) In the fourth quadrant, x is positive and y negative (see Figure 7.7). From the reference triangle we construct the following table:

Function:	$\sin \theta$	$\cos \theta$	$\tan \theta$	$\csc \theta$	$\sec \theta$	$\cot \theta$
Sign:	$\dfrac{-}{+} = -$	$\dfrac{+}{+} = +$	$\dfrac{-}{+} = -$	$\dfrac{+}{-} = -$	$\dfrac{+}{+} = +$	$\dfrac{+}{-} = -$

If a function value and the quadrant of an angle are known, the values of the remaining functions can be determined, as shown in the next example.

Example 4 Determine $\sin \theta$, given that $\cos \theta = \frac{1}{3}$ and $\tan \theta < 0$.

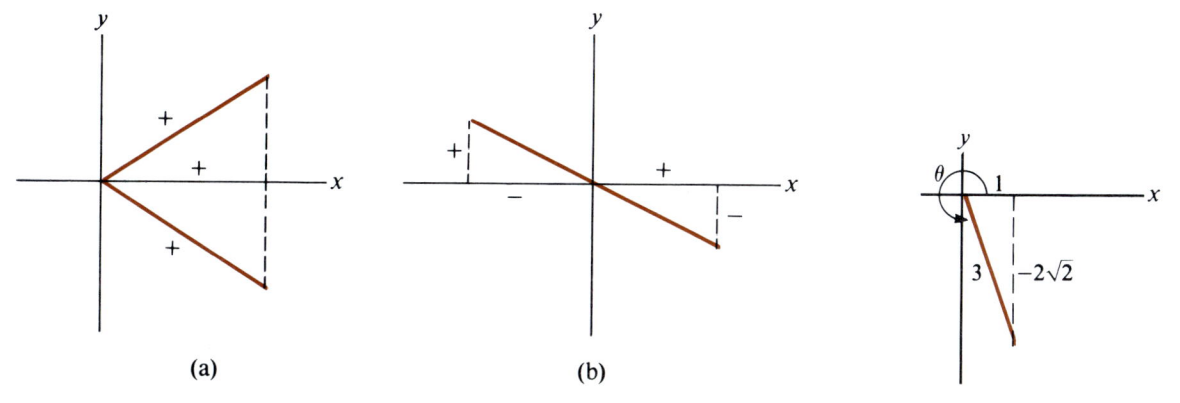

Figure 7.8

Figure 7.9

Solution. The first step is to determine the quadrant in which θ lies. Since $\cos \theta > 0$, θ is either in the first or fourth quadrant, as can be seen from Figure 7.8(a). Since $\tan \theta < 0$, θ must be in either the second or the fourth quadrant (see Figure 7.8(b)). To satisfy both conditions, θ must be in the fourth quadrant.

From $\cos \theta = \frac{1}{3}$, we have $x = 1$ and $r = 3$. The numerical value of y is found by using the Pythagorean theorem:
$$y^2 + 1^2 = 3^2 \quad \text{or} \quad y = \pm\sqrt{8} = \pm 2\sqrt{2}$$
and since θ is in the fourth quadrant, $y = -2\sqrt{2}$. Now draw the reference triangle (Figure 7.9). It follows that
$$\sin \theta = \frac{-2\sqrt{2}}{3} = -\frac{2\sqrt{2}}{3}$$

Example 5 Given that $\csc \theta = -\sqrt{13}/2$ and $\cos \theta < 0$, find $\sec \theta$.

Solution. Since $\csc \theta < 0$, θ is either in the third or fourth quadrant (Figure 7.10(a)); since $\cos \theta < 0$, θ is either in the second or third quadrant (Figure 7.10(b)). It follows that θ must be in the third quadrant.
$$\text{From } \csc \theta = \frac{r}{y} = -\frac{\sqrt{13}}{2} = \frac{\sqrt{13}}{-2}, \text{ we get } y = -2 \text{ and } r = \sqrt{13}.$$

Also,
$$x^2 + (-2)^2 = (\sqrt{13})^2 \quad \text{or} \quad x = \pm 3$$

Since θ is in the third quadrant, $x = -3$. Now draw the reference triangle in Figure 7.11. Thus
$$\sec \theta = \frac{\sqrt{13}}{-3} = -\frac{\sqrt{13}}{3}$$

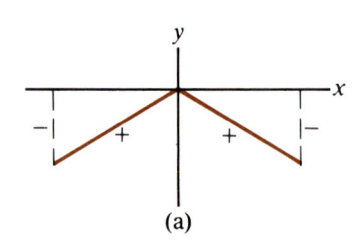

Figure 7.10

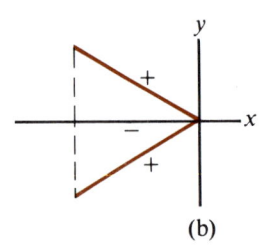

Figure 7.11

Exercises / Section 7.1

In Exercises 1–12, find the values of the six trigonometric functions for the angle whose terminal side passes through the given point.

1. $(-4, 3)$
2. $(-3, -4)$
3. $(5, -12)$
4. $(-5, -12)$
5. $(1, 2\sqrt{2})$
6. $(-3, 2)$
7. $(-2, -6)$
8. $(3, -5)$
9. $(-1, \sqrt{15})$
10. $(2, \sqrt{5})$
11. $(-3, 6)$
12. $(-4, -7)$

13. If $\tan \theta = \frac{5}{2}$ and $\cos \theta > 0$, find $\sin \theta$.
14. If $\sin \theta = \frac{1}{3}$ and $\tan \theta < 0$, find $\cot \theta$.
15. If $\sec \theta = -\sqrt{6}$ and $\cot \theta > 0$, find $\csc \theta$.
16. If $\cot \theta = -\frac{3}{2}$ and $\sin \theta < 0$, find $\cos \theta$.
17. If $\sin \theta = -\sqrt{7}/4$ and $\sec \theta > 0$, find $\tan \theta$.
18. If $\cos \theta = -\sqrt{7}/3$ and $\sin \theta > 0$, find $\sin \theta$.
19. If $\csc \theta = -\sqrt{3}$ and $\cot \theta < 0$, find $\sec \theta$.
20. If $\cos \theta = -3/\sqrt{21}$ and $\sin \theta > 0$, find $\tan \theta$.
21. If $\cos \theta = -\sqrt{3}/4$ and $\csc \theta < 0$, find $\sin \theta$.
22. If $\tan \theta = -\frac{7}{2}$ and $\cos \theta > 0$, find $\sec \theta$.
23. If $\cot \theta = \frac{1}{3}$ and $\cos \theta < 0$, find $\cos \theta$.
24. If $\sec \theta = \frac{7}{3}$ and $\cot \theta > 0$, find $\sin \theta$.
25. If $\sec \theta = -\sqrt{6}/2$ and $\tan \theta < 0$, find $\csc \theta$.
26. If $\sin \theta = \sqrt{2}/2$ and $\cot \theta < 0$, find $\tan \theta$.
27. If $\cos \theta = -\sqrt{15}/5$ and $\sin \theta < 0$, find $\cot \theta$.
28. If $\sin \theta = \sqrt{6}/3$ and $\cos \theta > 0$, find $\sec \theta$.

7.2 Special Angles

We shall now learn to find the trigonometric functions of special angles. Because these angles are used so frequently in the examples and exercises of more advanced courses as well as trigonometry, you must be thoroughly familiar with them.

Special angles

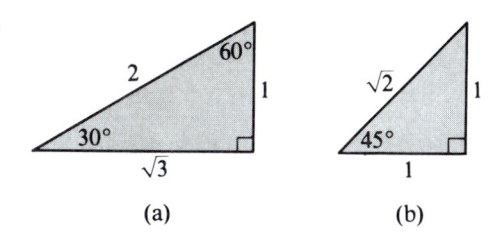

Figure 7.12

Recall that the triangles involving special angles are those illustrated in Figure 7.12. As we saw in Chapter 4, these triangles are adequate for discussing angles in the first quadrant. Since we want to consider special angles in any quadrant, the triangles in Figure 7.12 will serve as reference triangles.

The first three examples illustrate the technique for finding the function values of special angles by means of diagrams.

Example 1 Find tan 135° and sec 135°.

Solution. The angle is shown in Figure 7.13. Note that the reference angle α in the figure is $180° - 135° = 45°$. If we now drop a perpendicular from a point on the terminal side to the x-axis and use the numbers in the triangle in Figure 7.12(b), we obtain the reference triangle in Figure 7.14.

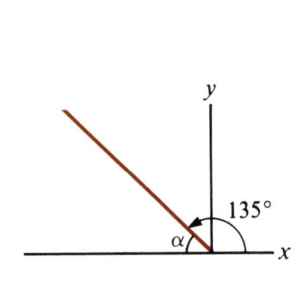

Figure 7.13

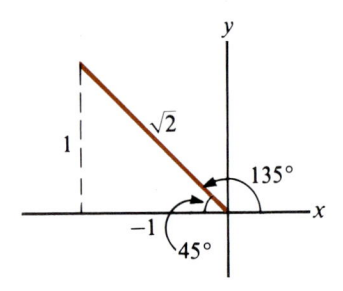

Figure 7.14

It follows that

$$\tan 135° = \frac{1}{-1} = -1 \quad \text{and} \quad \sec 135° = \frac{\sqrt{2}}{-1} = -\sqrt{2}$$

Example 2 Find sin 240° and cos 240°.

Solution. The angle is in the third quadrant, since $180° < 240° < 270°$, and the reference angle is $240° - 180° = 60°$. The resulting reference triangle, shown in Figure 7.15, is a 30°–60° right triangle. (Refer to Figure 7.12(a).) It follows that

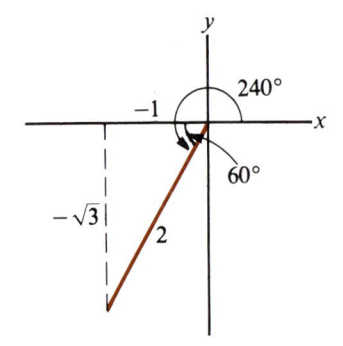

$$\sin 240° = -\frac{\sqrt{3}}{2} \quad \text{and} \quad \cos 240° = -\frac{1}{2}$$

Figure 7.15

7.2 SPECIAL ANGLES

Example 3 Find sec 330° and cot 330°.

Solution. The angle is in the fourth quadrant. Since 360° − 330° = 30°, we conclude that 30° is the reference angle. The resulting reference triangle is shown in Figure 7.16. Hence

$$\sec 330° = \frac{2}{\sqrt{3}} = \frac{2\sqrt{3}}{3} \quad \text{and} \quad \cot 330° = \frac{\sqrt{3}}{-1} = -\sqrt{3}$$

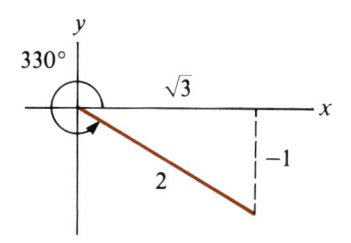

Figure 7.16

Quadrantal angle

As noted in Chapter 4, if the terminal side coincides with an axis, the angle is called a **quadrantal angle.** The values of trigonometric functions of quadrantal angles can also be found by using diagrams, as shown in Examples 4 and 5.

Example 4 Find cos 180° and csc 180°.

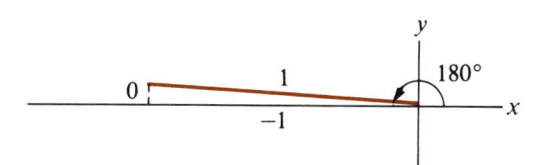

Figure 7.17

Solution. Since the terminal side coincides with the negative x-axis, we do not get a triangle. However, as we did in Chapter 4, we can pretend that such a triangle exists by separating the coincident sides. This means that two sides have the same length of 1 unit and the third side has a length of 0. The resulting diagram (Figure 7.17) shows that

$$\cos 180° = \frac{-1}{1} = -1$$

and

$$\csc 180° = \frac{1}{0} \quad \text{(undefined)}$$

Alternatively, let $(-1, 0)$ be a point on the terminal side. Then by the definitions of the trigonometric functions, we have

$$\cos 180° = \frac{x}{r} = \frac{-1}{1} = -1$$

and

$$\csc 180° = \frac{r}{y} = \frac{1}{0} \quad \text{(undefined)}$$

Example 5 Find $\cot 270°$, $\tan 270°$, and $\sin 270°$.

Solution. Let $(0, -1)$ be a point on the terminal side. Then by the definitions of the trigonometric functions, we get

$$\cot 270° = \frac{x}{y} = \frac{0}{-1} = 0$$

$$\tan 270° = \frac{y}{x} = \frac{-1}{0} \quad \text{(undefined)}$$

and

$$\sin 270° = \frac{y}{r} = \frac{-1}{1} = -1$$

Alternatively, we may draw a reference triangle by separating the coincident sides and denoting the length of the third side by 0. Care must be taken to let one side be perpendicular to the x-axis (see Figure 7.18). Thus

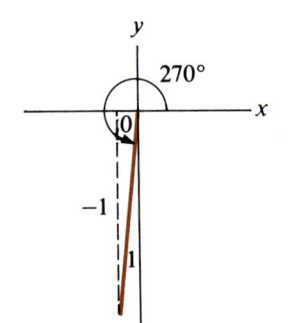

$$\cot 270° = \frac{0}{-1} = 0$$

$$\tan 270° = \frac{-1}{0} \quad \text{(undefined)}$$

$$\sin 270° = \frac{-1}{1} = -1$$

Figure 7.18

In the remaining examples we return to the problem of finding special angles, given the value of a trigonometric function.

Example 6 Find all angles θ such that $0° \leq \theta < 360°$ if $\cos \theta = \frac{1}{2}$.

Solution. Since $\cos \theta > 0$ in quadrants I and IV, we draw the reference triangles shown in Figure 7.19 with $x = 1$ and $r = 2$. Since the reference angle is $60°$, it follows that $\theta = 60°$ and $\theta = 300°$.

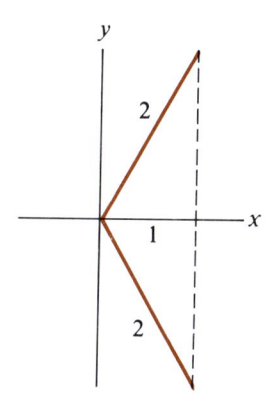

Figure 7.19

Example 7 Find all angles such that $0° \leq \theta < 360°$ if $\sin \theta = -\sqrt{2}/2$.

Solution. First note that $\sin \theta < 0$ in quadrants III and IV. Since

$$\sin \theta = -\frac{\sqrt{2}}{2} = \frac{-\sqrt{2}}{2} = \frac{y}{r}$$

we have in either instance $y = -\sqrt{2}$ and $r = 2$. The numerical value of x is obtained from the equation

$$x^2 + (-\sqrt{2})^2 = 2^2, \quad \text{which yields} \quad x = \pm\sqrt{2}.$$

Now draw Figure 7.20. Since the reference angle is $45°$, it follows that $\theta = 225°$ and $\theta = 315°$.

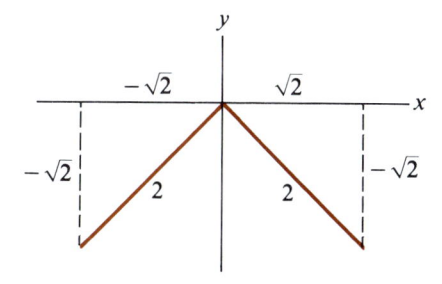

Figure 7.20

Exercises / Section 7.2

In Exercises 1–40, find the value of each trigonometric function by means of a diagram. (Do not use a table or a calculator.)

1. $\sin 30°$
2. $\cos 120°$
3. $\tan 330°$
4. $\sec 240°$
5. $\cos 135°$
6. $\sec 225°$
7. $\csc 30°$
8. $\cot 315°$
9. $\sin(-30°)$
10. $\cos 0°$
11. $\csc(-270°)$
12. $\sin 180°$
13. $\sec(-60°)$
14. $\sec 180°$
15. $\cot 90°$
16. $\cos 240°$
17. $\csc 270°$
18. $\cos(-90°)$
19. $\cot 120°$
20. $\cot(-120°)$
21. $\csc 180°$
22. $\tan 240°$
23. $\sin 300°$
24. $\cos 150°$
25. $\sec(-150°)$
26. $\tan(-180°)$
27. $\cot 300°$
28. $\sin 0°$
29. $\sec 210°$
30. $\cos(-45°)$
31. $\cos(-270°)$
32. $\cot 210°$
33. $\sin 315°$
34. $\tan 300°$
35. $\tan 150°$
36. $\csc 210°$
37. $\sec 390°$
38. $\sin 405°$
39. $\cos 420°$
40. $\tan 450°$

41. Show that $\sin^2 150° + \cos^2 150° = 1$.
42. Show that $1 + \tan^2 210° = \sec^2 210°$.
43. Show that $(\cos 300°)/(\sin 300°) = \cot 300°$.
44. Show that $\sec^2(-30°) = \tan^2(-30°) + 1$.

In Exercises 45–75, find all angles θ such that $0° \leq \theta < 360°$ for the given value of the trigonometric function. (Do not use a table or a calculator.)

45. $\sin \theta = \dfrac{1}{2}$
46. $\cos \theta = -\dfrac{1}{2}$
47. $\csc \theta = 2$
48. $\sec \theta = -2$
49. $\sin \theta = -\dfrac{\sqrt{3}}{2}$
50. $\sec \theta = \sqrt{2}$
51. $\cos \theta = -1$
52. $\sin \theta = 0$
53. $\tan \theta = \sqrt{3}$
54. $\cot \theta = -\sqrt{3}$
55. $\sec \theta = 2$
56. $\sin \theta = -\dfrac{1}{2}$
57. $\sin \theta = -\dfrac{\sqrt{2}}{2}$
58. $\cos \theta = -\dfrac{\sqrt{3}}{2}$
59. $\cot \theta = 0$
60. $\csc \theta = 1$
61. $\tan \theta = -1$
62. $\cot \theta = 1$
63. $\cos \theta = -\dfrac{\sqrt{2}}{2}$
64. $\tan \theta = \dfrac{\sqrt{3}}{3}$
65. $\csc \theta = \dfrac{2\sqrt{3}}{3}$
66. $\sec \theta = -\dfrac{2\sqrt{3}}{3}$
67. $\tan \theta = -\sqrt{3}$
68. $\csc \theta$ undefined
69. $\tan \theta$ undefined
70. $\cos \theta = \dfrac{\sqrt{2}}{2}$
71. $\cot \theta = -\dfrac{\sqrt{3}}{3}$
72. $\cos \theta = 0$
73. $\sin \theta = \dfrac{\sqrt{2}}{2}$
74. $\tan \theta = -\dfrac{\sqrt{3}}{3}$
75. $\sec \theta = \dfrac{2\sqrt{3}}{3}$

7.3 Trigonometric Functions of Arbitrary Angles

In this section we will (1) determine the values of trigonometric functions by means of tables and calculators and (2) determine an angle when given the value of a trigonometric function.

The first two examples illustrate how to find the value of a trigonometric function of a given angle.

Example 1 Use Table 1 of Appendix B to find sin 125°10′.

Solution. Since the angle is in the second quadrant, we find the reference angle by subtracting 125°10′ from 180°:

$$\begin{array}{r} 179°60' \\ \underline{125°10'} \\ 54°50' \end{array}$$

(See Figure 7.21.) Since $\sin \theta > 0$ for θ in the second quadrant, it follows that

$$\sin 125°10' = \sin 54°50' = 0.8175$$

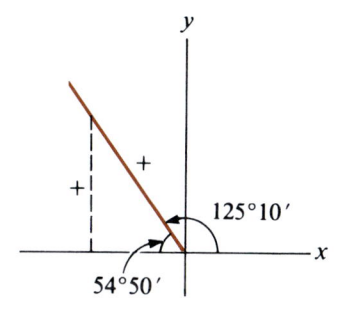

Figure 7.21

Example 2 Find cos 220°16′.

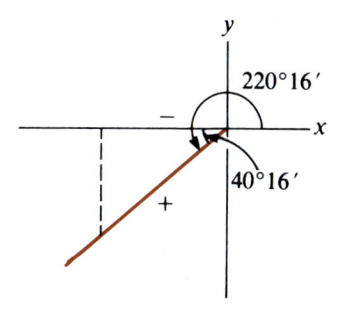

Figure 7.22

Solution. Since the angle is in the third quadrant, the reference angle is $220°16' - 180° = 40°16'$ (Figure 7.22 on page 235). Since $\cos\theta < 0$ for θ in the third quadrant, it follows that:

$$\cos 220°16' = -\cos 40°16' = -0.7630$$

If the trigonometric functions are obtained with a calculator, the procedure of Chapter 4 carries over directly: *Enter the angle measurement and press the appropriate function key.*

However, if the value of the function is given and the angle is to be found, then special problems arise both with the table and the calculator.

Example 3 Find θ, given that $\tan\theta = -1.312$ and θ is in quadrant II by using (a) Table 1 of Appendix B; (b) a calculator.

Solution. (a) Since Table 1 of Appendix B lists function values only for angles between 0° and 90°, the most we can get is the reference angle. (Recall that reference angles are always acute.) From

$$\tan\alpha = 1.312$$

we get, by interpolation, $\alpha = 52°41'$. (See Figure 7.23.) To obtain θ, we subtract α from 180°:

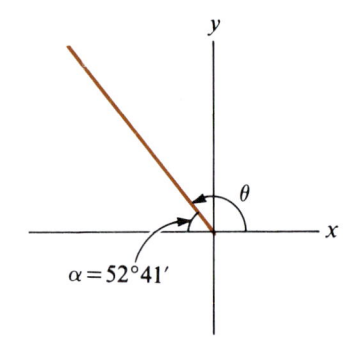

$$\begin{array}{r} 179°60' \\ \underline{52°41'} \\ \theta = 127°19' \end{array}$$

Figure 7.23

CALCULATOR COMMENT

(b) When finding θ with a calculator, it is important to observe that θ is always given in some definite range. For the tangent function, the range generally agreed on is $-90° < \theta < 90°$. So if $\tan\theta = -1.312$, enter -1.312 and press $\boxed{\text{INV}}\boxed{\text{TAN}}$. The result is $-52°41'$. Hence the desired reference angle is $\alpha = 52°41'$ and $\theta = 127°19'$ from part (a).

Most if not all calculators are programmed to yield θ in the following range:

7.3 TRIGONOMETRIC FUNCTIONS OF ARBITRARY ANGLES   237

Function	Angle
sin θ	$-90° \leq \theta \leq 90°$
cos θ	$0° \leq \theta \leq 180°$
tan θ	$-90° < \theta < 90°$

Example 4 Use a calculator to find θ, given that sec $\theta = -2.987$ and θ is in quadrant III.

Solution. Let sec $\beta = 2.987$. Then cos $\beta = \frac{1}{2.987}$ and, using a calculator, we find that $\beta = 70°26'$. The sequence is

2.987 $\boxed{1/x}$ $\boxed{\text{INV}}$ $\boxed{\text{COS}}$

and it yields $70.440592° = 70°26'$, which is the reference angle for θ. Thus

$$\theta = 180° + 70°26' = 250°26'.$$

Example 5 Use a calculator to find all angles θ such that $0° \leq \theta < 360°$ if sin $\theta = -0.2745$.

Solution. The sequence is

0.2745 $\boxed{\text{INV}}$ $\boxed{\text{SIN}}$

and yields $15.93222° = 15°56'$, which is the reference angle. Since sin $\theta < 0$ in quadrants III and IV, we get

$$\theta = 180° + 15°56' = 195°56'$$

and

$$\theta = 360° - 15°56' = 344°4'$$

Example 6 If a projectile is hurled with velocity v (in feet per second) along an inclined plane making a constant angle α with the horizontal, then the projectile's range up the plane is given by

$$R = \frac{2v^2 \cos \theta \sin (\theta - \alpha)}{32 \cos^2 \alpha}$$

where θ is the angle with the horizontal at which the projectile is aimed. If $v = 100$ ft/sec, $\alpha = 10.4°$, and $\theta = 16.7°$, find the range.

Solution. From the given formula we get

$$R = \frac{2(100)^2 \cos 16.7° \sin (16.7° - 10.4°)}{32 \cos^2 10.4°}$$

The sequence is

2 × 100 x^2 × 16.7 COS × (16.7
− 10.4) SIN ÷ 32 ÷ 10.4 COS
x^2 =

Display: 67.904045

So the projectile lands 67.9 ft away along the plane.

Exercises / Section 7.3

In Exercises 1–30, use Table 1 of Appendix B or a calculator to find the value of each of the given trigonometric functions.

1. sin 60°10′
2. cos 129°30′
3. sec 259°40′
4. tan (−30°20′)
5. csc 318°15′
6. sin 280°18′
7. cot 215°54′
8. cos 155°44′
9. sin 140°32′
10. cos 320°9′
11. sec 32°5′
12. tan 18°16′
13. cot 19°39′
14. sin 24°56′
15. cos 25°56′
16. sec 333°33′
17. csc 251°48′
18. tan 140°25′
19. cot (−142°18′)
20. sin (−256°20′)
21. cot (−75°36′)
22. cos 218°47′
23. tan 229°23′
24. sec 400°
25. sin 480°16′
26. cos 390°51′
27. cot 740°10′
28. sec (−410°)
29. sin (−370°20′)
30. tan 385°5′

In Exercises 31–50, use Table 1 of Appendix B or a calculator to find all angles θ such that $0° \leq \theta < 360°$.

31. sin θ = 0.2924
32. tan θ = 0.1492
33. cos θ = 0.1776
34. cot θ = 0.1812
35. sin θ = −0.1933
36. cos θ = −0.1945
37. csc θ = −1.130
38. sec θ = 2.316
39. cot θ = −3.859
40. tan θ = 2.893
41. csc θ = 2.754
42. sin θ = −0.5163
43. cos θ = −0.7318
44. cot θ = −3.896
45. sec θ = 2.374
46. csc θ = −1.478
47. tan θ = 1.179
48. sin θ = 0.3146
49. cos θ = −0.6438
50. cot θ = 0.3842

51. The range R (in meters) along the ground of a projectile fired at velocity v (in meters per second) at an angle θ with the horizontal is given by

$$R = \frac{v^2}{9.8} \sin 2\theta$$

Find the range if θ = 32.1° and v = 24.2 m/sec.

52. Repeat Exercise 51 for θ = 53.0° and v = 40.3 m/sec.

53. The largest weight that can be pulled up a plane inclined at an angle θ with the horizontal by a force F is

$$W = \frac{F}{\mu} (\cos \theta + \mu \sin \theta)$$

where μ is the coefficient of friction. If θ = 22°20′, F = 30.3 lb, and μ = 0.350, find W.

54. A weight W is to be dragged along a horizontal plane by a force whose line of action makes an angle θ with the plane. The force required to move the weight is given by

$$F = \frac{\mu W}{\mu \sin \theta + \cos \theta}$$

where μ is the coefficient of friction. If $W = 55.2$ lb, $\theta = 29.3°$, and $\mu = 0.200$, find F.

55. The formula for the magnetic intensity is

$$B = \frac{F}{qv \sin \theta}$$

where q is the magnitude of the charge, v its velocity, θ the angle between the direction of motion and the direction of the magnetic field, and F the force acting on the moving charge. If $B = 10$ webers/m², $F = 3.2 \times 10^{-11}$ N, $v = 2.0 \times 10^7$ m/sec, and $q = 1.6 \times 10^{-19}$ C, find θ.

56. Repeat Exercise 55 with $q = 1.6 \times 10^{-19}$ C, $F = 4.2 \times 10^{-11}$ N, $v = 3.2 \times 10^7$ m/sec, and $B = 10$ webers/m².

7.4 Radians

An alternate unit of angle measurement is the **radian.**

As noted in Chapter 4, degree measure is so old that its origins are obscure. Why 360° was chosen for a whole angle is not clear. Apart from the fact that the number 360 has many divisors, it is a quite arbitrary choice. For someone accustomed to the decimal system, even 100 units for a whole angle seems more natural. However, an even more natural unit of measure was eventually found. This unit, called the *radian,* is based on certain properties of the circle and is ideally suited for theoretical work. In fact, when trigonometric functions are discussed in calculus, radian measure is used almost exclusively.

Consider the circle in Figure 7.24. As the length of the radius r is measured off along the circumference, a central angle is created. The measure of this angle is called 1 *radian* (abbreviated "rad"), and this measure is independent of the size of the circle.

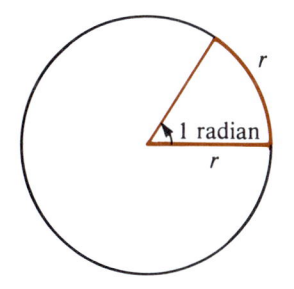

Figure 7.24

> **Definition of radian measure**: One **radian** is the size of an angle whose vertex is at the center of a circle and whose intercepted arc on the circumference is equal in length to the radius of the circle.

To translate familiar angle measurements into radian form, we need a relation between degree and radian measure. Suppose we measure off r two times along the circumference (Figure 7.25(a)) and then three times (Figure 7.25(b)). (See page 240.) In the first case the size of the angle is 2 rad, and in the second case 3 rad.

In general, the size of an angle is equal to the number of times that r can be measured off along the circumference. Consequently, the size of the *whole angle* (360°) in radian measure is equal to the number of times that *r* can be measured off along the entire circumference. Judging from Figure 7.25(b), this appears to be just over 6 times. How can we determine the precise number? Although the diagrams do not seem to offer much help, the number can be determined directly from the formula $C = 2\pi r$, where *C* is the circumference. Since the number we are looking for is now seen to be 2π, it follows that **2π rad = 360°**. This leads to the following equivalence:

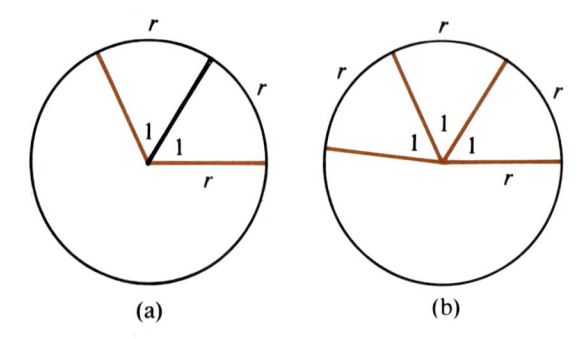

$$\pi \text{ rad} = 180° \tag{7.3}$$

This relationship, in turn, gives us some of the special angles. For example, $\pi/2$ rad = 90°, $\pi/3$ rad = 60°, $\pi/4$ rad = 45°, and so on.

Degrees:	0°	30°	45°	60°	90°	180°	270°	360°
Radians:	0	$\frac{\pi}{6}$	$\frac{\pi}{4}$	$\frac{\pi}{3}$	$\frac{\pi}{2}$	π	$\frac{3\pi}{2}$	2π

These measurements may appear strange at first, but most people do not find radian measure any more peculiar than degree measure once they become used to it. For example, just as 90° brings to mind a right angle, so does $\pi/2$ if it is seen often enough.

To convert degree measure to radian measure and vice versa, we need two other relationships, which follow directly from the fact that π rad = 180°:

$$1° = \frac{\pi}{180} \text{ rad} \approx 0.01745 \text{ rad} \tag{7.4}$$

$$1 \text{ rad} = \frac{180°}{\pi} = 57°17'45'' \approx 57.3° \tag{7.5}$$

For example,

$$15° = 15° \frac{\pi}{180°} = \frac{\pi}{12} \text{ rad}$$

Conversely,

$$\frac{2\pi}{9} \text{ rad} = \frac{2\pi}{9} \times \mathbf{1\ rad} = \frac{2\pi}{9}\left(\frac{180°}{\pi}\right) = 40°$$

> To convert degree measure to radian measure, multiply by $\pi/180°$.
>
> To convert radian measure to degree measure, multiply by $180°/\pi$.

In a pure conversion problem, it is customary to express radian measure in terms of π. However, if the problem contains physical data, the measure may have to be expressed in decimal form so that it can be rounded off to the proper number of significant figures.

One final point about units in radian measure: As already observed, the central angle in radians is equal to the number of times that r can be measured off along the circumference. Since the arc length is kr for some $k \geq 0$, the size of the central angle is k radians (see Figure 7.26). Expressed in another way.

$$\frac{kr}{r} = k \text{ rad}$$

In general, if s is the arc length and θ the central angle (Figure 7.27), then $\theta = s/r$.

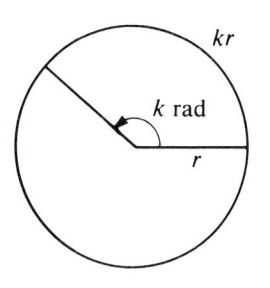

Figure 7.26

> **Radian measure of central angles:**
>
> $$\theta = \frac{s}{r} \qquad (7.6)$$
>
> **Figure 7.27**

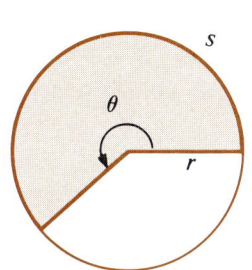

This relationship shows that a radian measure is the ratio of two lengths and is therefore free of units.

> When an angle is expressed in radians, no units are indicated.

Thus 30° is expressed simply as $\pi/6$.

Example 1 Convert 12° and 150° to radian measure.

Solution. Multiplying each angle by $\pi/180°$, we obtain

$$12° = 12\pi/180 = \pi/15$$
$$150° = 150\pi/180 = 5\pi/6$$

Note that both measures are expressed in terms of π.

Example 2 Convert $\pi/9$ and $7\pi/6$ to degree measure.

Solution. We multiply by $180°/\pi$ in each case:

$$\frac{\pi}{9} = \frac{\pi}{9} \times \frac{180°}{\pi} = 20°$$

$$\frac{7\pi}{6} = \frac{7\pi}{6} \times \frac{180°}{\pi} = 210°$$

CALCULATOR COMMENT   A calculator can convert radian measure to decimal form easily, especially since scientific calculators have a special key for π. On the other hand, if the value of a trigonometric function of an angle in radians has to be found, care must be taken to *set the calculator in the radian mode*.

Caution. Once put into the radian mode, most calculators remain in this mode until changed back to degree mode or turned off.

Example 3 Find sin 1.921 and cos 2.135 with a calculator.

Solution. Set the calculator in the **radian mode.** Now enter 1.921 and press [SIN]. Thus

$$\sin 1.921 = 0.9393$$

Similarly,

$$\cos 2.135 = -0.5347$$

Exercises / Section 7.4

In Exercises 1–24, convert each degree measure to radian measure expressed in terms of π.

1. 30°
2. 45°
3. 60°
4. 15°
5. 20°
6. 32°
7. 72°
8. −45°
9. −60°
10. 150°
11. 210°
12. 225°
13. 99°
14. 135°
15. 220°
16. 315°
17. 108°
18. 144°
19. 38°
20. −112°
21. −336°
22. 276°
23. 117°
24. 189°

In Exercises 25–40, convert each radian measure to degree measure.

25. $\dfrac{\pi}{4}$
26. $\dfrac{4\pi}{3}$
27. $-\dfrac{7\pi}{6}$
28. $\dfrac{5\pi}{3}$
29. $\dfrac{5\pi}{36}$
30. $\dfrac{7\pi}{12}$
31. $\dfrac{11\pi}{10}$
32. $\dfrac{25\pi}{18}$
33. $\dfrac{16\pi}{9}$
34. $-\dfrac{17\pi}{12}$
35. $\dfrac{21\pi}{10}$
36. $\dfrac{17\pi}{18}$
37. $\dfrac{\pi}{60}$
38. $-\dfrac{17\pi}{90}$
39. $\dfrac{7\pi}{9}$
40. $\dfrac{5\pi}{12}$

In Exercises 41–48, use a calculator to convert the degree measures to radian measures accurate to four decimal places.

41. 17°35′
42. 74°10′
43. 132°48′
44. 218°53′
45. 320°24′
46. −25°5′
47. −100°10′
48. 170°51′

In Exercises 49–56, use a calculator to convert the radian measures to degree measures.

49. 0.4160
50. 1.7320
51. 2.6810
52. 2.9460
53. 3.2572
54. −4.8417
55. −0.9026
56. 2.1136

In Exercises 57–64, use a calculator to find the values of the given trigonometric functions. (Remember to set your calculator in the **radian mode.**)

57. sin (0.8642)
58. cos (1.3246)
59. sec (−0.9174)
60. csc (2.1385)
61. tan (−1.7392)
62. cot (1.9127)
63. cos (2.0176)
64. tan (2.4014)

7.5 Applications of Radian Measure

Although radian measure is of primary importance in more advanced theoretical work, some basic applications can be considered now. For example, it follows directly from the definition of radian measure, formula (7.6), that

$$s = r\theta \tag{7.7}$$

where θ is the central angle shown in Figure 7.28. This formula enables us to find the length of a circular arc, provided, of course, that θ is expressed in radians.

Length of a circular arc:

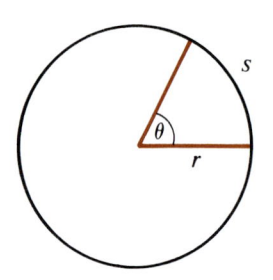

$$s = r\theta, \quad \theta \text{ in radians}$$

Figure 7.28

Example 1 Find the arc length s in Figure 7.29.

Solution. First we need to change $20°$ to radians:

$$20° = \frac{20\pi}{180} = \frac{\pi}{9}$$

Then by formula (7.7)

$$s = (3.0 \text{ in.}) \left(\frac{\pi}{9}\right) = 1.0 \text{ in.}$$

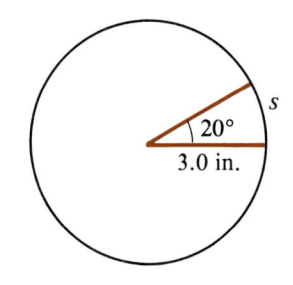

Figure 7.29

Radian measure also enables us to calculate the area of a sector of a circle. Since the area of a circular sector is proportional to the central angle, we have for the sectors in Figure 7.30

$$\frac{A}{\theta} = \frac{A'}{\theta'} \qquad (7.8)$$

Now suppose sector A' consists of the entire circle. Then $A' = \pi r^2$ and $\theta' = 2\pi$, and equation (7.8) becomes

$$\frac{A}{\theta} = \frac{\pi r^2}{2\pi}$$

or

$$A = \frac{1}{2} r^2 \theta \qquad (7.9)$$

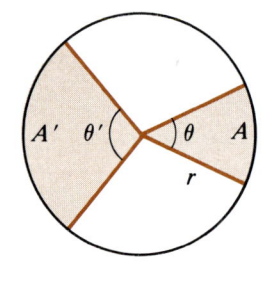

Figure 7.30

7.5 APPLICATIONS OF RADIAN MEASURE

The sector is shown in Figure 7.31.

Area of a circular sector:

$$A = \frac{1}{2} r^2 \theta, \quad \theta \text{ in radians}$$

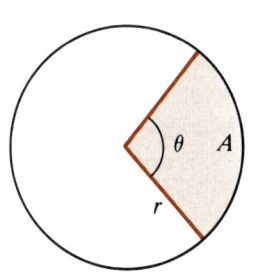

Figure 7.31

Example 2 Find the area of the circular sector in Figure 7.29.

Solution. From Example 1, $\theta = \pi/9$. So by formula (7.9),

$$A = \frac{1}{2} (3.0 \text{ in.})^2 \frac{\pi}{9} = 1.6 \text{ in.}^2$$

Example 3 A tree 193 ft from an observer on the ground intercepts an angle of 6.24°. (See Figure 7.32.) Find the approximate height of the tree.

Solution. The height of the tree is approximately equal to the length of the intercepted arc. (See Figure 7.32.) To find the length of the arc, we use the formula $s = r\theta$. Since

$$\theta = 6.24° = 6.24 \left(\frac{\pi}{180}\right) \approx 0.1089$$

it follows that the height is approximately equal to

$$(193 \text{ ft})(0.1089) = 21.0 \text{ ft}$$

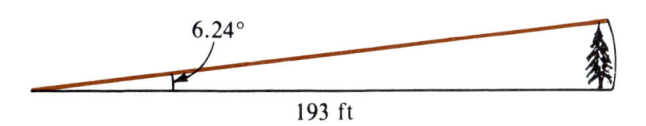

Figure 7.32

Angular velocity
Linear velocity

A particularly interesting application of radian measure involves rotational motion. Recall that for motion along a path, distance = rate × time, where the rate is assumed to be constant. Suppose that a particle is moving around a circle at a constant rate. If we denote the distance along the circle by s and the rate by v (for velocity), then $v = s/t$. Since $s = r\theta$, we have $v = (r\theta)/t = r(\theta/t)$. The ratio θ/t is called the **angular velocity** and is usually denoted by ω (omega); v is called the **linear velocity.** The relationship between linear and angular velocity is given next.

$$v = \omega r \qquad (7.10)$$

(The units for ω are radians per unit time.)

Example 4   A particle is moving about a circle of radius 3.0 in. with an angular velocity of 2.0 rad/min. Find the linear velocity v.

Solution. By formula (7.10)

$$v = 2.0 \text{ rad/min} \times 3.0 \text{ in.} = 6.0 \text{ in./min}$$

Note that since a radian is a unit-free number, it is not included in the final result. Consequently, v can be expressed simply as 6.0 in./min, which is the velocity of the particle along the rim.

In many problems ω is expressed in terms of revolutions per unit time. In such a case ω has to be converted to radians per unit time.

Example 5   An old phonograph record with a 5.0-in. radius rotates at the rate of 78 rev/min (revolutions per minute). Find the linear velocity of a point on the rim in feet per second.

Solution. Since 1 rev/min = 2π rad/min,

$$78 \text{ rev/min} = 156\pi \text{ rad/min}$$

Hence by formula (7.10)

$$v = 156\pi \, \frac{\text{rad}}{\text{min}} \times 5.0 \text{ in.}$$

$$= 780\pi \, \frac{\text{in.}}{\text{min}} \quad \text{(omitting radians)}$$

$$= 780\pi \, \frac{\text{in.}}{\text{min}} \times \frac{1 \text{ ft}}{12 \text{ in.}} \times \frac{1 \text{ min}}{60 \text{ sec}} = 3.4 \, \frac{\text{ft}}{\text{sec}}$$

(For more details on converting units, see Appendix A.)

Exercises / Section 7.5

1. A circle has a radius of 5.00 in. Find the length of the arc intercepted by a central angle of 0.524 (in radians).
2. A circle has a radius of 4.00 cm. Find the length of the arc intercepted by a central angle of 25.1°.
3. Find the area of the circular sector in Exercise 1.
4. Find the area of the circular sector in Exercise 2.
5. Find the length of the circular arc intercepted by an angle of 55°10′ if the radius of the circle is 10.0 cm.
6. Find the area of the circular sector in Exercise 5.
7. A tree 520 ft away intercepts an angle of 1.5°. Find the approximate height of the tree. (See Example 3.)
8. A building 1,010 ft away intercepts an angle of 3.0°. Find the approximate height.
9. Find the degree measure of the central angle that intercepts an arc of length 6.5 ft on a circle of radius 6.0 ft.
10. Find the degree measure of the central angle that intercepts an arc of length 20.0 m on a circle of radius 8.0 m.
11. The full moon intercepts an angle of $\frac{1}{2}°$. Find the diameter of the moon, given that the distance from the earth to the moon is about 240,000 mi.
12. The mean distance from the earth to the sun is 93 million miles. The diameter of the sun is 866,000 mi. Find its angular size, that is, the intercepted angle. Why do the sun and moon appear to be the same size to the naked eye (referring to Exercise 11)?
13. A more accurate measure of the distance from the earth to the moon is 239,000 mi (refer to Exercise 11). Given that the diameter of the moon is 2,163 mi, find the angular size of the moon to the nearest minute.
14. A pendulum of length 15.0 in. swings through an arc of 5°10′. Find the distance covered by the end of the pendulum as it swings from one end of the arc to the other.
15. Two circles are concentric if they have the same center. Suppose two circles have radii 8.00 cm and 10.0 cm, respectively. Find the area of the portion of the sector inside the larger circle and outside the smaller circle if the central angle is 125°50′.
16. The first reasonably accurate estimate of the radius of the earth was obtained by Eratosthenes (276–198 B.C.) in Egypt. He observed that whenever the sun was directly overhead in the town of Aswan, it made an angle of 7.2° with a line perpendicular to the ground in Alexandria 495 mi away (see Figure 7.33). Use Eratosthenes' method to estimate the radius.

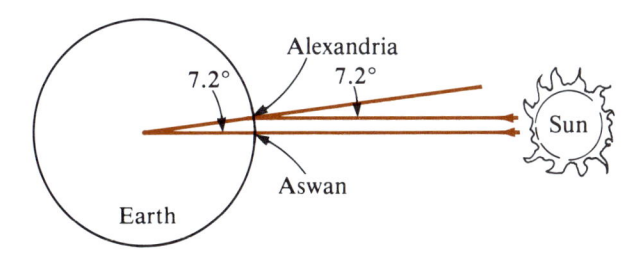

Figure 7.33

17. An object is moving about a circle of radius 10.0 cm with an angular velocity of 0.50 rad/sec. Find its linear velocity.

18. A circular disk 8.0 in. in diameter rotates at the rate of 5.2 rad/sec. Find the linear velocity of a point on the rim.

19. A wheel with a radius of 3.50 in. rotates at the rate of 5.70 rev/sec. Find the linear velocity of a point on the rim in feet per second.

20. A flywheel spins at the rate of 312 rev/min. If the radius of the flywheel is 8.15 cm, find the velocity of a point on the rim in meters per minute.

21. Find the velocity in miles per hour of a point on the equator due to the earth's rotation. (The radius of the earth is approximately 4,000 mi.)

22. A bicycle is traveling at the rate of 1,600 ft/min. Find the angular velocity of the wheels in radians per second, given that each wheel has a diameter of 32 in.

23. Determine the angular velocity of the wheels of a car in revolutions per second if each wheel has a radius of 14 in. and the car is traveling at 30 mi/hr. (Recall that 88 ft/sec = 60 mi/hr.)

24. A pulley belt is 10 ft long and takes 45 sec to make a complete revolution. If the radius of the pulley is 12 in., determine its angular velocity in radians per second.

25. Find the velocity in inches per minute of the tip of an 8-in. hour hand.

26. Determine the velocity of the moon in miles per hour relative to the earth. (Assume that the moon takes 28 days to make one revolution about the earth and that its distance from the earth is 240,000 mi.)

27. Determine the velocity in miles per second of the earth relative to the sun. (Assume that the distance from the earth to the sun is 93 million miles.)

Review Exercises / Chapter 7

In Exercises 1–4, find the values of the six trigonometric functions for the angle whose terminal side passes through the point indicated.

1. $(-2, \sqrt{5})$
2. $(-3, -1)$
3. $(2, -4)$
4. $(-5, 12)$

5. If $\sin \theta = -\frac{1}{3}$ and $\cot \theta > 0$, find $\cos \theta$.
6. If $\csc \theta = -\sqrt{6}/2$ and $\cos \theta < 0$, find $\cot \theta$.
7. If $\cos \theta = \sqrt{3}/4$ and $\tan \theta < 0$, find $\sin \theta$.
8. If $\tan \theta = \frac{1}{4}$ and $\sec \theta < 0$, find $\csc \theta$.

In Exercises 9–20, find the values of the trigonometric functions by means of diagrams. (Do not use a table or a calculator.)

9. $\sin 135°$
10. $\tan 180°$
11. $\cos 240°$
12. $\sec 315°$
13. $\tan 225°$
14. $\csc 150°$
15. $\sec 210°$
16. $\cot 330°$
17. $\csc 90°$
18. $\tan 270°$
19. $\cos 150°$
20. $\csc 300°$

In Exercises 21–28, use a diagram to find all angles θ such that $0° \leq \theta < 360°$. (Do not use a table or a calculator.)

21. $\cos \theta = \dfrac{1}{2}$
22. $\csc \theta = -2$
23. $\cot \theta = -\dfrac{\sqrt{3}}{3}$
24. $\tan \theta = \dfrac{\sqrt{3}}{3}$
25. $\tan \theta$ undefined
26. $\cos \theta = -1$
27. $\sec \theta = -\dfrac{2\sqrt{3}}{3}$
28. $\sin \theta = -\dfrac{\sqrt{2}}{2}$

In Exercises 29–32, use Table 1 of Appendix B or a calculator to find each value.

29. tan 123°15'   **30.** csc 250°18'   **31.** cos 318°50'   **32.** sec 107°5'

In Exercises 33–38, use Table 1 of Appendix B or a calculator to find θ such that $0° \leq \theta < 360°$.

33. $\cos \theta = -0.8107$ **34.** $\csc \theta = 1.397$ **35.** $\tan \theta = -0.4170$ **36.** $\sin \theta = 0.2713$
37. $\cot \theta = 1.786$ **38.** $\sec \theta = -3.162$

In Exercises 39–44, convert the given degree measures to radian measures expressed in terms of π.

39. 40°   **40.** 160°   **41.** −100°   **42.** −36°
43. 54°   **44.** 236°

In Exercises 45–50, convert the given radian measures to degree measures.

45. $\dfrac{5\pi}{6}$ **46.** $\dfrac{7\pi}{36}$ **47.** $\dfrac{13\pi}{10}$ **48.** $\dfrac{5\pi}{12}$
49. $\dfrac{17\pi}{9}$ **50.** $\dfrac{23\pi}{18}$

In Exercises 51–58, use a calculator to convert the given degree measures to radian measures. (Give answers to four decimal places.)

51. 25°18'   **52.** 79°40'   **53.** 144°6'   **54.** −23°51'
55. −257°10'   **56.** 340°26'   **57.** 390°   **58.** 400°

In Exercises 59–62, use a calculator to convert the given radian measures to degree measures.

59. 0.7162   **60.** 1.438   **61.** 2.605   **62.** 3.864

In Exercises 63–68, find the value of each trigonometric function. (Set your calculator in the radian mode.)

63. tan 2.161   **64.** sin (−0.3680)   **65.** cos (1.783)   **66.** tan (2.349)
67. csc 1.507   **68.** sec 5.458

69. A circle has a radius of 6.00 in. Find the length of the arc intercepted by a central angle of 32.0°.
70. Find the area of the circular sector in Exercise 69.
71. A tower 860 ft away intercepts an angle of 3.2°. Find the approximate height of the tower.
72. Find the degree measure of the central angle that intercepts an arc length of 13.3 ft on a circle of radius 7.80 ft.
73. The wheels of a bicycle have a radius of 16.0 in. and rotate at the rate of 2.20 rev/sec. Find the speed of the bicycle in miles per hour.
74. An artificial satellite travels around the earth in a circular orbit at an altitude of 300 mi. Find its velocity (in miles per hour) if it makes one revolution every 90 min. (The radius of the earth is approximately 4,000 mi.)
75. Find the velocity in miles per hour of a communications satellite that remains 22,300 mi above a point on the equator at all times. (See Exercise 74.)

CHAPTER 8

Graphs of Trigonometric Functions

Objectives Upon completion of this chapter, you should be able to:

1. Determine the period and amplitude of a given function of the form $y = a \sin bx$ and $y = a \cos bx$ and sketch the graph.
2. Determine the phase shift of a sinusoidal function and sketch the graph.
3. Sketch the graph of a function representing simple harmonic motion.
4. Sketch the graphs of the tangent, cotangent, secant, and cosecant functions.
5. Sketch certain graphs by adding coordinates.

8.1 Graphs of Sine and Cosine Functions

The brief discussion in Chapter 2 showed that a graph illustrates the behavior of a given function. In this chapter we will study the graphs of trigonometric functions, especially the sine and cosine functions, which are particularly useful in physical applications. Our first task is to draw the graphs of $y = \sin x$ and $y = \cos x$ by constructing a table of values and plotting enough points to obtain a smooth curve.

Since it is customary to use radian measure when graphing trigonometric functions, let us first recall the radian measure of certain special angles. From

$$\pi = 180°$$

we get $\pi/6 = 30°$. This relationship yields all special angles that are multiples of 30°. For example,

$$150° = 5 \cdot 30° = \frac{5\pi}{6} \quad \text{and} \quad 330° = 11 \cdot 30° = \frac{11\pi}{6}$$

Similarly, since $60° = \pi/3$,

$$120° = 2 \cdot 60° = \frac{2\pi}{3} \quad \text{and} \quad 300° = 5 \cdot 60° = \frac{5\pi}{3}$$

From $45° = \pi/4$, we get $225° = 5 \cdot 45° = 5\pi/4$, and so forth.

Using these special angles, we construct a table of values for the function $y = \sin x$. Plotting these points on the rectangular coordinate system, we obtain the graph shown in Figure 8.1.

$y = \sin x$

x:	0	$\frac{\pi}{6}$	$\frac{\pi}{4}$	$\frac{\pi}{3}$	$\frac{\pi}{2}$	$\frac{2\pi}{3}$	$\frac{5\pi}{6}$	π	$\frac{7\pi}{6}$	$\frac{4\pi}{3}$	$\frac{3\pi}{2}$	$\frac{11\pi}{6}$	2π	$2\pi + \frac{\pi}{6}$
y (exact):	0	$\frac{1}{2}$	$\frac{1}{\sqrt{2}}$	$\frac{\sqrt{3}}{2}$	1	$\frac{\sqrt{3}}{2}$	$\frac{1}{2}$	0	$-\frac{1}{2}$	$-\frac{\sqrt{3}}{2}$	-1	$-\frac{1}{2}$	0	$\frac{1}{2}$
y (decimal):	0	0.5	0.7	0.87	1	0.87	0.5	0	-0.5	-0.87	-1	-0.5	0	0.5

$y = \sin x$

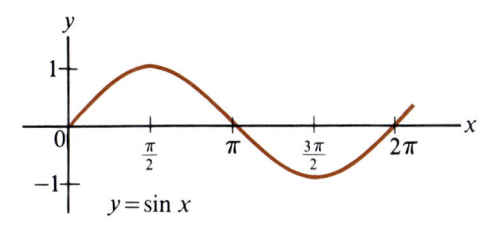

Figure 8.1

Observe that as x increases from 0 to $\pi/2$, the values of sin x increase from 0 to 1. As x continues to increase from $\pi/2$ to π, the values of sin x repeat in reverse order from 1 to 0. As x increases from π to $3\pi/2$, the values of sin x become negative, decreasing from 0 to -1. Finally, as x increases from $3\pi/2$ to 2π, the values of sin x increase once again, from -1 back to 0. *Starting at 2π, the values of sin x repeat.*

Note especially the zero values of sin x as well as the largest and smallest values:

x:	0	$\frac{1}{4}(2\pi) = \frac{\pi}{2}$	$\frac{1}{2}(2\pi) = \pi$	$\frac{3}{4}(2\pi) = \frac{3\pi}{2}$	$1 \cdot (2\pi) = 2\pi$
sin x:	0	1	0	-1	0

To obtain the graph of $y = \cos x$, we construct the following table:

$y = \cos x$

x:	0	$\frac{\pi}{6}$	$\frac{\pi}{4}$	$\frac{\pi}{3}$	$\frac{\pi}{2}$	$\frac{2\pi}{3}$	$\frac{5\pi}{6}$	π	$\frac{7\pi}{6}$	$\frac{4\pi}{3}$	$\frac{3\pi}{2}$	$\frac{5\pi}{3}$	$\frac{11\pi}{6}$	2π
y (exact):	1	$\frac{\sqrt{3}}{2}$	$\frac{1}{\sqrt{2}}$	$\frac{1}{2}$	0	$-\frac{1}{2}$	$-\frac{\sqrt{3}}{2}$	-1	$-\frac{\sqrt{3}}{2}$	$-\frac{1}{2}$	0	$\frac{1}{2}$	$\frac{\sqrt{3}}{2}$	1
y (decimal):	1	0.87	0.7	0.5	0	-0.5	-0.87	-1	-0.87	-0.5	0	0.5	0.87	1

Plotting these points, we get the graph shown in Figure 8.2.

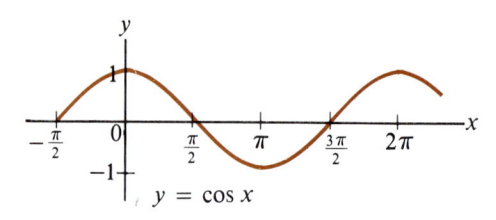

Figure 8.2

Note especially the zero values of $\cos x$ as well as the largest and smallest values:

x:	0	$\frac{1}{4}(2\pi) = \frac{\pi}{2}$	$\frac{1}{2}(2\pi) = \pi$	$\frac{3}{4}(2\pi) = \frac{3\pi}{2}$	$1 \cdot (2\pi) = 2\pi$
$\cos x$:	1	0	-1	0	1

A closer inspection of the tables for $y = \sin x$ and $y = \cos x$ suggests that the values of the sine and cosine functions and hence the shapes of their graphs are identical except for their positions. Indeed, if the graph of the cosine function is moved $\pi/2$ units to the right, it becomes the graph of the sine function.

Given these basic shapes, our real goal in this section is to sketch the graphs of $y = a \sin bx$ and $y = a \cos bx$ without plotting points. To this end, observe that the graph of $y = a \cos x$ can be obtained from the graph of $y = \cos x$ by multiplying each value of $\cos x$ by a. The number $|a|$ is called the *amplitude*. Consider the following examples.

Example 1 Sketch the graph of $y = 2 \cos x$.

Solution. First note that the coefficient 2 doubles the values of $y = \cos x$; otherwise, the shape is essentially the same as that of the basic cosine function. Therefore only the few points in the following table will need to be plotted:

x:	0	$\frac{\pi}{2}$	π	$\frac{3\pi}{2}$	2π
y:	2	0	-2	0	2

The graph, shown in Figure 8.3, looks like a tall version of the graph of $y = \cos x$.

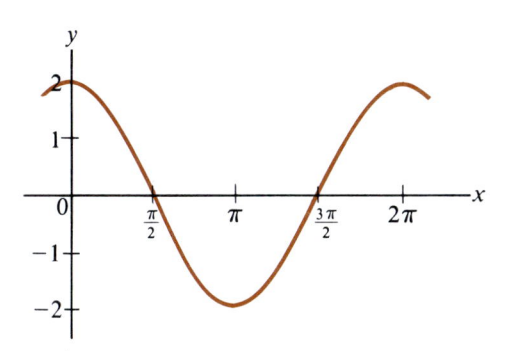

Figure 8.3

Example 2 Sketch the graph of $y = -3 \sin x$.

Solution. The effect of the coefficient -3 is twofold: It multiplies the values of the basic sine function by 3 and changes the sign of each value. The given

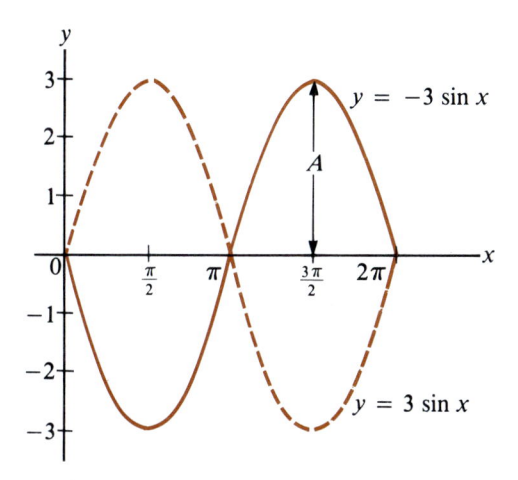

Figure 8.4

curve is therefore the "mirror image" of $y = 3 \sin x$ with respect to the *x*-axis. Since the shape of the basic sine curve is already known, we can sketch the curve by plotting only the highest and lowest points on the curve and the points where the curve crosses the *x*-axis (called the *x-intercepts*). The graph of $y = 3 \sin x$ is the dashed curve in Figure 8.4. The reflection of the dashed curve is the graph of $y = -3 \sin x$, shown by the solid curve in Figure 8.4.

These examples show that in the graphs of $y = a \sin x$ and $y = a \cos x$ the numerical value of the coefficient a is the maximum distance from the curve to the *x*-axis. This distance is called the **amplitude,** which we will denote by A. Thus $A = |a|$. (See Figure 8.4.)

Another important property of trigonometric functions is the **periodicity**. This term refers to the fact that the function values eventually repeat. In the case of the sine and cosine functions, the *y*-values repeat every 2π radians; this interval is called the **period**. Thus $y = \sin x$ and $y = \cos x$ are said to have a period of 2π. The general case is given next.

Period and amplitude of $y = a \sin bx$ and $y = a \cos bx$:

$$\textbf{Period: } P = \frac{2\pi}{b} \qquad \textbf{Amplitude: } A = |a| \tag{8.1}$$

To see why the period is $2\pi/b$, note first that the graph of $y = \sin x$ passes through the origin and crosses the *x*-axis at

$$x = \pi, 2\pi, 3\pi, \ldots \tag{8.2}$$

The second intercept, 2π, is equal to the period, denoted by P. To obtain the period of $y = a \sin bx$, we observe that

$$a \sin bx = 0$$

at the origin and whenever

$$bx = \pi, 2\pi, 3\pi, \ldots$$

or

$$x = \frac{\pi}{b}, \frac{2\pi}{b}, \frac{3\pi}{b}, \ldots \tag{8.3}$$

Comparing the intercepts given in statements (8.3) and (8.2) shows that $P = 2\pi/b$. By a similar argument, $2\pi/b$ is also the period of the cosine function. (Note that $b = 1$ for $y = \sin x$ and $y = \cos x$.)

For convenience, the functions $y = a \sin bx$ and $y = a \cos bx$ are both called **sinusoidal functions** and their graphs **sinusoidal curves.** (Slightly more general forms of the sinusoidal curves will be discussed in the next section.)

Sinusoidal function

8.1 GRAPHS OF SINE AND COSINE FUNCTIONS

> To graph a sinusoidal curve, determine the period and amplitude and sketch the curve from the basic shape.

The remaining examples illustrate the technique for graphing sinusoidal curves.

Example 3 Find the amplitude and period of $y = 2 \sin 3x$ and sketch the curve.

Solution. By statement (8.1) we have $P = 2\pi/b$ and $A = |a|$. It follows that

$$A = 2 \quad \text{and} \quad P = \frac{2\pi}{3}$$

To sketch the curve, we need only to mark its highest and lowest points and the end of the period starting at the origin. The sketch is shown in Figure 8.5 over two periods.

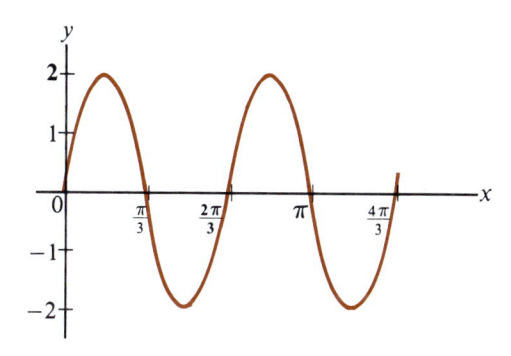

Figure 8.5

Example 4 Find the amplitude and period of $y = \frac{1}{2} \cos 2x$ and sketch the curve.

Solution. $A = \frac{1}{2}$ and $P = 2\pi/2 = \pi$. We now mark the highest and lowest points on the curve and the end of the period starting at the origin. Keeping

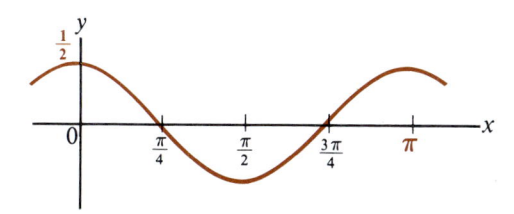

Figure 8.6

CHAPTER 8 GRAPHS OF TRIGONOMETRIC FUNCTIONS

in mind the shape of the basic cosine curve, the graph can be sketched directly (Figure 8.6).

Example 5 Sketch the graph of $y = 5 \sin \frac{1}{2}x$.

Solution. $A = 5$, $P = 2\pi/\frac{1}{2} = 4\pi$. See Figure 8.7.

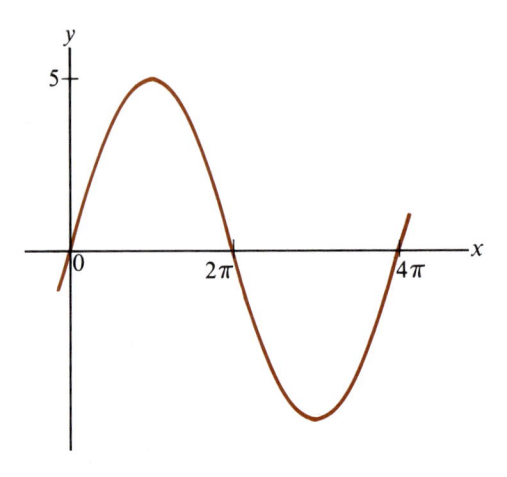

Figure 8.7

Example 6 Sketch the graph of $y = -6 \cos \frac{1}{3}x$.

Solution. $A = 6$, $P = 2\pi/\frac{1}{3} = 6\pi$. Because of the negative coefficient, the graph is the reflection of the graph of $y = 6 \cos \frac{1}{3}x$ (Figure 8.8).

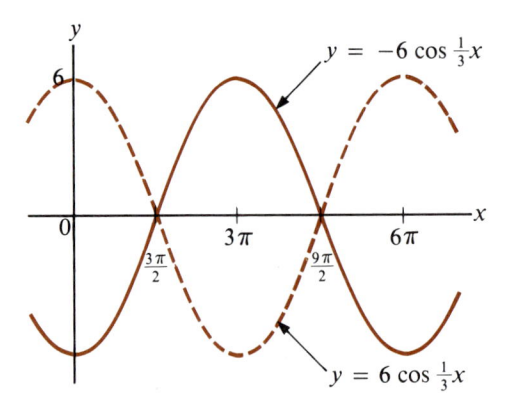

Figure 8.8

Exercises / Section 8.1

State the amplitude and period of each function; sketch the curve.

1. $y = 2 \sin x$
2. $y = 3 \cos x$
3. $y = -\sin x$
4. $y = -\cos x$
5. $y = \frac{1}{2} \sin x$
6. $y = \frac{1}{3} \cos 2x$
7. $y = 2 \cos 3x$
8. $y = \frac{1}{2} \sin \frac{1}{2} x$
9. $y = -5 \cos 2x$
10. $y = 6 \sin \frac{1}{2} x$
11. $y = \frac{1}{2} \cos 3x$
12. $y = 2 \sin 3x$
13. $y = 4 \sin \frac{2}{3} x$
14. $y = 6 \cos \frac{3}{4} x$
15. $y = -10 \sin \frac{1}{4} x$
16. $y = -12 \cos \frac{1}{8} x$
17. $y = 5 \sin \frac{1}{5} x$
18. $y = -\frac{3}{4} \sin \frac{1}{2} x$
19. $y = -4 \cos \frac{7}{3} x$
20. $y = 6 \sin \frac{3}{5} x$
21. $y = \sin \pi x$
22. $y = \cos \pi x$
23. $y = \frac{1}{3} \cos 3\pi x$
24. $y = 2 \sin 2\pi x$

8.2 Phase Shifts

The sinusoidal curves discussed in the last section are not the most general possible curves of this type. In addition to having a definite period and amplitude, the curves may be shifted to the left or right.

To see this behavior, consider the function

$$y = 2 \sin \left(x - \frac{\pi}{4} \right)$$

Suppose we compare the given curve to that of the function $y = 2 \sin x$, which has amplitude 2 and period 2π. The y-values of $y = 2 \sin (x - \pi/4)$ are identical to the values of $y = 2 \sin x$, but they correspond to different x-

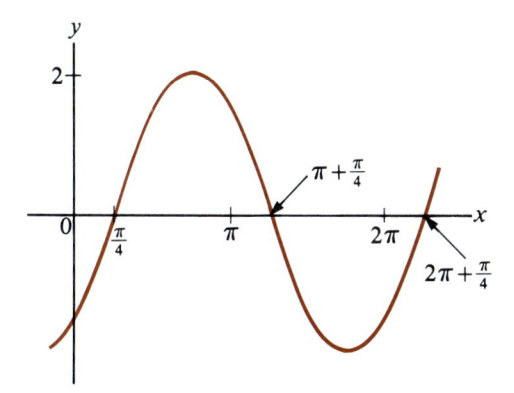

Figure 8.9

values. Note especially that to get sin 0, we must let $x = \pi/4$. So the point $(\pi/4, 0)$ corresponds to the origin. The other points are shifted similarly, so that the y-values start repeating at $x = 2\pi + \pi/4$ (instead of $x = 2\pi$ for the basic sine function). The graph is shown in Figure 8.9 on page 257.

Let's consider another case.

Example 1 Sketch the graph of $y = 2 \sin(x + \pi/4)$.

Solution. This function is similar to the function above, but this time we need to let $x = -\pi/4$ to obtain sin 0. Consequently, the point $(-\pi/4, 0)$ corresponds to the origin, and the entire curve is shifted to the left (Figure 8.10).

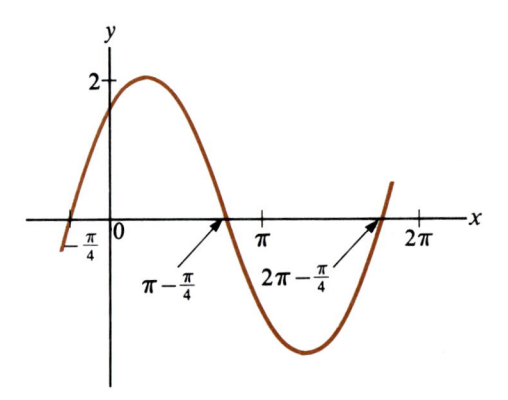

Figure 8.10

More generally, it is shown in analytic geometry that the graph of $y = f(x - h)$, $h > 0$, is the graph of $y = f(x)$ shifted h units to the right, while the graph of $y = f(x + h)$ is the graph of $y = f(x)$ shifted h units to the left.

To sketch the graph of $y = a \sin(bx + c)$, we first factor b and write $y = a \sin b(x + c/b)$, which shows that the shift is given by c/b. Similarly, from $y = a \cos b(x + c/b)$ we see that the shift is also given by c/b. The number c/b is commonly referred to as the **phase shift.**

We shall now summarize the basic features of sinusoidal functions.

Period, amplitude, and phase shift of $y = a \sin(bx \pm c)$ and $y = a \cos(bx \pm c)$:

Period: $P = \dfrac{2\pi}{b}$ **Amplitude:** $A = |a|$ **Phase Shift:** $\dfrac{c}{b}$

To sketch the graph of $y = a \sin(bx \pm c)$, we first sketch the graph of $y = a \sin bx$ using the techniques discussed in the previous section. We then shift this curve by c/b units to obtain the graph of $y = a \sin(bx \pm c)$. The procedure for sketching a cosine function is similar.

These ideas are illustrated in the remaining examples.

Example 2 Sketch the graph of $y = 3 \sin(2x - \pi)$.

Solution. Writing the function in the form $y = 3 \sin 2(x - \pi/2)$, we see that the phase shift is $\pi/2$. It follows that the desired graph is the graph of $y = 3 \sin 2x$ shifted $\pi/2$ units *to the right*. Consequently, it is actually simpler to sketch the graph of $y = 3 \sin 2x$ first and then translate the resulting curve to the right.

For $y = 3 \sin 2x$, we have $A = 3$ and $P = \pi$. The graph is the dashed curve in Figure 8.11. This graph is then shifted $\pi/2$ units to the right, as shown by the solid curve in Figure 8.11.

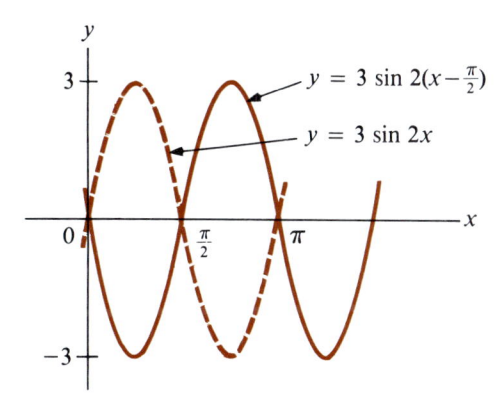

Figure 8.11

Example 3 Sketch the graph of

$$y = 4 \cos\left(\frac{1}{2}x + \frac{\pi}{16}\right)$$

Solution. By writing the function as

$$y = 4 \cos \frac{1}{2}\left(x + \frac{\pi}{8}\right)$$

we see that the graph can be obtained by shifting $y = 4 \cos \frac{1}{2}x$ to the left by $\pi/8$ units.

For $y = 4 \cos \frac{1}{2}x$ we have $A = 4$ and $P = 4\pi$. The graph is the dashed curve in Figure 8.12. Shifting the curve $\pi/8$ units *to the left* yields the solid curve shown in the same figure.

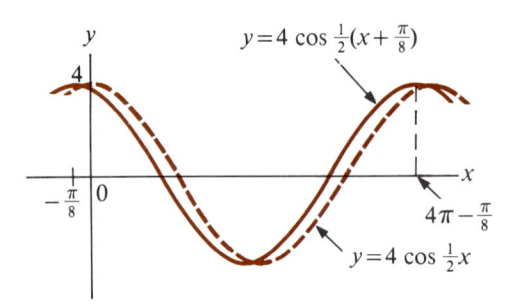

Figure 8.12

Exercises / Section 8.2

State the amplitude, period, and phase shift of each function. Sketch each curve over one period.

1. $y = 2 \sin \left(x - \frac{\pi}{4} \right)$
2. $y = 2 \cos \left(x + \frac{\pi}{4} \right)$
3. $y = \frac{1}{2} \sin \left(x + \frac{\pi}{8} \right)$
4. $y = \frac{1}{3} \cos \left(x - \frac{\pi}{4} \right)$
5. $y = \frac{1}{2} \sin \left(\frac{1}{2} x - \frac{\pi}{8} \right)$
6. $y = \frac{1}{2} \cos \left(\frac{1}{2} x + \frac{\pi}{8} \right)$
7. $y = 3 \sin \left(x + \frac{3\pi}{2} \right)$
8. $y = \frac{1}{2} \cos \left(2x - \frac{\pi}{4} \right)$
9. $y = 2 \cos \left(x - \frac{\pi}{3} \right)$
10. $y = 4 \sin (2x - 1)$
11. $y = -3 \cos (x - 2)$
12. $y = -4 \sin (3x - 3)$
13. $y = 3 \sin \left(2x + \frac{\pi}{2} \right)$
14. $y = 3 \cos \left(3x + \frac{3\pi}{4} \right)$
15. $y = 10 \cos \left(\frac{1}{3} x - \frac{4\pi}{5} \right)$
16. $y = 2 \sin \left(3x + \frac{\pi}{4} \right)$
17. $y = 2 \cos (\pi x + \pi)$
18. $y = \sin (\pi x - \pi)$
19. $y = 3 \sin \left(\frac{1}{2} \pi x - 1 \right)$
20. $y = 2 \sin (2\pi x + 1)$

8.3 Applications of Sinusoidal Functions

Many natural phenomena are sinusoidal, even such diverse phenomena as water waves, sound waves, and alternating current. To understand why this is so, we must study a phenomenon called **simple harmonic motion.**

Simple Harmonic Motion

Consider a wheel of radius r with a handle on the rim rotating at a constant rate. If a light source is placed some distance away, we can study the movement of the shadow of the handle on the wall (Figure 8.13). The problem is equivalent to finding the projection of the point P onto the y-axis in Figure 8.14. In fact, observe that the distance d from P to the horizontal axis can be found from the relation $\sin \theta = d/r$, so that

$$d = r \sin \theta$$

Now assume that the motion starts at a and proceeds in a counterclockwise direction with angular velocity ω. By definition, $\omega = \theta/t$, so that $\theta = \omega t$.

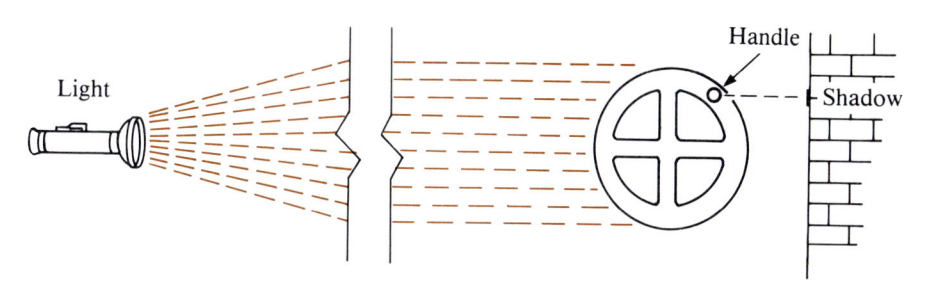

Figure 8.13

Hence

$$d = r \sin \omega t \tag{8.4}$$

This equation expresses the distance d from the horizontal axis as a function of time. Moreover, the projection onto the y-axis in Figure 8.14 is the y-coordinate of the graph of $y = r \sin \omega t$. Consequently, the graph of $y = r \sin \omega t$ gives a pictorial representation of the variable distance from P to the t-axis. Observe also that $A = r$ and $P = 2\pi/\omega$; that is, the amplitude is the radius of the circle, and the period corresponds to one complete revolution.

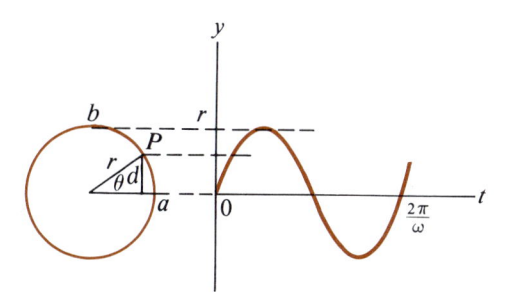

Figure 8.14

The motion just described is called **simple harmonic motion**. If the motion begins at b, then it is described by $y = r \cos \omega t$.

Simple harmonic motion is described by

$$y = r \sin \omega t \quad \text{or} \quad y = r \cos \omega t \tag{8.5}$$

This type of motion turns out to be the common trait among the phenomena mentioned at the beginning of this section. For example, the motion of a weight hanging on a spring and oscillating vertically is simple harmonic motion; that is, the motion of the weight is identical to the motion of the shadow in Figure 8.13. Other examples are the vertical motion of a floating object caused by water waves and the sound-producing vibration of the end of a tuning fork.

Example 1 An object is moving around a circle of radius **5.0** cm with a constant angular velocity of **2** rad/sec. If the motion starts at point *a* in Figure 8.14, sketch the graph of the function describing the motion of the vertical projection.

Solution. By formula (8.5), $y = 5.0 \sin 2t$. Note that $A = 5.0$ cm and $P = \pi$ sec. The graph is shown in Figure 8.15 over two periods.

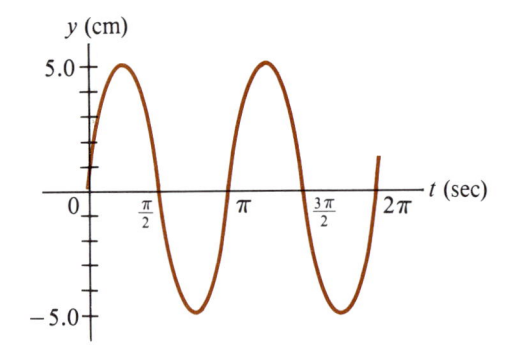

Figure 8.15

Alternating Current

Equations similar to those for simple harmonic motion arise in the study of alternating current. If a wire is moved rapidly through a magnetic field so that it "cuts" the lines of force, then a current is induced in the wire. More precisely, if B is the flux density of the magnetic field (in webers per square meter), L the length of the wire (in meters), and v the velocity of the wire (in meters per second), and if the wire cuts the lines of force at right angles, then the induced emf (electromotive force in volts) is given by

$$E = BLv \tag{8.6}$$

The velocity comes into play because it is proportional to the number of lines of force that the wire is able to cut in a given time interval.

Formula (8.6) is precisely the principle used in the operation of a generator. Consider the schematic diagram in Figure 8.16. The rectangular coil is rotated so that the wires cut the lines of force of the magnetic field, and the terminals of the coil are connected to the rings on the left. Brushes bearing against the rings connect the coil to an external circuit.

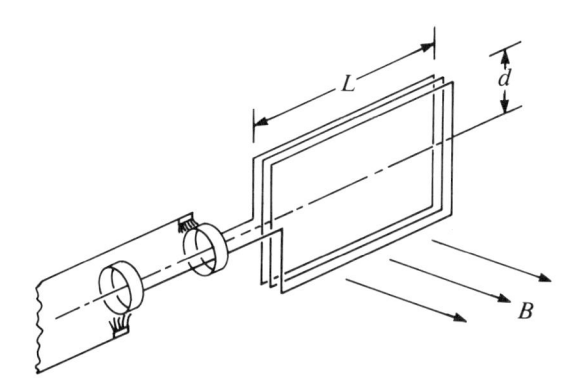

Figure 8.16

Suppose that the angular velocity of the coil is ω. Then by formula (7.10) in Section 7.5,

$$v = \omega d \tag{8.7}$$

At the instant that the plane of the coil is parallel to the lines of force, we have

$$E = BL\omega d$$

by formula (8.6), since the lines of force are cut at right angles. If the angle that the plane of the coil makes with the lines of force is $\pi/2 - \omega/t$, then E is reduced, since the wires now cut fewer lines of force during a given time interval. Suppose we think of B as the number of lines of force cut in the parallel position during the time interval; now examine Figure 8.17. A little reflection shows that the number of lines of force cut is reduced to $B \sin \omega t$ during the same time interval. Thus E varies according to

$$E = BL\omega d \sin \omega t$$

As already noted, if $\omega t = \pi/2$, then $\sin \omega t = 1$ and E attains its maximum value $BL\omega d$, denoted by $E_{\max}$. It follows that the emf is given by

$$E = E_{\max} \sin \omega t \tag{8.8}$$

If the motion starts at $t = -\alpha/\omega$ sec, then equation (8.8) becomes

$$E = E_{\max} \sin (\omega t + \alpha) \tag{8.9}$$

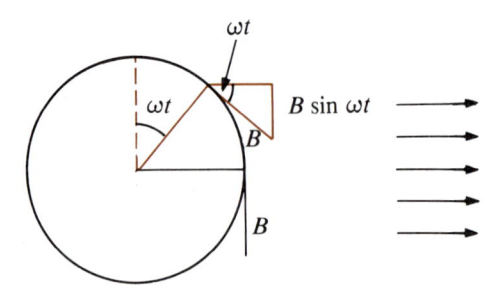

264 CHAPTER 8 GRAPHS OF TRIGONOMETRIC FUNCTIONS

Figure 8.17

From the relation $i = E/R$, it follows that the current in the external circuit is given by

$$i = I_{max} \sin(\omega t + \alpha) \tag{8.10}$$

Example 2 If $I_{max} = 5.00$ A, $\omega = 100\pi$ rad/sec, and $\alpha = \pi/3$, sketch the graph of the current.

Solution. By equation (8.10),

$$i = 5.00 \sin\left(100\pi t + \frac{\pi}{3}\right)$$

Since $i = 5.00 \sin 100\pi(t + \frac{1}{300})$, the amplitude is $I_{max} = 5.00$ A, $P = 2\pi/100\pi = \frac{1}{50}$ sec, and the shift is $\frac{1}{300}$ sec to the left. The graph is shown in Figure 8.18. At

$$t = \frac{1}{100} - \frac{1}{300} = \frac{1}{150} \text{ sec}$$

i becomes negative, indicating that the current has reversed direction. This corresponds to the instant when the coil in Figure 8.16 passes the horizontal position. The current is said to be **alternating.**

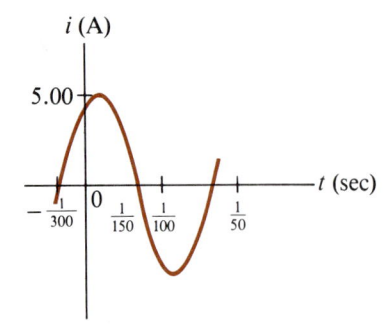

Figure 8.18

Finally, you should note that the reciprocal of the period is called the **frequency.** This is a natural idea, for if a wave has a period of $\frac{1}{20}$ sec, then one cycle is completed in $\frac{1}{20}$ sec. Thus 20 cycles are completed in 1 sec. We now say that the frequency is 20 cycles/sec, or 20 **hertz** (abbreviated "Hz"). This unit was named after the German physicist Heinrich Hertz (1857–1894), a pioneer in the study of electromagnetic phenomena.

Example 3
Sound

The sound of a tuning fork may be expressed in the form $y = A \sin 2\pi f t$, where f is the frequency. If a tuning fork vibrates at 190 Hz and has an amplitude of 0.001 in., then $y = 0.001 \sin 2\pi(\mathbf{190})t$. Note that the period is $1/f = \frac{1}{190}$ sec. This waveform can actually be observed on an oscilloscope.

Exercises / Section 8.3

1. An object moves around a circle of radius 3.0 in. starting at point (3.0, 0) with angular velocity 2 rad/sec. Sketch the graph of the resulting motion of the vertical projection.
2. Repeat Exercise 1 with the motion starting at (0, 3.0).
3. An object moves around a circle of radius 5 cm at the rate of 4 rev/min. Sketch the motion of the vertical projection, which is assumed to start at (0, 5).
4. Repeat Exercise 3 with the motion starting at (5, 0).
5. If $E_{max} = 6.00$ V, $\omega = 80\pi$ rad/sec, and $\alpha = \pi/4$, sketch the graph of E as a function of time.
6. If $E_{max} = 10.0$ V, $\omega = 120\pi$ rad/sec, and $\alpha = \pi/6$, sketch the graph of E as a function of time.
7. If $I_{max} = 6.0$ A, $\omega = 120\pi$ rad/sec, and $\alpha = \pi/4$, sketch the current as a function of time.
8. If $I_{max} = 9.0$ A, $\omega = 150\pi$ rad/sec, and $\alpha = \pi/6$, sketch the current as a function of time.
9. A weight is hanging on a spring and oscillating vertically. The displacement of the spring from the equilibrium position as a function of time is given by $y = 5 \cos 2t$. Sketch the curve of this motion.
10. Under certain conditions the vertical displacement of a horizontal string as a function of position and time is given by $y = A \sin(t - x/v)$. If $A = 2$ cm and $v = -10$ cm/sec, sketch the graph of the string at the instant when $t = 1$ sec.
11. A tuning fork vibrates at 224 Hz and has an amplitude of 0.002 in. Sketch the curve produced on an oscilloscope. (See Example 3.)
12. The form of a water wave is given by $y = A \sin 2\pi(ft - r/\lambda)$, where f is the frequency, r the distance from the source, and λ the wavelength, which is approximately $5.12T^2$ ft, where T is the period. Find the equation for the vertical motion of a floating body as a function of time if the body is 10.0 ft from the source and the period and amplitude are 8.0 sec and 2.8 ft, respectively.
13. Repeat Exercise 12 if the body is 20.0 ft from the source and if the wave has a period of 6.0 sec and an amplitude of 2.1 ft. Sketch the curve.

8.4 Graphs of Other Trigonometric Functions

The graphs of the remaining trigonometric functions occur less often than the sinusoidal curves in technical applications but play an important role in more advanced mathematics. Thus we should at least become familiar with them.

To obtain the graph of $y = \tan x$, we make up a table of values.

$y = \tan x$

x:	0	$\dfrac{\pi}{6}$	$\dfrac{\pi}{4}$	$\dfrac{\pi}{3}$	$\dfrac{\pi}{2}$	$-\dfrac{\pi}{6}$	$-\dfrac{\pi}{4}$	$-\dfrac{\pi}{3}$	$-\dfrac{\pi}{2}$
y:	0	0.6	1	1.7	undefined	-0.6	-1	-1.7	undefined

In the neighborhood of $x = \pi/2$ and $x = -\pi/2$, the values of tan x become numerically large, as a calculator will readily confirm. (See Figure 8.19.) The dashed vertical lines at

$$x = \pm \frac{\pi}{2}, \pm \frac{3\pi}{2}, \ldots$$

are approached by the graph and are called **vertical asymptotes.** Although the table of values does not extend beyond $x = |\pi/2|$, the values can be seen to repeat themselves, indicating that the function $y = \tan x$ has a period of π.

$y = \tan x$

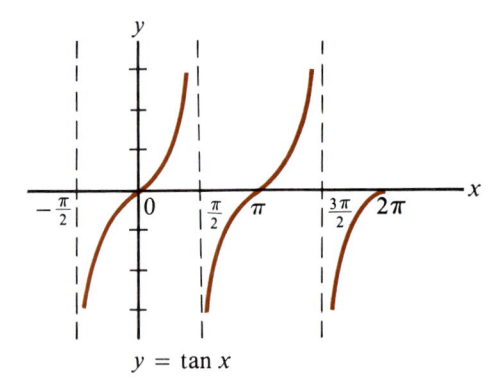

Figure 8.19

The concept of amplitude is not defined since the curve extends indefinitely in the vertical direction.

The graphs of $y = \cot x$, $y = \sec x$, and $y = \csc x$ are shown in Figures 8.20 through 8.22, respectively. The graphs of $y = \sec x$ and $y = \csc x$ can be obtained from the graphs of $y = \cos x$ and $y = \sin x$ by using the reciprocal relations

$$\sec x = \frac{1}{\cos x} \quad \text{and} \quad \csc x = \frac{1}{\sin x}$$

(See Example 1.) In other words, the graph of $y = \sec x$ is obtained by starting with $y = \cos x$ and plotting the reciprocals of the values of cos x.

8.4 GRAPHS OF OTHER TRIGONOMETRIC FUNCTIONS

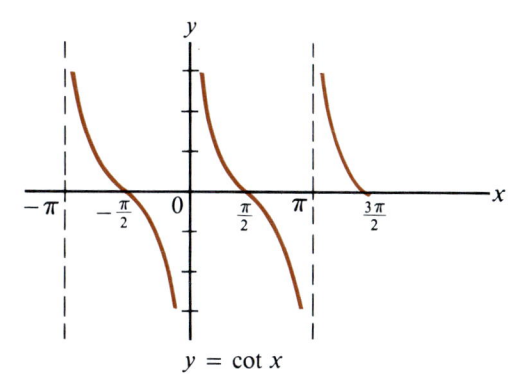

$y = \cot x$

Figure 8.20

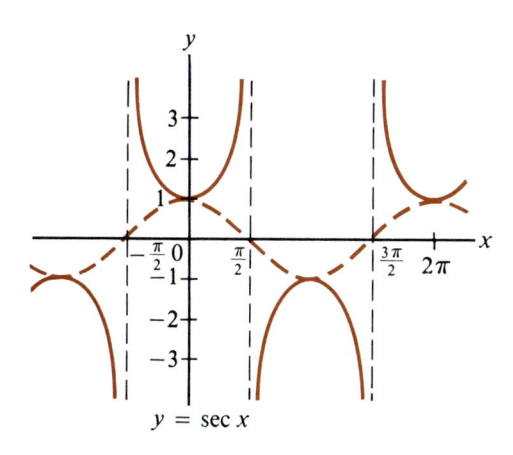

$y = \sec x$

Figure 8.21

$y = \csc x$

Figure 8.22

When $\cos x = 1$, we have $\sec x = 1$; when $0 < \cos x < 1$, then $\sec x > 1$; and when $\cos x = 0$, then $\sec x = \frac{1}{0}$, which is undefined. To show how the curves are related, the graphs of $y = \cos x$ and $y = \sin x$ are shown as dashed curves in Figures 8.21 and 8.22, respectively. Note that $y = \sec x$ and $y = \csc x$ have period 2π, the same as for $\cos x$ and $\sin x$.

Example 1 Sketch the graph of $y = 2 \csc x$.

Solution. The most direct way to sketch this graph is to start with Figure 8.22 and then multiply the values by 2.

Alternatively, we can start with the graph of $y = 2 \sin x$ (the dashed curve in Figure 8.23 on page 268) and obtain some of the reciprocal values. For example, when $2 \sin x = 2$, then $2 \csc x = 2$ also. When $\sin x = 0$, then $\csc x$ is undefined, confirming the existence of asymptotes at $x = 0, \pm\pi, \pm 2\pi, \ldots$ (see Figure 8.23.)

x:	0	$\frac{\pi}{6}$	$\frac{\pi}{4}$	$\frac{\pi}{3}$	$\frac{\pi}{2}$	$\frac{5\pi}{6}$	π
$\frac{2}{\sin x}$:	undefined	4	$2\sqrt{2}$	$\frac{4}{\sqrt{3}}$	2	4	undefined

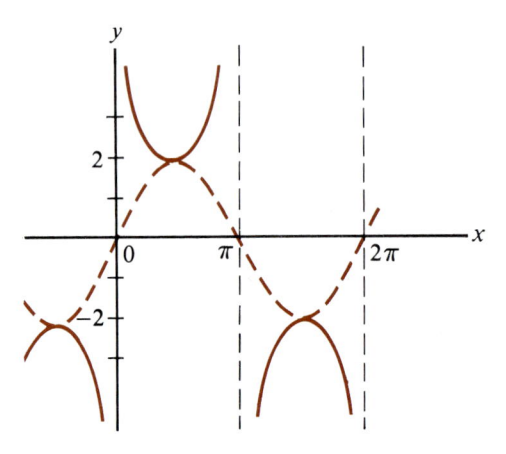

Figure 8.23

General Forms

To sketch the graph of $y = a \tan bx$, recall that $y = \tan x$ has period π and that the first asymptote to the right of the origin is at $x = \pi/2$. The distance from the origin to this asymptote is therefore half a period. For $y = \tan bx$, the first asymptote is at $x = \pi/2b$; so the period must be

$$P = \frac{2\pi}{2b} = \frac{\pi}{b}$$

The **period** of $y = a \tan bx$ and $y = a \cot bx$ is

$$P = \frac{\pi}{b} \qquad (8.11)$$

The **period** of $y = a \sec bx$ and $y = a \csc bx$ is

$$P = \frac{2\pi}{b} \qquad (8.12)$$

Example 2 Sketch the graph of $y = 3 \cot 2x$.

Solution. The period is $\pi/2$. Since the basic shape of the cotangent function is known from Figure 8.20, we readily obtain the desired sketch (Figure 8.24).

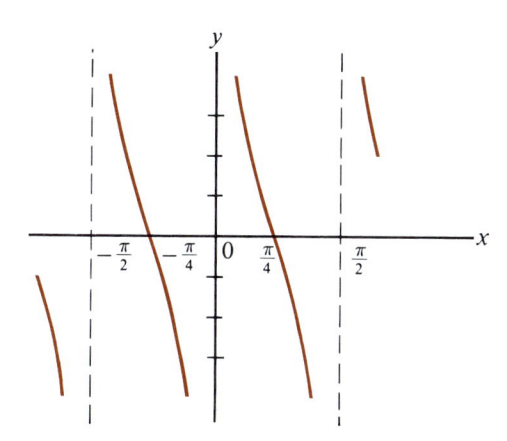

Figure 8.24

Exercises / Section 8.4

In Exercises 1–20, state the period of each function and sketch the curve.

1. $y = 2 \tan x$
2. $y = 3 \sec x$
3. $y = 4 \cot x$
4. $y = 2 \csc x$
5. $y = \sec 2x$
6. $y = \tan 3x$
7. $y = \csc 2x$
8. $y = 2 \cot 2x$
9. $y = \frac{1}{2} \tan 3x$
10. $y = \frac{1}{3} \sec x$
11. $y = -\csc 2x$
12. $y = -\tan 2x$
13. $y = 3 \sec \frac{1}{2} x$
14. $y = 4 \tan \frac{1}{3} x$
15. $y = 5 \csc \frac{2}{3} x$
16. $y = 6 \cot \frac{4}{3} x$
17. $y = \tan \left(x + \frac{\pi}{6}\right)$
18. $y = \sec \left(2x - \frac{\pi}{6}\right)$
19. $y = \frac{1}{2} \cot 3\pi x$
20. $y = 3 \csc 2\pi x$

21. A dam has a V-shaped notch with angle θ. If the water reaches a level 1 m above the bottom of the notch, then the rate of flow in meters per second across the notch is given by $Q = 2.506 \tan (\theta/2)$. Sketch the graph for $0 \leq \theta < \pi$.

22. If friction is taken into account, then the heaviest weight that can be pulled up an inclined plane depends only on the angle θ that the plane makes with the horizontal. The methods of differential calculus show that the maximum is attained when $\mu = \tan \theta$, where μ is the coefficient of friction. Sketch the graph of $\mu = \tan \theta$ for $0 \leq \theta < \pi/2$.

8.5 Addition of Coordinates

Many applications of trigonometric functions involve combinations of two or more functions. Therefore, in this section we will study the **addition of coordinates.** To graph a function that is the sum of two other functions, we graph the individual functions and add their *y*-values graphically. Because of their importance, we shall confine ourselves to sinusoidal and simple algebraic functions.

Example 1 Sketch the graph of $y = x + \sin x$.

Solution. We first graph the functions **y = x** and **y = sin x**, shown as dashed curves in Figure 8.25. When adding the coordinates, care must be taken to treat the *y*-coordinates of the points below the *x*-axis as negative. Certain key points are particularly useful. For example, the graph of $y = \sin x$ crosses the *x*-axis at $x = 0, \pi, 2\pi, \ldots$. Since the corresponding *y*-values are 0, the points on the graph to be drawn lie on $y = x$. (See Figure 8.25.) Point P_1 is a typical point between the zeros of $y = \sin x$; its coordinates are $(\pi/2, \pi/2 + 1)$. Similarly, the coordinates of P_2 are $(3\pi/2, 3\pi/2 - 1)$. We locate as many points as necessary to obtain a smooth graph, shown as the solid curve in Figure 8.25.

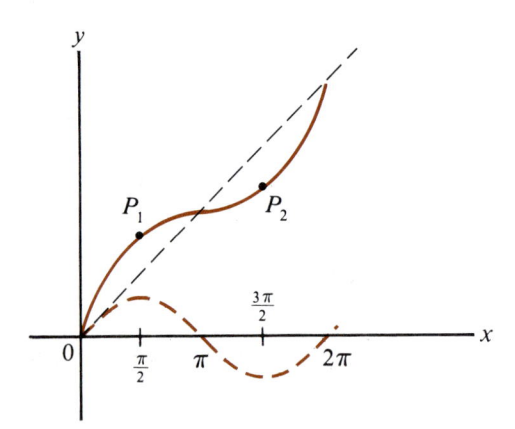

Figure 8.25

As pointed out in Section 8.3, the sound of a tuning fork can be seen as a wave on an oscilloscope; this wave takes the form $y = A \sin 2\pi ft$. It is interesting to note that, theoretically, complex sounds can be reproduced by a proper combination of tuning forks. Therefore, the functions describing such sounds are necessarily combinations of sines and cosines. One such combination is a *Fourier series,* useful in the study of vibration, heat flow, electrical circuits, and so on. An example of a Fourier series is

$$y = \sin t + \frac{1}{2} \sin 2t + \frac{1}{3} \sin 3t + \frac{1}{4} \sin 4t + \cdots \qquad (8.13)$$

The three dots indicate that the sum continues indefinitely.

Example 2 Graph the first two terms of equation (8.13) by addition of coordinates.

Solution. The graphs of $y = \sin t$ and $y = \frac{1}{2} \sin 2t$ are shown as dashed curves in Figure 8.26. Note that the graph of $y = \frac{1}{2} \sin 2t$ crosses the t-axis at

$$t = \frac{\pi}{2}, \pi, \frac{3\pi}{2}, 2\pi, \ldots$$

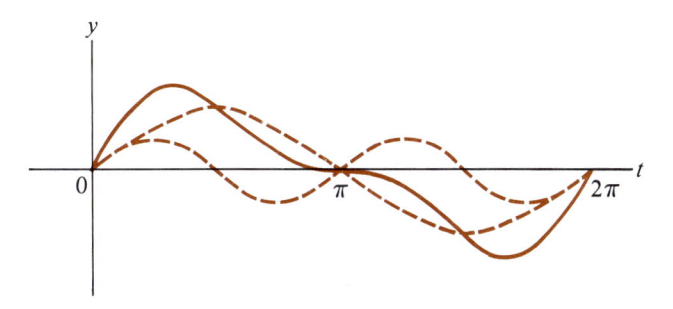

Figure 8.26

At these t-values the points on the curve to be sketched lie on the curve $y = \sin t$. Using this information, we obtain the solid curve in Figure 8.26.

Graphing more and more terms of equation (8.13) results in a graph that gets ever closer to the "sawtooth function" shown in Figure 8.27.

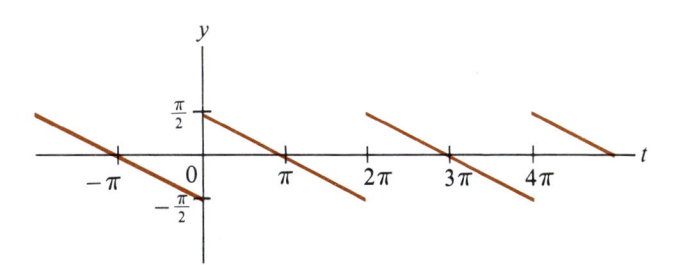

Figure 8.27

Exercises / Section 8.5

In Exercises 1–10, sketch the graph of each function by addition of coordinates.

1. $y = x + \cos x$
2. $y = 2x + \sin x$
3. $y = 2 \sin x + \cos x$
4. $y = 2 \sin x + 3 \cos x$
5. $y = 2 \cos x + \sin 2x$
6. $y = 3 \cos x + \cos 2x$
7. $y = \frac{1}{2} x - \sin x$
8. $y = \sin 2x - \cos x$
9. $y = \cos x + 2 \sin 2x$
10. $y = \cos \frac{1}{2} x + 2 \sin x$

11. Graph the first two terms of the Fourier series

$$f(t) = \sin \pi t + \frac{1}{3} \sin 3\pi t + \frac{1}{5} \sin 5\pi t + \cdots$$

12. Graph the first two terms of the Fourier series

$$f(t) = \cos t + \frac{1}{9} \cos 3t + \frac{1}{25} \cos 5t + \cdots$$

13. The current in a certain circuit is given by $i = 4.0 \cos 60\pi t + 2.0 \sin 60\pi t$. Sketch the graph of the current (in amperes) as a function of time (in seconds).

14. A weight of mass 1 slug is oscillating on a spring with spring constant 64 lb/ft. If the weight has an initial velocity of 4 ft/sec and an initial displacement of 2 ft, its displacement y (in feet) as a function of t (in seconds) is given by $y = 2 \cos 8t + \frac{1}{2} \sin 8t$. Sketch the curve.

Review Exercises / Chapter 8

In Exercises 1–8, state the amplitude and period and sketch the curves.

1. $y = 2 \cos x$
2. $y = \frac{1}{2} \sin 2x$
3. $y = 4 \cos 4x$
4. $y = -2 \cos x$
5. $y = -\frac{1}{2} \sin 2x$
6. $y = 3 \sin \frac{1}{2} x$
7. $y = 4 \sin \frac{1}{2} x$
8. $y = 8 \cos \frac{1}{4} x$

In Exercises 9–16, state the amplitude, period, and shift. Sketch the curves.

9. $y = 2 \sin \left(x + \frac{\pi}{4} \right)$
10. $y = 2 \cos \left(x - \frac{\pi}{4} \right)$
11. $y = \frac{1}{2} \cos \left(x + \frac{\pi}{8} \right)$
12. $y = \frac{1}{3} \sin \left(x - \frac{\pi}{4} \right)$
13. $y = 4 \cos \left(\frac{1}{2} x - \frac{\pi}{8} \right)$
14. $y = 2 \sin \left(\frac{1}{2} x + \frac{\pi}{8} \right)$
15. $y = \frac{1}{2} \sin \left(2x - \frac{\pi}{4} \right)$
16. $y = 4 \sin (2x - 2)$

In Exercises 17–20, sketch the given curves.

17. $y = \sec 2x$

18. $y = 3 \tan 2x$

19. $y = 4 \cot \frac{1}{2} x$

20. $y = \frac{1}{4} \csc 3x$

21. An object moving around a circle of radius 4 cm starts at (4, 0) with an angular velocity of 2 rad/sec. Sketch the graph of the motion of the vertical projection.

22. Repeat Exercise 21 if the motion starts at (0, 4).

23. An object moves around a circle of radius 6 cm at the rate of 5 rev/min. If the motion starts at (0, 6), sketch the graph of the motion of the vertical projection.

24. Repeat Exercise 23 if the motion starts at (6, 0).

25. If $E_{max} = 5.00$ V, $\omega = 60\pi$ rad/sec, and $\alpha = \pi/3$, sketch the graph of E as a function of time.

26. If $I_{max} = 4.00$ A, $\omega = 90\pi$ rad/sec, and $\alpha = \pi/6$, sketch the current as a function of time.

27. A tuning fork vibrates at 236 Hz and has an amplitude of 0.003 in. Sketch the corresponding curve appearing on an oscilloscope.

28. Recall that the equation of motion of a water wave is given by $y = A \sin 2\pi(ft - r/\lambda)$. Suppose the motion is frozen at $t = 1$ sec. Sketch the wave at this instant if the period is 8 sec and the amplitude 3 ft ($\lambda = 5.12\ T^2$).

29. The current (in amperes) is given by $i = 2.0 \sin 120\pi t + 3.0 \cos 120\pi t$. (Note that the frequency is $1/P = 60$ Hz.) Sketch the curve.

30. A weight of mass 1 slug is oscillating on a spring with spring constant 16 lb/ft. If the weight has an initial velocity of 8 ft/sec and an initial displacement of 1 ft, its displacement y (in feet) as a function of time t (in seconds) is given by $y = \cos 4t + 2 \sin 4t$. Sketch the curve.

31. Graph the first two terms of the Fourier series

$$f(t) = \sin t - \frac{1}{2} \sin 2t + \frac{1}{3} \sin 3t - \frac{1}{4} \sin 4t + \cdots$$

by adding the coordinates between $-\pi$ and π.

CHAPTER 9

Vectors and Oblique Triangles

Objectives Upon completion of this chapter, you should be able to:

1. Add and subtract vectors graphically.
2. Find the magnitude and direction of a given vector.
3. Resolve a vector into its components.
4. Solve certain physical problems by vector methods.
5. Solve oblique triangles by the:
   a. Law of sines.
   b. Law of cosines.

9.1 Vectors

Scalar

Vector

Initial and terminal point

Most quantities discussed so far can be fully described by a number. Examples are areas, volumes, and temperature. Quantities that have only magnitude are called **scalar** quantities. Other physical phenomena have both **magnitude and direction.** For example, the description of the velocity of a moving body must account for both the magnitude of the velocity and its direction. To describe a force, we need to know both the magnitude of the force and the direction in which the force is acting. Such entities are called **vectors.**

An arrow is a convenient way to represent a vector graphically. The arrow points in the direction of the vector, and the length of the arrow represents the magnitude. A vector is denoted by a boldfaced letter, while the same letter in italic represents the magnitude. Thus **A** represents a vector and A its magnitude. (In handwriting, a common notation for **A** is $\vec{A}$.) The base of the arrow is called the **initial point,** and the tip of the arrow the **terminal point.** (See Figure 9.1.) The magnitude is denoted by $|\mathbf{A}|$.

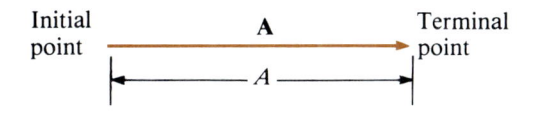

Figure 9.1

Addition of Vectors

If two vectors are to be added, both their magnitude and direction must be taken into account. Let **A** and **B** represent two vectors with the same initial point O (Figure 9.2). In the parallelogram determined by **A** and **B**, the vector sum **A** + **B**, called the **resultant,** is the vector with initial point O coinciding with the diagonal of the parallelogram. The initial point was introduced here only for convenience, for *the initial point of a vector can be placed anywhere in the plane*. For this reason the sum is also given by Figure 9.3.

Definition of vector addition:

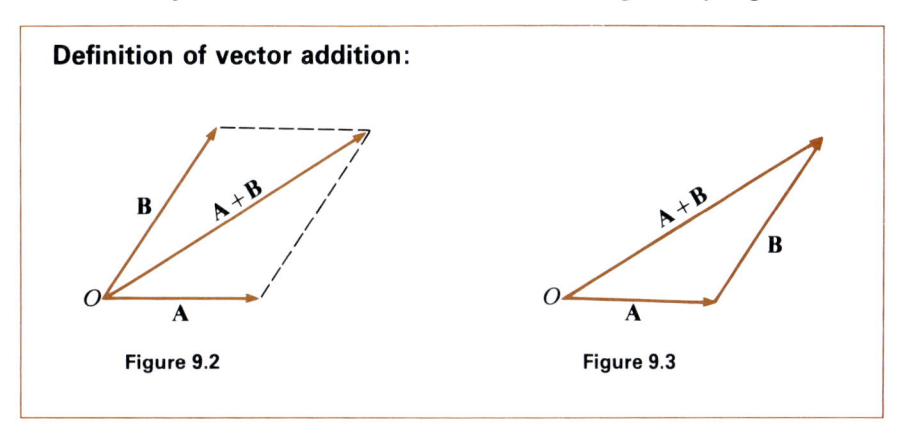

Figure 9.2

Figure 9.3

If more than two vectors have to be added, the procedure shown in Figure 9.3 is by far the more convenient. Consider the example below.

Example 1 Find the sum of the vectors **A, B, C,** and **D** shown in Figure 9.4.

Solution. Draw vector **A**. Place the initial point of **B** at the terminal point of **A**. Now place the initial point of **C** at the terminal point of **B**, and the initial

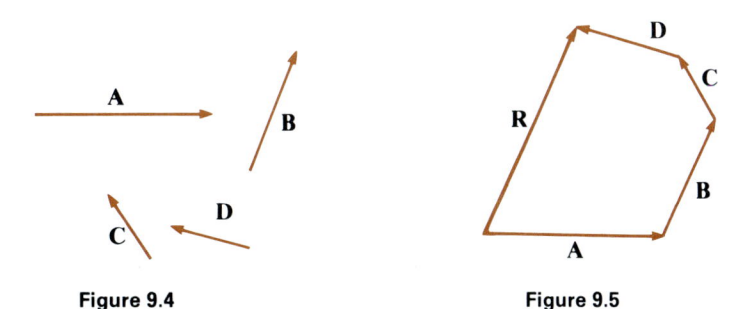

Figure 9.4

Figure 9.5

276 CHAPTER 9 VECTORS AND OBLIQUE TRIANGLES

Commutativity

point of **D** at the terminal point of **C**. The final resultant **R** is the vector from the initial point of **A** to the terminal point of **D**, as shown in Figure 9.5. (Since vector addition is commutative, the vectors can be added in any order.)

Scalar Multiplication

Scalar product

If a vector **A** is multiplied by a number c, the resulting vector $c\mathbf{A}$, called the **scalar product**, has the same direction as **A** and magnitude cA. For example, the vector $3\mathbf{A}$ is three times as long as **A** and has the same direction as **A**.

Example 2 Given the vectors **A** and **B** in Figure 9.6, draw $2\mathbf{A} + 3\mathbf{B}$.

Solution. From vector **A** construct the vector $2\mathbf{A}$, which has the same direction as **A** but twice the magnitude. (See Figure 9.7.) In a similar manner we construct the vector $3\mathbf{B}$ from **B**. The resultant $\mathbf{R} = 2\mathbf{A} + 3\mathbf{B}$ is shown in Figure 9.7.

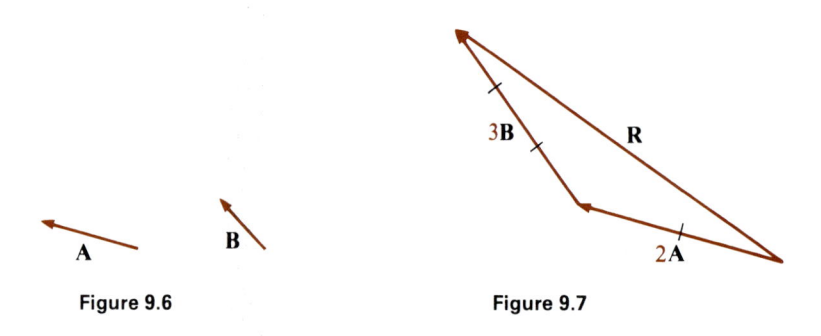

Figure 9.6

Figure 9.7

Subtraction of Vectors

To subtract vectors, define $-\mathbf{B}$ as equal to $(-1)\mathbf{B}$, which has the same magnitude as **B** but is oppositely directed. Thus $\mathbf{A} - \mathbf{B} = \mathbf{A} + (-\mathbf{B})$, which can be added in the usual way (Figure 9.8). It also follows that $\mathbf{A} - \mathbf{B}$ is the

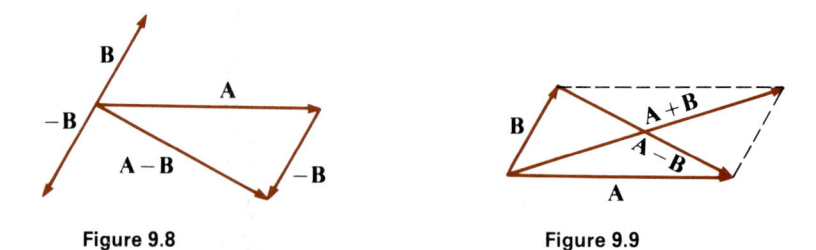

Figure 9.8

Figure 9.9

second diagonal in the parallelogram determined by **A** and **B**, as shown in Figure 9.9.

Exercises / Section 9.1

Carry out each of the indicated operations with the vectors given in Figure 9.10 by means of a diagram.

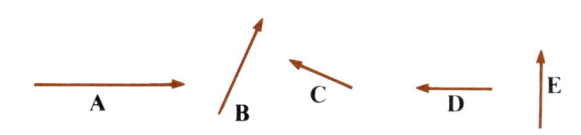

Figure 9.10

1. **A** + **C**
2. **C** + **E**
3. **A** + **B** + **C**
4. **B** + **C** + **D**
5. 2**D** + **E**
6. 2**B** + 2**C** + **D**
7. **A** − **C**
8. 2**B** − **C**
9. 2**A** − 3**B**
10. **A** + 2**B** − 2**E**
11. 2**D** − 4**C**
12. 2**C** − 3**E**
13. 2**C** + **E**
14. **A** + 2**E**
15. **A** + **E** + 2**C**
16. 2**D** − **E**
17. **A** + 2**B** + **C**
18. 2**B** + **C** − **A**

9.2 Vector Components

While vector addition by means of diagrams is conceptually important, it is impractical for computations. In this section we shall see how the resultant can be determined exactly. To this end we need to combine the vector concept with that of the coordinate system: We place the initial point of the vector at the origin so that the terminal point uniquely describes the vector. For example, the vector in Figure 9.11 can be specified by the point $(-2, 2)$.

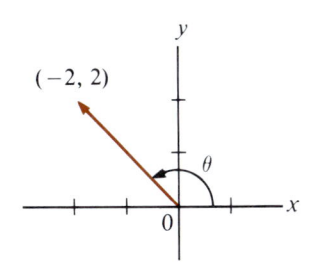

Figure 9.11

The magnitude, which can be computed directly by the Pythagorean theorem, turns out to be $\sqrt{(-2)^2 + 2^2} = \sqrt{8} = 2\sqrt{2}$. The direction is given by the angle $\theta = 135°$ in standard position.

278 CHAPTER 9 VECTORS AND OBLIQUE TRIANGLES

Unit vectors   A common alternative is to define two **unit vectors i** and **j** (magnitude of 1 unit) along the positive x- and y-axes, respectively, as shown in Figure 9.12. It follows that the vector **(−2, 2)** can be written $\mathbf{A} = -2\mathbf{i} + 2\mathbf{j}$. (See Figure 9.13.)

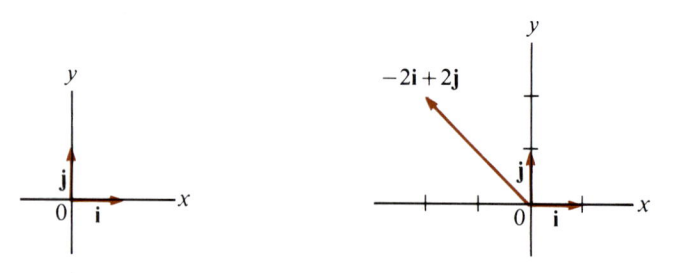

Figure 9.12

Figure 9.13

In general, if the terminal point of vector **A** is denoted by (A_x, A_y), then

$$\mathbf{A} = A_x\mathbf{i} + A_y\mathbf{j} \tag{9.1}$$

The magnitude A is now denoted by

$$|\mathbf{A}| = \sqrt{A_x^2 + A_y^2} = A \tag{9.2}$$

The direction θ is the angle determined by **A** and the positive x-axis, so that

$$\tan \theta = \frac{A_y}{A_x} \tag{9.3}$$

Magnitude and direction of $\mathbf{A} = A_x\mathbf{i} + A_y\mathbf{j}$:

Magnitude: $|\mathbf{A}| = \sqrt{A_x^2 + A_y^2}$

Direction: The direction is obtained from the relation

$$\tan \theta = \frac{A_y}{A_x}$$

These ideas are illustrated in the examples below.

Example 1   Find the magnitude and direction of the vector **A** whose terminal point is $(-3, -5)$.

Solution. The vector is shown in Figure 9.14. The magnitude is

$$|\mathbf{A}| = \sqrt{(-3)^2 + (-5)^2} = \sqrt{9 + 25} = \sqrt{34}$$

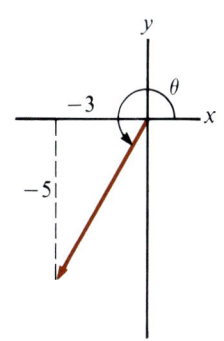

To find the direction, note that

$$\tan \theta = \frac{-5}{-3} = 1.667$$

Using a calculator, we find that the reference angle is 59°2'. Hence $\theta = 239°2'$.

Figure 9.14

Example 2 Find the magnitude and direction of the vector $\mathbf{A} = \sqrt{7}\mathbf{i} - 3\mathbf{j}$.

Solution. $|\mathbf{A}| = \sqrt{7 + 9} = 4$

and

$$\tan \theta = \frac{-3}{\sqrt{7}}$$

Using a calculator, the reference angle is found to be 48°35', so that $\theta = 311°25'$.

It is sometimes desirable to work in reverse: Starting with the magnitude and direction of a vector, we express the vector in the form $\mathbf{A} = A_x\mathbf{i} + A_y\mathbf{j}$. This process is called **resolving the vector into its components.**

Example 3 A vector $\mathbf{A}$ has magnitude **3** and direction **320°**. Resolve the vector into its components.

Solution. The vector is shown in Figure 9.15. To find A_x, we note that $\cos \theta = A_x/3$. Hence $A_x = 3 \cos 320° = 2.30$. Similarly, $A_y = 3 \sin 320° = -1.93$. It follows that

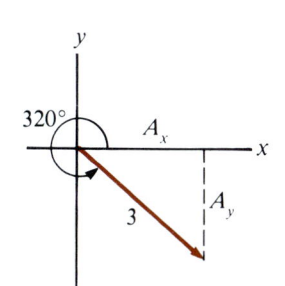

$$\mathbf{A} = 2.30\mathbf{i} - 1.93\mathbf{j}$$

Figure 9.15

> If the direction of **A** is θ, then
> $$\mathbf{A} = (|\mathbf{A}|\cos\theta)\mathbf{i} + (|\mathbf{A}|\sin\theta)\mathbf{j} \qquad (9.4)$$

Exercises / Section 9.2

In Exercises 1–16, find the magnitude and direction of each vector.

1. $\mathbf{i} + \mathbf{j}$
2. $-\mathbf{i} + \sqrt{3}\mathbf{j}$
3. $-3\mathbf{i} - 3\mathbf{j}$
4. $\sqrt{3}\mathbf{i} - \mathbf{j}$
5. $2\mathbf{i} + 3\mathbf{j}$
6. $-4\mathbf{i} + 5\mathbf{j}$
7. $-\sqrt{3}\mathbf{i} - \sqrt{6}\mathbf{j}$
8. $-\mathbf{i} + \sqrt{15}\mathbf{j}$
9. $-\sqrt{7}\mathbf{i} - 3\mathbf{j}$
10. $\mathbf{i} - 5\mathbf{j}$
11. $-\sqrt{13}\mathbf{i} + \sqrt{3}\mathbf{j}$
12. $2\mathbf{i} + \sqrt{2}\mathbf{j}$
13. $\mathbf{i} - 2\mathbf{j}$
14. $-2\mathbf{i} + 5\mathbf{j}$
15. $-\mathbf{i} - 4\mathbf{j}$
16. $-2\mathbf{i} - 7\mathbf{j}$

In Exercises 17–26, resolve each vector into its components. (Give answers to two decimal places.)

17. $|\mathbf{A}| = 3, \theta = 80°$
18. $|\mathbf{A}| = 2, \theta = 110°$
19. $|\mathbf{A}| = \sqrt{6}, \theta = 145°$
20. $|\mathbf{A}| = \sqrt{3}, \theta = 185°$
21. $|\mathbf{A}| = 4, \theta = 261°$
22. $|\mathbf{A}| = \sqrt{5}, \theta = 312°$
23. $|\mathbf{A}| = 3, \theta = 350°$
24. $|\mathbf{A}| = 7, \theta = 73°$
25. $|\mathbf{A}| = \sqrt{2}, \theta = 112°$
26. $|\mathbf{A}| = \sqrt{3}, \theta = 121°$

9.3 Basic Applications of Vectors

The vector concepts introduced so far lend themselves particularly well to applications involving right triangles. First we need to recall the rules of significant figures for triangles.

Degree measurement to nearest	Significant figures for measurement of a side
1°	2
10' or 0.1°	3
1' or 0.01°	4

The examples in this section all involve vectors representing forces or velocities acting at right angles to each other. The first three examples illustrate the problem of finding the resultant from two given vectors.

Example 1 Two forces act on the same object at right angles to each other. One force is 25.0 N (newtons) and the other is 16.1 N. Find the resultant.

9.3 BASIC APPLICATIONS OF VECTORS

Solution. This problem can be solved without referring to a coordinate system. From Figure 9.16, the magnitude of the resultant is $\sqrt{(25.0)^2 + (16.1)^2} = 29.7$ N. To obtain the direction, note that $\tan \theta = \frac{16.1}{25.0}$, so that $\theta = 32°50'$, rounded off to the nearest 10'.

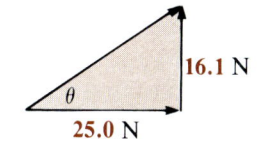

Figure 9.16

Example 2 A small plane is headed east at 150.0 mi/hr. The wind is from the north at 20.10 mi/hr. What is the resulting velocity and direction of the plane?

Solution. From the diagram (Figure 9.17), the resulting velocity is $\sqrt{(150.0)^2 + (20.10)^2} = 151.3$ mi/hr. Also, $\tan \theta = \frac{20.10}{150.0}$, so that $\theta = 7°38'$. The direction is said to be $7°38'$ *south of east*.

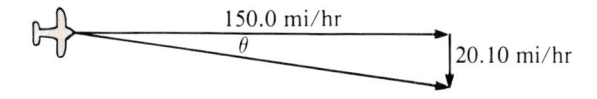

Figure 9.17

Example 3 A 30.0-lb weight hanging on a rope is pulled sideways by a force of 10.5 lb. Determine the resulting tension on the rope and the angle that the rope makes with the vertical. (Assume the system to be in equilibrium.)

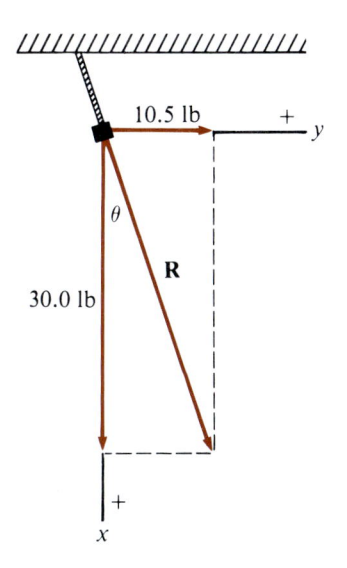

Figure 9.18

Solution. In the diagram (Figure 9.18 on page 281) the tension must be equal to the resultant **R** of the two forces, the 30.0-lb downward force and the 10.5-lb pull to the side. Relative to the coordinate system shown in Figure 9.18,

$$\mathbf{R} = 30.0\mathbf{i} + 10.5\mathbf{j}$$

So the tension on the rope is

$$|\mathbf{R}| = \sqrt{(30.0)^2 + (10.5)^2} = 31.8 \text{ lb}$$

The angle θ with the vertical is found from $\tan \theta = \frac{10.5}{30.0}$, or $\theta = 19.3°$ to the nearest tenth of a degree.

The last example shows how resolving a vector into its components can be of interest for physical reasons.

Example 4 A block of ice weighing 95.0 lb is resting on a plane inclined 15.0° to the horizontal. What is the force required to keep the block from sliding down the plane? (See Figure 9.19; the weight of an object always acts vertically downward.)

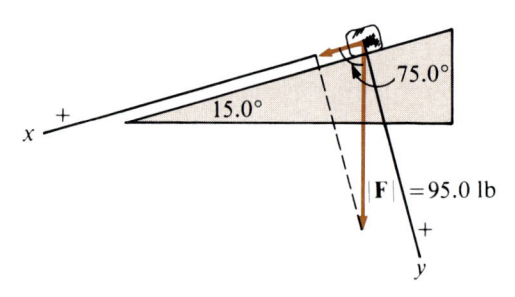

Figure 9.19

Solution. In this problem the downward force **F**, where $|\mathbf{F}| = 95.0$ lb, has to be split into two components, one directed down the plane and the other perpendicular to the plane. (The components must obey the parallelogram law.) The component along the plane represents the tendency of the ice to slide downward. In other words

$$F_x = 95.0 \cos 75.0° = 24.6 \text{ lb}$$

So a force of 24.6 lb is required to keep the block from slipping.

Exercises / Section 9.3

In Exercises 1–4, two forces with respective magnitudes F_1 and F_2 are acting on the same object at right angles. Determine the resultant.

1. $F_1 = 10.2$ lb, $F_2 = 5.70$ lb
2. $F_1 = 24.3$ lb, $F_2 = 15.6$ lb
3. $F_1 = 5.800$ N, $F_2 = 19.30$ N
4. $F_1 = 30.00$ N, $F_2 = 21.20$ N

5. A boat is heading directly across a river at 15.7 mi/hr. If the river flows at 4.10 mi/hr, find the actual speed and direction of the boat.

6. A boat is heading directly across a river at 14.3 km/hr. The river is flowing at 6.20 km/hr. What is the direction and speed of the boat?

7. A boy and a girl are pulling a cart in mutually perpendicular directions with forces of 32 lb and 45 lb, respectively. What single force will have the same effect?

8. Two tractors are pulling on a tree stump with forces of 12,000 lb and 14,000 lb, respectively, in mutually perpendicular directions. What single force will have the same effect?

9. With the sun directly overhead, a jet is taking off at 112 mi/hr at an angle of 42.0° with the ground. How fast is its shadow moving along the ground?

10. Two forces of 75.0 lb and 32.0 lb, respectively, are acting on an object in mutually perpendicular directions. What is the magnitude of the resultant?

11. A boat crosses a river flowing at 4.30 mi/hr and lands at a point directly across. If the velocity of the boat is 10.0 mi/hr in still water, find its velocity with respect to the bank and the direction in which the boat must head.

12. A boat crossing a river landed at a point directly across from where it started. Due to the flow of the river, it headed in a direction 25° upstream from the line directly across. Its velocity in still water was 17 mi/hr. Find the rate at which the river was flowing.

13. A jet is heading due north at 400 km/hr with the wind from the west at 29 km/hr. Find the resultant velocity and direction. (See Example 2.)

14. A small plane is heading due east at 121 mi/hr with the wind from the north at 25.3 mi/hr. Find the resultant velocity and direction.

15. A pilot wishes to fly directly south. His air speed is 225 km/hr, and the wind blows from the east at 30.1 km/hr. In what direction should he orient his plane, and what is the resultant speed relative to the ground?

16. A 107-lb weight is supported as illustrated in Figure 9.20. Find the tension on the rope.

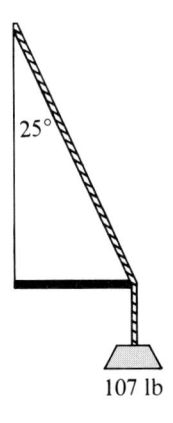

Figure 9.20

17. Find the force required to keep a 3,125-lb car parked on a hill that makes an angle of 11°31′ with the horizontal.
18. What is the force required to keep a 295-lb cart on a ramp inclined at an angle of 15.0° to the horizontal?
19. A force of 473 lb is required to pull a boat up a ramp inclined at 19.0° to the horizontal. Find the weight of the boat.
20. A woman is dragging a crate across the floor by means of a rope. She is able to pull with a force of 79 lb at an angle of 41° with the ground. What force parallel to the floor would have the same effect?
21. A 98-lb weight hanging on a rope is pulled sideways by a force of 25 lb. Assuming the system to be in equilibrium, what is the tension on the rope and the angle of the rope with the vertical?
22. A 31-lb weight hanging on a rope is to be pulled sideways so that the rope makes an angle of 17° with the vertical. What is the force required to accomplish this?
23. Determine the force against the horizontal support in Figure 9.21.

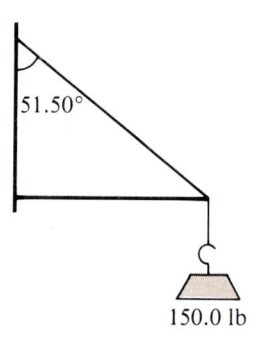

Figure 9.21

24. A 51-lb weight hanging on a rope is pulled sideways by a force of 15 lb. If the system is in equilibrium, determine the tension on the rope and the angle that the rope makes with the vertical.

9.4 The Law of Sines

So far all of our applications have involved right triangles. We will now consider oblique triangles (triangles not containing a right angle) and thereby greatly increase the usefulness of our previous methods. We shall begin by studying the *law of sines* in this section and then the *law of cosines* in the next.

To derive the law of sines, consider the triangles in Figure 9.22. Now pick an arbitrary vertex such as B and drop a perpendicular to the side opposite. (The triangles in the figure illustrate the different cases.) Making use of the resulting right triangles, we get from Figure 9.22(a)

$$\sin A = \frac{h}{c} \quad \text{and} \quad \sin C = \frac{h}{a}$$

9.4 THE LAW OF SINES

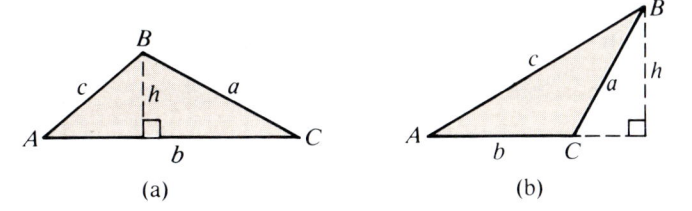

Figure 9.22

or $h = c \sin A$ and $h = a \sin C$. From Figure 9.22(b),

$$h = c \sin A \quad \text{and} \quad h = a \sin(180° - C) = a \sin C$$

In both cases

$$c \sin A = a \sin C$$

and

$$\frac{c \sin A}{\sin A \sin C} = \frac{a \sin C}{\sin A \sin C} \quad \text{dividing by } \sin A \sin C$$

$$\frac{a}{\sin A} = \frac{c}{\sin C} \tag{9.5}$$

If we drop a perpendicular from C to c, we get by the same argument

$$\frac{a}{\sin A} = \frac{b}{\sin B} \tag{9.6}$$

Combining equations (9.5) and (9.6), we get

$$\frac{a}{\sin A} = \frac{b}{\sin B} = \frac{c}{\sin C} \tag{9.7}$$

called the **law of sines** or simply the **sine law**.

Law of sines:

$$\frac{\sin A}{a} = \frac{\sin B}{b} = \frac{\sin C}{c}$$

or

$$\frac{a}{\sin A} = \frac{b}{\sin B} = \frac{c}{\sin C}$$

The law of sines enables us to solve many oblique triangles. **Solving a triangle** means finding the measures of all sides and angles of the triangle.

Note: Examples 1 and 2 refer to Figure 9.22.

Example 1 If $A = 45°41'$, $C = 130°26'$, and $b = 5.000$, find B, a, and c.

Solution. Since the sum of the angles of a triangle is 180°, $B = 180° - (45°41' + 130°26') = 179°60' - 176°7' = 3°53'$. From the law of sines

$$\frac{a}{\sin 45°41'} = \frac{5.000}{\sin 3°53'} \quad \text{or} \quad a = \frac{5.000 \sin 45°41'}{\sin 3°53'} = 52.82$$

$$\frac{c}{\sin 130°26'} = \frac{5.000}{\sin 3°53'} \quad \text{or} \quad c = \frac{5.000 \sin 130°26'}{\sin 3°53'} = 56.19$$

Thus $B = 3°53'$, $a = 52.82$, and $c = 56.19$.

Example 2 Given that $B = 54°10'$, $b = 15.1$, and $c = 10.0$, find the remaining parts (Figure 9.23).

Solution. Since only one angle is known, one of the ratios must involve this angle. We now find angle C:

$$\frac{\sin C}{10.0} = \frac{\sin 54°10'}{15.1}$$

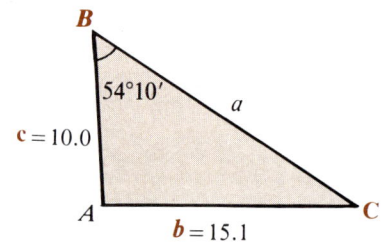

Figure 9.23

Thus

$$\sin C = \frac{10.0 \sin 54°10'}{15.1} = 0.5369$$

and $C = 32°30'$ (to the nearest 10'). It follows that $A = 180° - (32°30' + 54°10') = 179°60' - (86°40') = 93°20'$.

Using angle A, we now find a:

$$\frac{a}{\sin 93°20'} = \frac{15.1}{\sin 54°10'} \quad \text{or} \quad a = \frac{15.1 \sin 93°20'}{\sin 54°10'} = 18.6$$

Thus $A = 93°20'$, $C = 32°30'$, and $a = 18.6$.

If three sides of a triangle are known, then the sine law is of no help: All the ratios in the sine law involve an angle, at least one of which has to be known. We run into a similar problem if $a = 5$, $b = 6$, and $C = 20°$, even though one angle is known:

$$\frac{c}{\sin 20°} = \frac{6}{\sin B} \quad \text{and} \quad \frac{c}{\sin 20°} = \frac{5}{\sin A}$$

9.4 THE LAW OF SINES

Both equations involve two unknowns. These cases will be studied in the next section. The cases to which the sine law does apply are summarized below.

> To solve an oblique triangle by the **law of sines**, we need either:
> 1. Two angles and one side, or
> 2. Two sides and the angle opposite one of them.

The sine law can be applied to certain problems involving vectors, as shown in the next example.

Example 3 A plane has an air speed of 525 mi/hr and wants to fly on a course 20.0° east of north. If the wind is out of the north at 40.1 mi/hr, determine the direction in which the plane must head to stay on the proper course. What is the resultant velocity relative to the ground?

Solution. Denote the desired velocity vector by **v**. We can see from the diagram in Figure 9.24 that the angle x opposite **v** has to be known in order to

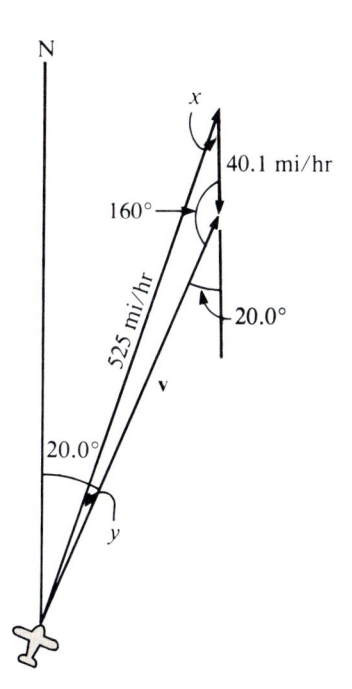

Figure 9.24

determine $|\mathbf{v}| = v$. To this end we first find angle y:

$$\frac{\sin y}{40.1} = \frac{\sin 160°}{525} \quad \text{or} \quad \sin y = \frac{40.1 \sin 160°}{525} = 0.0261$$

Thus $y = 1.5°$. It follows that $x = 180° - 161.5 = 18.5°$. Finally,

$$\frac{v}{\sin 18.5°} = \frac{525}{\sin 160°} \quad \text{or} \quad v = \frac{525 \sin 18.5°}{\sin 160°}$$

The calculator sequence is

CALCULATOR COMMENT

525 $\boxed{\times}$ 18.5 $\boxed{\text{SIN}}$ $\boxed{\div}$ 160 $\boxed{\text{SIN}}$ $\boxed{=}$

Display: 487.06179
If this is rounded off to three significant figures, we get 487.

Thus the plane must head 18.5° east of north, and it will fly at the rate of 487 mi/hr relative to the ground.

Caution. Suppose we want to find the obtuse angle C (greater than 90°) in Figure 9.25. By the sine law

$$\frac{\sin C}{4} = \frac{\sin 35°}{3}$$

or $\sin C = 0.7648$. If we now use a calculator to find C, we might conclude that $C = 49.9°$. However, since angle C is obtuse, the correct value is $180° - 49.9° = 130.1°$. Consequently, *we have to know in advance whether an angle to be found by the sine law is obtuse or acute.* (Recall that an angle in a triangle is *obtuse* if its measure is more than 90° and *acute* if its measure is less than 90°.)

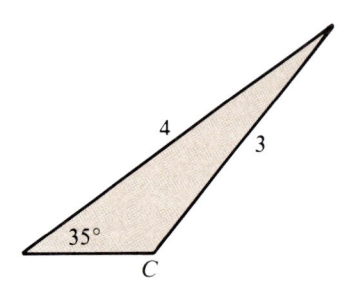

Figure 9.25

Ambiguous case

The case just described is called the *ambiguous case*. If for some reason the relevant information is withheld, then there may exist two solutions or no solution at all. The different cases are summarized in the chart on the next page.

Number of possible triangles	Sketch	Condition
0	Figure 9.26	$a < h$ or $a < b \sin A$
1	Figure 9.27	$a = h$ or $a = b \sin A$
2	Figure 9.28	$b > a > h$ or $b > a > b \sin A$

Exercises / Section 9.4

In Exercises 1–20, solve the triangles from the given information. (Refer to Figure 9.22 on page 285.)

1. $A = 20°10'$, $C = 50°40'$, $b = 4.00$
2. $A = 40.3°$, $B = 80.5°$, $a = 5.32$
3. $B = 25°50'$, $C = 130°20'$, $c = 15.1$
4. $A = 112.1°$, $C = 10.5°$, $c = 36.0$
5. $A = 19°50'$, $a = 102$, $b = 46.5$
6. $C = 47°38'$, $a = 0.7980$, $c = 1.320$
7. $C = 31.5°$, $a = 13.3$, $c = 6.82$
8. $A = 56°$, $a = 8.1$, $c = 10$
9. $A = 29.3°$, $a = 71.6$, $c = 136$; angle C acute
10. Same as Exercise 9 with angle C obtuse
11. $B = 46.0°$, $C = 54.0°$, $a = 236$
12. $A = 10.1°$, $C = 22.7°$, $c = 0.450$

13. $C = 63.6°$, $a = 12.4$, $c = 11.6$; angle A acute
14. Same as Exercise 13 with angle A obtuse
15. $B = 33°45'$, $a = 1.146$, $b = 2.805$
16. $B = 33°45'$, $a = 1.146$, $b = 0.6200$
17. $B = 33°45'$, $a = 1.146$, $b = 1.000$; angle C obtuse
18. $A = 72.3°$, $a = 86.0$, $c = 73.0$
19. $A = 126.50°$, $C = 10.40°$, $a = 136.4$
20. $B = 28.0°$, $C = 16.7°$, $c = 1.03$
21. A plane maintaining an air speed of 560 mi/hr is heading 10° west of north. A north wind causes the actual course to be 11° west of north. Find the velocity of the plane with respect to the ground.
22. Two forces act on the same object in directions that are 40.0° apart. If one force is 48.0 lb, what must the other force be so that the combined effect is equivalent to a force of 65.0 lb?
23. A surveyor wants to find the width of a river from a certain point on the bank. Since no other points on the bank nearby are accessible, he takes the measurements shown in Figure 9.29. Find the width of the river.

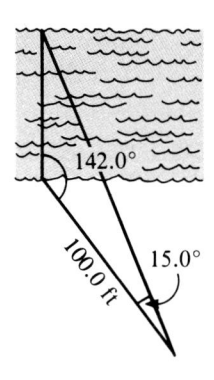

Figure 9.29

24. A building 81 ft tall stands on top of a hill. From a point at the foot of the hill, the angles of elevation of the top and bottom of the building are 39°10' and 37°40', respectively. What is the distance from the point to the bottom of the building?
25. From a point on the ground, the angle of elevation of a balloon is 49.0°. From a second point 1,250 ft away on the opposite side of the balloon and in the same vertical plane as the balloon and the first point, the angle of elevation is 33.0°. Find the distance from the second point to the balloon.
26. Find the area of the triangle in Figure 9.30.

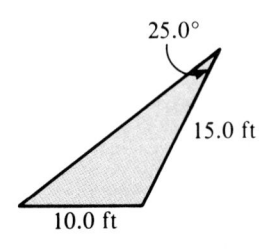

Figure 9.30

9.5 The Law of Cosines

We saw in the previous section that not all triangles can be solved by the law of sines. In this section we shall consider the solution of these triangles by means of the *law of cosines*.

Use the law of cosines to solve an oblique triangle given:
1. Two sides and the included angle, or
2. Three sides.

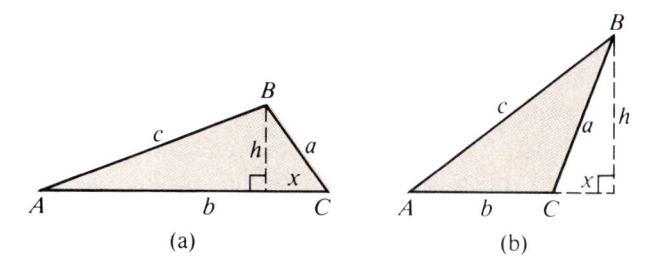

Figure 9.31

To derive the law of cosines, consider the triangles in Figure 9.31. Drop a perpendicular from B to the opposite side and let x denote the distance from C to the foot of the perpendicular. Since $h = c \sin A$, we get from the Pythagorean theorem

$$a^2 = c^2 \sin^2 A + x^2 \tag{9.8}$$

In Figure 9.31(a), $b - x = c \cos A$, or $x = b - c \cos A$. In Figure 9.31(b), $x + b = c \cos A$, or $x = c \cos A - b$. Substituting the respective expressions in equation (9.8), we get

$$a^2 = c^2 \sin^2 A + (b - c \cos A)^2$$

and

$$a^2 = c^2 \sin^2 A + (c \cos A - b)^2$$

If we multiply the expressions on the right, we find that in both cases

$$a^2 = c^2 \sin^2 A + c^2 \cos^2 A - 2bc \cos A + b^2$$

or

$$a^2 = c^2(\sin^2 A + \cos^2 A) + b^2 - 2bc \cos A \tag{9.9}$$

Now recall that for an angle θ in standard position, $\sin \theta = y/r$ and $\cos \theta = x/r$. Hence

$$\sin^2 \theta + \cos^2 \theta = \frac{y^2}{r^2} + \frac{x^2}{r^2} = \frac{x^2 + y^2}{r^2} = \frac{r^2}{r^2} = 1$$

Thus $\sin^2 A + \cos^2 A = 1$ for any angle A, so that equation (9.9) becomes

$$a^2 = b^2 + c^2 - 2bc \cos A \tag{9.10}$$

This formula is known as the **law of cosines** or simply the **cosine law.**

By dropping the perpendicular from the other vertices, we get two other forms of the cosine law, as summarized below.

Law of cosines:

$$a^2 = b^2 + c^2 - 2bc \cos A$$
$$b^2 = a^2 + c^2 - 2ac \cos B$$
$$c^2 = a^2 + b^2 - 2ab \cos C$$

The law of cosines can also be stated verbally.

Law of cosines (verbal form): The square of any side of a triangle equals the sum of the squares of the other two sides minus twice the product of those two sides and the cosine of the angle between them.

Note especially that if $A = 90°$, then formula (9.10) reduces to

$$a^2 = b^2 + c^2 - 2bc \cdot 0 = b^2 + c^2$$

It follows that the cosine law is a generalization of the Pythagorean theorem.

The first two examples illustrate how the law of cosines is used to solve oblique triangles.

Example 1 Given that $C = 20.4°$, $a = 7.60$, and $b = 10.0$, find the remaining parts (Figure 9.32).

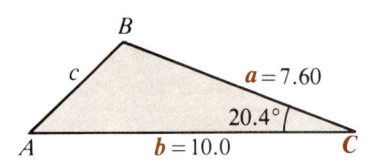

Figure 9.32

Solution. By the law of cosines

$$c^2 = (7.60)^2 + (10.0)^2 - 2(7.60)(10.0) \cos 20.4°$$

Using a calculator we find that $c = 3.91$.

Having obtained c, we can now find A. Suppose we use the cosine law again. Then

$$(7.60)^2 = (3.91)^2 + (10.0)^2 - 2(3.91)(10.0) \cos A$$
$$(7.60)^2 - (3.91)^2 - (10.0)^2 = -2(3.91)(10.0) \cos A$$
$$\cos A = \frac{(7.60)^2 - (3.91)^2 - (10.0)^2}{-2(3.91)(10.0)} = 0.7357$$

and $A = 42.6°$. Finally, $B = 117.0°$.

The calculator sequence is

CALCULATOR COMMENT

7.6 $\boxed{x^2}$ $\boxed{-}$ 3.91 $\boxed{x^2}$ $\boxed{-}$ 10 $\boxed{x^2}$ $\boxed{=}$ $\boxed{\div}$ 2 $\boxed{+/-}$ $\boxed{\div}$ 3.91 $\boxed{\div}$ 10 $\boxed{=}$ $\boxed{\text{INV}}$ $\boxed{\text{COS}}$

Display: 42.6 (to the nearest tenth of a degree)

Note that once we have obtained c in Example 1, we can find angle A by the sine law:

$$\frac{\sin A}{7.60} = \frac{\sin 20.4°}{3.91} \quad \text{or} \quad \sin A = \frac{7.60 \sin 20.4°}{3.91}$$

which yields $A = 42.7°$. The difference in the values is due to round-off errors.

Besides disagreements due to round-off errors, we are actually faced with a much more serious problem. As noted in the previous section, whenever the sine law is used, we need to know in advance whether an angle is acute or obtuse. The next example demonstrates the problem that can arise.

Example 2 Given that $a = 4.3$, $b = 5.2$, and $c = 8.2$, find the three angles (Figure 9.33).

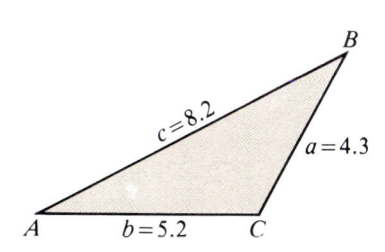

Figure 9.33

Solution. Suppose we first find angle B:

$$(5.2)^2 = (4.3)^2 + (8.2)^2 - 2(4.3)(8.2) \cos B$$

or

$$\cos B = \frac{(5.2)^2 - (4.3)^2 - (8.2)^2}{-2(4.3)(8.2)} = 0.8322$$

and $B = 34°$ (to the nearest degree).

Let us now find angle C by the cosine law:

$$\cos C = \frac{(8.2)^2 - (5.2)^2 - (4.3)^2}{-2(5.2)(4.3)} = -0.4855.$$

It follows that $C = 119°$ and $A = 180° - (119° + 34°) = 27°$.

However, suppose without examining the situation more closely, we had decided to use the sine law to get C. We now step into

$$\frac{\sin C}{8.2} = \frac{\sin 34°}{5.2}$$

so that $C = 62°$. Yet the correct value is $180° - 62° = 118°$, since C is obtuse. (Notice again the difference due to round-off errors.)

One way to avoid this problem is to find the second angle by the cosine law, especially if a calculator is used: Since $\cos C$ is negative, the calculator yields the correct value automatically. A good alternative is given next.

> First use the cosine law to find the angle opposite the longest side. Then use the sine law to find the remaining angles.

The law of cosines can also be used to solve problems involving vectors, as shown in the remaining examples.

Example 3 Two forces of 55.0 lb and 37.0 lb, respectively, are acting on the same object. If the angle between their directions is 23.4°, what single force would produce the same effect?

Solution. Denote the resultant force by **F** and consider the diagram in Figure 9.34. By the law of cosines

$$|\mathbf{F}|^2 = (55.0)^2 + (37.0)^2 - 2(55.0)(37.0) \cos 156.6°$$

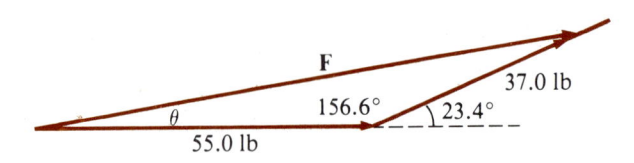

Figure 9.34

Using a calculator we get $|\mathbf{F}| = 90.2$ lb. Since the angle opposite the longest side is already known, we can safely apply the sine law to find angle θ, the direction of $\mathbf{F}$ relative to the 55.0-lb force. By the sine law, $\theta = 9.4°$. (If the cosine law is used, the calculated value turns out to be 9.3°.)

Example 4 A ship sailing at the rate of 21 knots in still water is heading 16° west of south. It runs into a strong current of 5.3 knots from the south. Find the resultant velocity vector.

Solution. In Figure 9.35, denote the desired velocity vector by $\mathbf{v}$. Then

$$|\mathbf{v}|^2 = (5.3)^2 + (21)^2 - 2(5.3)(21) \cos 16°$$

Using a calculator, we find that $|\mathbf{v}| = 16$ knots. To obtain the direction, we need to find θ in Figure 9.35. By the cosine law:

$$(5.3)^2 = (21)^2 + (16)^2 - 2(21)(16) \cos \theta$$

from which $\theta = 5°$. (The angle θ can also be found by the sine law.) So the direction is $16° + 5° = 21°$ west of south.

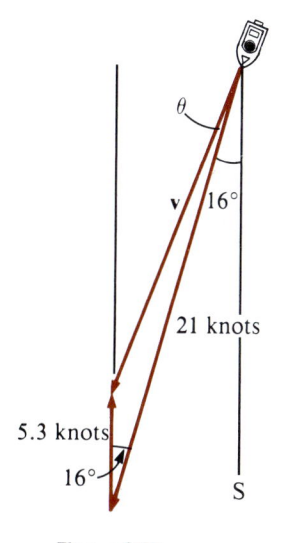

Figure 9.35

Exercises / Section 9.5

In Exercises 1–12, solve the given triangles. (See Examples 1 and 2.)

1. $A = 46.3°$, $b = 1.00$, $c = 2.30$
2. $B = 62.7°$, $a = 7.00$, $c = 10.0$
3. $C = 125°10'$, $a = 178$, $b = 137$
4. $A = 100.0°$, $b = 2.36$, $c = 1.97$
5. $a = 20.1$, $b = 30.3$, $c = 25.7$
6. $a = 2.46$, $b = 1.97$, $c = 4.10$
7. $A = 14°40'$, $b = 11.7$, $c = 7.80$
8. $a = 0.471$, $b = 0.846$, $c = 0.239$
9. $C = 39.4°$, $a = 126$, $b = 80.1$
10. $A = 63.0°$, $b = 35.1$, $c = 86.1$
11. $a = 12.85$, $b = 21.46$, $c = 9.179$
12. $a = 20$, $b = 23$, $c = 18$

13. A civil engineer wants to find the length of a proposed tunnel. From a distant point he observes that the respective distances to the ends of the tunnel are 585 ft and 624 ft. The angle between the lines of sight is 33.4°. Find the length of the proposed tunnel.

14. Two forces of 150.0 lb and 80.0 lb produce a resultant force of 209 lb. Find the angle between the forces.

15. A ship sails 16.0 km due east, turns 20.0° north of east, and then continues for another 11.5 km. Find its distance from the starting point.

296 CHAPTER 9 VECTORS AND OBLIQUE TRIANGLES

16. A car travels 85 km due west of point A and then northwest for another 50 km. Find its distance from point A.
17. A small plane is heading 5.0° north of east with an air speed of 151 mi/hr. The wind is from the south at 35.3 mi/hr. Find the actual course and the velocity with respect to the ground.
18. In Figure 9.36 the point P is moving around the circle. Find d as a function of θ.
19. Find the perimeter of the triangle in Figure 9.37.

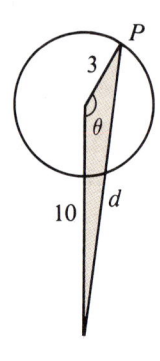

Figure 9.36

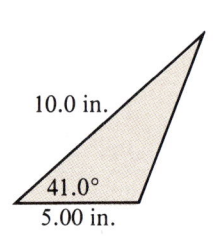

Figure 9.37

20. A river is flowing east at 4.10 mi/hr. A boat crosses the river in the direction 29.0° east of north. If the speedometer reads 10.0 mi/hr, what is the boat's actual speed and direction?

Review Exercises / Chapter 9

In Exercises 1–8, find the magnitude and direction of each vector.

1. $-\mathbf{i} + \mathbf{j}$
2. $-\mathbf{i} - \sqrt{3}\mathbf{j}$
3. $\mathbf{i} + 2\mathbf{j}$
4. $\mathbf{i} - 3\mathbf{j}$
5. $\sqrt{15}\mathbf{i} + \mathbf{j}$
6. $-3\mathbf{i} + \sqrt{7}\mathbf{j}$
7. $-\sqrt{2}\mathbf{i} - 2\mathbf{j}$
8. $\sqrt{2}\mathbf{i} + 3\mathbf{j}$

In Exercises 9–14, resolve each vector into its components. (Give answers to two decimals.)

9. $|\mathbf{A}| = 5$, $\theta = 75°$
10. $|\mathbf{A}| = 3$, $\theta = 115°$
11. $|\mathbf{A}| = \sqrt{7}$, $\theta = 220°$
12. $|\mathbf{A}| = \sqrt{3}$, $\theta = 328°$
13. $|\mathbf{A}| = 6$, $\theta = 216°$
14. $|\mathbf{A}| = \sqrt{10}$, $\theta = 153°$

In Exercises 15–24, solve each triangle from the given information.

15. $A = 46.3°$, $C = 53.7°$, $b = 5.26$
16. $A = 29.4°$, $B = 115.2°$, $c = 63.5$
17. $B = 10°40'$, $C = 130°20'$, $c = 236$
18. $A = 41°$, $B = 52°$, $b = 22$
19. $A = 26°$, $a = 25$, $c = 37$ (angle C is obtuse)
20. $C = 31.6°$, $a = 38.4$, $b = 62.0$
21. $A = 19°23'$, $b = 11.23$, $c = 30.04$
22. $B = 51°$, $a = 1.9$, $c = 1.4$
23. $a = 3.74$, $b = 5.86$, $c = 5.50$
24. $a = 321.7$, $b = 276.4$, $c = 248.2$

25. A railroad car weighing 9.8 tons is resting on a track inclined 7° with the horizontal. What is the force needed to keep it from rolling downhill?
26. A 320-lb cart is resting on a ramp inclined 12.0° to the horizontal. How much of the weight does the ramp support?
27. A surveyor wants to determine the width of a river. He can reach one point on the bank, but no other point nearby is accessible. Instead, he takes the measurements shown in Figure 9.38. Find the width of the river.
28. Find the perimeter of the triangle in Figure 9.39.
29. A 12.0-lb weight hanging from a rope is pushed sideways so that the rope makes a 15.0° angle with the vertical. If the system is in equilibrium, find the tension on the rope.
30. A 463-lb weight is supported as shown in Figure 9.40. Find the compression force on the angled bracket PQ.

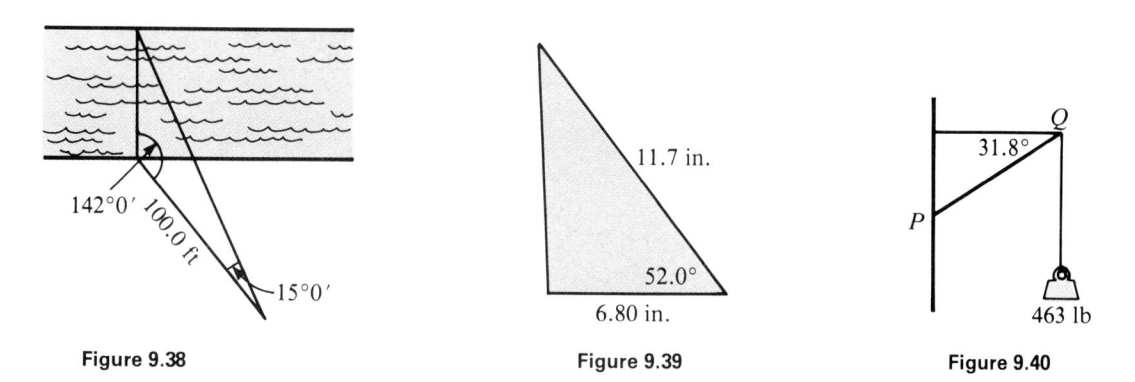

Figure 9.38

Figure 9.39

Figure 9.40

31. A force of 25.1 lb and a force of 39.4 lb produce a resultant force of 59.9 lb. Find the angle between the two forces.
32. From a point on the ground, the angle of elevation of the top of a tower is 26.3°. From a second point 48.3 ft closer to the tower, the angle of elevation of the top is 37.4°. Find the height of the tower.
33. A racing car is traveling at 90.0 mi/hr along a straight road inclined at 7.0° to the horizontal. Determine the rate of ascent.
34. A plane is heading 20°0′ south of west at 311 mi/hr. If the wind is from the east at 20.0 mi/hr, determine its velocity with respect to the ground and the resulting direction.

Cumulative Review Exercises / Chapters 7–9

1. If $\sec \theta = -\sqrt{5}$ and $\cot \theta < 0$, find $\csc \theta$.
2. Use a diagram to find (a) $\sec 210°$; (b) $\cot 300°$.
3. Find all angles between $0°$ and $360°$ such that (a) $\cos \theta = \dfrac{\sqrt{2}}{2}$; (b) $\tan \theta = -\dfrac{\sqrt{3}}{3}$
4. Use a calculator to find $\cot 215.78°$.
5. Use a calculator to find all angles between $0°$ and $360°$, given that $\csc \theta = -1.341$.
6. Change $112°$ to radian measure.
7. Change $\dfrac{25\pi}{9}$ to degree measure.
8. A circle has radius 6.00 cm. Find the length of the arc intercepted by a central angle of $10.6°$.
9. Find the area of the circular sector whose central angle is $78.0°$ if the radius of the circle is 12.0 in.
10. The range R (in meters) along the ground of a projectile fired at velocity v (m/sec) at angle θ with the horizontal is

    $$R = \dfrac{v^2}{9.8} \sin 2\theta$$

    Find the range if $\theta = 25.6°$ and $v = 32.7$ m/sec.
11. The cylindrical shaft on a motor rotates at the rate of 35 rev/sec. If the radius of the shaft is 8.5 cm, find the velocity of a point on the rim in meters per second.
12. Sketch the graph of $y = -10 \sin \dfrac{1}{4} x$.
13. Sketch the graph of $y = 2 \cos \left(x - \dfrac{\pi}{3}\right)$.
14. Find the magnitude and direction of the vector $-2\sqrt{3}\mathbf{i} + 2\mathbf{j}$.
15. A block of ice weighing 50.0 lb is resting on an inclined plane which makes an angle of $26.8°$ with the horizontal. What force is required to keep the block from slipping?
16. A small plane is heading west at 145 mi/hr. The wind is from the south at 20.3 mi/hr. Find the resulting direction and the velocity with respect to the ground.
17. Two tractors are pulling a heavy machine in directions that are $20.0°$ apart. If one force is 619 lb, what must the other force be so that the combined effect is a force of 1150 lb?
18. A ship sailing at the rate of 28 knots in still water is heading in the direction $15°$ east of south. Find its actual velocity and direction if it runs into a strong current of 4.6 knots from the south.

CHAPTER 10

Exponents and Radicals

Objectives Upon completion of this chapter, you should be able to:

1. Use the laws of exponents to perform the basic operations with positive, negative, zero, and fractional exponents.
2. Simplify expressions containing exponents.
3. Write a given expression in simplest radical form by:
   a. Removing perfect nth powers from the radicand.
   b. Rationalizing the denominator.
   c. Reducing the order of the radical.
4. Perform addition, subtraction, multiplication, and division of radical expressions.

10.1 Review of Exponents

The purpose of this section is to review the basic operations involving exponents, which are listed below.

Laws of exponents:

$$a^m a^n = a^{m+n} \tag{10.1}$$

$$\frac{a^m}{a^n} = a^{m-n}, \quad a \neq 0, \quad m > n \tag{10.2}$$

$$(a^m)^n = a^{mn} \tag{10.3}$$

$$(ab)^n = a^n b^n \tag{10.4}$$

$$\left(\frac{a}{b}\right)^n = \frac{a^n}{b^n}, \quad b \neq 0 \tag{10.5}$$

300 CHAPTER 10 EXPONENTS AND RADICALS

The following examples illustrate the basic laws:

Example 1 $x^3 x^6 x^{10} = x^{3+6+10} = x^{19}$ adding exponents

Example 2 $\dfrac{a^{12}}{a^8} = a^{12-8} = a^4$ subtracting exponents

Example 3 $(a^3)^4 = a^{3 \cdot 4} = a^{12}$ multiplying exponents

Example 4 $(2x)^3 = 2^3 x^3 = 8x^3$ by rule (10.4)

The basic laws are frequently used in combination, as illustrated in the remaining examples.

Example 5 Simplify:
$$\frac{(2x^3 a^4)^2}{3x^2 a^3}$$

Solution.
$$\frac{(2x^3 a^4)^2}{3x^2 a^3} = \frac{2^2 (x^3)^2 (a^4)^2}{3x^2 a^3} \qquad (ab)^n = a^n b^n$$
$$= \frac{4 x^6 a^8}{3 x^2 a^3} \qquad \text{multiplying exponents; } 2^2 = 4$$
$$= \frac{4 x^4 a^5}{3} \qquad \text{subtracting exponents}$$

Example 6 Simplify:
$$(3x^2 y^3)^2 (4x^3 y)$$

Solution. $(3x^2 y^3)^2 (4x^3 y) = 3^2 (x^2)^2 (y^3)^2 (4x^3 y)$ $(ab)^n = a^n b^n$
$\qquad\qquad = 9 x^4 y^6 (4 x^3 y)$ multiplying exponents
$\qquad\qquad = 9 \cdot 4 x^4 x^3 y^6 y = 36 x^7 y^7$

since $9 \cdot 4 = 36$, $x^4 x^3 = x^7$, and $y^6 y = y^7$.

Example 7 Perform the following multiplication and simplify:
$$\left(\frac{6 v^7 L^6 c^4}{15 v^4 L^4 c} \right)^4 \cdot \left(\frac{5a}{v^8 L^4 c^2} \right)^2$$

10.1 REVIEW OF EXPONENTS   **301**

Solution.

$$\left(\frac{6v^7L^6c^4}{15v^4L^4c}\right)^4 \cdot \left(\frac{5a}{v^8L^4c^2}\right)^2 = \left(\frac{2v^3L^2c^3}{5}\right)^4 \cdot \left(\frac{5a}{v^8L^4c^2}\right)^2$$

$$= \frac{(2v^3L^2c^3)^4}{5^4} \cdot \frac{(5a)^2}{(v^8L^4c^2)^2} \qquad \left(\frac{a}{b}\right)^n = \frac{a^n}{b^n}$$

$$= \frac{16v^{12}L^8c^{12}}{5^4} \cdot \frac{5^2a^2}{v^{16}L^8c^4} = \frac{16a^2c^8}{5^2v^4} = \frac{16a^2c^8}{25v^4}$$

Example 8 Simplify:

$$\frac{x^{3a+1}y^{a+3}}{x^{3a}y^3}$$

Solution. Even though the exponents are letters, the basic laws can be applied: Subtracting exponents, we get

$$x^{(3a+1)-3a}y^{(a+3)-3} = xy^a$$

Exercises / Section 10.1

Perform the indicated operations and simplify.

1. $3^2 \cdot 2^3$
2. $\dfrac{5^6}{5^4}$
3. $\left(\dfrac{3}{4}\right)^3$
4. $(2^2 3^2)^2$
5. $(-3x^4y^6)(-2xy^3)$
6. $(-4a^3b^2c^5)(3abc^4)$
7. $\dfrac{-24a^7b^4}{-12a^3b}$
8. $\dfrac{25c^8d^4}{-15c^6d^3}$
9. $(-RS^2)^4$
10. $(-2v_1^2v_2^3)^3$
11. $(-4C_1^2C_2)^3(2C_1C_2^3)^3$
12. $\dfrac{3u^3v^2}{5w^4} \cdot \dfrac{10v^2w^5}{9u^4}$
13. $\dfrac{9s^2t^3}{3v} \cdot \dfrac{3v^2}{21st^4}$
14. $\dfrac{7a^3b^4}{8c^5d^2} \div \dfrac{21a^7b}{16c^3d^7}$
15. $\dfrac{9mn^3}{16s^5} \div \dfrac{27mn^2}{16s^6}$
16. $(2xy)^3 \div xy$
17. $\dfrac{(3RV^2)^2}{3R^2V}$
18. $\dfrac{9w^4x^6y^8}{(3wx^2y^2)^3}$
19. $\dfrac{8x^{10}y^5z^6}{(-2x^2y^2z^3)^4}$
20. $\dfrac{18a^2bc^2}{(-3a^2bc^2)^3}$
21. $\dfrac{(3abc)^3}{(3abc)^2}$
22. $\dfrac{(5p^2q^3r)^2}{(3pqr^5)^3}$
23. $\dfrac{(7v_0s_1^2s_2)^3}{(7v_0s_1^3s_2^4)^2}$
24. $\dfrac{(3x^2y^3z)^4}{(3x^2y^3z^2)^4}$
25. $\left(\dfrac{6x^3y^6}{9x^2y^7}\right)^3$
26. $\left(\dfrac{8v_1^4v_2^2v_3}{12v_1^3v_2^4v_3^2}\right)^2$
27. $\left(\dfrac{14p^3q^4r^5}{21p^5qr^6}\right)^4$
28. $\left(\dfrac{27a^{10}c^8x^5}{45a^{15}c^3x^4}\right)^3$
29. $(-3x^2y^2z^3)^2(5xy^3z)^4$
30. $(-4\alpha^2\beta^3)^2(-\alpha^2\beta\omega)^5$
31. $\left(\dfrac{2ab^2}{3xy}\right)^2\left(\dfrac{3xy^2}{2a^2b}\right)^3$
32. $\left(\dfrac{5a^2b}{m^2n^2}\right)^3\left(\dfrac{m^3n^2}{10ab}\right)^2$
33. $\left(\dfrac{2R^2V^2}{C^3}\right)^3\left(\dfrac{C^2}{4RV}\right)^4$

34. $\left(\dfrac{4xy}{3st}\right)^2\left(\dfrac{3st^2}{8x^2y}\right)^4$

35. $\dfrac{x^{a+1}}{x^a}$

36. $\dfrac{x^a}{x^{a+2}}$

37. $\dfrac{y^{a+2}}{y^{a+1}}$

38. $\dfrac{y^{2a}}{y^a}$

39. $\dfrac{x^{a+1}y^{3a}}{x^a y^{2a}}$

40. $\dfrac{a^{x+2}b^x}{a^{x+1}b^{2x}}$

41. $\dfrac{a^{x-2}}{a^{x-3}}$

42. $\left(\dfrac{y^{b-1}}{y^{b-3}}\right)^2$

43. $\left(\dfrac{x^{a+4}}{x^{a-1}}\right)^b$

44. $\left(\dfrac{y^{2b+1}}{y^{b+1}}\right)^2$

45. $\left(\dfrac{x^{3b+2}}{x^{b+2}}\right)^3$

10.2 Zero and Negative Integral Exponents

So far exponents were assumed to be positive integers. In this section we will introduce zero exponents and exponents that are negative integers.

Let us first consider the zero exponent a^0, $a \neq 0$. There are several ways to make sense out of this expression. For example, note that

$$\dfrac{a^n}{a^n} = 1, \quad a \neq 0$$

If the law of division is to hold, then

$$\dfrac{a^n}{a^n} = a^{n-n} = a^0$$

We conclude that $a^0 = 1$. As a check, observe that

$$a^n a^0 = a^{n+0} = a^n$$

by the rule for multiplication. This product makes sense only if we define $a^0 = 1$.

Similar checks for consistency can be made for negative exponents. Consider, for example,

$$\dfrac{a^3}{a^5} = \dfrac{1}{a^2}$$

If we decide to carry out the division by subtracting the exponents, we get

$$\dfrac{a^3}{a^5} = a^{3-5} = a^{-2}$$

We conclude that $a^{-2} = 1/a^2$. (In general, we define $a^{-n} = 1/a^n$). As another check, note that

$$a^{-2}a^6 = a^{-2+6} = a^4$$

by the rule for multiplication. This result agrees with

$$a^{-2}a^6 = \dfrac{a^6}{a^2} = a^4$$

obtained from the definition. In fact, rules (10.1) through (10.5) in Section 10.1 hold for zero and negative exponents as well. Let us now summarize the new definitions.

Zero and negative exponents:

$$a^0 = 1, \quad a \neq 0 \tag{10.6}$$

$$a^{-n} = \frac{1}{a^n}, \quad a \neq 0 \tag{10.7}$$

The following examples illustrate how the basic operations can be performed with zero and negative exponents.

Example 1 $\quad \dfrac{x^{-4}}{x^{-2}} = x^{-4-(-2)} = x^{-2} \qquad \dfrac{a^m}{a^n} = a^{m-n}$

Example 2 $\quad (2x^{-3})^4 = 2^4(x^{-3})^4 = 2^4 x^{(-3)(4)} = 16x^{-12} \qquad (ab)^n = a^n b^n;\ (a^m)^n = a^{mn}$

Example 3 $\quad (x^{-2})^{-4} = x^{(-2)(-4)} = x^8 \qquad (a^m)^n = a^{mn}$

Example 4 $\quad x^3 x^{-4} x^{-6} = x^{3-4-6} = x^{-7} \qquad a^m a^n = a^{m+n}$

Example 5 $\quad \dfrac{a^0 x^4}{x^9} = 1 \cdot x^{4-9} = x^{-5} \qquad a^0 = 1;$ subtracting exponents

Before attempting more involved simplifications, observe that

$$\frac{1}{a^{-n}} = \frac{1}{\frac{1}{a^n}} = a^n$$

or $a^n = 1/a^{-n}$.

When a **factor** is moved from the numerator of a fraction to the denominator or from the denominator to the numerator, the sign on the exponent is changed:

$$a^n = \frac{1}{a^{-n}} \quad \text{and} \quad a^{-n} = \frac{1}{a^n} \tag{10.8}$$

These rules are illustrated in the examples that follow.

Example 6 $\dfrac{a^{-3}b}{c^{-4}} = \dfrac{bc^4}{a^3}$ since $a^{-3} = \dfrac{1}{a^3}$ and $\dfrac{1}{c^{-4}} = c^4$

Example 7 $\dfrac{2^{-1}x^{-2}}{y^4} = \dfrac{1}{2x^2y^4}$ by rule (10.8)

Example 8 $\dfrac{x^2y^3}{3^{-1}} = \dfrac{3}{x^{-2}y^{-3}}$ by rule (10.8)

Example 9 Evaluate: **a.** $(2^{-1})^{-3}$ **b.** $\dfrac{3^{-2}5^0}{2^{-3}}$ **c.** $\dfrac{9^{-2}}{3^{-6}}$

Solution. **a.** Multiplying exponents, we get
$$(2^{-1})^{-3} = 2^3 = 8$$

b. $\dfrac{3^{-2}5^0}{2^{-3}} = \dfrac{3^{-2}\cdot 1}{2^{-3}} = \dfrac{2^3}{3^2} = \dfrac{8}{9}$

c. Since $9 = 3^2$, the expression can be written
$$\dfrac{(3^2)^{-2}}{3^{-6}} = \dfrac{3^{-4}}{3^{-6}}$$
$$= \dfrac{3^6}{3^4} = 3^2 = 9$$

Although negative exponents are useful in making calculations, final results are ordinarily written without them.

Example 10 Simplify and write without negative exponents:
$$\dfrac{2^{-3}T_a^{-3}T_b^{-4}}{4^{-4}T_a^{2}T_b^{-6}}$$

Solution. Since $4 = 2^2$, $4^{-4} = (2^2)^{-4} = 2^{-8}$. Thus
$$\dfrac{2^{-3}T_a^{-3}T_b^{-4}}{4^{-4}T_a^{2}T_b^{-6}} = \dfrac{2^{-3}T_a^{-3}T_b^{-4}}{2^{-8}T_a^{2}T_b^{-6}}$$
$$= \dfrac{2^8 T_b^6}{2^3 T_a^2 T_a^3 T_b^4} \qquad \dfrac{1}{a^{-n}} = a^n;\ a^{-n} = \dfrac{1}{a^n}$$

$$= \frac{2^8 T_b^6}{2^3 T_a^5 T_b^4} \qquad T_a^2 T_a^3 = T_a^5$$

$$= \frac{2^5 T_b^2}{T_a^5} \qquad \text{subtracting exponents}$$

$$= \frac{32 T_b^2}{T_a^5} \qquad 2^5 = 32$$

The same result could be obtained by subtracting exponents in the original expression. For example

$$\frac{T_b^{-4}}{T_b^{-6}} = T_b^{-4-(-6)} = T_b^2$$

Example 11 Simplify:

$$(2^{-1} R^{-3} r^{-4} V^2)^{-2}$$

Solution. The expression can be written

$$\left(\frac{V^2}{2R^3 r^4}\right)^{-2}$$

However, because so many of the exponents are negative, it is better to multiply the exponents directly. Thus

$$(2^{-1} R^{-3} r^{-4} V^2)^{-2} = 2^2 R^6 r^8 V^{-4} \qquad \text{multiplying exponents}$$

$$= \frac{4 R^6 r^8}{V^4}$$

Fractions containing sums of terms with negative exponents have to be changed to complex fractions and then simplified, as shown in the next example.

Example 12 Simplify:

$$\frac{x^{-1}}{x^{-1} + y^{-1}}$$

Solution. Since the denominator is a sum of two terms (not a product!), the terms in the denominator cannot be "moved up." In other words,

$$\frac{x^{-1}}{x^{-1} + y^{-1}} \neq \frac{x^{-1}(x + y)}{1}$$

Instead, we write the given fraction as a complex fraction:

$$\frac{x^{-1}}{x^{-1}+y^{-1}} = \frac{\dfrac{1}{x}}{\dfrac{1}{x}+\dfrac{1}{y}}$$

$$= \frac{\dfrac{1}{x}}{\dfrac{y}{xy}+\dfrac{x}{xy}} = \frac{\dfrac{1}{x}}{\dfrac{x+y}{xy}}$$

$$= \frac{1}{x} \cdot \frac{xy}{x+y} = \frac{y}{x+y}$$

Common errors

a. Forgetting that

$$a^{-1} \neq -\frac{1}{a}$$

Instead,

$$a^{-1} = \frac{1}{a}$$

b. Forgetting that it is not correct to write

$$\frac{1}{a^{-1}+b^{-1}} \text{ as } a+b$$

since a^{-1} and b^{-1} are *not factors*. Instead,

$$\frac{1}{a^{-1}+b^{-1}} = \frac{1}{\dfrac{1}{a}+\dfrac{1}{b}} = \frac{1}{\dfrac{a+b}{ab}} = \frac{ab}{a+b}$$

Example 13 In the mathematics of finance, the *present value* is the amount of money that must be invested in order to accumulate to a prescribed amount in a given time interval. An *annuity* is a sequence of equal payments R made at equal time intervals. The present value A of an annuity, which is the total value of all payments before the first payment is made, can be written

$$A = R\frac{(1+i)^{-n-1}-(1+i)^{-1}}{(1+i)^{-1}-1}$$

where i is the interest rate. Simplify the expression for A.

Solution.
$$A = R\frac{(1 + i)^{-n-1} - (1 + i)^{-1}}{(1 + i)^{-1} - 1}$$
$$= R\frac{(1 + i)^{-n-1} - (1 + i)^{-1}}{(1 + i)^{-1} - 1} \cdot \frac{(1 + i)^{n+1}}{(1 + i)^{n+1}}$$
$$= R\frac{1 - (1 + i)^n}{(1 + i)^n - (1 + i)^{n+1}} \quad \text{adding exponents}$$
$$= R\frac{1 - (1 + i)^n}{(1 + i)^n[1 - (1 + i)]} \quad \text{common factor } (1 + i)^n$$
$$= R\frac{1 - (1 + i)^n}{(1 + i)^n(-i)} \quad \text{simplifying denominator}$$
$$= R\frac{(1 + i)^n - 1}{i(1 + i)^n} \quad \text{multiplying numerator and denominator by } -1$$

Exercises / Section 10.2

In Exercises 1–8, evaluate the given expressions.

1. 4^{-2} **2.** $3^{-1} \cdot 3^0$ **3.** $(2^{-1})^{-4}$ **4.** $(2^{-2})^3$

5. $\dfrac{6^{-2}}{6^{-4}}$ **6.** $\dfrac{9^{-2}}{3^{-2}}$ **7.** $\dfrac{4^{-4}}{16^{-2}}$ **8.** $\dfrac{2^0 \cdot 2^2}{8^{-2}}$

In Exercises 9–58, combine the expressions and write the results without zero or negative exponents.

9. $\dfrac{x^2}{x^{-3}}$ **10.** $\dfrac{a^{-3}}{a^{-4}}$ **11.** $\dfrac{a^{-3}b^2}{b^{-4}}$ **12.** $\dfrac{2^0 z^2 w^{-3}}{2^{-1} z^{-2} w^{-4}}$

13. $2^{-2} x^0 z^{-1}$ **14.** $\dfrac{5^{-1} V^2 W^{-2}}{2 V^0 W^3}$ **15.** $\dfrac{2^{-4} f_1^{-2} f_2^4 f_3^{-1}}{4^{-3} f_1^{-3} f_2^{-1} f_3^4}$ **16.** $\dfrac{3 a^{-2} b^4 t^{-7}}{9^{-1} a^3 b^{-5} t^{-5}}$

17. $\dfrac{3^{-4} s^{-3} t^4 u^{-7}}{9^{-2} s^{-2} t^2}$ **18.** $\dfrac{5^0 e^0 n^{-3} m^{-1}}{5^{-2} e^{-2} n^{-7} m}$ **19.** $\dfrac{3^{-1} q^3 r^{-2} s^{-5}}{4^{-2} q^{-4} s^2}$ **20.** $\dfrac{7^{-2} a^0 x^{-4} w^3}{3^0 a^{-6} x^0 w^{-1}}$

21. $\left(\dfrac{a}{b}\right)^{-1}$ **22.** $\left(\dfrac{x}{y}\right)^{-2}$ **23.** $(2x^{-2}y^2)^{-3}$ **24.** $(3^{-1}a^{-2}b^{-6})^{-3}$

25. $(2^{-2}x^{-1}b^2)^4$ **26.** $(2^{-3}x^{-12}y^{16})^{-1}$ **27.** $(2x^{-7}y^0)^0$ **28.** $(2V^{-2}R^{-3})^{-2}$

29. $(4^{-1}\pi^2\omega^{-2})^{-3}$ **30.** $(5e^{-3}\pi^{-4})^{-1}$ **31.** $\left(\dfrac{a^{-2}}{2b^2}\right)^4$ **32.** $\left(\dfrac{2^{-1}x^{-2}y^4}{x^{-6}y^{-1}}\right)^{-2}$

33. $\left(\dfrac{2^{-3}L^{-2}R^3C^{-2}}{4^{-1}L^2R^{-1}C^3}\right)^{-3}$ **34.** $\left(\dfrac{a^2p^{-4}e^{-1}}{a^{-2}p^{-3}e^0}\right)^{-2}$ **35.** $\left(\dfrac{2a^3c_1^{-1}c_2^{-2}}{3^{-1}a^{-1}c_1^{-2}c_2^{-1}}\right)^3$ **36.** $\left(\dfrac{4R^{-3}r^2}{2^{-1}R^{-7}r^{-6}}\right)^{-1}$

37. $\left(\dfrac{s^{-5}r^{-6}t^{-2}}{s^{-3}r^5t}\right)^{-2}$ **38.** $\left(\dfrac{2^{-1}f^{-3}g^3h^{-4}}{f^3g^{-1}h^{-2}}\right)^4$ **39.** $\left(\dfrac{a^2b^{-2}y^{-3}}{3^{-2}a^{-2}b^{-3}y^{-4}}\right)^{-2}$ **40.** $\left(\dfrac{7^{-1}x^{-2}y^3w^{-1}}{2^0x^0y^{-3}w^2}\right)^{-2}$

41. $(\pi^{-e})^{-2}$ **42.** $(2x^{-a})^{-1}$ **43.** $(3a^xb^{-y})^{-1}$ **44.** $(2^{-1}q^{-2r})^{-3}$

45. $(a^bx^c)^b$ **46.** $(c^{-x}d^y)^{-t}$ **47.** $x^2 + \dfrac{1}{x^{-2}}$ **48.** $2a^{-1} + \dfrac{1}{a}$

308   CHAPTER 10   EXPONENTS AND RADICALS

49. $x^{-1} + y^{-1}$

50. $2^{-1} - 3^{-1}$

51. $\dfrac{a^{-1} + b}{a^{-1}}$

52. $\dfrac{1 + x^{-2}}{x}$

53. $(1 + b^{-1})^{-1}$

54. $\dfrac{a^{-1}}{1 + a^{-1}}$

55. $\dfrac{a^{-1}b^{-1}}{a^{-1} - b^{-1}}$

56. $\left(\dfrac{x^{-1}y^{-1}}{x^{-1} + y^{-1}}\right)^{-1}$

57. $\dfrac{a^{-2}b^{-2}}{a^{-2} - b^{-2}}$

58. $\left(\dfrac{x^{-2}y^{-2}}{x^{-2} + y^{-2}}\right)^{-2}$

59. Show that the combined resistance of two resistors R and r in parallel can be written $(R^{-1} + r^{-1})^{-1}$.

60. The relationship between the object distance q, the image distance p, and the focal length f of a lens can be written
$$f^{-1} = p^{-1} + q^{-1}$$
Find f if $q = \tfrac{3}{2}$ cm and $p = 3$ cm.

61. In determining the period of a pendulum, the expression $(1 - n^{-1})^{-1}$ has to be simplified. Carry out this simplification.

62. The total resistance of the circuit in Figure 10.1 is given by
$$R = [R_1^{-1} + (R_2 + R_3)^{-1}]^{-1}$$
Simplify this expression.

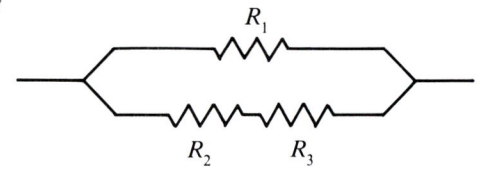

Figure 10.1

63. The *present value* of an investment is the amount of money P that needs to be invested in order to yield a certain amount S at the end of a prescribed time period. The equation for the relationship is
$$P = S(1 + i)^{-n}$$
where i is the interest rate compounded annually and n the number of years. How much has to be invested in order to obtain $10,000 at the end of eight years if the interest rate is 12% compounded annually?

64. Determine the size of an investment that will yield $15,000 at the end of 10 years if the interest rate is 10% compounded annually. (Refer to Exercise 63.)

10.3   Fractional Exponents

In this section we shall consider **fractional exponents**. As in the case of zero and negative integral exponents, the definitions will be formulated to preserve the basic laws of exponents. In particular, to define $a^{1/2}$, we recall that $(a^n)^m = a^{nm}$ and require that
$$(a^{1/2})^2 = a$$

Now let $b = a^{1/2}$. Then $b^2 = (a^{1/2})^2 = a$, so that $b = \sqrt{a}$. It follows that $a^{1/2} = \sqrt{a}$. In general, since $(a^{1/n})^n = a$, $a^{1/n} = \sqrt[n]{a}$. To define $a^{m/n}$, observe that
$$a^{m/n} = (a^{1/n})^m = (a^m)^{1/n}$$

These results are summarized on the following page.

10.3 FRACTIONAL EXPONENTS

Fractional exponents:

$$a^{1/n} = \sqrt[n]{a} \tag{10.9}$$

$$a^{m/n} = \sqrt[n]{a^m} = (\sqrt[n]{a})^m \tag{10.10}$$

Since fractional exponents are really radicals in disguise, numerical expressions with fractional exponents are evaluated by changing the expression to the radical form. It is usually easier to find the root first and then the power. For example, $8^{2/3} = (\sqrt[3]{8})^2 = 2^2 = 4$. Consider another example.

Example 1 Evaluate:

    a. $16^{3/2}$ **b.** $32^{4/5}$ **c.** $8^{-1/3}$ **d.** $27^{-2/3}$

Solution. **a.** $16^{3/2} = (\sqrt{16})^3 = 4^3 = 64$
             b. $32^{4/5} = (\sqrt[5]{32})^4 = 2^4 = 16$
             c. $8^{-1/3} = \dfrac{1}{8^{1/3}} = \dfrac{1}{2}$
             d. $27^{-2/3} = \dfrac{1}{27^{2/3}} = \dfrac{1}{(\sqrt[3]{27})^2} = \dfrac{1}{3^2} = \dfrac{1}{9}$

Algebraic expressions with fractional exponents are simplified by using the laws of exponents, as shown in the remaining examples.

Example 2 Simplify:

    a. $x^{3/2}x^{1/3}$ **b.** $\dfrac{a^{3/4}}{a^{5/2}}$ **c.** $(16x^6y^2)^{1/2}$

Solution. **a.** $x^{3/2}x^{1/3} = x^{3/2 + 1/3} = x^{11/6}$ adding exponents

            b. $\dfrac{a^{3/4}}{a^{5/2}} = a^{3/4 - 5/2} = a^{-7/4} = \dfrac{1}{a^{7/4}}$ subtracting exponents

            c. $(16x^6y^2)^{1/2} = (16)^{1/2}(x^6)^{1/2}(y^2)^{1/2} = 4x^3y$ since $(ab)^n = a^n b^n$

Example 3 Simplify and write without zero or negative exponents:

$$\dfrac{27^{-2/3}a^{-1/7}b^{-1/5}c^{-3/8}d^0}{8^{1/3}a^{3/7}b^{-2/3}c^{-1/4}}$$

Solution. As in the case of integral exponents, the various terms can be combined by subtracting exponents. For example,

$$\frac{a^{-1/7}}{a^{3/7}} = a^{-1/7-3/7} = a^{-4/7} = \frac{1}{a^{4/7}}$$

A good alternative is to write the expression without negative exponents first. Since $d^0 = 1$, we get

$$\frac{b^{2/3}c^{1/4}}{8^{1/3}27^{2/3}a^{3/7}a^{1/7}b^{1/5}c^{3/8}} \qquad \frac{1}{a^n} = a^{-n}, \frac{1}{a^{-n}} = a^n$$

$$= \frac{b^{2/3-1/5}c^{1/4-3/8}}{2 \cdot 9a^{4/7}} = \frac{b^{7/15}c^{-1/8}}{18a^{4/7}} = \frac{b^{7/15}}{18a^{4/7}c^{1/8}}$$

Example 4 Simplify and write without negative exponents:

$$\left(\frac{8^{-1}c_1^{-3/4}c_2^{-6}}{27^{-3}c_1^{-3/2}c_2^{9/4}}\right)^{-2/3}$$

Solution. The terms inside the parentheses could be combined first. However, because of the large number of negative exponents, it may be easier to multiply exponents first. Then we get

$$\frac{8^{2/3}c_1^{1/2}c_2^4}{27^2c_1c_2^{-3/2}} = \frac{4c_2^4c_2^{3/2}}{729c_1c_1^{-1/2}} = \frac{4c_2^{11/2}}{729c_1^{1/2}}$$

Expressions like the one in the next example occur frequently in calculus.

Example 5 If the distance $x(t)$ from the origin of a particle moving along the x-axis as a function of time is $x(t) = 2t^2(1 + t)^{1/3}$, we get from calculus that the velocity is

$$v = \frac{2}{3}t^2(1 + t)^{-2/3} + 4t(1 + t)^{1/3}$$

Write v as a single fraction.

Solution. The expression for v can be written

$$v = \frac{2t^2}{3(1 + t)^{2/3}} + 4t(1 + t)^{1/3}$$

$$= \frac{2t^2}{3(1 + t)^{2/3}} + \frac{4t(1 + t)^{1/3}}{1} \cdot \frac{3(1 + t)^{2/3}}{3(1 + t)^{2/3}}$$

Now note that

$$4t(1 + t)^{1/3} \cdot 3(1 + t)^{2/3} = 12t(1 + t)^{1/3+2/3}$$
$$= 12t(1 + t) = 12t + 12t^2$$

10.3 FRACTIONAL EXPONENTS

Thus

$$v = \frac{2t^2 + 12t + 12t^2}{3(1+t)^{2/3}} = \frac{14t^2 + 12t}{3(1+t)^{2/3}}$$

Exercises / Section 10.3

In Exercises 1–20, evaluate the given expressions.

1. $81^{1/2}$
2. $64^{1/3}$
3. $27^{2/3}$
4. $16^{3/4}$
5. $49^{-1/2}$
6. $27^{-1/3}$
7. $(0.09)^{1/2}$
8. $\left(\frac{1}{4}\right)^{1/2}$
9. $\left(\frac{1}{9}\right)^{-1/2}$
10. $32^{2/5}$
11. $(-64)^{1/3}$
12. $(-8)^{1/3}$
13. $(-32)^{3/5}$
14. $(-32)^{4/5}$
15. $49^{-3/2}$
16. $27^{-4/3}$
17. $125^{-1/3}$
18. $(-125)^{-1/3}$
19. $(-125)^{-2/3}$
20. $(0.064)^{1/3}$

In Exercises 21–68, simplify the given expressions and write the results without zero or negative exponents.

21. $a^{1/2}a^{-1/2}$
22. $x^{3/4}x^{1/4}$
23. $m^{1/7}m^{2/7}$
24. $p^{-1/3}p^{4/3}$
25. $\dfrac{x^{7/3}}{x^{1/4}}$
26. $\dfrac{Q^{1/5}}{Q^{-2/5}}$
27. $\dfrac{\omega^{-1/4}}{\omega^{-3/4}}$
28. $\dfrac{y^{-1/3}}{y^{1/6}}$
29. $(-5x^{1/2})(-2x^{-3/4})$
30. $(3a^{-5/9})(10a^{1/3})$
31. $\dfrac{2a^{1/3}b^{-1/4}}{4a^{-1/6}b^{-1/5}}$
32. $\dfrac{6V^{1/5}I^{1/12}}{9V^{1/10}I^{1/6}}$
33. $\dfrac{15R_1^{-1/3}R_2^{2/3}}{25R_1^{1/4}R_2^{3/4}}$
34. $\dfrac{21v_1^{1/2}v_2^{-3/4}}{18v_1^{-2/9}v_2^{2/5}}$
35. $\dfrac{36^{1/2}a^{-3/5}w^{-1/8}}{27^{1/3}a^{-5/6}w^{3/4}}$
36. $\dfrac{6^{1/3}s^{-3/7}t^{2/3}}{3^{1/3}s^{-1/7}t^{-2/5}}$
37. $(2^{1/2}x^{3/2}y^8)^2$
38. $(2^{2/3}x^{-1/3}y^{1/3})^3$
39. $(9a^4y^{-6})^{-1/2}$
40. $(27L^{-6}R^9)^{-1/3}$
41. $(32^{-1}x^5y^{-10}z^{-15})^{1/5}$
42. $(2^{-3}v^{-6}q^{-12})^{-1/3}$
43. $\left(\dfrac{2^0a^2c^{-3}}{2^{-1}a^{1/3}c^{-1/4}}\right)^{-1}$
44. $\left(\dfrac{27R^{-9}C}{8R^3C^{-3}}\right)^{-2/3}$

45. $\left(\dfrac{64^{-1}m^9d^0}{m^{-6/5}d^{-12}}\right)^{-1/6}$

46. $\left(\dfrac{16a^{-2}c^{2/5}}{25a^{4/3}c^{-4}}\right)^{-1/2}$

47. $\left(\dfrac{81m^{-8}v^{-4}s^0}{16^{-1}m^4v^{4/5}}\right)^{-1/4}$

48. $\left(\dfrac{4^{-1}x^{-2/3}y^{4/5}t^0}{16^{-1}x^{-2}y^{-4}}\right)^{3/2}$

49. $\left(\dfrac{L^{-4/3}T^{-8/3}}{16^{-1}L^{-16/3}T^4}\right)^{3/4}$

50. $\left(\dfrac{32r^{-10}s^{-2/5}}{r^{5/8}s^{1/5}}\right)^{-4/5}$

51. $x(x+2)^{-1/2} + (x+2)^{1/2}$

52. $2x(x+1)^{1/2} + (x+1)^{-1/2}$

53. $(x+2)(x-3)^{-1/2} + (x-3)^{1/2}$

54. $(x-4)(x+3)^{-1/2} + (x+3)^{1/2}$

55. $x(x+1)^{-2/3} + (x+1)^{1/3}$

56. $(x+2)(x-3)^{-2/3} + (x-3)^{1/3}$

57. $(x+6)(x-1)^{-2/3} - (x-1)^{1/3}$

58. $x(x+2)^{-2/3} - (x+2)^{1/3}$

59. $(x-3)(x-1)^{-1/3} - 2(x-1)^{2/3}$

60. $x(x-2)^{-3/4} + 2(x-2)^{1/4}$

61. $x(x+2)^{-4/5} - 3(x+2)^{1/5}$

62. $(x+3)(x-1)^{-5/6} - (x-1)^{1/6}$

63. $(x-1)(x+2)^{-6/7} - (x+2)^{1/7}$

64. $[x^{1/(a-1)}]^{a-1}$

65. $[x^{1/(a-1)}]^{a^2-1}$

66. $\left(\dfrac{x^{3a-b}}{x^a}\right)^{1/(2a-b)}$

67. $\left(\dfrac{x^{1/a}}{x^{-1}}\right)^{-a/(1+a)}$

68. $\left(\dfrac{x^{4a-1}}{x^{2a}}\right)^{1/(4a^2-1)}$

69. When a gas is compressed adiabatically (with no gain or loss of heat), the pressure and volume of the gas are related by the formula $p = kv^{-7/5}$, where k is a constant. Express the formula in the form of a radical.

70. Two hallways of width a and b intersect at right angles (Figure 10.2). The longest rod that can be carried horizontally around the corner is of length $L = (a^{2/3} + b^{2/3})^{3/2}$. If $a = 27$ ft and $b = 8$ ft, show that the length of the longest rod is $\sqrt{2{,}197}$ ft.

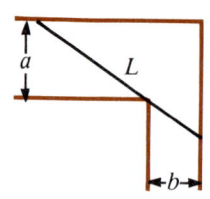

Figure 10.2

71. The distance between the centers of two spheres of radii a and b, respectively, is d. From a point P on the line of centers AB, the greatest amount of surface area is visible if the distance D from P to A is

$$D = da^{3/2}(a^{3/2} + b^{3/2})^{-1}$$

Find D if $a = 4$ cm, $b = 9$ cm, and $d = 20$ cm.

72. The equation $E = T/(x^2 + a^2)^{3/2}$ gives the electric-field intensity on the axis of a uniformly charged ring, where T is the total charge on the ring and a the radius of the ring. Write E in the form of a radical.

73. The rate of runoff of rainfall from an area of land to an inlet reaches its maximum after t minutes, where

$$t = C\sqrt[3]{L}i^{-2/3}(\tan\theta)^{-1/3}$$

Here L is the distance from the most remote point in the area, θ the angle the area makes with the horizontal, i the rain intensity, and C a constant. Show that t can be written

$$t = C \left(\frac{L}{i^2 \tan \theta}\right)^{1/3}$$

74. If V_1 and V_2 are the respective maximum and minimum volumes of air in the cylinder of an internal combustion engine, then V_1/V_2 is called the *compression ratio*. The *efficiency* Eff (expressed as a percentage) is given by

$$\text{Eff} = 100 \left(1 - \frac{1}{\left(\frac{V_1}{V_2}\right)^{k-1}}\right)$$

where k is a constant. Show that the formula is equivalent to

$$\text{Eff} = 100 \left(\frac{V_1^{k-1} - V_2^{k-1}}{V_1^{k-1}}\right)$$

75. If the position of a particle moving along the x-axis as a function of time is $x(t) = 2t(1 + t)^{1/2}$, then its velocity v is given by

$$v = t(1 + t)^{-1/2} + 2(1 + t)^{1/2}$$

Write v as a single fraction.

10.4 The Simplest Radical Form and Addition/Subtraction of Radicals

In this section we shall learn how to simplify radicals. Also, the addition and subtraction of radicals, which are natural by-products of simplification, are considered briefly.

First recall the laws of radicals from Chapter 1.

Laws of radicals:

$$\sqrt[n]{a^n} = (\sqrt[n]{a})^n = a \qquad (10.11)$$

$$\sqrt[n]{a}\,\sqrt[n]{b} = \sqrt[n]{ab} \qquad (10.12)$$

$$\frac{\sqrt[n]{a}}{\sqrt[n]{b}} = \sqrt[n]{\frac{a}{b}} \qquad (10.13)$$

$$\sqrt[m]{\sqrt[n]{a}} = \sqrt[mn]{a} \qquad (10.14)$$

Note: In the radical expression $\sqrt[n]{a}$, the number under the radical sign is called the **radicand** and the index is called the **order** of the radical.

The laws of radicals follow directly from the laws of exponents, as shown next.

(10.11): $\sqrt[n]{a^n} = (a^n)^{1/n} = a$ and $(\sqrt[n]{a})^n = (a^{1/n})^n = a$

(10.12): $\sqrt[n]{a}\,\sqrt[n]{b} = a^{1/n}b^{1/n} = (ab)^{1/n} = \sqrt[n]{ab}$

(10.13): $\dfrac{\sqrt[n]{a}}{\sqrt[n]{b}} = \dfrac{a^{1/n}}{b^{1/n}} = \left(\dfrac{a}{b}\right)^{1/n} = \sqrt[n]{\dfrac{a}{b}}$

(10.14): $\sqrt[m]{\sqrt[n]{a}} = (a^{1/n})^{1/m} = a^{1/mn} = \sqrt[mn]{a}$

In Chapter 1 we used law (10.12) to simplify certain radicals. Let's recall the procedure.

Example 1 To simplify $\sqrt{50}$, first note that $50 = 25 \cdot 2$ and $\sqrt{25} = 5$. So $\sqrt{50} = \sqrt{25 \cdot 2} = \sqrt{25}\sqrt{2} = 5\sqrt{2}$.

Similarly,

$$\sqrt[5]{64} = \sqrt[5]{32 \cdot 2} = \sqrt[5]{32}\sqrt[5]{2} = 2\sqrt[5]{2}$$

> To simplify a radical of the nth order, remove all perfect nth-power factors from the radical.

Example 2 Simplify: **a.** $\sqrt{81x^5y^3}$ **b.** $\sqrt[4]{32a^5b^{10}}$ **c.** $\sqrt[6]{128a^{11}b^{20}}$

Solution.

a. $\sqrt{81x^5y^3} = \sqrt{9^2x^4xy^2y} = \sqrt{(9^2x^4y^2)(xy)} = \sqrt{9^2x^4y^2}\sqrt{xy} = 9x^2y\sqrt{xy}$

b. First observe that $\sqrt[4]{a^4} = a$ and $\sqrt[4]{b^8} = (b^8)^{1/4} = b^2$. Hence $\sqrt[4]{32a^5b^{10}} = \sqrt[4]{16 \cdot 2a^4ab^8b^2} = \sqrt[4]{16a^4b^8}\sqrt[4]{2ab^2} = 2ab^2\sqrt[4]{2ab^2}$

c. Here we use the fact that $\sqrt[6]{a^6} = a$ and $\sqrt[6]{b^{18}} = (b^{18})^{1/6} = b^3$. Thus $\sqrt[6]{128a^{11}b^{20}} = \sqrt[6]{64 \cdot 2a^6a^5b^{18}b^2} = \sqrt[6]{64a^6b^{18}}\sqrt[6]{2a^5b^2} = 2ab^3\sqrt[6]{2a^5b^2}$

Now recall from Chapter 1 that $1/\sqrt{a}$ can be written

$$\dfrac{1}{\sqrt{a}} = \dfrac{1}{\sqrt{a}} \cdot \dfrac{\sqrt{a}}{\sqrt{a}} = \dfrac{\sqrt{a}}{a}$$

Rationalizing the denominator This operation is called **rationalizing the denominator.** To generalize this operation to radicals of any order, it is best to change the radicand to an equivalent fraction whose denominator is a perfect square, as shown in the following example.

Example 3 Rationalize the denominators: **a.** $\sqrt{\dfrac{3}{2}}$ **b.** $\sqrt{\dfrac{2a}{b}}$

10.4 THE SIMPLEST RADICAL FORM AND ADDITION/SUBTRACTION OF RADICALS **315**

Solution. a. To simplify $\sqrt{3/2}$, we multiply the numerator and denominator inside the radical sign by **2**:

$$\sqrt{\frac{3}{2}} = \sqrt{\frac{3 \cdot 2}{2 \cdot 2}} = \sqrt{\frac{6}{2^2}} = \frac{\sqrt{6}}{\sqrt{2^2}} = \frac{\sqrt{6}}{2}$$

b. Here we multiply the numerator and denominator inside the radical by **b**:

$$\sqrt{\frac{2a}{b}} = \sqrt{\frac{2ab}{bb}} = \sqrt{\frac{2ab}{b^2}} = \frac{\sqrt{2ab}}{\sqrt{b^2}} = \frac{\sqrt{2ab}}{b}$$

The method of Example 3 works with radicals of any order: We change the radicand to an equivalent fraction whose denominator is a perfect nth root.

Example 4 Rationalize the denominators: **a.** $\sqrt[3]{\frac{3}{2}}$ **b.** $\sqrt[3]{\frac{2a}{b}}$ **c.** $\sqrt[4]{\frac{x}{y^6}}$

Solution. a. We multiply the numerator and denominator inside the radical by 2^2:

$$\sqrt[3]{\frac{3}{2}} = \sqrt[3]{\frac{3 \cdot 2^2}{2 \cdot 2^2}} = \frac{\sqrt[3]{3 \cdot 2^2}}{\sqrt[3]{2^3}} \qquad \text{by rule (10.13)}$$

$$= \frac{\sqrt[3]{12}}{2} \qquad \text{by rule (10.11)}$$

b. $\sqrt[3]{\frac{2a}{b}} = \sqrt[3]{\frac{2ab^2}{bb^2}} = \frac{\sqrt[3]{2ab^2}}{\sqrt[3]{b^3}} \qquad \text{by rule (10.13)}$

$$= \frac{\sqrt[3]{2ab^2}}{b} \qquad \text{by rule (10.11)}$$

c. $\sqrt[4]{\frac{x}{y^6}} = \sqrt[4]{\frac{xy^2}{y^6 y^2}} = \frac{\sqrt[4]{xy^2}}{\sqrt[4]{y^8}} = \frac{\sqrt[4]{xy^2}}{y^2}$

If the radical contains negative exponents, we use rule (10.8) to rewrite the expression so it contains only positive exponents. We then rationalize the denominator and remove all perfect nth powers from the radical in the numerator.

Example 5 Simplify:

$$\sqrt{\frac{3^{-2}u^{-1}v^{-4}w^3}{27u^0v^{-1}}}$$

Solution.
$$\sqrt{\frac{3^{-2}u^{-1}v^{-4}w^3}{27u^0v^{-1}}} = \sqrt{\frac{3^{-2}u^{-1}v^{-4}w^3}{3^3v^{-1}}}$$
$$= \sqrt{\frac{w^3}{3^5uv^3}}$$
$$= \sqrt{\frac{w^3(3uv)}{3^5uv^3(3uv)}}$$
$$= \sqrt{\frac{3uvw^3}{3^6u^2v^4}}$$
$$= \frac{\sqrt{3uvw^3}}{\sqrt{3^6u^2v^4}}$$
$$= \frac{w\sqrt{3uvw}}{27uv^2}$$

Example 6 Simplify:
$$\sqrt[4]{\frac{3^{-1}R^{-1}C}{8C^3}}$$

Solution.
$$\sqrt[4]{\frac{3^{-1}R^{-1}C}{8C^3}} = \sqrt[4]{\frac{3^{-1}R^{-1}C}{2^3C^3}}$$
$$= \sqrt[4]{\frac{C}{2^3 \cdot 3RC^3}}$$
$$= \sqrt[4]{\frac{C(2 \cdot 3^3R^3C)}{2^3 \cdot 3RC^3(2 \cdot 3^3R^3C)}}$$
$$= \frac{\sqrt[4]{54R^3C^2}}{\sqrt[4]{2^4 \cdot 3^4R^4C^4}}$$
$$= \frac{\sqrt[4]{54R^3C^2}}{6RC}$$

It is sometimes possible to **reduce the order of a radical** by means of fractional exponents: Write the radicand as a single positive power (if possible), write the entire radical as a fractional power, multiply the exponents, and reduce the resulting fractional exponent (if possible). The result is a radical with a smaller index. Consider the next example.

Example 7
a. $\sqrt[4]{9} = \sqrt[4]{3^2} = (3^2)^{1/4} = 3^{2/4} = 3^{1/2} = \sqrt{3}$
b. $\sqrt[6]{16} = \sqrt[6]{2^4} = (2^4)^{1/6} = 2^{4/6} = 2^{2/3} = \sqrt[3]{2^2} = \sqrt[3]{4}$
c. $\sqrt[10]{x^6} = (x^6)^{1/10} = x^{6/10} = x^{3/5} = \sqrt[5]{x^3}$
d. $\sqrt[15]{x^3y^9} = \sqrt[15]{(xy^3)^3} = [(xy^3)^3]^{1/15} = (xy^3)^{1/5} = \sqrt[5]{xy^3}$

Example 8 Reduce the order of the radical
$$\sqrt[6]{x^4y^8z^{16}}$$

Solution. The reduction is best accomplished by means of fractional exponents:
$$\sqrt[6]{x^4y^8z^{16}} = (x^4y^8z^{16})^{1/6} = [(xy^2z^4)^4]^{1/6} = (xy^2z^4)^{4/6}$$
$$= (xy^2z^4)^{2/3} = \sqrt[3]{(xy^2z^4)^2} = \sqrt[3]{x^2y^4z^8}$$

We shall now summarize the procedures for simplifying radicals that have been considered so far.

A radical is in simplest form if the following conditions are met:

1. No power in the radical exceeds the order of the radical.
2. No radical appears in the denominator.
3. No fractional or negative powers appear in the radical.
4. The power of the radicand and the index of the radical have no common factors.

If any of the conditions are not met, proceed as follows:

To simplify a radical:

1. Remove all perfect nth powers from the radical of order n.
2. Rationalize the denominator.
3. Reduce the order of the radical if possible.

Addition and Subtraction of Radicals

In adding and subtracting radicals, recall that *only similar terms can be combined*.

To add or subtract radicals, we express each radical in its simplest form and combine similar terms.

For example,
$$9\sqrt{2} + 5\sqrt{2} = (9 + 5)\sqrt{2} = 14\sqrt{2}$$
On the other hand, $9\sqrt{2} + 5\sqrt[3]{2}$ cannot be combined since the radicals have different orders.

Example 9 Combine the following radicals:

$$\sqrt[3]{54} + \sqrt{25y} - \sqrt[3]{16} - \sqrt{36y}$$

Solution. The radicals can be written as follows:

$$\sqrt[3]{2 \cdot 27} + \sqrt{25y} - \sqrt[3]{2 \cdot 8} - \sqrt{36y}$$

Removing perfect nth powers from each radicand, we get

$$3\sqrt[3]{2} + 5\sqrt{y} - 2\sqrt[3]{2} - 6\sqrt{y}$$

Combining similar terms, we have

$$(3\sqrt[3]{2} - 2\sqrt[3]{2}) + (5\sqrt{y} - 6\sqrt{y}) = \sqrt[3]{2} - \sqrt{y}$$

Example 10 Combine the following radicals:

$$\sqrt{8s^3t} - \sqrt{2s^5t^7} + \sqrt{32s^3t^5}$$

Solution.

$$\sqrt{8s^3t} - \sqrt{2s^5t^7} + \sqrt{32s^3t^5} = \sqrt{2 \cdot 4s^2st} - \sqrt{2s^4st^6t} + \sqrt{2 \cdot 16s^2st^4t}$$
$$= 2s\sqrt{2st} - s^2t^3\sqrt{2st} + 4st^2\sqrt{2st}$$
$$= (2s - s^2t^3 + 4st^2)\sqrt{2st}$$

Common error Forgetting that

$$\sqrt{a} + \sqrt{b} \neq \sqrt{a+b}$$

$\sqrt{a} + \sqrt{b}$ is already in simplest form. Also,

$$\sqrt{x^2 + y^2} \neq x + y$$

$\sqrt{x^2 + y^2}$ is already in simplest form and should not be written as $x + y$.

CALCULATOR COMMENT To find the approximate value of a radical with a calculator, use the key $\boxed{x^y}$ (or a similar key). For example, to evaluate

$$(\sqrt[5]{2^7 + 6})^2$$

write the expression in the form $(2^7 + 6)^{2/5}$. The sequence is

$$\boxed{(}\ 2\ \boxed{x^y}\ 7\ \boxed{+}\ 6\ \boxed{)}\ \boxed{x^y}\ \boxed{(}\ 2\ \boxed{\div}\ 5\ \boxed{)}\ \boxed{=}$$

Display: 7.0931953
or

$$2\ \boxed{x^y}\ 7\ \boxed{=}\ \boxed{+}\ 6\ \boxed{=}\ \boxed{x^y}\ \boxed{(}\ 2\ \boxed{\div}\ 5\ \boxed{)}\ \boxed{=}$$

10.4 THE SIMPLEST RADICAL FORM AND ADDITION/SUBTRACTION OF RADICALS 319

Example 11 The amount of water A flowing across a dam with a V-shaped notch is given by

$$A = 2.506 \left(\tan \frac{\theta}{2}\right) h^{2.47}$$

where A is the volume of the flow in cubic meters per second, h the height in meters above the bottom of the notch, and θ the angle of the notch. If $h = 2.30$ m and $\theta = 20.3°$, find A.

Solution. From the given formula,

$$A = 2.506 \left(\tan \frac{20.3°}{2}\right) (2.30)^{2.47}$$

The sequence is

2.506 $\boxed{\times}$ $\boxed{(}$ 20.3 $\boxed{\div}$ 2 $\boxed{)}$ $\boxed{\text{TAN}}$ $\boxed{\times}$ 2.30 $\boxed{x^y}$ 2.47 $\boxed{=}$

Display: 3.5104991
So the water flows across at the rate of 3.51 m³/sec.

Exercises / Section 10.4

In Exercises 1–4, write each expression as a single radical. (See formula (10.14).)

1. $\sqrt{\sqrt{x}}$
2. $\sqrt{\sqrt[3]{a}}$
3. $\sqrt[3]{\sqrt[4]{b}}$
4. $\sqrt[5]{\sqrt{c}}$

In Exercises 5–12, reduce the order of the given radicals. (See Example 8.)

5. $\sqrt[4]{a^2b^2c^6}$
6. $\sqrt[6]{s^6t^9u^{12}}$
7. $\sqrt[8]{u^4v^6w^8}$
8. $\sqrt[12]{f^3g^6h^9}$
9. $\sqrt[9]{v^6m^6g^9}$
10. $\sqrt[6]{8x^3y^6z^9}$
11. $\sqrt[8]{25p^4q^8r^{10}}$
12. $\sqrt[15]{27x^9y^{12}v^{15}}$

In Exercises 13–20, simplify the given radicals. (See Examples 1 and 2.)

13. $\sqrt{4x^3}$
14. $\sqrt{8x^3y^5}$
15. $\sqrt[3]{54a^4c^8}$
16. $\sqrt[4]{48V^5W^{10}}$
17. $\sqrt[4]{32u^{13}v^{18}}$
18. $\sqrt[5]{64L^7R^{13}C^{18}}$
19. $\sqrt{18m^2n^3s^7t^{11}}$
20. $\sqrt{24v^8x^3y^4z^7}$

In Exercises 21–40, rationalize the denominator in each expression. (See Examples 3 and 4.)

21. $\dfrac{1}{\sqrt{5}}$
22. $\sqrt{\dfrac{2}{7}}$

23. $\sqrt{\dfrac{3}{x}}$ 24. $\sqrt{\dfrac{4}{b}}$

25. $\sqrt{\dfrac{a}{bc}}$ 26. $\sqrt{\dfrac{k}{mn}}$

27. $\sqrt{\dfrac{v}{3pq}}$ 28. $\sqrt{\dfrac{s}{5uv}}$

29. $\sqrt[3]{\dfrac{2}{3}}$ 30. $\sqrt[4]{\dfrac{1}{5}}$

31. $\sqrt[3]{\dfrac{v}{4\pi}}$ 32. $\sqrt[3]{\dfrac{B}{9A}}$

33. $\sqrt[4]{\dfrac{3}{8L^5V^2}}$ 34. $\sqrt[3]{\dfrac{3}{2ab}}$

35. $\sqrt{\dfrac{7}{3x^3y^5}}$ 36. $\sqrt[5]{\dfrac{1}{6}}$

37. $\sqrt[3]{\dfrac{aR}{9C}}$ 38. $\sqrt[4]{\dfrac{a^2c}{8b^2}}$

39. $\dfrac{1}{\sqrt{2pqr^2}}$ 40. $\sqrt[3]{\dfrac{2q}{9c^2d}}$

In Exercises 41–52, combine the given radicals. (See Examples 9 and 10.)

41. $\sqrt{24} + \sqrt{54}$ 42. $\sqrt{12} - \sqrt{27} + \sqrt{48}$

43. $2\sqrt{5} + \sqrt{80} - \sqrt{45}$ 44. $\sqrt{28} + \sqrt{63} - \sqrt{112}$

45. $\sqrt[3]{24} - \sqrt[3]{81}$ 46. $\sqrt{20} + \sqrt{4x} - \sqrt{80} + \sqrt{9x}$

47. $\sqrt{20} - \sqrt[3]{16} + \sqrt{125} - \sqrt[3]{54}$ 48. $\sqrt[3]{250} + \sqrt[3]{8x} - \sqrt[3]{54} - \sqrt[3]{27x}$

49. $3b\sqrt{a^3b} - 4a\sqrt{ab^3}$ 50. $\sqrt{8uv^4} - \sqrt{2u^3} - \sqrt{2u^3v^2}$

51. $\sqrt{25m^3n^3} - \sqrt{m^5n^3} + \sqrt{4m^5n^5}$ 52. $\sqrt{16p^4q^3} + \sqrt{4p^2q^3} + \sqrt{p^8q^5}$

In Exercises 53–68, simplify the given expressions. Write without zero or negative exponents and rationalize denominators. (See Examples 5 and 6.)

53. $\sqrt{\dfrac{4a^4b^{-2}}{3a^0b}}$ 54. $\sqrt{\dfrac{2^{-1}c^{-4}d^2}{2c^{-2}d^{-1}}}$

55. $\sqrt{\dfrac{w^{-4}z^{-1}}{9^{-1}w}}$ 56. $\sqrt{\dfrac{3^2T^{-1}W^{-2}}{9^{-2}T^2W}}$

57. $\sqrt{\dfrac{2^{-3}t^0xz}{4^{-3}x^{-1}z^4}}$ 58. $\sqrt{\dfrac{3V^0R^{-2}C}{9^2R^3C^5}}$

59. $\sqrt{\dfrac{3^{-1}u^{-2}v^2w^{-2}}{uv^{-1}w^5}}$ 60. $\sqrt{\dfrac{5^{-3}aL^{-6}}{2a^{-4}c^0L}}$

61. $\sqrt[3]{\dfrac{2^{-1}a^{-1}}{a^2c}}$ 62. $\sqrt[3]{\dfrac{16^{-1}cx^{-1}}{c^3x^4}}$

63. $\sqrt[3]{\dfrac{3u^{-4}v^{-1}}{4u^{-2}}}$ 64. $\sqrt[3]{\dfrac{st^{-1}z^{-2}}{9s^6t}}$

10.4 THE SIMPLEST RADICAL FORM AND ADDITION/SUBTRACTION OF RADICALS 321

65. $\sqrt[4]{\dfrac{x^{-1}y^{-2}}{8x^2y^5z^{-1}}}$

66. $\sqrt[5]{\dfrac{2^{-6}A^{-2}B^{-6}}{4AB^{-1}}}$

67. $\sqrt{\dfrac{7^{-2}s^2t^{-3}u^{-7}}{7s^5t^2u^{-2}}}$

68. $\sqrt[4]{\dfrac{2^{-1}x^{-3}y^2}{y^{-1}}}$

69. The average speed of a molecule of an ideal gas is $\sqrt{8kT/\pi m}$, where m is the mass, T the absolute temperature, and k the Boltzmann constant. Simplify this expression by rationalizing the denominator.

70. According to the special theory of relativity, the mass m of a body moving at velocity v relative to a stationary object is

$$m = \dfrac{m_0}{\sqrt{1 - \dfrac{v^2}{c^2}}}$$

where m_0 is the mass of the body at rest and c the velocity of light. Simplify this expression.

71. The velocity in meters per second of sound in air at temperature T (in degrees Celsius) is given by

$$v = 331.7\sqrt{1 + \dfrac{T}{273}}$$

Simplify this formula by combining terms and rationalizing the denominator.

72. Simplify the expression

$$\dfrac{1}{2L}\sqrt{R^2 - \dfrac{4L}{C}}$$

which arises in the study of electrical circuits.

73. The flow of water around an obstacle often results in *stagnation points* (where the velocity is zero). For a cylindrical obstacle placed at the origin, the x-coordinates of the stagnation points are

$$x = \pm\sqrt{1 - \dfrac{k^2}{16\pi^2}}$$

where k depends on the nature of the unimpeded flow. Simplify the expression for x.

74. A manufacturing process produces $100p\%$ defectives on the average. The process is *out of control* if a sample of size n contains more than

$$3\sqrt{\dfrac{p(1-p)}{n}}$$

defectives above the average. Simplify this expression by rationalizing the denominator.

75. By rationalizing the denominator, simplify the formula

$$t = C\sqrt[3]{\dfrac{L}{mi^2}}$$

which arises in the study of soil mechanics.

76. Evaluate $\sqrt[6]{(2.71)^4 + 3.12}$.

77. Evaluate $\sqrt[9]{(3.84)^6 + (2.56)^3}$.

10.5 Multiplication of Radicals

In the last section we briefly considered the addition and subtraction of radicals. We still need to discuss multiplication and division. Since these operations are somewhat more involved, we will discuss them in separate sections.

Formula (10.12), $\sqrt[n]{a}\sqrt[n]{b} = \sqrt[n]{ab}$, was used in the last section to simplify certain radicals. The same formula can be used to multiply single radicals, as shown in the next example.

Example 1

a. $\sqrt{2x}\sqrt{3x} = \sqrt{(2x)(3x)} = \sqrt{6x^2} = x\sqrt{6}$

b. $\sqrt[3]{6a^2}\sqrt[3]{4a^2} = \sqrt[3]{24a^4} = \sqrt[3]{3 \cdot 8a^3 a} = 2a\sqrt[3]{3a}$

c. $\sqrt[4]{4u^3v^2w}\sqrt[4]{8uv^3w^2} = \sqrt[4]{32u^4v^5w^3} = \sqrt[4]{2 \cdot 16u^4v^4vw^3} = 2uv\sqrt[4]{2vw^3}$

The next three examples illustrate the multiplication of multinomials containing radical expressions.

Example 2 Multiply $(\sqrt{3} - \sqrt{2})(\sqrt{3} + \sqrt{2})$.

Solution. Since the product has the form $(a - b)(a + b) = a^2 - b^2$, we get

$$(\sqrt{3} - \sqrt{2})(\sqrt{3} + \sqrt{2}) = (\sqrt{3})^2 - (\sqrt{2})^2 = 3 - 2 = 1$$

Example 3 Show that $a - b = (\sqrt{a} - \sqrt{b})(\sqrt{a} + \sqrt{b})$.

Solution. $(\sqrt{a} - \sqrt{b})(\sqrt{a} + \sqrt{b}) = (\sqrt{a})^2 - (\sqrt{b})^2 = a - b$

Note that this product shows that $a - b$ can be factored as a difference of two squares.

Example 4 Multiply $(\sqrt{a} - \sqrt{b})(a + \sqrt{ab} + b)$.

Solution.
$$\begin{array}{r} a + \sqrt{ab} + b \\ \sqrt{a} - \sqrt{b} \\ \hline a\sqrt{a} + a\sqrt{b} + b\sqrt{a} \\ - a\sqrt{b} - b\sqrt{a} - b\sqrt{b} \\ \hline a\sqrt{a} - b\sqrt{b} \end{array}$$

Thus $(\sqrt{a} - \sqrt{b})(a + \sqrt{ab} + b) = a\sqrt{a} - b\sqrt{b}$.

Note that the product can be written

$$a^{3/2} - b^{3/2} = (a^{1/2} - b^{1/2})[a + (ab)^{1/2} + b]$$

which has the form of a difference of two cubes.

10.5 MULTIPLICATION OF RADICALS

If radicals of different orders are to be multiplied and combined into a single radical, then the order of one or both radicals must be changed. This change can be accomplished by using fractional exponents, as shown in the next example.

Example 5 Multiply: a. $\sqrt[3]{3} \sqrt[5]{3}$ b. $\sqrt[3]{a} \sqrt[4]{b}$

Solution.

a. $\sqrt[3]{3} \sqrt[5]{3} = 3^{1/3} 3^{1/5} = 3^{5/15} 3^{3/15} = 3^{(5/15 + 3/15)} = 3^{8/15} = (3^8)^{1/15} = \sqrt[15]{3^8}$
$= \sqrt[15]{6{,}561}$

b. $\sqrt[3]{a} \sqrt[4]{b} = a^{1/3} b^{1/4} = a^{4/12} b^{3/12} = (a^4 b^3)^{1/12} = \sqrt[12]{a^4 b^3}$

Example 6 A topic in calculus is finding the center of mass of a plate. Suppose a plate of uniform density k (weight per unit area) is bounded by the graphs of $y = \sqrt{2x}$ and $y = \sqrt{x}$ and by a vertical line. In determining the y-coordinate of its center of mass, the following expression arises:

$$k \frac{\sqrt{2x} + \sqrt{x}}{2} (\sqrt{2x} - \sqrt{x})$$

Simplify this expression.

Solution. The expression can be written

$$\frac{k}{2} (\sqrt{2x} + \sqrt{x})(\sqrt{2x} - \sqrt{x}) = \frac{k}{2} [(\sqrt{2x})^2 - (\sqrt{x})^2]$$

$$= \frac{k}{2} (2x - x) = \frac{1}{2} kx$$

Exercises / Section 10.5

In Exercises 1–20, perform the multiplications and simplify.

1. $\sqrt{2y} \sqrt{6y}$
2. $\sqrt{16x} \sqrt{2y}$
3. $\sqrt{3ab} \sqrt{2ab}$
4. $\sqrt{5uv} \sqrt{6u}$
5. $\sqrt[3]{2x^2y} \sqrt[3]{4xy}$
6. $\sqrt[4]{4c^3d} \sqrt[4]{4c^2d^5}$
7. $\sqrt[5]{8m^3c^2} \sqrt[5]{8m^2c^4}$
8. $\sqrt[5]{4R^4C^3} \sqrt[5]{8RC^4}$
9. $\sqrt{2st} \sqrt{2s} \sqrt{5t}$
10. $\sqrt[3]{18a^2b} \sqrt[3]{3a^2b} \sqrt[3]{ab^2}$
11. $\sqrt[4]{9\pi^3\omega^2} \sqrt[4]{3\pi\omega^2} \sqrt[4]{3\pi^3\omega}$
12. $\sqrt[3]{4e^2f^2} \sqrt[3]{4ef^2} \sqrt[3]{ef}$
13. $\sqrt{3} \sqrt[3]{3}$
14. $\sqrt{3} \sqrt[4]{4}$
15. $\sqrt{x} \sqrt[4]{y}$
16. $\sqrt[3]{a} \sqrt[5]{b}$
17. $\sqrt{c} \sqrt[5]{d}$
18. $\sqrt{m} \sqrt[6]{n}$
19. $\sqrt[4]{\pi} \sqrt[6]{e}$
20. $\sqrt[6]{f} \sqrt[8]{g}$

324 CHAPTER 10 EXPONENTS AND RADICALS

In Exercises 21–46, multiply the given multinomials and simplify.

21. $(\sqrt{3} + \sqrt{7})(\sqrt{3} - \sqrt{7})$
22. $(\sqrt{5} - \sqrt{3})(\sqrt{5} + \sqrt{3})$
23. $(\sqrt{7} - \sqrt{2})(\sqrt{7} + \sqrt{2})$
24. $(\sqrt{11} - \sqrt{5})(\sqrt{11} + \sqrt{5})$
25. $(\sqrt{2} + \sqrt{3})^2$
26. $(1 - \sqrt{7})^2$
27. $(2 + \sqrt{13})^2$
28. $(\sqrt{5} - \sqrt{2})^2$
29. $(2\sqrt{3} - 4\sqrt{5})(2\sqrt{3} + 4\sqrt{5})$
30. $(3\sqrt{5} - \sqrt{11})(3\sqrt{5} + \sqrt{11})$
31. $(2\sqrt{13} - 2\sqrt{2})(2\sqrt{13} + 2\sqrt{2})$
32. $(\sqrt{17} - 2\sqrt{3})(\sqrt{17} + 2\sqrt{3})$
33. $(x - \sqrt{y})(x + \sqrt{y})$
34. $(\sqrt{a} - a)(\sqrt{a} + a)$
35. $(1 - \sqrt{k})(1 + \sqrt{k})$
36. $(2 - 2\sqrt{b})(2 + 2\sqrt{b})$
37. $(2 + \sqrt{q})^2$
38. $(3 - \sqrt{L})^2$
39. $(\sqrt{a} + 2\sqrt{b})(2\sqrt{a} - 3\sqrt{b})$
40. $(\sqrt{s} - 3\sqrt{t})(3\sqrt{s} - 4\sqrt{t})$
41. $(2\sqrt{5} - 3\sqrt{7})(\sqrt{5} + 4\sqrt{7})$
42. $(3\sqrt{6} + 2\sqrt{5})(2\sqrt{6} - 4\sqrt{5})$
43. $(\sqrt{a} + \sqrt{b})(a - \sqrt{ab} + b)$
44. $(\sqrt{x} - \sqrt{y})(x - \sqrt{xy} - y)$
45. $(\sqrt{a} - \sqrt{b})(a + \sqrt{ab} - b)$
46. $(\sqrt{a} + \sqrt{b} - \sqrt{ab})(\sqrt{a} - \sqrt{b} + \sqrt{ab})$

47. One root of the quadratic equation $x^2 + 2x - 2 = 0$ is $x = -1 + \sqrt{3}$, obtained by using the quadratic formula. Check this root by substituting the value of x into the equation.

48. Repeat Exercise 47 for the other root, $x = -1 - \sqrt{3}$.

49. The roots of the equation $x^2 - 4x + 2 = 0$ are $x = 2 \pm \sqrt{2}$, obtained by using the quadratic formula. Check the roots by substituting the values for x into the equation.

10.6 Division of Radicals

In this section we take up division involving radical expressions. We shall consider two basic types, division by a single radical and division by a multinomial.

Recall that

$$\frac{\sqrt[n]{a}}{\sqrt[n]{b}} = \sqrt[n]{\frac{a}{b}}$$

Thus, to divide two radicals of the same index, combine the two radicals and rationalize the denominator.

Example 1

a. $\dfrac{\sqrt{3ab}}{\sqrt{2b}} = \sqrt{\dfrac{3ab}{2b}} = \sqrt{\dfrac{3a}{2}} = \sqrt{\dfrac{3a(2)}{2(2)}} = \dfrac{\sqrt{6a}}{2}$

b. $\dfrac{\sqrt[4]{5\pi^3}}{\sqrt[4]{8r}} = \sqrt[4]{\dfrac{5\pi^3}{8r}} = \sqrt[4]{\dfrac{5\pi^3(2r^3)}{2^3 r(2r^3)}} = \dfrac{\sqrt[4]{10\pi^3 r^3}}{2r}$

10.6 DIVISION OF RADICALS

Dividing by a multinomial, such as $2 \div (\sqrt{3} - 4)$, is essentially a problem in rationalizing the denominator. Suppose we multiply the numerator and denominator of the fraction $2/(\sqrt{3} - 4)$ by $\sqrt{3} + 4$:

$$\frac{2}{\sqrt{3} - 4} = \frac{2}{\sqrt{3} - 4} \cdot \frac{\sqrt{3} + 4}{\sqrt{3} + 4} = \frac{2(\sqrt{3} + 4)}{3 - 16} = -\frac{2}{13}(\sqrt{3} + 4)$$

The expression $\sqrt{3} + 4$ is called the **conjugate** of $\sqrt{3} - 4$.

This procedure suggests the definition below.

Expressions of the form

$$a + b \quad \text{and} \quad a - b$$

are called **conjugates.**

This definition can be used in stating the rule for rationalizing binomial denominators containing square roots.

To **rationalize** a fraction in which the denominator contains a sum or difference of two terms, at least one of which is a square root, multiply the numerator and denominator by the conjugate of the denominator.

This technique is illustrated in the remaining examples.

Example 2 Simplify the fraction

$$\frac{1}{\sqrt{2} + 1}$$

Solution. We multiply the numerator and denominator by $\sqrt{2} - 1$, the conjugate of $\sqrt{2} + 1$:

$$\frac{1}{\sqrt{2} + 1} = \frac{1}{\sqrt{2} + 1} \cdot \frac{\sqrt{2} - 1}{\sqrt{2} - 1} = \frac{\sqrt{2} - 1}{2 - 1} = \sqrt{2} - 1$$

Example 3 Simplify the fraction

$$\frac{1}{\sqrt{x} - 2\sqrt{y}}$$

Solution.

$$\frac{1}{\sqrt{x} - 2\sqrt{y}} = \frac{1}{\sqrt{x} - 2\sqrt{y}} \cdot \frac{\sqrt{x} + 2\sqrt{y}}{\sqrt{x} + 2\sqrt{y}} = \frac{\sqrt{x} + 2\sqrt{y}}{x - 4y}$$

Example 4
Simplify the fraction

$$\frac{\sqrt{15} - 2}{\sqrt{5} + \sqrt{3}}$$

Solution.
$$\frac{\sqrt{15} - 2}{\sqrt{5} + \sqrt{3}} \cdot \frac{\sqrt{5} - \sqrt{3}}{\sqrt{5} - \sqrt{3}} = \frac{\sqrt{75} - \sqrt{45} - 2\sqrt{5} + 2\sqrt{3}}{5 - 3}$$

$$= \frac{\sqrt{3 \cdot 25} - \sqrt{5 \cdot 9} - 2\sqrt{5} + 2\sqrt{3}}{2}$$

$$= \frac{5\sqrt{3} - 3\sqrt{5} - 2\sqrt{5} + 2\sqrt{3}}{2}$$

$$= \frac{7\sqrt{3} - 5\sqrt{5}}{2}$$

Example 5
The time required to drain the contents of a vessel through an opening in the bottom from level L_1 to level L_2 is given by

$$t = \frac{k(L_1 - L_2)}{\sqrt{2g}(\sqrt{L_1} + \sqrt{L_2})}$$

where g is the acceleration due to gravity and k is a constant. Simplify the expression for t.

Solution. Multiplying the numerator and denominator of the fraction by the conjugate of the denominator, we get

$$t = \frac{k(L_1 - L_2)}{\sqrt{2g}(\sqrt{L_1} + \sqrt{L_2})} \cdot \frac{\sqrt{L_1} - \sqrt{L_2}}{\sqrt{L_1} - \sqrt{L_2}}$$

$$= \frac{k(L_1 - L_2)(\sqrt{L_1} - \sqrt{L_2})}{\sqrt{2g}(L_1 - L_2)} = \frac{k(\sqrt{L_1} - \sqrt{L_2})}{\sqrt{2g}}$$

$$= \frac{k\sqrt{2g}(\sqrt{L_1} - \sqrt{L_2})}{2g}$$

Exercises / Section 10.6

In Exercises 1–36, carry out the indicated divisions and simplify.

1. $\dfrac{\sqrt{2a}}{\sqrt{b}}$

2. $\dfrac{\sqrt{x}}{\sqrt{3y}}$

3. $\dfrac{\sqrt{4c}}{\sqrt{d}}$

4. $\dfrac{\sqrt{9ac}}{\sqrt{c}}$

5. $\dfrac{\sqrt{2xy}}{\sqrt{3x}}$

6. $\dfrac{\sqrt[3]{7x^2z^2}}{\sqrt[3]{4xy^2}}$

7. $\dfrac{\sqrt[3]{2v_1^2 v_2}}{\sqrt[3]{9v_1^2 v_3^2}}$

8. $\dfrac{\sqrt[3]{C_1^2 C_2^2}}{\sqrt[3]{6C_1 C_2}}$

9. $\dfrac{\sqrt[4]{3\pi^3 r}}{\sqrt[4]{8\pi^2}}$

10. $\dfrac{\sqrt[5]{m^3 v^2}}{\sqrt[5]{8vw^3}}$

11. $\dfrac{\sqrt{a}}{\sqrt[3]{a}}$

12. $\dfrac{\sqrt[4]{a}}{\sqrt{a}}$

13. $\dfrac{1}{1-\sqrt{2}}$

14. $\dfrac{2}{\sqrt{3}-1}$

15. $\dfrac{4}{\sqrt{5}+2}$

16. $\dfrac{2}{2+\sqrt{6}}$

17. $\dfrac{\sqrt{2}+2}{\sqrt{2}+1}$

18. $\dfrac{3-\sqrt{5}}{3+\sqrt{5}}$

19. $\dfrac{1+\sqrt{2}}{3-\sqrt{2}}$

20. $\dfrac{1-\sqrt{5}}{4+\sqrt{5}}$

21. $\dfrac{\sqrt{3}-\sqrt{2}}{\sqrt{3}+\sqrt{2}}$

22. $\dfrac{\sqrt{5}+2\sqrt{3}}{\sqrt{5}+\sqrt{3}}$

23. $\dfrac{\sqrt{10}-2}{\sqrt{5}-\sqrt{2}}$

24. $\dfrac{1-\sqrt{21}}{\sqrt{3}-\sqrt{7}}$

25. $\dfrac{\sqrt{18}+3}{\sqrt{6}-\sqrt{3}}$

26. $\dfrac{2\sqrt{3}+2}{\sqrt{2}-\sqrt{6}}$

27. $\dfrac{2}{1-\sqrt{a}}$

28. $\dfrac{5}{\sqrt{x}+2}$

29. $\dfrac{3}{\sqrt{b}-2}$

30. $\dfrac{1}{\sqrt{a}-\sqrt{b}}$

31. $\dfrac{\sqrt{a}+\sqrt{b}}{\sqrt{a}-\sqrt{b}}$

32. $\dfrac{\sqrt{c}-\sqrt{d}}{\sqrt{c}+\sqrt{d}}$

33. $\dfrac{1}{\sqrt{a}-\sqrt{a-b}}$

34. $\dfrac{1}{\sqrt{R}+\sqrt{R-C}}$

35. $\dfrac{1}{\sqrt[3]{a}-\sqrt[3]{b}}$ (*Hint:* $a - b = (a^{1/3})^3 - (b^{1/3})^3 = (a^{1/3} - b^{1/3})(a^{2/3} + a^{1/3}b^{1/3} + b^{2/3})$)

36. $\dfrac{1}{\sqrt[3]{a}+\sqrt[3]{b}}$

37. By rationalizing the numerator, show that the function $f(x) = x - \sqrt{x^2 - 2}$ approaches zero as x gets large.

38. Suppose the position of a particle moving along the x-axis as a function of time is $x(t) = \sqrt{t}$. From calculus, the velocity is

$$v = \lim_{h \to 0} \frac{\sqrt{t+h} - \sqrt{t}}{h}$$

where the symbol $\lim_{h \to 0}$ means "h approaches 0." Show that $v = 1/(2\sqrt{t})$ by multiplying the fraction by

$$\frac{\sqrt{t+h} + \sqrt{t}}{\sqrt{t+h} + \sqrt{t}}$$

and letting h be equal to 0.

39. The *coefficient of reflection* R is defined as the ratio of the amplitude of the reflected wave to the amplitude of the incident wave. If two ropes whose masses per unit length are μ_1 and μ_2, respectively, lie along the x-axis, and are joined at the origin, then

$$R = \frac{\sqrt{\mu_1} - \sqrt{\mu_2}}{\sqrt{\mu_1} + \sqrt{\mu_2}}$$

Simplify the expression for R.

40. Simplify the following expression from a problem in hydrostatics:

$$\frac{I}{1 - \sqrt{V\rho}}$$

328 CHAPTER 10 EXPONENTS AND RADICALS

41. Two heat sources A and B having intensities a and b, respectively, are 1 unit apart. From a point P between A and B, the temperature is lowest if the distance from P to A is

$$\frac{\sqrt[3]{a}}{\sqrt[3]{a} + \sqrt[3]{b}}$$

Simplify this expression. (See Exercise 36.)

42. Simplify the following expression from a problem in mechanics:

$$\frac{\sqrt{v}}{\sqrt{v} - \sqrt{m}}$$

Review Exercises / Chapter 10

In Exercises 1–6, evaluate the expressions.

1. $5^0 5^{-2}$
2. $(3^{-2})^{-1}$
3. $\dfrac{4^{-3}}{16^{-3}}$
4. $64^{-1/3}$
5. $(0.008)^{1/3}$
6. $(-27)^{-1/3}$

In Exercises 7–30, combine the terms and write the results without zero or negative exponents.

7. $\dfrac{3^{-2}V^3W^{-6}}{2^{-1}V^0W^4}$
8. $\dfrac{4^{-1}\pi^{-6}r^2}{4\pi^{-4}r^{-1}}$
9. $\left(\dfrac{2C_1}{C_2}\right)^{-1}$
10. $(3^{-1}p^{-6}q^{-3}c^{-4})^{-3}$
11. $(2^{-2}v^{-3}w^2z^{-1})^4$
12. $\left(\dfrac{2^{-2}f^{-3}g^4}{3^{-1}f^{-4}g^{-2}}\right)^{-2}$
13. $\left(\dfrac{3^{-2}R^{-4}r^3v^0}{3^{-1}R^2r^{-6}}\right)^2$
14. $\left(\dfrac{2v_1^{-2}v_2^3v_3^4}{7^{-2}v_1^{-1}v_2^{-1}v_3^6}\right)^{-1}$
15. $\left(\dfrac{x^{a+1}}{x^{2a+1}}\right)^{-1}$
16. $\dfrac{V^{-1} + I}{V^{-1}}$
17. $(1 - C^{-1})^{-1}$
18. $\left(\dfrac{a^{-1}b^{-1}}{a^{-1} + b^{-1}}\right)^{-2}$
19. $\dfrac{4^{1/2}v^{3/2}w^{-1/2}}{v^{1/2}}$
20. $(4w^4v^{-1})^{3/2}$
21. $(-5x^{3/2})(-2x^{-3/4})(3x)^0$
22. $\dfrac{2^{-1}A^{1/4}B^{-2/3}}{A^{-1/6}B^{-1/5}}$
23. $\dfrac{3^{-1/3}F_1^{-1/7}F_2^{1/3}}{9^{-1/3}F_1^{4/7}F_2^{1/4}}$
24. $(16a^0b^4c^{-8})^{-1/2}$
25. $\left(\dfrac{m^{12}n^{-9}}{64m^{-3/4}n^{-8}}\right)^{-1/6}$
26. $x(x-3)^{-1/2} + (x-3)^{1/2}$
27. $(x+1)(x-1)^{-2/3} + (x-1)^{1/3}$
28. $x(x+2)^{-3/4} + 2(x+2)^{1/4}$

29. $(x + 1)(x - 2)^{-4/5} - 4(x - 2)^{1/5}$
30. $[x^{1/(a+1)}]^{(a^2-1)}$
31. Simplify $\sqrt[3]{\sqrt{x}}$.
32. Simplify $\sqrt[3]{\sqrt[5]{b}}$.

In Exercises 33–34, reduce the order of the given radicals.

33. $\sqrt[6]{64a^6b^9c^{15}}$
34. $\sqrt[8]{16x^4y^6z^{12}}$

In Exercises 35–38, simplify the given radicals.

35. $\sqrt{12R^3r^7}$
36. $\sqrt[3]{24\pi^{10}q^8}$
37. $\sqrt[4]{16a^5b^{11}}$
38. $\sqrt[4]{32x^6y^8z^{13}}$

In Exercises 39–42, combine the radicals.

39. $\sqrt{8} + \sqrt{18} - \sqrt{50}$
40. $\sqrt{20} - \sqrt{80} + \sqrt{180}$
41. $\sqrt{45} - \sqrt[3]{24} + \sqrt{125} - \sqrt[3]{81}$
42. $\sqrt{36x^3y^3} - \sqrt{x^5y^5} + \sqrt{16xy^3}$

In Exercises 43–52, simplify each expression. (Write without zero or negative exponents and rationalize the denominator.)

43. $\sqrt{\dfrac{4a^{-4}b^{-3}}{5a^{-2}b}}$
44. $\sqrt{\dfrac{3^{-1}R^{-6}r^3}{3R^{-3}r^{-1}}}$
45. $\sqrt{\dfrac{3^{-1}s^{-1}tc^0}{2^{-1}t^2}}$
46. $\sqrt{\dfrac{3ab^2c^0}{5a^{-2}b}}$
47. $\sqrt[3]{\dfrac{3^{-1}R^{-1}}{R^2s}}$
48. $\sqrt[3]{\dfrac{2x^{-2}y^{-1}}{x^{-4}}}$
49. $\sqrt[3]{\dfrac{ab^{-1}}{9a^5b}}$
50. $\sqrt[4]{\dfrac{3u^{-1}v^{-1}}{8}}$
51. $\sqrt[5]{\dfrac{3^{-4}\pi\omega^{-2}}{\pi^4}}$
52. $\sqrt[4]{\dfrac{3^{-1}a^{-3}b^{-2}}{b}}$

In Exercises 53–60, carry out the multiplications and simplify.

53. $\sqrt[3]{3xy}\ \sqrt[3]{9x^2y}$
54. $\sqrt[5]{4R^2V^2}\ \sqrt[5]{8R^3V}$
55. $\sqrt[3]{a}\ \sqrt[4]{a}$
56. $\sqrt[4]{a}\ \sqrt[3]{b}$
57. $(\sqrt{6} - \sqrt{2})(\sqrt{6} + \sqrt{2})$
58. $(\sqrt{5} - \sqrt{3})^2$
59. $(1 - 3\sqrt{x})(1 + 3\sqrt{x})$
60. $(\sqrt{a} - 2\sqrt{b})(3\sqrt{a} + 5\sqrt{b})$

In Exercises 61–66, perform each indicated division by rationalizing the denominator.

61. $\dfrac{4}{3 - \sqrt{6}}$
62. $\dfrac{4}{3 + \sqrt{5}}$
63. $\dfrac{1 - \sqrt{10}}{\sqrt{2} - \sqrt{5}}$
64. $\dfrac{1}{\sqrt{R} + \sqrt{C}}$
65. $\dfrac{1}{\sqrt{\pi - 1} + \sqrt{\pi}}$
66. $\dfrac{1}{\sqrt{x - 2} - \sqrt{x}}$

67. In the mathematics of finance, the *present value* P is the amount that must be invested at interest rate i to yield S dollars at the end of n years. If the interest is compounded annually, then

$$P = S(1 + i)^{-n}$$

How much money must be invested at present to yield $10,000 after 5 years at 8% interest compounded annually?

68. The volume of a sphere is given by $V = \frac{4}{3}\pi r^3$. Solve for r and express the result in the simplest radical form.

69. A Borda's pendulum consists of a spherical bob of radius r and a long, thin wire of negligible weight (Figure 10.3). If the length of the wire from the point of suspension to the center of the bob is L, then the period P (in seconds) for one cycle of the pendulum for small oscillations is

$$P = 2\pi \sqrt{\frac{L}{g}\left(1 + \frac{2r^2}{5L^2}\right)}$$

where g is the acceleration due to gravity. Simplify this expression by combining the terms and rationalizing the denominator.

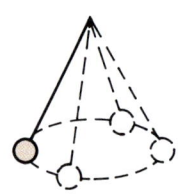

Figure 10.3

CHAPTER **11**

Complex Numbers

Objectives Upon completion of this chapter, you should be able to:

1. Represent a given complex number as a vector.
2. Add and subtract complex numbers graphically.
3. Determine when two complex numbers are equal.
4. Perform the four fundamental operations with complex numbers in rectangular form.
5. Change a complex number from rectangular to polar and exponential forms.
6. Change a complex number from polar to rectangular form.
7. Multiply and divide complex numbers in polar form.
8. Find powers and roots of complex numbers by using De Moivre's theorem.
9. Use complex numbers to study the phase relations between voltage and current. (Optional)

11.1 Basic Concepts and Definitions

When complex numbers arose in our discussion of quadratic equations in Chapter 6, we concentrated mostly on notation. In this chapter we shall study the meaning and properties of complex numbers in greater detail. A knowledge of these properties is essential in more advanced technical applications. (One such application is discussed in Section 11.6.)

Let us first recall that the square root of a negative number is called an *imaginary number*. This term, introduced by Descartes, suggests that such numbers cannot exist. Indeed, we cannot find a *real* number a such that $a = \sqrt{-4}$. (If $a = \sqrt{-4}$, then $a^2 = -4$, but the square of a real number cannot be

331

negative.) However, square roots of negative numbers can be extremely useful in certain applications. To realize this advantage, we have to *extend* the real number system in such a way that square roots of negative numbers are included. This can be done by introducing a number that is *not a real number*. This number, denoted by j, has the property

$$j^2 = -1 \tag{11.1}$$

and is called the **basic imaginary unit.**

Basic imaginary unit:

$$j = \sqrt{-1} \quad \text{or} \quad j^2 = -1 \tag{11.2}$$

With this number, the square root of a negative number can be written as a product of j and a real number. For example,

$$\sqrt{-4} = \sqrt{4(-1)} = \sqrt{4}\sqrt{-1} = 2j$$

In general,

$$\sqrt{-a}, \quad a > 0, \quad \text{is written} \quad \sqrt{a}\,j \tag{11.3}$$

To repeat: *j is not a real number* and must therefore be viewed as a new kind of number. By introducing j, we can extend the real number system to include square roots of negative numbers.

Even more general than imaginary numbers are **complex numbers,** which have the form

$$a + bj, \; a \text{ and } b \text{ real numbers} \tag{11.4}$$

Complex numbers include real numbers as special cases.

Definition of a complex number:

If a and b are real numbers and $j = \sqrt{-1}$, then $a + bj$ is called a **complex number**; a is called the **real part** and b the **imaginary part**.

If $a = 0$, then the complex number has the form bj and is called a **pure imaginary number.** For example, $2j = 0 + 2j$ is a pure imaginary number. If $b = 0$, then the complex number is real. For example, $3 = 3 + 0j$ is real. From this point of view *a real number is a complex number whose imaginary part is zero.*

A complex number can be given a geometric interpretation as a vector, as shown in Figure 11.1. The initial point of the arrow representing $a + bj$ is

11.1 BASIC CONCEPTS AND DEFINITIONS

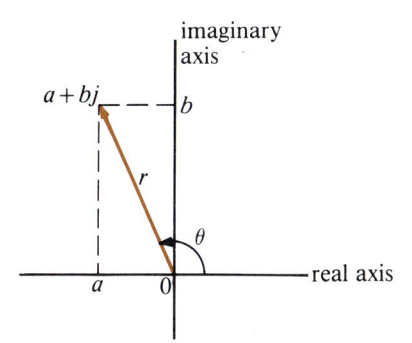

Figure 11.1

placed at the origin and the tip at the point (a, b). The horizontal axis is then called the **real axis** and the vertical axis the **imaginary axis,** although the terms *x-axis* and *y-axis* are also commonly used. This coordinate system used for plotting complex numbers is called the **complex plane.**

Real axis
Imaginary axis
Complex plane
Absolute value
Argument
Conjugate

The length of the arrow is $r = \sqrt{a^2 + b^2}$, called the **absolute value** or **modulus.** The angle θ made with the positive real axis is called the **argument.** (See Figure 11.1.) The **conjugate** of the complex number $a + bj$, denoted by $\overline{a + bj}$, is the reflection of $a + bj$ in the real axis. (See Figure 11.2.) Thus $\overline{a + bj} = a - bj$.

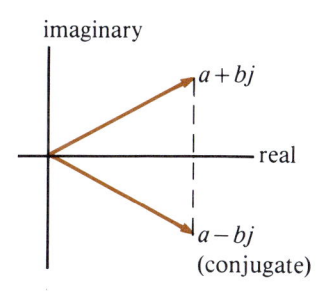

Figure 11.2

A complex number as a vector: The complex number $a + bj$ is represented as a **vector** extending from 0 to (a, b) in the complex plane (Figure 11.1).

    Absolute value $r = \sqrt{a^2 + b^2}$
    Argument   angle θ (Figure 11.1)
    Conjugate $\overline{a + bj} = a - bj$ (Figure 11.2)

Example 1 Find the absolute value and conjugate of the complex number $-3 - 4j$.

Solution. The absolute value r is given by

$$r = \sqrt{(-3)^2 + (-4)^2} = \sqrt{9 + 16} = 5$$

The conjugate of $-3 - 4j$ is $-3 + 4j$, shown in Figure 11.3. (As already noted, the axes are often labeled x-axis and y-axis respectively.)

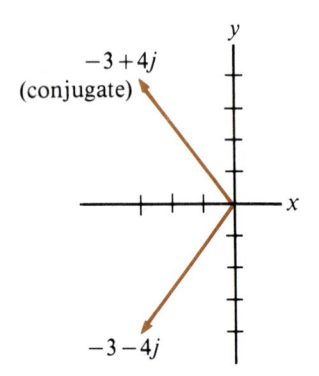

Figure 11.3

Addition

Addition of complex numbers is defined so that it is identical to vector addition. In other words, the sum of $z_1 = a + bj$ and $z_2 = c + dj$ is the resultant obtained from the parallelogram determined by z_1 and z_2. The resultant can also be obtained algebraically by adding the "components" of z_1 and z_2; that is,

$$z_1 + z_2 = (a + bj) + (c + dj) = (a + c) + (b + d)j \qquad (11.5)$$

The sum is shown in Figure 11.4.

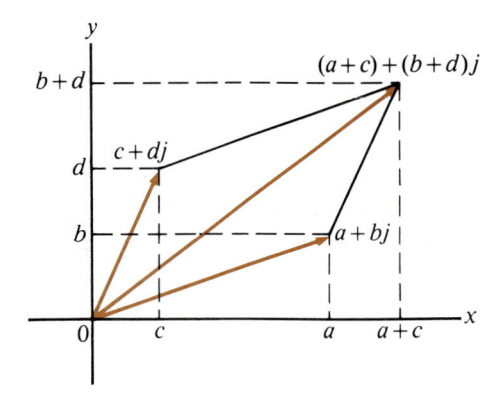

Figure 11.4

Subtraction   As would be the case with ordinary vectors, the difference $z_1 - z_2$ is the second diagonal of the parallelogram in Figure 11.4. Thus

$$z_1 - z_2 = z_1 + (-z_2) = (a + bj) + (-c - dj)$$

or

$$z_1 - z_2 = (a - c) + (b - d)j \tag{11.6}$$

(See Figure 11.5.) Note, however, that the vector representing $z_1 - z_2$ has to be drawn with the initial point at the origin to conform to the usual convention.

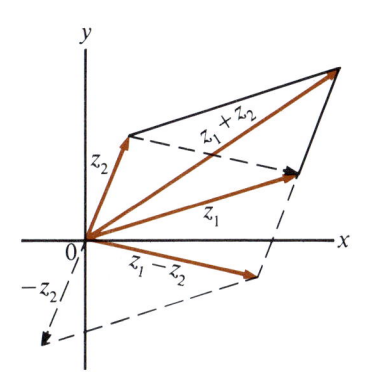

Figure 11.5

Example 2

Add the complex numbers $2 - 3j$ and $-3 - 4j$ graphically.

Solution. The sum is shown in Figure 11.6. Note that the sum is $(2 - 3j) + (-3 - 4j) = (2 - 3) + (-3 - 4)j = -1 - 7j$.

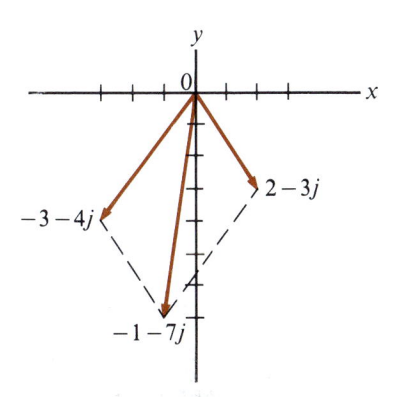

Figure 11.6

Example 3 Subtract $1 - 3j$ from $4 + 2j$ graphically.

Solution. Subtracting $1 - 3j$ is equivalent to adding $-1 + 3j$, shown in Figure 11.7. Note that the result is $(4 + 2j) + (-1 + 3j) = (4 - 1) + (2 + 3)j = 3 + 5j$.

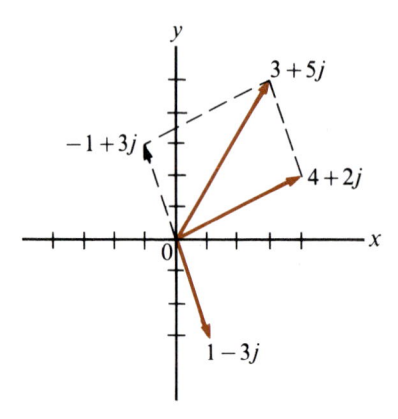

Figure 11.7

To complete this section, we should define the equality of complex numbers.

> **Equality of two complex numbers**: Two complex numbers are equal if, and only if, their real parts are equal and their imaginary parts are equal:
>
> $$a + bj = c + dj \quad \text{if, and only if,} \quad a = c \quad \text{and} \quad b = d \quad (11.7)$$

Example 4 Determine the values of x and y if

$$(-x + 3) + (x + 4)j = 2y - yj$$

Solution. By definition (11.7) we get the following system of equations:

$$-x + 3 = 2y \quad \text{equating real parts}$$
$$\underline{x + 4 = -y} \quad \text{equating imaginary parts}$$
$$7 = y \quad \text{adding}$$

Substituting $y = 7$ into the second equation, we get $x + 4 = -7$, or $x = -11$. So the two complex numbers are equal if $x = -11$ and $y = 7$.

Remark. As mentioned in Chapter 6, i is another common designation for $\sqrt{-1}$. However, when complex numbers were first applied to the study of electricity, i had already become the standard notation for instantaneous current. As a result, j came to be used for the basic imaginary unit in physics and technology.

Exercises / Section 11.1

In Exercises 1–4, draw each complex number and its conjugate in the complex plane.

1. $-4 - 6j$
2. $5 - 10j$
3. $-2 + 7j$
4. $-3 - j$

In Exercises 5–20, perform the indicated operations graphically and check algebraically.

5. $(1 - 3j) + (4 + 2j)$
6. $(2 + j) + (3 - 4j)$
7. $(-4 - 2j) + (6 + 5j)$
8. $(-3 + 7j) + (-6 + 2j)$
9. $(7 + 5j) - (2 - j)$
10. $(3 + 2j) - (1 - 2j)$
11. $(-2 + 3j) - (2 - 3j)$
12. $(1 - 4j) - (-2 - 4j)$
13. $(5 + 4j) - (5 - 3j)$
14. $(-4 - j) + (-3 + 2j)$
15. $(6 + 7j) + (4 - 7j)$
16. $(10 - 6j) + (-5 - 3j)$
17. $(-12 + 4j) - (-5 + 5j)$
18. $(7 + 12j) + (1 - 11j)$
19. $(-5 - 9j) - (-3 + 5j)$
20. $(-1 - 4j) - (10 + 2j)$

In Exercises 21–30, determine the values of x and y for which equality holds. (See Example 4.)

21. $2 + 3j = x + yj$
22. $x + 4j = 2 + yj$
23. $x + 3j = 2 - yj$
24. $-x + 2j = -3 - yj$
25. $(x + 1) + yj = 4 - j$
26. $(x + y) + (2x + y)j = 2 + 3j$
27. $(x - 2y) + (x - y)j = 4 + 5j$
28. $(x + 1) + (x + 3)j = y - yj$
29. $(x - 3) + (y + 2)j = -y + xj$
30. $(2x + y) + (-x + y)j = x - yj$

11.2 The Fundamental Operations

To study the four fundamental operations involving complex numbers, we shall examine several examples and summarize the results at the end.

Addition and subtraction

Let us return to the operations of **addition** and **subtraction**, already considered in Section 11.1.

Example 1 Perform the indicated operations:

a. $(-5 + 2j) + (-4 + 7j)$
b. $(-5 + 2j) - (-4 + 7j)$
c. $(2 - 3\sqrt{2}j) - (5 - 2\sqrt{2}j) - (-6 + 10\sqrt{2}j)$

Solution.   a. $(-5 + 2j) + (-4 + 7j) = (-5 - 4) + (2j + 7j) = -9 + 9j$
b. $(-5 + 2j) - (-4 + 7j) = -5 + 2j + 4 - 7j = -1 - 5j$
c. $(2 - 3\sqrt{2}j) - (5 - 2\sqrt{2}j) - (-6 + 10\sqrt{2}j)$
$= 2 - 3\sqrt{2}j - 5 + 2\sqrt{2}j + 6 - 10\sqrt{2}j = 3 - 11\sqrt{2}j$

Multiplication

The usual procedure for multiplying binomials can be applied to the **multiplication** of complex numbers, but j^2 has to be changed to -1.

Example 2   Multiply: a. $(2 - 3j)(4 + 2j)$ b. $(-5 + 7j)(2 + 6j)$

Solution.
a. $(2 - 3j)(4 + 2j) = 8 - 8j - 6j^2$
$= 8 - 8j + 6 = 14 - 8j,$ since $j^2 = -1$
b. $(-5 + 7j)(2 + 6j) = -10 - 16j + 42j^2$
$= -10 - 16j - 42 = -52 - 16j$

A special case of multiplying complex numbers involves powers of j. Since $j^2 = -1$, we have $j^3 = j^2 \cdot j = -j$. Similarly, $j^4 = j^3 \cdot j = -j \cdot j = -j^2 = 1$. Now the cycle repeats: $j^5 = j^4 \cdot j = j$, $j^6 = j^5 \cdot j = j \cdot j = -1$, and so forth.

Example 3   Simplify   a. j^{10};   b. j^{20};   c. j^{31}.

Solution.   a. $j^{10} = j^4 j^4 j^2 = (1)(1)(-1) = -1$
b. $j^{20} = (j^4)^5 = 1^5 = 1$
c. $j^{31} = j^{28} j^3 = (j^4)^7 j^3 = 1^7(-j) = -j$

Division

Since a complex number is a radical expression in disguise, **division** can be carried out by the same method used in Section 10.6 to divide radicals. In other words, *we rationalize the denominator by multiplying numerator and denominator by the conjugate of the denominator.*

Example 4   Perform the following divisions: a. $\dfrac{1}{1 - 2j}$ b. $\dfrac{2 - j}{3 + 4j}$

Solution.   a. $\dfrac{1}{1 - 2j} = \dfrac{1}{1 - 2j} \cdot \dfrac{1 + 2j}{1 + 2j}$
$= \dfrac{1 + 2j}{1 - 4j^2}$

$$= \frac{1 + 2j}{5}$$

$$= \frac{1}{5} + \frac{2}{5}j$$

b. $\dfrac{2 - j}{3 + 4j} = \dfrac{2 - j}{3 + 4j} \cdot \dfrac{3 - 4j}{3 - 4j}$

$$= \frac{6 - 11j + 4j^2}{9 - 16j^2} = \frac{6 - 11j - 4}{9 + 16}$$

$$= \frac{2 - 11j}{25}$$

$$= \frac{2}{25} - \frac{11}{25}j$$

Let us now summarize the four fundamental operations.

Fundamental operations for complex numbers:

Addition	$(a + bj) + (c + dj) = (a + c) + (b + d)j$
Subtraction	$(a + bj) - (c + dj) = (a - c) + (b - d)j$
Multiplication	$(a + bj)(c + dj) = (ac - bd) + (ad + bc)j$
Division (procedure)	$\dfrac{a + bj}{c + dj} = \dfrac{(a + bj)\overline{(c + dj)}}{(c + dj)\overline{(c + dj)}}$

As has been emphasized repeatedly, the symbol $\sqrt[n]{}$ denotes the principal nth root. In particular, $\sqrt{16} = 4$, not -4. Because of this convention, the rule

$$\sqrt{a}\,\sqrt{b} = \sqrt{ab}$$

holds only for positive a and b. For example, while

$$\sqrt{-4}\,\sqrt{-9} = (2j)(3j) = 6j^2 = -6$$

the law of radicals, if applied here, would yield $\sqrt{36} = 6$, not -6. This observation leads to the rule below.

$\sqrt{-a}$, $a > 0$, should always be written $\sqrt{a}\,j$ before performing any algebraic operation.

By using $\sqrt{a}\,j$ instead of $\sqrt{-a}$, problems with signs are avoided.

Exercises / Section 11.2

In Exercises 1–47, perform the indicated operations and express the results in the form $a + bj$.

1. $(5 + 3j) + (4 + 7j)$
2. $(-2 + j) - (10 + 2j)$
3. $(\sqrt{5} - j) + (2\sqrt{5} - 3j)$
4. $(\sqrt{2} + j) + 2j$
5. $2 + (4 - 6j)$
6. $(2 - 5j) - (7 + 2j) - (8 + 6j)$
7. $(-1 - 7j) + (2 - 8j) - (-3 - 9j)$
8. $(\sqrt{2} + 2\sqrt{3}j) - (2\sqrt{2} - 3\sqrt{3}j)$
9. $(-3 - j) + (-2 + 3j) - (4 - 11j)$
10. $(\sqrt{3} + j) + (\sqrt{3} - 3j) - (2\sqrt{3} + 5j)$
11. $(-1 - 2j) - (6 - 4j) - (5 + j)$
12. $-3j + (1 - 9j) - (11 + 15j)$
13. $(2 + j)(3 - j)$
14. $(1 - 3j)(2 + 2j)$
15. $(2 - j)(3 - 2j)$
16. $(3 - 4j)(1 + j)$
17. $(3 - 4j)(3 + 4j)$
18. $(4 - 6j)(4 + 6j)$
19. $(4 - 5j)(1 - 3j)$
20. $(6 + j)(-5 + 2j)$
21. $(1 + j)^2$
22. $(3 - 4j)^2$
23. $(\sqrt{3} - 2j)^2$
24. $(2 + \sqrt{3}j)^2$
25. $(\sqrt{2} - j)^2$
26. $(2 - 3j)(1 + 7j)$
27. $(5 - 7j)(2 + 4j)$
28. $(4 + 6j)(5 + 10j)$
29. $\dfrac{1}{1 - 3j}$
30. $\dfrac{1}{2 - j}$
31. $\dfrac{2}{2 + j}$
32. $\dfrac{3}{2 + 3j}$
33. $\dfrac{1 - 2j}{1 + 2j}$
34. $\dfrac{2 + 3j}{2 - 3j}$
35. $\dfrac{3 - 4j}{3 + 4j}$
36. $\dfrac{1 - 6j}{1 + 6j}$
37. $\dfrac{2 + 3j}{1 - 4j}$
38. $\dfrac{1 + j}{4 + 5j}$
39. $\dfrac{3 - 2j}{6 + 10j}$
40. $\dfrac{2 - 3j}{4 + 2j}$
41. $\dfrac{(1 - j)(1 + 2j)}{2 + j}$
42. $\dfrac{(2 - 3j)(2 - j)}{1 + 2j}$
43. $\dfrac{(3 - j)(2 - 3j)}{2 + 2j}$
44. $\dfrac{(2 - j)(3 + 2j)}{2 - 3j}$
45. $\dfrac{(1 - 2j)(1 + j)}{(1 - j)(3 + j)}$
46. $\dfrac{(1 - 3j)(3 - 2j)}{(1 - j)(2 + j)}$
47. $\dfrac{(2 - 3j)(2 - j)}{(2 + j)(1 - 4j)}$

In Exercises 48–60, simplify the given powers of j.

48. j^7
49. j^{11}
50. j^{17}
51. j^{21}
52. j^{22}
53. j^{26}
54. j^{69}
55. j^{71}
56. j^{100}
57. j^{102}
58. j^{500}
59. j^{48}
60. j^{55}

11.3 Polar and Exponential Forms

Complex numbers have been successfully applied to problems in science and technology, in part because of their geometric interpretation. To exploit this interpretation fully, we need to be able to represent a complex number in terms of its absolute value and argument. This representation leads to the **polar** and **exponential forms** of complex numbers.

Polar Form

Consider the complex number $x + yj$ in Figure 11.8. If we denote the absolute value by r and the argument by θ, then $\cos\theta = x/r$ and $\sin\theta = y/r$. It follows that $x = r\cos\theta$ and $y = r\sin\theta$. Also, $x^2 + y^2 = r^2$ and $\tan\theta = y/x$. These ideas are summarized below.

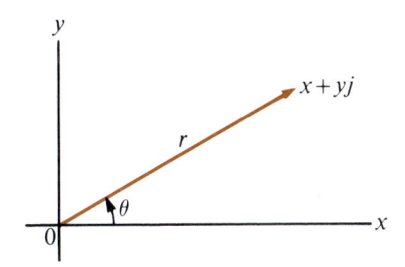

Figure 11.8

$$x = r\cos\theta \quad \text{and} \quad y = r\sin\theta \tag{11.8}$$

$$r = \sqrt{x^2 + y^2} \quad \text{and} \quad \tan\theta = \frac{y}{x} \tag{11.9}$$

where the signs of x and y determine the quadrant of the argument θ.

342 CHAPTER 11 COMPLEX NUMBERS

If we substitute the expressions for x and y into the form $x + yj$, we get by (11.8)

$$x + yj = r \cos \theta + rj \sin \theta$$

or, by factoring r,

$$x + yj = r(\cos \theta + j \sin \theta) \tag{11.10}$$

The right side of equation (11.10) is called the **polar form** of the complex number $x + yj$ and is often abbreviated "r cis θ." The abbreviation "cis θ" stands for $\cos \theta + i \sin \theta$, referring to the older form $i = \sqrt{-1}$. (Another common notation is $r \angle \theta$.) To distinguish between the two forms, $x + yj$ is called the **rectangular form** of the complex number.

Rectangular Form	$x + yj$
Polar Form	$r(\cos \theta + j \sin \theta) = r$ cis θ

Converting from the rectangular to the polar form requires only basic trigonometry, as illustrated in the first three examples. (In the examples and exercises, we shall make frequent use of special angles.)

Example 1 Change the complex number $-1 + \sqrt{3}j$ to polar form.

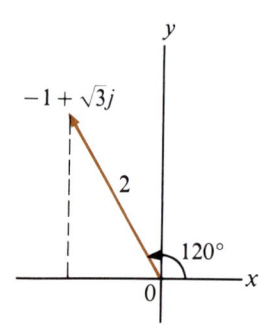

Figure 11.9

Solution. The number is shown in Figure 11.9.
After dropping a perpendicular from $(-1, \sqrt{3})$ to the x-axis, we get from the resulting 30°–60° reference triangle $r = \mathbf{2}$ and $\theta = \mathbf{120°}$. Hence

$$-1 + \sqrt{3}j = \mathbf{2}(\cos \mathbf{120°} + j \sin \mathbf{120°}) = 2 \text{ cis } 120°$$

Example 2 Change $-2 - 2j$ to polar form.

Solution. Draw Figure 11.10. It follows from the reference triangle that $\theta = 225°$. The absolute value is

$$r = \sqrt{(-2)^2 + (-2)^2} = \sqrt{8} = 2\sqrt{2}$$

Thus

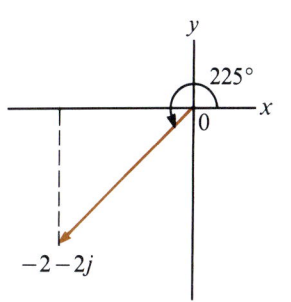

$$-2 - 2j = 2\sqrt{2} \text{ cis } 225°$$

Figure 11.10

Example 3 Change $2 - 5j$ to polar form.

Solution. The number is shown in Figure 11.11.
The absolute value is

$$r = \sqrt{2^2 + (-5)^2} = \sqrt{29}$$

Since θ is not a special angle, we need to use a calculator (or Table 1). From $\tan \theta = -\frac{5}{2}$, we obtain $\theta = -68.2°$. The polar form is therefore given by

$$\sqrt{29} \text{ cis } (-68.2°) = \sqrt{29} \text{ cis } 291.8°$$

In fact,

$$2 - 5j = \sqrt{29} \text{ cis } (291.8° + k \cdot 360°), \quad k = 0, \pm 1, \pm 2, \ldots \quad (11.11)$$

(See remark below.)

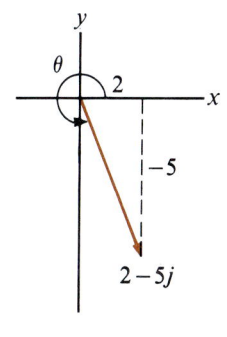

Figure 11.11

Remark. The polar form (11.11) lists all possible arguments of the complex number $2 - 5j$. In this and the next section, there is no compelling reason to specify all arguments. For uniformity, we choose θ such that $0° \leq \theta < 360°$. However, we shall see in Section 11.5 that in some problems the general polar form

$$r \text{ cis } (\theta + k \cdot 360°), \quad k = 0, \pm 1, \pm 2, \ldots$$

is needed.

The next example illustrates the conversion of a complex number in polar form to rectangular form.

Example 4 Change 2 cis 150° to rectangular form.

Solution.
$$2 \text{ cis } 150° = 2(\cos 150° + j \sin 150°)$$
$$= 2\left(-\frac{\sqrt{3}}{2} + \frac{1}{2}j\right)$$
$$= -\sqrt{3} + j$$

We can see from Example 4 that the polar form of a complex number can be changed to rectangular form by replacing $\cos \theta$ and $\sin \theta$ by their values. If θ is a special angle, then the rectangular form can be expressed in exact form by using radicals. If θ is not a special angle, then we shall express the function values of θ to four decimal places. For uniformity, other values will be rounded off to three significant figures.

Example 5 Change 3 cis 230° to polar form.

Solution.
$$3 \text{ cis } 230° = 3(\cos 230° + j \sin 230°)$$
$$= 3(-0.6428 - 0.7660j)$$
$$= -1.93 - 2.30j$$

Exponential Form

A third form of a complex number, called the **exponential form,** is particularly important in applications to electronics. This form is based on a relationship called *Euler's identity*:

$$e^{j\theta} = \cos \theta + j \sin \theta \tag{11.12}$$

The letter e represents an irrational number whose approximate value is 2.71828. The origin and significance of this number are explored in the study of calculus.

Comparing the polar and exponential forms, it follows that the exponential form can be obtained by replacing cis θ by $e^{j\theta}$. Thus

$$r \text{ cis } \theta = re^{j\theta} \tag{11.13}$$

The forms are summarized next.

> **Three forms of a complex number:**
>
> | | |
> |---|---|
> | **Rectangular form** | $x + yj$ |
> | **Polar form** | r cis θ |
> | **Exponential form** | $re^{j\theta}$ |

11.3 POLAR AND EXPONENTIAL FORMS **345**

In the exponential form, θ is usually expressed in radians. For θ in radians, $j\theta$ is a complex power for which the laws of exponents hold. It is precisely this property that makes the exponential form so useful.

Example 6 Write $-1 + j$ in exponential form.

Solution. $r = \sqrt{(-1)^2 + 1^2} = \sqrt{2}$ and $\theta = 135°$

In radian measure

$$\theta = 135° = \frac{135\pi}{180} = \frac{3\pi}{4}$$

Thus

$$-1 + j = \sqrt{2}e^{(3\pi/4)j}$$

Exercises / Section 11.3

In Exercises 1–16, express the given complex numbers in polar form ($0° \leq \theta < 360°$).

1. 1
2. -2
3. -4
4. j
5. $2j$
6. $-3j$
7. $-5j$
8. 5
9. $1 + j$
10. $-\sqrt{3} + j$
11. $1 - \sqrt{3}j$
12. $\sqrt{3} + j$
13. $-3 - 3j$
14. $-2 + 2\sqrt{3}j$
15. $-3\sqrt{3} + 3j$
16. $4 - 4j$

In Exercises 17–24, change the given complex numbers to exponential form ($0 \leq \theta < 2\pi$). (See Example 6.)

17. $2j$
18. $-4j$
19. $1 - j$
20. $-3 + 3\sqrt{3}j$
21. $\sqrt{3} - j$
22. $-1 - \sqrt{3}j$
23. $-2 + 2j$
24. $6\sqrt{3} + 6j$

In Exercises 25–36, express the given complex numbers in rectangular form. (Use a diagram and express each result in exact form.)

25. 2 cis 45°
26. 4 cis 210°
27. 2 cis 120°
28. $\sqrt{2}$ cis 135°
29. 3 cis 240°
30. 6 cis 300°
31. 5 cis 150°
32. $\sqrt{2}$ cis 225°
33. 3 cis 0°
34. 4 cis 180°
35. 5 cis 90°
36. 6 cis 270°

In Exercises 37–42, use a calculator to convert each complex number to polar form. Express the argument to the nearest tenth of a degree ($0° \leq \theta < 360°$).

37. $2 + 5j$
38. $-1 + 4j$
39. $-2 - 4j$
40. $-1 - \sqrt{11}j$
41. $\sqrt{14} - 2j$
42. $\sqrt{5} + \sqrt{3}j$

In Exercises 43–50, use a calculator to convert the given complex numbers to rectangular form. (Use three significant figures.)

43. 2.00 cis 37.1°
44. 3.12 cis 136.0°
45. 0.361 cis 221.3°
46. 4.96 cis 310.8°
47. 6.34 cis 274.6°
48. 10.3 cis 162.9°
49. 3.03 cis 25.6°
50. 4.96 cis 173.4°

11.4 Products and Quotients of Complex Numbers

As we saw in Section 11.1, vector interpretation of complex numbers yields simple and natural geometric interpretations of addition and subtraction. Unfortunately, geometric interpretations of multiplication and division are not given as easily. However, the polar form is particularly convenient for these operations: It yields the desired geometric interpretation and even paves the way for the calculation of powers and roots.

Consider the following arbitrary complex numbers in polar and rectangular forms:

$$r_1(\cos \theta_1 + j \sin \theta_1) = r_1 e^{j\theta_1}$$

and

$$r_2(\cos \theta_2 + j \sin \theta_2) = r_2 e^{j\theta_2}$$

Their product is given by

$$r_1(\cos \theta_1 + j \sin \theta_1) \cdot r_2(\cos \theta_2 + j \sin \theta_2)$$
$$= r_1 e^{j\theta_1} \cdot r_2 e^{j\theta_2} = r_1 r_2 e^{j(\theta_1 + \theta_2)} \quad \text{adding exponents}$$
$$= r_1 r_2 [\cos(\theta_1 + \theta_2) + j \sin(\theta_1 + \theta_2)]$$

Product of two complex numbers: The absolute value of the product of two complex numbers is the product of their absolute values. The argument of the product is the sum of their arguments:

$$(r_1 \operatorname{cis} \theta_1)(r_2 \operatorname{cis} \theta_2) = r_1 r_2 \operatorname{cis}(\theta_1 + \theta_2) \tag{11.14}$$

This delightfully terse and simple derivation illustrates the great power of the exponential form.

For the quotient of two complex numbers we get

$$r_1 e^{j\theta_1} \div r_2 e^{j\theta_2} = \frac{r_1 e^{j\theta_1}}{r_2 e^{j\theta_2}} = \frac{r_1}{r_2} e^{j(\theta_1 - \theta_2)} \quad \text{subtracting exponents}$$

Quotient of two complex numbers: The absolute value of the quotient of two complex numbers is the quotient of their absolute values. The argument of the quotient is the difference of their arguments in the proper order:

$$\frac{r_1 \operatorname{cis} \theta_1}{r_2 \operatorname{cis} \theta_2} = \frac{r_1}{r_2} \operatorname{cis}(\theta_1 - \theta_2) \tag{11.15}$$

Example 1 Multiply $(2 \text{ cis } 20°)(2 \text{ cis } 30°)$.

Solution. By formula (11.14)

$$(2 \text{ cis } 20°)(2 \text{ cis } 30°) = (2 \cdot 2) \text{ cis}(20° + 30°)$$
$$= 4 \text{ cis } 50°$$

(See Figure 11.12.)

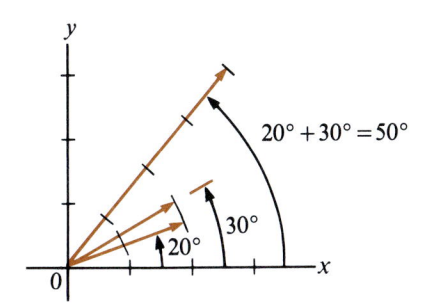

Figure 11.12

Example 2 Carry out the following division:

$$\frac{12 \text{ cis } 150°}{4 \text{ cis } 220°}$$

Solution. By formula (11.15) we get

$$\frac{12 \text{ cis } 150°}{4 \text{ cis } 220°} = \frac{12}{4} \text{ cis}(150° - 220°) = 3 \text{ cis}(-70°)$$
$$= 3 \text{ cis}(360° - 70°) = 3 \text{ cis } 290°$$

(See Figure 11.13.)

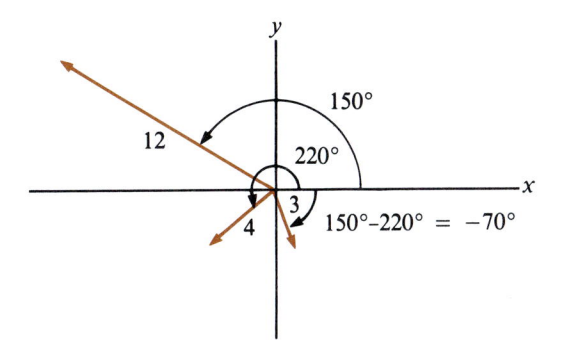

Figure 11.13

Exercises / Section 11.4

Carry out the indicated operations. (Leave answers in polar form.)

1. (2 cis 31°)(3 cis 62°)
2. (4 cis 81°)(8 cis 22°)
3. (6 cis 130°)(3 cis 220°)
4. (10 cis 265°)(4 cis 10°)
5. (15 cis 230°)(2 cis 300°)
6. (20 cis 328°)(30 cis 275°)
7. (18 cis 139°)(3 cis 46°)
8. (24 cis 318°)(4 cis 342°)
9. $\dfrac{36 \text{ cis } 100°}{9 \text{ cis } 40°}$
10. $\dfrac{50 \text{ cis } 280°}{10 \text{ cis } 150°}$
11. $\dfrac{39 \text{ cis } 160°}{13 \text{ cis } 260°}$
12. $\dfrac{8 \text{ cis } 70°}{24 \text{ cis } 90°}$
13. $\dfrac{3 \text{ cis } 172°}{2 \text{ cis } 321°}$
14. $\dfrac{5 \text{ cis } 216°}{8 \text{ cis } 284°}$
15. $\dfrac{(25 \text{ cis } 102°)(\text{cis } 87°)}{15 \text{ cis } 36°}$
16. $\dfrac{(4 \text{ cis } 25°)(6 \text{ cis } 230°)}{8 \text{ cis } 125°}$
17. $\dfrac{(6 \text{ cis } 60°)(4 \text{ cis } 148°)}{12 \text{ cis } 342°}$
18. $\dfrac{10 \text{ cis } 119°}{(4 \text{ cis } 73°)(5 \text{ cis } 156°)}$
19. $\dfrac{2 \text{ cis } 165°}{(7 \text{ cis } 43°)(4 \text{ cis } 301°)}$
20. $\dfrac{3 \text{ cis } 20°}{(8 \text{ cis } 70°)(10 \text{ cis } 117°)}$

11.5 Powers and Roots of Complex Numbers

We now have enough information to find the powers and roots of complex numbers. In addition, we shall study a method for finding all the roots of a number by means of *De Moivre's theorem*.

Raising a complex number to an integral power involves only repeated multiplication. Thus

$$[r(\cos \theta + j \sin \theta)]^3$$
$$= [r(\cos \theta + j \sin \theta)][r(\cos \theta + j \sin \theta)][r(\cos \theta + j \sin \theta)]$$
$$= r^3[\text{cis}(\theta + \theta + \theta)]$$
$$= r^3(\cos 3\theta + j \sin 3\theta)$$

So in general,

$$[r(\cos \theta + j \sin \theta)]^n = r^n(\cos n\theta + j \sin n\theta) \tag{11.16}$$

or

$$(r \text{ cis } \theta)^n = r^n \text{ cis } n\theta \tag{11.17}$$

Formula (11.17) is known as **De Moivre's theorem,** after Abraham De Moivre (1667–1754). De Moivre, a contemporary of Isaac Newton, was a major figure in the development of probability theory. His theorem is actually valid for all real numbers. If n is rational, it can be used to find roots, as we shall see later in this section.

Abraham De Moivre

11.5 POWERS AND ROOTS OF COMPLEX NUMBERS

De Moivre's theorem:

$$(r \text{ cis } \theta)^n = r^n \text{ cis } n\theta, \quad \text{for all real numbers } n$$

The first two examples illustrate the use of De Moivre's theorem for finding the nth power of a complex number.

Example 1 Use De Moivre's theorem to find $(1 + \sqrt{3}j)^{10}$.

Solution. $(1 + \sqrt{3}j)^{10} = (2 \text{ cis } 60°)^{10} = 2^{10} \text{ cis}(10 \cdot 60°)$

by De Moivre's theorem. It follows that

$$(1 + \sqrt{3}j)^{10} = 2^{10} \text{ cis } 600° = 2^{10} \text{ cis}(600° - 360°)$$
$$= 2^{10} \text{ cis } 240° = 2^{10}(\cos 240° + j \sin 240°)$$
$$= 2^{10}\left(-\frac{1}{2} - \frac{\sqrt{3}}{2}j\right) = 2^9(-1 - \sqrt{3}j) = -512 - 512\sqrt{3}j$$

Example 2 Use De Moivre's theorem to find $(-2 + 3j)^5$.

Solution. First we need to convert $-2 + 3j$ to polar form: $r = \sqrt{(-2)^2 + 3^2} = \sqrt{13}$; since $\tan \theta = -\frac{3}{2}$ and θ is in the second quadrant, we get $\theta = 123.69°$ to two decimal places. Then

$$(-2 + 3j)^5 = (13^{1/2} \text{ cis } 123.69°)^5$$
$$= 13^{5/2} \text{ cis}(5 \cdot 123.69°)$$
$$= 13^{5/2}(\cos 618.45° + j \sin 618.45°)$$
$$= 13^{5/2}(\cos 258.45° + j \sin 258.45°)$$
$$= -122 - 597j$$

carried out to three significant figures.

Roots

To use De Moivre's theorem for finding roots of complex numbers, we need to list all the arguments in order to obtain all the roots. In other words, we write r cis θ in the form

$$r \text{ cis}(\theta + k \cdot 360°)$$

Roots of a complex number: The n nth roots of the complex number r cis θ are given by

$$[r \text{ cis}(\theta + k \cdot 360°)]^{1/n} = r^{1/n} \text{ cis } \frac{1}{n}(\theta + k \cdot 360°),$$
$$k = 0, 1, 2, \ldots, n - 1$$

In the remaining examples, De Moivre's theorem is used to find all the roots of a complex number.

Example 3 Find all the roots of the equation $x^6 - 1 = 0$.

Solution. From $x^6 - 1 = 0$, we get

$$x^6 = 1$$

So we need to find the 6 sixth roots of 1. (Every number has n nth roots.)

To express 1 in polar form, note that the absolute value is 1 and the argument $0°$. Now recall that all the arguments differ by a multiple of $360°$. So the complete polar form is

$$1 = 1 \text{ cis } (0° + k \cdot 360°), \quad k = 0, 1, 2, \ldots$$

(Listing only positive multiples turns out to be sufficient, as we shall see.) Next, by De Moivre's theorem

$$[1 \text{ cis } (0° + k \cdot 360°)]^{1/6}, \quad k = 0, 1, 2, \ldots$$

$$= 1^{1/6} \text{ cis } \frac{0° + k \cdot 360°}{6}, \quad k = 0, 1, 2, \ldots$$

$$= 1^{1/6} \text{ cis } (k \cdot 60°), \quad k = 0, 1, 2, \ldots$$

Since the absolute value is a positive real number, it is understood that $1^{1/6}$ is the principal sixth root, that is, $1^{1/6} = 1$. So the sixth roots are

$$\text{cis } (k \cdot 60°), k = 0, 1, 2, \ldots$$

Converting back to the rectangular form, we get:

$k = 0$: $\cos 0° + j \sin 0° = 1$

$k = 1$: $\cos 60° + j \sin 60° = \frac{1}{2} + \frac{\sqrt{3}}{2} j$

$k = 2$: $\cos 120° + j \sin 120° = -\frac{1}{2} + \frac{\sqrt{3}}{2} j$

$k = 3$: $\cos 180° + j \sin 180° = -1$

$k = 4$: $\cos 240° + j \sin 240° = -\frac{1}{2} - \frac{\sqrt{3}}{2} j$

$k = 5$: $\cos 300° + j \sin 300° = \frac{1}{2} - \frac{\sqrt{3}}{2} j$

Starting with $k = 6$, the cycle begins again: $\cos 360° + j \sin 360° = 1$, which corresponds to $k = 0$. If $k = -1$, we get $\cos (-60°) + j \sin (-60°) = \cos 300° + j \sin 300°$, corresponding to $k = 5$. In other words, the values $k = 0, 1, \ldots, 5$ generate all six roots.

Note that only 1 and -1 could have been obtained directly by inspection. That the remaining numbers are also roots of unity can be checked by

direct multiplication. For example,
$$\left(\frac{1}{2} + \frac{\sqrt{3}}{2}j\right)^6 = 1$$

Example 4 Find the 5 fifth roots of $1 - j$.

Solution. Since $r = \sqrt{2}$ and $\theta = 315°$, we obtain
$$1 - j = \sqrt{2} \text{ cis } (315° + k \cdot 360°), \quad k = 0, 1, 2, 3, 4$$
By De Moivre's theorem the roots are given by
$$(1 - j)^{1/5} = [2^{1/2} \text{ cis } (315° + k \cdot 360°)]^{1/5} = (2^{1/2})^{1/5} \text{ cis } \frac{315° + k \cdot 360°}{5}$$
$$= 2^{1/10} \text{ cis } (63° + k \cdot 72°), \quad k = 0, 1, 2, 3, 4$$

Since $2^{1/10} = \sqrt[10]{2}$, the principal tenth root, the roots of $1 - j$ can now be listed:

$k = 0$: $\sqrt[10]{2}(\text{cis } 63° + j \sin 63°)$
$k = 1$: $\sqrt[10]{2}(\cos 135° + j \sin 135°)$
$k = 2$: $\sqrt[10]{2}(\cos 207° + j \sin 207°)$
$k = 3$: $\sqrt[10]{2}(\cos 279° + j \sin 279°)$
$k = 4$: $\sqrt[10]{2}(\cos 351° + j \sin 351°)$

If $k = 5$, we get $\sqrt[10]{2}$ cis $423° = \sqrt[10]{2}$ cis $63°$, which corresponds to $k = 0$.

These examples show that the n nth roots of $a + bj$ are equally spaced along the circumference of a circle of radius $\sqrt[n]{r}$ centered at the origin and $360°/n$ apart. Thus the tips of the vectors form the vertices of a regular polygon of n sides. For example, the 5 fifth roots of $1 - j$ (Example 4) are shown in Figure 11.14.

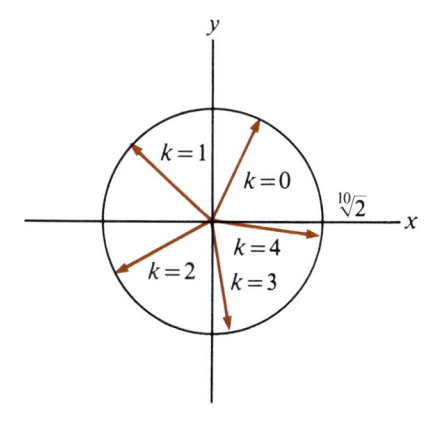

Figure 11.14

Example 5 Find the 7 seventh roots of $2j$ in polar form.

Solution. $2j = 2 \text{ cis } (90° + k \cdot 360°)$

The seven roots are therefore given by

$$[2 \text{ cis } (90° + k \cdot 360°)]^{1/7} = \sqrt[7]{2} \text{ cis } \frac{90° + k \cdot 360°}{7},$$
$$k = 0, 1, 2, 3, 4, 5, 6$$

Common error Writing

$$[2 \text{ cis } (20° + k \cdot 360°)]^{1/2}$$

as

$$2^{1/2} \frac{\text{cis } (20° + k \cdot 360°)}{2}$$

The correct form is

$$[2 \text{ cis } (20° + k \cdot 360°)]^{1/2} = 2^{1/2} \text{ cis } \frac{20° + k \cdot 360°}{2}$$

Exercises / Section 11.5

In Exercises 1–12, use De Moivre's theorem to find the indicated powers. Express the answers in rectangular form.

1. $(1 + j)^5$
2. $(-1 + j)^6$
3. $(2 - 2j)^8$
4. $(-3 - 3j)^4$
5. $(-1 + \sqrt{3}j)^5$
6. $(\sqrt{3} + j)^6$
7. $(-\sqrt{3} - j)^6$
8. $(1 - \sqrt{3}j)^7$
9. $(-2 + 2\sqrt{3}j)^4$
10. $(-3\sqrt{3} - 3j)^5$
11. $\left(\frac{1}{2} - \frac{\sqrt{3}}{2}j\right)^6$
12. $\left(\frac{\sqrt{3}}{2} + \frac{1}{2}j\right)^8$

In Exercises 13–16, use a calculator to find the indicated powers. Round off the values to three significant figures. (See Example 2.)

13. $(1 + 2j)^4$
14. $(2 - 3j)^4$
15. $(-1 + 4j)^5$
16. $(-2 - 4j)^5$

In Exercises 17–20, find the powers in polar form.

17. $(3 \text{ cis } 75.3°)^{10}$
18. $(4 \text{ cis } 118.4°)^{12}$
19. $(3 \text{ cis } 137.4°)^{15}$
20. $(3 \text{ cis } 236.1°)^{16}$

In Exercises 21–24, find the indicated roots. Express the results in rectangular form. (See Example 3.)

21. Cube roots of 1.
22. Fourth roots of -16.
23. Square roots of $-j$.
24. Cube roots of j.

In Exercises 25–40, find the indicated roots in polar form. (See Examples 4 and 5.)

25. Fifth roots of $32j$.
26. Fourth roots of $1 + \sqrt{3}j$.
27. Sixth roots of $1 + \sqrt{3}j$.
28. Fifth roots of $-1 + j$.
29. Fifth roots of $-32j$.
30. Sixth roots of $-2 + 2\sqrt{3}j$.
31. Eighth roots of $-1 - j$.
32. Seventh roots of $1 - j$.
33. Sixth roots of $-2\sqrt{3} - 2j$.
34. Ninth roots of $2j$.
35. Eighth roots of $3 - 3\sqrt{3}j$.
36. Fifth roots of $1 + j$.
37. Ninth roots of $-3j$.
38. Fifth roots of $-\sqrt{3} - j$.
39. Tenth roots of $-1 - \sqrt{3}j$.
40. Tenth roots of $\sqrt{3} - j$.

11.6 Phasors (Optional)

A particularly interesting application of complex numbers is the study of phase relations between voltage and current in a simple alternating current circuit. The problem arises because, as we saw in Chapter 8, the current produced by a generator is sinusoidal. While the current and voltage reach their peak periodically, they do not ordinarily peak at the same time. Since the phase relations are known for the individual elements, they can be determined for the circuit as a whole.

Consider the circuit in Figure 11.15 containing a resistor R, an inductor L, and a capacitor C in series with a generator E. Each of the elements offers

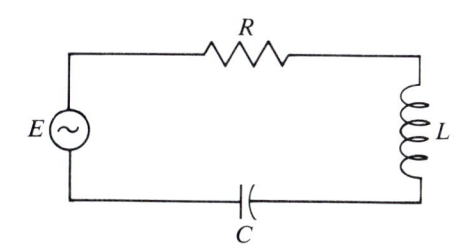

Figure 11.15

a type of resistance to the current flow, called the **reactance**. The reactance is denoted by X and is measured in ohms. We already know that $X = R$ for the resistor. The voltage across the resistor is given by $V = IR$. An analogous form for the voltage exists for all three elements: The voltages across the resistor, capacitor, and inductor are given by

$$V_R = IX_R = IR, \quad V_C = IX_C, \quad \text{and} \quad V_L = IX_L \qquad (11.18)$$

respectively.

Now consider the following facts: The voltage across the resistor reaches its peak at the same time as the current. The voltage and current are said to be **in phase**. The voltage across the inductor reaches its peak before the current and is said to **lead** the current (by 90°). Similarly, the voltage across the capacitor **lags** the current (by 90°). To obtain the phase relation for the combination, we use complex numbers in the following way: V_R is represented by a positive real number, V_L by the pure imaginary number $V_L j$, and V_C by the pure imaginary number $-V_C j$, as shown in Figure 11.16. The total voltage across the combination of elements is represented by

$$V = IR + IX_L j - IX_C j = I[R + j(X_L - X_C)] \tag{11.19}$$

If we let

$$Z = R + (X_L - X_C)j \tag{11.20}$$

(Figure 11.17), we get

$$V = IZ \tag{11.21}$$

which has the same form as $V = IX$. Z, called the **impedance,** is the total effective resistance of the elements, which takes into account the phase relation between the voltage and the current. More precisely, the magnitude of Z,

$$|Z| = \sqrt{R^2 + (X_L - X_C)^2}$$

is the *total effective resistance measured in ohms*. The argument of Z is the phase angle between the voltage and the current (Figure 11.17). If θ is positive, then the voltage leads the current; if θ is negative, then the voltage lags the current.

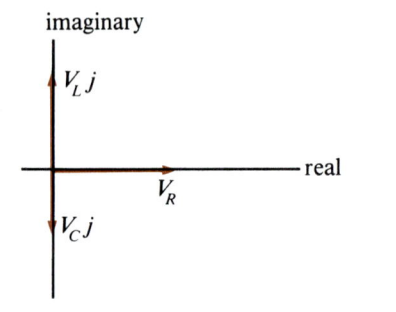

Figure 11.16

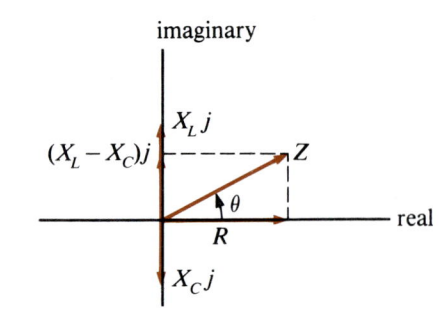

Figure 11.17

Example 1 If $R = 10.0\ \Omega$, $X_L = 5.00\ \Omega$, and $X_C = 20.0\ \Omega$, find **(a)** the impedance; and **(b)** the magnitude of the voltage across the combination at the instant when $I = 4.21$ A.

Solution. a. Draw the diagram in Figure 11.18. By formula (11.20)

$$Z = 10.0 + (5.00 - 20.0)j = 10.0 - 15.0j$$

So the magnitude of Z is

$$|Z| = \sqrt{(10.0)^2 + (15.0)^2} = 18.0\ \Omega$$

imaginary

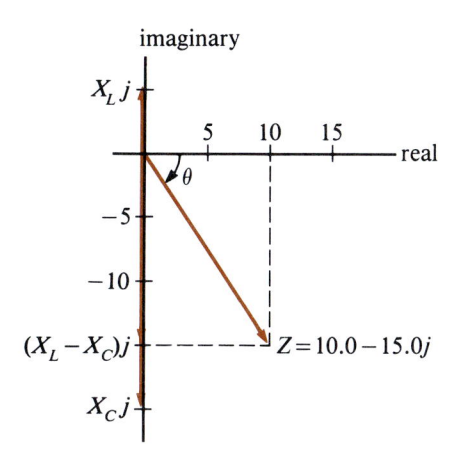

Figure 11.18

From $\tan \theta = -\frac{15.0}{10.0}$, we get $\theta = -56.3°$. So the voltage lags the current by 56.3°.

b. It follows from formula (11.21) that

$$|V| = I|Z| = (4.21)(18.0) = 75.8 \text{ V}$$

Exercises / Section 11.6

Find Z, $|Z|$, and θ for the given values of R, X_L, and X_C. Refer to the circuit in Figure 11.15.

1. $R = 3.0 \, \Omega$, $X_L = 8.0 \, \Omega$, $X_C = 4.0 \, \Omega$
2. $R = 4.0 \, \Omega$, $X_L = 7.0 \, \Omega$, $X_C = 4.0 \, \Omega$
3. $R = 12.0 \, \Omega$, $X_L = 17.0 \, \Omega$, $X_C = 8.00 \, \Omega$
4. $R = 13.1 \, \Omega$, $X_L = 15.3 \, \Omega$, $X_C = 10.2 \, \Omega$
5. $R = 20.0 \, \Omega$, $X_L = 12.3 \, \Omega$, $X_C = 29.7 \, \Omega$
6. $R = 55.3 \, \Omega$, $X_L = 25.7 \, \Omega$, $X_C = 42.6 \, \Omega$

Review Exercises / Chapter 11

In Exercises 1–4, perform the indicated operations graphically. Check your answers algebraically.

1. $(2 + 3j) + (1 - 2j)$
2. $(-2 + j) - (1 - 4j)$
3. $(2 - j) - (1 + 2j)$
4. $(1 - 3j) + (2 + 4j)$

In Exercises 5–8, determine the values of x and y for which equality holds.

5. $(x - 1) + yj = 3 + 2j$
6. $(x + 2) + (2x + 1)j = -y + yj$
7. $(x + y) + (x - y)j = 2 + 4j$
8. $(x - 4) + (2y + 2)j = y + xj$

In Exercises 9–18, perform the indicated operations and express the results in rectangular form.

9. $(\sqrt{2} - j) + (2\sqrt{2} + j)$
10. $(4\sqrt{5} + 2j) - (2\sqrt{5} - j)$
11. $(3 - 4j)(3 + 4j)$
12. $(1 - 7j)(1 + 7j)$
13. $(1 - 3j)^2$
14. $(2 + 3j)(7 - 4j)$

15. $\dfrac{3}{4-3j}$ **16.** $\dfrac{2}{4-j}$

17. $\dfrac{1-2j}{2+j}$ **18.** $\dfrac{2-3j}{1+2j}$

In Exercises 19–26, express the given complex numbers in polar form ($0° \le \theta < 360°$).

19. 2   **20.** $2j$
21. -3 **22.** $-4j$
23. $2+2j$ **24.** $2-2j$
25. $-1+\sqrt{3}j$ **26.** $-2-2\sqrt{3}j$

In Exercises 27–32, express the given complex numbers in exponential form ($0 \le \theta < 2\pi$).

27. $3j$ **28.** -2
29. $-3+3j$ **30.** $-2\sqrt{3}+2j$
31. $3-3\sqrt{3}j$ **32.** $-\sqrt{3}-j$

In Exercises 33–38, write the given complex numbers in rectangular form. (Use a diagram and express the results in terms of radicals.)

33. 2 cis 135°   **34.** 2 cis 240°
35. 4 cis 225°   **36.** 6 cis 330°
37. 3 cis 210°   **38.** 4 cis 300°

In Exercises 39–42, use a calculator to convert the given complex numbers to rectangular form. (Use three significant figures.)

39. 3.20 cis 136.0°   **40.** 1.32 cis 250.1°
41. 6.03 cis 25.1°   **42.** 4.96 cis 317.0°

In Exercises 43–46, use a calculator to convert each complex number to polar form. Express the argument to the nearest tenth of a degree.

43. $3-5j$ **44.** $-2-7j$
45. $-3+j$ **46.** $4-3j$

In Exercises 47–54, carry out each indicated operation. Leave the answer in polar form.

47. (2 cis 110°)(3 cis 32°)   **48.** (6 cis 151°)(2 cis 32°)

49. (2 cis 300°)($\sqrt{5}$ cis 72°)   **50.** $\dfrac{7 \text{ cis } 280°}{2 \text{ cis } 142°}$

51. $\dfrac{21 \text{ cis } 312°}{3 \text{ cis } 76°}$ **52.** $\dfrac{(6 \text{ cis } 281°)(12 \text{ cis } 39°)}{24 \text{ cis } 124°}$

53. $\dfrac{(25 \text{ cis } 53°)(2 \text{ cis } 132°)}{100 \text{ cis } 334°}$ **54.** $\dfrac{(3\sqrt{2} \text{ cis } 17°)(2 \text{ cis } 44°)}{\sqrt{2} \text{ cis } 216°}$

In Exercises 55–58, use De Moivre's theorem to find the indicated powers. Express the results in exact rectangular form.

55. $(1 + j)^7$

56. $(-2 + 2j)^8$

57. $(-\sqrt{3} - j)^6$

58. $(-1 + \sqrt{3}j)^7$

In Exercises 59–62, use a calculator to find the indicated powers. Round off to three significant figures. (See Example 2, Section 11.5.)

59. $(2 + j)^5$

60. $(-3 - 2j)^5$

61. $(-1 + 3j)^4$

62. $(3 - j)^4$

63. Find the cube roots of $8j$ in rectangular form.

64. Find the fourth roots of -1 in rectangular form.

In Exercises 65–67, find the roots in polar form.

65. Fifth roots of 1.

66. Square roots of $1 + j$.

67. Sixth roots of $\sqrt{3} + j$.

CHAPTER 12

Logarithmic and Exponential Functions

Objectives Upon completion of this chapter, you should be able to:
1. Change a given exponential form to a logarithmic form.
2. Change a given logarithmic form to an exponential form.
3. Sketch the graph of a given exponential or logarithmic function.
4. Use the laws of logarithms to write a given expression as a sum, difference, or multiple of logarithms.
5. Use the laws of logarithms to combine certain logarithmic expressions into a single logarithm.
6. Find the common and natural logarithm of a number.
7. Find the antilogarithm of a number.
8. Perform arithmetic operations using logarithms. (Optional)
9. Change the base of a logarithm.
10. Solve exponential and logarithmic equations.
11. Graph a given function on logarithmic paper. (Optional)

12.1 The Definition of Logarithm

Logarithms were first introduced by John Napier (1550–1617), a Scottish nobleman, as a device for carrying out lengthy arithmetic computations in a relatively fast way. At a time when arithmetic had to be done by hand, this was an enormous breakthrough that was a blessing for the whole scientific community, particularly astronomy. Even the slide rule was an offshoot of logarithms. In the days of the electronic calculator, the importance of logarithms as a computational device has all but vanished, but the importance in theoretical work and in advanced mathematics is greater than ever.

To see how logarithms are defined, let us consider a typical expression containing an exponent:

$$2^3 = 8$$

John Napier

358

Recall that 2 is called the *base* and 3 the *exponent*, and we say "2 cubed equals 8." If we start with the number 8, then we could also say "8 is written as a power of 2" or "the exponent corresponding to 8 is 3, provided that 2 is the base." Now, a **logarithm** is merely an exponent. Using this term, we could say "the logarithm of 8 is 3, provided that the base is 2." More simply, "the logarithm of 8 to the base 2 is 3." This statement is abbreviated

$$\log_2 8 = 3$$

where *log* stands for *logarithm*. In summary,

$$2^3 = 8 \quad \text{and} \quad \log_2 8 = 3$$

mean exactly the same thing. In general,

$$x = b^y \quad \text{means the same as} \quad y = \log_b x \qquad (12.1)$$

Definition of logarithm:

$$y = \log_b x \text{ means the same as } b^y = x, \quad b > 0, \quad b \neq 1$$

The equation $y = \log_b x$ is read *y is equal to the logarithm of x to the base b*, or *y equals log x to the base b*.

Example 1

a. $\log_3 9 = 2$ means $3^2 = 9$.
b. $\log_4 2 = \frac{1}{2}$ means $4^{1/2} = 2$.
c. $\log_8 \frac{1}{2} = -\frac{1}{3}$ means $8^{-1/3} = \frac{1}{2}$.

As we shall see throughout this chapter, logarithms have many applications. In this section, however, we are concerned only with the basic definitions. In the examples and exercises, we shall convert exponential forms ($x = b^y$) to logarithmic forms ($y = \log_b x$) and vice versa. Again for practice, we shall solve simple equations involving logarithms. More complicated equations will be taken up in Section 12.7.

Example 2 Write $2^{-3} = \frac{1}{8}$ in logarithmic form.

Solution. Since the base is 2 and the exponent -3, we write

$$\log_2 \frac{1}{8} = -3$$

Example 3 Write $\log_9 3 = \frac{1}{2}$ in exponential form.

Solution. Since 9 is the base and $\frac{1}{2}$ the exponent, we get

$$9^{1/2} = 3$$

Example 4 Find the value of the unknown:

  a. $\log_{27} x = -\dfrac{1}{3}$ **b.** $\log_b \dfrac{1}{16} = -2$

Solution.   **a.** If we change $\log_{27} x = -\frac{1}{3}$ to the exponential form, we get

$$27^{-1/3} = x$$

Thus

$$x = \dfrac{1}{27^{1/3}} = \dfrac{1}{\sqrt[3]{27}} = \dfrac{1}{3}$$

b. By writing

$$\log_b \dfrac{1}{16} = -2 \quad \text{as} \quad b^{-2} = \dfrac{1}{16}$$

we get

$$\dfrac{1}{b^2} = \dfrac{1}{16}$$

$$\dfrac{1}{b^2} = \dfrac{1}{4^2}$$

$$b = 4$$

(From $b^2 = 16$ we actually get $b = \pm 4$, but since $b > 0$, we choose the positive square root.)

Example 5 Find x: **a.** $\log_3 x = 1$ **b.** $\log_3 x = 0$

Solution.   **a.** $\log_3 x = 1$ can be written $3^1 = x$; thus $x = 3$.
  b. $\log_3 x = 0$ is equivalent to $3^0 = x$; thus $x = 1$.

The results of Example 5 can be generalized.

$$\log_b b = 1, \quad \log_b 1 = 0 \tag{12.2}$$

Exercises / Section 12.1

In Exercises 1–16, change each exponential form to the logarithmic form.

1. $3^4 = 81$ **2.** $2^6 = 64$ **3.** $10^3 = 1{,}000$ **4.** $7^2 = 49$

5. $4^4 = 256$ **6.** $5^{-2} = \dfrac{1}{25}$ **7.** $2^{-4} = \dfrac{1}{16}$ **8.** $49^{1/2} = 7$

9. $3^0 = 1$ **10.** $2^0 = 1$ **11.** $\left(\dfrac{3}{4}\right)^2 = \dfrac{9}{16}$ **12.** $\left(\dfrac{1}{2}\right)^{-2} = 4$

13. $\left(\dfrac{3}{2}\right)^{-3} = \dfrac{8}{27}$ **14.** $10^{-4} = 0.0001$ **15.** $3^{-4} = \dfrac{1}{81}$ **16.** $\left(\dfrac{1}{10}\right)^{-2} = 100$

In Exercises 17–32, change each logarithmic form to the exponential form.

17. $\log_5 125 = 3$ **18.** $\log_{10} 100 = 2$ **19.** $\log_5 1 = 0$ **20.** $\log_6 6 = 1$

21. $\log_2 2 = 1$ **22.** $\log_2 \dfrac{1}{64} = -6$ **23.** $\log_3 \dfrac{1}{27} = -3$ **24.** $\log_6 216 = 3$

25. $\log_{1/2} \dfrac{1}{4} = 2$ **26.** $\log_{1/3} 27 = -3$ **27.** $\log_{10} \dfrac{1}{1,000} = -3$ **28.** $\log_4 \dfrac{1}{256} = -4$

29. $\log_6 \dfrac{1}{36} = -2$ **30.** $\log_{15} 225 = 2$ **31.** $\log_{25} \dfrac{1}{25} = -1$ **32.** $\log_7 \dfrac{1}{7} = -1$

In Exercises 33–50, find the value of the unknown.

33. $\log_3 x = 4$ **34.** $\log_4 x = 2$ **35.** $\log_3 27 = a$ **36.** $\log_5 125 = a$

37. $\log_{10} x = 4$ **38.** $\log_b 49 = 2$ **39.** $\log_b 32 = 5$ **40.** $\log_{3/2} x = 2$

41. $\log_{25} x = \dfrac{1}{2}$ **42.** $\log_5 x = -2$ **43.** $\log_4 x = -3$ **44.** $\log_6 x = -1$

45. $\log_{3/4} x = -2$ **46.** $\log_b 16 = -2$ **47.** $\log_b \dfrac{27}{8} = -3$ **48.** $\log_b \dfrac{1}{64} = -6$

49. $\log_{49} \dfrac{1}{7} = a$ **50.** $\log_{27} \dfrac{1}{3} = a$

12.2 Graphs of Exponential and Logarithmic Functions

We have seen repeatedly that the behavior of a function can be observed from its graph. In this section we shall discuss two new functions, the exponential and logarithmic functions, with the aid of their graphs.

> **Exponential function**: An exponential function has the form
> $$y = b^x, \quad b > 0, \quad b \neq 1 \tag{12.3}$$

The exponential function appears to have a simple, even familiar, form and yet is quite different from the exponents discussed in Chapter 10. The reason is that the variable x in equation (12.3) need not be an integer or even an arbitrary rational number—it can take on any real value. So we need to assume without proof that $y = b^x$ is defined for all real x.

As usual, we shall construct our graphs by plotting points from a table of values. Consider the next example.

Example 1 Sketch the graph of $y = 2^x$.

Solution. To graph the function, we construct the following table of values:

x:	-3	-2	-1	0	1	2	3
y:	$\frac{1}{8}$	$\frac{1}{4}$	$\frac{1}{2}$	1	2	4	8

By plotting the points from the values in the table and connecting them by a smooth curve, we obtain the graph in Figure 12.1.

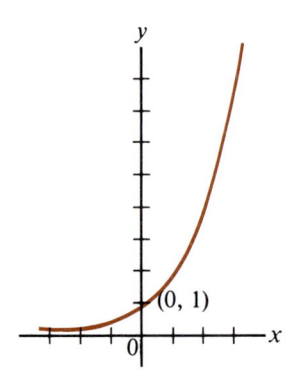

Figure 12.1

Since $b^0 = 1$, all graphs of the form $y = b^x$ have a single y-intercept at $(0, 1)$. If $b > 1$, then the graph is always similar to Figure 12.1, *approaching the negative x-axis* and *rising rapidly through positive x-values*. For example, the y-values of $y = 3^x$ increase rapidly as x increases.

We now turn to the logarithmic function.

Logarithmic function: A logarithmic function has the form

$$y = \log_b x, \quad b > 0, \quad b \neq 1 \tag{12.4}$$

A logarithmic function can be graphed by converting it to the form

$$x = b^y$$

Using this form, a table of values can readily be constructed, as shown in the next example.

Example 2
Sketch the graph of $y = \log_2 x$.

Solution. The function $y = \log_2 x$ can be written

$$x = 2^y$$

To obtain the table of values, we assign various values to y and compute x. Note that the resulting table is identical to the table in Example 1 with x and y interchanged.

y:	-3	-2	-1	0	1	2	3
x:	$\frac{1}{8}$	$\frac{1}{4}$	$\frac{1}{2}$	1	2	4	8

The graph is shown in Figure 12.2.

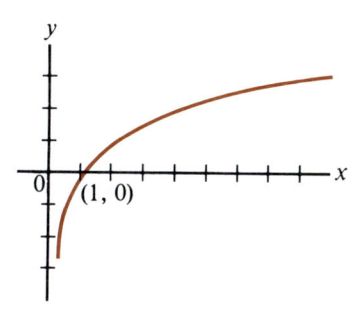

Figure 12.2

All graphs of the form $y = \log_b x$ have a single x-intercept at $(1, 0)$. If $b > 1$, the graph will be similar to Figure 12.2, *approaching the negative y-axis and rising as x increases*. Also, since $b^y > 0$, x is always positive. Since the graph crosses the x-axis at $(1, 0)$, $\log_b x < 0$ for $0 < x < 1$, and $\log_b x > 0$ for $x > 1$. For example, $\log_2 \frac{1}{2}$ is negative, while $\log_2 3$ is positive. (Although far from obvious, $\log_b x$ is complex for $x < 0$. Consequently, we shall exclude the case $x < 0$ from our discussion.)

Properties of the logarithmic function $y = \log_b x$:

1. If $0 < x < 1$, $\log_b x < 0$. $(b > 1)$
2. If $x > 1$, $\log_b x > 0$. $(b > 1)$
3. $\log_b 1 = 0$, $\log_b b = 1$.

CHAPTER 12 LOGARITHMIC AND EXPONENTIAL FUNCTIONS

For the exponential function, the case $0 < b < 1$ arises occasionally. For this case $y = b^x$ differs from the shape in Figure 12.1, as shown in the next example.

Example 3
Half-life

The *half-life* of a radioactive element is the time required for half of a given amount of the element to decay. If N_0 is the given amount and H the half-life, then the amount left as a function of time t is given by

$$N = N_0 \left(\frac{1}{2}\right)^{t/H}, \quad t \geq 0$$

For example, the half-life of radium is 1,590 years, so that

$$N = N_0 \left(\frac{1}{2}\right)^{t/1,590}$$

As another example, suppose a radioactive isotope has a half-life of 1.0 day. If the initial amount is 1.0 kg, then the amount after t days is

$$N = 1.0 \left(\frac{1}{2}\right)^{t/1.0} \quad \text{(in kilograms)}$$

Sketch the curve.

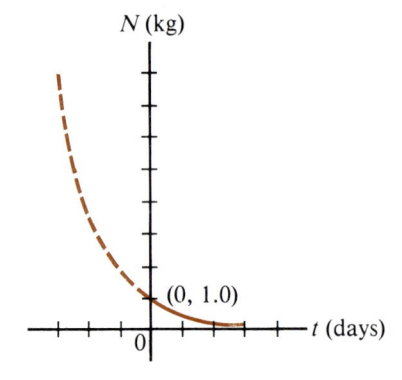

Figure 12.3

Solution. The curve (Figure 12.3) is plotted from the following table of values:

t (days):	0	1	2	3
N (kg):	1.0	0.5	0.25	0.125

Note that the curve approaches the horizontal axis on the positive side. To show the behavior of this exponential function, the curve in Figure 12.3 is drawn for negative as well as positive values. However, since the original amount of 1.0 kg corresponds to $t = 0$, the N-values are meaningful only for $t \geq 0$. Consequently, the portion of the curve for negative values is drawn as a dashed curve.

Exercises / Section 12.2

In Exercises 1–20, sketch the graph of each function.

1. $y = 3^x$
2. $y = 4^x$
3. $y = 5^x$
4. $y = 6^x$
5. $y = \left(\frac{1}{3}\right)^x$
6. $y = \left(\frac{1}{4}\right)^x$
7. $y = 2^{-x}$
8. $y = 3^{-x}$
9. $y = 2^{2x}$
10. $y = \left(\frac{1}{2}\right)^{3x}$
11. $y = \log_3 x$
12. $y = \log_4 x$
13. $y = \log_5 x$
14. $y = \log_6 x$
15. $y = \log_{1/2} x$
16. $y = \log_{1/3} x$
17. $y = \log_{1/4} x$
18. $y = 2 \log_2 x$
19. $y = 3 \log_4 x$
20. $y = 3 \log_5 x$

21. A certain radioactive isotope has a half-life of 2.0 years. If 10.0 g are present initially, then the amount left after t years is
$$N = 10.0 \left(\frac{1}{2}\right)^{t/2.0}, \quad t \geq 0$$
Sketch N as a function of t.

22. The number of bacteria in a culture doubles every hour. If there are 1,000 bacteria initially, then the number of bacteria after t hours is given by $N = 1{,}000(2^t)$, $t \geq 0$. Sketch the curve.

23. In a certain chemical reaction, a substance is converted into another substance. Starting with 1 kg of unconverted substance, the time required for the amount of unconverted substance to shrink to N ($0 < N \leq 1$) is given by
$$t = -10 \log_{10} N, \quad 0 < N \leq 1$$
where t is in seconds. Sketch t as a function of N.

24. An object at room temperature (20°C) is placed in an oven kept at constant temperature 100°C. If after 1 min the temperature has risen to 40°C, it can be shown that the temperature at any time is given by
$$T = 100 - 80 \left(\frac{3}{4}\right)^t, \quad t \geq 0$$
Sketch T as a function of t.

25. The tension T (in pounds) on a certain rope around a cylindrical post at a point P is related to the central angle θ determined by P by the formula $\log_{10} T = 0.5\theta$, where $\theta \geq 0$ is measured in radians. Write this formula in exponential form and sketch the graph.

26. In a certain chemical reaction, the amount N (in kilograms) of unconverted substance varies with time t (in seconds) according to the relation
$$\log_{10} N = -0.5t, \quad t \geq 0$$
Write this formula in exponential form and sketch the graph.

12.3 Properties of Logarithms

In this section we shall study the basic properties of logarithms. A knowledge of these properties is essential in using logarithms effectively.

Let us first recall the laws of exponents, expressed in the forms that we shall need below:

$$b^M b^N = b^{M+N} \tag{12.5}$$

$$\frac{b^M}{b^N} = b^{M-N} \tag{12.6}$$

$$(b^M)^r = b^{Mr} \tag{12.7}$$

Since logarithms are exponents, the laws of logarithms are just the laws of exponents stated in a different form. For example, formula (12.5) corresponds to

$$\log_b xy = \log_b x + \log_b y \tag{12.8}$$

To see why, let $M = \log_b x$ and $N = \log_b y$. Then

$$b^M = x \quad \text{and} \quad b^N = y$$

and

$$xy = b^M b^N = b^{M+N}$$

Converting the last expression to a logarithmic form, we get

$$\log_b xy = M + N$$

Substituting back, we get

$$\log_b xy = \log_b x + \log_b y$$

To obtain the form corresponding to rule (12.6), we take the quotient of x and y to get

$$\frac{x}{y} = \frac{b^M}{b^N} = b^{M-N}$$

In logarithmic form

$$\log_b \frac{x}{y} = M - N = \log_b x - \log_b y \tag{12.9}$$

The form corresponding to rule (12.7) can be obtained similarly. Let $M = \log_b x$, so that $b^M = x$. The rth power of x is

$$x^r = (b^M)^r = b^{Mr}$$

Converting to logarithmic form again, we have

$$\log_b x^r = Mr = r \log_b x \tag{12.10}$$

Thus we obtain the three formulas listed in the box below.

> **Properties of logarithms**: For any positive real number b, $b \neq 1$, for any *positive* real numbers x and y, and for any real number r:
> 1. $\log_b x + \log_b y = \log_b xy$. (12.11)
> 2. $\log_b x - \log_b y = \log_b (x/y)$. (12.12)
> 3. $\log_b x^r = r \log_b x$. (12.13)

These properties are illustrated in the next example.

Example 1
a. $\log_3 35 = \log_3 5 \cdot 7 = \log_3 5 + \log_3 7$ by property (12.11)
b. $\log_5 \dfrac{7}{11} = \log_5 7 - \log_5 11$ by property (12.12)
c. $\log_{10} 16 = \log_{10} 2^4 = \mathbf{4} \log_{10} 2$ by property (12.13)

Certain special cases involve the following properties of logarithms: $\log_b b = 1$ and $\log_b 1 = 0$. (See statement (12.2) in Section 12.1.)

Example 2 Evaluate: a. $\log_5 25$;   b. $\log_4 \frac{1}{4}$.

Solution.   a. Since $25 = 5^2$, we may write
$$\log_5 25 = \log_5 5^2$$
Since $\log_b x^r = r \log_b x$, we have
$$\log_5 5^2 = \mathbf{2} \log_5 5$$
Finally, since $\log_b b = 1$, it follows that
$$2 \log_5 5 = 2 \cdot \mathbf{1} = 2$$

b. $\log_4 \dfrac{1}{4} = \log_4 1 - \log_4 4$ by property (12.12)
         $= 0 - 1 = -1$ since $\log_b 1 = 0$

Now that we have seen how the properties of logarithms can be used to simplify certain elementary expressions, we can break up more complicated logarithmic expressions into sums, differences, and multiples of logarithms. Consider the next example.

Example 3 Break up the following logarithms into sums, differences, or multiples of logarithms. If possible, use the fact that $\log_b b = 1$ and $\log_b 1 = 0$.

 a. $\log_3 3x^2$ **b.** $\log_7 \dfrac{1}{x^3}$ **c.** $\log_3 \sqrt[5]{3}$ **d.** $\log_2 \dfrac{1}{4x^2}$

Solution.

a. $\log_3 3x^2 = \log_3 3 + \log_3 x^2$ log of a product
$= 1 + 2 \log_3 x$ $\log_b b = 1$ and (12.13)

b. $\log_7 \dfrac{1}{x^3} = \log_7 1 - \log_7 x^3$ log of a quotient
$= 0 - \log_7 x^3$ $\log_b 1 = 0$
$= -3 \log_7 x$ property (12.13)

c. $\log_3 \sqrt[5]{3} = \log_3 3^{1/5}$
$= \dfrac{1}{5} \log_3 3$
$= \dfrac{1}{5}$

d. $\log_2 \dfrac{1}{4x^2} = \log_2 1 - \log_2 4x^2$
$= 0 - (\log_2 4 + \log_2 x^2)$
$= -\log_2 2^2 - \log_2 x^2$
$= -2 \log_2 2 - 2 \log_2 x$
$= -2 - 2 \log_2 x$

The properties of logarithms can also be used to combine expressions involving logarithms. Consider the next example.

Example 4 Combine $\frac{1}{2} \log_5 x - 2 \log_5 x$ into a single logarithm.

Solution. $\dfrac{1}{2} \log_5 x - 2 \log_5 x = \log_5 x^{1/2} - \log_5 x^2$ $\log_b x^r = r \log_b x$

$= \log_5 \dfrac{x^{1/2}}{x^2}$ log of a quotient

$= \log_5 \dfrac{1}{x^{3/2}}$

Example 5 Given that $\log_{10} 7 = 0.8451$, find $\log_{10} 49$.

Solution. $\log_{10} 49 = \log_{10} 7^2 = 2 \log_{10} 7$
$= 2(0.8451)$
$= 1.6902$

Common error   Writing

$$\log_b (x + y) \quad \text{as} \quad \log_b x + \log_b y$$

By the first property of logarithms,

$$\log_b x + \log_b y = \log_b xy$$

The expression $\log_b (x + y)$ cannot be written as a sum of two logarithms.

Example 6   The loudness β of a sound (in decibels) is given by

$$\beta = 10(\log_{10} I - \log_{10} I_0)$$

where I is the intensity of the sound in watts per square meter and I_0 the intensity of the faintest audible sound. Write β as a single logarithm.

Solution.
$$\beta = 10 (\log_{10} I - \log_{10} I_0)$$
$$= 10 \log_{10} \frac{I}{I_0}$$
$$= \log_{10} \left(\frac{I}{I_0}\right)^{10}$$

Exercises / Section 12.3

In Exercises 1–40, write each given expression as the sum, difference or multiple of logarithms. Whenever possible, use $\log_b b = 1$ and $\log_b 1 = 0$ to simplify the result. (See Example 3.)

1. $\log_{10} 81$
2. $\log_2 27$
3. $\log_3 27$
4. $\log_5 21$
5. $\log_7 36$
6. $\log_7 49$
7. $\log_{10} 100$
8. $\log_5 125$
9. $\log_2 16$
10. $\log_3 81x^3$
11. $\log_3 \sqrt{27}$
12. $\log_{10} 1{,}000$
13. $\log_5 25x^2$
14. $\log_7 \sqrt{7x}$
15. $\log_3 15x^2$
16. $\log_4 4x^3$
17. $\log_4 20\sqrt{a}$
18. $\log_2 \dfrac{1}{2x}$
19. $\log_3 \dfrac{1}{9a^2}$
20. $\log_4 \dfrac{1}{16V^2}$
21. $\log_3 \dfrac{1}{\sqrt{3C}}$
22. $\log_2 \dfrac{1}{\sqrt[3]{4v}}$
23. $\log_5 \dfrac{1}{30\sqrt{u}}$
24. $\log_5 \dfrac{1}{50r^2}$
25. $\log_a \dfrac{1}{a^2}$
26. $\log_c \dfrac{1}{\sqrt{c}}$
27. $\log_e \dfrac{1}{\sqrt[3]{e}}$
28. $\log_e 36e^2$
29. $\log_e \dfrac{25}{e^3}$
30. $\log_e \dfrac{49}{\sqrt{e}}$
31. $\log_e \dfrac{1}{\sqrt{\pi e}}$
32. $\log_e 9\sqrt{e}$
33. $\log_e 2\sqrt[3]{e}$
34. $\log_{10} 900$
35. $\log_{10} \dfrac{1}{400}$
36. $\log_{10} \dfrac{1}{16{,}000}$

37. $\log_{10} 0.009$ **38.** $\log_{10} 0.00025$ **39.** $\log_{10} \dfrac{x}{300}$

40. $\log_{10} \dfrac{x^2}{75}$

In Exercises 41–56, write each expression as a single logarithm. (See Example 4.)

41. $\log_2 3 + \log_2 5$ **42.** $\log_5 10 - \log_5 5$ **43.** $\log_4 30 - \log_4 3$
44. $\log_6 14 - \log_6 7$ **45.** $\log_2 7 + \log_2 5 - \log_2 3$ **46.** $\log_e 2 + \log_e 5 - \log_e 7$
47. $2 \log_2 5 - \log_2 3$ **48.** $3 \log_{10} 4 - \log_{10} 32$ **49.** $\dfrac{1}{2} \log_{10} 16$
50. $\dfrac{1}{3} \log_6 27$ **51.** $\dfrac{1}{2} \log_2 x - \log_2 3$ **52.** $\dfrac{1}{2} \log_5 V + \log_5 L$
53. $\dfrac{1}{2} \log_e m + \dfrac{1}{2} \log_e v$ **54.** $\dfrac{1}{3} \log_e s + \dfrac{1}{4} \log_e 16$ **55.** $\dfrac{1}{2} \log_7 s + \dfrac{1}{2} \log_7 t - \log_7 2$
56. $\dfrac{1}{3} \log_5 a^6 - \dfrac{1}{3} \log_5 64$

In Exercises 57–64, evaluate the logarithms. Use the values $\log_{10} 2 = 0.3010$ and $\log_{10} 3 = 0.4771$.

57. $\log_{10} 6$ **58.** $\log_{10} 9$ **59.** $\log_{10} 4$
60. $\log_{10} 8$ **61.** $\log_{10} 12$ **62.** $\log_{10} 18$
63. $\log_{10} 20$ **64.** $\log_{10} 40$

65. If p_0 is the pressure at sea level and p the pressure at the top of a column of air h meters in height and having a uniform temperature T (in degrees Kelvin), then

$$\log_{10} p - \log_{10} p_0 = -0.0149 \dfrac{k}{T}$$

where k is a constant. Combine the terms on the left side and write the equation in exponential form.

66. The fallout from a nuclear explosion can be written $\log_{10} R - \log_{10} R_0 = kt$, where R is the amount of radiation after the explosion, R_0 the amount before the explosion, and k a constant. Combine the terms on the left side and write R as a function of time t.

67. If p (in millimeters of mercury) is the vapor pressure of carbon tetrachloride, then

$$\log_{10} p + \log_{10} (5.97 \times 10^7) = -\dfrac{1706.4}{T}$$

where T is the temperature in degrees Kelvin. Combine the terms on the left side and write the equation in exponential form.

12.4 Common Logarithms

Although logarithms were defined for an arbitrary base $b > 0$, $b \neq 1$, in practice only two bases are generally used. Logarithms to base 10, called **common logarithms,** are suitable for numerical work, since our number system is in base 10. They also arise in certain applications. Logarithms to base e are called **natural logarithms,** since they arose quite naturally in the development of the calculus. In this section we shall discuss common logarithms.

To see the practical importance of base 10, consider the folowing number, expressed in scientific notation:

$$u = 2.36 \times 10^6$$

Taking logarithms to base 10 of both sides, we get

$$\log_{10} u = \log_{10} (2.36 \times 10^6) = \log_{10} 2.36 + \log_{10} 10^6$$
$$= \log_{10} 2.36 + 6 \log_{10} 10$$
$$= \log_{10} 2.36 + 6 \qquad \text{since } \log_b b = 1$$

The number $\log_{10} 2.36$ is called the **mantissa** and the number 6 is called the **characteristic.** In general, for the number

$$u = M \times 10^c, \quad 0 < M < 10$$

the number $\log_{10} M$ is the *mantissa* and the number c the *characteristic*.

To see how these ideas can be used to find logarithms, let us first suppose that we are given the logarithm of 2:

$$\log_{10} 2 = 0.3010$$

Now consider the following logarithms involving powers of 10:

$$\log_{10} 20 = \log_{10} (2 \times 10^1) = 1 + 0.3010 = 1.3010$$
$$\log_{10} 200 = \log_{10} (2 \times 10^2) = 2 + 0.3010 = 2.3010$$
$$\log_{10} 2{,}000 = \log_{10} (2 \times 10^3) = 3 + 0.3010 = 3.3010$$
$$\log_{10} 20{,}000 = \log_{10} (2 \times 10^4) = 4 + 0.3010 = 4.3010$$
$$\log_{10} 0.2 = \log_{10} (2 \times 10^{-1}) = -1 + 0.3010 = -0.6990$$
$$\log_{10} 0.02 = \log_{10} (2 \times 10^{-2}) = -2 + 0.3010 = -1.6990$$

Note the emerging pattern: The mantissa is 0.3010 in all cases, while the characteristic depends only on the position of the decimal point. So given $\log_{10} 2 = 0.3010$, we can find $\log_{10} 2 \times 10^n$ for any integer n.

In general, to find the logarithm of a number, we need a table of logarithms only for the numbers between 1 and 10. (Such a table is given in Appendix B.) To find the logarithm of a number, we write the number in scientific notation, determine the characteristic, and obtain the mantissa from the table.

It is customary to omit the number 10 indicating the base. Thus $\log N$ is understood to mean $\log_{10} N$.

$$\log N \quad \text{means} \quad \log_{10} N$$

Example 1 Find log 3,610.

Solution. 3,610 = **3.61** × 10³. So the mantissa is **log 3.61** and the characteristic **3**. From Table 2 of Appendix B, log 3.61 = **0.5575**. Hence

$$\log 3{,}610 = 3.5575$$

Example 2 Find log 74,600.

Solution. 74,600 = **7.46** × 10⁴

From the table, **log 7.46 = 0.8727**. Since the characteristic is **4**,

$$\log 74{,}600 = 4.8727$$

Example 3 Find log 0.00813.

Solution. 0.00813 = 8.13 × 10⁻³

From the table, log 8.13 = 0.9101; since the characteristic is −3, we get

$$\log 0.00813 = -3 + 0.9101 = -2.0899$$

To preserve the mantissa obtained from the table, the logarithm is also written **−3** + 0.9101 = **7 − 10** + 0.9101 = (7 + 0.9101) − 10, or

$$\log 0.00813 = 7.9101 - 10$$

CALCULATOR COMMENT   Logarithms can be found using a calculator. For example, to find log 0.000581, enter 0.000581 and press the $\boxed{\text{LOG}}$ key.

Example 4 Find log 0.000371 by using the table. Check the result with a calculator.

Solution. 0.000371 = 3.71 × 10⁻⁴

From the table, log 3.71 = 0.5694. Hence

$$\log 0.000371 = -4 + 0.5694 = -3.4306$$

or, since −4 = 6 − 10,

$$\log 0.000371 = 6.5694 - 10$$

To check the result with a calculator, enter 0.000371 and press $\boxed{\text{LOG}}$. The result is −3.4306 to four decimal places.

Antilogarithm   If log N is known, we can use the table to find N. The number N is called the **antilogarithm.**

12.4 COMMON LOGARITHMS

Example 5 If $\log N = 3.8943$, find N.

Solution. We look up the number 0.8943 in the table and find that it corresponds to 7.84. Since the characteristic is 3,

$$N = 7.84 \times 10^3 = 7,840$$

Example 6 If $\log N = 0.9560$, find N.

Solution. Since the number 0.9560 is not listed, we need to interpolate by the same method used in Chapter 4:

$$1 \left[x \left[\begin{array}{l} \log 9.03 = 0.9557 \\ \log N\ \ \ = 0.9560 \\ \log 9.04 = 0.9562 \end{array} \right] 3 \right] 5$$

Thus

$$\frac{x}{1} = \frac{3}{5} \quad \text{and} \quad x = 0.6$$

We conclude that $N = 9.036$.

CALCULATOR COMMENT As with most calculations, it is simpler to obtain the antilogarithm with a calculator: enter 0.9560, press the $\boxed{\text{INV}}$ key, and then press the $\boxed{\text{LOG}}$ key. The sequence is

0.9560 $\boxed{\text{INV}}$ $\boxed{\text{LOG}}$

Display: 9.0364947

On some calculators the proper sequence is

0.9560 $\boxed{10^x}$

Example 7 If $\log N = -3.6747$, find N.

Solution. The sequence is

-3.6747 $\boxed{\text{INV}}$ $\boxed{\text{LOG}}$

or

-3.6747 $\boxed{10^x}$

The result is $N = 2.1149 \times 10^{-4}$.

Example 8 The atmospheric pressure P (in pounds per square inch) can be obtained from the equation

$$\log_{10} P = 0.434(2.69 - 0.21h)$$

where h is the altitude in miles above sea level. Find the atmospheric pressure at an altitude of **45** mi.

Solution. From the given formula,

$$\log_{10} P = 0.434(2.69 - 0.21 \cdot 45)$$

The sequence is

2.69 $\boxed{-}$ 0.21 $\boxed{\times}$ 45 $\boxed{=}$ $\boxed{\times}$ 0.434 $\boxed{=}$ $\boxed{\text{INV}}$ $\boxed{\text{LOG}}$

Display: 0.001164555

So the atmospheric pressure is 0.0012 lb/in² to 2 significant figures.

Exercises / Section 12.4

In Exercises 1–10, find the common logarithm of each number by using Table 2 of Appendix B. Check the result with a calculator.

1. log 5.59
2. log 559
3. log 83,400
4. log 613,000
5. log 0.00524
6. log 1,290,000
7. log 2,478 (interpolate)
8. log 72,870
9. log 853,200
10. log 0.002463

In Exercises 11–20, use Table 2 of Appendix B to find the antilogarithm in each case. Check the result with a calculator.

11. 1.8639
12. 2.5599
13. 3.7300
14. 8.7745 − 10
15. 9.9325 − 10
16. 2.8497
17. 3.6472
18. 8.7221 − 10
19. 7.9308 − 10
20. 6.9615 − 10

21. The acidity of a chemical solution is determined by the concentration of the hydrogen ion H⁺, written [H⁺], and is measured in moles per liter (mol/L). The hydrogen potential pH is defined by

$$\text{pH} = -\log_{10} [\text{H}^+]$$

Given that the acid concentration of water is 10^{-7} mol/L, determine the pH of water.

22. The acid concentration of blood is 3.98×10^{-8} mol/L. Determine the pH value.

23. A certain satellite has a power supply whose output P (in watts) is given by $\log_{10} P = -t/901$, where t is the number of days that the battery has operated. What will the output be after 379 days?

24. A record company estimates that its monthly profit (in thousands of dollars) from a new hit record is $P = 12 - 15 \log_{10} (1 + t)$, where t is the number of months after release. Calculate the profit (or loss) for the first six months and determine when the record should be withdrawn.

12.5 Computations with Logarithms (Optional)

In this section we shall briefly see how logarithms are used in numerical calculations. As indicated earlier, this application of logarithms has lost much of its importance since the advent of the scientific calculator.

12.5 COMPUTATIONS WITH LOGARITHMS

Example 1 Multiply $396{,}100 \times 0.0005686$ by means of logarithms.

Solution. Let

$$N = (396{,}100)(0.0005686)$$

Then

$$\log N = \log 396{,}100 + \log 0.0005686$$

by property (12.11). From Table 2 of Appendix B,

$$\begin{aligned} \log 396{,}100 &= 5.5978 \\ \log 0.0005686 &= 6.7548 - 10 \\ \log N &= 12.3526 - 10 \quad \text{(adding)} \end{aligned}$$

which gives $\log N = 2.3526$. The desired product is the antilogarithm $N = 225.2$.

Example 2 Calculate $\sqrt[6]{2{,}746}$

Solution. The extraction of a root requires the third law of logarithms (12.13),

$$\log_b x^r = r \log_b x$$

Let $N = \sqrt[6]{2{,}746} = (2{,}746)^{1/6}$. Then $\log N = \frac{1}{6} \log 2{,}746 = \frac{1}{6}(3.4387) = 0.5731$. The desired root is the antilogarithm $N = 3.742$.

Example 3 Calculate $\sqrt[4]{0.007140}$.

Solution. Let $N = (0.007140)^{1/4}$. From Table 2,

$$\log 0.007140 = 7.8537 - 10$$

This is the usual form for negative characteristics. To obtain this form *after* the division by 4, we need to write

$$7.8537 - 10 = 37.8537 - 40$$

Now dividing by 4 leaves -10. Thus

$$\log N = \frac{1}{4}(37.8537 - 40) = 9.4634 - 10$$

and $N = 0.2907$.

Logarithms are particularly useful when several operations are combined, as in the next example.

Example 4 Calculate

$$\frac{3.87\sqrt[3]{7{,}462}}{\sqrt{0.0321}}$$

Solution. Let

$$N = \frac{3.87\sqrt[3]{7{,}462}}{\sqrt{0.0321}}$$

Then $\log N = \log 3.87 + \frac{1}{3} \log 7{,}462 - \frac{1}{2} \log 0.0321$

$\log 3.87 = 0.5877$	
$\frac{1}{3} \log 7{,}462 = 1.2909$	$\log 7{,}462 = 3.8728$ by interpolation
1.8786	adding
$\frac{1}{2} \log 0.0321 = 9.2533 - 10$	$\log 0.0321 = 8.5065 - 10 = 18.5065 - 20$
2.6253	subtracting

Thus $\log N = 2.6253$ and $N = 422$ to three significant figures.

Exercises / Section 12.5

Use logarithms to perform the indicated calculations.

1. $(0.00962)(8{,}464)$
2. $(0.6416)(72.11)$
3. $\dfrac{79.60}{24.94}$
4. $\dfrac{6.477}{0.03420}$
5. $\sqrt[3]{8{,}642}$
6. $\sqrt[3]{0.0005128}$
7. $\sqrt[10]{0.000000176}$
8. $\sqrt[5]{6{,}416}$
9. $\dfrac{\sqrt[3]{46.71}}{\sqrt[4]{3.068}}$
10. $\dfrac{\sqrt[3]{1.36}}{\sqrt[5]{0.0846}}$
11. $47.62\sqrt{5.620}\;\sqrt[3]{7.301}$
12. $\sqrt[4]{(0.001230)(0.009647)}$
13. $\dfrac{2{,}765\sqrt[3]{0.0003620}}{5.247}$
14. $(0.007648\sqrt[3]{2313})^{1/2}$
15. $\dfrac{9.623\sqrt[3]{5.128}}{\sqrt{0.07225}}$
16. $\dfrac{0.0000123}{\sqrt[6]{0.000926}}$

12.6 Natural Logarithms, Change of Base, and Powers of e

The last two sections were devoted to common logarithms. In this section we shall briefly discuss logarithms to base e, which are called **natural logarithms** and play an important role in theoretical work.

We already saw that $e \approx 2.71828$ is used in the exponential form of a complex number. Although the origin of this number is clarified in the study of calculus, the examples and exercises that follow will indicate how such logarithms arise.

To avoid having to write the base each time, $\log_e x$ is denoted by "ln x."

Notation for natural logarithm:

$\log_e x$ is denoted by ln x

where $e \approx 2.71828$.

The natural logarithm of a number can be obtained from the common logarithm. This method is a special case of the more general problem of changing from one base to another, which is stated next.

Formula for changing the base of a logarithm:

$$\log_b x = \frac{\log_a x}{\log_a b} \qquad (12.14)$$

where a is any new base.

To derive this formula, let $M = \log_b x$, so that $b^M = x$. Take the logarithm to base a of both sides to obtain

$$\log_a b^M = \log_a x$$
$$M \log_a b = \log_a x \qquad \log_b x^r = r \log_b x$$
$$M = \frac{\log_a x}{\log_a b} \qquad \text{dividing by } \log_a b$$

Since $M = \log_b x$, we get

$$\log_b x = \frac{\log_a x}{\log_a b}$$

Example 1 Find ln 3.21 by means of common logarithms.

Solution. By formula (12.14)

$$\ln 3.21 = \log_e 3.21 = \frac{\log_{10} 3.21}{\log_{10} e}$$
$$= \frac{\log_{10} 3.21}{\log_{10} 2.718}$$
$$= \frac{0.5065}{0.4343} = 1.166 \quad \text{(using Table 2 of Appendix B)}$$

It is true in general that

$$\ln x = \frac{\log_{10} x}{\log_{10} e} = \frac{\log x}{0.4343}$$

Since $1/0.4343 = 2.3026$, we now have the following relationships:

$$\ln x = 2.3026 \log x \qquad (12.15)$$
$$\log x = 0.4343 \ln x \qquad (12.16)$$

CALCULATOR COMMENT   Natural logarithms can also be obtained by using a scientific calculator, as shown in the next example.

Example 2   Use a calculator to find   **a.** $\ln 4.7125$;   **b.** $\log_6 3.8926$.

Solution.   **a.** Enter 4.7125 and press $\boxed{\text{LN}}$ or $\boxed{\ln x}$ to get $\ln 4.7125 = 1.5502$.

b. By (12.14),

$$\log_6 3.8926 = \frac{\ln 3.8926}{\ln 6} = 0.7585$$

Example 3
Half-life
The *half-life* of a radioactive substance is the time required for half of a given amount to decay. If a radioactive substance has a half-life H, then the time taken for the initial amount N_0 to shrink to N is given by

$$t = \frac{H}{\ln 2} \ln \frac{N_0}{N} \qquad (12.17)$$

How long will **10** g of polonium, which has a half-life of **140** days, take to shrink to **7.0** g?

Solution. By formula (12.17),

$$t = \frac{140}{\ln 2} \ln \frac{10}{7.0}$$

The sequence is

140 $\boxed{\div}$ 2 $\boxed{\text{LN}}$ $\boxed{\times}$ $\boxed{(}$ 10 $\boxed{\div}$ 7.0 $\boxed{)}$ $\boxed{\text{LN}}$ $\boxed{=}$

The final result is 72 days.

CALCULATOR COMMENT   Because of the close relationship between logarithms and exponents, the exponential form $y = e^{bx}$ also occurs frequently in applications. The function values can be obtained with a scientific calculator by pressing $\boxed{e^x}$ or $\boxed{\text{INV}}$ $\boxed{\ln x}$.

12.6 NATURAL LOGARITHMS, CHANGE OF BASE, AND POWERS OF e

Example 4 Just as interest can be compounded quarterly or daily, it can also be compounded continuously. The formula is

$$P = P_0 e^{rt} \qquad (12.18)$$

where P_0 is the initial amount invested, t the time in years, and r the rate of interest. To what amount will $1,000 accumulate after 9.5 years at 10.25% interest compounded continuously?

Solution. Since $10.25\% = 0.1025$, we get by formula (12.18)

$$P = 1,000 e^{(0.1025)(9.5)} = \$2,647.86$$

The sequence is

$$0.1025 \; \boxed{\times} \; 9.5 \; \boxed{=} \; \boxed{e^x} \; \boxed{\times} \; 1,000 \; \boxed{=}$$

or

$$\boxed{(} \; 0.1025 \; \boxed{\times} \; 9.5 \; \boxed{)} \; \boxed{e^x} \; \boxed{\times} \; 1,000 \; \boxed{=}$$

Exercises / Section 12.6

In Exercises 1–4, use a calculator to find the natural logarithms to four decimal places.

1. ln 7.32
2. ln 27.601
3. ln 0.5173
4. ln 0.00123

In Exercises 5–10, use a calculator to find the indicated logarithms to four decimal places. (See Example 2.)

5. $\log_5 3.864$
6. $\log_3 27.164$
7. $\log_4 0.00713$
8. $\log_7 0.03926$
9. $\log_6 126.77$
10. $\log_2 0.000127$

In Exercises 11–20, use a calculator to find the exponential values to four decimal places.

11. e^3
12. $e^{1.5}$
13. e^{-2}
14. $e^{0.162}$
15. $e^{-0.92}$
16. $e^{2.68}$
17. $e^{0.013}$
18. $e^{-1.79}$
19. $e^{-0.03}$
20. $e^{-0.056}$

21. A radioactive isotope has a half-life of 50 years. Use formula (12.17) to find the time required for 50 g of the substance to shrink to 40 g.

22. A woman deposits $2,150 into a long-term savings account that pays 11% compounded continuously if left for 8 years. How much will she have at the end of this period? [Use formula (12.18).]

23. The present value is the amount of money that has to be invested in order to receive a specified amount at the end of a certain time period. The equation is given by $P_0 = P e^{-rt}$. If you wish to have $10,000 at the end of 6 years, how much will you have to invest at 10.2% compounded continuously? (Compare this problem with Example 4.)

24. The quantity $n! = n \cdot (n-1) \cdot (n-2) \cdots 2 \cdot 1$, read "$n$ factorial" can be approximated by Stirling's formula

$$n! \approx \sqrt{2\pi n} \left(\frac{n}{e}\right)^n$$

Approximate 30!.

25. For a certain gas, an enclosed volume of 10 cm³ is gradually increased to v. The average pressure (in atmospheres) is given by

$$P_{av} = \frac{4}{v-10} \ln \frac{v}{10}$$

Find the average pressure if the volume is increased to 50 cm³.

26. An electric circuit consists of a resistor of $R = 10.0 \, \Omega$ and an inductor of $L = 0.0123$ H. When the current is $I_0 = 3.50$ A, the current source is removed and the current dies out quickly according to the equation

$$I = I_0 e^{-Rt/L}$$

Determine the current after 1.00 millisecond.

27. The distance traveled by a motorboat after the engine is shut off is

$$x = \frac{1}{k} \ln(v_0 kt + 1)$$

where v_0 is the velocity when the motor is running and k a constant. If $k = 0.00300$, a typical value, determine how far a motorboat traveling at 20.0 ft/sec will continue in the first 15.0 sec after the motor is shut off.

28. An object of weight w moving around a circle of radius a is subject to a retarding force. If the initial velocity is v_0, then the angular displacement as a function of time t is given by

$$\theta = \frac{w}{kga} \ln\left(1 + \frac{kgv_0 t}{w}\right)$$

where k is a constant. Suppose $k = 0.0040$. Determine the angular displacement of an object after 8.0 sec if $v_0 = 25$ ft/sec, $w = 10$ lb, and $a = 50$ ft ($g = 32$ ft/sec²).

29. A certain satellite has a power supply whose output in watts is given by $P = 30.0 e^{-t/700.0}$, where t is the number of days that the battery has operated. (Note that $P = 30.0$ W when $t = 0$.) What is the output after 1 year?

30. The radioactive element strontium-90 decays according to the formula

$$N = N_0 e^{-0.025t}$$

where N_0 is the initial amount and t the time in years. If $N_0 = 100$ g, determine the amount left after
a. 1 year; **b.** 10 years; **c.** 100 years.

12.7 Special Exponential and Logarithmic Equations

One of the most interesting applications of logarithms is the solution of equations in which the unknown is an exponent. To solve such equations, we write the equation in the form $a^x = b$ and take the logarithm of both sides:

$$\log a^x = \log b$$

By the third law of logarithms,

$$x \log a = \log b$$

and

$$x = \frac{\log b}{\log a}$$

12.7 SPECIAL EXPONENTIAL AND LOGARITHMIC EQUATIONS

Example 1 Solve the equation $3^x = 12$.

Solution. First we take common logarithms of both sides:

$$\log 3^x = \log 12$$

By (12.13), the third property of logarithms, we get

$$x \log 3 = \log 12$$

Then, dividing by log 3,

$$x = \frac{\log 12}{\log 3} = 2.262$$

The sequence is

12 $\boxed{\text{LOG}}$ $\boxed{\div}$ 3 $\boxed{\text{LOG}}$ $\boxed{=}$

Display: 2.2618595

Natural logarithms can be used equally well, as shown in the next example.

Example 2 Solve the equation

$$3(4^{x+1}) = 20$$

Solution. The given equation is equivalent to

$$4^{x+1} = \frac{20}{3} \quad \text{dividing both sides by 3}$$

Taking natural logarithms, we get

$$\ln 4^{x+1} = \ln \frac{20}{3}$$

$$(x + 1) \ln 4 = \ln \frac{20}{3} \quad \log_b x^r = r \log_b x$$

$$x + 1 = \frac{\ln \frac{20}{3}}{\ln 4} \quad \text{dividing by ln 4}$$

$$x = -1 + \frac{\ln \frac{20}{3}}{\ln 4}$$

$$x = 0.3685$$

by using a calculator.

If two or more logarithms are contained in an equation, they must first be combined by using the properties of logarithms, as shown in the next example.

Example 3 Solve the equation

$$2 \ln x - \ln (x + 1) = 0$$

Solution. The left side must be written as a single logarithm:

$$2 \ln x - \ln (x + 1) = 0$$
$$\ln x^2 - \ln (x + 1) = 0 \qquad \log_b x^r = r \log_b x$$
$$\ln \frac{x^2}{x + 1} = \log_e \frac{x^2}{x + 1} = 0 \qquad \text{difference of two logs}$$

By using the definition of a logarithm, this equation can be written in exponential form. Since e is the base and 0 the exponent, we get

$$\frac{x^2}{x + 1} = e^0 = 1$$
$$x^2 = x + 1$$
$$x^2 - x - 1 = 0$$
$$x = \frac{1 \pm \sqrt{1 + 4}}{2} = \frac{1 \pm \sqrt{5}}{2}$$

Taking the positive root, we conclude that $x = \frac{1}{2}(1 + \sqrt{5})$ is the solution of the equation.

Logarithmic equations arise in various technical fields. Some of these applications are illustrated in the examples and exercises that follow.

Example 4 Stars are classified according to their visual brightness, which is called *magnitude*. Stars of the first magnitude are the brightest. Let b_n and b_m denote the measured (actual) brightness of two stars having magnitudes n and m, respectively. They are related by the formula

$$m - n = 2.5 \log \frac{b_n}{b_m} \qquad (12.19)$$

  a. If one star is 10 times brighter than another, what is the difference in their magnitudes?
  b. Sirius has magnitude 1.4. How much brighter is Sirius than a star of magnitude 12?

Solution.   a. Let $b_n/b_m = 10$. Then by formula (12.19), the difference in magnitudes is

$$m - n = 2.5 \log 10 = 2.5$$

b. Let $m = 12$ and $n = 1.4$. Now use formula (12.19) to find the ratio b_n/b_m:

$$12 - 1.4 = 2.5 \log \frac{b_n}{b_m}$$

$$10.6 = 2.5 \log \frac{b_n}{b_m}$$

$$\log_{10} \frac{b_n}{b_m} = \frac{10.6}{2.5} = 4.24$$

Since 10 is the base and 4.24 the exponent, we now get from the definition of logarithm

$$\frac{b_n}{b_m} = 10^{4.24} = 17{,}000$$

We conclude that Sirius is 17,000 times brighter than the other star.

Example 5 In certain chemical reactions, the amount of unconverted substance is given by $x = x_0 e^{kt}$, where x_0 is the initial amount, t the time, and k a negative constant that depends on the substance. Solve this equation for t.

Solution. From the given equation,

$$\frac{x}{x_0} = e^{kt}$$

Taking natural logarithms of both sides, we get

$$\ln \frac{x}{x_0} = \ln e^{kt}$$

Recalling that $\log_b b = 1$, it follows that

$$\ln \frac{x}{x_0} = kt \ln e$$

$$\ln \frac{x}{x_0} = kt \cdot 1$$

$$t = \frac{1}{k} \ln \frac{x}{x_0}$$

Exercises / Section 12.7

In Exercises 1–24, solve each equation for x.

1. $2^x = 5$
2. $3^x = 10$
3. $4^x = 20$
4. $6^x = 5$
5. $3^x = \frac{1}{2}$
6. $3^{x+2} = \frac{1}{4}$

7. $5^{x-1} = 2$

8. $2(3^x) = 7$

9. $2(6^x) = 21$

10. $3(7^{x-2}) = 59$

11. $9^{2x+3} = 5$

12. $15^{3x-1} = 6$

13. $2 \log_3 x = 4$

14. $\log_5 (x + 2) = 2$

15. $\log_2 x = 0.64590$

16. $\log_{3/2} x = 5$

17. $\log_2 x + \log_2 (x - 3) = 2$

18. $2 \log x - \log (x + 2) = 0$

19. $\frac{1}{2} \ln x + \ln 2 = 1$

20. $2 \log x = 1 - \log 5$

21. $\log x + \log x^2 = \log 8$

22. $2 \ln x - \ln (x + 3) = 0$

23. $2 \ln x - 2 \ln (x - 1) = 0$

24. $2 \ln x - 2 \ln (x + 1) = 0$

In Exercises 25–28, solve for the indicated letter.

25. $i = \frac{E}{R} e^{-t/RC}$, for t

26. $\log \frac{I}{I_0} = -\beta x$, for I

27. $Q = P_1 V_1 \log \frac{V_2}{V_1}$, for V_2

28. $T = T_2 + (T_1 - T_2)e^{-at}$, for a

29. What is the difference in magnitudes if one star is brighter than another by **a.** 20 times? **b.** 100 times? (Refer to Example 4.)

30. How much brighter is a star of magnitude 1 than a star of magnitude 10? (Refer to Example 4.)

31. The subjective impression of sound, measured in decibels, is related to the intensity of sound (measured in watts per square meter) by the formula

$$\beta = 10 \log \frac{I}{I_0}$$

where I_0 is an arbitrary reference level corresponding roughly to the faintest sound that can be heard (10^{-12} W/m²). Rustling leaves have an intensity of 10^{-11} W/m². What is the loudness in decibels?

32. An elevated train has a loudness of 90 decibels. How many times more intense is the sound of an elevated train than the sound of rustling leaves (10 decibels)? (Refer to Exercise 31.)

33. Determine the intensity level of riveting, which has a loudness of 95 decibels. (Refer to Exercise 31.)

34. A certain body cools according to the equation

$$T = 20 + 60e^{-t/100}$$

where T is the temperature in degrees Celsius and t the time in minutes. Find an expression for t as a function of T.

35. Another logarithmic scale, used for measuring the magnitude of an earthquake, is the Richter scale

$$R = \log \frac{I}{I_0}$$

where I_0 is some arbitrary reference level. How much more intense was the famous San Francisco earthquake of 1906, which measured 8.25 on the Richter scale, than a minor tremor measuring 3.0 on the Richter scale?

36. How much more intense was the 1964 Alaska earthquake, which registered 7.5 on the Richter scale, than an earthquake measuring only 5.0 on the Richter scale? (Refer to Exercise 35.)

37. A certain bacteria population doubles every hour. If there are 1,000 bacteria initially, then the number of bacteria after t hours is given by $N = 1,000(2^t)$.
   a. Determine the bacteria population after 3.5 hr.
   b. Determine when the population will reach 20,000.

38. In a body of water, the light intensity I diminishes with the depth. If I_0 is the intensity level at the surface, then at depth d (in feet)

$$I = I_0 e^{-kd}$$

where k is a constant. For a chlorinated swimming pool, a typical value for k is 0.0080. At what depth is the intensity 95% of the intensity at the surface? (*Hint:* Find d such that $I = 0.95 I_0$.)

39. Logarithms may be used to determine the age of a fossil in a procedure called *carbon dating*. This method is based on the fact that carbon-14 is found in all organisms in a fixed percentage. When an organism dies, the carbon-14 decays according to the formula

$$P = (P_0)(2^{-t/5,580})$$

where P_0 is the initial amount. Suppose a fossil contains only 25% of the original amount. How long has the organism been dead? (Find t such that $P = 0.25 P_0$.)

40. If a fossil contains 42% of the original amount of carbon-14, how long ago did the organism die? (Refer to Exercise 39.)

41. The decay of uranium-238, which has a half-life of 4.5 billion years, is

$$N = N_0 e^{-(1.54 \times 10^{-10})t}$$

Determine how long 0.1% of the initial amount N_0 takes to decay. (*Hint:* If 0.1% has decayed, then $0.999 N_0$ remains.)

42. From the formula $\text{pH} = -\log_{10} [\text{H}^+]$, find the concentration of hydrogen ions (in moles per liter) of milk whose hydrogen potential pH is 6.39. (Refer to Exercise 21 of Section 12.4.)

43. The atmospheric pressure (in millimeters of mercury) is given by

$$P = 760 e^{-0.00013h}$$

where h is the height (in meters) above sea level. At what height above sea level is the atmospheric pressure equal to 605 mm of mercury?

12.8 Graphs on Logarithmic Paper (Optional)

So far all the graphs we have studied were drawn using the **linear scale**, on which the integers are all equally spaced. For some equations it is more convenient to use graph paper on which one axis (usually the y-axis) or both axes use a **logarithmic scale**. To see what is meant by a logarithmic scale, consider the following table:

Logarithmic scale

x:	1	2	3	4	5	6	7	8	9	10	20	30	100
$\log x$:	0	0.301	0.477	0.602	0.699	0.778	0.845	0.903	0.954	1	1.301	1.477	2

386 CHAPTER 12 LOGARITHMIC AND EXPONENTIAL FUNCTIONS

Figure 12.4

The resulting scale, using log x instead of x, is shown in Figure 12.4. Since the logarithmic scale has cycles, the first number in each cycle must be labeled as a power of 10: 0.01, 0.1, 1, 10, 100, 1,000, and so on. (See Figure 12.5.)

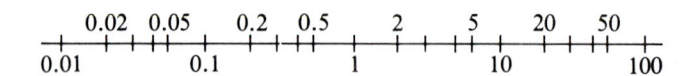

Figure 12.5

Log paper

If one scale is linear and one logarithmic (usually the y-axis), it is called *semilog paper*. If both axes have a logarithmic scale, it is called *log-log paper*. (Graph paper using logarithmic scales is available in many bookstores.) When the y-variable has a range that is much larger than that of the x-variable, it is often advantageous to use semilog paper. Log-log paper is frequently used if both variables have large scales.

We shall concentrate mostly on the important special case where the graph to be drawn appears as a straight line on semilog or log-log paper. To this end, let us recall from Chapter 3 that the graph of

$$ax + by = c \tag{12.20}$$

is a straight line. This equation can also be written in the form

$$y = mx + b \tag{12.21}$$

To graph an equation on logarithmic paper, we take the logarithm of both sides and graph the resulting equation. Consider the next example.

Example 1 Graph $y = 5^x$ on semilog paper.

Solution. To graph an equation on either type of log paper, we first take the logarithm of both sides of the equation. Thus, in this case,

$$\log y = x \log 5$$

If we let $u = \log y$, the equation becomes

$$u = (\log 5)x$$

which has the same form as equation (12.21). The equation will therefore appear as a line when graphed on semilog paper. (The reason semilog paper is used is that u is a logarithm but x is not.) Next, note that when $x = 0$, then $y = 1$. To obtain the graph, we need only one other point, but it is best to include a check point. So if $x = 3$, then $y = 125$, and if $x = 4$, $y = 625$. The graph is shown in Figure 12.6.

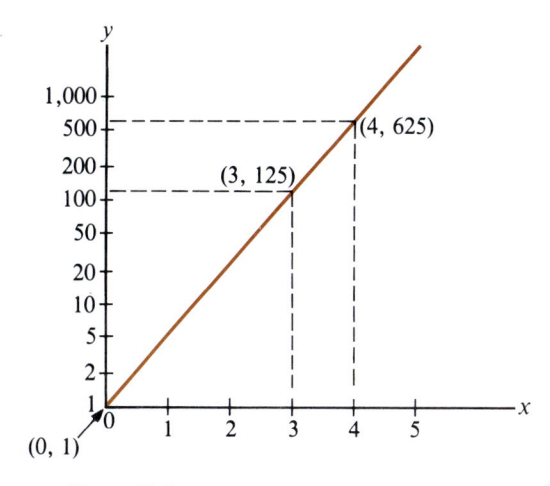

Figure 12.6

Example 2 Graph $y = 2(3^{-x})$ on semilog paper.

Solution. Taking logarithms again, we get

$$\log y = \log 2 - (\log 3)x$$

If we let $u = \log y$, then the resulting equation

$$u = -(\log 3)x + \log 2$$

has the same form as equation (12.21). The graph, which is a line, is constructed from the following short table of values:

x:	-1	0	1	2	3	4
y:	6	2	0.67	0.22	0.074	0.025

See Figure 12.7 on page 388.

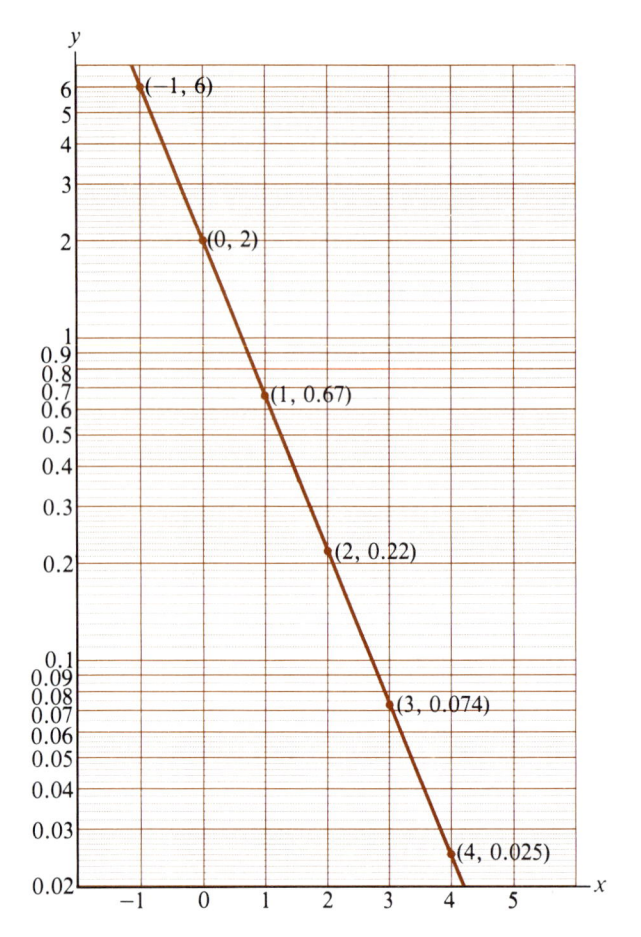

388 CHAPTER 12 LOGARITHMIC AND EXPONENTIAL FUNCTIONS

Figure 12.7

Example 3 Graph $y^{1/2} = 3x^{1/3}$ on log-log paper.

Solution. From the given equation

$$\frac{1}{2} \log y = \log 3 + \frac{1}{3} \log x$$

We now let $u = \log y$ and $v = \log x$. The equation becomes

$$\frac{1}{2} u = \frac{1}{3} v + \log 3$$

which has the form of the equation of a line, equation (12.20). Since both variables are logarithms, log-log paper is appropriate. The graph, shown in

Figure 12.8, is constructed from the following short table:

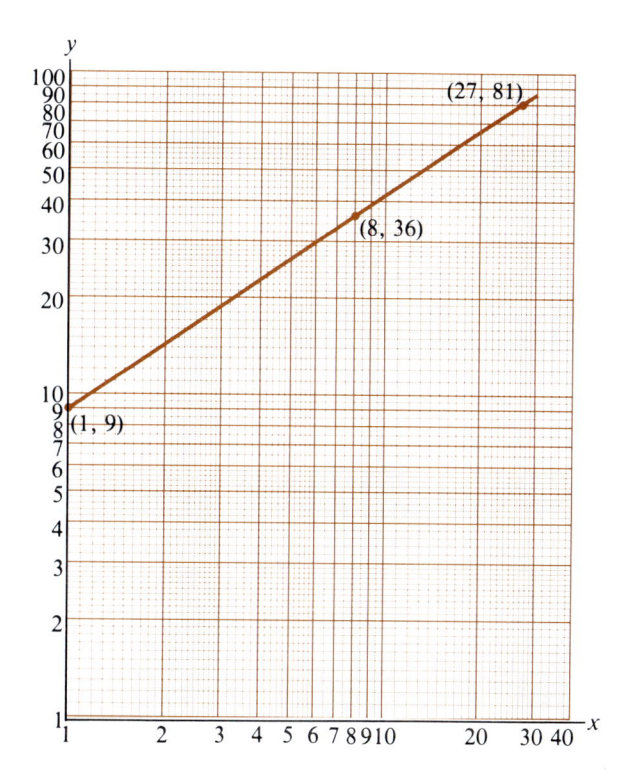

x:	1	8	27
y:	9	36	81

Figure 12.8

Example 4 Determine the type of graph paper for which the following equations are straight lines:

a. $x^4 y^3 = 1$ **b.** $3^x y = 2$

Solution.   a. Taking logarithms, we get

$$4 \log x + 3 \log y = 0$$

Letting $u = \log y$ and $v = \log x$, we get $4v + 3u = 0$, which has the form of a line. Since both u and v are logarithms, log-log paper is required.

b. From $3^x y = 2$, we get

$$x \log 3 + \log y = \log 2$$

Letting $u = \log y$, we get $u + (\log 3)x = \log 2$, which fits form (12.20). Since u is a logarithm and x is not, semilog paper is required.

Occasionally experimental data are plotted on logarithmic paper to give a clearer picture of relationships. (See Exercise 13.)

Exercises / Section 12.8

In Exercises 1–4, graph each equation on semilog paper.

1. $y = 3^x$ **2.** $y = 3(2^x)$ **3.** $y = 3(4^{-x})$ **4.** $y = 2^{-x}$

In Exercises 5–8, graph each equation on log-log paper.

5. $y = 3x^4$ **6.** $y = 2x^{2/3}$ **7.** $y^3 = 3x^{2/3}$ **8.** $x^2 y^3 = 1$

In Exercises 9–12, determine the type of graph paper for which the graph of each given equation will be a straight line; sketch the graph.

9. $y(3^x) = 1$ **10.** $y^2 = 2^x$ **11.** $\sqrt{xy^3} = 1$ **12.** $x\sqrt{y} = 2$

13. Experimenters determined the relationship between the pressure P (in atmospheres) and the volume V (in cubic inches) of a certain gas. The following results were obtained:

P:	1	3	5	7
V:	2.69	1.23	0.85	0.67

Plot the resulting curve on log-log paper.

14. The period P of a pendulum is directly proportional to the square root of its length L, a fact first discovered by Galileo. An experiment to test this relationship yielded the following results:

L (feet):	1	2	3	4	5
P (seconds):	1.13	1.59	1.95	2.24	2.50

Plot the resulting curve on log-log paper. Do the data confirm the relationship $P = k\sqrt{L}$?

Review Exercises / Chapter 12

In Exercises 1–4, change each exponential form to the logarithmic form.

1. $3^{-3} = \dfrac{1}{27}$

2. $\left(\dfrac{1}{2}\right)^2 = \dfrac{1}{4}$

3. $7^{-2} = \dfrac{1}{49}$

4. $a^3 = b$

In Exercises 5–8, change each logarithmic form to the exponential form.

5. $\log_4 4 = 1$

6. $\log_3 \dfrac{1}{9} = -2$

7. $\log_{1/4} \dfrac{1}{16} = 2$

8. $\log_{1/2} 2 = -1$

In Exercises 9–16, find the value of the unknown.

9. $\log_5 x = -2$

10. $\log_{1/2} x = -3$

11. $\log_2 \dfrac{1}{16} = a$

12. $\log_b 36 = 2$

13. $\log_{3/2} x = 3$

14. $\log_b 9 = -2$

15. $\log_b \dfrac{1}{32} = -5$

16. $\log_8 \dfrac{1}{2} = a$

In Exercises 17–22, sketch each function.

17. $y = 3^x$

18. $y = e^x$

19. $y = e^{-x}$

20. $y = \log_5 x$

21. $y = \log_{1/2} x$

22. $y = \left(\dfrac{1}{3}\right)^x$

In Exercises 23–36, write the given expressions as sums, differences, or multiples of logarithms. Whenever possible, simplify the result by using $\log_b 1 = 0$ and $\log_b b = 1$.

23. $\log_2 32$

24. $\log_5 35$

25. $\log_{10} 1{,}000$

26. $\log_3 81$

27. $\log_5 125x^3$

28. $\log_b b^3$

29. $\log_6 \sqrt{6x}$

30. $\log_3 \dfrac{1}{3b}$

31. $\log_2 \dfrac{1}{16a^2}$

32. $\log_{10} \dfrac{1}{\sqrt{10a}}$

33. $\ln \dfrac{16}{e^2}$

34. $\ln \sqrt{\pi e}$

35. $\ln \dfrac{1}{\sqrt[3]{e}}$

36. $\log_{10} 0.004$

In Exercises 37–40, write each expression as a single logarithm.

37. $\log_5 26 - \log_5 13$

38. $\ln a + \frac{1}{2} \ln b$

39. $\frac{1}{2} \log_6 x - 2 \log_6 b$

40. $\frac{1}{3} \ln 27$

In Exercises 41–46, use Table 2 of Appendix B to find the common logarithm. Check with a calculator.

41. $\log 2.58$

42. $\log 2{,}580$

43. $\log 9{,}142$

44. $\log 73.28$

45. $\log 0.005765$

46. $\log 0.0004517$

In Exercises 47–50, use Table 2 of Appendix B to find the antilogarithm. Check with a calculator.

47. 2.6517

48. 3.8324

49. $9.5327 - 10$

50. $7.4385 - 10$

In Exercises 51–66, use a calculator to find the values to four decimal places.

51. $\log 2.64$

52. $\ln 2.64$

53. $\ln 37.42$

54. $\log 37.42$

55. $\ln 10.74$

56. $\ln 8.03$

57. $e^{-1.2}$

58. $e^{-2.3}$

59. $e^{2.7}$

60. $e^{0.365}$

61. $e^{-1.6}$

62. $e^{-0.342}$

63. $\log_5 3.92$

64. $\log_2 35.6$

65. $\log_7 0.728$

66. $\log_6 0.00312$

In Exercises 67–71, solve each equation for x.

67. $2^x = 6$

68. $3^{x-1} = \frac{1}{2}$

69. $2(3^{-x}) = 5$

70. $\log x + 2 \log x = \log 27$

71. $\log_3 x + \log_3 (x + 1) = 1$

72. If $L = R \ln (V_1/V_2)$, show that $V_1 = V_2 e^{L/R}$.

73. Recall that the formula relating the loudness of a sound in decibels to its intensity is

$$\beta = 10 \log \frac{I}{I_0}$$

A quiet automobile has a loudness of 50 decibels, and busy street traffic 70 decibels. How many times more intense is the latter?

74. If one star is 30 times brighter than another, what is their difference in magnitude? (Refer to formula (12.19).)

75. Newton's law of cooling states that the rate of change of the temperature of a body is directly proportional to the difference of the temperature of the body and the temperature of the surrounding medium. If T_0 is the initial temperature and M_T the temperature of the medium, then the temperature of the body as a function of time is

$$T = M_T + (T_0 - M_T)e^{-kt}$$

where k is a constant. For a certain body, $M_T = 10°C$, $T_0 = 50°C$, and $k = 0.13$. Find the temperature of the body after 5.0 min.

76. The current i as a function of time is given by $i = 3.0e^{-2t}$. Graph i (in amperes) as a function of time (in seconds).

77. Solve the equation in Exercise 76 for t.

78. Determine the present value of $5,000 if the money is to be invested for 7.5 years at 9.75% compounded continuously.

79. If P dollars is compounded n times a year, then the value of the investment at the end of t years is $A = P(1 + r/n)^{nt}$, where r is the interest rate. Determine how long an initial investment will take to double if invested at 10% compounded quarterly.

80. A body cools according to the formula $T = 10(1 + e^{-0.11t})$, where T is in degrees Celsius and t in minutes. Since $T = 20$ when $t = 0$, the initial temperature is 20°C. Determine how long the temperature will take to drop to 18°C.

81. A bacteria culture grows according to $N = 10,000(1.75)^t$, where t is measured in hours. Find how long the bacteria population will take to reach 50,000.

82. Carbon-14 decays according to the formula

$$P = P_0(2^{-t/5,580})$$

If a fossil contains 30% of the original carbon, how long has the organism been dead?

83. From the formula $pH = -\log_{10} [H^+]$, determine the concentration of hydrogen ions (in moles per liter) of wine grapes (pH = 3.15).

84. The atmospheric pressure (in millimeters of mercury) is given by

$$P = 760e^{-0.00013h}$$

where h is the height (in meters) above sea level. At what height is the atmospheric pressure equal to 550 mm of mercury?

Cumulative Review Exercises / Chapters 10–12

In Exercises 1–6, simplify the given expressions and write the results without zero or negative exponents.

1. $\dfrac{2^0 x^2 y^{-1/2}}{x^{-2} y^{3/2}}$

2. $\dfrac{3^{-1} g_1^{-2} g_2^2 g_3^{-5}}{6 g_1^4 g_2^{-3} g_3^{-3}}$

3. $\dfrac{a^{-1} - b^{-1}}{a^{-1}}$

4. $(64 L^{12} R^{-9})^{-1/3}$

5. $\dfrac{27 x^0 x^{-1/2} y^{1/5}}{9 x^{1/3} y^{-1/15}}$

6. $(a-2)(a+3)^{-1/2} - (a+3)^{1/2}$

7. Rationalize the denominator:

$$\sqrt[4]{\dfrac{1}{ab^3}}$$

8. Simplify and rationalize the denominator:

$$\sqrt[3]{\dfrac{4^{-1} G^{-1} H^{-2}}{G^3 H^3}}$$

9. Rationalize the denominator:

$$\dfrac{\sqrt{2} - 1}{\sqrt{5} - \sqrt{2}}$$

10. Combine the following complex numbers:

$$(-2 + 3j) + (6 - 2j) - (2 - j)$$

11. Perform the indicated multiplication:

$$(2 - 3j)(-1 + j)$$

12. Divide $1 - j$ by $2 + 3j$.

13. Change the complex number $-2\sqrt{3} - 2j$ to

   a. polar form   b. exponential form.

14. Divide 3 cis 10° by 6 cis 150°.

15. Find the four fourth roots of $-1 + \sqrt{3} j$. (Leave answer in polar form.)

16. Combine into a single logarithm:

$$\tfrac{1}{2} \log_6 P + \log_6 Q - 2 \log_6 R$$

17. Solve for x:

$$3^x = 7$$

18. Solve the formula

$$S = S_1 - S_2 e^{-kt}$$

   for t.

19. Recall that the equation relating the object distance q and the image distance p to the focal length f of a thin lens is

$$f^{-1} = p^{-1} + q^{-1}.$$

   Solve this equation for p.

394

20. Use the formula $P = S(1 + i)^{-n}$ (for obtaining the present value) to find the amount of money that has to be invested to yield $3000 in 6 years at 8.5% interest compounded annually.

21. The amount N of a certain radioactive isotope varies with time according to the relation

$$\log_{10} N = -0.4t, \quad t \geq 0.$$

Write this formula in exponential form.

22. The power supply P (in watts) of a certain satellite after t days is $P = 20.0(e^{-0.00341t})$. Note that $P = 20.0$ W when $t = 0$. How many days will it take for the power supply to diminish to 10.0 W?

CHAPTER 13

More on Equations and Systems of Equations

Objectives Upon completion of this chapter, you should be able to:
1. Solve a given system of equations graphically.
2. Solve a given system of equations algebraically by:
   a. Addition or subtraction.
   b. Substitution.
3. Solve equations in quadratic form.
4. Solve equations containing radicals.

13.1 Solving Systems of Equations Graphically

The main purpose of this section is to learn how to solve certain systems of equations graphically. Recall that we have already studied how to solve systems of two linear equations in Chapter 3. In this section we shall move on to systems of two quadratic equations, as well as a few other types. However, first we need to look at the graphs of second-order equations in two variables x and y, which are defined next.

> **General quadratic equation in x and y:**
> $$Ax^2 + Bxy + Cy^2 + Dx + Ey + F = 0 \qquad (13.1)$$
> where A, B, and C are not all zero.

Conic sections

The graphs of equations of form (13.1) are collectively known as **conic sections,** which include the **circle, ellipse, parabola,** and **hyperbola.** Our purpose here is to classify the different conic sections according to their equations; we shall take up the details in Chapter 19. As a starting point, consider the example below of a quadratic equation.

Example 1 Graph the equation $x^2 + y^2 = 25$.

Solution. As usual, we shall obtain the graph by plotting enough points to get a smooth curve. So let us solve the given equation for y in terms of x:

$$x^2 + y^2 = 25$$
$$y^2 = 25 - x^2$$
$$y = \pm\sqrt{25 - x^2}$$

Now observe that whenever $x > 5$ or $x < -5$, the radicand is negative and the corresponding y-values imaginary. Since complex numbers cannot be plotted on the Cartesian coordinate system, we must exclude them here. The x-values must therefore lie between -5 and $+5$. Observe also that for both $x = 4$ and $x = -4$, we get $y = \pm 3$, written compactly as $x = \pm 4$, $y = \pm 3$. Some of the other values are listed in the following table:

x:	0	± 1	± 2	± 3	± 4	± 5
y:	± 5	± 4.9	± 4.6	± 4	± 3	0

The graph, shown in Figure 13.1, is a *circle*.

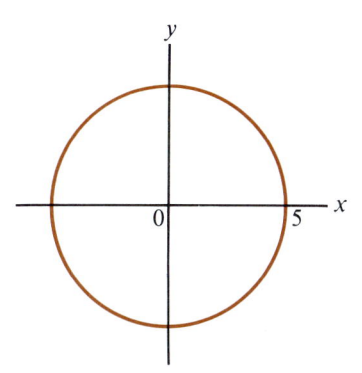

Figure 13.1

Example 1 has shown that $x^2 + y^2 = 25$ is a circle. The general case is given next.

> A **circle** of radius r centered at the origin is represented by the equation
> $$x^2 + y^2 = r^2 \qquad (13.2)$$

If the coefficients of x^2 and y^2 (in the equation of the circle) are positive but have different values, we get a curve called an *ellipse*.

Example 2 Graph the equation $x^2 + 2y^2 = 16$.

Solution. As in the case of the circle in Example 1, we solve the equation for y in terms of x and construct a table of values:

$$x^2 + 2y^2 = 16$$
$$2y^2 = 16 - x^2$$
$$y^2 = \frac{16 - x^2}{2}$$
$$y = \pm \sqrt{\frac{16 - x^2}{2}}$$

x:	0	± 1	± 2	± 3	± 4
y:	± 2.8	± 2.7	± 2.4	± 1.9	0

Note that the values of x numerically greater than 4 lead to imaginary values for y and must be excluded. The graph, shown in Figure 13.2, is an ellipse.

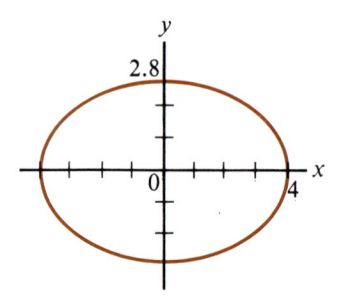

Figure 13.2

> **Ellipse:** An equation of the form
> $$ax^2 + by^2 = c, \quad a \neq b \quad \text{and} \quad a, b, c > 0 \qquad (13.3)$$
> represents an ellipse centered at the origin.

13.1 SOLVING SYSTEMS OF EQUATIONS GRAPHICALLY

The next example illustrates the graph of another conic section, called a *parabola*.

Example 3 Graph the equation $y = 4x - 2x^2$.

Solution. For this equation, y is defined for all values of x. The graph, shown in Figure 13.3, is a parabola.

x:	-2	-1	0	1	2	3	4
y:	-16	-6	0	2	0	-6	-16

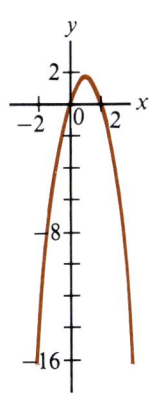

Figure 13.3

Parabola: An equation of the form
$$y = ax^2 + bx + c \quad \text{or} \quad x = ay^2 + by + c, \quad a \neq 0 \tag{13.4}$$
represents a parabola.

Our final example of a conic section is a *hyperbola*.

Example 4 Graph the equation $2x^2 - y^2 = 8$.

Solution. Once again we solve the equation for y in terms of x:
$$-y^2 = -2x^2 + 8$$
$$y^2 = 2x^2 - 8$$
$$y = \pm\sqrt{2x^2 - 8}$$

To avoid imaginary values, we must have $x \geq 2$ or $x \leq -2$.

x:	± 2	± 2.1	± 2.5	± 3	± 4	± 5
y:	0	± 0.9	± 2.1	± 3.2	± 4.9	± 6.5

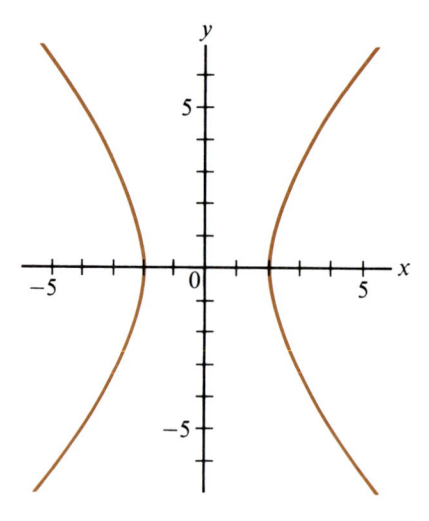

[Figure: hyperbola graph]

Figure 13.4

The curve, shown in Figure 13.4, is a hyperbola.

Hyperbola: An equation of the form
$$ax^2 - by^2 = c, \quad ay^2 - bx^2 = c, \quad \text{or} \quad xy = k \quad (a,b > 0) \quad (13.5)$$
represents a hyperbola.

Solving Systems of Equations Graphically

Our brief discussion of conic sections has provided enough information so that we can solve certain systems of equations graphically. To solve such systems, we graph the given equations and estimate the points of intersection.

Example 5 Graphically solve the system

$$xy = 4$$
$$y = 4x^2$$

Solution. By form (13.5) the first equation is a hyperbola, plotted from the following table of values:

x:	-4	-3	-2	-1	0	1	2	3	4
y:	-1	$-\frac{4}{3}$	-2	-4	undefined	4	2	$\frac{4}{3}$	1

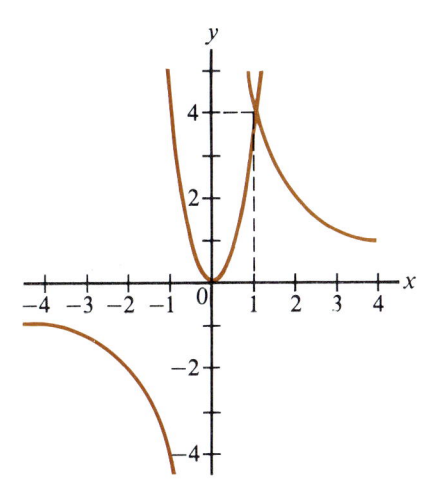

Figure 13.5

(See Figure 13.5.) The second equation is a parabola by form (13.4).

x:	-2	-1	$-\frac{1}{2}$	0	$\frac{1}{2}$	1	2
y:	16	4	1	0	1	4	16

According to Figure 13.5, the graphs intersect at (1.0, 4.0). So the approximate solution of the system is $x = 1.0$, $y = 4.0$.

Example 6 Graphically solve the system

$$x^2 - y^2 = 4$$
$$x^2 + 2y^2 = 64$$

Solution. The hyperbola $x^2 - y^2 = 4$ (Figure 13.6 on page 402) is plotted from the following table of values:

x:	2	2.5	3	4	5
y:	0	± 1.5	± 2.2	± 3.5	± 4.6

The second equation represents an ellipse by form (13.3).

x:	0	± 2	± 4	± 6	± 8
y:	± 5.7	± 5.5	± 4.9	± 3.7	0

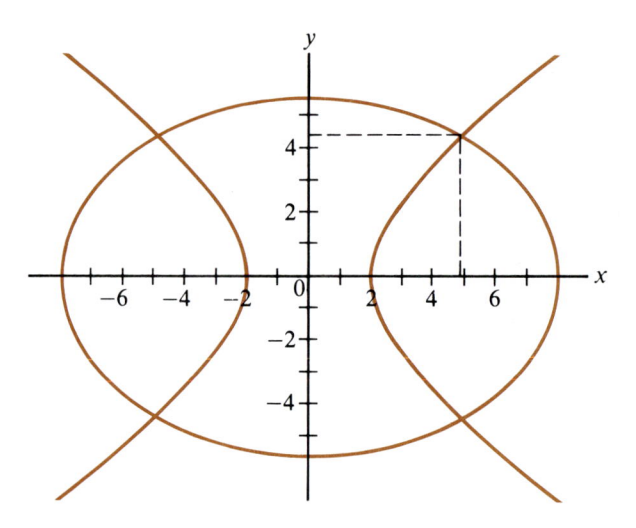

Figure 13.6

According to Figure 13.6, there are four points of intersection and hence four solutions to the system: (4.9, 4.5), (4.9, −4.5), (−4.9, 4.5), and (−4.9, −4.5).

After discussing graphical methods of solution in Chapter 3, we found that algebraic methods were far more convenient. For linear equations and, to a large extent, quadratic equations, this is true, as we shall see in the next section. However, many systems have no algebraic methods of solution. An example of such a system is given next.

Example 7 Solve the following system graphically:

$$y - 4x = 4$$
$$y = e^{-x}$$

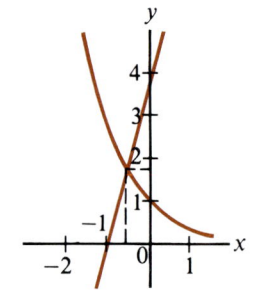

Solution. The equation $y - 4x = 4$ is a line passing through $(-1, 0)$ and $(0, 4)$, as shown in Figure 13.7. The exponential function is plotted from the following table of values:

x:	−2	−1	−0.8	−0.6	−0.4	0	1
y:	7.4	2.7	2.2	1.8	1.5	1	0.4

Figure 13.7

From the graphs, the approximate solution is $(-0.6, 1.8)$.

Exercises / Section 13.1

Solve each system of equations graphically. Estimate the solutions to the nearest tenth of a unit.

1. $x^2 + y^2 = 8$
   $x - y = 0$
2. $2x + y = 3$
   $y = x^2$
3. $y^2 - 3x + 1 = 0$
   $3x - 2y = 0$
4. $x - 2y = 1$
   $y^2 - x^2 = 4$
5. $2y - x = 2$
   $x^2 - y^2 = 2$
6. $2x^2 + y^2 = 8$
   $x - y = 1$
7. $xy = 6$
   $y^2 = x$
8. $x^2 - 2y^2 = 4$
   $xy = 1$
9. $3x + 2y = 6$
   $xy = 1$
10. $2x^2 + y^2 = 32$
    $y = x^2$
11. $x^2 + 4y^2 = 16$
    $x^2 + y^2 = 9$
12. $x^2 - 2y^2 = 4$
    $2x^2 + y^2 = 18$
13. $3y^2 - 5x^2 = 12$
    $2y + x^2 = 0$
14. $2x^2 - 4y^2 = 9$
    $4y^2 - x^2 = 12$
15. $y^2 = x$
    $y = e^{-x}$
16. $y = \sin x$
    $y = x^2$
17. $y = \log_3 x$
    $y = 2^{-x}$
18. $y = \cos x$
    $y = \sqrt{x}$

13.2 Solving Systems of Equations Algebraically

In the previous section we saw that graphical methods can be used to obtain approximate solutions to systems of equations. However, if both equations are algebraic, the exact solution can often be obtained algebraically. In fact, the methods for solving systems of quadratic equations are identical to the methods used in Chapter 3 for solving systems of linear equations, namely, the method of **substitution** and the method of **addition or subtraction**.

To solve the system

$$y = 4 - x^2$$
$$y - x = 1$$

Substitution by the *method of substitution,* we solve the second equation for y in terms of x to get $y = x + 1$ and then substitute this expression for y in the first equation. Thus

$$y = 4 - x^2 \qquad \text{first equation}$$
$$x + 1 = 4 - x^2 \qquad \text{substituting } x + 1 \text{ for } y$$
$$x^2 + x - 3 = 0$$
$$x = \frac{-1 \pm \sqrt{13}}{2}$$

From $y = x + 1$, we now find the corresponding values for y:

$$x = \frac{-1 + \sqrt{13}}{2}: \qquad x = \frac{-1 - \sqrt{13}}{2}:$$
$$y = \frac{-1 + \sqrt{13}}{2} + 1 \qquad y = \frac{-1 - \sqrt{13}}{2} + 1$$

$$= \frac{-1 + \sqrt{13}}{2} + \frac{2}{2} \qquad\qquad = \frac{-1 - \sqrt{13}}{2} + \frac{2}{2}$$

$$= \frac{1 + \sqrt{13}}{2} \qquad\qquad\qquad = \frac{1 - \sqrt{13}}{2}$$

So the solution is given by

$$\left(\frac{-1 + \sqrt{13}}{2}, \frac{1 + \sqrt{13}}{2}\right), \ \left(\frac{-1 - \sqrt{13}}{2}, \frac{1 - \sqrt{13}}{2}\right)$$

As a check, we substitute $x = \frac{1}{2}(-1 + \sqrt{13})$ in the equation $y = 4 - x^2$:

$$y = 4 - \left(\frac{-1 + \sqrt{13}}{2}\right)^2$$

$$= 4 - \frac{1 - 2\sqrt{13} + 13}{4}$$

$$= \frac{16}{4} - \frac{14 - 2\sqrt{13}}{4}$$

$$= \frac{2 + 2\sqrt{13}}{4}$$

$$= \frac{1 + \sqrt{13}}{2} \quad \text{(solution checks)}$$

The other pair is checked similarly. (Note that the solution gives us the coordinates of the intersection of a parabola and a straight line.)

Example 1 Solve the system

$$2x - y = 2$$
$$xy = 4$$

Solution. As before, we shall use the method of substitution. From the second equation, $y = 4/x$. Substituting into the first equation, we get

$$2x - y = 2 \qquad\qquad \text{first equation}$$
$$2x - \frac{4}{x} = 2 \qquad\qquad \text{substituting } \frac{4}{x} \text{ for } y$$
$$2x^2 - 4 = 2x \qquad\qquad \text{clearing fractions}$$
$$2x^2 - 2x - 4 = 0 \qquad\qquad \text{collecting terms on left side}$$
$$x^2 - x - 2 = 0 \qquad\qquad \text{dividing by 2}$$
$$(x - 2)(x + 1) = 0 \qquad\qquad \text{factoring}$$
$$x = 2, -1$$

From $xy = 4$, we find $y = 2$ and $y = -4$, respectively. So the solution is given by $(2, 2)$ and $(-1, -4)$.

As a check, let us substitute the values obtained into the first equation:

$$2(2) - 2 = 2; \quad 2(-1) - (-4) = 2$$

The solution checks. (The solution yields the coordinates of the points of intersection of a hyperbola and a straight line.)

Addition or subtraction Sometimes one of the unknowns can be eliminated by *addition or subtraction*, as shown in the next example.

Example 2 Solve the system

$$2x^2 - 3y^2 = 19$$
$$3x^2 + y^2 = 1$$

Solution. We multiply both sides of the second equation by 3 and add:

$$\begin{aligned} 2x^2 - 3y^2 &= 19 \\ 9x^2 + 3y^2 &= 3 \quad &&3(3x^2 + y^2) = 3(1) \\ \hline 11x^2 &= 22 \quad &&\text{adding} \\ x^2 &= 2 \quad &&\text{dividing by 11} \\ x &= \pm\sqrt{2} \end{aligned}$$

To find the corresponding y-values, both x-values may be substituted simultaneously since the exponents are even. Using the second equation, we get

$$3(\pm\sqrt{2})^2 + y^2 = 1$$
$$6 + y^2 = 1$$
$$y^2 = -5$$
$$y = \pm\sqrt{5}j$$

It is important to note that since the y-values were found at once, both $\sqrt{5}j$ and $-\sqrt{5}j$ correspond to each x-value, thereby yielding four sets of solutions:

$$(\sqrt{2}, \sqrt{5}j), \quad (\sqrt{2}, -\sqrt{5}j), \quad (-\sqrt{2}, \sqrt{5}j), \quad (-\sqrt{2}, -\sqrt{5}j)$$

Since all members of this solution set involve pure imaginary numbers, the graphs of these equations, a hyperbola and an ellipse, do not intersect at any point. However, algebraically speaking, all four pairs satisfy the equation.

Example 3 Recall that two resistances R_1 and R_2 in series have a total resistance of $R_1 + R_2$ (in ohms). If the same resistances are connected in parallel, then the combined resistance R_T (in ohms) is such that $1/R_T = 1/R_1 + 1/R_2$. Suppose $R_1 + R_2 = 15\ \Omega$ and $R_T = \frac{10}{3}\ \Omega$. Find the two resistances.

Solution. From the given information

$$R_1 + R_2 = 15$$
$$\frac{1}{R_1} + \frac{1}{R_2} = \frac{3}{10}$$

The second equation can be written

$$\frac{R_1 + R_2}{R_1 R_2} = \frac{3}{10}$$
$$10(R_1 + R_2) = 3R_1 R_2$$

From the first equation, $R_2 = 15 - R_1$. Substituting into the second equation, we get

$$10R_1 + 10(15 - R_1) = 3R_1(15 - R_1) \qquad \text{substitution}$$
$$10R_1 + 150 - 10R_1 = 45R_1 - 3R_1^2$$
$$3R_1^2 - 45R_1 + 150 = 0$$
$$R_1^2 - 15R_1 + 50 = 0$$
$$(R_1 - 10)(R_1 - 5) = 0$$
$$R_1 = 10, 5$$

From $R_2 = 15 - R_1$ we have the following: If $R_1 = 10$, then $R_2 = 5$; if $R_1 = 5$, then $R_2 = 10$. We conclude that the two resistances are 5 Ω and 10 Ω.

Exercises / Section 13.2

In Exercises 1–28, solve the systems of equations.

1. $y = x + 2$
   $y = x^2$

2. $y = x - 3$
   $y^2 = x - 1$

3. $y - x^2 = 1$
   $y - x = 3$

4. $y = 4 - x$
   $y = x^2 - 4x + 4$

5. $y + x^2 - 4 = 0$
   $y - x - 2 = 0$

6. $y = x - 2$
   $y^2 - x^2 = 3$

7. $x^2 - x + y = 0$
   $x + y = 0$

8. $y = x + 2$
   $x^2 + y^2 = 4$

9. $y^2 + x = 3$
   $y - x + 1 = 0$

10. $4x - 3y - 4 = 0$
    $4x - y^2 = 0$

11. $y = x^2 - 4$
    $2y = x + 2$

12. $x^2 + y = 3$
    $x + y = 1$

13. $x + y^2 - 4y + 2 = 0$
    $x - y^2 + 4y - 4 = 0$

14. $y = x + 1$
    $y^2 = 3x + 2$

15. $x + 2y = 3$
    $x^2 + 2y^2 = 2$

16. $2x + y = 1$
    $3x^2 - 2y^2 = 2$

17. $x + y + 2 = 0$
    $xy = 1$

18. $x + y = 5$
    $xy = 6$

19. $2y^2 + xy + x^2 = 7$
    $x - 2y = 5$

20. $2x - 3y = 2$
    $xy = 4$

21. $x^2 + y^2 = 5$
    $3x^2 - 2y^2 = -5$

22. $2x^2 + y^2 = 4$
    $x^2 + y^2 = 3$

23. $3x^2 + 3y^2 = 2$
    $7x^2 + 3y^2 = 10$

24. $x^2 - 2y^2 = 1$
    $4x^2 + y^2 = 5$

25. $4y^2 - 3x^2 = 29$
    $y^2 + 7x^2 = 46$

26. $5x^2 + 2y^2 = 37$
    $x^2 + y^2 = 5$

27. $x^2 + y^2 = 16$
    $y^2 - 2x^2 = 10$

28. $4x^2 + 3y^2 = 24$
    $3x^2 - 2y^2 = 35$

29. The area of a metal plate is 25 cm², while the length is 4 times the width. Find its dimensions.

30. Recall that the focal length f of a lens is related to the object distance q and the image distance p by

$$\frac{1}{f} = \frac{1}{p} + \frac{1}{q}$$

If the focal length is 2.0 cm and $p + q = 9.0$ cm, find p and q.

31. Starting at $t = 0$, two particles travel along a line. The distances s from the origin are given by $s = 80 - 16t^2$ and $s = 16t^2 - 100t$, respectively, where s is measured in feet and t in seconds. At what time do the particles meet?

32. Two resistors have a combined resistance of 18 Ω if connected in series and of 4 Ω if connected in parallel. Determine the two resistances.

33. Repeat Exercise 32 if the combined resistance is 9.00 Ω when connected in series and 1.82 Ω when connected in parallel.

34. A rectangular pasture is to be divided into two equal parts by a fence parallel to the smaller sides. The area of each subdivision is 2,500 ft² and the total amount of fence used is 350 ft. Find the dimensions of the pasture.

35. The kinetic energy of a body of mass m and velocity v is given by $\frac{1}{2}mv^2$ and its momentum by mv. Both are conserved in perfectly elastic collisions. Two billiard balls of mass 500 g each collide with respective velocities v_1 and v_2 (in centimeters per second). The following relationships were found experimentally:

$$\frac{1}{2}(500)v_1^2 + \frac{1}{2}(500)v_2^2 = 325{,}000 \quad \text{(ergs)}$$

$$500v_1 + 500v_2 = 25{,}000 \quad \text{(g} \cdot \text{cm/sec)}$$

Find the two velocities.

36. The laws of conservation of energy and momentum give rise to the following system of equations for two colliding bodies:

$$m_1v_1 + m_2v_2 = m_1V_1 + m_2V_2$$

$$\frac{1}{2}m_1v_1^2 + \frac{1}{2}m_2v_2^2 = \frac{1}{2}m_1V_1^2 + \frac{1}{2}m_2V_2^2.$$

Show that $V_1 - V_2 = v_2 - v_1$.

13.3 Equations in Quadratic Form

Some equations that don't appear to be quadratic really are quadratic equations in disguise. The reason is that the unknown in a quadratic equation may be any quantity. If we denote the unknown quantity by $f(x)$, we obtain

$$a[f(x)]^2 + b[f(x)] + c = 0$$

Quadratic form which is called an **equation in quadratic form.** Such equations can be solved by letting $z = f(x)$ and then solving the resulting quadratic equation

$$az^2 + bz + c = 0$$

for z. The roots of the given equation are then found by solving the equation $z = f(x)$ for x.

The examples given next illustrate how to solve a variety of equations in quadratic form, starting with $z = x^2$.

Example 1 Solve the equation $x^4 - 4x^2 + 4 = 0$.

Solution. This equation is not quadratic, but suppose we let $z = x^2$. Then $x^4 = z^2$ and we get

$$z^2 - 4z + 4 = 0$$

which is quadratic, after all! Solving for z, we now have

$$(z - 2)^2 = 0$$
$$z = 2, 2$$

Since $z = x^2$, it follows that $x^2 = 2$ and $x = \pm\sqrt{2}, \pm\sqrt{2}$ (repeating roots).
Check: $(\pm\sqrt{2})^4 - 4(\pm\sqrt{2})^2 + 4 = 4 - 8 + 4 = 0$.

Example 2 Solve the equation $2x + 7\sqrt{x} - 15 = 0$.

Solution. As in Example 1, we let z be the literal part of the middle term. Thus $z = \sqrt{x}$ and $z^2 = x$. After substituting, we get the following quadratic equation:

$$2z^2 + 7z - 15 = 0$$
$$(2z - 3)(z + 5) = 0$$
$$z = \frac{3}{2}, -5$$

From $z = \sqrt{x}$, it follows that $\sqrt{x} = \frac{3}{2}$ and $\sqrt{x} = -5$. However, since $\sqrt{x}$ is the principal square root, which is always positive, the value $\sqrt{x} = -5$ must be excluded. This leaves $\sqrt{x} = \frac{3}{2}$, or $x = \frac{9}{4}$.
Check: If $x = \frac{9}{4}$,

$$2x + 7\sqrt{x} - 15 = 2\left(\frac{9}{4}\right) + 7\sqrt{\frac{9}{4}} - 15$$

$$= \frac{9}{2} + 7\left(\frac{3}{2}\right) - 15$$

$$= \frac{9}{2} + \frac{21}{2} - \frac{30}{2} = 0$$

The solution checks.

Example 3 Solve the equation $x^6 + 7x^3 - 8 = 0$.

Solution. Let $z = x^3$, so that $z^2 = x^6$. Then we get

$$z^2 + 7z - 8 = 0$$
$$(z + 8)(z - 1) = 0$$
$$z = 1, -8$$

Since $z = x^3$, we get $x = \sqrt[3]{z}$. Hence $x = \sqrt[3]{1}$ and $x = \sqrt[3]{-8}$.

To find all the roots of this equation, we need De Moivre's theorem. For $x = \sqrt[3]{1}$,

$$1 = 1 \text{ cis } (0° + k \cdot 360°)$$

The cube roots are

$$[1 \text{ cis } (0° + k \cdot 360°)]^{1/3} = \text{cis } k \cdot 120°, \quad k = 0, 1, 2$$

In rectangular form,

$$x = 1, -\frac{1}{2} + \frac{\sqrt{3}}{2}j, -\frac{1}{2} - \frac{\sqrt{3}}{2}j$$

For $\sqrt[3]{-8}$, we have

$$-8 = 8 \text{ cis } (180° + k \cdot 360°)$$

The cube roots are

$$[8 \text{ cis } (180° + k \cdot 360°)]^{1/3} = 2 \text{ cis } (60° + k \cdot 120°), \quad k = 0, 1, 2$$

In rectangular form

$$x = -2, 1 + \sqrt{3}j, 1 - \sqrt{3}j$$

The roots can be checked in the usual way.

Example 4 Solve the equation

$$2x^{2/3} + x^{1/3} - 6 = 0.$$

Solution. Letting $z = x^{1/3}$, $z^2 = (x^{1/3})^2 = x^{2/3}$, and we have

$$2z^2 + z - 6 = 0$$
$$(2z - 3)(z + 2) = 0$$
$$z = \frac{3}{2}, -2$$

Since $x = z^3$, we get $x = (\frac{3}{2})^3 = \frac{27}{8}$ and $x = (-2)^3 = -8$.

Example 5 Solve the equation

$$(x^2 - x)^2 - 4(x^2 - x) - 12 = 0$$

Solution. Let $z = x^2 - x$, so that $z^2 = (x^2 - x)^2$:

$$z^2 - 4z - 12 = 0$$
$$(z - 6)(z + 2) = 0$$
$$z = 6, -2$$

Since $z = x^2 - x$, we obtain the following equations:

$$x^2 - x = 6 \qquad\qquad x^2 - x = -2$$
$$x^2 - x - 6 = 0 \qquad\qquad x^2 - x + 2 = 0$$
$$(x - 3)(x + 2) = 0 \qquad\qquad x = \frac{1 \pm \sqrt{1 - 8}}{2}$$
$$x = 3, -2 \qquad\qquad x = \frac{1 \pm \sqrt{7}j}{2}$$
$$\qquad\qquad x = \frac{1}{2} \pm \frac{\sqrt{7}}{2}j.$$

Systems of quadratic equations sometimes lead to quadratic forms if the solution is carried out by the method of substitution. (See Exercise 26.)

Quadratic forms may also occur in equations containing trigonometric, exponential, or logarithmic functions, as shown in the next example.

Example 6 Solve the equation

$$e^{-2x} - e^{-x} - 2 = 0$$

Solution. Let $z = e^{-x}$, $z^2 = (e^{-x})^2 = e^{-2x}$:

$$z^2 - z - 2 = 0$$
$$(z - 2)(z + 1) = 0$$
$$z = 2, -1$$

From $z = e^{-x}$, we get

$$e^{-x} = 2$$

Taking natural logarithms of both sides, we have

$$\ln e^{-x} = \ln 2$$
$$-x \ln e = \ln 2 \qquad\qquad \ln x^r = r \ln x$$
$$-x = \ln 2 \qquad\qquad \ln e = 1$$
$$x = -\ln 2 \approx -0.693$$

The other root, $z = -1$, leads to $e^{-x} = -1$, which is impossible since e^{-x} is always positive.

Exercises / Section 13.3

In Exercises 1–25, solve the given equations.

1. $x^4 - 5x^2 + 4 = 0$
2. $x^4 - 7x^2 + 12 = 0$
3. $x^4 - 3x^2 - 4 = 0$
4. $x^4 - 4x^2 - 12 = 0$
5. $x^6 - 7x^3 - 8 = 0$
6. $x^6 - 28x^3 + 27 = 0$
7. $x^{-2} - x^{-1} - 6 = 0$
8. $x^{-2} + x^{-1} - 12 = 0$
9. $x^{-4} + 2x^{-2} - 24 = 0$
10. $2x^{-4} + x^{-2} - 6 = 0$
11. $4x^{-4} + 7x^{-2} - 2 = 0$
12. $6x^{-4} - 7x^{-2} + 2 = 0$
13. $x - 6\sqrt{x} + 5 = 0$
14. $x + 2\sqrt{x} - 48 = 0$
15. $6x + 7\sqrt{x} - 20 = 0$
16. $3x - 13\sqrt{x} + 14 = 0$
17. $(x - 1) - 7\sqrt{x - 1} + 10 = 0$
18. $2(x + 1) - \sqrt{x + 1} - 10 = 0$
19. $x^{2/3} + x^{1/3} - 6 = 0$
20. $6x^{2/3} + 13x^{1/3} - 5 = 0$
21. $(x^2 - 4x)^2 + 2(x^2 - 4x) - 8 = 0$
22. $(x^2 + 2x)^2 + 4(x^2 + 2x) + 3 = 0$
23. $24x^4 + 10x^2 - 21 = 0$
24. $24x^4 - 41x^2 + 12 = 0$
25. $\sqrt{x} + \sqrt[4]{x} = 6$

26. In designing a machine, a technician needs a rectangular metal plate whose area is 2 ft². If the diagonal is to be 2 ft in length, what must the dimensions be?

27. A rectangular metal plate has an area of 120 in² and a diagonal of 20 in. Find the dimensions.

In Exercises 28–31, solve each equation.

28. $e^{2y} - e^y = 0$
29. $\sin^2 \theta - \sin \theta = 0$, $0° \leq \theta < 360°$
30. $2 \sin^2 \theta + \sin \theta - 1 = 0$, $0° \leq \theta < 360°$
31. $2 \cos^2 \theta - \cos \theta - 1 = 0$, $0° \leq \theta < 360°$

13.4 Radical Equations

Our final topic in this chapter is the solution of equations containing radicals, which we shall call **radical equations.**

Except for certain quadratic forms discussed in the previous section, radical equations are generally solved by eliminating the radical. If all the radicals are square roots, we square both sides of the equation. If the radicals have index n, we raise both sides to the nth power.

Unfortunately, when both sides of an equation are raised to a power, the solutions obtained may not all satisfy the given equation and must therefore be checked. Solutions that do not check are called **extraneous solutions** or **roots.** Extraneous roots may also occur when solving fractional equations, as we already saw in Section 5.10.

Extraneous roots

The first three examples illustrate the solution of equations containing a single square root.

Example 1 Solve the equation $\sqrt{x + 4} = x - 2$.

Solution. We can eliminate the radical by squaring both sides of the equation:

$$\sqrt{x + 4} = x - 2 \qquad \text{original equation}$$
$$(\sqrt{x + 4})^2 = (x - 2)^2 \qquad \text{squaring both sides}$$
$$x + 4 = x^2 - 4x + 4 \qquad (a - b)^2 = a^2 - 2ab + b^2$$
$$0 = x^2 - 5x$$
$$x(x - 5) = 0$$
$$x = 0, 5$$

Check: Let $x = $ **5**; then

$$\sqrt{x + 4} = \sqrt{5 + 4} = 3; \quad x - 2 = 5 - 2 = 3 \quad \text{(solution checks)}$$

Let $x = $ **0**; then

$$\sqrt{x + 4} = \sqrt{4} = 2; \quad x - 2 = -2$$

The two values do not agree, so that $x = 0$ is an extraneous root. Therefore, the only solution is $x = 5$.

Example 2 Solve the equation $\sqrt{3x + 4} = -1$.

Solution. Since the principal square root is positive, this equation cannot have a solution. If this observation is overlooked, we get

$$(\sqrt{3x + 4})^2 = (-1)^2$$
$$3x + 4 = 1$$
$$x = -1$$

Check: If $x = -1$, $\sqrt{3x + 4} = \sqrt{1} = 1$, which does not agree with the right side. The only (apparent) root is therefore *extraneous*.

Example 3 Solve the equation $\sqrt{35 - 2x} = x$.

Solution. Squaring both sides, we get

$$(\sqrt{35 - 2x})^2 = x^2$$
$$35 - 2x = x^2$$
$$x^2 + 2x - 35 = 0$$
$$(x + 7)(x - 5) = 0$$
$$x = 5, -7$$

Check: Let $x = 5$; then $\sqrt{35 - 2x} = \sqrt{35 - 10} = \sqrt{25} = 5$, which agrees with the right side. Now let $x = -7$; then $\sqrt{35 - 2x} = \sqrt{35 - 2(-7)} = \sqrt{49} = 7$, not -7. The root is extraneous.

The solution is therefore given by $x = 5$.

Isolating the radical

If an equation contains a radical as well as other terms on the same side, we first **isolate** the radical, that is, we write the radical on one side of the equation and collect the remaining terms on the opposite side. Squaring both sides of the equation then eliminates the radical.

Example 4 Solve the equation $\sqrt{x + 3} - x - 3 = 0$.

Solution. Squaring both sides of this equation directly would lead to a complicated expression containing $\sqrt{x + 3}$. Thus, we must first *isolate* the radical by keeping the radical on the left side and collecting the remaining terms on the right side:

$$\sqrt{x + 3} = x + 3 \qquad \text{isolating the radical}$$

Then we get

$$(\sqrt{x + 3})^2 = (x + 3)^2 \qquad \text{squaring both sides}$$
$$x + 3 = x^2 + 6x + 9 \qquad (a + b)^2 = a^2 + 2ab + b^2$$
$$x^2 + 5x + 6 = 0$$
$$(x + 2)(x + 3) = 0$$
$$x = -2, -3$$

Check: Let $x = -2$. Then

$$\sqrt{x + 3} - x - 3 = \sqrt{-2 + 3} - (-2) - 3 = \sqrt{1} + 2 - 3 = 0$$

Let $x = -3$. Then

$$\sqrt{x + 3} - x - 3 = \sqrt{-3 + 3} - (-3) - 3 = 0$$

Since both values check, this equation has no extraneous roots.

The equation in the next example contains a sum of two radicals, of which only one can initially be isolated.

Example 5 Solve the equation $\sqrt{x + 2} + \sqrt{3x + 4} = 2$.

Solution. In this equation, eliminating both radicals at once is impossible. The best procedure is placing the radicals on opposite sides and then squaring. Thus

$$\sqrt{x+2} = 2 - \sqrt{3x+4} \quad \text{isolating one radical}$$
$$(\sqrt{x+2})^2 = (2 - \sqrt{3x+4})^2 \quad \text{squaring both sides}$$
$$x + 2 = 2^2 - 2 \cdot 2\sqrt{3x+4} + (\sqrt{3x+4})^2 \quad (a-b)^2 = a^2 - 2ab + b^2$$
$$x + 2 = 4 - 4\sqrt{3x+4} + 3x + 4$$

Although one radical is now gone, the other remains. We now isolate this radical by collecting all other terms on the opposite side:

$$x + 2 - 4 - 3x - 4 = -4\sqrt{3x+4} \quad \text{isolating the remaining radical}$$
$$-2x - 6 = -4\sqrt{3x+4} \quad \text{simplifying the left side}$$
$$x + 3 = 2\sqrt{3x+4} \quad \text{dividing by } -2$$
$$(x+3)^2 = (2\sqrt{3x+4})^2 \quad \text{squaring both sides}$$
$$x^2 + 6x + 9 = 4(3x+4) \quad (a+b)^2 = a^2 + 2ab + b^2$$
$$x^2 - 6x - 7 = 0$$
$$(x-7)(x+1) = 0$$
$$x = 7, -1$$

The root $x = -1$ checks, but $x = 7$ is extraneous.

Common error Forgetting the middle term when squaring. In Example 5

$$(2 - \sqrt{3x+4})^2 \quad \text{is not} \quad 4 + (3x+4)$$

The correct expression is obtained by recalling that

$$(a-b)^2 = a^2 - 2ab + b^2$$

so that

$$(2 - \sqrt{3x+4})^2 = 4 - 4\sqrt{3x+4} + (3x+4)$$

Example 6 The base of a right triangle is 3 ft shorter than the height, and the perimeter is 36 ft. Find the dimensions.

Solution. In Figure 13.8, x is the length of the base and $x + 3$ the height. By the Pythagorean theorem, the length of the hypotenuse is $\sqrt{x^2 + (x+3)^2}$. Since the perimeter is 36 ft, we obtain the equation

$$x + (x+3) + \sqrt{x^2 + (x+3)^2} = 36$$
$$2x + 3 + \sqrt{2x^2 + 6x + 9} = 36$$

Isolating the radical, we get

$$\sqrt{2x^2 + 6x + 9} = 33 - 2x$$
$$2x^2 + 6x + 9 = 1,089 - 132x + 4x^2$$
$$2x^2 - 138x + 1,080 = 0$$

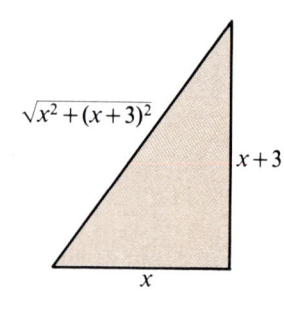

Figure 13.8

$$x^2 - 69x + 540 = 0$$
$$(x - 9)(x - 60) = 0$$
$$x = 9, 60$$

The root $x = 60$ is extraneous. The base is therefore 9 ft and the height 12 ft.

Exercises / Section 13.4

In Exercises 1–24, solve the given equations. Check for possible extraneous roots.

1. $\sqrt{x - 4} = 2$
2. $\sqrt{2x - 1} = 3$
3. $\sqrt{2x + 6} = -2$
4. $\sqrt{2x + 6} = 2$
5. $\sqrt{x + 9} = x + 3$
6. $\sqrt{2x + 4} = x - 2$
7. $\sqrt{x + 6} = 2\sqrt{x}$
8. $\sqrt{x + 4} = \sqrt{2x + 3}$
9. $\sqrt{3x + 4} = \sqrt{4x - 5}$
10. $\sqrt[3]{x + 3} = 2$
11. $\sqrt{15 - 2x} = x$
12. $\sqrt{48 - 2x} = x$
13. $\sqrt{2x - 3} = x - 3$
14. $\sqrt{4 - x} = x - 2$
15. $\sqrt{x} + \sqrt{x - 1} = 1$
16. $12 + 2x - 11\sqrt{x} = 0$
17. $\sqrt{x - 1} - \sqrt{6 - x} = 1$
18. $\sqrt{4 - x} + \sqrt{7 - x} = 3$
19. $\sqrt{2x - 1} - \sqrt{x - 4} = 2$
20. $\sqrt{3x + 4} - \sqrt{2x + 1} = 1$
21. $\sqrt{x - 2} - \sqrt{x - 5} = 1$
22. $\sqrt{5x - 1} - \sqrt{x + 2} = 1$
23. $\sqrt[3]{2x - 5} = \sqrt[3]{6x + 7}$
24. $\sqrt[4]{x + 4} = \sqrt{x - 2}$

25. Solve the equation $L = \sqrt{v} - \sqrt{v - 1}$ for v.

26. Recall that the magnitude of the impedance in an alternating current circuit is $|Z| = \sqrt{R^2 + (X_L - X_C)^2}$. Solve this equation for R.

27. The base of a right triangle is 14 in. longer than the height, while the perimeter is 60 in. Find the lengths of the base and the height.

28. According to the special theory of relativity, the kinetic energy K of a body moving at velocity v relative to an observer is

$$K = \frac{m_0 c^2}{\sqrt{1 - \frac{v^2}{c^2}}} - m_0 c^2$$

where c is the velocity of light and m_0 the mass of the body at rest (relative to the observer). Show that

$$v = c\sqrt{1 - \left(\frac{m_0 c^2}{K + m_0 c^2}\right)^2}$$

29. In an experiment on the refraction of light (Figure 13.9 on page 416), a light ray from a point A in medium M_1 to a point B in medium M_2 was observed to follow the path shown, a distance of 15 cm. Determine where the light enters M_2. (Find x in Figure 13.9; there are two possible solutions, depending on which medium has the larger index of refraction.)

30. A lighthouse is situated 3 mi from a straight shore. The lighthouse keeper wants to travel to a point A on the shore 10 mi from P, the point on the shore nearest the lighthouse (Figure 13.10 on page 416). Since A is not accessible by boat, he lands at a point between A and P and walks the rest of the way, traveling a total of 11 mi. Where does he land?

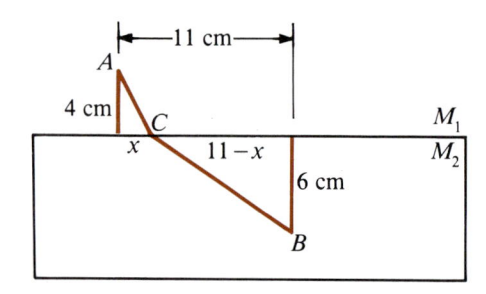

Figure 13.9

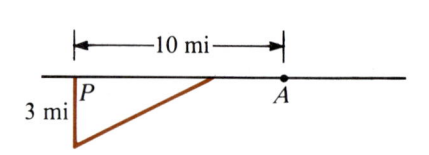

Figure 13.10

31. Under certain conditions, the impedance in a circuit is given by

$$Z = \frac{1}{\sqrt{R + \left(\frac{1}{2\pi fC}\right)^2}}$$

Solve this formula for C.

32. One form of the equation of the parabola is $\sqrt{x} + \sqrt{y} = \sqrt{a}$. Solve this equation for y.

Review Exercises / Chapter 13

In Exercises 1–2, solve each system graphically to the nearest tenth of a unit.

1. $x^2 - y^2 = 2$
   $y + 1 = x$

2. $y = \cos x$
   $y = \sqrt{2x}$

In Exercises 3–10, solve the systems of equations algebraically.

3. $y = x + 2$
   $y = 2x^2 + 1$

4. $y = x^2 - 2$
   $y - x + 1 = 0$

5. $y = x^2 - 4$
   $y - 3x + 6 = 0$

6. $x + y + 3 = 0$
   $xy = 2$

7. $x^2 + 3y^2 = 4$
   $2x^2 + 4y^2 = 5$

8. $x^2 - 2y^2 = 4$
   $3x^2 + 4y^2 = 6$

9. $x^2 + y^2 = 4$
   $xy = 2$

10. $x^2 - y^2 = 3$
    $xy = 2$

In Exercises 11–30, solve each equation.

11. $x^4 - 20x^2 + 64 = 0$

12. $4x^4 - 5x^2 - 9 = 0$

13. $4x^4 + 13x^2 - 12 = 0$

14. $10x^{-2} - 13x^{-1} - 3 = 0$

15. $2x^{-2} + x^{-1} - 15 = 0$

16. $x^{-4} + 12x^{-2} - 64 = 0$

17. $4x^{-4} - 17x^{-2} + 4 = 0$

18. $2x + \sqrt{x} - 6 = 0$

19. $6x + 13\sqrt{x} - 28 = 0$

20. $15x - 7\sqrt{x} - 2 = 0$

21. $(x + 1) - 6\sqrt{x + 1} + 8 = 0$

22. $x^{2/3} + 2x^{1/3} - 8 = 0$

23. $(x^2 + 2x)^2 + 3(x^2 + 2x) + 2 = 0$
24. $\sqrt{x + 10} = -3$
25. $\sqrt{2 - x} = x + 4$
26. $\sqrt{x} + \sqrt{x - 2} = 2$
27. $\sqrt[3]{x - 3} = -2$
28. $\sqrt{x + 11} - \sqrt{x} = 1$
29. $\sqrt{4x - 3} - \sqrt{x + 1} = 1$
30. $\sqrt{x + 3} + \sqrt{x + 6} = \sqrt{5 - 2x}$

31. Two resistors connected in series have a combined resistance of 10.0 Ω. The same resistors connected in parallel have a combined resistance of 2.172 Ω. Find the two resistances.

32. Two bodies weighing 100 g and 80 g collide with velocities v_1 and v_2 (in centimeters per second), respectively. From the laws of conservation of energy and momentum, the following relationships are obtained:

$$100v_1^2 + 80v_2^2 = 11{,}600$$
$$50v_1 + 40v_2 = 700$$

Find v_1 and v_2.

33. A technician wants to design a rectangular metal plate having an area of 540 in.² and a diagonal 39 in. long. What must the dimensions be?

34. The base of a right triangle is 4 in. longer than the height and the perimeter is 48 in. Find the dimensions.

35. According to the theory of relativity, the mass of a body moving at velocity v relative to an observer is given by

$$m = \frac{m_0}{\sqrt{1 - \dfrac{v^2}{c^2}}}$$

where c is the velocity of light and m_0 the mass of the body at rest (relative to the observer). Solve for $v > 0$.

CHAPTER **14**

Higher-Order Equations in One Variable

Objectives Upon completion of this chapter, you should be able to:
1. Perform synthetic division.
2. Use synthetic division to:
   a. Determine whether $x - r$ is a factor of a given polynomial function $f(x)$.
   b. Determine whether a particular value is a zero of a given polynomial function $f(x)$.
   c. Find $f(r)$ for a given value of r.
3. Given one or more roots of a polynomial equation, find the remaining roots.
4. Find the rational roots of a given polynomial equation.
5. Find the irrational roots of a polynomial equation by the method of successive linear approximations.

14.1 Introduction

So far we have confined ourselves to equations of first and second degree except for some special quadratic forms. In this chapter we shall turn our attention to equations of higher degree, which are defined next.

> **A polynomial equation of nth degree** has the form
> $$a_0x^n + a_1x^{n-1} + \cdots + a_{n-1}x + a_n = 0, \quad a_0 \neq 0 \qquad (14.1)$$

Geronimo Cardano

Karl F. Gauss

To see how far we have come, let us first note that equations of third degree were studied by the Persian poet Omar Khayyám in the twelfth century and equations of fourth degree by the Italian mathematician and physician Geronimo Cardano (1501–1576) in 1545. Although there is some doubt about the true originator, a publication attributed to Cardano had such an impact on the development of algebra that the year 1545 is frequently considered the beginning of the modern period in mathematics.

The next great step was taken by the German mathematician Karl F. Gauss (1777–1855), who proved in his doctoral dissertation of 1799 that every polynomial equation has a root. This statement, known as the *fundamental theorem of algebra,* will be discussed in greater detail in Section 14.4.

Many attempts were made to apply Cardano's method to equations of degree greater than four, but up to the time of Gauss's famous theorem and afterwards, none of these efforts was successful. Imagine the shock when in 1824 a young Norwegian named Niels Abel (1802–1829) proved that no such method exists. In other words, there is no formula that will yield the roots of an equation of degree larger than four in terms of radicals (unlike the situation with quadratic equations, for which such a formula does exist).

In spite of this general problem, many equations can be solved by special methods. Techniques also exist for finding approximate solutions in decimal form. These are the topics to be discussed in this chapter.

14.2 The Remainder and Factor Theorems

As noted in the introduction, the purpose of this chapter is to develop special techniques for solving certain higher-order equations. These techniques depend on two theorems, the **remainder theorem** and the **factor theorem,** which will be discussed in this section.

Given the form (14.1) of a polynomial equation, we can define a **polynomial function** $f(x)$.

A polynomial function has the form

$$f(x) = a_0 x^n + a_1 x^{n-1} + \cdots + a_{n-1} x + a_n \tag{14.2}$$

where $a_0, a_1, \ldots, a_n$ are real or complex constants.

If $f(x)$ is divided by $x - r$, we obtain a polynomial $Q(x)$ and remainder R. These are related by the form

$$f(x) = Q(x)(x - r) + R \tag{14.3}$$

as illustrated in the next two examples.

Example 1 Divide $f(x) = 2x^2 - 4x + 4$ by $x - 3$ and write $f(x)$ in the form $f(x) = Q(x)(x - r) + R$.

Solution.
$$\begin{array}{r} 2x + 2 \\ x - 3 \overline{\smash{\big)}\, 2x^2 - 4x + 4} \\ \underline{2x^2 - 6x } \\ 2x + 4 \\ \underline{2x - 6} \\ 10 \end{array}$$

Thus,
$$f(x) = (2x + 2)(x - 3) + 10$$
which can be checked by direct multiplication.

Example 2 Divide $f(x) = 5x^2 + 3x - 16$ by $x + 2$ and write the function in the form $f(x) = Q(x)(x - r) + R$.

Solution.
$$\begin{array}{r} 5x - 7 \\ x + 2 \overline{\smash{\big)}\, 5x^2 + 3x - 16} \\ \underline{5x^2 + 10x } \\ -7x - 16 \\ \underline{-7x - 14} \\ -2 \end{array}$$

Thus,
$$f(x) = (5x - 7)(x + 2) - 2$$

An immediate consequence of formula (14.3) is the **remainder theorem**, which will be used in Section 14.6 to find approximate roots of a given polynomial equation $f(x) = 0$. Let **$x = r$** in the formula $f(x) = Q(x)(x - r) + R$. Then
$$f(r) = Q(r)(r - r) + R = Q(r) \cdot 0 + R$$
or

$$f(r) = R \qquad (14.4)$$

In other words, R is equal to the function value of $f(x)$ at $x = r$.

14.2 THE REMAINDER AND FACTOR THEOREMS

> **Remainder theorem**: If a polynomial function is divided by $x - r$, then the remainder is equal to $f(r)$.

Example 3 If $f(x) = x^3 - 4x^2 + 5x - 3$, evaluate $f(2)$ by using the remainder theorem.

Solution. To get the proper remainder, we must divide $f(x)$ by $x - 2$. The division yields

$$f(x) = (x^2 - 2x + 1)(x - 2) - 1$$

Thus $f(2) = -1$.

Example 4 Use the remainder theorem to find R when $f(x) = 2x^3 - x^2 + 2x - 1$ is divided by $x + 3$.

Solution. Since the divisor is $x + 3$, we have $r = -3$. Hence

$$R = f(-3) = 2(-3)^3 - (-3)^2 + 2(-3) - 1 = -70$$

Another immediate consequence of formula (14.3) is the **factor theorem**: If $R = 0$ in the formula $f(x) = Q(x)(x - r) + R$, then

$$f(x) = Q(x)(x - r)$$

> **Factor theorem**: If r is a root of $f(x) = 0$, then $x - r$ is a factor of $f(x)$. Conversely, if $x - r$ is a factor of $f(x)$, then r is a root of $f(x) = 0$. In symbols,
>
> $$f(x) = Q(x)(x - r)$$

(The factor theorem will be used in Sections 14.4 and 14.5 to find one or more factors of a given polynomial equation $f(x) = 0$. Finding such factors greatly simplifies the computations involved in solving the equation.)

Example 5 Determine whether $x + 5$ is a factor of $f(x) = x^4 + 5x^3 + 2x^2 + 11x + 5$ by means of the factor theorem.

Solution. Since $x + 5$ is the divisor, $r = -5$. Since

$$f(-5) = (-5)^4 + 5(-5)^3 + 2(-5)^2 + 11(-5) + 5 = 0$$

it follows that $x + 5$ is a factor of $f(x)$.

Example 6 Determine whether $x - 4$ is a factor of $f(x) = 3x^4 - 2x^3 + 3x - 5$.

Solution. Since

$$f(4) = 3(4)^4 - 2(4)^3 + 3(4) - 5 = 647 \neq 0$$

$x - 4$ is not a factor.

Summary. The remainder theorem says that whenever a polynomial function $f(x)$ is divided by $x - r$, the remainder R is equal to $f(r)$, the value of $f(x)$ at $x = r$. If the remainder R is zero, then $f(r) = 0$ and r is therefore a root of $f(x) = 0$. By the factor theorem, if r is a root of $f(x) = 0$, then $x - r$ is a factor of $f(x)$. These observations will be used in later sections to solve a given polynomial equation $f(x) = 0$.

Exercises / Section 14.2

In Exercises 1–20, find the remainder in each division problem by the remainder theorem. Check by long division. (See Example 4.)

1. $(x^3 + 2x^2 - x + 1) \div (x - 1)$
2. $(x^3 + 3x^2 + x - 3) \div (x + 2)$
3. $(x^3 - 2x^2 - 2x + 3) \div (x + 3)$
4. $(x^3 - x^2 + x - 2) \div (x - 1)$
5. $(x^3 - 4x^2 - x + 2) \div (x - 3)$
6. $(2x^3 - x^2 + x + 3) \div (x - 2)$
7. $(2x^3 + 3x - 100) \div (x - 4)$
8. $(2x^3 - x^2 + 10) \div (x + 3)$
9. $(x^4 + x^3 - x^2 + x + 1) \div (x - 1)$
10. $(x^4 + x^3 - 3x - 1) \div (x + 2)$
11. $(x^4 - 2x^3 + x^2 + x - 3) \div (x - 3)$
12. $(x^3 - x^2 - x - 50) \div (x - 5)$
13. $(2x^4 + 3x^3 + x - 38) \div (x + 3)$
14. $(4x^3 - 10x^2 + 5x + 3) \div (x - 2)$
15. $(4x^3 + 20x^2 - 18x - 30) \div (x + 5)$
16. $(5x^3 - 32x^2 + 20x - 40) \div (x - 6)$
17. $(5x^4 - 28x^3 - 20x - 300) \div (x - 6)$
18. $(6x^4 - 25x^3 + 4x^2 + 7x - 25) \div (x - 4)$
19. $(3x^5 + 5x^4 - 12x^3 + 4x^2 + 10x - 6) \div (x + 3)$
20. $(4x^5 + 20x^4 + 8x^2 + 35x - 25) \div (x + 5)$

In Exercises 21–30, use the factor theorem to determine whether the second expression is a factor of the first. (See Examples 5 and 6.)

21. $x^3 - 5x^2 + 8x - 4, \ x - 2$
22. $x^3 - 2x^2 - 11x - 5, \ x + 2$
23. $x^4 - 2x^3 - 9x^2 + x + 12, \ x + 2$
24. $x^4 - 5x^2 + 15x + 9, \ x + 3$
25. $2x^4 - 7x^3 - 19x^2 + 23x - 15, \ x - 5$
26. $2x^4 + 4x^3 - 13x^2 + 10x - 10, \ x + 4$
27. $3x^5 - 7x^4 - 6x^3 + x^2 - 2x - 3, \ x - 3$
28. $6x^5 - 36x^4 - x^3 + 6x^2 + x - 6, \ x - 6$
29. $3x^5 + 17x^4 + 10x^3 + x^2 + 7x + 5, \ x + 5$
30. $5x^5 + 16x^4 + 2x^3 - 2x^2 + 4x + 3, \ x + 3$

14.3 Synthetic Division

The remainder and factor theorems point out the close relationship between the value of a polynomial function $f(x)$ at r and the remainder due to division by $x - r$. To use these theorems effectively later on, a fast and efficient means of dividing $f(x)$ by $x - r$ is extremely helpful. **Synthetic division** provides us with this means. After we study synthetic division, we shall return to the factor and remainder theorems.

Synthetic Division

To derive the method of synthetic division, let us consider a regular division problem and then see what steps can be eliminated:

$$
\begin{array}{r}
x^3 - x^2 - 4x - 7 \\
x - 2 \overline{\smash{)}x^4 - 3x^3 - 2x^2 + x + 5} \\
\underline{x^4 - 2x^3} \\
-x^3 - 2x^2 \\
\underline{-x^3 + 2x^2} \\
-4x^2 + x \\
\underline{-4x^2 + 8x} \\
-7x + 5 \\
\underline{-7x + 14} \\
-9
\end{array}
$$

First, observe that the entire division procedure depends only on the coefficients. The powers of x serve as placeholders, provided that the given polynomial is written in descending powers of x. Let's see what the division problem looks like without the powers of x:

$$
\begin{array}{r}
1 - 1 - 4 - 7 \\
1 - 2 \overline{\smash{)}1 - 3 - 2 + 1 + 5} \\
\underline{1 - 2} \\
-1 - 2 \\
\underline{-1 + 2} \\
-4 + 1 \\
\underline{-4 + 8} \\
-7 + 5 \\
\underline{-7 + 14} \\
-9
\end{array}
$$

Since the division procedure we are developing applies only to the division by $x - r$, the number 1 in the divisor $1 - 2$ can be omitted. Even more significantly, the division process contains a number of repeated terms (shown in color) that can also be eliminated. The division now has the following form:

$$\begin{array}{r} 1-1-4-7 \\ -2\overline{)1-3-2+1+5} \\ \underline{-2} \\ -1 \\ \underline{+2} \\ -4 \\ \underline{+8} \\ -7 \\ \underline{+14} \\ -9 \end{array}$$

This format can now be condensed by eliminating the empty spaces:

$$\begin{array}{r} 1-1-4-7 \\ -2\overline{)1-3-2+1+5} \\ \underline{-2+2+8+14} \\ -1-4-7-9 \end{array}$$

Observe next that the numbers in the first row are contained in the fourth row, except for the first term. By repeating this term in the fourth row, the first row can be eliminated:

$$\begin{array}{r} -2\overline{)1-3-2+1+5} \\ \underline{-2+2+8+14} \\ 1-1-4-7-9 \end{array}$$

To simplify even further, note that the division can actually be carried out by the last scheme: Bring down the 1 in the first row and multiply by -2. Then place the product -2 in the second row under the -3. Subtract -2 from -3 and write the resulting -1 in the third row, and so on. Now, in subtraction we change the sign of the subtrahend and add. This change in sign becomes automatic if we replace the -2 in the divisor by 2.

We now have a greatly simplified method of division. All we need to do in addition is observe the custom of placing the 2 on the right:

Synthetic division

$$\begin{array}{r} 1-3-2+1+5\underline{)2} \\ 2-2-8-14 \\ \hline 1-1-4-7-9 \end{array}$$

The steps can be described as follows: Bring down the first 1, multiply by 2, place the product 2 under the -3, add 2 and -3, and place the sum -1 below the 2. Then repeat the process.

In general, to divide a polynomial function $f(x)$ by $x - r$ using synthetic division, follow the procedure outlined next.

14.3 SYNTHETIC DIVISION

Synthetic division: To divide a polynomial $f(x)$ by $x - r$:

1. Write $f(x)$ in descending powers of x and supply a zero for each missing power.
2. Omit the powers of x and set up the division scheme in the following manner:
$$a_0 + a_1 + a_2 + \cdots + a_n \overline{)r}$$
3. Bring down a_0, multiply a_0 by r, place the product beneath a_1, and add the product to a_1. Multiply the sum by r, place the product beneath a_2, and add the product to a_2. Continue the process to the end of the row.
4. The numbers in the bottom row are, reading from left to right, the coefficients of the quotient, which is of degree one less than $f(x)$. The last number in the bottom row is $R = f(r)$.

Example 1 Use synthetic division to divide $2x^3 - 3x + 2$ by $x + 3$.

Solution. Since $x - r = x + 3$, we have $r = -3$. Now supply a zero for the missing power x^2 and proceed as follows:

$$
\begin{array}{r}
2 + 0 - 3 + 2 \overline{)-3} \\
\underline{-6 + 18 - 45} \\
2 - 6 + 15 - 43
\end{array}
$$

The quotient, obtained from the bottom row, is therefore $2x^2 - 6x + 15$, while the remainder is $f(-3) = -43$.

Example 2 Determine whether $x - 4$ is a factor of $3x^3 - 14x^2 + 13x - 20$.

Solution. By synthetic division, we have

$$
\begin{array}{r}
3 - 14 + 13 - 20 \overline{)4} \\
\underline{12 - 8 + 20} \\
3 - 2 + 5 + 0
\end{array}
$$

Since the remainder is 0, $x - 4$ is a factor.

Function Values, Zeros, and Factors

We learned in Section 14.2 that $x - r$ is a factor of $f(x)$ if $f(r) = 0$. Consequently, computing $f(r)$ appeared to be the quickest way to determine whether $x - r$ was a factor. However, as we have seen, synthetic division is usually simpler than carrying out the necessary arithmetic and is actually a convenient method for computing $f(r)$. Moreover, if $f(r) = 0$, then r is called a **zero** of $f(x)$. So synthetic division can be used to determine whether a number is a zero of $f(x)$.

> **Function Values, Zeros, and Factors:**
> 1. To compute $f(r)$, divide $f(x)$ by $x - r$ by means of synthetic division. The remainder R is equal to $f(r)$.
> 2. If $f(r) = 0$, then r is called a **zero of $f(x)$**.
> 3. If r is a zero of $f(x)$, then $x - r$ is a factor of $f(x)$.

These ideas are illustrated in the examples below.

Example 3 Given that $f(x) = 9x^4 - 7x^2 + x + 4$, find $f(-\frac{1}{3})$ by synthetic division.

Solution. Since $f(-\frac{1}{3})$ is the remainder when $f(x)$ is divided by $x + \frac{1}{3}$, we proceed by synthetic division as follows:

$$
\begin{array}{r}
9 + 0 - 7 + 1 + 4 \underline{\big) -\frac{1}{3}} \\
\underline{-3 + 1 + 2 - 1} \\
9 - 3 - 6 + 3 + \mathbf{3}
\end{array}
$$

Thus $f(-\frac{1}{3}) = \mathbf{3}$.

Example 4 Determine whether the number $\mathbf{-2}$ is a zero of $f(x) = 2x^5 - x^4 - 20x + 40$.

Solution. Supplying zeros for the missing powers x^3 and x^2, we get the following scheme:

$$
\begin{array}{r}
2 - 1 + 0 + 0 - 20 + 40 \underline{\big) -\mathbf{2}} \\
\underline{-4 + 10 - 20 + 40 - 40} \\
2 - 5 + 10 - 20 + 20 + \mathbf{0}
\end{array}
$$

Since the remainder is $\mathbf{0}$, we have shown that $f(-2) = \mathbf{0}$. (We have also shown that $x + 2$ is a factor of $f(x)$.)

Example 5 Determine whether $x = \frac{1}{2}$ is a zero of $f(x) = 6x^5 - 3x^4 - 4x^2 + 8x - 3$.

Solution. Supplying a 0 for the missing power x^3, we get

$$
\begin{array}{r}
6 - 3 + 0 - 4 + 8 - 3 \underline{\big) \frac{1}{2}} \\
\underline{3 + 0 + 0 - 2 + 3} \\
6 + 0 + 0 - 4 + 6 + 0
\end{array}
$$

Since the remainder is 0, $f(\frac{1}{2}) = 0$.

Exercises / Section 14.3

In Exercises 1–20, find the quotient and remainder by means of synthetic division.

1. $(x^3 + 2x^2 - x + 1) \div (x - 1)$
2. $(x^3 + 3x^2 + x - 3) \div (x + 2)$
3. $(x^3 - 4x^2 - x + 2) \div (x - 3)$
4. $(2x^3 - x^2 + x + 3) \div (x - 2)$
5. $(x^4 + x^3 - 3x - 1) \div (x + 2)$
6. $(2x^4 + 3x^3 + x - 38) \div (x + 3)$
7. $(4x^3 + 20x^2 - 18x - 30) \div (x + 5)$
8. $(5x^3 - 32x^2 + 20x - 40) \div (x - 6)$
9. $(4x^5 + 20x^4 + 8x^2 + 35x - 25) \div (x + 5)$
10. $(3x^5 + 5x^4 - 12x^3 + 4x^2 + 10x - 6) \div (x + 3)$
11. $(2x^4 + 8x^3 + 3x^2 + 15x + 8) \div (x + 4)$
12. $(3x^4 - 25x^2 - 5x + 2) \div (x - 3)$
13. $(4x^4 + 2x^3 - 4x^2 + x + 2) \div \left(x - \frac{1}{2}\right)$
14. $(4x^5 - x^3 + 6x^2 + 5x - 3) \div \left(x + \frac{1}{2}\right)$
15. $(x^5 - 8x^2 - 4x + 6) \div (x - 2)$
16. $(x^5 - 30x^3 - 35x^2 - 6x + 1) \div (x - 6)$
17. $(9x^5 + 2x^3 - 4x^2 - 5x + 2) \div \left(x - \frac{1}{3}\right)$
18. $(x^5 - 3x^4 - 15x^3 - 20x^2 + 14x - 10) \div (x - 6)$
19. $(x^4 + 10x^3 + 20x^2 + 35x + 20) \div (x + 8)$
20. $(8x^5 - x^2 + 4x - 3) \div \left(x - \frac{1}{2}\right)$

In Exercises 21–32, use synthetic division to determine whether the second expression is a factor of the first.

21. $2x^3 - 8x^2 + 5x + 3, \; x - 3$
22. $5x^3 + 12x^2 - x - 10, \; x + 2$
23. $x^4 + 4x^3 + 2x^2 - x - 20, \; x + 4$
24. $5x^3 + 20x^2 + 30x + 24, \; x + 5$
25. $2x^4 + 4x^3 - 2x - 4, \; x + 2$
26. $3x^3 - 10x^2 + 2x - 40, \; x - 4$
27. $2x^5 - 20x^3 + 4x^2 + 5x + 3, \; x - 3$
28. $3x^4 - 15x^3 + x^2 - 6x + 4, \; x - 5$
29. $2x^4 - 12x^3 - x + 6, \; x - 6$
30. $3x^5 + 11x^4 - 4x^3 + x^2 - 16, \; x + 4$
31. $x^5 + 32, \; x + 2$
32. $x^6 - 1, \; x - 1$

In Exercises 33–38, determine whether the indicated value is a zero of the polynomial.

33. $4x^3 - 10x^2 - 6x, \; 3$
34. $3x^4 + 9x^3 - 10x^2 + 10x + 8, \; -4$
35. $4x^4 + 4x^3 + x^2 - 6x + 1, \; \frac{1}{2}$
36. $6x^3 + 15x^2 + 9x + 7, \; -2$
37. $4x^4 + 5x^3 - 6x^2 + 4x - 3, \; \frac{3}{4}$
38. $8x^5 - 6x^4 + 10x^3 - x^2 + 3x + 1, \; -\frac{1}{4}$

In Exercises 39–45, find $f(r)$ for the indicated value of r.

39. $f(x) = 2x^3 - 4x^2 + x - 2; \; r = 3$
40. $f(x) = 3x^3 + 5x^2 - 11x + 2; \; r = \frac{1}{3}$
41. $f(x) = 4x^3 - 3x^2 - 7x + 1; \; r = -\frac{1}{4}$
42. $f(x) = x^3 - x^2 - 30x + 10; \; r = -6$
43. $f(x) = x^4 + x^3 - 10x^2 - 20x + 30; \; r = 5$
44. $f(x) = 4x^4 - 7x^3 + 2x^2 - x + 1; \; r = -\frac{1}{4}$
45. $f(x) = 6x^4 - 8x^3 - x^2 - 2x - 6; \; r = \frac{1}{3}$

14.4 The Fundamental Theorem of Algebra

To solve higher-order equations, we need to employ special tools such as the factor theorem and synthetic division. Another useful tool, discussed in this section, is the criterion for determining the number of roots of a polynomial equation. This criterion follows from the **fundamental theorem of algebra**, first mentioned in Section 14.1, and enables us to solve polynomial equations under certain special conditions.

> **Fundamental theorem of algebra**: Every polynomial equation
> $$f(x) = a_0 x^n + a_1 x^{n-1} + \cdots + a_n = 0$$
> has at least one real or complex root.

The fundamental theorem of algebra leads to a much more general statement by repeated application of the factor theorem. Let r_1 be the root of $f(x) = 0$ guaranteed by the fundamental theorem. Then by the factor theorem

$$f(x) = Q_1(x)(x - r_1)$$

where $Q_1(x)$ is of degree $n - 1$. Since $Q_1(x) = 0$ is also a polynomial equation, it has a root r_2 by the fundamental theorem. Thus

$$Q_1(x) = Q_2(x)(x - r_2)$$

where $Q_2(x)$ is of degree $n - 2$. It follows that

$$f(x) = Q_2(x)(x - r_1)(x - r_2)$$

Continuing this process, we obtain n linear factors of $f(x)$ and hence n roots of the equation $f(x) = 0$. Thus

$$f(x) = (x - r_1)(x - r_2) \cdots (x - r_n) \tag{14.5}$$

where $r_1, r_2, \ldots, r_n$ are the roots of $f(x) = 0$. Unfortunately, the numbers $r_1, r_2, \ldots, r_n$ may not all be distinct; thus, the number of distinct roots may be less than n. For this reason, we need to introduce the following definition:

> **Multiplicity of a root**: If a root r_k occurs k times ($1 \leq k \leq n$), then r_k is said to be a **root of multiplicity k**.

With this definition, we can make the general statements below.

> 1. A polynomial of degree n can be factored into n linear factors.
> 2. A polynomial equation of degree n has n roots if a root of multiplicity k is counted as k roots.

14.4 THE FUNDAMENTAL THEOREM OF ALGEBRA

Example 1 Consider the following equation of degree 11:

$$(x - 1)(x + 2)^4(x - 5)^6 = 0$$

The roots of the equation are: $x = 1$ (single root), $x = -2$ (multiplicity 4), and $x = 5$ (multiplicity 6). The total number of roots, counting their multiplicities, is $1 + 4 + 6 = 11$, which is equal to the degree of the equation.

Recall from Section 14.1 that no general method exists for solving by algebraic means an equation of degree 5 or higher. However, if one or more of the roots is already known, the remaining roots may possibly be found by the factor theorem, as shown in the remaining examples.

Example 2 One zero of the polynomial function $f(x) = 2x^3 - 7x^2 + x + 1$ is $\frac{1}{2}$. Determine the remaining zeros.

Solution. By synthetic division

$$\begin{array}{r} 2 - 7 + 1 + 1 \underline{)\frac{1}{2}} \\ \underline{1 - 3 - 1} \\ 2 - 6 - 2 + 0 \end{array}$$

Hence

$$f(x) = (2x^2 - 6x - 2)\left(x - \frac{1}{2}\right)$$

$$= 2(x^2 - 3x - 1)\left(x - \frac{1}{2}\right)$$

It follows that $f(x) = 0$ if $x - \frac{1}{2} = 0$ or $x^2 - 3x - 1 = 0$. The roots of $x^2 - 3x - 1 = 0$ are

$$x = \frac{3 \pm \sqrt{9 + 4}}{2} = \frac{3 \pm \sqrt{13}}{2}$$

So the zeros of $f(x)$ are $\frac{1}{2}$, $\frac{1}{2}(3 + \sqrt{13})$, and $\frac{1}{2}(3 - \sqrt{13})$.

Example 3 Solve the equation

$$x^4 - x^3 - 26x^2 + 84x - 72 = 0$$

Double root given that 2 is a root of multiplicity 2, also called a **double root**.

Solution. By synthetic division

$$\begin{array}{r} 1 - 1 - 26 + 84 - 72 \underline{)2} \\ \underline{2 + 2 - 48 + 72} \\ 1 + 1 - 24 + 36 + 0 \end{array}$$

This division shows that
$$x^4 - x^3 - 26x^2 + 84x - 72 = (x^3 + x^2 - 24x + 36)(x - 2)$$

Since 2 is a double root, it must also be a root of the equation $x^3 + x^2 - 24x + 36 = 0$. Thus we repeat the synthetic division process:

$$\begin{array}{r} 1 + 1 - 24 + 36 \underline{)2} \\ \underline{2 + 6 - 36} \\ 1 + 3 - 18 + 0 \end{array}$$

Dividing by a known root always leaves a remainder of 0. If we avoid writing down the zero remainder, then synthetic divisions can be carried out successively:

$$\begin{array}{r} 1 - 1 - 26 + 84 - 72\underline{)2} \\ \underline{2 + 2 - 48 + 72} \\ 1 + 1 - 24 + 36 \underline{)2} \\ \underline{2 + 6 - 36} \\ 1 + 3 - 18 \end{array}$$

The resulting quotient $x^2 + 3x - 18$ factors into $(x + 6)(x - 3)$. It follows that
$$x^4 - x^3 - 26x^2 + 84x - 72 = (x - 2)^2(x + 6)(x - 3)$$
$$= 0$$

The roots are therefore given by $x = 2, 2, -6, 3$.

Complex roots

This procedure is by no means restricted to real roots. Moreover, if the coefficients of a polynomial equation are real and if $a + bj$ is a root, then so is $a - bj$. This observation follows from the quadratic formula:

$$x = \frac{-b \pm \sqrt{b^2 - 4ac}}{2a} = -\frac{b}{2a} \pm \frac{\sqrt{b^2 - 4ac}}{2a}$$

In other words, any complex roots will always occur as a pair of complex conjugates.

Example 4 Solve the equation
$$x^4 - 6x^3 + 15x^2 - 18x + 10 = 0$$
given that $2 - j$ is a root.

Solution. Since $2 - j$ is a root, so is $2 + j$. By synthetic division we get

$$\begin{array}{rrrrr} 1 & -6 & +15 & -18 & +10\,\underline{)2-j} \\ & 2-j & -9+2j & 14-2j & -10 \\ \hline 1 & -4-j & 6+2j & -4-2j & \underline{)2+j} \\ & 2+j & -4-2j & +4+2j & \\ \hline 1 & -2 & +2 & & \end{array}$$

The zeros of the resulting quadratic factor $x^2 - 2x + 2$ can be found by the quadratic formula

$$x = \frac{2 \pm \sqrt{4 - 8}}{2} = 1 \pm j$$

The roots are therefore $x = 2 \pm j, 1 \pm j$.

Exercises / Section 14.4

In Exercises 1–8, state the degree of the equation, each root, and its multiplicity.

1. $(x + 2)^2(x - 3)^2(x + 1) = 0$
2. $(x - 4)^3(x + 1)^2(x + 2) = 0$
3. $x^2(x + 4)^3(x - 5)^2 = 0$
4. $x^2(x - 7)^3(x + 4)^4 = 0$
5. $(2x + 1)^4(x - 10)^5 = 0$
6. $(3x + 2)^5(x - 3)^2 = 0$
7. $x^2(4x - 3)^3(x + 7)^4 = 0$
8. $(x - 10)^3(x + 9)^2(x - 3)^4 = 0$

In Exercises 9–32, find the roots not given.

9. $x^3 - 7x + 6 = 0$; one root is 2
10. $x^3 + 4x^2 + x - 6 = 0$; one root is -3
11. $x^3 + x^2 - 16x - 16 = 0$; one root is -4
12. $x^3 - x^2 - 17x - 15 = 0$; one root is 5
13. $x^3 - 4x + 3 = 0$; one root is 1
14. $2x^3 + 6x^2 + 3x - 2 = 0$; one root is -2
15. $2x^3 - 2x^2 - 9x - 9 = 0$; one root is 3
16. $x^3 - 14x - 8 = 0$; one root is 4
17. $x^4 - 8x^3 + 21x^2 - 22x + 8 = 0$; 1 is a double root
18. $x^4 + 6x^3 + 9x^2 - 4x - 12 = 0$; -2 is a double root
19. $x^4 + 3x^3 + 4x^2 + 3x + 1 = 0$; -1 is a root of multiplicity 2
20. $3x^4 - 13x^3 - 2x^2 + 39x + 9 = 0$; 3 is a root of multiplicity 2
21. $x^4 - x^3 - x^2 - x - 2 = 0$; one root is j
22. $x^4 - 2x^3 - 7x^2 - 2x - 8 = 0$; one root is $-j$
23. $x^4 + x^3 - 2x^2 + 4x - 24 = 0$; one root is $-2j$
24. $x^4 + 2x^3 - 15x^2 + 18x - 216 = 0$; one root is $3j$
25. $x^4 + 2x^3 + 5x^2 + 2x + 4 = 0$; one root is j
26. $x^4 - x^3 + 13x^2 - 16x - 48 = 0$; one root is $4j$
27. $x^4 - 5x^3 + 10x^2 - 10x + 4 = 0$; one root is $1 + j$
28. $x^4 - 23x^2 + 68x - 60 = 0$; one root is $2 - j$
29. $4x^4 - 16x^3 + 21x^2 - 11x + 2 = 0$; $\frac{1}{2}$ is a root of multiplicity 2
30. $9x^4 - 21x^3 - 107x^2 - 63x - 10 = 0$; $-\frac{1}{3}$ is a root of multiplicity 2
31. $x^4 - 7x^3 + 13x^2 - 2x - 8 = 0$; two roots are 2 and 4
32. $x^4 + 10x^3 - 31x^2 - 280x + 300 = 0$; two roots are 5 and -6

14.5 Rational Roots

The technique for solving equations discussed in the last section has one serious limitation: At least one root has to be known in advance. In this section we shall learn how to determine one or more rational roots from a list of numbers that includes the rational roots. The remaining roots can then be found by the factor theorem. If no rational roots exist, the method of this section cannot be used. This case will be discussed in Section 14.6.

Let us first consider a special type of polynomial equation (with integral coefficients) for which the coefficient of the nth power is unity:

$$f(x) = x^n + a_1 x^{n-1} + \cdots + a_n = 0 \tag{14.6}$$

By the discussion in Section 14.4

$$f(x) = (x - r_1)(x - r_2) \cdots (x - r_n) = 0 \tag{14.7}$$

Multiplying out the middle expression shows that the constant a_n is numerically equal to a product of integers:

$$a_n = r_1 r_2 \cdots r_n \tag{14.8}$$

So a_n is divisible by each r_i ($i = 1, 2, \ldots, n$). Furthermore, since each r_i is a root of $f(x) = 0$, a_n is divisible by each root. It follows, then, that writing down all the factors of a_n gives a list of all possible integral roots. However, these factors do not indicate the sign of a root or give any irrational root.

Example 1 Solve the equation $x^3 - 7x + 6 = 0$.

Solution. The integral divisors of 6 are ± 1, ± 2, ± 3, and ± 6. This is a list of all possible integral roots. Suppose we try $x = 1$ by synthetic division:

$$
\begin{array}{r}
1 + 0 - 7 + 6 \underline{)1} \\
\underline{1 + 1 - 6} \\
1 + 1 - 6 + 0
\end{array}
$$

Since the remainder is 0, $x = 1$ is a root. The factor $x^2 + x - 6$ can be written $(x + 3)(x - 2)$. So the roots are -3, 2, and 1.

If the coefficient of the highest power is different from 1, then the list of all possible rational roots may be considerably larger. Note that any polynomial equation

$$f(x) = a_0 x^n + a_1 x^{n-1} + \cdots + a_n = 0, \quad a_0 \neq 0$$

can be written

$$f(x) = a_0 \left(x^n + \frac{a_1}{a_0} x^{n-1} + \cdots + \frac{a_n}{a_0} \right) = 0 \tag{14.9}$$

By a similar argument we can now obtain a much more general result.

14.5 RATIONAL ROOTS

If the coefficients of
$$a_0 x^n + a_1 x^{n-1} + \cdots + a_{n-1} x + a_n = 0$$
are integers, then each **rational** root, after being reduced to lowest terms, has a factor of a_n for its numerator and a factor of a_0 for its denominator.

This criterion can be used to obtain a list of all possible rational roots. Consider the example below.

Example 2 Solve the equation $4x^3 - 24x^2 + 35x - 12 = 0$.

Solution. To construct a list of all possible rational roots, we first list the factors of the numerator and denominator separately:

Factors of numerator: $\pm 1, \pm 2, \pm 3, \pm 4, \pm 6, \pm 12$

Factors of denominator: $\pm 1, \pm 2, \pm 4$

Now we form all possible ratios. Omitting repetitions, we get

$$\pm 1, \pm 2, \pm 3, \pm 4, \pm 6, \pm 12, \pm \frac{1}{4}, \pm \frac{1}{2}, \pm \frac{3}{2}, \pm \frac{3}{4} \qquad (14.10)$$

This long list would take quite a while to check, even with synthetic division. Fortunately, the number of possibilities can be reduced considerably. Let us postpone solving this equation and see how we can further limit the possible roots.

One way to reduce the number of possible roots is by **Descartes' rule of signs.**

Descartes' rule of signs: The number of positive roots of a polynomial equation $f(x) = 0$ with real coefficients is equal to the number of sign changes of $f(x)$ or less than this number by an even integer. The number of negative roots is equal to the number of sign changes of $f(-x)$ or less than this number by an even integer.

To see what is meant by sign changes, consider the polynomial equation $3x^4 - 2x^3 + 3x^2 - x - 1 = 0$. Since the sign changes from positive to negative, then to positive, and then back to negative, there are three sign changes. So the equation has either three positive roots or only $3 - 2 = 1$ positive root. The equation $4x^3 + 10x^2 - 5x + 3 = 0$ has two sign changes and hence either two positive roots or none. The equation $7x^3 + 3x + 2 = 0$ has no variations in signs and hence no positive roots.

Even more useful than Descartes' rule of signs is the criterion for restricting the range in which the real roots can lie. Any number that is greater than or equal to the greatest root of an equation is called an **upper bound** of the roots. Any number that is less than or equal to the least root of an equation is called a **lower bound** of the roots.

> **Upper and lower bounds:** If the coefficient of x^n in a polynomial equation $f(x) = 0$ is positive, and if no negative terms are in the third row of the synthetic division of $f(x)$ by $x - k$, $k > 0$, then no roots are larger than k.
>
> If the signs of the third row of the synthetic division of $f(x)$ by $x - (-k) = x + k$ are alternating plus and minus (zero can be called either plus or minus), then no roots are less than $-k$.

To check the first part of this criterion, recall that

$$f(x) = Q(x)(x - k) + R$$

Now, the coefficients of $Q(x)$ and the number R are precisely the entries in the third row of the synthetic division. If these numbers are positive and if $x > k > 0$, then $Q(x)(x - k) + R > 0$ (since $x - k > 0$). So $f(x) > 0$ for $x > k$. The argument for the second part is similar.

Let us now see how these ideas can be used to finish solving the equation in Example 2.

Example 2 (Continuation) Recall that the list of all positive rational roots is given in statement (14.10). Note next that the equation $4x^3 - 24x^2 + 35x - 12 = 0$ has three sign changes. By Descartes' rule of signs, the total number of positive roots must be either three or one.

$$f(-x) = 4(-x)^3 - 24(-x)^2 + 35(-x) - 12$$
$$= -4x^3 - 24x^2 - 35x - 12$$

we see that $f(-x)$ has no variations in sign. Consequently, there cannot be any negative roots at all, and we need to check only positive roots.

Of the positive roots, $x = 1$ is the easiest to check:

$$\begin{array}{r} 4 - 24 + 35 - 12 \,\underline{)1} \\ 4 - 20 + 15 \\ \hline 4 - 20 + 15 + 3 \end{array}$$

Since $R = 3$, $x = 1$ is not a root, but the signs in the third row are not all positive, which means that we may have a root larger than 1. Let us try $x = 4$:

$$\begin{array}{r} 4 - 24 + 35 - 12 \,\underline{)4} \\ 16 - 32 + 12 \\ \hline 4 - 8 + 3 + 0 \end{array}$$

14.5 RATIONAL ROOTS

Since $R = 0$, $x = 4$ is a root. The resulting factor $4x^2 - 8x + 3$ can be written $(2x - 1)(2x - 3)$, so that the remaining roots are $x = \frac{3}{2}$ and $x = \frac{1}{2}$.

Before summarizing the procedure in this section, let us recall that if r is a root of $f(x) = 0$, then by the factor theorem we have

$$f(x) = Q(x)(x - r)$$

where $Q(x)$ is of degree one less than $f(x)$. The equation

$$Q(x) = 0$$

Depressed equation   is called the **depressed equation.**

To find the rational roots of $f(x) = 0$:

1. Use Descartes' rule of signs to determine the possible number of positive and negative roots.
2. List all possible rational roots.
3. Test the positive integers.
   a. If one of the values is an upper bound, disregard all larger values.
   b. If one of the values is a root, continue by working with the depressed equation. Remember to check for multiple roots.
4. Test the fractions that remain after the upper bound is obtained. If a value is a root, continue by working with the depressed equation. Check for multiple roots.
5. Repeat Steps 3 and 4 for negative roots using lower bounds.
6. Continue the process until a quadratic equation is obtained; solve the quadratic equation.

Remark. While the procedure under discussion is designed to find rational roots, the quadratic equation obtained in the last step may have irrational or complex roots..

The remaining examples illustrate the procedure for finding the rational roots of $f(x) = 0$.

Example 3 Solve the equation $4x^4 - 19x^2 + 3x + 18 = 0$.

Solution. Since $f(x)$ has two sign changes, there exist either two positive roots or none. Also,

$$\begin{aligned} f(-x) &= 4(-x)^4 - 19(-x)^2 + 3(-x) + 18 \\ &= 4x^4 - 19x^2 - 3x + 18 \end{aligned}$$

which has two sign changes. So there are either two negative roots or none.

Factors of numerator: $\pm 1, \pm 2, \pm 3, \pm 6, \pm 9, \pm 18$

Factors of denominator: $\pm 1, \pm 2, \pm 4$

All possible distinct rational roots:

$$\pm 1, \pm 2, \pm 3, \pm 6, \pm 9, \pm 18, \pm \frac{1}{2}, \pm \frac{1}{4}, \pm \frac{3}{2}, \pm \frac{3}{4}, \pm \frac{9}{2}, \pm \frac{9}{4}$$

Let us avoid fractions for now and try $x = 2$ to see if any roots are larger than 2:

$$\begin{array}{r} 4 + 0 - 19 + 3 + 18 \underline{)2} \\ \underline{8 + 16 - 6 - 6} \\ 4 + 8 - 3 - 3 + 12 \end{array}$$

The value $x = 2$ is not a root. Since the signs in the last row are not all positive, we may have a root larger than 2. Let's try $x = 3$:

$$\begin{array}{r} 4 + 0 - 19 + 3 + 18 \underline{)3} \\ \underline{12 + 36 + 51 + 162} \\ 4 + 12 + 17 + 54 + 180 \end{array}$$

Even though 3 is not a root, at least we know that no root can be larger than 3 (since the numbers in the third row are all positive). Since 2 also failed, let's try $x = \frac{3}{2}$:

$$\begin{array}{r} 4 + 0 - 19 + 3 + 18 \underline{)\tfrac{3}{2}} \\ \underline{6 + 9 - 15 - 18} \\ 4 + 6 - 10 - 12 \underline{)\tfrac{3}{2}} \\ \underline{6 + 18 + 12} \\ 4 + 12 + 8 \end{array}$$

We see that $x = \frac{3}{2}$ is a double root. (*A root should always be checked for such repetitions.*) From

$$4x^2 + 12x + 8 = 4(x^2 + 3x + 2) = 0$$

we get

$$4(x + 2)(x + 1) = 0$$
$$x = -1, -2$$

The roots are therefore $x = \frac{3}{2}, \frac{3}{2}, -1, -2$.

Example 4 Solve the equation $2x^3 - 7x^2 + 12x - 9 = 0$.

Solution. Since $f(x)$ has three sign changes, there are either three positive roots or one. Also, $f(-x) = -2x^3 - 7x^2 - 12x - 9$. Since $f(-x)$ has no sign changes, there are no negative roots.

Factors of numerator: ±1, ±3, ±9
Factors of denominator: ±1, ±2

All possible rational roots:
$$\pm 1,\ \pm 3,\ \pm 9,\ \pm\frac{1}{2},\ \pm\frac{3}{2},\ \pm\frac{9}{2}$$

Avoiding fractions for now, let us try $x = 3$:

$$\begin{array}{r}2 - 7 + 12 - 9\ \underline{|3}\\ \underline{\quad\ \ 6 - \ \ 3 + 27}\\ 2 - 1 + \ \ 9 + 18\end{array}$$

While 3 is not a root, a root larger than 3 may exist, such as $\frac{9}{2}$:

$$\begin{array}{r}2 - 7 + 12 - 9\ \underline{\big|\frac{9}{2}}\\ \underline{\qquad\ \ 9 + \ \ 9 + \frac{189}{2}}\\ 2 + 2 + 21 + \frac{171}{2}\end{array}$$

Since the numbers in the third row are positive, no roots are bigger than $\frac{9}{2}$. So we try $x = \frac{3}{2}$:

$$\begin{array}{r}2 - 7 + 12 - 9\ \underline{\big|\frac{3}{2}}\\ \underline{\qquad\ \ 3 - \ \ 6 + 9}\\ 2 - 4 + \ \ 6 + 0\end{array}$$

Hence $x = \frac{3}{2}$ is a root. From the resulting factor, we have

$$2x^2 - 4x + 6 = 0$$
$$2(x^2 - 2x + 3) = 0$$
$$x = \frac{2 \pm \sqrt{4 - 12}}{2} = \frac{2 \pm 2\sqrt{2}j}{2}$$
$$= 1 \pm \sqrt{2}j$$

So the roots are $x = \frac{3}{2}$ and $x = 1 \pm \sqrt{2}j$.

Example 5 Solve the equation $4x^4 + 4x^3 + 9x^2 + 8x + 2 = 0$.

Solution. Since there are no sign changes, the equation has no positive roots. Since

$$f(-x) = 4x^4 - 4x^3 + 9x^2 - 8x + 2$$

$f(x)$ has either four negative roots, two negative roots, or none. (If none, then all roots would be complex.)

Factors of numerator: ±1, ±2

Factors of denominator: ±1, ±2, ±4

All possible rational roots:

$$\pm 1, \pm 2, \pm \frac{1}{2}, \pm \frac{1}{4}$$

Let us try $x = -1$:

$$4 + 4 + 9 + 8 + 2 \overline{)-1}$$
$$\underline{-4 + 0 - 9 + 1}$$
$$4 - 0 + 9 - 1 + 3$$

Reminder. Zero can be regarded as either positive or negative. As written, the signs in the last row alternate, indicating that no roots can be less than -1. So we try $x = -\frac{1}{2}$:

$$4 + 4 + 9 + 8 + 2 \overline{)-\tfrac{1}{2}}$$
$$\underline{-2 - 1 - 4 - 2}$$
$$4 + 2 + 8 + 4 \overline{)-\tfrac{1}{2}}$$
$$\underline{-2 + 0 - 4}$$
$$4 + 0 + 8$$

(Note that $x = -\frac{1}{2}$ was tried twice.)

From the resulting factor, we get

$$4x^2 + 8 = 0$$
$$4(x^2 + 2) = 0$$
$$x = \pm \sqrt{2}j$$

So the roots are $x = -\frac{1}{2}, -\frac{1}{2}, \pm\sqrt{2}j$.

Exercises / Section 14.5

In Exercises 1–26, solve each equation.

1. $x^3 - 2x^2 - x + 2 = 0$
2. $x^3 - 7x + 6 = 0$
3. $x^3 + 3x^2 - 4x - 12 = 0$
4. $x^3 + x^2 - 14x - 24 = 0$
5. $x^4 + x^3 - 7x^2 - x + 6 = 0$
6. $x^4 - 2x^3 - 7x^2 + 20x - 12 = 0$
7. $x^4 + 5x^3 + x^2 - 21x - 18 = 0$
8. $x^4 - 8x^3 + 17x^2 + 2x - 24 = 0$
9. $x^4 - 3x^3 - 8x^2 + 21x + 9 = 0$
10. $x^4 + 5x^3 - 27x - 27 = 0$
11. $x^4 - 10x^3 + 35x^2 - 56x + 48 = 0$
12. $x^4 - 7x^3 + 16x^2 - 15x + 9 = 0$
13. $2x^3 - 5x^2 + 9x - 9 = 0$
14. $3x^3 + 10x^2 + 20x + 16 = 0$

15. $4x^4 + 8x^3 - 3x^2 - 5x + 2 = 0$

16. $2x^4 - 7x^3 + 5x^2 + 3x - 2 = 0$

17. $4x^4 + 8x^3 - 3x^2 - 7x - 2 = 0$

18. $4x^4 - 8x^3 - 19x^2 + 23x - 6 = 0$

19. $6x^4 - 11x^3 - 8x^2 + 25x - 12 = 0$

20. $9x^4 - 15x^3 - 17x^2 + 40x - 16 = 0$

21. $6x^4 + 5x^3 - 22x^2 - 11x + 10 = 0$

22. $3x^4 + 13x^3 + 14x^2 - 4x - 8 = 0$

23. $4x^4 - 8x^3 - 11x^2 + 33x - 18 = 0$

24. $4x^4 - 8x^3 - 15x^2 + 29x - 10 = 0$

25. $6x^4 + 7x^3 - 57x^2 + 42x - 8 = 0$

26. $6x^4 - 19x^3 - 37x^2 + 62x + 24 = 0$

27. A sphere of radius r and specific gravity c will sink in water to a depth h according to the formula $h^3 - 3rh^2 + 4r^3c = 0$. Find the depth to which a wooden ball having a radius of 3 in. and specific gravity 0.5 will sink.

28. If $c = \tfrac{1}{2}$ in the formula $h^3 - 3rh^2 + 4r^3c = 0$ in Exercise 27, show that r is a root. Why is $h = r$ the only root that is physically realistic?

29. A box with a capacity of 32 ft³ has a square base and a height that is 2 ft less than the length of the base. Find the dimensions.

30. The length of one side of a rectangle is equal to the cube of the length of the side adjacent. If the perimeter is $\tfrac{20}{27}$ cm, find the dimensions.

31. If three capacitors C_1, C_2, and C_3 are connected in series, then the capacitance C_T of the combination satisfies the relationship

$$\frac{1}{C_T} = \frac{1}{C_1} + \frac{1}{C_2} + \frac{1}{C_3}$$

The capacitance of the second is 3 μF (microfarads) more than that of the first. The capacitance of the third is 6 μF more than that of the second. If $C_T = \tfrac{12}{7}$ μF, find the capacitance of each.

14.6 Irrational Roots

If all the roots of an equation are irrational, we have no way of making a list of possible roots. Fortunately, there are numerical procedures by which an irrational root can be approximated to any desired degree of accuracy.

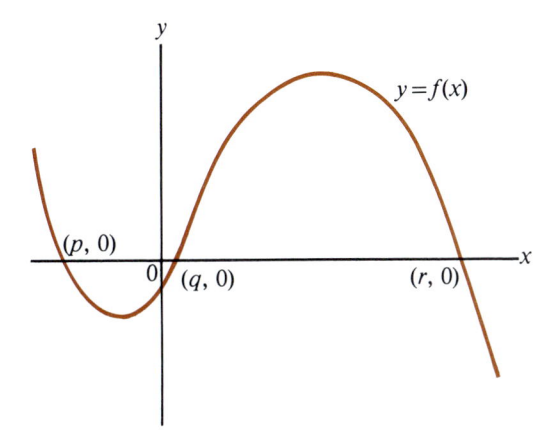

Figure 14.1

The method we shall study in this section is called the **method of successive linear approximations.** The idea behind the approximation scheme can be seen geometrically. Consider the graph of the function $y = f(x)$. Any value of x for which $y = 0$ is an *x-intercept,* that is, a point where the graph crosses the *x*-axis. For example, the graph in Figure 14.1 crosses the *x*-axis at $(p, 0)$, $(q, 0)$, and $(r, 0)$. So $x = p$, $x = q$, and $x = r$ are the real roots of the equation $f(x) = 0$. Now let (x_1, y_1) and (x_2, y_2) be two points on the graph close to an intercept but on opposite sides of the *x*-axis. (See Figure 14.2.)

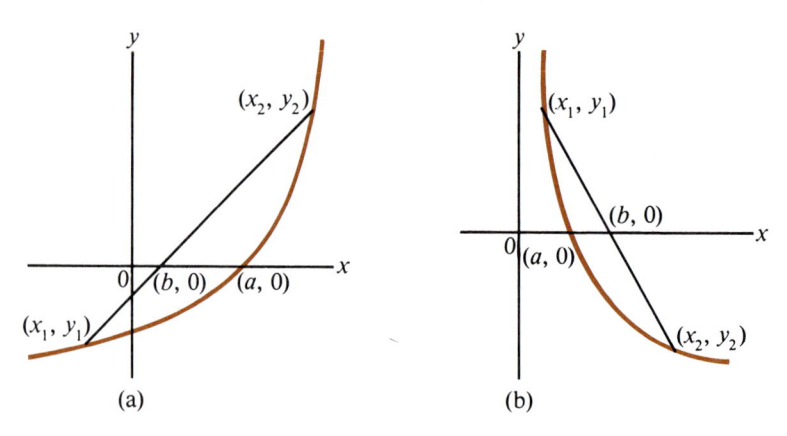

Figure 14.2

Then the line segment joining (x_1, y_1) and (x_2, y_2) will cross the *x*-axis at a point that is very close to the intercept. In other words, in Figure 14.2, $x = b$ is approximately equal to the root $x = a$. As we shall see, this approximation can be found by repeated use of synthetic division. This process can be facilitated by a formula expressing b in terms of the coordinates of (x_1, y_1) and (x_2, y_2), which is given below.

$$b = x_1 - \frac{y_1(x_2 - x_1)}{y_2 - y_1} \quad (14.11)$$

Formula (14.11) is merely a convenience since b can also be found from a scale drawing.

Example 1 Find the irrational root of $x^3 - 6x + 2 = 0$ that lies between 2 and 3.

Solution. First we compute the function values for $x = 2$ and $x = 3$ by synthetic division:

```
1 + 0 − 6 + 2 ⌋2            1 + 0 − 6 +  2 ⌋3
    2 + 4 − 4                    3 + 9 +  9
─────────────────           ─────────────────
1 + 2 − 2 − 2               1 + 3 + 3 + 11
```

Since $f(2) = -2$ and $f(3) = 11$, the graph does indeed cross the x-axis between $(2, 0)$ and $(3, 0)$. As indicated earlier, we can estimate the intersection by a scale drawing, as shown in Figure 14.3. According to the figure, the

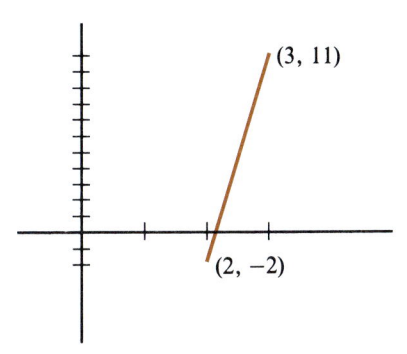

Figure 14.3

root is approximately equal to 2.2. The same estimate can be obtained from formula (14.11), letting $(x_1, y_1) = (2, -2)$ and $(x_2, y_2) = (3, 11)$:

$$b = 2 - \frac{-2(3 - 2)}{11 - (-2)} = 2 + \frac{2}{13} = 2.2$$

The next step is to place the root between two values close to 2.2. To this end, we compute the function values for 2.1, 2.2, and 2.3. Carrying two significant figures for now, we obtain the following:

```
1 + 0   − 6   + 2  )2.1          1 + 0   − 6   + 2  )2.2
       2.1 + 4.4 − 3.4                   2.2 + 4.8 − 2.6
    ─────────────────                 ─────────────────
    1 + 2.1 − 1.6 − 1.4              1 + 2.2 − 1.2 − 0.6

                 1 + 0   − 6   + 2  )2.3
                        2.3 + 5.3 − 1.6
                     ─────────────────
                     1 + 2.3 − 0.7 + 0.4
```

So $f(2.1) = -1.4$, $f(2.2) = -0.6$, and $f(2.3) = 0.4$. It follows that the graph crosses the x-axis between 2.2 and 2.3. To estimate the location, another scale drawing could be used, but it is easier to use formula (14.11) with $(x_1, y_1) = (2.2, -0.6)$ and $(x_2, y_2) = (2.3, 0.4)$:

$$b = 2.2 - \frac{-0.6(2.3 - 2.2)}{0.4 - (-0.6)} = 2.2 + 0.06 = 2.26$$

As always, this value is only an estimate. To refine the estimate, let us locate the root between two values close to 2.26. Suppose we try 2.26 and 2.27, carrying the calculations to three significant figures this time:

```
1 + 0    − 6    + 2   )2.26         1 + 0    − 6    + 2   )2.27
       2.26 + 5.11 − 2.01                  2.27 + 5.15 − 1.93
    ──────────────────────              ──────────────────────
    1 + 2.26 − 0.89 − 0.01              1 + 2.27 − 0.85 + 0.07
```

Since $f(2.26) = -0.01$ and $f(2.27) = 0.07$, the root lies between 2.26 and 2.27. Since the function value (remainder) is *numerically smaller* for $x = 2.26$, the root must be closer to 2.26 than to 2.27. We conclude that $x = 2.26$ is the root accurate to two decimal places.

The process can be continued to obtain even greater accuracy.

Example 2 Find the irrational root of $2x^3 - 5x^2 + 4 = 0$ located between -1 and 0.

Solution. Since $f(-1) = -3$ and $f(0) = 4$, the graph appears to be rising and does indeed cross the x-axis between the values indicated. By formula (11.14), for $(x_1, y_1) = (-1, -3)$ and $(x_2, y_2) = (0, 4)$,

$$b = -1 - \frac{-3[0 - (-1)]}{4 - (-3)} = -1 + 0.4 = -0.6$$

As before, we try to locate the root between two values close to -0.6. For this value we get

$$\begin{array}{r} 2 - 5 + 0 + 4 \;)\overline{-0.6} \\ \underline{-1.2 + 3.72 - 2.2} \\ 2 - 6.2 + 3.72 + \mathbf{1.8} \end{array}$$

Thus $f(-0.6) = \mathbf{1.8}$. Since the graph is rising as we move to the right, the function values should decrease as we move to the left. So let's try -0.7:

$$\begin{array}{r} 2 - 5 + 0 + 4 \;)\overline{-0.7} \\ \underline{-1.4 + 4.48 - 3.14} \\ 2 - 6.4 + 4.48 + \mathbf{0.86} \end{array}$$

Since the remainder is still positive, let's move further to the left to -0.8:

$$\begin{array}{r} 2 - 5 + 0 + 4 \;)\overline{-0.8} \\ \underline{-1.6 + 5.28 - 4.22} \\ 2 - 6.6 + 5.28 - \mathbf{0.22} \end{array}$$

Now the remainder is negative, so the root must lie between -0.8 and -0.7. By formula (14.11)

$$b = -0.8 - \frac{-0.22[-0.7 - (-0.8)]}{0.86 - (-0.22)} = -0.8 + 0.02 = -0.78$$

The remaining trials are

$$\begin{array}{r} 2 - 5 + 0 + 4 \;)\overline{-0.78} \\ \underline{-1.56 + 5.12 - 3.99} \\ 2 - 6.56 + 5.12 + \mathbf{0.01} \end{array} \qquad \begin{array}{r} 2 - 5 + 0 + 4 \;)\overline{-0.79} \\ \underline{-1.58 + 5.20 - 4.11} \\ 2 - 6.58 + 5.20 - \mathbf{0.11} \end{array}$$

Since the function value corresponding to -0.78 is the smaller numerically, we conclude that $x = \mathbf{-0.78}$ is the desired root to two decimal places.

14.6 IRRATIONAL ROOTS

Example 3 Find the irrational root of $x^4 - 8x^3 + 12x^2 - 8x + 8$ located between 1 and 2.

Solution.

$$
\begin{array}{r}
1 - 8 + 12 - 8 + 8 \,\overline{)1} \\
1 - 7 + 5 - 3 \\
\hline
1 - 7 + 5 - 3 + \mathbf{5}
\end{array}
\qquad
\begin{array}{r}
1 - 8 + 12 - 8 + 8 \,\overline{)2} \\
2 - 12 + 0 - 16 \\
\hline
1 - 6 - 0 - 8 - \mathbf{8}
\end{array}
$$

By formula (14.11)

$$b = 1 - \frac{5(2-1)}{-8-5} = 1 + \frac{5}{13} = 1.4$$

The trials beginning with 1.4 are given next.

$$
\begin{array}{r}
1 - 8 \;\; + 12 \;\; - 8 \;\; + 8 \,\overline{)1.4} \\
1.4 - 9.2 + 3.9 - 5.7 \\
\hline
1 - 6.6 + 2.8 - 4.1 + \mathbf{2.3} \\
1 - 8 \;\; + 12 \;\; - 8 \;\; + 8 \,\overline{)1.5} \\
1.5 - 9.8 + 3.3 - 7.1 \\
\hline
1 - 6.5 + 2.2 - 4.7 + \mathbf{0.9} \\
1 - 8 \;\; + 12 \;\; - 8 \;\; + 8 \,\overline{)1.6} \\
1.6 - 10.2 + 2.9 - 8.16 \\
\hline
1 - 6.4 + 1.8 - 5.1 - \mathbf{0.16}
\end{array}
$$

Note that $f(1.5) = \mathbf{0.9}$ and $f(1.6) = \mathbf{-0.16}$, so that the root lies between 1.5 and 1.6. Again by formula (14.11)

$$b = 1.5 - \frac{0.9(0.1)}{-0.16 - 0.9} = 1.5 + 0.08 = 1.58$$

The remaining trials are

$$
\begin{array}{r}
1 - 8 \;\; + 12 \;\; - 8 \;\; + 8 \,\overline{)1.58} \\
1.58 - 10.144 + 2.93 - 8.01 \\
\hline
1 - 6.42 + 1.856 - 5.07 - \mathbf{0.01} \\
1 - 8 \;\; + 12 \;\; - 8 \;\; + 8 \,\overline{)1.57} \\
1.57 - 10.10 + 2.98 - 7.88 \\
\hline
1 - 6.43 + 1.90 - 5.02 + \mathbf{0.12}
\end{array}
$$

Because of the small remainder, the desired root is $x = 1.58$ to two decimal places.

If the location of the root is not indicated or if more than one root must be found, the initial estimate must be obtained from a graph. In most cases plotting points for integral values of x is sufficient.

CHAPTER 14 HIGHER-ORDER EQUATIONS IN ONE VARIABLE

Example 4 Solve the equation $x^3 - 2x - 5 = 0$.

Solution. First we construct the following short table for $y = x^3 - 2x - 5$:

x:	-2	-1	0	1	2	3
y:	-9	-4	-5	-6	-1	16

The graph is shown in Figure 14.4. The root evidently lies between 2 and 3. By formula (14.11)

$$b = 2 - \frac{-1(1)}{16 + 1} = 2 + \frac{1}{17} = 2.1$$

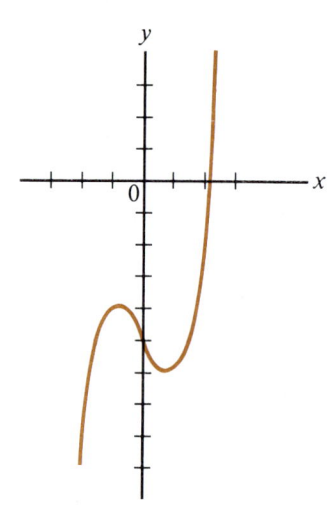

Figure 14.4

We now proceed as before:

$$\begin{array}{r} 1 + 0 - 2 - 5 \overline{)2.0} \\ 2 + 4 + 4 \\ \hline 1 + 2 + 2 - 1 \end{array} \qquad \begin{array}{r} 1 + 0 - 2 - 5 \overline{)2.1} \\ 2.1 + 4.4 + 5.04 \\ \hline 1 + 2.1 + 2.4 + 0.04 \end{array}$$

$$b = 2.0 - \frac{-1(0.1)}{0.04 + 1} = 2.10$$

$$\begin{array}{r} 1 + 0 - 2 - 5 \overline{)2.09} \\ 2.09 + 4.37 + 4.95 \\ \hline 1 + 2.09 + 2.37 - 0.05 \end{array} \qquad \begin{array}{r} 1 + 0 - 2 - 5 \overline{)2.10} \\ 2.10 + 4.41 + 5.06 \\ \hline 1 + 2.10 + 2.41 + 0.06 \end{array}$$

Thus $x = 2.09$ to two decimal places.

Exercises / Section 14.6

In Exercises 1–12, find the irrational root between the indicated values to two decimal places.

1. $x^3 - 4x - 2 = 0$; $[2, 3]$
2. $x^3 - 8x + 1 = 0$; $[2, 3]$
3. $2x^3 + 3x - 6 = 0$; $[1, 2]$
4. $3x^3 + x - 1 = 0$; $[0, 1]$
5. $x^3 - 3x^2 - x + 2 = 0$; $[0, 1]$
6. $x^3 - 5x^2 + 2x + 3 = 0$; $[1, 2]$
7. $2x^3 - 3x^2 - 10x - 5 = 0$; $[3, 4]$
8. $2x^3 - 9x^2 + x + 10 = 0$; $[1, 2]$
9. $x^3 + 4x + 14 = 0$; $[-2, -1]$
10. $4x^3 + 2x^2 + 2x + 2 = 0$; $[-1, 0]$
11. $x^4 + 2x^3 + x - 3 = 0$; $[0, 1]$
12. $x^4 - x^2 - 3x - 2 = 0$; $[-1, 0]$

13. Determine the root of the equation $x^3 - 3x^2 + 2x - 3 = 0$ located between 2 and 3 to three decimal places.
14. Determine the root of the equation $x^3 - 4x^2 + 4 = 0$ located between 1 and 2 to three decimal places.
15. Find the negative root of the equation $x^3 - 4x^2 + 3x + 2 = 0$ to three decimal places.
16. Solve the equation $x^3 + 2x + 6 = 0$ to two decimal places.
17. Find the largest root of the equation $x^4 - 4x^3 - 2x^2 + 12x + 8 = 0$ to two decimal places.
18. A sphere of radius r and specific gravity c will sink in water to a depth h according to the equation $h^3 - 3rh^2 + 4r^3c = 0$. Find to two decimal places the depth to which a ball having a radius of 1.00 ft and specific gravity 0.75 will sink.
19. A box is to be constructed from a square piece of cardboard 10 in. on each side by cutting equal squares from each corner and bending up the sides. (See Figure 14.5.) Determine to two decimal places the size of the square that must be cut out if the volume is to be 73 in.^{3}.

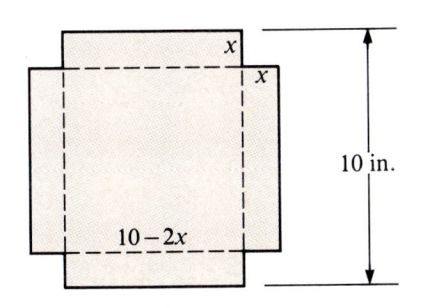

Figure 14.5

20. Use the method for finding irrational roots to determine the approximate value of $\sqrt[3]{6}$ to two decimal places.

Review Exercises / Chapter 14

In Exercises 1–4, determine whether the second expression is a factor of the first.

1. $x^3 - x^2 - 4x + 4$, $x - 2$
2. $x^4 + 2x^3 - x^2 + 7x + 3$, $x + 3$
3. $x^5 + 3x^4 + 15x^2 - 4x + 1$, $x + 4$
4. $x^5 - 5x^4 + x^3 - 5x^2 + 4x - 20$, $x - 5$

In Exercises 5–10, perform the indicated divisions by synthetic division.

5. $(x^3 - 2x^2 + 3x + 1) \div (x - 1)$
6. $(x^4 - 3x^2 + 2x - 10) \div (x - 3)$
7. $(2x^5 + 10x^4 - 2x^3 - 20x^2 + 60x - 60) \div (x + 4)$
8. $(3x^4 - 15x^3 + x^2 - 7x + 12) \div (x - 5)$
9. $(4x^5 - 2x^3 - 2x^2 + 3x - 4) \div (x - 1)$
10. $(3x^5 + 8x^4 + 10x^2 + 5x + 4) \div (x + 3)$

In Exercises 11–16, determine whether the given value is a zero of the function.

11. $x^4 + 4x^3 - 5x + 6$, -2
12. $x^4 + 3x^3 - 6x + 4$, -2
13. $4x^3 + 10x^2 - 8x + 6$, -3
14. $3x^4 - 15x^3 + 10x^2 + 7x + 4$, 4
15. $2x^4 + 10x^3 + x^2 + 6x - 5$, -5
16. $x^5 + 5x^4 - 12x^3 - 30x^2 + 40x + 24$, -6

In Exercises 17–20, use synthetic division to find $f(r)$ for the given value of r.

17. $f(x) = 2x^3 - 3x^2 + 7x - 5$, $r = \frac{1}{2}$
18. $f(x) = 6x^4 + 5x^3 + 6x^2 + 10x - 10$, $r = -\frac{3}{2}$
19. $f(x) = 6x^4 - 2x^3 + 3x^2 - 4x + 5$, $r = \frac{1}{3}$
20. $f(x) = 3x^4 + 5x^3 - x^2 + 4x - 2$, $r = -\frac{2}{3}$

In Exercises 21–26, find the roots not given.

21. $x^4 + 2x^3 - 11x^2 - 12x + 36 = 0$; -3 is a double root
22. $2x^3 + 3x^2 + 5x + 2 = 0$; $-\frac{1}{2}$ is a root
23. $9x^4 - 24x^3 + 4x^2 + 4x - 1 = 0$; $\frac{1}{3}$ is a double root
24. $x^3 + 4x^2 + x + 4 = 0$; $-j$ is a root
25. $x^4 + 2x^3 - 14x^2 + 2x - 15 = 0$; j is a root
26. $x^3 + x^2 - 4x + 6 = 0$; $1 + j$ is a root

In Exercises 27–34, solve the given equations by the method described in Section 14.5.

27. $x^3 + 7x^2 + 4x - 12 = 0$
28. $x^3 - 5x^2 + 6x - 8 = 0$
29. $x^4 - 9x^2 - 4x + 12 = 0$
30. $x^4 - 4x^3 + x^2 - 6x + 36 = 0$
31. $4x^3 + 8x^2 - 11x + 3 = 0$
32. $9x^3 - 6x^2 - 20x - 8 = 0$
33. $9x^4 - 15x^3 - 20x^2 + 17x - 3 = 0$
34. $4x^4 - 24x^3 + 21x^2 + 41x + 12 = 0$

In Exercises 35–38, find the irrational root between the indicated values to two decimal places.

35. $2x^3 + x - 1 = 0$; $[0, 1]$
36. $2x^3 - 9x^2 + 2x + 12 = 0$; $[1, 2]$
37. $x^4 - x^2 - 2x - 1 = 0$; $[-1, 0]$
38. $2x^4 + 3x^3 + 2x - 6 = 0$; $[0, 1]$

39. Find all the real roots of the equation $x^3 - 3x - 4 = 0$ accurate to three decimal places.
40. Find all the real roots of the equation $x^3 - 6x^2 - 4x + 6 = 0$ accurate to two decimal places.

CHAPTER **15**

Determinants and Matrices

Objectives Upon completion of this chapter, you should be able to:
1. Expand a determinant by minors.
2. Use the properties of determinants to find the value of a determinant of any order.
3. Solve a given system of equations by Cramer's rule.
4. Perform algebraic operations with matrices.
5. Use row operations to:
    a. Find the inverse of a given matrix.
    b. Solve a given system of equations.
6. Use the inverse of the coefficient matrix to solve a given system of equations.

15.1 Introduction

In the first part of this chapter we shall continue our study of determinants, which we began in Chapter 3. Our goal is to develop various techniques for evaluating higher-order determinants and to use determinants for solving larger systems of equations. The second part of the chapter is devoted to a study of matrix algebra and the application of matrix methods to solving equations.

15.2 Review of Determinants

Recall from Chapter 3 that a 2×2 determinant is defined by

$$\begin{vmatrix} a_1 & b_1 \\ a_2 & b_2 \end{vmatrix} = a_1 b_2 - a_2 b_1 \tag{15.1}$$

Example 1
$$\begin{vmatrix} -4 & -3 \\ 4 & 1 \end{vmatrix} = (-4)(1) - (4)(-3) = -4 + 12 = 8$$

A third-order determinant

$$\begin{vmatrix} a_1 & b_1 & c_1 \\ a_2 & b_2 & c_2 \\ a_3 & b_3 & c_3 \end{vmatrix} \tag{15.2}$$

Minor — is expanded by minors. A **minor** of a given element is the determinant found by deleting all the elements in the row and column in which that element lies. One possible expansion of determinant (15.2) by minors is

$$\begin{vmatrix} a_1 & b_1 & c_1 \\ a_2 & b_2 & c_2 \\ a_3 & b_3 & c_3 \end{vmatrix} = a_1 \begin{vmatrix} b_2 & c_2 \\ b_3 & c_3 \end{vmatrix} - b_1 \begin{vmatrix} a_2 & c_2 \\ a_3 & c_3 \end{vmatrix} + c_1 \begin{vmatrix} a_2 & b_2 \\ a_3 & b_3 \end{vmatrix} \tag{15.3}$$

Rule for signs — The sign depends on the position of the element: If the sum of the number of the row of the element and the number of the column is even, affix a plus sign; if the sum is odd, affix a minus sign. For example, in expansion (15.3), the element b_1 lies in the first row, second column; since $1 + 2 = 3$, b_1 is preceded by a minus sign. A determinant can be expanded along any row or column.

Example 2 Expand the determinant

$$\begin{vmatrix} -2 & 4 & 3 \\ 1 & 3 & -2 \\ -1 & 5 & -3 \end{vmatrix}$$

Solution. Arbitrarily choosing the third column, we get

$$\begin{vmatrix} -2 & 4 & 3 \\ 1 & 3 & -2 \\ -1 & 5 & -3 \end{vmatrix} = 3 \begin{vmatrix} 1 & 3 \\ -1 & 5 \end{vmatrix} - (-2) \begin{vmatrix} -2 & 4 \\ -1 & 5 \end{vmatrix} + (-3) \begin{vmatrix} -2 & 4 \\ 1 & 3 \end{vmatrix}$$

Note that the element 3 lies in row 1, column 3; since $1 + 3 = 4$, we affix a plus sign. The element -2 lies in row 2, column 3, so we affix a minus sign. Finally, -3 lies in row 3, column 3, and $3 + 3 = 6$, which is again even. So we affix a plus sign. Continuing the expansion, we get

$$3 \begin{vmatrix} 1 & 3 \\ -1 & 5 \end{vmatrix} + 2 \begin{vmatrix} -2 & 4 \\ -1 & 5 \end{vmatrix} - 3 \begin{vmatrix} -2 & 4 \\ 1 & 3 \end{vmatrix}$$
$$= 3(5 + 3) + 2(-10 + 4) - 3(-6 - 4) = 42$$

Example 3 Expand the determinant
$$\begin{vmatrix} -5 & 3 & 6 \\ -2 & 0 & 0 \\ 7 & 1 & 4 \end{vmatrix}$$

Solution. Because of the zeros, it is best to expand the determinant along the second row. Thus
$$\begin{vmatrix} -5 & 3 & 6 \\ -2 & 0 & 0 \\ 7 & 1 & 4 \end{vmatrix} = -(-2)\begin{vmatrix} 3 & 6 \\ 1 & 4 \end{vmatrix} + 0 + 0 = 2(12 - 6) = 12$$

Exercises / Section 15.2

Evaluate each determinant.

1. $\begin{vmatrix} 2 & 3 \\ -4 & 0 \end{vmatrix}$ 2. $\begin{vmatrix} 0 & 1 \\ -3 & 2 \end{vmatrix}$ 3. $\begin{vmatrix} 1 & -4 \\ 7 & -8 \end{vmatrix}$ 4. $\begin{vmatrix} -3 & -5 \\ -4 & 6 \end{vmatrix}$

5. $\begin{vmatrix} 1 & 0 & 0 \\ 2 & -3 & 4 \\ 3 & -2 & 1 \end{vmatrix}$ 6. $\begin{vmatrix} 2 & 2 & 1 \\ 0 & -3 & 4 \\ 0 & 6 & 0 \end{vmatrix}$ 7. $\begin{vmatrix} 0 & 1 & 6 \\ 0 & 0 & -3 \\ -7 & -4 & 3 \end{vmatrix}$ 8. $\begin{vmatrix} 3 & 4 & 2 \\ 0 & -1 & 0 \\ 8 & 3 & -5 \end{vmatrix}$

9. $\begin{vmatrix} 2 & -7 & 10 \\ 3 & 0 & 1 \\ -2 & 4 & 3 \end{vmatrix}$ 10. $\begin{vmatrix} 2 & 8 & 9 \\ 0 & 3 & 6 \\ -2 & -4 & 0 \end{vmatrix}$ 11. $\begin{vmatrix} 4 & 6 & 4 \\ -7 & 0 & 3 \\ 11 & -5 & 1 \end{vmatrix}$ 12. $\begin{vmatrix} -3 & -2 & 1 \\ 6 & 10 & 3 \\ -4 & -3 & 11 \end{vmatrix}$

13. $\begin{vmatrix} 3 & 6 & 0 \\ -5 & 7 & 3 \\ -8 & 0 & 10 \end{vmatrix}$ 14. $\begin{vmatrix} 2 & -6 & 8 \\ -7 & 12 & 3 \\ 2 & 0 & 4 \end{vmatrix}$

15.3 Properties of Determinants

So far we have discussed expansion by minors only for third-order determinants. In this section we shall extend the expansion by minors to determinants of any order. To do this, however, we must first know several properties of determinants, which are discussed next.

To get an overview of the problem, consider the following expansion of a fourth-order determinant by minors:

$$\begin{vmatrix} a_1 & b_1 & c_1 & d_1 \\ a_2 & b_2 & c_2 & d_2 \\ a_3 & b_3 & c_3 & d_3 \\ a_4 & b_4 & c_4 & d_4 \end{vmatrix} = a_1 \begin{vmatrix} b_2 & c_2 & d_2 \\ b_3 & c_3 & d_3 \\ b_4 & c_4 & d_4 \end{vmatrix} - b_1 \begin{vmatrix} a_2 & c_2 & d_2 \\ a_3 & c_3 & d_3 \\ a_4 & c_4 & d_4 \end{vmatrix}$$
$$+ c_1 \begin{vmatrix} a_2 & b_2 & d_2 \\ a_3 & b_3 & d_3 \\ a_4 & b_4 & d_4 \end{vmatrix} - d_1 \begin{vmatrix} a_2 & b_2 & c_2 \\ a_3 & b_3 & c_3 \\ a_4 & b_4 & c_4 \end{vmatrix}$$

Given the amount of arithmetic required to expand a single third-order determinant, the expansion of a fourth-order determinant looks like a discouraging task. For even higher order determinants, the amount of work becomes prohibitive. Fortunately, the work can be greatly simplified by making use of several properties of determinants, which are stated next.

> **Properties of determinants:**
> 1. If all the entries in a row or column are 0, then the value of the determinant is 0.
> 2. If two rows (or columns) of a given determinant are interchanged, then the value of the resulting determinant is the negative of the value of the given determinant.
> 3. If a determinant has two identical rows or columns, then the value of the determinant is 0.
> 4. Whenever the elements of a row or column have a common factor, this factor can be removed from the row or column and placed in front of the determinant as a multiplicative constant.
> 5. If all the elements of a row (or column) are multiplied by k and the resulting numbers added to the corresponding elements of another row (or column), then the value of the determinant is unchanged.

A proof of these properties requires a more sophisticated study of determinants than is needed for our purposes. However, to illustrate the type of argument used, we shall check Property 2 for an arbitrary third-order determinant and then use Property 2 to prove Property 3. Thus, we start with the following:

$$\begin{vmatrix} a_1 & b_1 & c_1 \\ a_2 & b_2 & c_2 \\ a_3 & b_3 & c_3 \end{vmatrix} = a_1 \begin{vmatrix} b_2 & c_2 \\ b_3 & c_3 \end{vmatrix} - b_1 \begin{vmatrix} a_2 & c_2 \\ a_3 & c_3 \end{vmatrix} + c_1 \begin{vmatrix} a_2 & b_2 \\ a_3 & b_3 \end{vmatrix}$$

If the first two rows are interchanged, we get by expanding along the second row

$$\begin{vmatrix} a_2 & b_2 & c_2 \\ a_1 & b_1 & c_1 \\ a_3 & b_3 & c_3 \end{vmatrix} = -a_1 \begin{vmatrix} b_2 & c_2 \\ b_3 & c_3 \end{vmatrix} + b_1 \begin{vmatrix} a_2 & c_2 \\ a_3 & c_3 \end{vmatrix} - c_1 \begin{vmatrix} a_2 & b_2 \\ a_3 & b_3 \end{vmatrix}$$

$$= - \begin{vmatrix} a_1 & b_1 & c_1 \\ a_2 & b_2 & c_2 \\ a_3 & b_3 & c_3 \end{vmatrix}$$

This confirms Property 2.

Given Property 2, Property 3 can actually be proved. Let the value of the determinant be D. If two identical rows (or columns) are interchanged and the value of the new determinant denoted by D', then $D' = -D$ by

Property 2. However, since the rows (or columns) that were interchanged are identical, $D' = D$. Hence $D = -D$, which is possible only if $D = 0$.

Although all five properties have their uses, Property 5 is by far the most important.

> Use **Property 5** to change all the elements but one in a row (or column) to zeros. Expand along this row (or column) to obtain a determinant of order one less.

Let's illustrate Property 5 with an example.

Example 1 Change all the elements except the 1 in the first column to zeros and evaluate the resulting determinant:

$$\begin{vmatrix} 1 & 2 & -1 \\ 2 & 0 & 3 \\ -3 & -4 & -2 \end{vmatrix}$$

Solution. By **Property 5** we may multiply the first row by -2 and add the resulting numbers to the corresponding elements in the second row. Thus

$$\begin{vmatrix} 1 & 2 & -1 \\ 2 & 0 & 3 \\ -3 & -4 & -2 \end{vmatrix} = \begin{vmatrix} 1 & 2 & -1 \\ 2 + (-2)(1) & 0 + (-2)(2) & 3 + (-2)(-1) \\ -3 & -4 & -2 \end{vmatrix}$$

$$= \begin{vmatrix} 1 & 2 & -1 \\ 0 & -4 & 5 \\ -3 & -4 & -2 \end{vmatrix}$$

Similarly, we may multiply the first row by 3 and add the resulting numbers to the third row to obtain

$$\begin{vmatrix} 1 & 2 & -1 \\ 0 & -4 & 5 \\ -3 & -4 & -2 \end{vmatrix} = \begin{vmatrix} 1 & 2 & -1 \\ 0 & -4 & 5 \\ 0 & 2 & -5 \end{vmatrix} = 1 \begin{vmatrix} -4 & 5 \\ 2 & -5 \end{vmatrix}$$

$$= 20 - 10 = 10$$

Notation. To make the operation involving Property 5 easier to describe, we shall adopt the notation

> $kR_j + R_i$

452 CHAPTER 15 DETERMINANTS AND MATRICES

to indicate that the jth row is multiplied by k and the resulting elements added to the ith row. The analogous operation with columns will be denoted by

$$kC_j + C_i$$

With this notation, the operations in Example 1 can now be described compactly as follows:

$$\begin{vmatrix} 1 & 2 & -1 \\ 2 & 0 & 3 \\ -3 & -4 & -2 \end{vmatrix} \begin{array}{c} -2R_1 + R_2 \\ 3R_1 + R_3 \end{array} = \begin{vmatrix} 1 & 2 & -1 \\ 0 & -4 & 5 \\ 0 & 2 & -5 \end{vmatrix}$$

This notation, in conjunction with Property 5, is further illustrated in the next example.

Example 2 Evaluate the determinant

$$\begin{vmatrix} -2 & 5 & -2 & 1 \\ 1 & 0 & -3 & 2 \\ 2 & -1 & 3 & -4 \\ 3 & 6 & -2 & 0 \end{vmatrix}$$

Solution. Remember that *our immediate goal is to change all elements but one in a given row or column to zeros*. This can be done in several ways. For example, retaining the -2 in the upper left-hand corner, we can obtain a zero for the element directly below by the operation $\frac{1}{2}R_1 + R_2$:

$$\begin{vmatrix} -2 & 5 & -2 & 1 \\ 1 & 0 & -3 & 2 \\ 2 & -1 & 3 & -4 \\ 3 & 6 & -2 & 0 \end{vmatrix} \begin{array}{c} \frac{1}{2}R_1 + R_2 \end{array} = \begin{vmatrix} -2 & 5 & -2 & 1 \\ 0 & \frac{5}{2} & -4 & \frac{5}{2} \\ 2 & -1 & 3 & -4 \\ 3 & 6 & -2 & 0 \end{vmatrix}$$

In a similar manner, we can obtain zeros for the remaining elements in the first column.

Avoiding a fractional k

However, working with $k = \frac{1}{2}$, a fraction, is inconvenient. We can avoid a fractional k by working with 1 or -1 instead of -2. Suppose we retain the 1 in the first column and obtain zeros for the remaining elements in the first column as follows:

$$\begin{vmatrix} -2 & 5 & -2 & 1 \\ 1 & 0 & -3 & 2 \\ 2 & -1 & 3 & -4 \\ 3 & 6 & -2 & 0 \end{vmatrix} \begin{array}{c} 2R_2 + R_1 \\ \\ -2R_2 + R_3 \\ -3R_2 + R_4 \end{array} = \begin{vmatrix} 0 & 5 & -8 & 5 \\ 1 & 0 & -3 & 2 \\ 0 & -1 & 9 & -8 \\ 0 & 6 & 7 & -6 \end{vmatrix}$$

15.3 PROPERTIES OF DETERMINANTS

In expanding the new determinant along the first column, note that the element 1 is in row 2, column 1. The expansion yields

$$-(1)\begin{vmatrix} 5 & -8 & 5 \\ -1 & 9 & -8 \\ 6 & 7 & -6 \end{vmatrix}$$

Now retain the -1 in the first column and obtain zeros for the remaining elements in this column.

$$(-1)\begin{vmatrix} 5 & -8 & 5 \\ -1 & 9 & -8 \\ 6 & 7 & -6 \end{vmatrix} \begin{matrix} 5R_2 + R_1 \\ \\ 6R_2 + R_3 \end{matrix} = (-1)\begin{vmatrix} 0 & 37 & -35 \\ -1 & 9 & -8 \\ 0 & 61 & -54 \end{vmatrix}$$

$$= (-1)\left\{[-(-1)]\begin{vmatrix} 37 & -35 \\ 61 & -54 \end{vmatrix}\right\}$$

$$= -[(37)(-54) - (61)(-35)] = -137$$

Obtaining a 1 or −1

Example 2 shows that to obtain zeros with relative ease, we need either a 1 or a -1. (Otherwise, the constant k is a fraction.) If such a term is lacking, it may be possible to obtain a 1 or -1 by one of the operations. Consider the next example.

Example 3 Evaluate

$$\begin{vmatrix} 2 & 1 & -3 & 4 \\ 0 & 2 & -2 & 3 \\ -4 & 0 & 5 & -6 \\ 2 & -3 & 4 & 6 \end{vmatrix}$$

Solution. For the sake of variation, let us obtain zeros in the first row except for the 1. The operations are

$$-2C_2 + C_1, \quad 3C_2 + C_3, \quad \text{and} \quad -4C_2 + C_4$$

$$\begin{vmatrix} 2 & 1 & -3 & 4 \\ 0 & 2 & -2 & 3 \\ -4 & 0 & 5 & -6 \\ 2 & -3 & 4 & 6 \end{vmatrix} = \begin{vmatrix} 0 & 1 & 0 & 0 \\ -4 & 2 & 4 & -5 \\ -4 & 0 & 5 & -6 \\ 8 & -3 & -5 & 18 \end{vmatrix}$$

$$= -(1)\begin{vmatrix} -4 & 4 & -5 \\ -4 & 5 & -6 \\ 8 & -5 & 18 \end{vmatrix}$$

The simplest way to create a -1 (or 1) is to factor 4 (or -4) from each element in the first column (Property 4). Thus

$$(-1)\begin{vmatrix} -4 & 4 & -5 \\ -4 & 5 & -6 \\ 8 & -5 & 18 \end{vmatrix} = (-1)(4)\begin{vmatrix} -1 & 4 & -5 \\ -1 & 5 & -6 \\ 2 & -5 & 18 \end{vmatrix} \quad \begin{array}{c} -1R_1 + R_2 \\ 2R_1 + R_3 \end{array}$$

$$= (-4)\begin{vmatrix} -1 & 4 & -5 \\ 0 & 1 & -1 \\ 0 & 3 & 8 \end{vmatrix}$$

$$= (-4)(-1)\begin{vmatrix} 1 & -1 \\ 3 & 8 \end{vmatrix} = 44$$

Example 4 In some cases a 1 or -1 can be obtained by using Property 5. For example,

$$\begin{vmatrix} 3 & 7 & 7 \\ 13 & -5 & 6 \\ 14 & 0 & 20 \end{vmatrix} \quad -4R_1 + R_2 \quad = \begin{vmatrix} 3 & 7 & 7 \\ 1 & -33 & -22 \\ 14 & 0 & 20 \end{vmatrix}$$

or

$$\begin{vmatrix} 3 & 7 & 7 \\ 13 & -5 & 6 \\ 14 & 0 & 20 \end{vmatrix} \quad -3R_1 + R_3 \quad = \begin{vmatrix} 3 & 7 & 7 \\ 13 & -5 & 6 \\ 5 & -21 & -1 \end{vmatrix}$$

Example 5 Evaluate

$$\begin{vmatrix} 1 & -1 & 2 & 4 & -2 \\ 1 & 0 & -1 & -3 & 1 \\ 0 & -3 & 0 & 2 & 2 \\ 0 & 4 & 3 & 0 & -3 \\ 0 & -6 & 5 & 6 & 0 \end{vmatrix}$$

Solution. By $-1R_1 + R_2$ we get

$$\begin{vmatrix} 1 & -1 & 2 & 4 & -2 \\ 0 & 1 & -3 & -7 & 3 \\ 0 & -3 & 0 & 2 & 2 \\ 0 & 4 & 3 & 0 & -3 \\ 0 & -6 & 5 & 6 & 0 \end{vmatrix} = (1)\begin{vmatrix} 1 & -3 & -7 & 3 \\ -3 & 0 & 2 & 2 \\ 4 & 3 & 0 & -3 \\ -6 & 5 & 6 & 0 \end{vmatrix} \quad \begin{array}{c} 3R_1 + R_2 \\ -4R_1 + R_3 \\ 6R_1 + R_4 \end{array}$$

$$= \begin{vmatrix} 1 & -3 & -7 & 3 \\ 0 & -9 & -19 & 11 \\ 0 & 15 & 28 & -15 \\ 0 & -13 & -36 & 18 \end{vmatrix}$$

$$= (1)\begin{vmatrix} -9 & -19 & 11 \\ 15 & 28 & -15 \\ -13 & -36 & 18 \end{vmatrix}$$

15.3 PROPERTIES OF DETERMINANTS

To obtain a -1, we use the operation $-2C_1 + C_2$:

$$\begin{vmatrix} -9 & -19 & 11 \\ 15 & 28 & -15 \\ -13 & -36 & 18 \end{vmatrix} = \begin{vmatrix} -9 & -1 & 11 \\ 15 & -2 & -15 \\ -13 & -10 & 18 \end{vmatrix} \begin{matrix} \\ -2R_1 + R_2 \\ -10R_1 + R_3 \end{matrix}$$

$$= \begin{vmatrix} -9 & -1 & 11 \\ 33 & 0 & -37 \\ 77 & 0 & -92 \end{vmatrix}$$

$$= -(-1) \begin{vmatrix} 33 & -37 \\ 77 & -92 \end{vmatrix} = -187$$

Exercises / Section 15.3

Evaluate each determinant by using the properties of determinants.

1. $\begin{vmatrix} 1 & 3 & -5 \\ 2 & -4 & 0 \\ 0 & -2 & 2 \end{vmatrix}$

2. $\begin{vmatrix} 1 & 4 & 10 \\ 0 & -6 & -3 \\ -3 & 5 & -20 \end{vmatrix}$

3. $\begin{vmatrix} 3 & 0 & 1 \\ -4 & 7 & 2 \\ 8 & 6 & 5 \end{vmatrix}$

4. $\begin{vmatrix} 2 & 3 & 2 \\ 4 & 5 & 1 \\ -6 & -4 & -2 \end{vmatrix}$

5. $\begin{vmatrix} 1 & 4 & 8 \\ -2 & -7 & 3 \\ 3 & 6 & -5 \end{vmatrix}$

6. $\begin{vmatrix} 3 & 1 & -2 \\ -2 & 10 & 2 \\ 10 & -8 & 4 \end{vmatrix}$

7. $\begin{vmatrix} 1 & 12 & 7 \\ 6 & -5 & -2 \\ 5 & -12 & 6 \end{vmatrix}$

8. $\begin{vmatrix} -1 & 2 & -5 \\ 2 & 0 & 3 \\ -3 & 6 & 2 \end{vmatrix}$

9. $\begin{vmatrix} -8 & 3 & 5 \\ 7 & 2 & -1 \\ 0 & 6 & 4 \end{vmatrix}$

10. $\begin{vmatrix} 0 & -2 & 6 \\ 7 & 3 & -5 \\ 2 & -1 & -3 \end{vmatrix}$

11. $\begin{vmatrix} 9 & -13 & -6 \\ -4 & 5 & 3 \\ 7 & 2 & 5 \end{vmatrix}$

12. $\begin{vmatrix} -5 & 7 & 11 \\ -7 & 11 & 9 \\ 4 & 6 & 11 \end{vmatrix}$

(See Example 5.)

13. $\begin{vmatrix} 2 & 1 & -4 & 0 \\ 2 & -1 & 3 & -4 \\ 0 & -3 & 5 & 2 \\ 5 & 2 & 2 & 2 \end{vmatrix}$

14. $\begin{vmatrix} -3 & 0 & -1 & 3 \\ 1 & 2 & 0 & -1 \\ 3 & 4 & -2 & 5 \\ 2 & -2 & -4 & 6 \end{vmatrix}$

15. $\begin{vmatrix} 2 & -3 & 0 & 5 \\ 1 & 2 & -1 & 3 \\ -2 & 3 & -4 & 0 \\ 3 & -2 & 3 & 5 \end{vmatrix}$

16. $\begin{vmatrix} -1 & 2 & -3 & 0 \\ -5 & 0 & 2 & -2 \\ 0 & 4 & 0 & -1 \\ 2 & -1 & 6 & -3 \end{vmatrix}$

17. $\begin{vmatrix} 4 & 1 & 3 & 1 \\ 3 & -2 & -2 & 4 \\ 2 & 3 & 1 & -2 \\ -2 & 5 & 1 & 3 \end{vmatrix}$

18. $\begin{vmatrix} 3 & 1 & -2 & 1 \\ -2 & -1 & 1 & 2 \\ 1 & 2 & -3 & 1 \\ -1 & -3 & 3 & 2 \end{vmatrix}$

19. $\begin{vmatrix} -2 & -1 & 0 & 3 \\ 0 & 2 & 6 & 4 \\ 1 & 2 & 0 & 4 \\ 6 & 0 & 1 & 2 \end{vmatrix}$

20. $\begin{vmatrix} 4 & -1 & 0 & 3 \\ 3 & 0 & 2 & 0 \\ 3 & -1 & 2 & 1 \\ 1 & -3 & 0 & 2 \end{vmatrix}$

21. $\begin{vmatrix} 1 & 3 & -4 & 0 & -5 \\ 0 & -1 & 5 & 0 & -3 \\ 0 & -3 & 2 & 3 & 1 \\ 0 & 2 & -2 & 1 & 0 \\ 0 & 0 & 4 & -6 & 0 \end{vmatrix}$

22. $\begin{vmatrix} -2 & 1 & 0 & -3 & 1 \\ 1 & 1 & -2 & 2 & 0 \\ 0 & 1 & 1 & 1 & 0 \\ 3 & 2 & 0 & 4 & 0 \\ 0 & -4 & -1 & -2 & 0 \end{vmatrix}$
23. $\begin{vmatrix} 2 & 1 & 2 & 3 & 1 \\ 3 & 2 & 1 & 1 & 1 \\ 1 & 1 & 2 & 2 & 3 \\ 1 & 1 & 3 & 1 & 1 \\ 2 & 1 & 1 & 1 & 3 \end{vmatrix}$
24. $\begin{vmatrix} 1 & -1 & -2 & 0 & 2 \\ 1 & 1 & 0 & 2 & 0 \\ 0 & 0 & 1 & 2 & 1 \\ -1 & -1 & 1 & 0 & 1 \\ -2 & 2 & 1 & 0 & -1 \end{vmatrix}$

15.4 Cramer's Rule

In Section 3.6 we discussed solving systems of equations by **Cramer's rule**. Since we can now evaluate determinants of higher order, we can also use Cramer's rule to solve systems with more than three equations and three unknowns.

Even though Cramer's rule holds for systems of equations of any size, we shall state the rule for systems of three equations and three unknowns.

Cramer's rule: The solution of the system

$$a_1x + b_1y + c_1z = d_1$$
$$a_2x + b_2y + c_2z = d_2$$
$$a_3x + b_3y + c_3z = d_3$$

is given by

$$x = \frac{\begin{vmatrix} d_1 & b_1 & c_1 \\ d_2 & b_2 & c_2 \\ d_3 & b_3 & c_3 \end{vmatrix}}{\begin{vmatrix} a_1 & b_1 & c_1 \\ a_2 & b_2 & c_2 \\ a_3 & b_3 & c_3 \end{vmatrix}}, \quad y = \frac{\begin{vmatrix} a_1 & d_1 & c_1 \\ a_2 & d_2 & c_2 \\ a_3 & d_3 & c_3 \end{vmatrix}}{\begin{vmatrix} a_1 & b_1 & c_1 \\ a_2 & b_2 & c_2 \\ a_3 & b_3 & c_3 \end{vmatrix}},$$

$$z = \frac{\begin{vmatrix} a_1 & b_1 & d_1 \\ a_2 & b_2 & d_2 \\ a_3 & b_3 & d_3 \end{vmatrix}}{\begin{vmatrix} a_1 & b_1 & c_1 \\ a_2 & b_2 & c_2 \\ a_3 & b_3 & c_3 \end{vmatrix}}$$

Note that the denominator, which is the same for each of the unknowns, consists of the determinant whose elements are the coefficients of the unknowns. The respective numerators are formed by replacing the column of coefficients of the unknown to be found by the column of numbers on the right side.

15.4 CRAMER'S RULE

Cramer's rule can be stated more compactly by means of the following notation:

Notation for Cramer's rule

$$D = \begin{vmatrix} a_1 & b_1 & c_1 \\ a_2 & b_2 & c_2 \\ a_3 & b_3 & c_3 \end{vmatrix}$$

$$D_x = \begin{vmatrix} d_1 & b_1 & c_1 \\ d_2 & b_2 & c_2 \\ d_3 & b_3 & c_3 \end{vmatrix} \quad D_y = \begin{vmatrix} a_1 & d_1 & c_1 \\ a_2 & d_2 & c_2 \\ a_3 & d_3 & c_3 \end{vmatrix}$$

$$D_z = \begin{vmatrix} a_1 & b_1 & d_1 \\ a_2 & b_2 & d_2 \\ a_3 & b_3 & d_3 \end{vmatrix}$$

Then

$$x = \frac{D_x}{D}, \quad y = \frac{D_y}{D}, \quad z = \frac{D_z}{D}$$

Analogous notations are used for larger systems.

Example 1 Solve the following system of equations by using Cramer's rule:

$$\begin{aligned} 2x_1 - x_2 + 2x_3 - x_4 &= 1 \\ x_1 - 2x_3 + x_4 &= 0 \\ x_1 - 2x_2 + x_3 - 3x_4 &= 2 \\ -x_2 + x_3 - x_4 &= -1 \end{aligned}$$

Solution. The denominator D, common to all the unknowns, will be evaluated first:

$$D = \begin{vmatrix} 2 & -1 & 2 & -1 \\ 1 & 0 & -2 & 1 \\ 1 & -2 & 1 & -3 \\ 0 & -1 & 1 & -1 \end{vmatrix} \begin{matrix} -2R_2 + R_1 \\ \\ -1R_2 + R_3 \\ \\ \end{matrix} = \begin{vmatrix} 0 & -1 & 6 & -3 \\ 1 & 0 & -2 & 1 \\ 0 & -2 & 3 & -4 \\ 0 & -1 & 1 & -1 \end{vmatrix}$$

$$= - \begin{vmatrix} -1 & 6 & -3 \\ -2 & 3 & -4 \\ -1 & 1 & -1 \end{vmatrix} \begin{matrix} -2R_1 + R_2 \\ -1R_1 + R_3 \end{matrix}$$

$$= - \begin{vmatrix} -1 & 6 & -3 \\ 0 & -9 & 2 \\ 0 & -5 & 2 \end{vmatrix}$$

$$= -(-1) \begin{vmatrix} -9 & 2 \\ -5 & 2 \end{vmatrix} = -18 + 10 = -8$$

Now by Cramer's rule

$$x_1 = \frac{D_{x_1}}{D} = -\frac{1}{8}\begin{vmatrix} 1 & -1 & 2 & -1 \\ 0 & 0 & -2 & 1 \\ 2 & -2 & 1 & -3 \\ -1 & -1 & 1 & -1 \end{vmatrix} \quad \begin{matrix} -2R_1 + R_3 \\ 1R_1 + R_4 \end{matrix}$$

$$= -\frac{1}{8}\begin{vmatrix} 1 & -1 & 2 & -1 \\ 0 & 0 & -2 & 1 \\ 0 & 0 & -3 & -1 \\ 0 & -2 & 3 & -2 \end{vmatrix}$$

$$= -\frac{1}{8}\begin{vmatrix} 0 & -2 & 1 \\ 0 & -3 & -1 \\ -2 & 3 & -2 \end{vmatrix} = -\frac{1}{8}(-2)\begin{vmatrix} -2 & 1 \\ -3 & -1 \end{vmatrix}$$

$$= \frac{1}{4}(2+3) = \frac{5}{4}$$

The procedure for finding x_2 is similar:

$$x_2 = \frac{D_{x_2}}{D} = -\frac{1}{8}\begin{vmatrix} 2 & 1 & 2 & -1 \\ 1 & 0 & -2 & 1 \\ 1 & 2 & 1 & -3 \\ 0 & -1 & 1 & -1 \end{vmatrix} \quad \begin{matrix} -2R_2 + R_1 \\ -1R_2 + R_3 \end{matrix}$$

$$= -\frac{1}{8}\begin{vmatrix} 0 & 1 & 6 & -3 \\ 1 & 0 & -2 & 1 \\ 0 & 2 & 3 & -4 \\ 0 & -1 & 1 & -1 \end{vmatrix}$$

$$= -\frac{1}{8}(-1)\begin{vmatrix} 1 & 6 & -3 \\ 2 & 3 & -4 \\ -1 & 1 & -1 \end{vmatrix} \quad \begin{matrix} -2R_1 + R_2 \\ 1R_1 + R_3 \end{matrix}$$

$$= \frac{1}{8}\begin{vmatrix} 1 & 6 & -3 \\ 0 & -9 & 2 \\ 0 & 7 & -4 \end{vmatrix}$$

$$= \frac{1}{8}\begin{vmatrix} -9 & 2 \\ 7 & -4 \end{vmatrix} = \frac{1}{8}(36 - 14) = \frac{1}{8}(22) = \frac{11}{4}$$

For x_3 we have

$$x_3 = \frac{D_{x_3}}{D} = -\frac{1}{8}\begin{vmatrix} 2 & -1 & 1 & -1 \\ 1 & 0 & 0 & 1 \\ 1 & -2 & 2 & -3 \\ 0 & -1 & -1 & -1 \end{vmatrix} = -\frac{1}{2}$$

The simplest way to find x_4 is to substitute the values already found in one of the equations and to solve for x_4. From the fourth equation

$$-x_2 + x_3 - x_4 = -1$$

we get

$$-\frac{11}{4} - \frac{1}{2} - x_4 = -1$$

which gives $x_4 = -\frac{9}{4}$. As a check, let us substitute the values into the first equation $2x_1 - x_2 + 2x_3 - x_4 = 1$:

$$2\left(\frac{5}{4}\right) - \frac{11}{4} + 2\left(-\frac{1}{2}\right) - \left(-\frac{9}{4}\right) = \frac{10}{4} - \frac{11}{4} - \frac{4}{4} + \frac{9}{4} = \frac{4}{4} = 1$$

which checks.

Since every determinant has a unique value, Cramer's rule guarantees a unique solution for every linear system, provided that $D \neq 0$ (to avoid division by zero). Such a system is called **consistent.**

The system of equations

$$a_1 x + b_1 y + c_1 z = d_1$$
$$a_2 x + b_2 y + c_2 z = d_2$$
$$a_3 x + b_3 y + c_3 z = d_3$$

has a unique solution if, and only if, $D \neq 0$. An analogous statement holds for systems of n equations with n unknowns.

Exercises / Section 15.4

In Exercises 1–20, solve each system of equations by using Cramer's rule.

1. $x_1 + x_2 + 2x_3 = 9$
   $x_1 \quad\quad - x_3 = -2$
   $2x_1 - x_2 \quad\quad = 0$

2. $x_1 - x_2 \quad\quad = 2$
   $x_1 + 2x_2 - x_3 = -3$
   $\quad\quad -x_2 + x_3 = 3$

3. $3x_1 \quad\quad + 2x_3 = -1$
   $4x_1 - x_2 - 2x_3 = 7$
   $x_1 + x_2 \quad\quad = 2$

4. $2x_1 - 3x_2 \quad\quad = 8$
   $3x_1 - x_2 + 2x_3 = 8$
   $2x_1 + 3x_2 \quad\quad = -4$

5. $x_1 - 2x_2 - x_3 = 1$
   $2x_1 - x_2 - 2x_3 = -1$
   $3x_1 + x_2 + 2x_3 = 6$

6. $3x_1 - 3x_2 - 2x_3 = 1$
   $-x_1 + x_2 - 6x_3 = 3$
   $x_1 + x_2 + 2x_3 = 3$

7. $2x_1 - x_2 + 3x_3 = 16$
   $3x_1 + 4x_2 + 2x_3 = 7$
   $5x_1 - 6x_2 + 8x_3 = 47$

8. $x_1 - 2x_2 + x_3 = 1$
   $2x_1 + x_2 + 2x_3 = 2$
   $3x_1 + 3x_2 - 3x_3 = 2$

9. $3x_1 - 3x_2 + 7x_3 = 35$
   $6x_1 + 3x_2 - x_3 = -5$
   $9x_1 - 12x_2 + 3x_3 = 38$

10. $2x_1 - 5x_2 + 10x_3 = 21$
    $3x_1 + 5x_2 + 10x_3 = 6$
    $4x_1 - 10x_2 + 40x_3 = 64$

11. $2x_1 - x_2 + 2x_3 = 2$
$-4x_1 + 5x_2 - 3x_3 = 1$
$3x_1 - 4x_2 + x_3 = 3$

12. $2x_1 + x_2 - x_3 = 0$
$8x_1 - 2x_2 + x_3 = 1$
$4x_1 - 4x_2 + x_3 = 0$

13. $2x_2 + 3x_3 + x_4 = 1$
$2x_1 - 2x_2 + x_4 = 1$
$3x_1 - 2x_3 + 2x_4 = 13$
$x_1 + 3x_2 - x_3 = 9$

14. $x_1 + 2x_3 - 3x_4 = 12$
$3x_1 - 2x_2 + x_3 = -3$
$3x_2 - 3x_3 - 5x_4 = 12$
$2x_1 - x_2 - 2x_4 = 3$

15. $2x_1 - x_2 - 2x_3 - 2x_4 = -7$
$4x_1 - 2x_2 + 3x_3 = 3$
$6x_1 + 2x_2 + 4x_3 - 3x_4 = 16$
$8x_1 - 2x_3 - 4x_4 = -6$

16. $2x_1 + x_3 - x_4 = 0$
$x_1 - x_2 + x_4 = 1$
$x_1 + x_2 - x_4 = 0$
$2x_2 + x_3 + x_4 = -1$

17. $3x_1 - 2x_2 + x_3 - x_4 = 2$
$-x_1 - x_3 + 2x_4 = 1$
$x_1 - x_2 + x_3 = 0$
$x_1 + 2x_2 + 3x_3 - x_4 = -2$

18. $x_1 - x_2 - 3x_3 + x_4 = 2$
$2x_1 + x_3 - x_4 = 0$
$3x_1 - 2x_2 + 3x_3 + x_4 = -1$
$x_1 + x_2 - 2x_4 = 0$

19. $x_1 - x_2 + x_4 - x_5 = 1$
$x_2 - x_3 + x_4 + 2x_5 = 0$
$x_1 - x_2 + x_4 - 2x_5 = -1$
$x_2 + x_3 - x_4 = 0$
$2x_2 - x_3 + x_5 = 2$

20. $x_1 - 2x_2 + x_3 - x_5 = 0$
$2x_1 - x_3 + 2x_4 - x_5 = 1$
$x_2 - 2x_3 - x_4 + 2x_5 = 1$
$x_1 + x_3 + x_4 = 0$
$x_1 + x_3 + x_5 = 1$

21. The system of equations for finding the currents in the circuit in Figure 15.1 is

$I_1 + I_2 + I_3 - I_4 = 0$
$5I_1 - 4I_2 = -1$
$4I_2 - 3I_3 = -1$
$3I_3 + 6I_4 = 7$

Solve this system by using Cramer's rule.

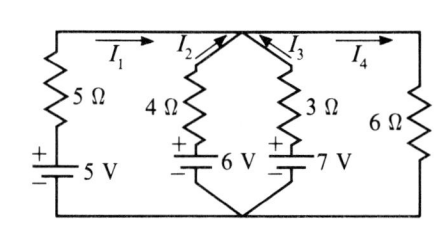

Figure 15.1

22. Use Cramer's rule to solve the system

$I_1 + I_2 + I_3 - I_4 = 0$
$2I_1 - I_2 = 1$
$I_2 - 3I_3 = 3$
$3I_3 + I_4 = -2$

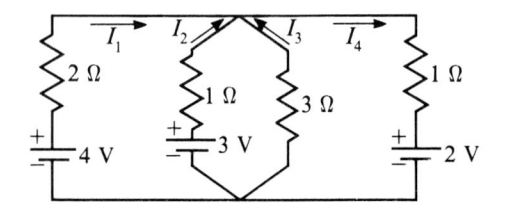

Figure 15.2

which is the system for finding the currents in the circuit shown in Figure 15.2. (See Exercise 35, Section 3.6.)

23. Four machine parts cost a total of $55. The first part costs $5 less than the other three combined, while the second part costs as much as the last two combined. Finally, the first part costs 5 times as much as the fourth part. Find the cost of each.

15.5 Matrix Algebra

We shall now turn to the other important concept in this chapter, that of a *matrix*. Matrices were first studied systematically by the English mathematician Arthur Cayley (1821–1895) and have grown in importance ever since. They are particularly well suited for solving systems of equations by means of computers, but matrix methods have been making inroads into many other fields as well, such as statistics, mechanics, and economics.

In this section we shall introduce the basic definitions and then study the algebra of matrices. The solution of equations by matrix methods will be left to a later section.

Arthur Cayley

Basic Definitions

A **matrix** is a rectangular array of numbers. The array is enclosed in brackets or parentheses to distinguish it from a determinant. For example, the array

$$\begin{bmatrix} 1 & 0 & -4 & 5 \\ 6 & -3 & 2 & 0 \end{bmatrix}$$

is a matrix. The distinction between a matrix and a determinant is an important one, for *a matrix does not have a numerical value attached to it*. A matrix with m **rows** and n **columns** is said to have **dimensions** $m \times n$. The numbers in the array are called the **elements** of the matrix. Capital letters such as A, B, and C will be used to denote a matrix.

$m \times n$ matrix

Double subscripts are customarily used to denote the individual elements of a matrix. For example, a general 3×3 and a general 2×4 matrix are written, respectively,

$$\begin{bmatrix} a_{11} & a_{12} & a_{13} \\ a_{21} & a_{22} & a_{23} \\ a_{31} & a_{32} & a_{33} \end{bmatrix} \quad \text{and} \quad \begin{bmatrix} a_{11} & a_{12} & a_{13} & a_{14} \\ a_{21} & a_{22} & a_{23} & a_{24} \end{bmatrix}$$

Meaning of a_{ij}

The subscript a_{ij} refers to the ***i*th row** and ***j*th column.** (This convention can be remembered by thinking of "RC," as in "Royal Crown."). Similarly, a 3×5 matrix has 3 rows and 5 columns.

A matrix can have a single row or column. Thus

$$\begin{bmatrix} a_{11} \\ a_{21} \\ a_{31} \end{bmatrix}$$

is a 3×1 matrix, while

$$[a_{11} \quad a_{12} \quad a_{13} \quad a_{14}]$$

Column matrix
Row matrix

is a 1×4 matrix. The first is called a **column matrix** and the second a **row matrix.**

Equality

To define equality in a natural way, note that a vector $\mathbf{A} = (a, b)$ can be viewed as a row or column matrix. Now recall that

$$(a, b) = (a', b')$$

if $a = a'$ and $b = b'$. So it seems reasonable to define the equality of matrices as follows:

> **Equality:** Two matrices are equal if, and only if, the corresponding elements are equal.

Example 1

$$\begin{bmatrix} a & b \\ c & 1 \end{bmatrix} = \begin{bmatrix} 3 & -4 \\ 2 & d \end{bmatrix}$$

if, and only if, $a = 3$, $b = -4$, $c = 2$, and $d = 1$.

Two matrices can be equal only if they have the same dimensions.

Example 2

$$\begin{bmatrix} 1 & -3 \end{bmatrix} \neq \begin{bmatrix} 1 & -3 \\ 2 & 0 \end{bmatrix}$$

since they have different dimensions.

Addition and Subtraction

The definitions of addition and subtraction of matrices can also be motivated by the corresponding operations for vectors:

$$(a, b) \pm (a', b') = (a \pm a', b \pm b')$$

> **Addition and subtraction:**
>
> The sum (difference) of two $m \times n$ matrices is obtained by adding (subtracting) the corresponding elements.

Example 3

$$\begin{bmatrix} 2 & 3 \\ -1 & 0 \\ 2 & 6 \end{bmatrix} + \begin{bmatrix} -2 & 5 \\ 6 & -4 \\ 0 & -3 \end{bmatrix} = \begin{bmatrix} 2-2 & 3+5 \\ -1+6 & 0-4 \\ 2+0 & 6-3 \end{bmatrix} = \begin{bmatrix} 0 & 8 \\ 5 & -4 \\ 2 & 3 \end{bmatrix}$$

Zero Matrix

In ordinary algebra

$$a + 0 = 0 \quad \text{for all } a$$

In matrix algebra, a matrix whose elements are all zero is called the **zero matrix** and is denoted by O. It has the analogous property

$$A + O = A$$

provided that A and O have the same dimensions.

Zero matrix:

$$O = \begin{bmatrix} 0 & 0 & \cdots & 0 \\ 0 & 0 & \cdots & 0 \\ \vdots & & & \\ 0 & 0 & \cdots & 0 \end{bmatrix}$$

Example 4 If

$$A = \begin{bmatrix} -2 & 3 & 0 \\ 1 & 4 & -3 \end{bmatrix}$$

then

$$A + O = \begin{bmatrix} -2 & 3 & 0 \\ 1 & 4 & -3 \end{bmatrix} + \begin{bmatrix} 0 & 0 & 0 \\ 0 & 0 & 0 \end{bmatrix} = \begin{bmatrix} -2 & 3 & 0 \\ 1 & 4 & -3 \end{bmatrix}$$

Scalar Product

A type of multiplication in matrix algebra involves a number k and a matrix A, which is defined next.

Scalar product: The product kA of a number k and a matrix A is called the **scalar product** and is obtained by multiplying each element of A by k.

Example 5 If $k = -2$ and

$$A = \begin{bmatrix} 1 & 2 \\ -3 & 4 \\ -7 & 8 \end{bmatrix}$$

then

$$kA = -2 \begin{bmatrix} 1 & 2 \\ -3 & 4 \\ -7 & 8 \end{bmatrix} = \begin{bmatrix} -2(1) & -2(2) \\ -2(-3) & -2(4) \\ -2(-7) & -2(8) \end{bmatrix} = \begin{bmatrix} -2 & -4 \\ 6 & -8 \\ 14 & -16 \end{bmatrix}$$

Example 6 If $k = 3$,

$$A = \begin{bmatrix} -3 & 4 & 6 \\ 0 & 1 & -2 \end{bmatrix}, \text{ and } B = \begin{bmatrix} 0 & 2 & 4 \\ -1 & -2 & 5 \end{bmatrix}$$

find $kA + B$.

Solution.
$$kA + B = 3 \begin{bmatrix} -3 & 4 & 6 \\ 0 & 1 & -2 \end{bmatrix} + \begin{bmatrix} 0 & 2 & 4 \\ -1 & -2 & 5 \end{bmatrix}$$

$$= \begin{bmatrix} -9 & 12 & 18 \\ 0 & 3 & -6 \end{bmatrix} + \begin{bmatrix} 0 & 2 & 4 \\ -1 & -2 & 5 \end{bmatrix}$$

$$= \begin{bmatrix} -9 & 14 & 22 \\ -1 & 1 & -1 \end{bmatrix}$$

Multiplication

Multiplying two matrices is harder to describe than the other operations. To see what leads to this operation, let us write the system of equations

$$\begin{aligned} a_{11}x_1 + a_{12}x_2 &= b_1 \\ a_{21}x_1 + a_{22}x_2 &= b_2 \end{aligned} \qquad (15.4)$$

in matrix form. Let

$$A = \begin{bmatrix} a_{11} & a_{12} \\ a_{21} & a_{22} \end{bmatrix}, \quad X = \begin{bmatrix} x_1 \\ x_2 \end{bmatrix}, \text{ and } B = \begin{bmatrix} b_1 \\ b_2 \end{bmatrix}$$

Then the left side of

$$AX = B$$

must be

$$AX = \begin{bmatrix} a_{11} & a_{12} \\ a_{21} & a_{22} \end{bmatrix} \begin{bmatrix} x_1 \\ x_2 \end{bmatrix} = \begin{bmatrix} a_{11}x_1 + a_{12}x_2 \\ a_{21}x_1 + a_{22}x_2 \end{bmatrix}$$

by equations (15.4). This means that the product of A and X is formed by multiplying the first (and then second) row of A by the corresponding elements in the column matrix X and adding the products.

This definition can be generalized. Let $AB = C$, where A has dimensions $m \times n$ and B has dimensions $n \times k$. The element c_{ij} in C is found by the scheme given next.

Matrix multiplication:

$$i\text{th row} \to \begin{bmatrix} a_{11} & a_{12} & \cdots & a_{1n} \\ \vdots & \vdots & & \vdots \\ a_{i1} & a_{i2} & \cdots & a_{in} \\ \vdots & \vdots & & \vdots \\ a_{m1} & a_{m2} & \cdots & a_{mn} \end{bmatrix} \begin{bmatrix} b_{11} & \cdots & b_{1j} & \cdots & b_{1k} \\ b_{21} & \cdots & b_{2j} & \cdots & b_{2k} \\ \vdots & & \vdots & & \vdots \\ b_{n1} & \cdots & b_{nj} & \cdots & b_{nk} \end{bmatrix} \overset{j\text{th column}}{=}$$

$$i \to \begin{bmatrix} & \vdots & \\ \cdots & c_{ij} & \cdots \\ & \vdots & \end{bmatrix}$$

where $c_{ij} = a_{i1}b_{1j} + a_{i2}b_{2j} + \cdots + a_{in}b_{nj}$.

To be able to form the product AB, the number of columns in A must be equal to the number of rows in B; otherwise, multiplication is undefined. Observe also that if A is an $m \times n$ matrix and B an $n \times k$ matrix, then $C = AB$ is an $m \times k$ matrix.

Example 7 If

$$A = \begin{bmatrix} 2 & -1 \\ -3 & 4 \\ 2 & -6 \end{bmatrix} \quad \text{and} \quad B = \begin{bmatrix} -1 & 2 & -3 & 1 \\ 5 & -6 & 2 & 2 \end{bmatrix}$$

find $C = AB$.

Solution. To find c_{11}, we multiply the elements of the first row of A by the corresponding elements in the first column of B and add the products:

$$\begin{bmatrix} \mathbf{2} & \mathbf{-1} \\ -3 & 4 \\ 2 & -6 \end{bmatrix} \begin{bmatrix} \mathbf{-1} & 2 & -3 & 1 \\ \mathbf{5} & -6 & 2 & 2 \end{bmatrix} = \begin{bmatrix} c_{11} & \cdots \\ \vdots & \end{bmatrix}$$

Thus $c_{11} = (\mathbf{2})(\mathbf{-1}) + (\mathbf{-1})(\mathbf{5}) = -7$.

The element c_{14} is found as follows:

$$\begin{bmatrix} \mathbf{2} & \mathbf{-1} \\ -3 & 4 \\ 2 & -6 \end{bmatrix} \begin{bmatrix} -1 & 2 & -3 & \mathbf{1} \\ 5 & -6 & 2 & \mathbf{2} \end{bmatrix} = \begin{bmatrix} \cdots & & & c_{14} \\ & & & \vdots \\ & & & \vdots \end{bmatrix}$$

Thus $c_{14} = (\mathbf{2})(\mathbf{1}) + (\mathbf{-1})(\mathbf{2}) = 0$.

For c_{32} we have

$$\begin{bmatrix} 2 & -1 \\ -3 & 4 \\ \mathbf{2} & \mathbf{-6} \end{bmatrix} \begin{bmatrix} -1 & \mathbf{2} & -3 & 1 \\ 5 & \mathbf{-6} & 2 & 2 \end{bmatrix} = \begin{bmatrix} & \vdots & & \\ & \vdots & & \\ & c_{32} & \cdots & \end{bmatrix}$$

Thus $c_{32} = (\mathbf{2})(\mathbf{2}) + (\mathbf{-6})(\mathbf{-6}) = 40$.

Continuing in this manner, we obtain the remaining elements of C:

$$\begin{bmatrix} 2 & -1 \\ -3 & 4 \\ 2 & -6 \end{bmatrix} \begin{bmatrix} -1 & 2 & -3 & 1 \\ 5 & -6 & 2 & 2 \end{bmatrix} = \begin{bmatrix} -7 & 10 & -8 & 0 \\ 23 & -30 & 17 & 5 \\ -32 & 40 & -18 & -10 \end{bmatrix}$$

One significant difference between matrix algebra and ordinary algebra can be seen in Example 7. We found the product $C = AB$. However,

$$BA = \begin{bmatrix} -1 & 2 & -3 & 1 \\ 5 & -6 & 2 & 2 \end{bmatrix} \begin{bmatrix} 2 & -1 \\ -3 & 4 \\ 2 & -6 \end{bmatrix}$$

Noncommutativity is undefined. Hence $AB \neq BA$, and we conclude that matrix multiplication is not commutative.

The Identity Matrix

In ordinary algebra, the number 1 has the property

$$1 \cdot a = a \cdot 1 = a \quad \text{for all } a$$

In matrix algebra the matrix I with the analogous property

$$IA = AI = A \quad \text{for all } A \tag{15.5}$$

is called the **multiplicative identity** or simply the **identity matrix.** The identity matrix is a **square matrix** ($n \times n$) with 1's along the diagonal starting in the upper left-hand corner and 0's elsewhere. For example, the 4×4 identity matrix is

$$I = \begin{bmatrix} 1 & 0 & 0 & 0 \\ 0 & 1 & 0 & 0 \\ 0 & 0 & 1 & 0 \\ 0 & 0 & 0 & 1 \end{bmatrix}$$

The diagonal referred to is called the **main diagonal.**

Identity matrix:

$$I = \begin{bmatrix} 1 & 0 & 0 & \cdots & 0 \\ 0 & 1 & 0 & \cdots & 0 \\ \vdots & & & & \\ 0 & 0 & 0 & \cdots & 1 \end{bmatrix}$$

Example 8 Show that $AI = IA = A$ if

$$A = \begin{bmatrix} -2 & 1 \\ -3 & 4 \end{bmatrix}$$

Solution.

$$\begin{bmatrix} -2 & 1 \\ -3 & 4 \end{bmatrix} \begin{bmatrix} 1 & 0 \\ 0 & 1 \end{bmatrix}$$
$$= \begin{bmatrix} (-2)(1) + (1)(0) & (-2)(0) + (1)(1) \\ (-3)(1) + (4)(0) & (-3)(0) + (4)(1) \end{bmatrix} = \begin{bmatrix} -2 & 1 \\ -3 & 4 \end{bmatrix}$$

$$\begin{bmatrix} 1 & 0 \\ 0 & 1 \end{bmatrix} \begin{bmatrix} -2 & 1 \\ -3 & 4 \end{bmatrix}$$
$$= \begin{bmatrix} (1)(-2) + (0)(-3) & (1)(1) + (0)(4) \\ (0)(-2) + (1)(-3) & (0)(1) + (1)(4) \end{bmatrix} = \begin{bmatrix} -2 & 1 \\ -3 & 4 \end{bmatrix}$$

According to equation (15.5), $AI = IA$, even though matrix multiplication is not commutative in general. The exercises will illustrate other differences between ordinary and matrix algebra.

Example 9 One form of the Lorentz transformation in special relativity is

$$x_1' = x_1$$
$$x_2' = x_2$$
$$x_3' = \gamma x_3 + j\beta\gamma x_4$$
$$x_4' = -j\beta\gamma x_3 + \gamma x_4$$

The transformation is frequently written in matrix form using the definition of multiplication:

$$\begin{bmatrix} x_1' \\ x_2' \\ x_3' \\ x_4' \end{bmatrix} = \begin{bmatrix} 1 & 0 & 0 & 0 \\ 0 & 1 & 0 & 0 \\ 0 & 0 & \gamma & j\beta\gamma \\ 0 & 0 & -j\beta\gamma & \gamma \end{bmatrix} \begin{bmatrix} x_1 \\ x_2 \\ x_3 \\ x_4 \end{bmatrix}$$

CHAPTER 15 DETERMINANTS AND MATRICES

Here $\gamma = 1/\sqrt{1 - \beta^2}$ and $\beta = v/c$, where v is the velocity of the body and c the velocity of light.

Exercises / Section 15.5

In Exercises 1–4, find the values of the unknowns in each case by using the definition of matrix equality.

1. $\begin{bmatrix} x & y \\ 1 & x \end{bmatrix} = \begin{bmatrix} -3 & 2 \\ 1 & -3 \end{bmatrix}$

2. $\begin{bmatrix} a & x \\ b & y \end{bmatrix} = \begin{bmatrix} 2 & 6 \\ -7 & 3 \end{bmatrix}$

3. $\begin{bmatrix} x & 0 & a+b \\ -4 & a & y \end{bmatrix} = \begin{bmatrix} 6 & 0 & -5 \\ -4 & 2 & 3 \end{bmatrix}$

4. $\begin{bmatrix} x+y & 1 & x-y \\ -2 & 2 & y \end{bmatrix} = \begin{bmatrix} 4 & 1 & 0 \\ -2 & 2 & 2 \end{bmatrix}$

In Exercises 5–12, refer to the following matrices:

$$A = \begin{bmatrix} 1 & -3 & -2 \\ 2 & 0 & 6 \end{bmatrix}, \quad B = \begin{bmatrix} 2 & -6 & 5 \\ 7 & 3 & 0 \end{bmatrix}, \quad C = \begin{bmatrix} -6 & -7 & 0 \\ 5 & 2 & 1 \end{bmatrix}$$

Perform the indicated operations.

5. $A + B$
6. $A - C$
7. $C - B$
8. $B + C$
9. $-3A$
10. $-2A - C$
11. $2B + C$
12. Verify:
    a. $A + B = B + A$
    b. $A + (B - C) = (A + B) - C$

In Exercises 13–24, perform the indicated multiplications.

13. $\begin{bmatrix} -1 & 2 \\ 0 & 1 \end{bmatrix} \begin{bmatrix} 3 \\ 4 \end{bmatrix}$

14. $\begin{bmatrix} 2 & 3 & 0 \\ 1 & -4 & 6 \end{bmatrix} \begin{bmatrix} -3 \\ 5 \\ 2 \end{bmatrix}$

15. $\begin{bmatrix} -1 & 2 \end{bmatrix} \begin{bmatrix} 3 & -4 \\ -6 & 0 \end{bmatrix}$

16. $\begin{bmatrix} 2 & 0 & -3 & 4 \end{bmatrix} \begin{bmatrix} 1 & 2 \\ -3 & 0 \\ 4 & -6 \\ 0 & 1 \end{bmatrix}$

17. $\begin{bmatrix} 10 & -2 \\ 0 & 5 \end{bmatrix} \begin{bmatrix} 6 & 8 \\ -1 & 0 \end{bmatrix}$

18. $\begin{bmatrix} 5 & -3 \\ 5 & 2 \\ 10 & 3 \end{bmatrix} \begin{bmatrix} 2 & 5 \\ -3 & -4 \end{bmatrix}$

19. $\begin{bmatrix} 0 & -1 \\ 3 & -8 \\ -6 & 7 \end{bmatrix} \begin{bmatrix} 1 & -5 & 0 \\ 2 & 4 & 3 \end{bmatrix}$

20. $\begin{bmatrix} -1 & 0 \\ 2 & 8 \end{bmatrix} \begin{bmatrix} 1 & 5 & -12 \\ -2 & 10 & 0 \end{bmatrix}$

21. $\begin{bmatrix} -5 & 10 & 3 \\ 2 & 0 & -6 \end{bmatrix} \begin{bmatrix} 6 & 7 \\ -10 & 3 \\ 12 & 6 \end{bmatrix}$

22. $\begin{bmatrix} 2 & 5 & 0 \\ 8 & 0 & 1 \\ -2 & 3 & 4 \end{bmatrix} \begin{bmatrix} 1 & 7 \\ -6 & 2 \\ 0 & 3 \end{bmatrix}$

23. $\begin{bmatrix} 2 & 0 & -5 & 6 \\ 7 & 3 & 0 & 2 \end{bmatrix} \begin{bmatrix} 1 & 2 \\ 0 & 3 \\ -5 & 6 \\ 10 & 12 \end{bmatrix}$

24. $\begin{bmatrix} -2 & 0 & 3 \\ 0 & -5 & 6 \\ 3 & -8 & 10 \end{bmatrix} \begin{bmatrix} 7 & 6 & 0 \\ -3 & 4 & 2 \\ -7 & 0 & 1 \end{bmatrix}$

In Exercises 25–28, find AB and, if possible, BA.

25. $A = \begin{bmatrix} 1 & -2 \\ 2 & 1 \end{bmatrix}$, $B = \begin{bmatrix} -2 & 3 \\ 4 & 0 \end{bmatrix}$

26. $A = \begin{bmatrix} 3 & -4 \\ 2 & 6 \\ 0 & 1 \end{bmatrix}$, $B = \begin{bmatrix} 6 & 0 & 2 \\ 2 & -1 & 3 \end{bmatrix}$

27. $A = \begin{bmatrix} 5 & -10 \\ 8 & 6 \end{bmatrix}$, $B = \begin{bmatrix} -4 \\ 2 \end{bmatrix}$

28. $A = \begin{bmatrix} -2 & 0 & 6 \\ 7 & 8 & 12 \end{bmatrix}$, $B = \begin{bmatrix} 7 \\ 2 \\ 0 \end{bmatrix}$

In Exercises 29–32, show that $AI = IA = A$.

29. $\begin{bmatrix} -2 & 0 \\ 1 & 8 \end{bmatrix}$

30. $\begin{bmatrix} 1 & 0 \\ 6 & 1 \end{bmatrix}$

31. $\begin{bmatrix} 0 & -3 & 4 \\ 1 & 3 & 7 \\ -4 & 0 & 1 \end{bmatrix}$

32. $\begin{bmatrix} 0 & 0 & -1 & 1 \\ 2 & 1 & 0 & 3 \\ 0 & 1 & 0 & 2 \\ 0 & 0 & 1 & 6 \end{bmatrix}$

33. If $ab = 0$, then $a = 0$ or $b = 0$ for any real or complex numbers a and b. This fundamental property of the number system does not carry over to matrix algebra. If

$$A = \begin{bmatrix} 1 & 1 & 1 \\ 1 & 3 & 2 \\ 2 & 0 & 1 \end{bmatrix} \quad \text{and} \quad B = \begin{bmatrix} -1 & 2 & -4 \\ -1 & 2 & -4 \\ 2 & -4 & 8 \end{bmatrix}$$

show that $AB = 0$ (even though $A \neq 0$ and $B \neq 0$).

34. If $ab = ac$, $a \neq 0$, then $b = c$ for real or complex numbers (law of cancellation). If

$$A = \begin{bmatrix} 2 & -1 & 3 \\ 3 & 2 & 1 \\ 1 & -4 & 5 \end{bmatrix}, \quad B = \begin{bmatrix} 1 & -1 & -3 & 2 \\ -1 & 1 & 4 & -1 \\ -1 & 2 & 3 & -1 \end{bmatrix}, \quad C = \begin{bmatrix} -2 & -3 & 2 & -2 \\ 2 & 3 & -1 & 3 \\ 2 & 4 & -2 & 3 \end{bmatrix}$$

show that $AB = AC$. This example shows that the law of cancellation does not necessarily hold.

35. If

$$A = \begin{bmatrix} 1 & -2 \\ 3 & -1 \end{bmatrix} \quad \text{and} \quad B = \begin{bmatrix} -1 & 3 \\ -2 & 1 \end{bmatrix}$$

show that $A^2 - B^2 \neq (A - B)(A + B)$. (Note: $A^n = A \cdot A \cdot \ldots \cdot A$ [n times].)

36. If

$$A = \begin{bmatrix} j & 0 \\ 0 & j \end{bmatrix}$$

show that $A^4 = I$.

37. In quantum mechanics, electron spin can be studied in terms of the *Pauli matrices*

$$s_x = \frac{1}{2}\begin{bmatrix} 0 & 1 \\ 1 & 0 \end{bmatrix}, \quad s_y = \frac{1}{2}\begin{bmatrix} 0 & -j \\ j & 0 \end{bmatrix}, \quad s_z = \frac{1}{2}\begin{bmatrix} 1 & 0 \\ 0 & -1 \end{bmatrix}$$

Verify that

$$2s_y s_z = js_x, \quad 2s_z s_x = js_y, \quad \text{and} \quad 2s_x s_y = js_z$$

15.6 Inverse of a Matrix and Row Operations

The purpose of this section is twofold: to introduce the *inverse* of a matrix and to find the inverse by a matrix technique called *row operations*. This technique can also be used to solve systems of equations, as we shall see in the next section.

The inverse matrix plays a role similar to division in algebra: Although division as such is not defined in matrix algebra, an operation does exist analogous to

$$a^{-1}a = 1, \quad a \neq 0 \tag{15.6}$$

This operation involves the **inverse matrix** of A, denoted by A^{-1}, which has the property

$$A^{-1}A = AA^{-1} = I \tag{15.7}$$

As we shall see shortly, not every matrix has an inverse.

> **Inverse of a matrix**: The **inverse matrix** of A, denoted by A^{-1}, is the matrix such that
>
> $$A^{-1}A = AA^{-1} = I$$

The definition of an inverse matrix is illustrated in the next example.

Example 1 Given

$$A = \begin{bmatrix} 1 & 1 \\ 1 & 0 \end{bmatrix}, \quad \text{show that} \quad B = \begin{bmatrix} 0 & 1 \\ 1 & -1 \end{bmatrix}$$

is the inverse of A.

Solution. $$AB = \begin{bmatrix} 1 & 1 \\ 1 & 0 \end{bmatrix}\begin{bmatrix} 0 & 1 \\ 1 & -1 \end{bmatrix} = \begin{bmatrix} 1 & 0 \\ 0 & 1 \end{bmatrix}$$

and

$$BA = \begin{bmatrix} 0 & 1 \\ 1 & -1 \end{bmatrix}\begin{bmatrix} 1 & 1 \\ 1 & 0 \end{bmatrix} = \begin{bmatrix} 1 & 0 \\ 0 & 1 \end{bmatrix}$$

Thus $B = A^{-1}$.

To show that $B = A^{-1}$ for a given matrix A, it is sufficient to check just one of the products.

Two methods are commonly employed for finding the inverse of a matrix, the proofs of which are too lengthy for us here. The first of these involves determinants. Let A be a 3×3 matrix, and denote by $|A|$ the determinant of A. Thus

$$A = \begin{bmatrix} a_{11} & a_{12} & a_{13} \\ a_{21} & a_{22} & a_{23} \\ a_{31} & a_{32} & a_{33} \end{bmatrix} \quad \text{and} \quad |A| = \begin{vmatrix} a_{11} & a_{12} & a_{13} \\ a_{21} & a_{22} & a_{23} \\ a_{31} & a_{32} & a_{33} \end{vmatrix}$$

Then the inverse matrix is given by

Inverse matrix

$$A^{-1} = \frac{1}{|A|} \begin{bmatrix} \begin{vmatrix} a_{22} & a_{23} \\ a_{32} & a_{33} \end{vmatrix} & -\begin{vmatrix} a_{12} & a_{13} \\ a_{32} & a_{33} \end{vmatrix} & \begin{vmatrix} a_{12} & a_{13} \\ a_{22} & a_{23} \end{vmatrix} \\ -\begin{vmatrix} a_{21} & a_{23} \\ a_{31} & a_{33} \end{vmatrix} & \begin{vmatrix} a_{11} & a_{13} \\ a_{31} & a_{33} \end{vmatrix} & -\begin{vmatrix} a_{11} & a_{13} \\ a_{21} & a_{23} \end{vmatrix} \\ \begin{vmatrix} a_{21} & a_{22} \\ a_{31} & a_{32} \end{vmatrix} & -\begin{vmatrix} a_{11} & a_{12} \\ a_{31} & a_{32} \end{vmatrix} & \begin{vmatrix} a_{11} & a_{12} \\ a_{21} & a_{22} \end{vmatrix} \end{bmatrix} \quad (15.8)$$

Since $1/|A|$ is a constant, A^{-1} has the form of a scalar product. Formula (15.8) can be extended to matrices of any size.

Since the elements of the matrix (15.8) are determinants, which always exist and have unique values, A^{-1} always exists, provided that $|A| \neq 0$ (to avoid division by 0). We now have the criterion below.

Existence of the inverse of a matrix: If A is a square matrix, then A^{-1} exists and is unique if, and only if, $|A| \neq 0$. If A is not a square matrix, then $|A|$ is undefined and A^{-1} does not exist.

Since matrix (15.8) takes too long to evaluate for larger matrices, we shall concentrate mostly on another method for determining the inverse of a matrix. This method is based on **row operations**, which are defined next.

Row operations on a matrix:
1. Interchange any two rows.
2. Multiply any row by a nonzero constant.
3. Add a nonzero multiple of any row to any other row.

Equivalence

If matrix A is changed to matrix B by one of the row operations, then matrix A is said to be **equivalent** to matrix B, designated by $A \sim B$. The term *equivalent* is borrowed from the theory of equations, for two systems of equations are also called equivalent if they have the same solution set. In fact, the row operations are precisely those operations used to solve a given system by addition or subtraction.

Note especially that Row Operation 3 is identical to Property 5 for determinants applied to rows. For that reason, we are going to employ the same notation we used earlier:

$$kR_j + R_i$$

(See Section 15.3.)

Row operations provide a convenient method for matrix inversion. The technique consists of transforming the given matrix by successive row operations to the identity matrix and simultaneously performing the same row operations on the identity matrix. The converted identity matrix is the inverse of the given matrix. In practice, these steps are best carried out by **augmenting** the given matrix with the identity matrix. For example, if the given matrix is

$$\begin{bmatrix} a_{11} & a_{12} & a_{13} \\ a_{21} & a_{22} & a_{23} \\ a_{31} & a_{32} & a_{33} \end{bmatrix}, \text{ then } \begin{bmatrix} a_{11} & a_{12} & a_{13} & 1 & 0 & 0 \\ a_{21} & a_{22} & a_{23} & 0 & 1 & 0 \\ a_{31} & a_{32} & a_{33} & 0 & 0 & 1 \end{bmatrix}$$

Augmented matrix

is called the **augmented matrix.** Loosely speaking, the two matrices are "put together."

Matrix inversion by row operations: Transform the augmented matrix

$$\begin{bmatrix} \overbrace{\begin{matrix} a_{11} & a_{12} & \cdots & a_{1n} \end{matrix}}^{A} & 1 & 0 & \cdots & 0 \\ a_{21} & a_{22} & \cdots & a_{2n} & 0 & 1 & \cdots & 0 \\ \vdots & & & & & & & \\ a_{n1} & a_{n2} & \cdots & a_{nn} & 0 & 0 & \cdots & 1 \end{bmatrix}$$

by successive row operations to

$$\begin{bmatrix} 1 & 0 & \cdots & 0 & \overbrace{\begin{matrix} b_{11} & b_{12} & \cdots & b_{1n} \end{matrix}}^{A^{-1}} \\ 0 & 1 & \cdots & 0 & b_{21} & b_{22} & \cdots & b_{2n} \\ \vdots & & & & & & & \\ 0 & 0 & \cdots & 1 & b_{n1} & b_{n2} & \cdots & b_{nn} \end{bmatrix}$$

The next example illustrates the technique.

15.6 INVERSE OF A MATRIX AND ROW OPERATIONS

Example 2 Find A^{-1} for

$$\begin{bmatrix} 1 & 1 & 0 \\ 2 & 3 & 3 \\ -1 & 1 & 1 \end{bmatrix}$$

Solution. We form the augmented matrix by putting the given matrix together with the identity matrix:

$$\begin{bmatrix} 1 & 1 & 0 & 1 & 0 & 0 \\ 2 & 3 & 3 & 0 & 1 & 0 \\ -1 & 1 & 1 & 0 & 0 & 1 \end{bmatrix}$$

We need to transform this matrix to the form

$$\begin{bmatrix} 1 & 0 & 0 & \cdots \\ 0 & 1 & 0 & \cdots \\ 0 & 0 & 1 & \cdots \end{bmatrix} \quad (15.9)$$

by successive row operations. By Row Operation 3, adding a nonzero multiple of one row to another row, we get

$$\begin{bmatrix} 1 & 1 & 0 & 1 & 0 & 0 \\ 2 & 3 & 3 & 0 & 1 & 0 \\ -1 & 1 & 1 & 0 & 0 & 1 \end{bmatrix} \begin{array}{l} \\ -2R_1 + R_2 \\ 1R_1 + R_3 \end{array}$$

$$\sim \begin{bmatrix} 1 & 1 & 0 & 1 & 0 & 0 \\ 0 & 1 & 3 & -2 & 1 & 0 \\ 0 & 2 & 1 & 1 & 0 & 1 \end{bmatrix}$$

The first column now has the desired form. To change the second column, we first obtain the 1 on the main diagonal and then the 0's. Since we already have a 1, we can proceed:

$$\begin{bmatrix} 1 & 1 & 0 & 1 & 0 & 0 \\ 0 & 1 & 3 & -2 & 1 & 0 \\ 0 & 2 & 1 & 1 & 0 & 1 \end{bmatrix} \begin{array}{l} -1R_2 + R_1 \\ \\ -2R_2 + R_3 \end{array}$$

$$\sim \begin{bmatrix} 1 & 0 & -3 & 3 & -1 & 0 \\ 0 & 1 & 3 & -2 & 1 & 0 \\ 0 & 0 & -5 & 5 & -2 & 1 \end{bmatrix}$$

These operations complete the second column. Moving on to the third column, we first obtain the 1 on the main diagonal and then the 0's. Multiplying the third row by $-\frac{1}{5}$ (Row Operation 2), we have

$$\begin{bmatrix} 1 & 0 & -3 & 3 & -1 & 0 \\ 0 & 1 & 3 & -2 & 1 & 0 \\ 0 & 0 & -5 & 5 & -2 & 1 \end{bmatrix}$$

$$\sim \begin{bmatrix} 1 & 0 & -3 & 3 & -1 & 0 \\ 0 & 1 & 3 & -2 & 1 & 0 \\ 0 & 0 & 1 & -1 & \frac{2}{5} & -\frac{1}{5} \end{bmatrix} \quad \begin{array}{c} 3R_3 + R_1 \\ -3R_3 + R_2 \end{array}$$

$$\sim \begin{bmatrix} 1 & 0 & 0 & 0 & \frac{1}{5} & -\frac{3}{5} \\ 0 & 1 & 0 & 1 & -\frac{1}{5} & \frac{3}{5} \\ 0 & 0 & 1 & -1 & \frac{2}{5} & -\frac{1}{5} \end{bmatrix}$$

Since the augmented matrix has now been transformed to form (15.9), we conclude that

$$A^{-1} = \begin{bmatrix} 0 & \frac{1}{5} & -\frac{3}{5} \\ 1 & -\frac{1}{5} & \frac{3}{5} \\ -1 & \frac{2}{5} & -\frac{1}{5} \end{bmatrix}$$

By the definition of a scalar product, we can remove the factor $\frac{1}{5}$ from all the elements and write

$$A^{-1} = \frac{1}{5} \begin{bmatrix} 0 & 1 & -3 \\ 5 & -1 & 3 \\ -5 & 2 & -1 \end{bmatrix}$$

Check:

$$A^{-1}A = \frac{1}{5} \begin{bmatrix} 0 & 1 & -3 \\ 5 & -1 & 3 \\ -5 & 2 & -1 \end{bmatrix} \begin{bmatrix} 1 & 1 & 0 \\ 2 & 3 & 3 \\ -1 & 1 & 1 \end{bmatrix}$$

$$= \frac{1}{5} \begin{bmatrix} 5 & 0 & 0 \\ 0 & 5 & 0 \\ 0 & 0 & 5 \end{bmatrix} = \begin{bmatrix} 1 & 0 & 0 \\ 0 & 1 & 0 \\ 0 & 0 & 1 \end{bmatrix}$$

In our next example we shall find the inverse of a 4×4 matrix.

Example 3 Find A^{-1} for

$$A = \begin{bmatrix} 0 & 1 & 0 & 2 \\ 1 & 2 & 0 & 0 \\ 0 & 0 & 3 & 0 \\ 0 & 1 & 2 & 5 \end{bmatrix}$$

15.6 INVERSE OF A MATRIX AND ROW OPERATIONS **475**

Solution. As before, we start with the augmented matrix:

$$\begin{bmatrix} 0 & 1 & 0 & 2 & 1 & 0 & 0 & 0 \\ 1 & 2 & 0 & 0 & 0 & 1 & 0 & 0 \\ 0 & 0 & 3 & 0 & 0 & 0 & 1 & 0 \\ 0 & 1 & 2 & 5 & 0 & 0 & 0 & 1 \end{bmatrix} \text{ (interchange } R_1 \text{ and } R_2\text{)}$$

$$\sim \begin{bmatrix} 1 & 2 & 0 & 0 & 0 & 1 & 0 & 0 \\ 0 & 1 & 0 & 2 & 1 & 0 & 0 & 0 \\ 0 & 0 & 3 & 0 & 0 & 0 & 1 & 0 \\ 0 & 1 & 2 & 5 & 0 & 0 & 0 & 1 \end{bmatrix} \begin{matrix} -2R_2 + R_1 \\ \\ \\ -1R_2 + R_4 \end{matrix}$$

$$\sim \begin{bmatrix} 1 & 0 & 0 & -4 & -2 & 1 & 0 & 0 \\ 0 & 1 & 0 & 2 & 1 & 0 & 0 & 0 \\ 0 & 0 & 3 & 0 & 0 & 0 & 1 & 0 \\ 0 & 0 & 2 & 3 & -1 & 0 & 0 & 1 \end{bmatrix} \left(\text{multiply } R_3 \text{ by } \tfrac{1}{3}\right)$$

$$\sim \begin{bmatrix} 1 & 0 & 0 & -4 & -2 & 1 & 0 & 0 \\ 0 & 1 & 0 & 2 & 1 & 0 & 0 & 0 \\ 0 & 0 & 1 & 0 & 0 & 0 & \tfrac{1}{3} & 0 \\ 0 & 0 & 2 & 3 & -1 & 0 & 0 & 1 \end{bmatrix} \begin{matrix} \\ \\ \\ -2R_3 + R_4 \end{matrix}$$

$$\sim \begin{bmatrix} 1 & 0 & 0 & -4 & -2 & 1 & 0 & 0 \\ 0 & 1 & 0 & 2 & 1 & 0 & 0 & 0 \\ 0 & 0 & 1 & 0 & 0 & 0 & \tfrac{1}{3} & 0 \\ 0 & 0 & 0 & 3 & -1 & 0 & -\tfrac{2}{3} & 1 \end{bmatrix} \left(\text{multiply } R_4 \text{ by } \tfrac{1}{3}\right)$$

$$\sim \begin{bmatrix} 1 & 0 & 0 & -4 & -2 & 1 & 0 & 0 \\ 0 & 1 & 0 & 2 & 1 & 0 & 0 & 0 \\ 0 & 0 & 1 & 0 & 0 & 0 & \tfrac{1}{3} & 0 \\ 0 & 0 & 0 & 1 & -\tfrac{1}{3} & 0 & -\tfrac{2}{9} & \tfrac{1}{3} \end{bmatrix} \begin{matrix} 4R_4 + R_1 \\ -2R_4 + R_2 \\ \\ \end{matrix}$$

$$\sim \begin{bmatrix} 1 & 0 & 0 & 0 & -\tfrac{10}{3} & 1 & -\tfrac{8}{9} & \tfrac{4}{3} \\ 0 & 1 & 0 & 0 & \tfrac{5}{3} & 0 & \tfrac{4}{9} & -\tfrac{2}{3} \\ 0 & 0 & 1 & 0 & 0 & 0 & \tfrac{1}{3} & 0 \\ 0 & 0 & 0 & 1 & -\tfrac{1}{3} & 0 & -\tfrac{2}{9} & \tfrac{1}{3} \end{bmatrix}$$

Thus

$$A^{-1} = \begin{bmatrix} -\dfrac{10}{3} & 1 & -\dfrac{8}{9} & \dfrac{4}{3} \\ \dfrac{5}{3} & 0 & \dfrac{4}{9} & -\dfrac{2}{3} \\ 0 & 0 & \dfrac{1}{3} & 0 \\ -\dfrac{1}{3} & 0 & -\dfrac{2}{9} & \dfrac{1}{3} \end{bmatrix} = \dfrac{1}{9}\begin{bmatrix} -30 & 9 & -8 & 12 \\ 15 & 0 & 4 & -6 \\ 0 & 0 & 3 & 0 \\ -3 & 0 & -2 & 3 \end{bmatrix}$$

Check:

$$A^{-1}A = \dfrac{1}{9}\begin{bmatrix} -30 & 9 & -8 & 12 \\ 15 & 0 & 4 & -6 \\ 0 & 0 & 3 & 0 \\ -3 & 0 & -2 & 3 \end{bmatrix}\begin{bmatrix} 0 & 1 & 0 & 2 \\ 1 & 2 & 0 & 0 \\ 0 & 0 & 3 & 0 \\ 0 & 1 & 2 & 5 \end{bmatrix}$$

$$= \dfrac{1}{9}\begin{bmatrix} 9 & 0 & 0 & 0 \\ 0 & 9 & 0 & 0 \\ 0 & 0 & 9 & 0 \\ 0 & 0 & 0 & 9 \end{bmatrix} = \begin{bmatrix} 1 & 0 & 0 & 0 \\ 0 & 1 & 0 & 0 \\ 0 & 0 & 1 & 0 \\ 0 & 0 & 0 & 1 \end{bmatrix}$$

Exercises / Section 15.6

In Exercises 1–24, find the inverse matrix of each given matrix by means of row operations.

1. $\begin{bmatrix} 1 & 3 \\ 2 & 4 \end{bmatrix}$

2. $\begin{bmatrix} 3 & -2 \\ 1 & 1 \end{bmatrix}$

3. $\begin{bmatrix} 2 & 4 \\ 3 & 2 \end{bmatrix}$

4. $\begin{bmatrix} 3 & -2 \\ 4 & 0 \end{bmatrix}$

5. $\begin{bmatrix} 1 & 0 & 0 \\ 0 & 1 & 0 \\ 1 & 0 & 1 \end{bmatrix}$

6. $\begin{bmatrix} 1 & 2 & 5 \\ 2 & 3 & 8 \\ -1 & 1 & 2 \end{bmatrix}$

7. $\begin{bmatrix} 1 & 0 & 2 \\ 0 & -1 & 1 \\ 1 & -2 & 1 \end{bmatrix}$

8. $\begin{bmatrix} 1 & 2 & 3 \\ -1 & 1 & 0 \\ 0 & 2 & 1 \end{bmatrix}$

9. $\begin{bmatrix} 0 & 1 & 2 \\ 2 & 0 & 4 \\ 0 & -1 & 1 \end{bmatrix}$

10. $\begin{bmatrix} 2 & -1 & 1 \\ 1 & 0 & -1 \\ 0 & -1 & 2 \end{bmatrix}$

11. $\begin{bmatrix} 3 & 3 & 2 \\ 2 & 2 & 1 \\ 0 & 1 & -1 \end{bmatrix}$

12. $\begin{bmatrix} 2 & 0 & 0 \\ 2 & 4 & 4 \\ 2 & 0 & -1 \end{bmatrix}$

13. $\begin{bmatrix} 0 & 2 & 1 \\ 3 & 1 & 3 \\ 4 & -2 & -1 \end{bmatrix}$

14. $\begin{bmatrix} 1 & 4 & 0 \\ 3 & 3 & 1 \\ 2 & -2 & 0 \end{bmatrix}$

15. $\begin{bmatrix} 1 & 2 & 3 \\ 0 & 1 & -2 \\ 3 & 2 & 1 \end{bmatrix}$

16. $\begin{bmatrix} 2 & 1 & 0 \\ 0 & -1 & 0 \\ 4 & 3 & 1 \end{bmatrix}$

17. $\begin{bmatrix} 1 & 2 & 0 & 0 \\ -2 & -3 & 1 & 2 \\ 0 & 1 & 0 & -1 \\ 0 & 0 & 1 & 2 \end{bmatrix}$

18. $\begin{bmatrix} 3 & 2 & 0 & 0 \\ 0 & 1 & -2 & 0 \\ -2 & 0 & -3 & 0 \\ 0 & 1 & 0 & 1 \end{bmatrix}$

19. $\begin{bmatrix} 1 & 0 & 1 & 2 \\ 0 & 1 & 2 & -1 \\ 0 & 0 & 1 & 3 \\ 2 & 3 & 0 & 1 \end{bmatrix}$
20. $\begin{bmatrix} 3 & -2 & 0 & 1 \\ 0 & 1 & 2 & 0 \\ 2 & 1 & -2 & 0 \\ 1 & 0 & 0 & 1 \end{bmatrix}$
21. $\begin{bmatrix} 2 & 0 & 1 & 2 \\ 2 & 1 & 3 & 0 \\ 4 & 4 & 1 & 2 \\ 0 & 1 & 0 & -1 \end{bmatrix}$
22. $\begin{bmatrix} 1 & 1 & 2 & 1 \\ 2 & -1 & -1 & 2 \\ 0 & 0 & -1 & 2 \\ 2 & 3 & 1 & 0 \end{bmatrix}$
23. $\begin{bmatrix} 2 & -1 & 0 & 1 \\ 3 & -2 & 2 & 1 \\ 0 & 3 & 4 & 1 \\ 2 & 0 & 0 & 2 \end{bmatrix}$
24. $\begin{bmatrix} 1 & -1 & 0 & 0 \\ 3 & -2 & -1 & 0 \\ 2 & 2 & 0 & 0 \\ 0 & 3 & 1 & 1 \end{bmatrix}$

25. Find A^{-1} for the matrix in Exercise 5 by using formula (15.8).
26. Find A^{-1} for the matrix in Exercise 6 by using formula (15.8).

15.7 Solving Systems of Equations by Matrix Methods

In this section we shall learn two methods for solving systems of equations by means of matrices. The first method uses the inverse of the coefficient matrix; the second yields the solution directly by row operations.

To describe the first method, consider a typical system of three equations and three unknowns.

$$a_{11}x_1 + a_{12}x_2 + a_{13}x_3 = b_1$$
$$a_{21}x_1 + a_{22}x_2 + a_{23}x_3 = b_2 \quad (15.10)$$
$$a_{31}x_1 + a_{32}x_2 + a_{33}x_3 = b_3$$

Let

$$A = \begin{bmatrix} a_{11} & a_{12} & a_{13} \\ a_{21} & a_{22} & a_{23} \\ a_{31} & a_{32} & a_{33} \end{bmatrix}, \quad X = \begin{bmatrix} x_1 \\ x_2 \\ x_3 \end{bmatrix}, \quad B = \begin{bmatrix} b_1 \\ b_2 \\ b_3 \end{bmatrix}$$

Coefficient matrix

where A is called the **coefficient matrix**. If we write the system in the form $AX = B$, the method of solution can be compared with that of the single linear equation $ax = b$.

$$\begin{array}{ll} ax = b & AX = B \\ a^{-1}ax = a^{-1}b & A^{-1}AX = A^{-1}B \\ 1x = a^{-1}b & IX = A^{-1}B \\ x = a^{-1}b & X = A^{-1}B \quad (15.11) \end{array}$$

Once A^{-1} is found, the solution can be obtained by multiplying A^{-1} by the column matrix B.

The solution of the system
$AX = B$ is given by $X = A^{-1}B$

Example 1

Solve the system

$$x_1 + x_2 = 0$$
$$2x_1 + 3x_2 + 3x_3 = 1$$
$$-x_1 + x_2 + x_3 = 1$$

by using A^{-1}.

Solution. For this system

$$A = \begin{bmatrix} 1 & 1 & 0 \\ 2 & 3 & 3 \\ -1 & 1 & 1 \end{bmatrix}, \quad X = \begin{bmatrix} x_1 \\ x_2 \\ x_3 \end{bmatrix}, \quad B = \begin{bmatrix} 0 \\ 1 \\ 1 \end{bmatrix}$$

By the method of Section 15.6

$$A^{-1} = \frac{1}{5} \begin{bmatrix} 0 & 1 & -3 \\ 5 & -1 & 3 \\ -5 & 2 & -1 \end{bmatrix}$$

So by formula (15.11)

$$X = A^{-1}B = \frac{1}{5} \begin{bmatrix} 0 & 1 & -3 \\ 5 & -1 & 3 \\ -5 & 2 & -1 \end{bmatrix} \begin{bmatrix} 0 \\ 1 \\ 1 \end{bmatrix} = \frac{1}{5} \begin{bmatrix} -2 \\ 2 \\ 1 \end{bmatrix}$$

and

$$X = \begin{bmatrix} -\frac{2}{5} \\ \frac{2}{5} \\ \frac{1}{5} \end{bmatrix}$$

Considering the number of steps necessary to find A^{-1}, the method used in solving Example 1 does not seem as practical as employing Cramer's rule. However, as pointed out earlier, matrix algebra has found its way into ever more areas of application beyond the solution of equations. Systems of equations merely provide a simple and natural way to introduce the concept.

To describe the method of solution by row operations, let us return to the system (15.10). By **augmenting the coefficient matrix** A with the column matrix B, we get the matrix

Augmented matrix
$$\begin{bmatrix} a_{11} & a_{12} & a_{13} & b_1 \\ a_{21} & a_{22} & a_{23} & b_2 \\ a_{31} & a_{32} & a_{33} & b_3 \end{bmatrix}$$

15.7 SOLVING SYSTEMS OF EQUATIONS BY MATRIX METHODS

Recall that any row operation transforms this matrix into an equivalent matrix. So if the matrix is transformed to

$$\begin{bmatrix} 1 & 0 & 0 & s_1 \\ 0 & 1 & 0 & s_2 \\ 0 & 0 & 1 & s_3 \end{bmatrix} \tag{15.12}$$

we get the equivalent system of equations

$$x_1 + 0 + 0 = s_1$$
$$0 + x_2 + 0 = s_2$$
$$0 + 0 + x_3 = s_3$$

Hence $x_1 = s_1$, $x_2 = s_2$, and $x_3 = s_3$.

Solution by row operations:

Transform the augmented matrix

$$\begin{bmatrix} a_{11} & a_{12} & \cdots & a_{1n} & b_1 \\ a_{21} & a_{22} & \cdots & a_{2n} & b_2 \\ \vdots & & & & \\ a_{n1} & a_{n2} & \cdots & a_{nn} & b_n \end{bmatrix}$$

by successive row operations to ↓ ──── solution column

$$\begin{bmatrix} 1 & 0 & \cdots & 0 & s_1 \\ 0 & 1 & \cdots & 0 & s_2 \\ \vdots & & & & \\ 0 & 0 & \cdots & 1 & s_n \end{bmatrix}$$

The remaining examples illustrate how to solve systems by row operations.

Example 2 Solve the system in Example 1,

$$x_1 + x_2 = 0$$
$$2x_1 + 3x_2 + 3x_3 = 1$$
$$-x_1 + x_2 + x_3 = 1$$

by means of row operations.

Solution. From

$$A = \begin{bmatrix} 1 & 1 & 0 \\ 2 & 3 & 3 \\ -1 & 1 & 1 \end{bmatrix} \quad \text{and} \quad B = \begin{bmatrix} 0 \\ 1 \\ 1 \end{bmatrix}$$

we form the augmented matrix and transform this matrix to form (15.12). Thus

$$\begin{bmatrix} 1 & 1 & 0 & 0 \\ 2 & 3 & 3 & 1 \\ -1 & 1 & 1 & 1 \end{bmatrix} \begin{matrix} \\ -2R_1 + R_2 \\ 1R_1 + R_3 \end{matrix}$$

$$\sim \begin{bmatrix} 1 & 1 & 0 & 0 \\ 0 & 1 & 3 & 1 \\ 0 & 2 & 1 & 1 \end{bmatrix} \begin{matrix} -1R_2 + R_1 \\ \\ -2R_2 + R_3 \end{matrix}$$

$$\sim \begin{bmatrix} 1 & 0 & -3 & -1 \\ 0 & 1 & 3 & 1 \\ 0 & 0 & -5 & -1 \end{bmatrix} \left(\text{multiply } R_3 \text{ by } -\frac{1}{5} \right)$$

$$\sim \begin{bmatrix} 1 & 0 & -3 & -1 \\ 0 & 1 & 3 & 1 \\ 0 & 0 & 1 & \frac{1}{5} \end{bmatrix} \begin{matrix} 3R_3 + R_1 \\ -3R_3 + R_2 \\ \end{matrix}$$

$$\sim \begin{bmatrix} 1 & 0 & 0 & -\frac{2}{5} \\ 0 & 1 & 0 & \frac{2}{5} \\ 0 & 0 & 1 & \frac{1}{5} \end{bmatrix}$$

The solution is therefore given by $x_1 = -\frac{2}{5}$, $x_2 = \frac{2}{5}$, and $x_3 = \frac{1}{5}$.

Example 3 Solve the following system of equations by means of row operations:

$$\begin{aligned} 2x_1 - x_2 + 2x_4 &= 0 \\ x_1 + 2x_2 - x_3 - x_4 &= 1 \\ x_1 + 2x_3 + x_4 &= -1 \\ x_1 - x_2 - x_3 + x_4 &= 2 \end{aligned}$$

Solution.

$$\begin{bmatrix} 2 & -1 & 0 & 2 & 0 \\ 1 & 2 & -1 & -1 & 1 \\ 1 & 0 & 2 & 1 & -1 \\ 1 & -1 & -1 & 1 & 2 \end{bmatrix} \text{ (interchange } R_1 \text{ and } R_2\text{)}$$

$$\sim \begin{bmatrix} 1 & 2 & -1 & -1 & 1 \\ 2 & -1 & 0 & 2 & 0 \\ 1 & 0 & 2 & 1 & -1 \\ 1 & -1 & -1 & 1 & 2 \end{bmatrix} \begin{matrix} \\ -2R_1 + R_2 \\ -1R_1 + R_3 \\ -1R_1 + R_4 \end{matrix}$$

15.7 SOLVING SYSTEMS OF EQUATIONS BY MATRIX METHODS

$$\sim \begin{bmatrix} 1 & 2 & -1 & -1 & 1 \\ 0 & -5 & 2 & 4 & -2 \\ 0 & -2 & 3 & 2 & -2 \\ 0 & -3 & 0 & 2 & 1 \end{bmatrix} \quad -2R_4 + R_2 \quad \text{(to obtain a 1)}$$

$$\sim \begin{bmatrix} 1 & 2 & -1 & -1 & 1 \\ 0 & 1 & 2 & 0 & -4 \\ 0 & -2 & 3 & 2 & -2 \\ 0 & -3 & 0 & 2 & 1 \end{bmatrix} \quad \begin{matrix} -2R_2 + R_1 \\ \\ 2R_2 + R_3 \\ 3R_2 + R_4 \end{matrix}$$

$$\sim \begin{bmatrix} 1 & 0 & -5 & -1 & 9 \\ 0 & 1 & 2 & 0 & -4 \\ 0 & 0 & 7 & 2 & -10 \\ 0 & 0 & 6 & 2 & -11 \end{bmatrix} \quad -1R_4 + R_3 \quad \text{(to obtain a 1)}$$

$$\sim \begin{bmatrix} 1 & 0 & -5 & -1 & 9 \\ 0 & 1 & 2 & 0 & -4 \\ 0 & 0 & 1 & 0 & 1 \\ 0 & 0 & 6 & 2 & -11 \end{bmatrix} \quad \begin{matrix} 5R_3 + R_1 \\ -2R_3 + R_2 \\ \\ -6R_3 + R_4 \end{matrix}$$

$$\sim \begin{bmatrix} 1 & 0 & 0 & -1 & 14 \\ 0 & 1 & 0 & 0 & -6 \\ 0 & 0 & 1 & 0 & 1 \\ 0 & 0 & 0 & 2 & -17 \end{bmatrix} \quad \text{(divide } R_4 \text{ by 2)}$$

$$\sim \begin{bmatrix} 1 & 0 & 0 & -1 & 14 \\ 0 & 1 & 0 & 0 & -6 \\ 0 & 0 & 1 & 0 & 1 \\ 0 & 0 & 0 & 1 & -\frac{17}{2} \end{bmatrix} \quad 1R_4 + R_1$$

$$\sim \begin{bmatrix} 1 & 0 & 0 & 0 & \frac{11}{2} \\ 0 & 1 & 0 & 0 & -6 \\ 0 & 0 & 1 & 0 & 1 \\ 0 & 0 & 0 & 1 & -\frac{17}{2} \end{bmatrix}$$

The solution is therefore given by $x_1 = \frac{11}{2}$, $x_2 = -6$, $x_3 = 1$, and $x_4 = -\frac{17}{2}$.

Since the method of row operations is systematic and has certain operations repeating, it is relatively easy to program. Consequently, it is the best method to use when solving systems of equations on a computer.

Exercises / Section 15.7

In Exercises 1–32: **a.** solve each of the given systems of equations by means of row operations; **b.** solve each system of equations by using A^{-1}. (In Exercises 1–24, A is identical to the matrix in the corresponding exercises in Section 15.6.)

1. $x_1 + 3x_2 = 2$
   $2x_1 + 4x_2 = 1$

2. $3x_1 - 2x_2 = 7$
   $x_1 + x_2 = 4$

3. $2x_1 + 4x_2 = 6$
   $3x_1 + 2x_2 = -1$

4. $3x_1 - 2x_2 = 2$
   $4x_1 = 5$

5. $x_1 = 2$
   $ x_2 = 3$
   $x_1 + x_3 = 4$

6. $x_1 + 2x_2 + 5x_3 = 1$
   $2x_1 + 3x_2 + 8x_3 = 2$
   $-x_1 + x_2 + 2x_3 = 3$

7. $x_1 - 2x_3 = 3$
   $ -x_2 + x_3 = -1$
   $x_1 - 2x_2 + x_3 = 7$

8. $x_1 + 2x_2 + 3x_3 = -2$
   $-x_1 + x_2 = -1$
   $ 2x_2 + x_3 = 1$

9. $ x_2 + 2x_3 = -1$
   $2x_1 + 4x_3 = 6$
   $-x_2 + x_3 = 1$

10. $2x_1 - x_2 + x_3 = 3$
    $x_1 - x_3 = 2$
    $ - x_2 + 2x_3 = 2$

11. $3x_1 + 3x_2 + 2x_3 = 1$
    $2x_1 + 2x_2 + x_3 = 2$
    $ x_2 - x_3 = 3$

12. $2x_1 = 1$
    $2x_1 + 4x_2 + 4x_3 = 1$
    $2x_1 - x_3 = -1$

13. $ 2x_2 + x_3 = 2$
    $3x_1 + x_2 + 3x_3 = -1$
    $4x_1 - 2x_2 - x_3 = 2$

14. $x_1 + 4x_2 = 4$
    $3x_1 + 3x_2 + x_3 = 7$
    $2x_1 - 2x_2 = 3$

15. $x_1 + 2x_2 + 3x_3 = 1$
    $ x_2 - 2x_3 = 1$
    $3x_1 + 2x_2 + x_3 = -1$

16. $2x_1 + x_2 = 2$
    $ -x_2 = 2$
    $4x_1 + 3x_2 + x_3 = 4$

17. $x_1 + 2x_2 = 1$
    $-2x_1 - 3x_2 + x_3 + 2x_4 = 0$
    $ x_2 - x_4 = 2$
    $ x_3 + 2x_4 = -3$

18. $3x_1 + 2x_2 = 1$
    $ x_2 - 2x_3 = 1$
    $-2x_1 - 3x_3 = 3$
    $ x_2 + x_4 = 2$

19. $x_1 + x_3 + 2x_4 = 2$
    $ x_2 + 2x_3 - x_4 = 1$
    $ x_3 + 3x_4 = 1$
    $2x_1 + 3x_2 + x_4 = -1$

20. $3x_1 - 2x_2 + x_4 = 1$
    $ x_2 + 2x_3 = 1$
    $2x_1 + x_2 - 2x_3 = 1$
    $x_1 + x_4 = 1$

21. $2x_1 + x_3 + 2x_4 = 1$
    $2x_1 + x_2 + 3x_3 = 0$
    $4x_1 + 4x_2 + x_3 + 2x_4 = 1$
    $ x_2 - x_4 = 0$

22. $x_1 + x_2 + 2x_3 + x_4 = -1$
    $2x_1 - x_2 - x_3 + 2x_4 = 1$
    $ -x_3 + 2x_4 = -1$
    $2x_1 + 3x_2 + x_3 = 1$

23. $2x_1 - x_2 + x_4 = 2$
    $3x_1 - 2x_2 + 2x_3 + x_4 = 0$
    $ 3x_2 + 4x_3 + x_4 = 0$
    $2x_1 + 2x_4 = 3$

24. $x_1 - x_2 = 1$
    $3x_1 - 2x_2 - x_3 = 0$
    $2x_1 + 2x_2 = 3$
    $ 3x_2 + x_3 + x_4 = 2$

25. $x_1 + x_2 = 1$
    $x_1 + 2x_2 + x_3 = 1$
    $x_1 - 2x_3 = 3$

26. $2x_1 - x_2 + x_3 = 1$
    $ - 2x_2 + 2x_3 = -2$
    $x_1 + x_3 = 3$

27. $x_1 + x_3 = 0$
    $2x_1 + 2x_2 + 2x_3 = 1$
    $3x_1 - x_2 + x_3 = 1$

28. $x_1 + x_2 = 2$
    $ x_2 - x_3 = 3$
    $x_1 + x_2 - x_3 = 4$

29. $x_1 - x_3 + x_4 = -1$
$x_1 - 2x_2 + 2x_4 = 0$
$x_1 + x_2 - x_3 = 0$
$x_1 - x_2 + 4x_4 = 1$

30. $x_1 - x_3 + 2x_4 = 1$
$2x_1 + x_2 - 2x_3 + 4x_4 = 1$
$ - 3x_2 + x_3 + x_4 = 2$
$x_1 - x_3 + 3x_4 = 0$

31. $x_1 - 2x_3 + 3x_4 = -2$
$2x_1 + x_2 - 4x_3 + 5x_4 = -1$
$ 2x_2 + 3x_4 = 1$
$2x_1 - 5x_3 + 4x_4 = 1$

32. $3x_1 - x_2 + 2x_3 + 2x_4 = 2$
$x_1 + x_3 + x_4 = 1$
$ - x_2 - x_4 = 1$
$-x_1 + x_2 + x_3 - x_4 = 1$

33. Solve the given system by means of row operations. (See Exercise 22, Section 15.4.)

$I_1 + I_2 + I_3 - I_4 = 0$
$2I_1 - I_2 = 1$
$ I_2 - 3I_3 = 3$
$ 3I_3 + I_4 = -2$

34. Three machine parts cost a total of $40. The first part costs as much as the other two together, and the cost of 6 times the second is $2 more than the total cost of the other two. Find the cost of each part.

35. An experimenter determined that three resistors satisfy the following relationships:

$R_1 + 2R_2 + 2R_3 = 14$
$2R_1 + R_2 + R_3 = 10$
$R_1 + 2R_2 + 3R_3 = 18$

Find the three resistances (in ohms).

Review Exercises / Chapter 15

In Exercises 1–6, evaluate the given determinants.

1. $\begin{vmatrix} -10 & 2 \\ -3 & 1 \end{vmatrix}$

2. $\begin{vmatrix} 2 & -2 & 0 \\ 3 & -4 & 2 \\ 2 & 2 & 3 \end{vmatrix}$

3. $\begin{vmatrix} 4 & 2 & 4 & 3 \\ 2 & 3 & 1 & 1 \\ 4 & 4 & 3 & 2 \\ 4 & 4 & 3 & 5 \end{vmatrix}$

4. $\begin{vmatrix} 1 & 1 & 3 & 3 \\ 1 & -3 & 2 & -1 \\ 3 & 1 & 6 & -3 \\ 3 & 2 & -3 & 10 \end{vmatrix}$

5. $\begin{vmatrix} 2 & 0 & 0 & 0 & 5 \\ 1 & 0 & 1 & 0 & 2 \\ 1 & 0 & 0 & 0 & 3 \\ 3 & 3 & 2 & 0 & 1 \\ 2 & 1 & 3 & 2 & 0 \end{vmatrix}$

6. $\begin{vmatrix} 1 & 0 & 3 & 2 & 1 \\ 3 & 1 & 3 & 2 & 1 \\ 2 & 2 & 2 & 4 & 2 \\ 2 & 2 & -2 & -2 & -1 \\ 4 & 1 & 6 & 3 & 3 \end{vmatrix}$

In Exercises 7–10, solve each system of equations by using Cramer's rule.

7. $2x_1 + x_2 - x_3 = 0$
$x_1 - x_2 + x_3 = 1$
$x_1 - 2x_2 - 3x_3 = 2$

8. $x_1 - x_2 + 2x_3 = 1$
$x_1 - 3x_3 = 2$
$2x_1 + 2x_2 = 3$

9. $5x_1 + 6x_2 + 4x_3 + 3x_4 = 0$
$3x_1 - 4x_2 + 2x_3 + 8x_4 = -9$
$2x_1 + 3x_2 + 5x_3 + 4x_4 = 4$
$3x_1 + 5x_2 - x_3 + x_4 = 2$

10. $2x_1 - x_2 - x_3 + 2x_4 = 1$
$4x_1 - 2x_2 + x_3 + 3x_4 = 6$
$x_1 + x_2 + 2x_3 + 2x_4 = 3$
$2x_1 + 2x_2 - 3x_3 + x_4 = 2$

In Exercises 11–12, use the definition of matrix equality to determine the value of each of the unknowns.

11. $\begin{bmatrix} x+y & 1 & y \\ z & 2 & 1 \end{bmatrix} = \begin{bmatrix} 3 & 1 & 1 \\ -3 & 2 & 1 \end{bmatrix}$

12. $\begin{bmatrix} a-b & a+b & c \\ c+d & 4 & 5 \end{bmatrix} = \begin{bmatrix} 2 & 6 & -1 \\ 2 & 4 & 5 \end{bmatrix}$

In Exercises 13–16, refer to the following matrices:

$$A = \begin{bmatrix} 1 & -1 \\ 0 & 2 \\ 4 & 5 \end{bmatrix}, \quad B = \begin{bmatrix} -2 & 6 \\ 0 & -3 \\ 2 & 7 \end{bmatrix}, \quad C = \begin{bmatrix} 0 & 1 \\ 2 & -4 \\ 3 & 5 \end{bmatrix}$$

Find

13. $A + B - C$

14. $C - B$

15. $-2A + C$

16. $A - 3C$

In Exercises 17–20, perform the indicated multiplications.

17. $\begin{bmatrix} 1 & -2 & 3 \\ 2 & -5 & 6 \end{bmatrix} \begin{bmatrix} -2 \\ 2 \\ -1 \end{bmatrix}$

18. $\begin{bmatrix} -1 & 3 & -2 & 0 \end{bmatrix} \begin{bmatrix} 2 & -2 \\ -1 & 5 \\ 0 & -2 \\ 1 & 3 \end{bmatrix}$

19. $\begin{bmatrix} -1 & 2 \\ 0 & 6 \\ 4 & -1 \end{bmatrix} \begin{bmatrix} -1 & 4 & -3 \\ 0 & -8 & 5 \end{bmatrix}$

20. $\begin{bmatrix} 2 & -4 \\ 0 & -2 \end{bmatrix} \begin{bmatrix} 4 & 3 \\ -1 & 2 \end{bmatrix}$

In Exercises 21–27, find the inverse of each matrix by means of row operations.

21. $\begin{bmatrix} 2 & 6 \\ 3 & 11 \end{bmatrix}$

22. $\begin{bmatrix} 5 & 6 \\ 3 & 9 \end{bmatrix}$

23. $\begin{bmatrix} -3 & -5 & 0 \\ 2 & 6 & 1 \\ 0 & 7 & 3 \end{bmatrix}$

24. $\begin{bmatrix} 3 & -1 & 2 \\ 2 & 0 & 1 \\ 1 & 2 & 0 \end{bmatrix}$

25. $\begin{bmatrix} 2 & -3 & 4 \\ 1 & 0 & -5 \\ 0 & -2 & 1 \end{bmatrix}$

26. $\begin{bmatrix} 1 & 0 & 0 & 0 \\ 0 & 2 & 0 & -1 \\ 2 & -1 & 0 & 0 \\ 0 & 0 & -1 & 1 \end{bmatrix}$

27. $\begin{bmatrix} 3 & -1 & 0 & 2 \\ 1 & 0 & -1 & 3 \\ 0 & 1 & 0 & 1 \\ 2 & 2 & 2 & 1 \end{bmatrix}$

28. Find the inverse of the matrix in Exercise 25 by using formula (15.8).

In Exercises 29–33, **a.** solve each of the given systems of equations by using row operations; **b.** solve each of the systems by using A^{-1} (the coefficient matrices are identical to the matrices in Exercises 23–27, respectively).

29. $-3x_1 - 5x_2 = 2$
$2x_1 + 6x_2 + x_3 = -1$
$ 7x_2 + 3x_3 = 1$

30. $3x_1 - x_2 + 2x_3 = 2$
$2x_1 + x_3 = 1$
$x_1 + 2x_2 = 4$

31. $2x_1 - 3x_2 + 4x_3 = 1$
$x_1 - 5x_3 = 2$
$ -2x_2 + x_3 = -1$

32. $x_1 = 1$
$ 2x_2 - x_4 = 3$
$2x_1 - x_2 = -5$
$ -x_3 + x_4 = -2$

33. $3x_1 - x_2 + 2x_4 = 2$
$x_1 - x_3 + 3x_4 = 2$
$ x_2 + x_4 = 0$
$2x_1 + 2x_2 + 2x_3 + x_4 = -1$

In Exercises 34–35, solve the systems by using row operations.

34. In a problem involving static forces, four forces (in pounds) are related by the following equations:

$2f_1 + f_2 + 3f_3 = 15$
$ 2f_2 + f_3 + 4f_4 = 23$
$f_1 + 2f_3 + 3f_4 = 20$
$2f_1 + 3f_2 + 3f_3 = 19$

Find the forces.

35. Find the currents (in amperes) satisfying the following relationships:

$I_1 + I_2 - I_3 - I_4 = 0$
$2I_1 + 3I_2 = 2$
$ -2I_2 + 2I_3 = -2$
$ -3I_3 - I_4 = 3$

Cumulative Review Exercises / Chapters 13–15

1. Solve the given system of equations graphically. Estimate the solution to the nearest tenth of a unit.

    $x + y = 1$
    $y = x^2$

In Exercises 2 and 3, solve the given systems algebraically.

2. $x + y = 6$
    $xy = 8$

3. $x^2 - 2y^2 = 10$
    $x^2 + y^2 = 16$

4. Solve the following equation in quadratic form:

    $x^{-4} - 3x^{-2} - 4 = 0$

5. Solve for x:

    $\sqrt{x - 5} - \sqrt{x - 2} + 1 = 0$

In Exercises 6–11, use synthetic division.

6. Determine whether $x - \frac{1}{2}$ is a factor of the polynomial

    $4x^4 + 4x^3 + x^2 - 6x + 1$

7. Determine whether $x = -2$ is a zero of $f(x) = 2x^4 - 10x^2 - 2x + 4$.

8. Given $f(x) = 2x^3 - x + 6$, find $f(-1)$.

9. Solve the equation

    $x^4 - 8x^3 + 21x^2 - 22x + 8 = 0$

    given that 1 is a double root.

10. Find the rational roots of the equation

    $6x^4 - 11x^3 - 8x^2 + 25x - 12 = 0$

11. Estimate the irrational root of

    $x^3 - 3x^2 - x + 2 = 0$

    located between 0 and 1 to two decimal places.

12. Evaluate

    $$\begin{vmatrix} 2 & 2 & 2 & 5 \\ -4 & -1 & 3 & 2 \\ 2 & -3 & 5 & 0 \\ 0 & 1 & -4 & 2 \end{vmatrix}$$

13. Solve the following system by Cramer's rule:

$$I_1 + 3I_2 - I_3 = 9$$
$$3I_1 - 2I_3 + 2I_4 = 13$$
$$2I_1 - 2I_2 + I_4 = 1$$
$$2I_2 + 3I_3 + I_4 = 1$$

14. Perform the following addition of matrices:

$$\begin{bmatrix} 0 & -3 & 4 \\ 7 & 0 & 5 \\ -9 & 6 & 5 \end{bmatrix} + \begin{bmatrix} 2 & 0 & -6 \\ -8 & 7 & 0 \\ 3 & 0 & 6 \end{bmatrix}$$

15. Perform the following multiplication of matrices:

$$\begin{bmatrix} 6 & 0 & -2 \\ 1 & 3 & 5 \end{bmatrix} \begin{bmatrix} 1 & 0 \\ -2 & 1 \\ 3 & -4 \end{bmatrix}$$

16. Find the inverse of the following matrix:

$$\begin{bmatrix} 3 & 2 & 0 \\ 3 & 2 & 1 \\ 2 & 1 & -1 \end{bmatrix}$$

17. An experimenter determined that three forces F_1, F_2, and F_3 on a structure satisfy the following relationships:

$$F_1 + 2F_2 + 3F_3 = 1$$
$$F_2 - 2F_3 = 1$$
$$3F_1 + 2F_2 + F_3 = -1$$

Use the method of row operations to solve this system.

18. Two resistors have a combined resistance of 10.0 Ω if connected in series and 2.14 Ω if connected in parallel. Determine the two resistances.

CHAPTER **16**

Additional Topics in Trigonometry

Objectives Upon completion of this chapter, you should be able to:
1. State the fundamental trigonometric identities.
2. Use the fundamental trigonometric identities to
   a. Simplify certain trigonometric expressions.
   b. Prove additional elementary identities.
3. State the sum, difference, half-angle, and double-angle formulas.
4. Use the identities in objective (3) to:
   a. Find certain function values.
   b. Transform certain given trigonometric expressions.
   c. Prove other identities.
5. Solve trigonometric equations.
6. Evaluate inverse trigonometric relations and functions.

16.1 Fundamental Identities

We saw in earlier chapters that solving triangles is an integral part of trigonometry. Another branch, called **analytic trigonometry,** deals mainly with **identities.** This aspect of the subject plays a major role in more advanced areas of mathematics, especially calculus.

Most of this chapter is devoted to the study of trigonometric identities. Identities are then used in Section 16.6 to help solve trigonometric equations. The chapter ends with a brief study of inverse trigonometric functions.

Identity   Recall that an **identity** is an equation that is satisfied for every value of the variable. For example, $x^2 - 1 = (x - 1)(x + 1)$ is an identity. In trigonometry, identities arise almost as soon as the basic definitions are given. For example, $\sin \theta = 1/\csc \theta$ is valid for every $\theta \neq 0 \pm n\pi$. To obtain other identities, let us recall the basic definitions of the trigonometric functions.

Definitions of trigonometric functions:

$$\sin \theta = \frac{y}{r} \qquad \csc \theta = \frac{r}{y}$$

$$\cos \theta = \frac{x}{r} \qquad \sec \theta = \frac{r}{x}$$

$$\tan \theta = \frac{y}{x} \qquad \cot \theta = \frac{x}{y}$$

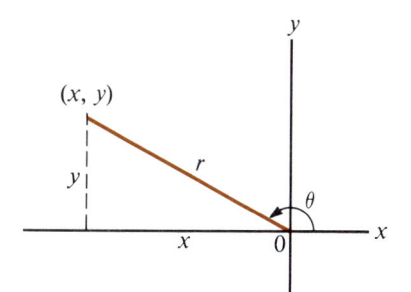

Figure 16.1

From these definitions we get the following reciprocal relations:

$$\sin \theta = \frac{1}{\csc \theta}, \quad \cos \theta = \frac{1}{\sec \theta}, \quad \tan \theta = \frac{1}{\cot \theta} \qquad (16.1)$$

Since $\tan \theta = y/x = (y/r)/(x/r)$, we get from the definitions of sine and cosine the identity

$$\tan \theta = \frac{y}{x} = \frac{\frac{y}{r}}{\frac{x}{r}} = \frac{\sin \theta}{\cos \theta} \qquad (16.2)$$

Finally, since $\cot \theta = 1/\tan \theta$, we have

$$\cot \theta = \frac{\cos \theta}{\sin \theta} \qquad (16.3)$$

Note especially that the secant, cosecant, tangent, and cotangent functions can be expressed in terms of sines and cosines:

$$\sec \theta = \frac{1}{\cos \theta}, \quad \csc \theta = \frac{1}{\sin \theta}, \quad \tan \theta = \frac{\sin \theta}{\cos \theta}, \quad \cot \theta = \frac{\cos \theta}{\sin \theta}$$

490 CHAPTER 16 ADDITIONAL TOPICS IN TRIGONOMETRY

It follows that any combination of these functions can be expressed in terms of sines and cosines. For example,

$$2 \tan \theta + \frac{1}{2} \cot \theta = \frac{2 \sin \theta}{\cos \theta} + \frac{\cos \theta}{2 \sin \theta}$$

Consider another example.

Example 1 Change the following expressions to equivalent expressions involving sines and cosines:

a. $\sec \theta + \tan \theta$ **b.** $\sin \theta + \dfrac{1}{\csc \theta}$

Solution.   **a.** $\sec \theta + \tan \theta = \dfrac{1}{\cos \theta} + \dfrac{\sin \theta}{\cos \theta} = \dfrac{1 + \sin \theta}{\cos \theta}$

by identities (16.1) and (16.2), respectively.

$$\textbf{b. } \sin \theta + \frac{1}{\csc \theta} = \sin \theta + \sin \theta = 2 \sin \theta$$

since $1/\csc \theta = \sin \theta$.

The definitions of the trigonometric functions yield other basic relationships, but the derivations can be carried out in a more interesting way. Consider the unit circle ($r = 1$) in Figure 16.2 and any point (a, b) on the circle. Note that

$$\sin \theta = \frac{a}{1} \quad \text{and} \quad \cos \theta = \frac{b}{1}$$

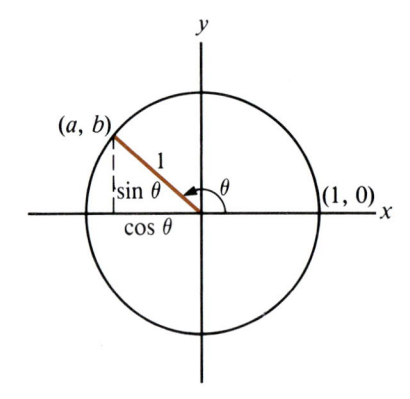

Figure 16.2

or

$$a = \sin \theta \quad \text{and} \quad b = \cos \theta$$

Since $a^2 + b^2 = 1$, it follows that

$$\sin^2 \theta + \cos^2 \theta = 1 \tag{16.4}$$

Similar identities hold for the remaining trigonometric functions. In Figure 16.3, the y-coordinate of P is equal to $\tan \theta$ and the length of PO is numerically equal to $\sec \theta$. By the Pythagorean theorem

$$1 + \tan^2 \theta = \sec^2 \theta \tag{16.5}$$

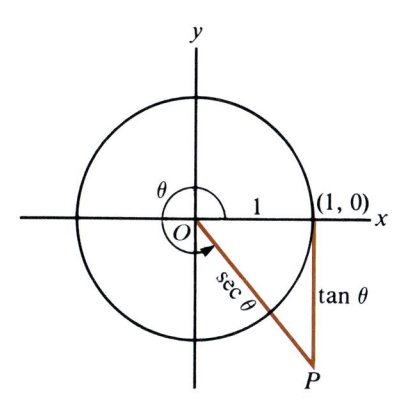

Figure 16.3

Of particular interest in Figure 16.3 is that the names *tangent* and *secant* suggest themselves quite naturally. (Recall that a secant line intersects a circle at two points and a tangent line at one point.)

The derivation of the remaining identity is similar and will be left as an exercise:

$$1 + \cot^2 \theta = \csc^2 \theta \tag{16.6}$$

Example 2 Use the fundamental identities to simplify the given expressions. Write the expressions in terms of sines and cosines, if necessary.

  a. $1 - \sin^2 \alpha$ b. $\dfrac{\cot \beta}{\csc \beta}$ c. $\csc x(1 - \cos^2 x)$

  d. $\dfrac{\sec^2 \theta - \tan^2 \theta}{\cot \theta}$

Solution.

  a. $1 - \sin^2 \alpha = \cos^2 \alpha$

by identity (16.4).

b. $\dfrac{\cot \beta}{\csc \beta} = \dfrac{\cos \beta}{\sin \beta} \cdot \dfrac{\sin \beta}{1} = \cos \beta$

by identities (16.3) and (16.1).

c. $\csc x (1 - \cos^2 x) = \csc x \sin^2 x$

by identity (16.4), and

$$\csc x \sin^2 x = \dfrac{1}{\sin x} \cdot \dfrac{\sin^2 x}{1} = \sin x \quad \left(\text{replacing } \csc x \text{ by } \dfrac{1}{\sin x}\right)$$

d. $\dfrac{\sec^2 \theta - \tan^2 \theta}{\cot \theta} = \dfrac{1}{\cot \theta}$

by identity (16.5), and

$$\dfrac{1}{\cot \theta} = \tan \theta$$

It follows that

$$\dfrac{\sec^2 \theta - \tan^2 \theta}{\cot \theta} = \tan \theta$$

For easy reference, the basic identities are given in the box below.

Fundamental trigonometric identities:

$$\sin \theta = \dfrac{1}{\csc \theta} \qquad \cos \theta = \dfrac{1}{\sec \theta} \qquad \tan \theta = \dfrac{1}{\cot \theta}$$

$$\tan \theta = \dfrac{\sin \theta}{\cos \theta} \qquad \cot \theta = \dfrac{\cos \theta}{\sin \theta}$$

$$\sin^2 \theta + \cos^2 \theta = 1$$
$$1 + \tan^2 \theta = \sec^2 \theta$$
$$1 + \cot^2 \theta = \csc^2 \theta$$

Alternate forms of the identities can be obtained by rearranging the terms. The following sets of identities are equivalent:

$$\sin^2 \theta + \cos^2 \theta = 1, \quad 1 - \cos^2 \theta = \sin^2 \theta, \quad 1 - \sin^2 \theta = \cos^2 \theta$$

$$1 + \tan^2 \theta = \sec^2 \theta, \quad \sec^2 \theta - \tan^2 \theta = 1$$

$$1 + \cot^2 \theta = \csc^2 \theta, \quad \csc^2 \theta - \cot^2 \theta = 1$$

Exercises / Section 16.1

In Exercises 1–14, change each expression to an equivalent expression involving sines and cosines. Simplify if possible.

1. $\cot \beta$
2. $\sin \alpha \cot \alpha$
3. $\cos \theta \tan \theta$
4. $\sec \theta$
5. $\cos \gamma \sec \gamma$
6. $1 - \tan \beta \cot \beta$
7. $\cot x + \dfrac{1}{\sin x}$
8. $\dfrac{1}{\sec \omega} + \cos \omega$
9. $\tan \theta \cos \theta \cot \theta$
10. $\cot^2 t \sin^2 t$
11. $\cot^2 s (1 + \tan^2 s)$
12. $\tan^2 x - \sec^2 x$
13. $1 - \sec^2 \theta$
14. $\dfrac{1 + \tan^2 x}{\sec^2 x}$

In Exercises 15–31, use the fundamental identities to simplify each given expression. Convert to an expression involving sines and cosines if necessary.

15. $\dfrac{\cos^2 x + \sin^2 x}{\sin x}$
16. $\dfrac{\tan \beta}{\sec \beta}$
17. $\dfrac{1}{1 - \cos^2 \theta}$
18. $\sin^2 \gamma (1 + \tan^2 \gamma)$
19. $\sin \theta (1 + \cot^2 \theta)$
20. $\csc \beta (1 - \cos^2 \beta)$
21. $\dfrac{\tan \theta \csc \theta}{\sec \theta}$
22. $\dfrac{\sec y}{\csc y}$
23. $\dfrac{\csc t}{\sec t}$
24. $\sin^2 x \sec^2 x$
25. $\dfrac{\sin \theta}{\tan \theta}$
26. $\dfrac{\tan^2 \omega - \sec^2 \omega}{\sec \omega}$
27. $\csc^2 \alpha - \cot^2 \alpha$
28. $\tan^2 y - \dfrac{\sec^2 y}{\csc^2 y}$
29. $\dfrac{1 + \tan^2 x}{\cos x}$
30. $\tan \theta \sin^2 \theta + \tan \theta \cos^2 \theta$
31. $\cot \theta \cos^2 \theta + \cot \theta \sin^2 \theta$

16.2 Proving Identities

In this section we shall use the fundamental identities to verify more complicated identities. Writing trigonometric expressions in alternate form is a skill required in more advanced work in mathematics.

In one respect, proving identities is similar to solving word problems: Each identity has its own features and must be verified in its own way. A facility for verifying identities can be developed only through practice. Although no general method can be given, the guidelines that follow will help you decide what approach to take.

Guidelines for proving identities

1. Memorize the fundamental identities and use them whenever possible.
2. Start with the more complicated side of the identity and try to reduce it to the simpler side.
3. Perform any algebraic operation indicated. For example, it may help to multiply out the terms in an expression, to factor an expression, to add fractions, and so on.
4. If everything else fails, try expressing all functions in terms of sines and cosines.
5. When working on one side of the identity, always keep in mind the other side for possible clues on how to proceed.

Caution. When proving an identity, the given relationship may not be treated as an equation—establishing equality is the very purpose of the verification. For this reason, it is not permissible to transpose terms, to multiply both sides by an expression, and so on. Instead, work on one side of the identity until the other side is obtained.

To see how to use the guidelines, consider the identity

$$\cot \theta \sin \theta = \cos \theta$$

In accordance with Guideline 2, start with the left side of the equation, since it is more complicated. While no algebraic operations come to mind (Guideline 3), the identity $\cot \theta = \cos \theta / \sin \theta$ may be useful (Guideline 1). This identity converts the left side to sines and cosines (Guideline 4), so that

$$\cot \theta \sin \theta = \frac{\cos \theta}{\sin \theta} \sin \theta = \cos \theta$$

The identity is thereby verified.

The examples below further illustrate these guidelines.

Example 1 Prove the identity

$$\cos^4 \theta - \sin^4 \theta = \cos^2 \theta - \sin^2 \theta$$

Solution. The left side, which is the more complicated side (Guideline 2), is factorable as a difference of two squares (Guideline 3). Thus

$$\cos^4 \theta - \sin^4 \theta = (\cos^2 \theta)^2 - (\sin^2 \theta)^2$$
$$= (\cos^2 \theta - \sin^2 \theta)(\cos^2 \theta + \sin^2 \theta) \qquad \text{difference of two squares}$$
$$= (\cos^2 \theta - \sin^2 \theta)(1) \qquad \text{replacing } \cos^2 \theta + \sin^2 \theta \text{ by } 1$$
$$= \cos^2 \theta - \sin^2 \theta$$

which is the right side. Note that Guideline 1 was also used.

16.2 PROVING IDENTITIES

Example 2 Prove the identity
$$\sin^3 \beta \cos \beta + \cos^3 \beta \sin \beta = \sin \beta \cos \beta$$

Solution. The left side, which is more complicated, contains a common factor $\sin \beta \cos \beta$ (Guideline 3). Thus

$\sin^3 \beta \cos \beta + \cos^3 \beta \sin \beta$
$\qquad = \sin \beta \cos \beta (\sin^2 \beta + \cos^2 \beta)$ common factor
$\qquad = \sin \beta \cos \beta (1)$ from Guideline 1
$\qquad = \sin \beta \cos \beta$

Example 3 Show that
$$\left(\csc \gamma + \frac{\cos^2 \gamma}{\sin^3 \gamma} \right) \sin \gamma = \csc^2 \gamma$$

Solution. We multiply the expression on the left side (Guideline 3) to obtain

$\left(\csc \gamma + \dfrac{\cos^2 \gamma}{\sin^3 \gamma} \right) \sin \gamma = \csc \gamma \sin \gamma + \dfrac{\cos^2 \gamma}{\sin^2 \gamma}$

$\qquad\qquad = \dfrac{1}{\sin \gamma} \sin \gamma + \left(\dfrac{\cos \gamma}{\sin \gamma} \right)^2$ $\quad \csc \gamma = 1/\sin \gamma$

$\qquad\qquad = 1 + \cot^2 \gamma$ $\quad \cos \gamma / \sin \gamma = \cot \gamma$

$\qquad\qquad = \csc^2 \gamma$ $\quad$ identity (16.6)

Example 4 Show that
$$\frac{1}{1 - \sin x} + \frac{1}{1 + \sin x} = 2 \sec^2 x$$

Solution. The left side is more complicated and contains two fractions, which should be combined:

$\dfrac{1 + \sin x}{(1 - \sin x)(1 + \sin x)} + \dfrac{1 - \sin x}{(1 + \sin x)(1 - \sin x)}$

$\qquad = \dfrac{(1 + \sin x) + (1 - \sin x)}{(1 - \sin x)(1 + \sin x)}$

$\qquad = \dfrac{2}{1 - \sin^2 x}$

$\qquad = \dfrac{2}{\cos^2 x}$ $\qquad$ since $\sin^2 x + \cos^2 x = 1$

$\qquad = 2 \sec^2 x$ $\qquad$ replacing $1/\cos x$ by $\sec x$

Example 5 Show that

$$\tan \theta + \cot \theta = \sec \theta \csc \theta$$

Solution. Since the left side contains two terms, it must be considered the more complicated side. Since no algebraic operations and no fundamental identities come to mind, we write both terms as expressions involving sines and cosines (Guideline 4). Thus

$$\tan \theta + \cot \theta = \frac{\sin \theta}{\cos \theta} + \frac{\cos \theta}{\sin \theta}$$

$$= \frac{\sin \theta \, \mathbf{\sin \theta}}{\cos \theta \, \mathbf{\sin \theta}} + \frac{\cos \theta \, \mathbf{\cos \theta}}{\sin \theta \, \mathbf{\cos \theta}}$$

$$= \frac{\sin^2 \theta + \cos^2 \theta}{\cos \theta \sin \theta} = \frac{1}{\cos \theta \sin \theta}$$

$$= \frac{1}{\cos \theta} \cdot \frac{1}{\sin \theta} = \sec \theta \csc \theta$$

Example 6 Show that

$$\frac{\sin^2 \theta}{1 + \cos \theta} = 1 - \cos \theta$$

Solution. Apart from the fact that the left side is more complicated than the right side, none of the guidelines we have used so far seem to apply. Thinking of Guideline 5, see if the right side offers a clue. It does suggest one possibility: Multiply the numerator and denominator of the left side by $1 - \cos \theta$, not only to introduce this expression but also to reduce the denominator. Thus

$$\frac{\sin^2 \theta}{1 + \cos \theta} \cdot \frac{1 - \cos \theta}{1 - \cos \theta} = \frac{\sin^2 \theta (1 - \cos \theta)}{1 - \cos^2 \theta}$$

$$= \frac{\cancel{\sin^2 \theta}(1 - \cos \theta)}{\cancel{\sin^2 \theta}} = 1 - \cos \theta$$

Guidelines 1 and 3 could actually be used to prove this identity: since $\sin^2 \theta + \cos^2 \theta = 1$, the numerator of the left side can be written $\sin^2 \theta = 1 - \cos^2 \theta = (1 - \cos \theta)(1 + \cos \theta)$, so that the factor $1 + \cos \theta$ cancels. However, the above technique using Guideline 5 is needed in Exercises 32 and 33.

Example 7 Show that

$$(\sec \beta - \tan \beta)^2 = \frac{1 - \sin \beta}{1 + \sin \beta}$$

Solution. The most promising approach is to multiply out the expression on the left side (Guideline 3):

$$(\sec \beta - \tan \beta)^2$$
$$= \sec^2 \beta - 2 \sec \beta \tan \beta + \tan^2 \beta \qquad (a-b)^2 = a^2 - 2ab + b^2$$
$$= \frac{1}{\cos^2 \beta} - \frac{2 \sin \beta}{\cos^2 \beta} + \frac{\sin^2 \beta}{\cos^2 \beta} \qquad \text{converting to sines and cosines}$$
$$= \frac{1 - 2 \sin \beta + \sin^2 \beta}{\cos^2 \beta} \qquad \text{combining fractions}$$
$$= \frac{(1 - \sin \beta)^2}{1 - \sin^2 \beta} \qquad \text{factoring and identity (16.4)}$$
$$= \frac{(1 - \sin \beta)^{\cancel{2}}}{\cancel{(1 - \sin \beta)}(1 + \sin \beta)} \qquad \text{difference of two squares}$$
$$= \frac{1 - \sin \beta}{1 + \sin \beta}$$

Example 8 The current in a certain circuit as a function of time is given by

$$i = \sqrt{0.04 \cos^2 \omega t - 0.04 + 2.0 \sin^2 \omega t}$$

Simplify this expression.

Solution.

$$i = \sqrt{0.04 \cos^2 \omega t - 0.04 + 2.0 \sin^2 \omega t}$$
$$= \sqrt{-0.04(1 - \cos^2 \omega t) + 2.0 \sin^2 \omega t} \qquad \text{common factor } -0.04$$
$$= \sqrt{-0.04 \sin^2 \omega t + 2.0 \sin^2 \omega t} \qquad \text{replacing } 1 - \cos^2 \omega t \text{ by } \sin^2 \omega t$$
$$= \sqrt{1.96 \sin^2 \omega t}$$
$$= 1.4 \sin \omega t$$

Exercises / Section 16.2

In Exercises 1–40, prove the given identities.

1. $\cot \theta \sin \theta = \cos \theta$
2. $\sin \theta \cot \theta \sec \theta = 1$
3. $\tan \theta \csc \theta = \sec \theta$
4. $\cos \theta + \sin \theta \tan \theta = \sec \theta$
5. $\dfrac{\cos^2 \beta}{\sin \beta} + \sin \beta = \csc \beta$
6. $\sin^3 x + \sin x \cos^2 x = \sin x$
7. $\dfrac{1 - \cos^2 \gamma}{\cos^2 \gamma} = \tan^2 \gamma$
8. $\sin^2 y + \tan^2 y + \cos^2 y = \sec^2 y$
9. $\dfrac{1 + \tan^2 \omega}{1 + \cot^2 \omega} = \tan^2 \omega$
10. $\tan \theta + \cot \theta = \dfrac{\tan \theta}{\sin^2 \theta}$

11. $(1 + \tan^2 x) \cos^2 x = 1$

12. $\dfrac{\csc \theta}{\tan \theta + \cot \theta} = \cos \theta$

13. $\dfrac{1}{\cot \gamma + \tan \gamma} = \sin \gamma \cos \gamma$

14. $(\tan \theta + \cot \theta)^2 = \sec^2 \theta + \csc^2 \theta$

15. $\dfrac{\sin \beta + \tan \beta}{1 + \cos \beta} = \tan \beta$

16. $\dfrac{\sin \theta}{1 + \cos \theta} + \dfrac{\sin \theta}{1 - \cos \theta} = 2 \csc \theta$

17. $\dfrac{\sin \theta}{1 - \cos \theta} - \dfrac{1 - \cos \theta}{\sin \theta} = 2 \cot \theta$

18. $\sec^2 x + \csc^2 x = \sec^2 x \csc^2 x$

19. $\cot^2 \theta - \cos^2 \theta = \cot^2 \theta \cos^2 \theta$

20. $\tan^4 \alpha + \tan^2 \alpha = \sin^2 \alpha \sec^4 \alpha$

21. $\dfrac{1 + \sin \theta}{\cos \theta} + \dfrac{\cos \theta}{1 + \sin \theta} = 2 \sec \theta$

22. $\dfrac{\cos \theta}{\tan \theta + \sec \theta} + \dfrac{\cos \theta}{\tan \theta - \sec \theta} = -2 \sin \theta$

23. $2 \csc x - \cot x \cos x = \sin x + \csc x$

24. $(1 - \cos \beta)(1 + \cos \beta) = \dfrac{1}{1 + \cot^2 \beta}$

25. $\dfrac{1 - \tan \gamma}{1 + \tan \gamma} = \dfrac{\cot \gamma - 1}{\cot \gamma + 1}$

26. $\dfrac{\cos \omega}{\cos \omega - \sin \omega} = \dfrac{1}{1 - \tan \omega}$

27. $\dfrac{\tan \theta}{\csc \theta - \cot \theta} - \dfrac{\sin \theta}{\csc \theta + \cot \theta} = \sec \theta + \cos \theta$

28. $\dfrac{1 - \tan^2 \alpha}{1 + \tan^2 \alpha} = 1 - 2 \sin^2 \alpha$

29. $\dfrac{1 + \tan^2 \theta}{\csc^2 \theta} = \tan^2 \theta$

30. $\dfrac{\cot \theta + \tan \theta}{\sec \theta} = \csc \theta$

31. $\dfrac{1 + \cos \gamma - \sin^2 \gamma}{\sin \gamma (1 + \cos \gamma)} = \cot \gamma$

32. $\dfrac{\cos \theta}{1 - \sin \theta} = \dfrac{1 + \sin \theta}{\cos \theta}$ (See Example 6.)

33. $\sec \theta + \tan \theta = \dfrac{\cos \theta}{1 - \sin \theta}$ (See Example 6.)

34. $\dfrac{1}{\sec \theta - \tan \theta} = \sec \theta + \tan \theta$

35. $\cos^4 x - \sin^4 x = 2 \cos^2 x - 1$

36. $\dfrac{\sin^4 \alpha - \cos^4 \alpha}{\sin^2 \alpha - \cos^2 \alpha} = 1$

37. $\dfrac{\tan \theta + \sec^2 \theta - 1}{\tan \theta - \sec^2 \theta + 1} = \dfrac{\cos \theta + \sin \theta}{\cos \theta - \sin \theta}$

38. $\left(\dfrac{\sec \beta + \csc \beta}{1 + \tan \beta}\right)^2 = \csc^2 \beta$

39. $\dfrac{\tan x - \sin x}{\sin^3 x} = \dfrac{\sec x}{1 + \cos x}$

40. $\dfrac{\tan \alpha + \cot \alpha}{\cos^2 \alpha} - \sin \alpha \sec^3 \alpha = \sec \alpha \csc \alpha$

41. An object traveling along a circle of radius a (in feet) at an angular velocity of $\omega/2\pi$ rev/sec has linear velocity
$$\sqrt{(a\omega \sin \omega t)^2 + (a\omega \cos \omega t)^2}$$
Simplify this expression. (See Example 8.)

42. Suppose a particle moves along a line with velocity $v = 2 \cos t + 2 \sin t$ (in meters per second). The methods of calculus show that the acceleration is given by $a = -2 \sin t + 2 \cos t$ (in meters per second per second). Show that $a = 0$ whenever $\tan t = 1$.

43. Neglecting air resistance, the equation of the path of a missile projected at velocity v_0 at an angle α with the horizontal is
$$y = x \tan \alpha - \dfrac{gx^2}{2v_0^2 \cos^2 \alpha}$$
Write y as a single fraction.

44. In the study of the motion of a pendulum, the expression $\sqrt{1 - k^2 \sin^2 \theta}$ arises. Show that

$$1 - k^2 \sin^2 \theta = k^2 \cos^2 \theta + 1 - k^2$$

45. A beam of length L (in inches) weighing w (pounds per inch) and clamped at the left end is subjected to a compressive force P at the free end. The minimum deflection is given by

$$y_{min} = \frac{wEI}{P^2}\left(1 - \frac{1}{2}\theta^2 - \sec\theta + \theta\tan\theta\right)$$

where $\theta = L\sqrt{P/EI}$. Show that

$$y_{min} = \frac{wEI}{P^2} \frac{2\cos\theta - \theta^2\cos\theta - 2 + 2\theta\sin\theta}{2\cos\theta}$$

46. If a vertical plate is partly submerged in a liquid, then the capillarity will cause the liquid to rise on the plate to a height of

$$h = c\sqrt{\frac{1 - \sin\theta}{2}}$$

where θ is the contact angle between the liquid and the plate and c a constant that depends on the surface tension and specific gravity of the liquid. Show that

$$h = \frac{c\cos\theta}{\sqrt{2(1 + \sin\theta)}}$$

47. In some problems on the motion of a pendulum, the expression $1/\sqrt{1 - \cos x}$ arises. Show that this expression is equivalent to $\sqrt{1 + \cos x}/\sin x$.

16.3 The Sum and Difference Formulas

It is sometimes useful to write a trigonometric function of the sum of two angles in terms of trigonometric functions of each angle. For example, $\sin(A + B)$ can be expressed in terms of $\sin A$, $\cos A$, $\sin B$, and $\cos B$.

To do so, let A and B be two acute angles. Then $A + B$ may be either acute (Figure 16.4) or obtuse (Figure 16.5). In both figures, PQ and MN are

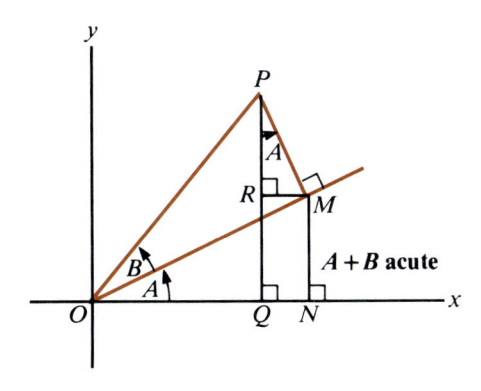

Figure 16.4

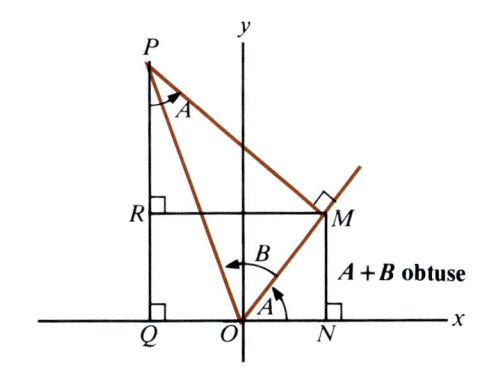

Figure 16.5

perpendicular to the *x*-axis, *PM* is perpendicular to *OM*, and *MR* is perpendicular to *PQ*. Note that $\angle MPQ = \angle A$, since the two angles have their sides perpendicular, right side to right side and left side to left side.

In both figures we have

$$\sin(A+B) = \frac{PQ}{OP} = \frac{PR + RQ}{OP} = \frac{PR}{OP} + \frac{RQ}{OP} = \frac{PR}{OP} + \frac{MN}{OP}$$

The last two fractions do not define functions of either *A* or *B*. However, if we multiply numerator and denominator of the first fraction by *PM* and the second by *OM*, each of the resulting ratios is a function of *A* or *B*:

$$\frac{PR}{OP} \cdot \frac{PM}{PM} + \frac{MN}{OP} \cdot \frac{OM}{OM}$$

$$= \frac{PR}{PM} \cdot \frac{PM}{OP} + \frac{MN}{OM} \cdot \frac{OM}{OP} = \cos A \sin B + \sin A \cos B$$

or

$$\sin(A+B) = \sin A \cos B + \cos A \sin B \qquad (16.7)$$

For the corresponding identity involving $\cos(A+B)$, we get

$$\cos(A+B) = \frac{OQ}{OP} = \frac{ON - QN}{OP} = \frac{ON}{OP} - \frac{QN}{OP} = \frac{ON}{OP} - \frac{RM}{OP}$$

$$= \frac{ON}{OM} \cdot \frac{OM}{OP} - \frac{RM}{PM} \cdot \frac{PM}{OP}$$

or

$$\cos(A+B) = \cos A \cos B - \sin A \sin B \qquad (16.8)$$

Since $\sin(-B) = -\sin B$ and $\cos(-B) = \cos B$, we also get

$$\sin(A-B) = \sin A \cos B - \cos A \sin B \qquad (16.9)$$

and

$$\cos(A-B) = \cos A \cos B + \sin A \sin B \qquad (16.10)$$

These four formulas can be written more compactly in the forms given below. (The combination "$\pm$" and "$\mp$" in formula (16.12) indicates that the terms have opposite signs.)

Sum and difference formulas:

$$\sin(A \pm B) = \sin A \cos B \pm \cos A \sin B \qquad (16.11)$$
$$\cos(A \pm B) = \cos A \cos B \mp \sin A \sin B \qquad (16.12)$$

As noted earlier, these identities enable us to express a function of a sum of two angles in terms of functions of the angles themselves. To illustrate these identities, let us find the values of certain trigonometric functions without tables or calculators.

Example 1 Find the exact value of cos 75° by means of the sum and difference formulas.

Solution. Since 75° is not a special angle, cos 75° cannot be found from a diagram. However, 75° = 30° + 45°, a sum of two special angles having known function values. So it follows from identity (16.8) that

$$\cos 75° = \cos(30° + 45°)$$
$$= \cos 30° \cos 45° - \sin 30° \sin 45°$$
$$= \frac{\sqrt{3}}{2} \cdot \frac{1}{\sqrt{2}} - \frac{1}{2} \cdot \frac{1}{\sqrt{2}} = \frac{\sqrt{3}-1}{2\sqrt{2}} \cdot \frac{\sqrt{2}}{\sqrt{2}} = \frac{\sqrt{6}-\sqrt{2}}{4}$$

Example 2 Find the exact value of

$$\sin 25° \cos 20° + \cos 25° \sin 20°$$

Solution. By identity (16.7)

$$\sin 25° \cos 20° + \cos 25° \sin 20° = \sin(25° + 20°)$$
$$= \sin 45° = \frac{\sqrt{2}}{2}$$

The sum and difference identities are sometimes used to combine certain expressions, as shown in the next example.

Example 3 Combine

$$\sin 3x \cos 2x - \cos 3x \sin 2x$$

into a single term.

Solution. By identity (16.9) we get directly

$$\sin(3x - 2x) = \sin x$$

In our study of the graphs of sinusoidal functions, we found that the graph of $y = \sin(x \pm c)$ can be obtained from the graph of $y = \sin x$ by translating the latter graph by c units. If c is a special angle, the relationship between the two functions can be obtained readily by means of the sum and difference identities.

Example 4 Simplify $\sin(x + \pi/2)$.

Solution. By identity (16.7)

$$\sin\left(x + \frac{\pi}{2}\right) = \sin x \cos \frac{\pi}{2} + \cos x \sin \frac{\pi}{2}$$
$$= (\sin x)(0) + (\cos x)(1)$$
$$= \cos x$$

Example 5 Simplify $\cos(2x - \pi)$.

Solution. By identity (16.10)

$$\cos(2x - \pi) = \cos 2x \cos \pi + \sin 2x \sin \pi$$
$$= (\cos 2x)(-1) + (\sin 2x)(0)$$
$$= -\cos 2x$$

The sum and difference identities for the tangent occur less frequently and are listed mainly for completeness.

By identities (16.7) and (16.8),

$$\tan(A + B) = \frac{\sin(A + B)}{\cos(A + B)} = \frac{\sin A \cos B + \cos A \sin B}{\cos A \cos B - \sin A \sin B}$$

Dividing numerator and denominator by $\cos A \cos B$, we get

$$\tan(A + B) = \frac{\dfrac{\sin A \cos B}{\cos A \cos B} + \dfrac{\cos A \sin B}{\cos A \cos B}}{\dfrac{\cos A \cos B}{\cos A \cos B} - \dfrac{\sin A \sin B}{\cos A \cos B}}$$

or

$$\tan(A + B) = \frac{\tan A + \tan B}{1 - \tan A \tan B} \qquad (16.13)$$

Similarly,

$$\tan(A - B) = \frac{\tan A - \tan B}{1 + \tan A \tan B} \qquad (16.14)$$

Example 6 Simplify $\tan(2x + \pi/4)$.

Solution. By identity (16.13)

$$\tan\left(2x + \frac{\pi}{4}\right) = \frac{\tan 2x + \tan \dfrac{\pi}{4}}{1 - \tan 2x \tan \dfrac{\pi}{4}} = \frac{1 + \tan 2x}{1 - \tan 2x}$$

16.3 THE SUM AND DIFFERENCE FORMULAS

Remark. The identities $\tan \theta = \sin \theta / \cos \theta$ and $\cot \theta = \cos \theta / \sin \theta$ were known to the Arabs. The Hindus knew the fundamental identity $\sin^2 \theta + \cos^2 \theta = 1$, while the formula for $\sin(A + B)$ was discovered by the Belgian mathematician Romanus (1561–1615).

Common error Writing

$$\sin(A + B) \quad \text{as} \quad \sin A + \sin B$$

Instead, $\sin(A + B)$ should be written as

$$\sin A \cos B + \cos A \sin B$$

Similarly,

$$\cos(A + B) \quad \text{should not be written} \quad \cos A + \cos B.$$

Example 7 If $i = 3 \sin(\omega t - \pi/2)$ is the current in a circuit and $e = 5 \sin \omega t$ the voltage, find an expression for the power $P = ei$ as a function of time and simplify the result.

Solution. The power is given by

$$P = ei = (5 \sin \omega t)\left[3 \sin\left(\omega t - \frac{\pi}{2}\right)\right]$$

$$= (5 \sin \omega t) \cdot 3\left[\left(\sin \omega t \cos \frac{\pi}{2} - \cos \omega t \sin \frac{\pi}{2}\right)\right]$$

$$= 15 \sin \omega t \,(\sin \omega t \cdot 0 - \cos \omega t \cdot 1)$$

$$= -15 \sin \omega t \cos \omega t$$

Exercises / Section 16.3

In Exercises 1–10, use the sum and difference identities to find each given value without using a table or a calculator. (See Examples 1 and 2.)

1. $\cos 15°$
2. $\sin 105°$
3. $\cos(-105°)$
4. $\sin 285°$
5. $\cos 16° \cos 29° - \sin 16° \sin 29°$
6. $\sin 50° \cos 10° + \cos 50° \sin 10°$
7. $\cos 55° \cos 10° + \sin 55° \sin 10°$
8. $\sin 76° \cos 16° - \cos 76° \sin 16°$
9. $\sin 39° \cos 6° + \cos 39° \sin 6°$
10. $\cos 18° \cos 12° - \sin 18° \sin 12°$

In Exercises 11–20, write each expression as a single term. (See Example 3.)

11. $\sin 4x \cos 2x - \cos 4x \sin 2x$
12. $\sin 6x \cos 3x - \sin 6x \cos 3x$
13. $\sin x \cos 2x + \cos x \sin 2x$
14. $\sin 3x \cos x + \cos 3x \sin x$
15. $\cos 3x \cos x + \sin 3x \sin x$
16. $\cos 5x \cos 3x + \sin 5x \sin 3x$
17. $\cos 5x \cos 4x - \sin 5x \sin 4x$
18. $\cos 2x \cos 3x - \sin 2x \sin 3x$
19. $\sin(x + y) \cos y - \cos(x + y) \sin y$
20. $\cos(2x - y) \cos(y - x) - \sin(2x - y) \sin(y - x)$

In Exercises 21–38, write each expression as a function of x or $2x$. (See Examples 4 and 5.)

21. $\cos(x + 30°)$
22. $\sin(x - 30°)$
23. $\cos\left(2x + \dfrac{\pi}{2}\right)$
24. $\sin(x - \pi)$
25. $\cos(x + \pi)$
26. $\cos(x - \pi)$
27. $\sin(2x - 2\pi)$
28. $\cos\left(x - \dfrac{\pi}{2}\right)$
29. $\sin\left(2x + \dfrac{\pi}{2}\right)$
30. $\cos\left(x - \dfrac{\pi}{3}\right)$
31. $\sin\left(x - \dfrac{\pi}{3}\right)$
32. $\sin\left(x + \dfrac{\pi}{6}\right)$
33. $\cos\left(x + \dfrac{\pi}{6}\right)$
34. $\cos\left(2x - \dfrac{\pi}{6}\right)$
35. $\sin\left(x + \dfrac{\pi}{4}\right)$
36. $\cos\left(x - \dfrac{\pi}{4}\right)$
37. $\tan\left(x + \dfrac{\pi}{4}\right)$
38. $\tan\left(2x - \dfrac{\pi}{4}\right)$

In Exercises 39–48, prove the given identities.

39. $\sin\left(\theta - \dfrac{\pi}{4}\right) = -\cos\left(\theta + \dfrac{\pi}{4}\right)$
40. $\sin 2\theta = \sin(\theta + \theta) = 2\sin\theta\cos\theta$
41. $\cos 2\theta = \cos(\theta + \theta) = \cos^2\theta - \sin^2\theta$
42. $\sin(x + y) + \sin(x - y) = 2\sin x\cos y$
43. $\sin(x + y) - \sin(x - y) = 2\cos x\sin y$
44. $\sin\left(x + \dfrac{\pi}{6}\right) + \cos\left(x + \dfrac{\pi}{3}\right) = \cos x$
45. $\sin(x + y)\sin(x - y) = \sin^2 x - \sin^2 y$
46. $\cos(x + y)\cos(x - y) = \cos^2 x - \sin^2 y$
47. $\tan(x - y) - \tan(y - x) = \dfrac{2(\tan x - \tan y)}{1 + \tan x \tan y}$
48. $\dfrac{\tan(x + y) - \tan y}{1 + \tan(x + y)\tan y} = \tan x$

In Exercises 49–51, we shall obtain a few other standard identities. The first four are known as the *product-to-sum formulas* and the last four as the *sum-to-product formulas*.

49. Derive the following **product-to-sum formulas** by adding and subtracting formulas (16.7) and (16.9):

$$\sin A \cos B = \dfrac{1}{2}[\sin(A + B) + \sin(A - B)] \qquad (16.15)$$

$$\cos A \sin B = \dfrac{1}{2}[\sin(A + B) - \sin(A - B)] \qquad (16.16)$$

50. Derive the following **product-to-sum formulas** by adding and subtracting formulas (16.8) and (16.10):

$$\cos A \cos B = \dfrac{1}{2}[\cos(A + B) + \cos(A - B)] \qquad (16.17)$$

$$\sin A \sin B = \dfrac{1}{2}[\cos(A - B) - \cos(A + B)] \qquad (16.18)$$

51. The **sum-to-product formulas** can be obtained from identities (16.15) through (16.18) by letting $A + B = x$ and $A - B = y$. Thus

$$A = \dfrac{x + y}{2} \quad \text{and} \quad B = \dfrac{x - y}{2}$$

16.3 THE SUM AND DIFFERENCE FORMULAS

By substituting show that

$$\sin x + \sin y = 2 \sin\left(\frac{x+y}{2}\right) \cos\left(\frac{x-y}{2}\right) \tag{16.19}$$

$$\sin x - \sin y = 2 \cos\left(\frac{x+y}{2}\right) \sin\left(\frac{x-y}{2}\right) \tag{16.20}$$

$$\cos x + \cos y = 2 \cos\left(\frac{x+y}{2}\right) \cos\left(\frac{x-y}{2}\right) \tag{16.21}$$

$$\cos x - \cos y = -2 \sin\left(\frac{x+y}{2}\right) \sin\left(\frac{x-y}{2}\right) \tag{16.22}$$

52. Prove the following identity occurring in the study of alpha particle scattering:

$$\sin \frac{1}{2}(\pi - \theta) = \cos \frac{\theta}{2}$$

53. The equation of a standing wave may be obtained by adding the displacements of two waves of equal amplitude and wavelength but traveling in opposite directions. Given that at some particular instant

$$y_1 = A \sin 2\left(x - \frac{\pi}{4}\right)$$

is the equation of a wave traveling in the positive x-direction and

$$y_2 = A \sin 2\left(x + \frac{\pi}{4}\right)$$

is the equation of the corresponding wave traveling in the negative x-direction, show that $y_1 + y_2 = 0$; that is, the waves cancel each other at the instant in question.

54. The current in a certain electric circuit is given by

$$i = A \sin\left(\omega t - \frac{\pi}{4}\right) + B \cos\left(\omega t + \frac{\pi}{4}\right)$$

Simplify this expression.

55. If a force $F_0 \cos \omega t$ is applied to a weight oscillating on a spring, then the energy supplied to the system can be written in the form

$$E = A\omega F_0 \cos(\omega t - \gamma) \cos \omega t$$

Show that

$$E = A\omega F_0(\cos^2 \omega t \cos \gamma + \cos \omega t \sin \omega t \sin \gamma)$$

56. A light ray strikes a glass plate of thickness a at an angle of incidence ϕ. If ϕ' is the angle of refraction within the glass, then the lateral displacement D of the emerging beam is given by

$$D = \frac{a \sin(\phi - \phi')}{\cos \phi'}$$

Show that $D = a(\sin \phi - \cos \phi \tan \phi')$.

57. Given that

$$y_1 = A \cos 2\pi \left(\frac{t}{T} - \frac{x}{\lambda}\right)$$

is the equation of a wave traveling in the positive x-direction and

$$y_2 = A \cos 2\pi \left(\frac{t}{T} + \frac{x}{\lambda}\right)$$

is the equation of the corresponding wave traveling in the negative x-direction, find $y = y_1 + y_2$, the equation of a standing wave. (Refer to Exercise 53.)

58. In the development of the theory of Fourier series (see Section 8.5) the product

$$\cos \frac{m\pi t}{p} \cos \frac{n\pi t}{p}$$

has to be written as a sum. Carry out this operation.

59. Show that the product of two complex numbers $r_1 \text{ cis } \theta_1$ and $r_2 \text{ cis } \theta_2$ is

$$r_1 r_2 [(\cos \theta_1 \cos \theta_2 - \sin \theta_1 \sin \theta_2) + j(\sin \theta_1 \cos \theta_2 + \cos \theta_1 \sin \theta_2)]$$

Simplify this expression to obtain the standard form $r_1 r_2 \text{ cis } (\theta_1 + \theta_2)$

16.4 Double-Angle Formulas

Some special cases of the sum and difference formulas occur often enough to warrant separate classification. One such classification includes the **double-angle formulas.**

Let $A = B$ in the identity

$$\sin (A + B) = \sin A \cos B + \cos A \sin B$$

Then

$$\sin (A + A) = \sin A \cos A + \cos A \sin A$$

or

$$\sin 2A = 2 \sin A \cos A$$

If $A = B$ in the identity

$$\cos (A + B) = \cos A \cos B - \sin A \sin B$$

then

$$\cos (A + A) = \cos A \cos A - \sin A \sin A$$

or

$$\cos 2A = \cos^2 A - \sin^2 A$$

If we let $\cos^2 A = 1 - \sin^2 A$, then $\cos 2A = 1 - \sin^2 A - \sin^2 A = 1 - 2 \sin^2 A$. Similarly, $\cos 2A = \cos^2 A - (1 - \cos^2 A) = 2 \cos^2 A - 1$.

16.4 DOUBLE-ANGLE FORMULAS

Double-angle formulas:

$$\sin 2A = 2 \sin A \cos A \qquad (16.23)$$

$$\cos 2A = \cos^2 A - \sin^2 A \qquad (16.24)$$

$$= 2 \cos^2 A - 1 \qquad (16.25)$$

$$= 1 - 2 \sin^2 A \qquad (16.26)$$

The double-angle formulas can be used to express the sine or cosine of twice an angle in terms of functions of a single angle. In particular, if $\sin \theta$ or $\cos \theta$ are known, we can use the identities to find $\sin 2\theta$ and $\cos 2\theta$. Consider the examples below.

Example 1 Use the double-angle formulas to find $\sin 2\theta$ and $\cos 2\theta$, given that $\sin \theta = \frac{5}{13}$, θ in quadrant II.

Solution. Since $\sin \theta = \frac{5}{13}$, θ in quadrant II, we obtain $\cos \theta = -\frac{12}{13}$ (Figure 16.6). Thus

$$\sin 2\theta = 2 \sin \theta \cos \theta = 2 \left(\frac{5}{13}\right)\left(-\frac{12}{13}\right) = -\frac{120}{169}$$

and

$$\cos 2\theta = \cos^2 \theta - \sin^2 \theta = \left(-\frac{12}{13}\right)^2 - \left(\frac{5}{13}\right)^2$$

$$= \frac{144}{169} - \frac{25}{169} = \frac{119}{169}$$

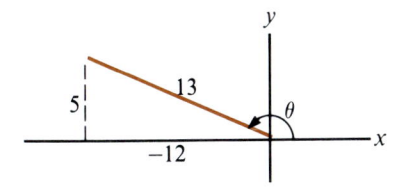

Figure 16.6

Example 2 Find $\sin 2\theta$ and $\cos 2\theta$, given that $\cos \theta = -\frac{2}{5}$, θ in quadrant III.

Solution. From the diagram (Figure 16.7 on page 508), we obtain $\sin \theta = -\sqrt{21}/5$. Hence

$$\sin 2\theta = 2 \sin \theta \cos \theta = 2 \left(-\frac{\sqrt{21}}{5}\right)\left(-\frac{2}{5}\right) = \frac{4\sqrt{21}}{25}$$

and

$$\cos 2\theta = \cos^2 \theta - \sin^2 \theta = \frac{4}{25} - \frac{21}{25} = -\frac{17}{25}$$

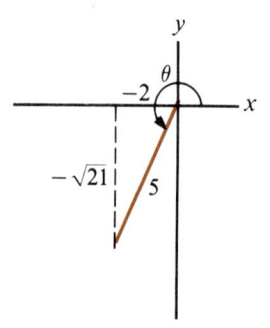

Figure 16.7

The double-angle formulas are also applicable to trigonometric functions of multiple angles. For example, from the identity $\sin 2A = 2 \sin A \cos A$, it follows that

$$\sin 16\theta = 2 \sin 8\theta \cos 8\theta$$

Similarly, since $\cos 2A = 2 \cos^2 A - 1$, we have

$$2 \cos^2 6x - 1 = \cos 12x$$

Example 3 Change $\cos^2 4x - \sin^2 4x$ to a single term.

Solution. By formula (16.24), $\cos 2A = \cos^2 A - \sin^2 A$, we get

$$\cos^2 4x - \sin^2 4x = \cos 8x \qquad A = 4x \text{ and } 2A = 8x$$

Example 4 Prove the identity

$$\cos^4 2\theta - \sin^4 2\theta = \cos 4\theta$$

Solution. Factoring the left side, we get

$$\cos^4 2\theta - \sin^4 2\theta = (\cos^2 \theta)^2 - (\sin^2 \theta)^2$$
$$= (\cos^2 2\theta - \sin^2 2\theta)(\cos^2 2\theta + \sin^2 2\theta)$$
$$= \cos^2 2\theta - \sin^2 2\theta = \cos 4\theta \qquad \cos^2 2\theta + \sin^2 2\theta = 1$$

by formula (16.24).

Common error Equating $\sin 2A$ with $2 \sin A$ and $\cos 2A$ with $2 \cos A$. As we have seen,

$$\sin 2A = 2 \sin A \cos A$$

and

$$\cos 2A = \cos^2 A - \sin^2 A$$

Example 5 The range R of a projectile fired with velocity v at an angle θ with the ground is given by

$$R = \frac{2v^2}{g} \sin \theta \cos \theta$$

Write R as a single trigonometric function of θ.

Solution.
$$R = \frac{2v^2}{g} \sin \theta \cos \theta$$
$$= \frac{v^2}{g} (2 \sin \theta \cos \theta)$$
$$= \frac{v^2}{g} \sin 2\theta$$

by the double-angle formula (16.23).

Exercises / Section 16.4

1. Find $\sin 2\theta$, given that $\sin \theta = \frac{3}{5}$, θ in quadrant I.
2. Find $\sin 2\theta$, given that $\sin \theta = \frac{4}{5}$, θ in quadrant II.
3. Find $\cos 2\theta$, given that $\sin \theta = -\frac{3}{5}$, θ in quadrant III.
4. Find $\cos 2\theta$, given that $\cos \theta = \frac{5}{13}$, θ in quadrant I.
5. Find $\sin 2\theta$, given that $\cos \theta = -\frac{12}{13}$, θ in quadrant II.
6. Find $\cos 2\theta$, given that $\sin \theta = -\frac{5}{13}$, θ in quadrant III.
7. Find $\sin 2\theta$, given that $\cos \theta = \frac{1}{2}$, θ in quadrant IV.
8. Find $\cos 2\theta$, given that $\sin \theta = -\frac{1}{2}$, θ in quadrant IV.
9. Find $\cos 2\theta$, given that $\sin \theta = \frac{2}{5}$, θ in quadrant II.
10. Find $\sin 2\theta$, given that $\sin \theta = \frac{1}{3}$, θ in quadrant I.
11. Find $\cos 2\theta$, given that $\cos \theta = -\frac{3}{7}$, θ in quadrant III.
12. Find $\sin 2\theta$, given that $\cos \theta = -\frac{2}{3}$, θ in quadrant II.

In Exercises 13–24, write each expression as a single trigonometric function. (See Examples 3 and 5.)

13. $\cos^2 3y - \sin^2 3y$
14. $\sin^2 x - \cos^2 x$
15. $2 \sin 3\theta \cos 3\theta$
16. $1 - 2 \sin^2 5x$

17. $2\cos^2 2\beta - 1$

18. $\sin 2x \cos 2x$

19. $1 - 2\cos^2 4y$

20. $2\sin^2 A - 1$

21. $\sin 4\omega \cos 4\omega$

22. $\sin 3\theta \cos 3\theta$

23. $4\sin 2x \cos 2x$

24. $6\sin 5x \cos 5x$

In Exercises 25–35, prove the given identities.

25. $\cos^4 x - \sin^4 x = \cos 2x$

26. $\sin 2\theta = \tan \theta (1 + \cos 2\theta)$

27. $1 - \cos 2\beta = \tan \beta \sin 2\beta$

28. $\sin 2\beta = \dfrac{2\tan \beta}{1 + \tan^2 \beta}$

29. $\dfrac{\cos 2\theta + \cos \theta + 1}{\sin 2\theta + \sin \theta} = \cot \theta$

30. $\sin 4x = 4\sin x \cos x \cos 2x$

31. $\dfrac{\cos^2 \gamma + 1}{2\cos^4 \gamma + \cos^2 \gamma - 1} = \sec 2\gamma$

32. $(\cos x + \sin x)^2 = 1 + \sin 2x$

33. $\dfrac{1 + \cos 2\omega}{\sin 2\omega} = \cot \omega$

34. $\dfrac{\cot^2 y - 1}{2\cot y} = \cot 2y$

35. $\dfrac{\csc^2 \theta - 2}{\csc^2 \theta} = \cos 2\theta$

36. By letting $A = B$ in identity (16.13), show that

$$\tan 2A = \dfrac{2\tan A}{1 - \tan^2 A}$$

which is the *double-angle formula for the tangent*.

37. Suppose a particle is traveling along a line according to the equation $s = 4\sin^2 t$, where s is measured in meters and t in seconds. Calculus shows that the velocity is given by $v = 8\sin t \cos t$. Write v as a single trigonometric function of t.

38. Prove the following identity from the derivation of Rutherford's scattering formula:

$$2\pi r^2 \sin \theta = 4\pi r^2 \sin \dfrac{\theta}{2} \cos \dfrac{\theta}{2}$$

39. An axle is placed through the center of a circular disk at an angle α. The magnitude T of the torque on the bearings holding the axle has the form $T = k\omega^2 \sin \alpha \cos \alpha$, where ω is the angular velocity. Show that

$$T = \dfrac{1}{2} k\omega^2 \sin 2\alpha$$

40. The equation of the path of a missile projected at velocity v at an angle θ with the ground is

$$y = x\tan \theta - \dfrac{gx^2}{2v^2 \cos^2 \theta}$$

Show that

$$y = \dfrac{x(v^2 \sin 2\theta - gx)}{2v^2 \cos^2 \theta}$$

16.5 Half-Angle Formulas

The identities in the previous section allow us to write a function of $2A$ in terms of functions of A. In this section we shall study the **half-angle formulas**, which enable us to express a function of $\frac{1}{2}A$ in terms of functions of A.

The half-angle formulas can be obtained from the double-angle formulas by properly rearranging the terms. If we start with

$$\cos 2x = 1 - 2 \sin^2 x$$

we get

$$2 \sin^2 x = 1 - \cos 2x$$

$$\sin^2 x = \frac{1 - \cos 2x}{2}$$

$$\sin x = \pm \sqrt{\frac{1 - \cos 2x}{2}}$$

Letting $x = A/2$, we have

$$\sin \frac{A}{2} = \pm \sqrt{\frac{1 - \cos A}{2}}$$

Similarly, from $\cos 2x = 2 \cos^2 x - 1$, we obtain

$$\cos \frac{A}{2} = \pm \sqrt{\frac{1 + \cos A}{2}}$$

The algebraic sign depends on the quadrant in which the terminal side of $A/2$ lies.

Half-angle formulas:

$$\sin \frac{A}{2} = \pm \sqrt{\frac{1 - \cos A}{2}} \qquad (16.27)$$

$$\cos \frac{A}{2} = \pm \sqrt{\frac{1 + \cos A}{2}} \qquad (16.28)$$

The half-angle identities can be used to express the sine and cosine of a given half-angle in terms of the cosine of the angle, as shown in the first two examples.

CHAPTER 16 ADDITIONAL TOPICS IN TRIGONOMETRY

Example 1 Use the appropriate half-angle formula to find the exact value of cos 165°.

Solution. By identity (16.28)

$$\cos 165° = \pm\sqrt{\frac{1 + \cos[(2)(165°)]}{2}} = \pm\sqrt{\frac{1 + \cos 330°}{2}}$$

$$= \pm\sqrt{\frac{1 + \frac{\sqrt{3}}{2}}{2}} = \pm\sqrt{\frac{1 + \frac{\sqrt{3}}{2}}{2} \cdot \frac{2}{2}}$$

$$= \pm\sqrt{\frac{2 + \sqrt{3}}{4}} = \pm\frac{\sqrt{2 + \sqrt{3}}}{2}$$

Since 165° is in the second quadrant, cos 165° is negative. So

$$\cos 165° = -\frac{\sqrt{2 + \sqrt{3}}}{2}$$

Example 2 Find $\cos \theta/2$, given that $\sin \theta = -\frac{12}{13}$, θ in quadrant IV.

Solution. If $\sin \theta = -\frac{12}{13}$, then $\cos \theta = \frac{5}{13}$ for θ in quadrant IV. Moreover, since $270° < \theta < 360°$, it follows that $135° < \theta/2 < 180°$. Thus $\cos \theta/2 < 0$ and

$$\cos \frac{\theta}{2} = -\sqrt{\frac{1 + \cos \theta}{2}} = -\sqrt{\frac{1 + \frac{5}{13}}{2}} = -\sqrt{\frac{1 + \frac{5}{13}}{2} \cdot \frac{13}{13}}$$

$$= -\sqrt{\frac{13 + 5}{26}} = -\sqrt{\frac{9}{13}}$$

$$= -\frac{3}{\sqrt{13}} = -\frac{3\sqrt{13}}{13}$$

Sometimes the half-angle formulas are used to simplify certain radical expressions.

Example 3 Simplify $\sqrt{6 - 6\cos 4\theta}$.

Solution. The expression resembles the right side of identity (16.27). To get the proper form, we need to remove 6 from the radical and obtain a 2 in the denominator. Thus

$$\sqrt{6 - 6\cos 4\theta} = \sqrt{6(1 - \cos 4\theta)} = \sqrt{6}\sqrt{1 - \cos 4\theta}$$

$$= \sqrt{6}\sqrt{\frac{2(1 - \cos 4\theta)}{2}}$$

$$= \sqrt{6}\sqrt{2}\sqrt{\frac{1-\cos 4\theta}{2}} = \sqrt{12}\sin 2\theta$$
$$= 2\sqrt{3}\sin 2\theta$$

In the study of calculus, the forms of the half-angle formulas stated below are sometimes more useful.

$$\sin^2 A = \frac{1 - \cos 2A}{2} \qquad (16.29)$$

$$\cos^2 A = \frac{1 + \cos 2A}{2} \qquad (16.30)$$

Note that these formulas are really the double-angle identities (16.25) and (16.26) slightly rewritten.

Example 4 Write $4\sin^2 3x$ without the square.

Solution. By identity (16.29) with $A = 3x$,

$$4\sin^2 3x = 4\left(\frac{1 - \cos 6x}{2}\right) = 2(1 - \cos 6x)$$

Example 5 Write $6\cos^2 2x$ without the square.

Solution. By identity (16.30) with $A = 2x$,

$$6\cos^2 2x = 6\left(\frac{1 + \cos 4x}{2}\right) = 3(1 + \cos 4x)$$

Example 6 Prove the identity

$$2\cos^2 \frac{\theta}{2} = \frac{\sin^2 \theta}{1 - \cos \theta}$$

Solution. Since $\sin^2 \theta = 1 - \cos^2 \theta$,

$$\frac{\sin^2 \theta}{1 - \cos \theta} = \frac{1 - \cos^2 \theta}{1 - \cos \theta} = \frac{(1 - \cos \theta)(1 + \cos \theta)}{1 - \cos \theta} \qquad \text{difference of two squares}$$

$$= 1 + \cos \theta = \frac{2(1 + \cos \theta)}{2}$$

$$= 2\cos^2 \frac{\theta}{2} \qquad \text{by (16.30)}$$

Example 7 The motion of a planet or comet about the sun can be described by an equation of the form

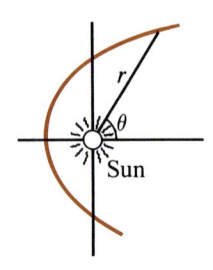

$$r = \frac{c}{a - b \cos \theta}$$

where a, b, and c are constants. (See Figure 16.8.) This equation represents a conic section. For example, the equation of the elliptic path of Mercury is

$$r = \frac{3.442 \times 10^7}{1 - 0.206 \cos \theta}$$

where r is measured in miles.

Figure 16.8

Some comets follow a path that is nearly parabolic:

$$r = \frac{A}{1 - \cos \theta}$$

Show that

$$r = \frac{A}{2} \csc^2 \frac{\theta}{2}$$

Solution. $r = \dfrac{A}{1 - \cos \theta} = \dfrac{A}{2} \cdot \dfrac{2}{1 - \cos \theta} = \dfrac{A}{2} \cdot \dfrac{1}{\dfrac{1 - \cos \theta}{2}}$

$= \dfrac{A}{2} \cdot \dfrac{1}{\sin^2 \dfrac{\theta}{2}} = \dfrac{A}{2} \csc^2 \dfrac{\theta}{2}$

Exercises / Section 16.5

In Exercises 1–5, find the exact value of each trigonometric function by means of the half-angle formulas.

1. sin 15°   **2.** cos 75°   **3.** cos 22.5°
4. cos 105°   **5.** sin 112.5°
6. Find $\sin(\theta/2)$, given that $\cos \theta = \frac{4}{5}$, θ in quadrant I.
7. Find $\sin(\theta/2)$, given that $\cos \theta = -\frac{24}{25}$, θ in quadrant III.
8. Find $\cos(\theta/2)$, given that $\sin \theta = \frac{3}{5}$, θ in quadrant II.
9. Find $\sin(\theta/2)$, given that $\cos \theta = \frac{5}{13}$, θ in quadrant IV.
10. Find $\cos(\theta/2)$, given that $\sin \theta = -\frac{7}{25}$, θ in quadrant IV.

In Exercises 11–16, simplify each given expression. (See Example 3.)

11. $\sqrt{\dfrac{1 - \cos 4\theta}{2}}$ **12.** $\sqrt{\dfrac{1 + \cos 6\theta}{2}}$ **13.** $\sqrt{1 + \cos 6\theta}$
14. $\sqrt{4 - 4 \cos 8\theta}$ **15.** $\sqrt{5 - 5 \cos 4\theta}$ **16.** $\sqrt{6 + 6 \cos 8\theta}$

In Exercises 17–24, eliminate the exponent. (See Examples 4 and 5.)

17. $\sin^2 4x$
18. $\sin^2 3x$
19. $\cos^2 2x$
20. $\cos^2 3x$
21. $2\sin^2 3x$
22. $4\cos^2 4x$
23. $12\sin^2 x$
24. $2\cos^2 x$

In Exercises 25–28, prove the given identities.

25. $\dfrac{\sin 2\theta}{2\sin\theta} = \cos^2 \dfrac{\theta}{2} - \sin^2 \dfrac{\theta}{2}$

26. $\cos^2 \dfrac{x}{4} = \dfrac{1 + \cos \dfrac{x}{2}}{2}$

27. $\csc^2 \theta = \dfrac{2}{1 - \cos 2\theta}$

28. $2\cos\dfrac{\beta}{2} = (1 + \cos\beta)\sec\dfrac{\beta}{2}$

29. Simplify the following expression from a problem in the study of the pendulum:

$$\dfrac{1}{\sqrt{1 - \cos x}}$$

30. In the study of the motion of a pendulum, the expression

$$A = \sqrt{\dfrac{1 - \cos\theta}{1 - \cos\alpha}}$$

needs to be simplified. Show that

$$A = \dfrac{\sin\dfrac{\theta}{2}}{\sin\dfrac{\alpha}{2}}$$

31. In determining the length of the path along which a particle will slide from a higher to a lower point in minimum time, the expression $\sqrt{2 - 2\cos\theta}$ needs to be simplified. Carry out this simplification.

32. A common exercise in calculus is determining the area under a curve. To find the area under one arch of the curve $y = \sin^2 x$, the equation must be written without the exponent. Rewrite this equation.

33. The index of refraction n of a prism with apex angle A whose minimum angle of refraction is δ is given by

$$n = \dfrac{\sin\dfrac{A + \delta}{2}}{\sin\dfrac{A}{2}}, \quad n \geq 0$$

Show that the expression is equivalent to

$$n = \sqrt{\dfrac{1 + \sin A \sin\delta - \cos A \cos\delta}{1 - \cos A}}$$

16.6 Trigonometric Equations

So far we have concentrated only on identities, equations that are valid for all values of the variable. Now we shall turn to **conditional equations,** which are valid only for certain values of the angle.

516 CHAPTER 16 ADDITIONAL TOPICS IN TRIGONOMETRY

For example, the equation

$$\sin \theta = 0$$

is not an identity, since equality holds only if

$$\theta = 0°, \pm 180°, \pm 360°, \text{ and so on}$$

To solve an equation containing a single trigonometric function, we solve the equation for this function and then determine the values of the angle for which equality holds. Consider the next example.

Example 1 Solve the equation $2 \cos x - 1 = 0$, $0 \leq x < 2\pi$.

Solution. The first step is to solve the given equation for $\cos x$. Thus

$$2 \cos x - 1 = 0 \qquad \text{given equation}$$
$$2 \cos x = 1 \qquad \text{transposing } -1$$
$$\cos x = \frac{1}{2} \qquad \text{dividing by 2}$$

The angles between 0 and 2π whose cosine is $\frac{1}{2}$ are

$$x = \frac{\pi}{3} \quad \text{and} \quad x = \frac{5\pi}{3}$$

Substituting into the given equation shows that the solutions check.

If an equation involves more than one function, we can often use the identities to convert it to an equation involving only one function, as shown in the next example.

Example 2 Solve the equation $\sec^2 x - 4 \tan^2 x = 0$, $0 \leq x < 2\pi$.

Solution. Since the equation involves two different functions, no direct solution is possible. However, if we recall that $1 + \tan^2 x = \sec^2 x$, we can convert one of the functions. Thus

$$\mathbf{\sec^2 x} - 4 \tan^2 x = 0$$
$$\mathbf{1 + \tan^2 x} - 4 \tan^2 x = 0$$
$$1 - 3 \tan^2 x = 0$$
$$\tan^2 x = \frac{1}{3}$$
$$\sqrt{\tan^2 x} = \pm \sqrt{\frac{1}{3}}$$

16.6 TRIGONOMETRIC EQUATIONS

$$\tan x = \pm \frac{\sqrt{1}}{\sqrt{3}}$$

$$\tan x = \pm \frac{1}{\sqrt{3}}$$

It follows that $x = 30°, 150°, 210°$, and $330°$. In radian measure

$$x = \frac{\pi}{6}, \frac{5\pi}{6}, \frac{7\pi}{6}, \frac{11\pi}{6}$$

Quadratic forms

Some trigonometric equations are actually in quadratic form, as shown in the next example.

Example 3 Solve the equation $2 \csc^2 x + 3 \csc x - 2 = 0, 0 \le x < 2\pi$.

Solution. Let $y = \csc x$. Then the equation becomes

$$2y^2 + 3y - 2 = 0$$
$$(2y - 1)(y + 2) = 0$$
$$y = -2, \frac{1}{2}$$

It follows from $y = \csc x$ that

$$\csc x = -2 \quad \text{and} \quad \csc x = \frac{1}{2}$$

Since a value of $\csc x$ cannot be less than unity, the equation $\csc x = \frac{1}{2}$ has no solution. From $\csc x = -2$, we obtain $x = 210°$ and $330°$. In radian measure

$$x = \frac{7\pi}{6}, \frac{11\pi}{6}$$

Example 4 Solve the equation $\sin 2x = 0, 0 \le x < 2\pi$.

Solution. Since $\sin 2x = 0$, we have $2x = 0°, 180°$, so that $x = 0°, 90°$. Because of the double angle, these are not the only solutions in the range $0 \le x < 360°$. From $2x = 360°, 540°$, we have $x = 180°, 270°$. In other words,

$$\sin 2x = 0$$

whenever

$$2x = 0, \pi, 2\pi, 3\pi$$

and

$$x = 0, \frac{\pi}{2}, \pi, \frac{3\pi}{2}$$

Note that the largest of the roots, $3\pi/2$, is still less than 2π, so that there are four solutions to the equation.

Example 5 Solve the equation $\cos 2x - \cos x = 0$, $0 \le x < 2\pi$.

Solution. Because of the double angle, $\cos 2x$ must first be changed to $2 \cos^2 x - 1$ by one of the double-angle formulas for the cosine function (reminder: $\cos 2x \ne 2 \cos x$). Then we obtain

$$\cos 2x - \cos x = 0$$
$$2 \cos^2 x - 1 - \cos x = 0$$
$$2 \cos^2 x - \cos x - 1 = 0$$
$$(2 \cos x + 1)(\cos x - 1) = 0 \qquad 2z^2 - z - 1 = (2z + 1)(z - 1)$$
$$\cos x = -\frac{1}{2}, 1$$

It follows that $x = 120°, 240°$, and $0°$. In radians

$$x = 0, \frac{2\pi}{3}, \frac{4\pi}{3}$$

Example 6 Use a calculator to solve the equation

$$2 \sin 2x - 3 \sin x = 0$$

to the nearest tenth of a degree ($0° \le x < 360°$).

Solution. By the double-angle formula for the sine function, $\sin 2x = 2 \sin x \cos x$, we get

$$2 \sin 2x - 3 \sin x = 0 \qquad \text{given equation}$$
$$2(2 \sin x \cos x) - 3 \sin x = 0$$
$$4 \sin x \cos x - 3 \sin x = 0$$
$$\sin x(4 \cos x - 3) = 0 \qquad \text{common factor } \sin x$$
$$\sin x = 0 \qquad\qquad\qquad\qquad\qquad\qquad 4 \cos x - 3 = 0$$
$$\cos x = \frac{3}{4}$$

From $\sin x = 0$, we get $x = 0°, 180°$. Using a calculator, $\cos x = \frac{3}{4}$ yields $x = 41.4°, 318.6°$.

Example 7 The range R (in feet) of a projectile fired at an angle θ with the horizontal at velocity v (in feet per second) is given by

$$R = \frac{2v^2 \cos \theta \sin \theta}{g}$$

where $g = 32$ ft/sec². (See Figure 16.9.) If $v = 40$ ft/sec, determine the angle θ at which the projectile has to be aimed to hit an object 45 ft away.

Figure 16.9

Solution. Substituting into the given equation, we get

$$\frac{2(40)^2 \cos \theta \sin \theta}{32} = 45$$

$$\frac{(40)^2}{32}(2 \sin \theta \cos \theta) = 45$$

$$2 \sin \theta \cos \theta = \frac{(45)(32)}{(40)^2}$$

$$\sin 2\theta = 0.9$$

$$2\theta = 64°, 116°$$

$$\theta = 32°, 58°$$

So the projectile can be aimed at either 32° or 58° to land 45 ft away.

Exercises / Section 16.6

In Exercises 1–31, solve the given equations for x, $0 \le x < 2\pi$.

1. $2 \sin x - 1 = 0$
2. $3 \sin x + 3 = 0$
3. $4 \tan x + 4 = 8$
4. $2 \sec x + 4 = 0$
5. $\cos^2 x - 1 = 0$
6. $2 \sin^2 x - 1 = 0$
7. $\sin^2 x - \sin x = 0$
8. $\tan x(\csc x + 1) = 0$
9. $4 \cos^2 x - 3 = 0$
10. $(\sec x - 1)(\tan x + 1) = 0$
11. $(\cot x - 1)(\cos x + 1) = 0$
12. $(2 \cos x - 1)(\csc x - 2) = 0$
13. $2 \sin^2 x - \sin x - 1 = 0$
14. $2 \sin^2 x - \sin x - 3 = 0$
15. $3 \cos^2 x - 7 \cos x + 4 = 0$
16. $2 \sec^2 x + 3 \sec x - 2 = 0$
17. $\sin x - \cos x = 0$
18. $\cot^2 x - \tan^2 x = 0$
19. $2 \sin^2 x + \cos x + 1 = 0$
20. $2 \tan^2 x - \sec^2 x = 0$
21. $\sin 2x = 1$

22. $\cos 2x = 1$ **23.** $\cos 2x = 0$ **24.** $\sin x + \sin 2x = 0$
25. $\sin^2 x + \cos 2x = 0$ **26.** $\sin 2x + \cos 2x = 0$ **27.** $\cos 2x - \cos x = 0$
28. $\cos 2x - \sin x = 0$ **29.** $\sin \frac{x}{2} = \cos \frac{x}{2}$ **30.** $1 - \sin x \cos x = 1$
31. $\sin x \cos x - \sin 2x = 0$

In Exercises 32–40, use a calculator to solve the given equations to the nearest tenth of a degree ($0 \le x < 360°$).

32. $3 \sin x \cos x - \cos x = 0$ **33.** $\tan^2 x - 2 = 0$ **34.** $2 \cos^2 x = 1 + \sin^2 x$
35. $2 \sin 2x + \cos x = 0$ **36.** $\sec^2 x - 2 \tan x - 4 = 0$ **37.** $2 \sin 2x = 3 \sin x$
38. $\cos^2 x - 2 \sin^2 x = 0$ **39.** $\csc^2 x - 3 \cot^2 x = 0$ **40.** $5 \sin^2 x + 8 \sin x - 4 = 0$

41. Certain problems in mechanics are simplified by rotating the coordinate axes. In the process, the following equation has to be solved:

$$2(C - A) \sin \theta \cos \theta + B(\cos^2 \theta - \sin^2 \theta) = 0$$

Solve this equation for θ ($0 \le \theta < 180°$), given that $A = B = 1$ and $C = 0$.

42. Suppose a projectile fired at a velocity of 80 ft/sec is to hit a target 100 ft away. At what angle with respect to the ground does the projectile have to be fired? (See Example 7.)

43. The current in a certain circuit is given by $i = e^{-5t}(\cos 4.0t - \sqrt{3} \sin 4.0t)$. Find the smallest positive value of t (in seconds) for which the current is zero.

44. For a certain mass oscillating on a spring, the vertical displacement is given by

$$x = 2.0 \cos 2t - 1.0 \sin 2t, \quad t \ge 0$$

where x is measured in centimeters and t in seconds. Find the smallest value of t for which the displacement is zero. (Set your calculator in the radian mode.)

45. Starting at $t = 0$, the current in a circuit is

$$i = 2 \sin^2 \omega t + 3 \sin \omega t$$

Find the smallest value of t (in seconds) for which $i = 2$ A.

16.7 Inverse Trigonometric Relations

We know from our study of equations that it is often desirable to solve a given equation for one of the variables in terms of the other variables. To solve a trigonometric equation $y = f(x)$ for the variable x, we need the concept of an *inverse trigonometric relation*.

Consider, for example, the function $y = \sin x$. To solve this equation for x in terms of y, we introduce the following notation:

$$x = \arcsin y$$

Expressed verbally, "x is a number (or angle measure) whose sine is y." Following the usual convention of placing y on the left side of the equal sign, we write

$$y = \arcsin x \qquad (16.31)$$

16.7 INVERSE TRIGONOMETRIC RELATIONS

The expression says that y is a number (or angle measure) whose sine is x. The equation $y = \arcsin x$ is called an **inverse trigonometric relation.**

To illustrate the meaning of this kind of notation, let us evaluate $y = \arcsin x$ for certain values of x.

Example 1 Find y if $y = \arcsin \frac{1}{2}$, $0 \leq y < 2\pi$.

Solution. By the definition of $\arcsin \frac{1}{2}$ we need to find a number whose sine is $\frac{1}{2}$. Two such numbers exist between 0 and 2π, namely,

$$y = \frac{\pi}{6} \quad \text{and} \quad y = \frac{5\pi}{6}$$

Example 2 Find y if $y = \arcsin 0$.

Solution. Since $\sin 0 = 0$ and $\sin \pi = 0$, it follows that

$$\sin(0 + k \cdot 2\pi) = 0 \quad \text{and} \quad \sin(\pi + k \cdot 2\pi) = 0,$$
$$k = 0, \pm 1, \pm 2, \ldots$$

or

$$\sin k\pi = 0, \quad k = 0, \pm 1, \pm 2, \ldots$$

So

$$y = k\pi, \quad k = 0, \pm 1, \pm 2, \ldots$$

Remark. Recall that a relation between two variables x and y is called a **function** if for every value of x there exists a unique value of y, denoted by $y = f(x)$. By this definition, $y = \arcsin x$ is *not* a function, as we can see from Example 2: The value $x = 0$ does not yield a unique value for y. To obtain a function, the values of y must be suitably restricted. That is the topic of the next section.

The other trigonometric functions have similar inverse relations, as shown in the next two examples.

Example 3 Find y if $y = \arccos(\sqrt{3}/2)$, $0 \leq y < 2\pi$.

Solution. The notation $\arccos(\sqrt{3}/2)$ has an analogous meaning as an angle whose cosine is $\sqrt{3}/2$. For $0 \leq y < 2\pi$, we have

$$y = \frac{\pi}{6} \quad \text{and} \quad y = \frac{11\pi}{6}$$

Example 4 Find y if $y = \arctan(-1)$, $0 \leq y < 2\pi$.

Solution. For y between 0 and 2π, the only angles whose tangents are equal to -1 are

$$y = \frac{3\pi}{4} \quad \text{and} \quad y = \frac{7\pi}{4}$$

As noted at the beginning of this section, the notation for the inverse relationship enables us to solve a trigonometric equation for x in terms of y, as shown in the next example.

Example 5 Solve the equation $y = 1 + \sin 2x$ for x in terms of y.

Solution. The equation $y = 1 + \sin 2x$ can also be written

$$\sin 2x = y - 1$$

Using the inverse relationship,

$$2x = \arcsin(y - 1)$$

we get

$$x = \frac{1}{2}\arcsin(y - 1)$$

Exercises / Section 16.7

In Exercises 1–16, find y ($0 \leq y < 2\pi$) without using a table or calculator.

1. $y = \arcsin \dfrac{\sqrt{3}}{2}$
2. $y = \arcsin(-1)$
3. $y = \arctan 1$
4. $y = \operatorname{arccot}(-1)$
5. $y = \operatorname{arccsc} 1$
6. $y = \arccos(-1)$
7. $y = \arcsin 0$
8. $y = \arcsin\left(-\dfrac{1}{2}\right)$
9. $y = \arccos \dfrac{1}{2}$
10. $y = \arcsin\left(-\dfrac{\sqrt{2}}{2}\right)$
11. $y = \arccos 0$
12. $y = \operatorname{arcsec} 1$
13. $y = \operatorname{arccsc}(-2)$
14. $y = \arctan\left(-\dfrac{1}{\sqrt{3}}\right)$
15. $y = \operatorname{arccot} \dfrac{1}{\sqrt{3}}$
16. $y = \arctan 0$

In Exercises 17–30, solve each equation for x in terms of y.

17. $y = \arctan x$
18. $y = \arccos 3x$
19. $y = 1 - \arcsin x$
20. $y = 2 + \operatorname{arcsec} x$
21. $y = \arcsin 2x - 1$
22. $y = \arccos(x - 2)$

23. $y = \text{arccsc}(x + 1)$ **24.** $y = \text{arccot } 2x$ **25.** $y = \text{arcsec } 3x + 1$
26. $y = \text{arcsin } 2(x - 2)$ **27.** $y = 3 \text{ arccot } 3x$ **28.** $y = 2 \text{ arctan } 5x + 1$
29. $y = 2 \text{ arcsin}(x + 1) + 3$ **30.** $y = 3 \text{ arccos}(x - 2) - 2$

16.8 Inverse Trigonometric Functions

We learned in our study of logarithms that the equation $y = b^x$ can be written $x = \log_b y$. While the two equations mean the same thing, the first expresses y as a function of x and the second expresses x as a function of y; $y = b^x$ and $y = \log_b x$ are called **inverse functions.** An analogous situation exists in trigonometry in the sense that every trigonometric function has an inverse function.

In the last section we introduced the customary notation for inverse trigonometric relations. We also noted that a relation such as $y = \arcsin x$ does not represent a function. Given the importance of the function concept, this state of affairs is unsatisfactory. The variable y must be suitably restricted so that every value of x yields a unique value of y. This restriction leads to the definition of an **inverse trigonometric function.**

First we need to consider the graph of the relation $y = \arcsin x$. Writing this equation in the form $x = \sin y$, we get the graph of the sine function with x and y interchanged, as shown in Figure 16.10. This graph shows why the

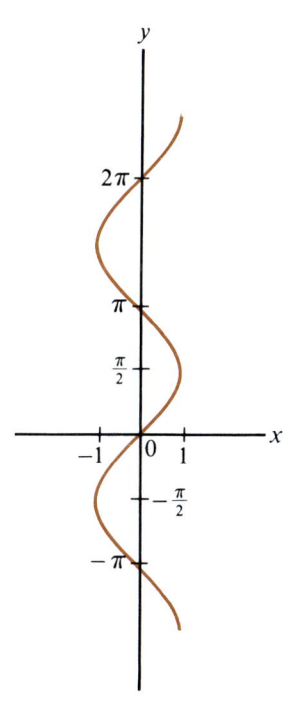

Figure 16.10

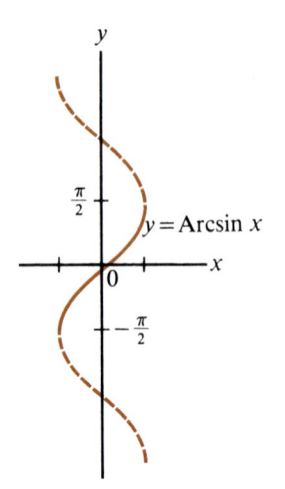

relation $y = \arcsin x$ is not a function: For every x such that $-1 \leq x \leq 1$, we get infinitely many values for y.

We can also see from the graph that y becomes unique if all but a small section of the graph is eliminated. This elimination can be done in several ways. The restriction that has become standard is $-\pi/2 \leq y \leq \pi/2$, which corresponds to the portion of the graph through the origin, drawn as the solid curve in Figure 16.11. To distinguish between the solid curve and the dashed curve, the equation of the solid curve is written

$$y = \text{Arcsin } x$$

using the capital letter A. Note especially that

$y = \text{Arcsin } x$ is a function.

Figure 16.11

Example 1 Find the exact value of y in each case:

a. $y = \arcsin \dfrac{1}{2} \quad (0 \leq y < 2\pi)$ **b.** $y = \text{Arcsin } \dfrac{1}{2}$

Solution. a. As we saw in the previous section,

$$y = \frac{\pi}{6} \quad \text{and} \quad y = \frac{5\pi}{6}$$

b. Since $-\pi/2 \leq y \leq \pi/2$, the only permissible value is

$$y = \frac{\pi}{6}$$

Thus $\text{Arcsin } \tfrac{1}{2} = \pi/6$, a unique value.

The inverse functions corresponding to $y = \arctan x$ and $y = \arccos x$ are obtained from the graphs of $x = \tan y$ and $x = \cos y$, shown in Figure 16.12 and Figure 16.13, respectively.

Following the usual conventions, $y = \text{Arctan } x$ is the solid curve in Figure 16.12 and $y = \text{Arccos } x$ the solid curve in Figure 16.13. Note that in all cases the angle y is in the first quadrant whenever x is positive. The different cases are summarized next.

Inverse trigonometric functions:

$y = \text{Arcsin } x, \quad -\pi/2 \leq y \leq \pi/2$ \hfill (16.32)

$y = \text{Arctan } x, \quad -\pi/2 < y < \pi/2$ \hfill (16.33)

$y = \text{Arccos } x, \quad 0 \leq y \leq \pi$ \hfill (16.34)

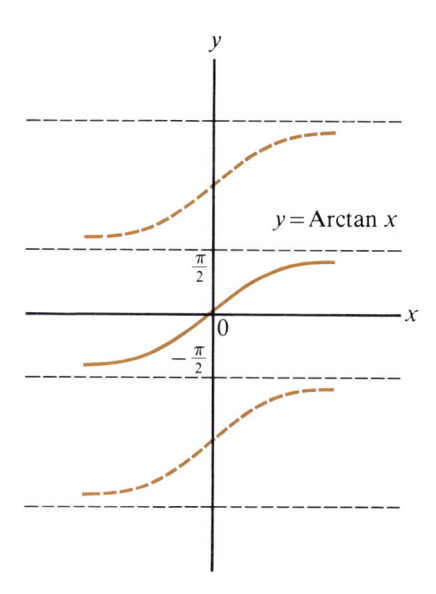

16.8 INVERSE TRIGONOMETRIC FUNCTIONS **525**

Figure 16.12 — $y = \text{Arctan } x$

Figure 16.13 — $y = \text{Arccos } x$

Although inverse trigonometric functions exist for the remaining functions, we shall confine ourselves to the cases already presented. One reason is that different authors define the other functions in different ways. For example, $y = \text{Arcsec } x$ is sometimes defined by using the restriction $0 \leq y < \pi/2$, $\pi/2 < y \leq \pi$ and sometimes by the restriction $0 \leq y < \pi/2$, $-\pi \leq y < -\pi/2$.

Example 2 Find the exact values of

    a. $\text{Arccos } \dfrac{1}{2}$ **b.** $\text{Arccos}\left(-\dfrac{1}{2}\right)$

Solution. a. Since x is positive, $\text{Arccos } \frac{1}{2}$ is in the first quadrant. Thus

$$\text{Arccos } \frac{1}{2} = \frac{\pi}{3}$$

(Don't forget that $\text{Arccos } \frac{1}{2}$ is an angle!)

    b. Since x is negative, the angle cannot be in the first quadrant. To find the proper quadrant, we must refer to the definition of $\text{Arccos } x$. By agreement, *the angle must lie between 0 and π.* Thus

$$\text{Arccos}\left(-\frac{1}{2}\right) = \frac{2\pi}{3}$$

Example 3 Find the exact value of Arctan (-1).

Solution. This is a problem that many students find troublesome. If we were looking merely for some angles between 0 and 2π, we would choose 135° and 315° ($3\pi/4$ and $7\pi/4$). Knowing that the value has to be unique, some students proceed to drop one of the values and keep only $7\pi/4$. Now, while this angle does lie in the fourth quadrant, this choice still violates the convention in statement (16.33). Since $-\pi/2 < y < \pi/2$, the angle chosen must be negative. Thus

$$\text{Arctan}\,(-1) = -\frac{\pi}{4}$$

Example 4 Find the exact value of Arcsin $\left(-\dfrac{\sqrt{3}}{2}\right)$.

Solution. Since x is negative, the angle cannot lie in the first quadrant. By the definition of Arcsin x, we have $-\pi/2 \leq y \leq \pi/2$, so that

$$\text{Arcsin}\,\left(-\frac{\sqrt{3}}{2}\right) = -\frac{\pi}{3}$$

CALCULATOR COMMENT

To show how strictly these conventions must be observed, let us find some of the values of the inverse trigonometric functions by using a calculator. (If the angles are not special angles, a calculator should be used anyway.)

Example 5 Use a calculator to find Arcsin (0.4278).

Radian mode

Solution. First set the calculator in the radian mode. Now enter 0.4278 and press the $\boxed{\text{INV}}$ key, followed by the $\boxed{\text{SIN}}$ key, to obtain 0.4421. As expected, the angle is in the first quadrant.

Degree mode

By setting the calculator in the degree mode, the same sequence yields Arcsin (0.4278) = 25.33°.

Example 6 Evaluate Arcsin (-0.6845).

Solution. Set the calculator in the radian mode and proceed as in Example 5. We obtain

$$\text{Arcsin}\,(-0.6845) = -0.7539$$

The result agrees with the convention in statement (16.32).

Inverse trigonometric functions can be used to solve certain trigonometric equations.

16.8 INVERSE TRIGONOMETRIC FUNCTIONS

Example 7 Solve the equation $y = 2 \tan 3x - 1$ for x in terms of y using the proper inverse function.

Solution.

$$y = 2 \tan 3x - 1 \qquad \text{given equation}$$
$$y + 1 = 2 \tan 3x \qquad \text{transposing } -1$$
$$\tan 3x = \frac{1}{2}(y + 1) \qquad \text{dividing by 2}$$
$$3x = \text{Arctan } \frac{1}{2}(y + 1) \qquad \text{inverse function}$$
$$x = \frac{1}{3} \text{Arctan } \frac{1}{2}(y + 1) \qquad \text{dividing by 3}$$

Example 8 Solve the equation $y = 3 \text{ Arccos } 2x$ for x in terms of y.

Solution.
$$y = 3 \text{ Arccos } 2x$$
$$\frac{y}{3} = \text{Arccos } 2x$$
$$2x = \cos \frac{y}{3}$$
$$x = \frac{1}{2} \cos \frac{y}{3}$$

The remaining examples involve trigonometric functions in a way that is particularly useful in calculus.

Example 9 Find the exact value of $\sin [\text{Arctan } (-\frac{1}{4})]$.

Solution. Recall that $\text{Arctan } (-\frac{1}{4})$ is an angle whose tangent is $-\frac{1}{4}$. Let $\theta = \text{Arctan } (-\frac{1}{4})$. To find $\sin \theta$, we draw the diagram in Figure 16.14. It follows that

$$\sin \left[\text{Arctan } \left(-\frac{1}{4}\right) \right] = \sin \theta = \frac{-1}{\sqrt{17}} = -\frac{\sqrt{17}}{17}$$

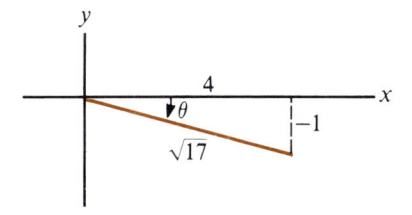

Figure 16.14

Example 10

Find an algebraic expression equivalent to tan (Arccos x).

Solution. Let θ = Arccos x. Thus θ is an angle whose cosine is $x/1$. Draw a right triangle with x on the adjacent side and 1 on the hypotenuse. (See Figure 16.15.) By the Pythagorean theorem, the length of the opposite side is $\sqrt{1 - x^2}$. It follows that

$$\tan (\text{Arccos } x) = \tan \theta = \frac{\sqrt{1 - x^2}}{x}$$

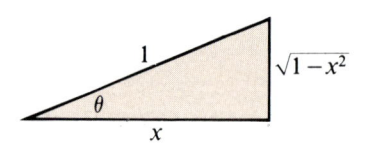

Figure 16.15

Example 11

The width w of a laser beam at a distance d from the source is given by

$$w = 2d \tan \frac{\alpha}{2}$$

where α is the angle of the beam. Solve this equation for α.

Solution.
$$w = 2d \tan \frac{\alpha}{2}$$

$$\frac{w}{2d} = \tan \frac{\alpha}{2}$$

$$\frac{\alpha}{2} = \text{Arctan } \frac{w}{2d}$$

$$\alpha = 2 \text{ Arctan } \frac{w}{2d}$$

Exercises / Section 16.8

In Exercises 1–17, find the exact value (in radian measure) of each expression without using a table or a calculator.

1. Arcsin $\frac{\sqrt{3}}{2}$
2. Arcsin (-1)
3. Arctan 1
4. Arccos (-1)
5. Arcsin 0
6. Arcsin $\left(-\frac{1}{2}\right)$
7. Arccos 0
8. Arctan $\left(-\frac{1}{\sqrt{3}}\right)$
9. Arctan 0

10. Arcsin 1 **11.** Arcsin $\left(-\dfrac{1}{\sqrt{2}}\right)$ **12.** Arcsin $\left(\dfrac{1}{\sqrt{2}}\right)$

13. Arctan $(-\sqrt{3})$ **14.** Arccos $\left(-\dfrac{1}{\sqrt{2}}\right)$ **15.** Arctan $\sqrt{3}$

16. Arccos $\dfrac{1}{\sqrt{2}}$ **17.** Arccos $\left(-\dfrac{\sqrt{3}}{2}\right)$

In Exercises 18–40, evaluate the given expressions without a table or a calculator. (See Example 9.)

18. sin (Arctan 2) **19.** tan $\left[\text{Arccos}\left(-\dfrac{1}{3}\right)\right]$ **20.** tan $\left[\text{Arcsin}\left(-\dfrac{1}{3}\right)\right]$

21. csc $\left[\text{Arcsin}\left(-\dfrac{3}{4}\right)\right]$ **22.** csc $\left[\text{Arccos}\left(-\dfrac{3}{4}\right)\right]$ **23.** cos [Arctan (-2)]

24. sec (Arctan 3) **25.** cos $\left(\text{Arcsin}\,\dfrac{2}{3}\right)$ **26.** sec $\left(\text{Arcsin}\,\dfrac{4}{5}\right)$

27. csc $\left[\text{Arctan}\left(-\dfrac{3}{4}\right)\right]$ **28.** tan $\left[\text{Arcsin}\left(-\dfrac{12}{13}\right)\right]$ **29.** cot $\left[\text{Arccos}\left(-\dfrac{5}{13}\right)\right]$

30. cot $\left[\text{Arctan}\left(-\dfrac{5}{6}\right)\right]$ **31.** sec $\left(\text{Arccos}\,\dfrac{1}{4}\right)$ **32.** csc $\left(\text{Arcsin}\,\dfrac{2}{5}\right)$

33. cot $\left[\text{Arcsin}\left(-\dfrac{1}{4}\right)\right]$ **34.** sec $\left[\text{Arcsin}\left(-\dfrac{3}{7}\right)\right]$ **35.** sin $\left[\text{Arccos}\left(-\dfrac{2}{5}\right)\right]$

36. csc (Arctan $\sqrt{5}$) **37.** sin $\left(\text{Arcsin}\,\dfrac{1}{5}\right)$ **38.** tan (Arctan 4)

39. cot (Arctan 4) **40.** cos $\left(\text{Arccos}\,\dfrac{2}{5}\right)$

In Exercises 41–50, for each expression find an equivalent algebraic expression. Use the positive square root in each case. (See Example 10.)

41. tan (Arcsin x) **42.** cos (Arctan x) **43.** sec (Arctan x)
44. sin (Arccos x) **45.** cot (Arcsin $2x$) **46.** sin (Arccos $2x$)
47. csc (Arctan $3x$) **48.** tan (Arccos $3x$) **49.** sin (Arccos $2x$)
50. tan (Arcsin $3x$)

In Exercises 51–58, use a calculator to evaluate each inverse function. (Set your calculator in the radian mode.)

51. Arctan 2 **52.** Arctan (-2) **53.** Arcsin $\left(-\dfrac{1}{3}\right)$

54. Arccos $\left(-\dfrac{2}{3}\right)$ **55.** Arctan (1.3142) **56.** Arcsin (-0.7418)

57. Arccos (-0.4915) **58.** Arctan (2.672)

In Exercises 59–68, solve the equations for x.

59. $y = 2\,\text{Arcsin}\,x$ **60.** $y = 3\,\text{Arccos}\,x$ **61.** $y = 2\sin x$
62. $y = 3\cos x$ **63.** $y = \text{Arctan}\,x + 3$ **64.** $y = \text{Arctan}\,(x + 3)$

65. $y = 2 \sin 3x$

66. $y = 4 \operatorname{Arcsin}(x + 4)$

67. $y = 4 \tan(x - 2)$

68. $y = \frac{1}{2} \operatorname{Arccos}(x + 1)$

69. A woman is walking toward a building 100 ft high. Show that when she is x feet from the base of the building, the angle of elevation of the top is given by $\theta = \operatorname{Arctan}(100/x)$.

70. Show that angle $A = \operatorname{Arctan}[(b/a) \tan B]$ in Figure 16.16.

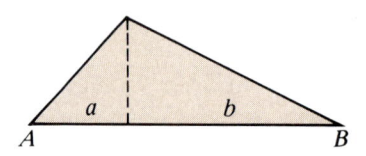

Figure 16.16

71. The formula $\phi = \operatorname{Arctan}(X/R)$ arises in the study of alternating current. Solve this formula for R.

72. A small body is revolving in a horizontal circle at the end of a cord of length L making an angle θ with the vertical (Figure 16.17). The time for one complete revolution is

$$T = 2\pi \sqrt{\frac{L \cos \theta}{g}}$$

Solve this equation for θ.

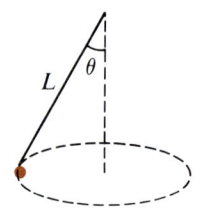

Figure 16.17

73. Recall that the equation of simple harmonic motion is

$$x = A \cos \sqrt{\frac{k}{m}}\, t$$

Find the formula for the time t required for the particle to move from its starting position $x = A$ (when $t = 0$) to a new position ($0 \leq t \leq \pi \sqrt{m/k}$).

74. The formula for *magnetic intensity* is

$$B = \frac{F}{qv \sin \phi}$$

where q is the magnitude of the charge, v its velocity, ϕ the angle between the direction of motion and the direction of the magnetic field, and F the force acting on the moving charge. Solve this formula for ϕ.

Review Exercises / Chapter 16

In Exercises 1–4, use the appropriate identity to evaluate each function without using a table or a calculator.

1. $\sin 22.5°$

2. $\cos 112.5°$

3. $\cos 12° \cos 18° - \sin 12° \sin 18°$

4. $\sin 110° \cos 20° - \cos 110° \sin 20°$

In Exercises 5–8, combine each expression into a single term.

5. $\sin 2x \cos 4x + \cos 2x \sin 4x$

6. $\cos 6x \cos x + \sin 6x \sin x$

7. $\sin 4x \cos x - \cos 4x \sin x$

8. $\cos(x - y) \cos y - \sin(x - y) \sin y$

In Exercises 9–14, write each expression as a function of x or $2x$.

9. $\cos(2x - \pi)$

10. $\sin\left(x - \dfrac{\pi}{2}\right)$

11. $\cos\left(2x + \dfrac{\pi}{2}\right)$

12. $\cos(x - 2\pi)$

13. $\sin\left(x - \dfrac{\pi}{6}\right)$

14. $\cos\left(x - \dfrac{\pi}{4}\right)$

15. Find $\sin 2\theta$, given that $\cos \theta = -\frac{4}{5}$, θ in quadrant II.

16. Find $\sin 2\theta$, given that $\sin \theta = -\frac{1}{3}$, θ in quadrant III.

17. Find $\cos 2\theta$, given that $\sin \theta = -\frac{5}{13}$, θ in quadrant IV.

18. Find $\sin(\theta/2)$, given that $\cos \theta = -\frac{4}{5}$, θ in quadrant II.

19. Find $\cos(\theta/2)$, given that $\sin \theta = -\frac{24}{25}$, θ in quadrant III.

20. Find $\cos(\theta/2)$, given that $\cos \theta = \frac{12}{13}$, θ in quadrant IV.

In Exercises 21–26, write each expression as a single trigonometric function.

21. $\cos^2 3x - \sin^2 3x$

22. $\sin^2 2x - \cos^2 2x$

23. $1 - 2\sin^2 4x$

24. $2\cos^2 3\beta - 1$

25. $2 \sin 3x \cos 3x$

26. $\sin 4x \cos 4x$

In Exercises 27–30, simplify the given expressions.

27. $\sqrt{\dfrac{1 - \cos 4\theta}{2}}$

28. $\sqrt{\dfrac{1 + \cos 4\theta}{2}}$

29. $\sqrt{1 + \cos 4\theta}$

30. $\sqrt{2 - 2\cos 8\theta}$

In Exercises 31–34, eliminate the exponent.

31. $\sin^2 3x$

32. $\cos^2 4x$

33. $2\cos^2 3x$

34. $4\sin^2 4x$

In Exercises 35–56, prove the given identities.

35. $\dfrac{\cos \beta \tan \beta + \sin \beta}{\tan \beta} = 2\cos \beta$

36. $(\sec \alpha - \tan \alpha)(\csc \alpha + 1) = \cot \alpha$

37. $\dfrac{1}{1 + \sin x} - \dfrac{1}{1 - \sin x} = -2 \tan x \sec x$

38. $\dfrac{\sec \theta}{\cot \theta + \tan \theta} = \sin \theta$

39. $\dfrac{1 + \sin^2 \theta \sec^2 \theta}{1 + \cos^2 \theta \csc^2 \theta} = \tan^2 \theta$

40. $\tan^2 \theta - \sin^2 \theta = \tan^2 \theta \sin^2 \theta$

41. $\dfrac{\cos^4 x - \sin^4 x}{1 - \tan^4 x} = \cos^4 x$

42. $\dfrac{1}{\csc x + \cot x} = \dfrac{1 - \cos x}{\sin x}$

43. $\cos y \sin (x - y) + \sin y \cos (x - y) = \sin x$

44. $\cos \left(x - \dfrac{\pi}{6}\right) - \cos \left(x + \dfrac{\pi}{6}\right) = \sin x$

45. $\cos (x + y) + \cos (x - y) = 2 \cos x \cos y$

46. $\cos 2x + 2 \sin^2 x = 1$

47. $\dfrac{2 - \sec^2 y}{\sec^2 y} = \cos 2y$

48. $\dfrac{2 \tan \theta}{\sin 2\theta} = \sec^2 \theta$

49. $\dfrac{1 - \tan^2 \theta}{1 + \tan^2 \theta} = \cos 2\theta$

50. $\cot 2\theta = \dfrac{\cot^2 \theta - 1}{2 \cot \theta}$

51. $\dfrac{1 - \cos 2\gamma + \sin 2\gamma}{1 + \cos 2\gamma + \sin 2\gamma} = \tan \gamma$

52. $\sin \theta = 2 \sin \dfrac{\theta}{2} \cos \dfrac{\theta}{2}$

53. $\csc^2 \theta = \dfrac{2}{1 - \cos 2\theta}$

54. $2 \sec^2 \theta = \dfrac{1}{1 + \cos 2\theta}$

55. $\left(\sin \dfrac{\theta}{2} - \cos \dfrac{\theta}{2}\right)^2 = 1 - \sin \theta$

56. $\sin^2 \alpha = \dfrac{\sin^2 2\alpha}{2(1 + \cos 2\alpha)}$

In Exercises 57–62, solve the given equations ($0 \le x < 2\pi$).

57. $\cos^2 x + 2 \sin x = 1$

58. $\sin 2x - 3 \sin x = 0$

59. $2 \cos^2 x + \cos x = 1$

60. $\tan x = \tan^2 x$

61. $\cos \dfrac{x}{2} + \cos x + 1 = 0$

62. $3 \cos^2 x - 14 \cos x + 8 = 0$

63. Suppose a projectile is fired along an inclined plane making an angle α to the horizontal from a gun making an angle θ to the horizontal. Calculus shows that the range of the projectile along the inclined plane is a maximum when θ satisfies the relation

$$\cos \theta \cos (\theta - \alpha) - \sin \theta \sin (\theta - \alpha) = 0$$

Find angle θ.

64. A body of weight W is dragged along a horizontal plane by a force whose line of action makes an angle θ with the plane. Calculus shows that the pull is least when θ satisfies the equation

$$\mu \cos \theta - \sin \theta = 0$$

where μ is the coefficient of friction. Show that the pull is least when $\theta = \text{Arctan } \mu$.

In Exercises 65–68, find y ($0 \le y < 2\pi$) in each case without using a table or a calculator.

65. $y = \arccos \dfrac{1}{2}$

66. $y = \text{arccsc } 1$

67. $y = \text{arccot} (-\sqrt{3})$

68. $y = \arctan \dfrac{1}{\sqrt{3}}$

In Exercises 69–72, evaluate each expression without using a table or a calculator.

69. $\text{Arccos} \left(-\dfrac{1}{2}\right)$

70. $\text{Arctan} \left(-\dfrac{1}{\sqrt{3}}\right)$

71. $\cot \left[\text{Arcsin} \left(-\dfrac{1}{6}\right)\right]$

72. $\sin [\text{Arctan} (-2)]$

73. Find an algebraic expression for sin (Arccos $2x$).

74. Solve for x: $y = 2$ Arctan $(x + 2)$.

75. Solve for x: $y = 2 \sin 4x$.

76. The expression $a \sin \theta + b \cos \theta$ can be written in simpler form by noting that

$$\sqrt{a^2 + b^2} \left(\frac{a}{\sqrt{a^2 + b^2}} \sin \theta + \frac{b}{\sqrt{a^2 + b^2}} \cos \theta \right) = a \sin \theta + b \cos \theta$$

and that

$$\sin \alpha = \frac{b}{\sqrt{a^2 + b^2}} \quad \text{and} \quad \cos \alpha = \frac{a}{\sqrt{a^2 + b^2}}$$

where α is an angle determined by a and b. (See Figure 16.18.) Use the identity for sin $(A + B)$ to show that

> $a \sin \theta + b \cos \theta = k \sin (\theta + \alpha)$
>
> where $k = \sqrt{a^2 + b^2}$ and α is any angle for which
>
> $$\sin \alpha = \frac{b}{\sqrt{a^2 + b^2}} \quad \text{and} \quad \cos \alpha = \frac{a}{\sqrt{a^2 + b^2}}$$

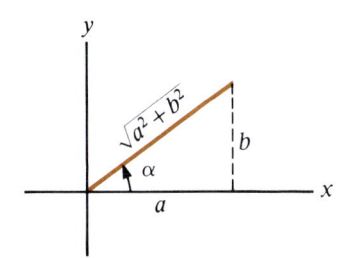

Figure 16.18

CHAPTER **17**

Inequalities

Objectives Upon completion of this chapter, you should be able to:
1. Graph inequalities.
2. Use the basic properties to solve linear inequalities algebraically.
3. Use the critical values to solve higher-degree inequalities.
4. Solve absolute inequalities.
5. Graph single inequalities and systems of inequalities.
6. Solve problems in linear programming. (Optional).

17.1 Basic Properties of Inequalities

A lot of time and effort in algebra and trigonometry is devoted to solving equations and verifying identities. Yet **inequalities** also play an important role, as we have seen from time to time. For example, we saw in the last chapter that if $y = $ Arctan x, then $-\pi/2 < y < \pi/2$. In this chapter we shall study inequalities in detail. This first section describes the basic properties of inequalities.

Sense of an inequality

First recall that $a < b$ means "a is less than b," and $a > b$ means "a is greater than b." The direction of the inequality is called the **sense** of the inequality. For example, the inequalities $a < b$ and $x < y$ have the same sense, while $a < b$ and $x > y$ have the opposite sense. To combine the notions of equality and inequality, we use the notation $a \leq b$, meaning "a is less than or equal to b," and $a \geq b$, meaning "a is greater than or equal to b."

Solution of an inequality

If an inequality contains a variable, then a **solution** of the inequality may exist. For example, the inequality $x + 1 < 2$ is satisfied when $x < 1$, while the inequality $x + 1 \leq 2$ is satisfied when $x \leq 1$.

Inequalities containing an unknown can be solved by a method similar to solving equations. Analogous geometric interpretations can also be given. To do so, however, we need to know the basic properties of inequalities.

Basic properties of inequalities:

1. For any real numbers a, b, and c, if $a < b$, then $a + c < b + c$ and $a - c < b - c$.
2. If $c > 0$ and $a < b$, then $ac < bc$ and $a/c < b/c$.
3. If $c < 0$ and $a < b$, then $ac > bc$ and $a/c > b/c$.
4. If a and b are positive and $a < b$, then

    $$a^n < b^n \quad \text{and} \quad \sqrt[n]{a} < \sqrt[n]{b}$$

5. If $a > 0$ and $b > 0$ (or if $a < 0$ and $b < 0$) and $a < b$, then

    $$\frac{1}{a} > \frac{1}{b}$$

Similar statements hold for the opposite sense, $a > b$.

It may be worthwhile now to summarize the basic properties of inequalities. The sense of an inequality is preserved whenever the same number is added to or subtracted from both sides. If both sides are multiplied or divided by c, then the sense is preserved whenever $c > 0$ and reversed whenever $c < 0$. With suitable restrictions, taking powers or roots preserves the sense of an inequality, while taking reciprocals reverses the sense.

Consider the example below.

Example 1 Since $4 < 7$,

$$\mathbf{2} \cdot 4 < \mathbf{2} \cdot 7 \quad \text{or} \quad 8 < 14 \qquad \text{by Property 2}$$

and

$$\mathbf{-3} \cdot 4 > \mathbf{-3} \cdot 7 \quad \text{or} \quad -12 > -21 \qquad \text{by Property 3}$$

Also,

$$\sqrt{4} < \sqrt{7} \quad \text{or} \quad 2 < \sqrt{7} \approx 2.6 \qquad \text{by Property 4}$$

and

$$\frac{1}{4} > \frac{1}{7} \qquad \text{by Property 5}$$

The next example illustrates how to solve a given inequality by using the basic properties. (Since the inequality involves a polynomial of first degree, the inequality is said to be **linear**.)

Example 2 Solve the inequality $2x - 4 < 5x + 6$.

Solution.
$$2x - 4 < 5x + 6$$
$$2x - 4 + 4 < 5x + 6 + 4 \qquad \text{by Property 1}$$
$$2x < 5x + 10$$
$$2x - 5x < 5x + 10 - 5x \qquad \text{by Property 1}$$
$$-3x < 10$$
$$x > \frac{10}{-3} = -\frac{10}{3} \qquad \text{by Property 3}$$

We conclude that the given inequality is satisfied whenever $x > -\frac{10}{3}$. Note that the sense of the inequality is reversed on the last step, since both sides are divided by -3. Otherwise, the procedure for solving (linear) inequalities is essentially the same as the procedure for solving equations.

To see how the solution of an inequality can also be shown graphically, consider the inequality $4x + 2y \geq 5$. Let us first graph the equation of the line $4x + 2y = 5$ by finding the intercepts. If $x = 0$, then $y = \frac{5}{2}$ and if $y = 0$, then $x = \frac{5}{4}$. So the intercepts are $(\frac{5}{4}, 0)$ and $(0, \frac{5}{2})$. (See Figure 17.1.)

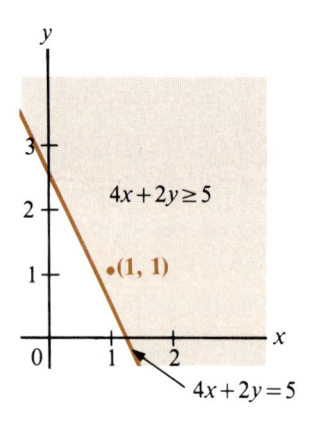

Figure 17.1

The coordinates of every point on the line satisfy the equation $4x + 5y = 5$. At all other points inequality holds. For example, if $x = 1$ and $y = 1$, then $4x + 2y = 6 > 5$; thus, the coordinates of **(1, 1)** satisfy the inequality. Checking one point is sufficient: The solution consists of the set of all points on the side of the line containing the point (1, 1), shown as the shaded region in Figure 17.1.

Let's consider another example.

Example 3 Graph the inequality $2x - 3y < 9$.

Solution. This time the coordinates of the points on the line $2x - 3y = 9$ do not satisfy the inequality, but the line is still needed to determine the boundary of the region. (See Figure 17.2.)

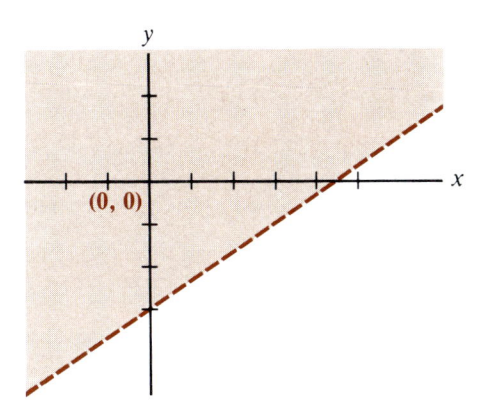

Figure 17.2

If $x = 0$ and $y = 0$, we have

$$0 - 0 < 9$$

So the point **(0, 0)** lies in the region. We conclude that the solution consists of the set of all points above the line (containing the origin). The line itself is *dashed* to indicate that the points on the boundary do not belong to the solution set.

Example 4 The current through two variable resistors in series is 2 A. Determine graphically the values of R_1 and R_2 for which the voltage across the two resistors is greater than 7 V.

Solution. By Ohm's law and the fact that the resistors are connected in series, we get

$$2R_1 + 2R_2 > 7$$

The graph of $2R_1 + 2R_2 = 7$ is the dashed line in Figure 17.3 on page 538. If $R_1 = 4$ and $R_2 = 0$, we get $8 + 0 > 7$. So the inequality is satisfied for any point in the shaded region.

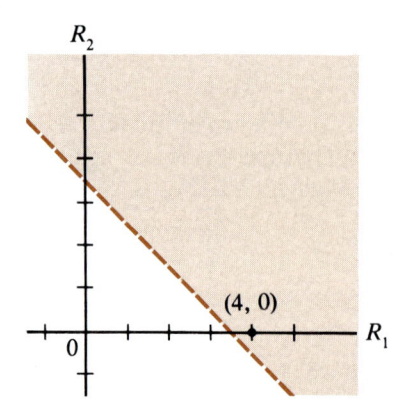

Figure 17.3

Exercises / Section 17.1

In Exercises 1–16, solve the given inequalities.

1. $x + 2 < 5$
2. $x - 3 > 4$
3. $2x + 4 \leq 8$
4. $3x - 1 \geq 5$
5. $2x + 1 < x - 5$
6. $3x + 2 > 5x - 6$
7. $5x - 2 \leq 6x + 3$
8. $2x - 2 < 7x - 3$
9. $-2x + 2 \geq x - 2$
10. $3 - 6x < -7x - 3$
11. $1 - 2x < 2x + 7$
12. $-1 - 3x > 5 + 2x$
13. $-2x - 1 \leq x - 8$
14. $2x + 1 \geq 6x - 3$
15. $5x - 3 > 3x + 2$
16. $2x + 2 < -x + 7$

In Exercises 17–26, graph the given inequalities.

17. $2x + 3y \leq 6$
18. $x + 4y \geq 4$
19. $x + 3y > 3$
20. $x - 2y < 3$
21. $2x - y > 0$
22. $-2x + y \leq 6$
23. $3x - 6y \geq 9$
24. $-2x - 5y \geq 5$
25. $-3x - 6y < -6$
26. $-x + 4y < 3$

27. The cost C (in dollars) of operating a certain machine is $C = 80 + 5t$ (t in hours). If the machine is started at $t = 0$, how long can the machine be run before the cost exceeds $118?

28. The relationship between degrees Fahrenheit and degrees Celsius is given by

$$C = \frac{5}{9}(F - 32)$$

Determine the values of C for which $F \leq 212°$, the boiling point of water.

29. The force F (in pounds) required to pull a certain spring x inches is given by $F = 10.0x$ within the limits $0.0 \leq x \leq 8.0$ (in inches). What values of F does this correspond to?

17.2 More Inequalities

The inequalities discussed in the last section involved polynomials of first degree. In this section we shall study inequalities containing polynomials of higher degree. The key to solving such inequalities is to systematically extend the method for solving linear inequalities: We change the given inequality to an equivalent inequality for which the right side is equal to zero. We then determine the values of x for which the left side is either positive or negative. To do this, we first find those values of x for which the left side is either equal to zero or undefined. Such values of x are called **critical values.**

> **Critical values of a function:** The **critical values** of a function $y = f(x)$ are the values of x for which $y = f(x)$ is either zero or undefined.

To see how critical values may be used to solve inequalities, consider the inequality

$$x^2 < 4 - 3x$$

By adding $3x - 4$ to both sides, we see that the given inequality is equivalent to

$$f(x) = x^2 + 3x - 4 < 0$$

To find the critical values, we set the left side equal to zero:

$$x^2 + 3x - 4 = 0$$
$$(x - 1)(x + 4) = 0$$
$$\boxed{x = 1, -4}$$

Thus $x = 1$ and $x = -4$ are the critical values. Note especially that the critical values are precisely those values at which the function $f(x) = (x - 1)(x + 4)$ changes signs.

To the left of -4, both factors are negative, so that the product is positive. Between -4 and 1 we have $x - 1 < 0$ and $x + 4 > 0$; the product is therefore negative. Finally, to the right of $x = 1$ both factors are positive, so that the product is positive. Consequently, the inequality is satisfied if $-4 < x < 1$. So the solution of the inequality is given by

$$-4 < x < 1$$

This procedure can be made more systematic by using a chart.

	Sign of		
x-values	$x - 1$	$x + 4$	$f(x)$
$x < -4$	−	−	+
$-4 < x < 1$	−	+	−
$x > 1$	+	+	+

540 CHAPTER 17 INEQUALITIES

Test value
While this procedure for solving inequalities is adequate, another shortcut is possible. Since $(x - 1)(x + 4) > 0$ whenever $x < -4$, it follows that $(x - 1)(x + 4) > 0$ for *any* value of x less than -4. For example, if $x = -5$, then $(x - 1)(x + 4) = (-6)(-1) = 6 > 0$. We shall call such a value a **test value**.

If we plot the critical values first, we can easily pick out some arbitrary test values between or beyond the critical values. (See Figure 17.4.)

Test values: -5, 0, 2

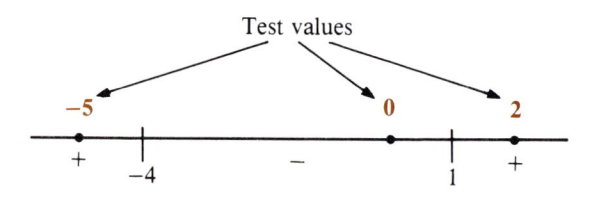

$+$ at -4, $-$ between, $+$ at 1

Figure 17.4

By using test values, the chart can be modified as follows:

x-values	Test values	$x - 1$	$x + 4$	$f(x)$
$x < -4$	-5	$-$	$-$	$+$
$-4 < x < 1$	0	$-$	$+$	$-$
$x > 1$	2	$+$	$+$	$+$

Before summarizing the procedure, let's consider the example below.

Example 1 Solve the inequality

$$(x - 2)(x + 3)(x + 5) < 0$$

Solution. From

$$f(x) = (x - 2)(x + 3)(x + 5) = 0$$

we see that the critical values are

$$\boxed{x = 2, -3, -5}$$

Test values: -6, -4, 0, 3

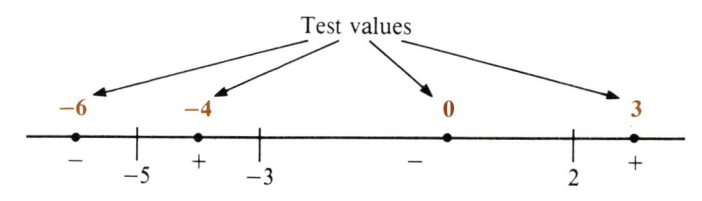

$-$ at -5, $+$ at -3, $-$ at 2, $+$

Figure 17.5

For the test values, we use *any* values between or beyond the critical values. (See Figure 17.5.)

We obtain the following chart:

		Sign of			
x-values	Test values	$x+5$	$x+3$	$x-2$	$f(x)$
$x < -5$	-6	$-$	$-$	$-$	$-$
$-5 < x < -3$	-4	$+$	$-$	$-$	$+$
$-3 < x < 2$	0	$+$	$+$	$-$	$-$
$x > 2$	3	$+$	$+$	$+$	$+$

Noting where $f(x)$ is negative, we conclude that the solution is given by

$$x < -5, \quad -3 < x < 2$$

The procedure for solving inequalities is summarized below.

Procedure for solving inequalities:

1. Change the given inequality to an equivalent inequality for which the right side is zero.
2. Find the critical values.
3. Choose arbitrary values between and beyond the critical values (test values).
4. Use the test values to determine the sign of the left side.

Example 2 Solve the inequality

$$(x+1)^2(x-3)(x-6) \geq 0$$

Solution. The critical values are $x = -1, 3, 6$ (Figure 17.6).

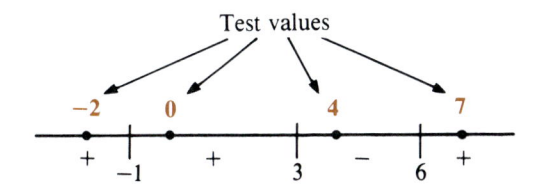

Figure 17.6

| | Test | Sign of | | | |
x-values	values	$(x+1)^2$	$x-3$	$x-6$	$f(x)$
$x < -1$	-2	+	$-$	$-$	+
$-1 < x < 3$	0	+	$-$	$-$	+
$3 < x < 6$	4	+	+	$-$	$-$
$x > 6$	7	+	+	+	+

Note that since $(x+1)^2 \geq 0$, $f(x)$ does not undergo a sign change at $x = -1$. Also, because of the equality part, the given inequality is satisfied at the critical values, including $x = -1$. The solution is now seen to be

$$x \leq 3, \quad x \geq 6$$

In the next example, one of the critical values is a value of x for which $f(x)$ is undefined.

Example 3 Solve the inequality

$$\frac{x^2 + 2x - 8}{x + 1} \leq 0$$

Solution. Factoring the numerator, the inequality becomes

$$f(x) = \frac{(x+4)(x-2)}{x+1} \leq 0$$

The critical values are $x = -4, 2, -1$. (The value $x = -1$ is a critical value because it leads to division by zero.) Note that even though $f(x)$ is undefined at $x = -1$, the sign of $f(x)$ changes at this value (Figure 17.7).

We obtain the following chart:

| | Test | Sign of | | | |
x-values	values	$x+4$	$x+1$	$x-2$	$f(x)$
$x < -4$	-5	$-$	$-$	$-$	$-$
$-4 < x < -1$	-2	+	$-$	$-$	+
$-1 < x < 2$	0	+	+	$-$	$-$
$x > 2$	3	+	+	+	+

The inequality is therefore satisfied whenever

$$x < -4 \quad \text{or} \quad -1 < x < 2$$

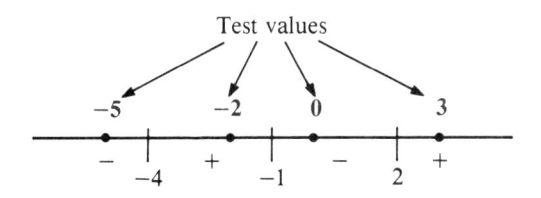

Figure 17.7

In addition, $f(x) = 0$ when $x = -4$ and $x = 2$. So the solution is

$$x \le -4, \quad -1 < x \le 2$$

Exercises / Section 17.2

In Exercises 1–32, solve the given inequalities.

1. $2x - 3 < 4x + 3$
2. $5x - 4 \ge 4x - 6$
3. $x^2 - 4 < 0$
4. $x^2 - x - 6 \le 0$
5. $x^2 + 5x \ge -6$
6. $x^2 > 6x - 8$
7. $x^3 < x^2 + 6x$
8. $x^3 + 3x^2 > 4x$
9. $(x + 2)(x - 3)(x + 5) \le 0$
10. $(x + 2)(x + 4)(x - 1) \ge 0$
11. $(x - 6)(x - 7)(x + 8) > 0$
12. $(x + 1)(x + 2)(x + 3) < 0$
13. $\dfrac{x - 1}{x - 2} < 0$
14. $\dfrac{x - 3}{x + 4} \le 0$
15. $\dfrac{x - 6}{x - 7} \ge 0$
16. $\dfrac{x + 3}{x + 5} \ge 0$
17. $\dfrac{x - 3}{4 - x} > 0$
18. $\dfrac{x + 6}{2 - x} < 0$
19. $\dfrac{(x - 3)(x + 4)}{x - 1} \le 0$
20. $\dfrac{(x + 2)(x - 1)}{x + 3} > 0$
21. $\dfrac{(x - 6)(x + 3)}{3 - x} < 0$
22. $\dfrac{(x + 5)(x + 2)}{(x - 3)(x + 4)} \le 0$
23. $\dfrac{(x - 1)(x + 6)}{(x - 2)(x - 3)} \ge 0$
24. $\dfrac{(x + 3)(x - 2)}{(x - 4)(x + 2)} < 0$
25. $(x - 2)^2(x - 4) > 0$
26. $(x + 3)(x - 1)^2 \le 0$
27. $(x - 3)^2(x - 7)^2 \ge 0$
28. $(x + 2)(x - 4)^2 < 0$
29. $(x - 4)^2(x + 1) < 0$
30. $(x - 3)^2(x + 2)^4 < 0$
31. $\dfrac{(x - 2)^2(x - 4)}{(x - 1)(x + 4)} \le 0$
32. $\dfrac{(x + 3)(x - 4)}{(x + 1)^2(x - 6)} \ge 0$

33. The current in a certain circuit is given by $i = 4t - 3t^2$, where i is measured in amperes and t in seconds. Determine the time interval for which $i \ge 0$.

34. The resistance R (in ohms) of a certain wire as a function of temperature T (in degrees Celsius) is given by $R = 10.0 + 0.10T + 0.0060T^2$. Find $T > 0$ such that $R > 20 \, \Omega$.

35. The deflection d (in feet) of a beam 20 ft long is given by $d = \frac{1}{100}(20 - x)x$, where x is the distance from one end of the beam. For what values of x is $d \le \frac{3}{4}$ ft?

36. The current in a circuit is given by $i = 2.0t^2 - 19.0t + 45.0$. In what time interval is $i \le 0$?

37. The weekly profit P of a company is $P = x^4 - 20x^3$, $x \ge 1$, where x is the week in the year. During what period is the company operating at a loss?

17.3 Absolute Inequalities

Many situations, particularly in calculus, lead to inequalities involving absolute values. In this section we shall consider some of the basic cases.

First consider the inequality $|x| < 1$. Any number between 0 and 1 would certainly satisfy this relation, but if we choose $x = -\frac{2}{3}$, then $|x| = |-\frac{2}{3}| = \frac{2}{3}$, which also satisfies the inequality. In fact, the inequality is satisfied for $-1 < x < 1$.

The general rule for solving absolute inequalities will now be stated.

Absolute inequalities:

1. If $|f(x)| < r$, then $-r < f(x) < r$. (17.1)
2. If $|f(x)| > r$, then $f(x) < -r$ or $f(x) > r$. (17.2)

Rule (17.1) is illustrated in the example below.

Example 1 Solve the inequality $|x - 2| < 3$.

Solution. By statement (17.1), $|x - 2| < 3$ is equivalent to

$$-3 < x - 2 < 3$$

Adding **2** to each member, we obtain

$$-3 + \mathbf{2} < x - 2 + \mathbf{2} < 3 + \mathbf{2}$$

or

$$-1 < x < 5$$

which is the solution (Figure 17.8).

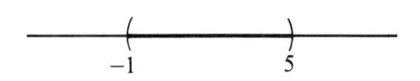

Figure 17.8

The inequality in the next example is solved with the help of rule (17.2).

Example 2 Solve the inequality $|3x + 2| \geq 6$.

Solution. By statement (17.2), the inequality $|3x + 2| \geq 6$ is equivalent to

$$3x + 2 \leq -6 \quad \text{or} \quad 3x + 2 \geq 6$$
$$3x \leq -8 \quad \text{or} \quad 3x \geq 4 \qquad \text{adding } -2$$
$$x \leq -\frac{8}{3} \quad \text{or} \quad x \geq \frac{4}{3} \qquad \text{dividing by 3}$$

So the given equality is satisfied whenever $x \leq -\frac{8}{3}$ or whenever $x \geq \frac{4}{3}$. The solution is therefore given by

$$x \leq -\frac{8}{3}, \quad x \geq \frac{4}{3}$$

In the remaining examples, the absolute inequalities lead to two quadratic inequalities.

Example 3 Solve the inequality $|x^2 - x + 1| > 2$.

Solution. By statement (17.2) the given inequality is equivalent to

$$x^2 - x + 1 > 2 \quad \text{or} \quad x^2 - x + 1 < -2$$

For the inequality on the left side, we have

$$x^2 - x + 1 > 2$$
$$x^2 - x - 1 > 0 \quad \text{adding } -2 \text{ to both sides}$$

The critical values are obtained by using the quadratic formula:

$$x^2 - x - 1 = 0$$
$$x = \frac{1 \pm \sqrt{1 + 4}}{2} = \frac{1 \pm \sqrt{5}}{2}$$

Hence $x^2 - x - 1 > 0$ whenever

$$x < \frac{1 - \sqrt{5}}{2} \quad \text{or} \quad x > \frac{1 + \sqrt{5}}{2} \tag{17.3}$$

by the method of the last section.

For the inequality on the right we have, by adding 2 to each side,

$$x^2 - x + 1 < -2$$
$$x^2 - x + 3 < 0 \tag{17.4}$$

By the quadratic formula, the critical values are

$$x = \frac{1 \pm \sqrt{1 - 12}}{2} = \frac{1 \pm \sqrt{11}j}{2}$$

which are complex numbers. Thus $x^2 - x + 3 \neq 0$. In other words, either $x^2 - x + 3 > 0$ for all x or $x^2 - x + 3 < 0$ for all x. Since $x^2 - x + 3 > 0$ when $x = 1$, the left side of inequality (17.4) is positive for all x. Thus, inequality (17.4) has no solution.

The solution of the given inequality, shown in Figure 17.9 on page 546, is therefore given by statement (17.3):

$$x < \frac{1 - \sqrt{5}}{2}, \quad x > \frac{1 + \sqrt{5}}{2}$$

546 CHAPTER 17 INEQUALITIES

$\frac{1}{2}(1-\sqrt{5})$ $\frac{1}{2}(1+\sqrt{5})$

Figure 17.9

Example 4 Solve the inequality $|x^2 - 2x - 2| < 1$.

Solution. By statement (17.1), we have

$$-1 < x^2 - 2x - 2 < 1$$

Both inequalities must be satisfied.
   Suppose we solve the right inequality first by subtracting 1 from each member. Thus

$$x^2 - 2x - 2 < 1$$
$$x^2 - 2x - 3 < 0 \qquad \text{adding } -1 \text{ to each side}$$
$$(x - 3)(x + 1) < 0 \qquad \text{factoring}$$

By the method of the previous section, the solution is

$$-1 < x < 3 \tag{17.5}$$

Solving the left inequality, we get

$$x^2 - 2x - 2 > -1$$
$$x^2 - 2x - 1 > 0$$

The critical values are

$$x = \frac{2 \pm \sqrt{4 + 4}}{2} = \frac{2 \pm 2\sqrt{2}}{2} = 1 \pm \sqrt{2}$$

The left inequality is therefore satisfied whenever

$$x < 1 - \sqrt{2} \quad (x < -0.4) \quad \text{or} \quad x > 1 + \sqrt{2} \quad (x > 2.4)$$

The two sets of solutions are shown in Figure 17.10. Since x must satisfy both sets of inequalities, we conclude from the diagram that the solution of the given inequality is given by

$$-1 < x < 1 - \sqrt{2}, \quad 1 + \sqrt{2} < x < 3$$

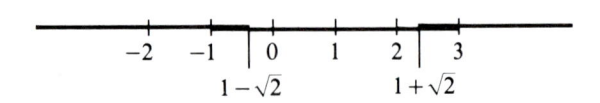

$-2 \quad -1 \quad 0 \quad 1 \quad 2 \quad 3$
$\qquad\qquad\quad 1-\sqrt{2} \qquad 1+\sqrt{2}$

Figure 17.10

Exercises / Section 17.3

In Exercises 1–20, solve the given inequalities.

1. $|x - 1| < 2$
2. $|x + 1| < 2$
3. $|x + 3| \leq 1$
4. $|x - 2| < 4$
5. $|2x + 1| < 1$
6. $|3x - 3| > 2$
7. $|2x - 4| \geq 3$
8. $|4x - 8| \leq 4$
9. $|2x + 7| < 6$
10. $|3x + 2| > 4$
11. $|2 - x| > 3$
12. $|1 - 2x| < 5$
13. $|x^2 - x - 9| > 3$
14. $|x^2 - x - 9| < 3$
15. $|x^2 - 4x + 2| > 2$
16. $|x^2 - x + 1| > 3$
17. $|x^2 - 2x - 7| < 1$
18. $|x^2 + 5x + 2| < 2$
19. $|x^2 + 6x + 2| < 3$
20. $|x^2 + 5x + 2| > 2$

21. The diameter d of a shaft is 2.5550 in., and the tolerance of the shaft is 0.0001 in. Express this statement as an inequality containing absolute values.

22. An athlete can run a mile in approximately 4 min 20 sec, expressed more precisely as $|T - (4 \text{ min } 20 \text{ sec})| \leq 5$ sec. What are his fastest and slowest times?

17.4 Graphing Inequalities

In this section we shall continue our study of the graphs of inequalities begun in Section 17.1. First we consider single inequalities, and then systems of inequalities.

To recall the basic ideas, consider the example of a linear inequality below.

Example 1 Graph the inequality $x - 2y < 4$.

Solution. First we need to graph the line $x - 2y = 4$. If $y = 0$, then $x = 4$; if $x = 0$, $y = -2$. So the intercepts are $(4, 0)$ and $(0, -2)$. The resulting graph is the dashed line in Figure 17.11. (Recall that the line is shown as a dashed line to indicate that the points on the line do not belong to the solution set.) Next we check a point not on the line. For example, at $(0, 0)$, we get from the

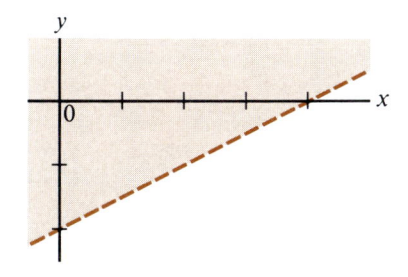

Figure 17.11

given inequality $0 - 0 < 4$, which satisfies the relation. So the coordinates of all the points on the side of the line containing the origin satisfy the inequality. (See Figure 17.11.)

The next example illustrates the graphical solution of a quadratic inequality.

Example 2 Graph the inequality $y \geq x^2 + 1$.

Solution. We shall first graph the curve $y = x^2 + 1$ from the following table:

x:	-3	-2	-1	0	1	2	3
y:	10	5	2	1	2	5	10

The graph is shown as a solid curve in Figure 17.12.
Now we check a point not on the curve. For the point **(0, 2)**, we get

$y \geq x^2 + 1$
2 > **0** + 1

which satisfies the inequality. The inequality is therefore satisfied by the coordinates of all the points in the shaded region shown in Figure 17.12.

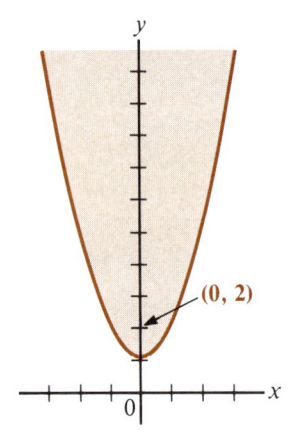

Figure 17.12

To graph a **system of inequalities,** we first graph the individual inequalities. In each case the solution is a region in which the coordinates of every

17.4 GRAPHING INEQUALITIES **549**

point satisfy the particular inequality. The common solution of the system is the resulting *intersection*, where all the regions overlap. These ideas are illustrated in the remaining examples.

Example 3 Solve the following system of inequalities graphically:

$$y > x$$
$$y < 2x$$

Solution. The points whose coordinates satisfy the inequality $y > x$ lie in the region above the line $y = x$, shown in Figure 17.13. Similarly, the points whose coordinates satisfy the inequality $y < 2x$ lie below the line $y = 2x$ (Figure 17.13). The solution must therefore be the intersection, where both regions overlap. The intersection is shown in Figure 17.14.

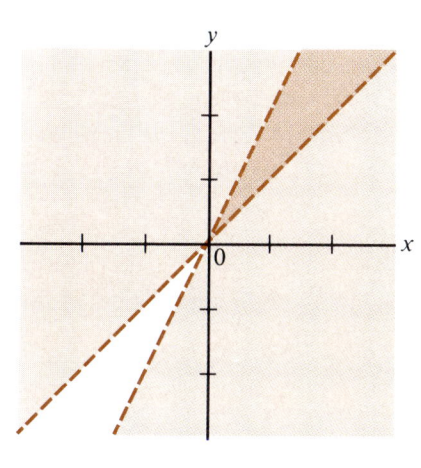

Figure 17.13

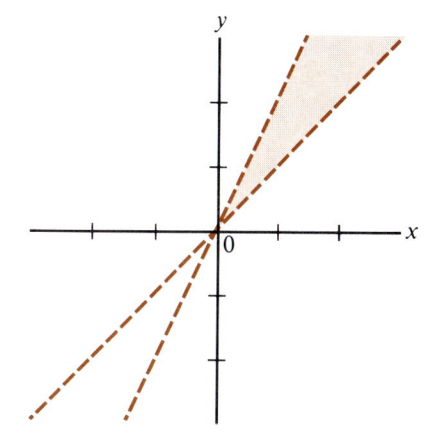

Figure 17.14

Example 4 Sketch the region in which the coordinates of the points satisfy the given inequalities:

$$x + y \leq 2$$
$$x \geq 0$$
$$y \geq 0$$

Solution. The region defined by $x \geq 0$ and $y \geq 0$ is the first quadrant (including the nonnegative coordinate axes). The points whose coordinates satisfy $x + y \leq 2$ lie below the line $x + y = 2$. The points whose coordinates satisfy all three inequalities lie in the shaded region shown in Figure 17.15.

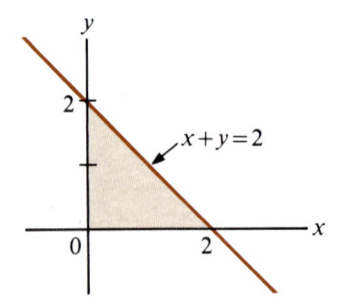

Figure 17.15

Example 5 Graph the following system of inequalities:

$$y > -3x$$
$$x + 4y < 11$$
$$y > \frac{2}{3}x$$

Solution. To find the proper region, we need to solve the corresponding equations in pairs. The first two equations,

$$y = -3x$$
$$x + 4y = 11$$

may be solved by the method of substitution:

$$x + 4(-3x) = 11$$
$$x - 12x = 11$$
$$x = -1, \quad y = 3$$

So the lines intersect at $(-1, 3)$.
  Similarly, the lines

$$x + 4y = 11$$
$$y = \frac{2}{3}x$$

intersect at $(3, 2)$. Finally, the lines

$$y = -3x$$
$$y = \frac{2}{3}x$$

intersect at the origin.
  The dashed lines and their intersections are shown in Figure 17.16. As can be readily checked, the shaded region in Figure 17.16 is the intersection

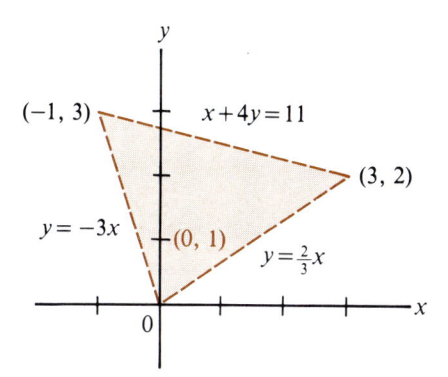

Figure 17.16

of the three solution sets. For example, the coordinates of the point **(0, 1)** satisfy all three inequalities.

Example 6 Graph the following system of inequalities:

$$y \leq x$$
$$y \geq x^2$$

Solution. Solving the corresponding system of equations, $y = x$ and $y = x^2$, by substitution, we get

$$x^2 = x$$
$$x^2 - x = 0$$
$$x(x - 1) = 0$$
$$x = 0, 1$$

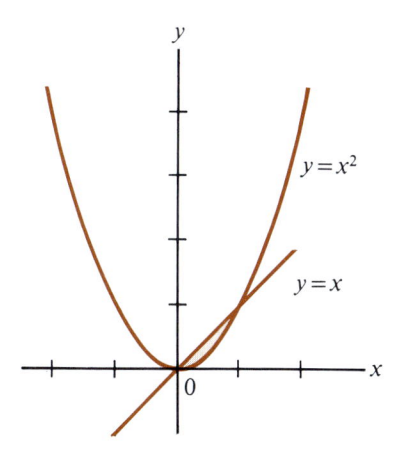

Figure 17.17

So the curves intersect at (0, 0) and (1, 1). The curve $y = x^2$ is sketched from the following table of values:

x:	-2	-1	0	1	2
y:	4	1	0	1	4

The shaded region between the graphs, shown in Figure 17.17, is the solution set.

Exercises / Section 17.4

In each of the following exercises, sketch the region in which the coordinates of the points satisfy the given inequalities.

1. $y \leq 3x + 6$

2. $y > 4 - 2x$

3. $y > x$
   $y < 3x$

4. $y > x^2$

5. $y > x^2 - 1$

6. $y < x^3$

7. $y \leq x$
   $y \leq 2 - x$
   $y \geq 0$

8. $1 < x < 2$

9. $0 \leq y \leq 1$
   $y \leq 2 - x$
   $x \geq 0$

10. $2y + x \leq 2$
    $x \geq 0$
    $y \geq 0$

11. $x + y \leq 4$
    $x + 2y \leq 6$
    $x \geq 0$
    $y \geq 0$

12. $x + 2y \leq 6$
    $2x + y \leq 6$
    $x \geq 0$
    $y \geq 0$

13. $2x + y \geq 6$
    $x + y \geq 5$
    $x \geq 0$
    $y \geq 0$

14. $3x + y \geq 4$
    $x + 2y \geq 3$
    $x \geq 0$
    $x \geq 0$

15. $y \leq x^2$
    $y \geq x^3$

16. $y > x^2 - 4$
    $y < 4 - 2x$

17. $y > x^2 - 2$
    $y < x$

18. $y \geq x^2$
    $y < x + 6$

17.5 Linear Programming (Optional)

Graphs of inequalities have important applications in a field called **linear programming.** Linear programming is a technique for determining which of several available options will yield the greatest benefit. For example, the method may be used to determine the maximum profit or the minimum cost, given certain constraints. (The term *linear programming* has nothing to do with computer programming.)

As we saw in the last section, a given set of inequalities determines a region. These inequalities are called the **constraints,** and the region is called the **set of feasible solutions,** or simply the **feasible region.** Linear programming is confined to regions that are **convex:** A line segment connecting two points in the region also lies in the region. For example, the region in Figure 17.18 is convex, but the region in Figure 17.19 is not.

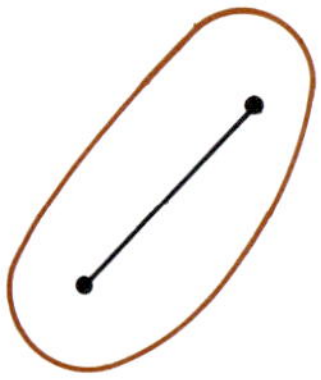

Figure 17.18

17.5 LINEAR PROGRAMMING 553

If the boundary of the region consists of straight lines, then the region is called **polygonal**. For example, the two regions shown in Figure 17.20 are convex polygonal regions. Each corner is called a **vertex.**

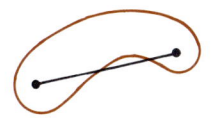

Figure 17.19

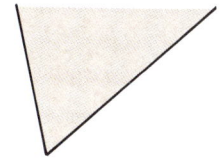

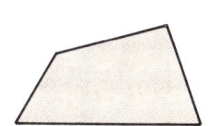

Figure 17.20

In linear programming, the quantity whose maximum or minimum value is to be determined is called the **objective function.** The objective function contains two variables and must have the form $P = ax + by + c$. Because of the given constraints, the problem is *to find the maximum or minimum value of P over the feasible region.* The theorem below is fundamental to linear programming.

> The maximum or minimum value of an objective function must occur at a vertex of a convex polygonal region.

The example below illustrates the technique.

Example 1 Given the following constraints:

$$x + 2y \geq 7$$
$$x + y \geq 5$$
$$x \geq 0$$
$$y \geq 0$$

find the minimum value of $P = 4x + 6y$.

Solution. To obtain the feasible region, we need to solve the system of equations

$$x + 2y = 7$$
$$x + y = 5$$

simultaneously. Subtracting the second equation from the first, we get $y = 2$. If $y = 2$, then $x = 3$, so that the lines intersect at (3, 2). By the method of the last section, we obtain the feasible region shown in Figure 17.21. (Recall that the feasible region is the solution set of the given system of inequalities.)

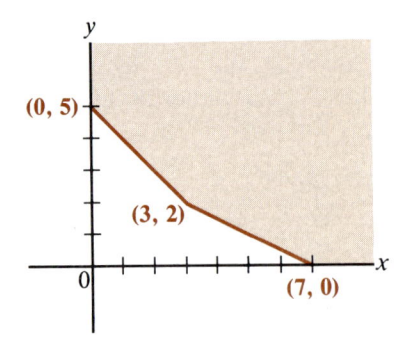

Figure 17.21

By the fundamental theorem, the minimum value of P occurs at one of the vertices. The value at each of the three vertices is given in the following chart:

Vertex	$P = 4x + 6y$
(0, 5)	$P = 4 \cdot 0 + 6 \cdot 5 = 30$
(3, 2)	$P = 4 \cdot 3 + 6 \cdot 2 = 24$
(7, 0)	$P = 4 \cdot 7 + 6 \cdot 0 = 28$

The minimum value of $P = 24$ occurs at the vertex (3, 2).

To see why the fundamental theorem guarantees a minimum or maximum value at a vertex, let P assume some arbitrary values. For example, if $P = 36$, we get

$$4x + 6y = 36$$

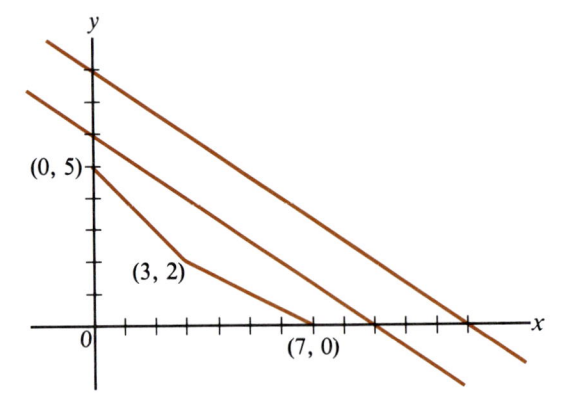

Figure 17.22

If $P = 48$, we get

$$4x + 6y = 48$$

The resulting parallel lines are shown in Figure 17.22.

We can see from the figure that the values of P get smaller as the lines move downward. The minimum value is obtained by moving the line downward as far as possible without leaving the feasible region, which occurs at the vertex $(3, 2)$. (Below this point, the values of x and y do not satisfy the constraints.)

Example 2 An electronics manufacturer produces two types of calculators, the basic model and the scientific model. Each model is produced by two sets of operations, electronic assembly (the first operation) and case assembly (the second operation). For the basic model, the time required for each operation is 5 min. For the scientific model, the first operation requires 10 min, but the second operation only 2 min, since some of the assembly is done in the first operation. The time available for the first operation is at most $133\frac{1}{3}$ hr (8,000 min) per month, and for the second at most $73\frac{1}{3}$ hr (4,400 min) per month. (During the remaining time, the equipment is used for manufacturing other products.) If the profit is $5 for each basic and $8 for each scientific calculator, how many of each should be produced per month to maximize the profit?

Solution. Let

$x = $ number of basic calculators

and

$y = $ number of scientific calculators

Then $5x + 10y$ is the total time in minutes required for the first operation and $5x + 2y$ the time required for the second operation. The constraints are therefore given by

$$5x + 10y \leq 8{,}000$$
$$5x + 2y \leq 4{,}400$$
$$x \geq 0$$
$$y \geq 0$$

(The last two conditions say that the number of calculators produced cannot be negative.)

The objective function representing the profit is

$$P = 5x + 8y$$

The feasible region is shown in Figure 17.23 on page 556.

556 CHAPTER 17 INEQUALITIES

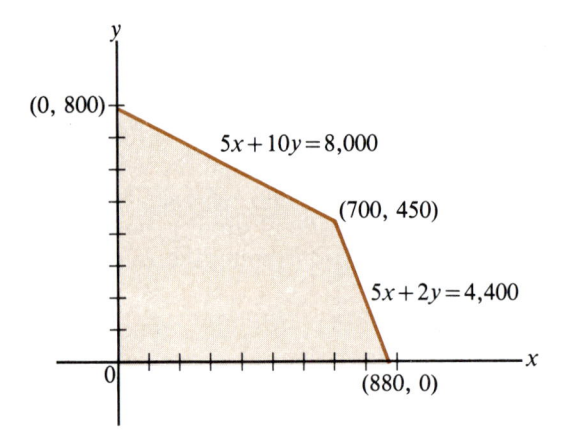

Figure 17.23

The profit evaluated at the different vertices is shown in the following chart:

Vertex	$P = 5x + 8y$
(0, 800)	$P = 5 \cdot 0 + 8 \cdot 800 = \$6,400$
(700, 450)	$P = 5 \cdot 700 + 8 \cdot 450 = \$7,100$
(880, 0)	$P = 5 \cdot 880 + 8 \cdot 0 = \$4,400$

Consequently, the manufacturer should produce 700 basic and 450 scientific calculators per month for a maximum profit of \$7,100.

Example 3 Suppose the management of the firm discussed in Example 2 decides that they cannot sell more than 1,000 calculators per month. How many calculators should then be produced to maximize the profit?

Solution. The extra constraint $x + y \leq 1,000$ leads to the following system of inequalities:

$$5x + 10y \leq 8,000$$
$$x + y \leq 1,000$$
$$5x + 2y \leq 4,400$$
$$x \geq 0$$
$$y \geq 0$$

To obtain the feasible region, the corresponding equations have to be solved in pairs. The lines

$$5x + 10y = 8,000$$
$$x + y = 1,000$$

intersect at (400, 600) and the lines

$$x + y = 1,000$$
$$5x + 2y = 4,400$$

intersect at (800, 200). The feasible region is shown in Figure 17.24. The values of P at the different vertices are shown in the following chart:

Vertex	$P = 5x + 8y$
(0, 800)	$P = 5 \cdot 0 + 8 \cdot 800 = \$6,400$
(400, 600)	$P = 5 \cdot 400 + 8 \cdot 600 = \$6,800$
(800, 200)	$P = 5 \cdot 800 + 8 \cdot 200 = \$5,600$
(880, 0)	$P = 5 \cdot 880 + 8 \cdot 0 = \$4,400$

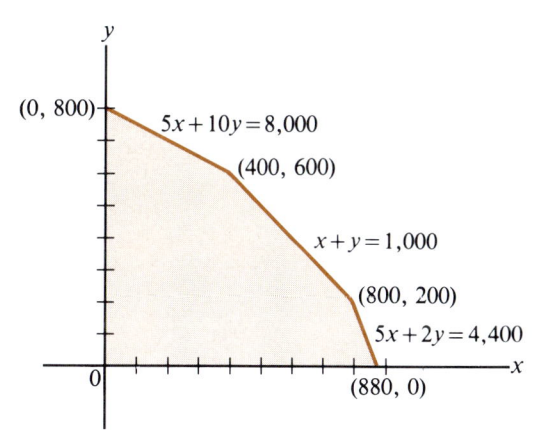

Figure 17.24

We conclude that the manufacturer should produce 400 basic and 600 scientific calculators per month for a maximum profit of $6,800.

If a linear programming problem contains three variables, then a three-dimensional feasible region is required. More than three variables require an abstract algebraic approach, which is commonly carried out with the aid of a computer.

Exercises / Section 17.5

In Exercises 1–9, find the minimum or maximum value for P, subject to the given constraints.

1. Maximum P
   Constraints:
   $x + y \leq 2$
   $x \geq 0$
   $y \geq 0$
   Objective function:
   $P = 2x + y$

2. Maximum P
   Constraints:
   $x + y \leq 2$
   $0 \leq y \leq 1$
   $x \geq 0$
   Objective function:
   $P = x + 3y$

3. Minimum P
   Constraints:
   $2x + y \geq 2$
   $x \geq 0$
   $y \geq 0$
   Objective function:
   $P = 3x + 2y$

4. Maximum P
   Constraints:
   $x + y \leq 3$
   $0 \leq x \leq 2$
   $y \geq 0$
   Objective function:
   $P = 2x + y$

5. Maximum P
   Constraints:
   $x + y \leq 4$
   $x + 2y \leq 5$
   $x \geq 0$
   $y \geq 0$
   Objective function:
   $P = 3x + 4y$

6. Minimum P
   Constraints:
   $2x + y \geq 6$
   $x + y \geq 5$
   $x \geq 0$
   $y \geq 0$
   Objective function:
   $P = 5x + 4y$

7. Minimum P
   Constraints:
   $2x + y \geq 7$
   $x + y \geq 5$
   $2x + 3y \geq 12$
   $x \geq 0$
   $y \geq 0$
   Objective function:
   $P = 3x + 4y$

8. Maximum P
   Constraints:
   $x + 3y \leq 15$
   $x + 2y \leq 11$
   $x + y \leq 8$
   $x \geq 0$
   $y \geq 0$
   Objective function:
   $P = 5x + 12y$

9. Maximum P
   Constraints:
   $x + 2y \leq 20$
   $x + y \leq 11$
   $2x + y \leq 14$
   $3x + y \leq 18$
   $x \geq 0$
   $y \geq 0$
   Objective function:
   $P = 3x + 2y$

10. Farmer Schmidt and his son wish to raise corn and soybeans on part of their farm. Each acre of corn requires 4 hr of labor for the father and 5 hr for the son during the season and yields a profit of $140. Each acre of soybeans requires 2 hr for the father and 3 hr for the son and yields a profit of $75. Because of other farm chores, the father does not wish to spend more than 200 hr during the season raising corn and soybeans, and the son not more than 270 hr. How many acres of each crop should they plant to maximize their profit? What is the maximum profit?

11. If in Exercise 10 the profit on each acre of corn is $120 and on each acre of soybeans $90, how many acres of each should Farmer Schmidt and his son plant, and what is the maximum profit?

12. A manufacturer produces widgets and gadgets. Each widget requires 4 hr of assembly time and 1 hr spraying time. Each gadget requires 3 hr assembly time and 2 hr spraying time. Assembling is done during the 8-hr day shift for a maximum of 40 hr per week. Spraying is done during the night shift for a maximum of 20 hr

per week. The profit on each widget is $15,000 and on each gadget $12,000. How many of each should be produced to maximize the profit? What is the maximum profit?

13. Farmer Armstrong has set aside a section of his farm for growing fruit trees. Each smaller fruit tree requires 150 ft² of space and 2 lb of insecticide. Each larger fruit tree requires 200 ft² of space but only 1 lb of insecticide. He can spare at most 5,000 ft² for the trees, and he cannot afford more than 50 lb of insecticide. His profit is $290 for each smaller tree and $380 for each larger tree. Determine how many of each kind of tree he should plant for the largest possible profit. What is the profit?

14. A corporation has at most 2,400 ft² available to store two types of motors. Each small motor requires 15 ft² of storage space, and each large motor 20 ft². Storage and maintenance costs are $4 for each small motor and $5 for each large motor. Suppose no more than $620 is to be spent on storage and maintenance. If the profit on the sale of each small motor is $150 and on each large motor $190, how many of each should be stored to obtain a maximum profit? What is the maximum profit?

15. As a sideline, the members of a family produce two types of high-quality lawn mowers, making some of the parts themselves and buying the rest. Type I requires 3 hr for making the parts, and Type II, 4 hr. Type I requires 3 hr for assembly, but Type II only 2 hr. Since they hold regular jobs, the members of the family cannot spend more than a total of 56 hr per month making parts and 39 hr on assembly. The profit is $100 for a Type I mower and $120 for a Type II, and the total number that can be sold each month is no more than 15. How many lawn mowers of each type should be made to maximize the profit? What is the maximum profit?

16. If in Exercise 15 the profit on the Type I lawn mower increases to $135, while the profit on the Type II remains the same, how many of each should be produced and what will the maximum profit be?

17. A certain kind of dog food sold in a bag is to be made from two types of products. Each pound of Product I contains 100 g of protein and 150 g of carbohydrates. Each pound of Product II contains 50 g of protein and 200 g of carbohydrates. The cost of Product I is 45¢ per pound, and of Product II, 30¢ per pound. If each bag must contain at least 400 g of protein and at least 1,100 g of carbohydrates, determine how many pounds of each product must be used per bag in order to minimize the cost. What is the minimum cost per bag?

18. In Exercise 17, how much of each product must be used to minimize the cost if the price of Product I is increased to 50¢ per pound and the price of Product II is decreased to 20¢ per pound? What is the minimum cost?

Review Exercises / Chapter 17

In Exercises 1–22, solve each inequality.

1. $2x + 2 < x - 3$
2. $5x - 1 \geq x + 4$
3. $x + 1 > 3x - 5$
4. $1 - 3x \geq 2x + 3$
5. $x^2 - 9 \leq 0$
6. $x^2 - x - 6 \leq 0$
7. $(x - 3)(x + 1)(x + 4) > 0$
8. $x^2 - x < 0$
9. $\dfrac{x - 1}{x + 2} \leq 0$
10. $\dfrac{x - 2}{x + 4} \leq 0$
11. $\dfrac{x - 6}{x - 1} > 0$
12. $\dfrac{x - 5}{2 - x} < 0$
13. $\dfrac{(x - 2)(x - 1)}{x + 4} \geq 0$
14. $\dfrac{(x + 3)(x + 5)}{x - 1} \leq 0$

15. $(x + 2)(x - 1)^2 \leq 0$
16. $(x - 1)^2(x + 3) \geq 0$
17. $|x - 4| \leq 2$
18. $|x - 1| > 3$
19. $|2x + 2| < 3$
20. $|1 - x| < 2$
21. $|x^2 - 4x - 4| < 1$
22. $|x^2 - 2x - 2| > 1$

In Exercises 23–28, sketch the region in which the coordinates of the points satisfy the given inequalities.

23. $y > 4x + 2$
24. $y \leq 3 - 6x$
25. $y \leq \sqrt{x}$
    $0 \leq x \leq 4$
    $y \geq 0$
26. $y < x^2$
27. $2y > x$
    $y < x$
28. $2x + y \leq 8$
    $x + y \leq 5$
    $x \geq 0$
    $y \geq 0$

29. The cost (in dollars) of operating a certain machine is $C = 100 + 4t$, where t is measured in hours. How long can the machine be run before the cost exceeds $126?

30. A technician receives $250 per week plus $18.75/hr for overtime, so that his pay can be expressed as $P = 250 + 18.75t$. For what value of t is $P \geq 350$?

31. If an object is hurled upward with a velocity of 80 ft/sec from a point 146 ft above the ground, its distance above the ground as a function of time (in seconds) is given by $s = 146 + 80t - 16t^2$. Find the time for which $s \leq 50$ ft. (The object is hurled upward when $t = 0$ sec.)

32. From past experience, the produce manager of a supermarket estimates that the number S of watermelons sold on July 3 will be $|S - 150| \leq 25$. Determine the anticipated sale.

In Exercises 33–34, find the minimum or maximum value for P, subject to the given constraints.

33. Maximum P
    Constraints:
    $x + 2y \leq 20$
    $x + y \leq 14$
    $x \geq 0$
    $y \geq 0$
    Objective function:
    $P = 3x + 4y$

34. Minimum P
    Constraints:
    $2x + y \geq 11$
    $x + 2y \geq 10$
    $x \geq 0$
    $y \geq 0$
    Objective function:
    $P = x + y$

35. Anticipating a price increase, the owner of a warehouse stocks up on two types of appliances. Each small appliance requires 10 ft² of space and is expected to sell for $100. Each large appliance requires 15 ft² of space and is expected to sell for $130. The storage cost is $10 for each small appliance and $12 for each large appliance. If at most 900 ft² are available for storage and at most $840 for the cost of storage, how many of each type should be stored to maximize the amount of money received? What is the maximum amount?

CHAPTER **18**

Geometric Progressions and the Binomial Theorem

Objectives Upon completion of this chapter, you should be able to:
1. Find the nth term of a geometric progression.
2. Find the sum of the terms of a geometric progression.
3. Find the sum of a geometric series.
4. Expand a given binomial by means of:
   a. Pascal's triangle.
   b. The binomial theorem.
5. Write a given expression of the form $(1 + b)^n$ as a binomial series.

18.1 Geometric Progressions

In science and technology, we occasionally encounter a problem involving a sequence of numbers that follow a definite pattern. One such sequence, called a **geometric progression,** will be studied in this section.

In a geometric progression, each number after the first in the sequence can be determined by multiplying the preceding term by a number called the **common ratio.** For example, the sequence

$$1, \frac{1}{3}, \frac{1}{9}, \frac{1}{27}, \cdots$$

has a common ratio $\frac{1}{3}$, while the sequence

$$3, \frac{3}{2}, \frac{3}{4}, \frac{3}{8}, \cdots$$

has a common ratio $\frac{1}{2}$. The general definition is stated next.

CHAPTER 18 GEOMETRIC PROGRESSIONS AND THE BINOMIAL THEOREM

> A **geometric progression** is a sequence of numbers of the form
> $$a, ar, ar^2, ar^3, \ldots, ar^{n-1} \quad (18.1)$$
> where r is called the **common ratio.** The nth term is
> $$a_n = ar^{n-1} \quad (18.2)$$

These definitions are illustrated in the first three examples.

Example 1 Write the first six terms of the geometric progression whose first term is 2 and whose common ratio is 3.

Solution. By the sequence (18.1), we have

$$2, 2(3), 2(3^2), 2(3^3), 2(3^4), 2(3^5)$$

Example 2 Write the tenth term of the geometric progression whose first term is 3 and whose common ratio is $\frac{1}{2}$.

Solution. By formula (18.2), the nth term is ar^{n-1}. Since $a = 3$ and $r = \frac{1}{2}$, we get

$$a_{10} = 3\left(\frac{1}{2}\right)^{10-1} = 3\left(\frac{1}{2}\right)^9 = \frac{3}{2^9} = \frac{3}{512}$$

Example 3 A certain radioactive substance has a half-life of 3 years, which means that the amount present is cut in half every 3 years. If there are 20 g of the substance initially, how many grams are left at the end of 24 years?

Solution. Since the amount is cut in half every 3 years, the common ratio is $\frac{1}{2}$. Since $a = 20$, the first term, the amount after 3 years is $20\left(\frac{1}{2}\right)$, corresponding to $n = 2$. After 24 years the initial amount is cut in half 8 times, corresponding to $n = 9$. So by formula (18.2), the amount is

$$ar^{n-1} = 20\left(\frac{1}{2}\right)^8 = \frac{20}{256} = \frac{5}{64} \text{ g}$$

It is possible to find the sum of the terms of a geometric progression. Let

$$s = a + ar + ar^2 + \cdots + ar^{n-1}$$

18.1 GEOMETRIC PROGRESSIONS

If we multiply s by r and subtract the resulting expression from s, we get

$$s = a + ar + ar^2 + \cdots + ar^{n-1}$$
$$rs = ar + ar^2 + \cdots + ar^{n-1} + ar^n$$
$$\overline{s - rs = a \phantom{+ ar + ar^2 + \cdots + ar^{n-1}} - ar^n}$$
$$s(1 - r) = a(1 - r^n) \qquad \text{factoring}$$
$$s = \frac{a(1 - r^n)}{1 - r} \qquad \text{dividing by } 1 - r$$

Sum of the first n terms of a geometric progression:

$$s = a + ar + ar^2 + \cdots + ar^{n-1} = \frac{a(1 - r^n)}{1 - r} \quad (r \neq 1) \qquad (18.3)$$

Example 4 Find the following sum, given that the terms form a geometric progression:

$$s = 3 + 6 + 12 + 24 + 48 + 96$$

Solution. The first term is $a = 3$ and the second term is $ar = 3r = 6$, so that $r = 2$. Since $96 = 3(32) = 3(2^5)$, we have $n - 1 = 5$ and $n = 6$. Since $ar^{n-1} = 3(2)^{6-1}$, the sum is

$$s = \frac{3(1 - 2^6)}{1 - 2} = \frac{3(-63)}{-1} = 189$$

An *annuity* is a sequence of equal payments at equal time intervals. Formula (18.3) can be used to compute the value of an annuity, as illustrated in the next example.

Example 5 One hundred dollars is invested each year at 12% compounded annually. Determine the value of the annuity at the end of 8 years (before the ninth payment is made).

Solution. Suppose we examine the growth of the first payment over the years. At the end of the first year, the first $100 has grown to $100 + 100(0.12) = 100(1 + 0.12) = 100(1.12)$. At the end of the second year, the amount is $100(1.12)(1.12) = 100(1.12)^2$; and at the end of the third year, $100(1.12)^3$. Continuing in this manner, at the end of the eighth year, the amount has grown to $100(1.12)^8$.

The payment made at the beginning of the second year has only 7 years in which to grow, leading to a final amount of $100(1.12)^7$. The third payment grows to $100(1.12)^6$, and so on.

The last payment draws interest for only 1 year, at which time the amount is 100(1.12).

The total amount A of the annuity is given by the sum of the terms of the geometric progression

$$A = 100(1.12) + 100(1.12)^2 + \cdots + 100(1.12)^8$$
$$= 100[1.12 + (1.12)^2 + \cdots + (1.12)^8]$$

Note that for the sum inside the brackets, we have $a = 1.12$, $r = 1.12$, and $n - 1 = 7$. By formula (18.3) and with the help of a calculator,

$$A = 100 \cdot \frac{1.12[1 - (1.12)^8]}{1 - 1.12} = \$1{,}377.57$$

Example 6 Determine the value of the annuity in Example 5 if the fund pays 12% interest compounded quarterly.

Solution. When interest is compounded quarterly, 3% interest is payed every 3 months. So at the end of 1 year, $100 accumulates to $100(1.03)^4$; and at the end of 2 years, to $100(1.03)^8$. At the end of 8 years, the first payment grows to $100(1.03)^{32}$. Thus the value of the annuity is given by

$$100[(1.03)^4 + (1.03)^8 + \cdots + (1.03)^{32}]$$
$$= 100\{(1.03)^4 + [(1.03)^4]^2 + \cdots + [(1.03)^4]^8\}$$

Since $a = (1.03)^4$, $r = (1.03)^4$, and $n - 1 = 7$, we have, using a calculator,

$$A = 100(1.03)^4 \cdot \frac{1 - [(1.03)^4]^8}{1 - (1.03)^4} = \$1{,}412.47$$

Exercises / Section 18.1

1. Find the sixth term of the geometric progression whose first term is 4 and whose common ratio is $\frac{2}{3}$.
2. Find the fourth term of the geometric progression whose first term is $\frac{1}{2}$ and whose common ratio is $\frac{3}{4}$.

In Exercises 3–6, find the nth term of the geometric progression.

3. $a = 5$, $r = -\frac{1}{2}$, $n = 10$
4. $a = 4$, $r = -\frac{2}{3}$, $n = 6$
5. $a = 2$, $r = -3$, $n = 5$
6. $a = 3$, $r = -4$, $n = 5$

In Exercises 7–18, the terms in each sum form a geometric progression. Find each sum by the method of Example 4.

7. $1 + 2 + 4 + 8 + 16 + 32$
8. $1 + 3 + 9 + 27 + 81$
9. $2 + 6 + 18 + 54 + 162$
10. $3 + 6 + 12 + 24 + 48 + 96 + 192$

11. $1 + \frac{1}{2} + \frac{1}{4} + \frac{1}{8} + \frac{1}{16} + \frac{1}{32} + \frac{1}{64}$

12. $1 + \frac{1}{3} + \frac{1}{9} + \frac{1}{27} + \frac{1}{81}$

13. $1 - \frac{1}{3} + \frac{1}{9} - \frac{1}{27}$ $\left(\text{Note: } r = -\frac{1}{3}.\right)$

14. $1 - \frac{1}{2} + \frac{1}{4} - \frac{1}{8} + \frac{1}{16} - \frac{1}{32}$

15. $2 + \frac{4}{3} + \frac{8}{9} + \frac{16}{27} + \frac{32}{81}$

16. $3 + \frac{9}{4} + \frac{27}{16} + \frac{81}{64} + \frac{243}{256}$

17. $3 - \frac{3}{2} + \frac{3}{4} - \frac{3}{8} + \frac{3}{16} - \frac{3}{32} + \frac{3}{64}$

18. $4 - \frac{4}{3} + \frac{4}{9} - \frac{4}{27} + \frac{4}{81}$

19. Five hundred dollars is invested at 8%. Find the value of the investment after one year if interest is compounded **a.** annually; **b.** semiannually; **c.** quarterly; **d.** monthly.

20. One thousand dollars is invested at 12%. Find the value of the investment after 1 year if interest is compounded **a.** annually; **b.** semiannually; **c.** quarterly; **d.** daily.

21. What is the value of an investment of $600 after 5 years if it earns 10% interest compounded semiannually?

22. What is the value of an $800 investment after 6 years if it earns 12% interest compounded quarterly?

23. Determine the value of the following annuity: $100 per year for 5 years (before the sixth payment is made). The interest is 10% compounded annually.

24. Determine the value of the following annuity: $200 per year for 6 years (before the seventh payment is made). Interest is 8% compounded annually.

25. What is the value of the annuity in Exercise 23 if the interest is compounded **a.** semiannually; **b.** quarterly?

26. Determine the value of the annuity in Exercise 24 if the interest is compounded semiannually.

27. How much will you accumulate at the end of 10 years (before the eleventh payment is made) by depositing $150 per year at 9% interest compounded annually?

28. Show that after n years (before payment $n + 1$ is made), the value of an annuity A_n consisting of P dollars deposited yearly at $100i\%$ interest compounded annually is given by

$$A_n = \frac{P(1 + i)[1 - (1 + i)^n]}{1 - (1 + i)} = \frac{P[(1 + i)^{n+1} - (1 + i)]}{i}$$

29. The number of bacteria in a culture doubles every 2 hr. If there are N bacteria initially, how many will there be 20 hr later?

30. The number of bacteria in a culture trebles every hour. If there are N bacteria initially, how many will there be after 8 hr?

31. A certain radioactive substance has a half-life of 1 hr. If there are N grams initially, how much will be left after 12 hr? (See Example 3.)

32. If a radioactive substance has a half-life of 2 hr, what fractional part of N grams will be left after 24 hr?

33. A ball is tossed upward from the ground to a height of 10 ft. If it rebounds one-half as far as it falls, find the distance traveled when it hits the ground for the fifth time.

34. A ball is tossed upward from the ground to a height of 15 ft. On each rebound it reaches $\frac{1}{3}$ of the height from which it fell. Determine the distance the ball has traveled when it hits the ground for the sixth time.

35. A ball dropped from a height of 12 ft rebounds $\frac{3}{4}$ as far as it falls. Find the distance the ball has traveled when it hits the ground for the seventh time.

18.2 Geometric Series

If a geometric progression is continued indefinitely, it is called an **infinite geometric progression.** If the common ratio is properly restricted, the sum of the terms can be defined, even though the number of terms is infinitely large. Such a sum is called a **geometric series.**

Recall formula (18.3):

$$s = a + ar + ar^2 + \cdots + ar^{n-1} = \frac{a(1 - r^n)}{1 - r}$$

Since the formula is valid for all n, the size of n need not be restricted. Suppose n gets ever larger, beyond any bound; does it still make sense to speak of the sum? It does if we note that r^n gets ever smaller if $|r| < 1$. For example, if $r = \frac{1}{2}$, then $(\frac{1}{2})^n$ approaches 0. Consequently, if $|r| < 1$, then

$$\frac{a(1 - r^n)}{1 - r} \quad \text{approaches} \quad \frac{a}{1 - r}$$

This observation suggests the definition below.

Sum of a geometric series: The sum of the geometric series

$$S = a + ar + ar^2 + \cdots + ar^{n-1} + \cdots \tag{18.4}$$

is defined as

$$S = \frac{a}{1 - r} \quad \text{if} \quad |r| < 1 \tag{18.5}$$

To show that formula (18.5) is indeed a reasonable definition, let us check the formula by changing a decimal to a common fraction.

Example 1   Change $0.121212\ldots$ to a common fraction.

Solution. The decimal fraction $N = 0.121212\ldots$ can be written

$$N = 0.12 + 0.0012 + 0.000012 + \cdots$$

$$= \frac{12}{10^2} + \frac{12}{10^4} + \frac{12}{10^6} + \cdots$$

Since $a = 12/10^2$ and $r = 1/10^2$, we get

$$N = \frac{\frac{12}{10^2}}{1 - \frac{1}{10^2}} = \frac{\left(\frac{12}{10^2}\right)(10^2)}{\left(1 - \frac{1}{10^2}\right)(10^2)} = \frac{12}{10^2 - 1} = \frac{12}{99} = \frac{4}{33}$$

As a check, long division gives $4 \div 33 = 0.121212\ldots$.

The remaining examples further illustrate the technique for finding the sum of a geometric series.

Example 2 Find the sum of the geometric series

$$S = \frac{2}{5} - \left(\frac{2}{5}\right)^2 + \left(\frac{2}{5}\right)^3 - \left(\frac{2}{5}\right)^4 + \cdots$$

Solution. Note that the series can be written

$$S = \frac{2}{5} + \frac{2}{5}\left(-\frac{2}{5}\right) + \frac{2}{5}\left(-\frac{2}{5}\right)^2 + \frac{2}{5}\left(-\frac{2}{5}\right)^3 + \cdots$$

So by formula (18.4), $a = \frac{2}{5}$ and $r = -\frac{2}{5}$. Thus

$$S = \frac{\frac{2}{5}}{1 - \left(-\frac{2}{5}\right)} = \frac{\frac{2}{5}}{1 + \frac{2}{5}} = \frac{\frac{2}{5}}{\frac{7}{5}} = \frac{2}{5} \cdot \frac{5}{7} = \frac{2}{7}$$

Example 3 A ball is tossed upward from the ground to a height of 6 ft. On each rebound it reaches $\frac{1}{2}$ its former height. How far will it travel before coming to rest?

Solution. The total distance d is given by

$$d = 2(6) + 2\left(\frac{6}{2}\right) + 2\left(\frac{6}{4}\right) + 2\left(\frac{6}{8}\right) + 2\left(\frac{6}{16}\right) + \cdots$$

Thus $a = 2(6) = 12$ and $r = \frac{1}{2}$. Hence

$$d = \frac{12}{1 - \frac{1}{2}} = 24 \text{ ft}$$

Exercises / Section 18.2

In Exercises 1–8, convert the decimal fractions to common fractions.

1. 0.4444 . . .   **2.** 0.6666 . . .
3. 0.131313 . . .   **4.** 0.212121 . . .
5. 0.312312 . . .   **6.** 0.613613 . . .
7. 0.1343434 . . .   **8.** 0.5727272 . . .

In Exercises 9–20, find the sum of the given geometric series.

9. $\frac{1}{2} + \frac{1}{4} + \frac{1}{8} + \frac{1}{16} + \cdots$ **10.** $\frac{1}{3} + \frac{1}{9} + \frac{1}{27} + \frac{1}{81} + \cdots$

11. $1 - \frac{1}{3} + \frac{1}{9} - \frac{1}{27} + \cdots$ **12.** $1 - \frac{1}{2} + \frac{1}{4} - \frac{1}{8} + \frac{1}{16} - \cdots$

13. $1 - \dfrac{1}{4} + \dfrac{1}{4^2} - \dfrac{1}{4^3} + \cdots$

14. $2 - 2\left(\dfrac{3}{5}\right) + 2\left(\dfrac{3}{5}\right)^2 - 2\left(\dfrac{3}{5}\right)^3 + \cdots$

15. $3 + 3\left(\dfrac{5}{9}\right) + 3\left(\dfrac{5}{9}\right)^2 + 3\left(\dfrac{5}{9}\right)^3 + \cdots$

16. $\dfrac{3}{4} + \dfrac{9}{16} + \dfrac{27}{64} + \cdots$

17. $\dfrac{3}{5} - \left(\dfrac{3}{5}\right)^2 + \left(\dfrac{3}{5}\right)^3 - \cdots$

18. $\dfrac{2}{3} - \dfrac{2}{9} + \dfrac{2}{27} - \dfrac{2}{81} + \cdots$

19. $4 - 3 + \dfrac{9}{4} - \dfrac{27}{16} + \dfrac{81}{64} - \cdots$

20. $5 - 2 + \dfrac{4}{5} - \dfrac{8}{25} + \dfrac{16}{125} - \cdots$

21. Each swing of a pendulum is 90% of its former swing. If the first swing is 10 in. long, find the distance that the tip of the pendulum travels before coming to rest.

22. A ball is dropped from a height of 9 m. If it reaches $\frac{2}{3}$ of its height on each rebound, how far does it travel before coming to rest?

23. A ball rebounds $\frac{3}{8}$ as far as it falls. If dropped from a height of 12 ft, how far does it travel before coming to rest?

24. A mass hanging on a spring oscillates up and down. Each oscillation is $\frac{3}{8}$ the preceding one. If the first oscillation is 4 in., how far does it travel before coming to rest?

25. A particle oscillates in a magnetic field. Each oscillation is $\frac{5}{8}$ the preceding one. If the first oscillation is 3 mm, how far does the particle travel before coming to rest?

18.3 Binomial Theorem and Series

In this section we shall consider another sequence of numbers arising in a formula called the **binomial theorem.** The purpose of this formula is to expand a binomial of the form $(a + b)^n$. This binomial form occurs often enough in certain technical applications and in advanced mathematics to warrant a separate section.

For small values of n, the binomial $(a + b)^n$ can be multiplied directly:

$$(a + b)^0 = 1$$
$$(a + b)^1 = a + b$$
$$(a + b)^2 = a^2 + 2ab + b^2$$
$$(a + b)^3 = a^3 + 3a^2b + 3ab^2 + b^3$$
$$(a + b)^4 = a^4 + 4a^3b + 6a^2b^2 + 4ab^3 + b^4$$

These forms have certain properties in common:

1. The first term is a^n, and the last term is b^n.
2. The exponent of a decreases by 1 from term to term, starting with a^n and ending with a^0.
3. The exponent of b increases by 1 from term to term, starting with b^0 and ending with b^n.
4. The exponents of a and b add up to n in each term.
5. The number of terms is $n + 1$.

In these expansions, only the coefficients do not seem to follow an obvious pattern. Yet a pattern does exist: Each coefficient is the sum of the coefficients immediately above it (when the expansion is laid out in pyramid fashion as above). Noted earlier by Chinese mathematicians and rediscovered by the French mathematician Blaise Pascal (1623–1662), the pattern can be described by means of **Pascal's triangle:**

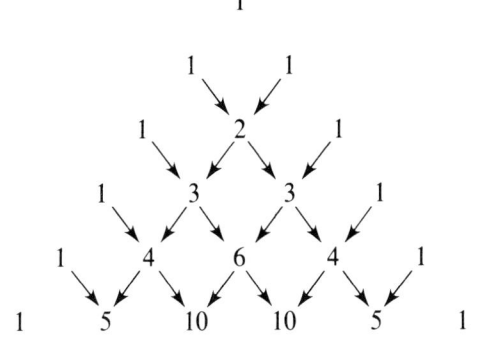

```
              1
           1     1
         1    2    1
       1    3    3    1
     1    4    6    4    1
   1    5   10   10    5    1
```

Example 1

Expand $(x + y)^6$ by means of Pascal's triangle.

Solution. The next row of the triangle, obtained by adding adjacent numbers, is

1 6 15 20 15 6 1

Thus

$$(x + y)^6 = x^6 + 6x^5y + 15x^4y^2 + 20x^3y^3 + 15x^2y^4 + 6xy^5 + y^6$$

The pattern emerging from Pascal's triangle can also be described mathematically. To do so, we need the definition of *factorial:*

Factorial:

The product

$$n(n - 1)(n - 2) \ldots (2)(1)$$

is called *n* **factorial** and is denoted by *n*!.

With the factorial notation, the binomial theorem can be stated in the form given next.

CHAPTER 18 GEOMETRIC PROGRESSIONS AND THE BINOMIAL THEOREM

> **Binomial theorem:**
>
> $$(a + b)^n = a^n + na^{n-1}b + \frac{n(n-1)}{2!}a^{n-2}b^2$$
> $$+ \frac{n(n-1)(n-2)}{3!}a^{n-3}b^3 + \cdots + b^n \qquad (18.6)$$

The binomial theorem is illustrated in the next two examples.

Example 2 Expand $(x + y)^4$ by the binomial theorem.

Solution. By the binomial theorem,

$$(x + y)^4 = x^4 + 4x^3y + \frac{(4)(3)}{2!}x^2y^2 + \frac{(4)(3)(2)}{3!}xy^3 + \frac{(4)(3)(2)(1)}{4!}y^4$$

Note that the coefficient of the last term reduces to $4!/4! = 1$. Thus

$$(x + y)^4 = x^4 + 4x^3y + 6x^2y^2 + 4xy^3 + y^4$$

Example 3 Expand $(2x + 3y)^5$.

Solution. In this binomial we have $a = 2x$ and $b = 3y$. So by the binomial theorem

$$[(2x) + (3y)]^5 = (2x)^5 + 5(2x)^4(3y) + \frac{(5)(4)}{2!}(2x)^3(3y)^2$$
$$+ \frac{(5)(4)(3)}{3!}(2x)^2(3y)^3 + \frac{(5)(4)(3)(2)}{4!}(2x)(3y)^4$$
$$+ \frac{(5)(4)(3)(2)(1)}{5!}(3y)^5$$
$$= 32x^5 + 240x^4y + 720x^3y^2 + 1080x^2y^3 + 810xy^4 + 243y^5$$

In some applications of the binomial theorem, only the first few terms may be of interest.

Example 4 Find the first five terms of the binomial $(v - 2t)^{10}$.

Solution. Writing the binomial in the form $[v + (-2t)]^{10}$, we get

$$[v + (-2t)]^{10} = v^{10} + 10v^9(-2t) + \frac{(10)(9)}{2!} v^8(-2t)^2$$

$$+ \frac{(10)(9)(8)}{3!} v^7(-2t)^3 + \frac{(10)(9)(8)(7)}{4!} v^6(-2t)^4 + \cdots$$

$$= v^{10} - 20v^9 t + 180v^8 t^2 - 960v^7 t^3 + 3{,}360v^6 t^4 - \cdots$$

It can be shown that the binomial theorem is valid for any rational n under certain conditions. Of particular interest is the special case $a = 1$ in formula (18.6), which is usually called the **binomial series.**

Binomial series:

$$(1 + b)^n = 1 + nb + \frac{n(n-1)}{2!} b^2 + \frac{n(n-1)(n-2)}{3!} b^3 + \cdots \quad (18.7)$$

where $|b| < 1$.

Note especially that whenever n is not a positive integer, the expansion does not have a last term. Such an expansion is called an **infinite series,** of which the geometric series is another example. The binomial series is valid only if $|b| < 1$.

Example 5 Use the binomial series to approximate $1/\sqrt[3]{0.894}$ to three decimal places.

Solution. To obtain the proper form, we write 0.894 as $1 - 0.106$. So by the binomial series

$$(1 - 0.106)^{-1/3} = 1 + \left(-\frac{1}{3}\right)(-0.106) + \frac{\left(-\frac{1}{3}\right)\left(-\frac{4}{3}\right)}{2!} (-0.106)^2$$

$$+ \frac{\left(-\frac{1}{3}\right)\left(-\frac{4}{3}\right)\left(-\frac{7}{3}\right)}{3!} (-0.106)^3 + \cdots$$

Using the first four terms, we get (with a calculator)

$$(0.894)^{-1/3} = 1 + 0.035333 + 0.0024969 + 0.00020585$$
$$= 1.0380$$

Since the next term is

$$\frac{\left(-\frac{1}{3}\right)\left(-\frac{4}{3}\right)\left(-\frac{7}{3}\right)\left(-\frac{10}{3}\right)}{4!}(-0.106)^4 = 0.000018$$

the use of additional terms does not affect the first three decimal places. Thus, $1/\sqrt[3]{0.894} = 1.038$.

If the evaluation of a radical requires more decimal places than a calculator can provide, the binomial series is a convenient method of evaluation.

Exercises / Section 18.3

In Exercises 1–4, use Pascal's triangle to expand the given binomials.

1. $(x - y)^4$ **2.** $(a - T)^3$ **3.** $(s + 2v)^4$ **4.** $(a + 3x)^3$

In Exercises 5–16, use the binomial theorem to expand the given binomials.

5. $(x - y)^3$ **6.** $(x - y)^4$ **7.** $(s + 2t)^4$ **8.** $(M + 3N)^3$
9. $(a - b^2)^5$ **10.** $(V^2 - T)^5$ **11.** $(S - 2W)^6$ **12.** $(2x + y)^6$
13. $(R + 2)^6$ **14.** $(x - y)^7$ **15.** $(R - 3)^6$ **16.** $(3a + b)^6$

In Exercises 17–22, write the first four terms of the binomial expansion.

17. $(x + 2)^{10}$ **18.** $(y + 3)^8$ **19.** $(k - m^2)^{12}$ **20.** $(r^2 - e)^9$
21. $(r + 2\pi)^8$ **22.** $(1 - \gamma)^{10}$

In Exercises 23–26, write the first four terms of the binomial series (18.7).

23. $(1 + x)^{-2}$ **24.** $(1 - R)^{-2}$ **25.** $(1 + s)^{1/2}$ **26.** $(1 + p)^{1/3}$

27. Write $(1 - x)^{-1}$ as a binomial series and compare it to the geometric series.

28. Write $(1 + x)^{-1}$ as a binomial series and compare it to the geometric series.

29. Approximate $\sqrt[3]{0.912}$ to four decimal places. (See Example 5.)

30. Approximate $\sqrt[3]{0.892}$ to three decimal places.

31. Approximate $1/\sqrt{0.832}$ to three decimal places.

32. Approximate $1/\sqrt{0.917}$ to four decimal places.

33. The period of a pendulum having a large swing is determined by means of so-called elliptic integrals. In the calculation, the expression

$$\frac{1}{\sqrt{1 - \sin^2 \frac{\theta}{2} \sin^2 \phi}}$$

has to be expanded. By letting $b = \sin^2(\theta/2) \sin^2 \phi$, show that

$$\frac{1}{\sqrt{1 - \sin^2 \frac{\theta}{2} \sin^2 \phi}} = 1 + \frac{1}{2}\sin^2\frac{\theta}{2}\sin^2\phi + \frac{3}{8}\sin^4\frac{\theta}{2}\sin^4\phi + \cdots$$

34. When finding the derivative of the function $f(x) = x^4$ in calculus, the following expression has to be simplified:

$$\frac{(x+h)^4 - x^4}{h}$$

Carry out this simplification.

35. In determining the length of the curve $y = \tfrac{1}{2}x^2$, $(-1 \le x \le 1)$, the expression $\sqrt{1 + x^2}$ arises. Find an approximation for this expression by writing the first four terms of the binomial series.

36. The deflection of a certain beam is given by $y = x(x - a/2)^4$. Expand this expression.

Review Exercises / Chapter 18

1. Find the sixth term of the geometric progression whose first term is 2 and common ratio is $\tfrac{1}{3}$.

2. Find the fifth term of the geometric progression whose first term is 3 and common ratio is $\tfrac{3}{4}$.

In Exercises 3–8, find each sum, given that the terms form a geometric progression.

3. $1 - \dfrac{1}{2} + \dfrac{1}{4} - \dfrac{1}{8} + \dfrac{1}{16} - \dfrac{1}{32} + \dfrac{1}{64}$

4. $1 - \dfrac{1}{3} + \dfrac{1}{9} - \dfrac{1}{27} + \dfrac{1}{81}$

5. $5 + \dfrac{10}{3} + \dfrac{20}{9} + \dfrac{40}{27} + \dfrac{80}{81}$

6. $2 + \dfrac{4}{5} + \dfrac{8}{25} + \dfrac{16}{125}$

7. $2 - \dfrac{4}{3} + \dfrac{8}{9} - \dfrac{16}{27} + \dfrac{32}{81} - \dfrac{64}{243}$

8. $3 - \dfrac{3}{5} + \dfrac{3}{25} - \dfrac{3}{125}$

In Exercises 9–14, find the sum of the given geometric series.

9. $1 - \dfrac{1}{3} + \dfrac{1}{3^2} - \dfrac{1}{3^3} + \cdots$

10. $1 - \dfrac{2}{3} + \left(\dfrac{2}{3}\right)^2 - \left(\dfrac{2}{3}\right)^3 + \cdots$

11. $2 + \dfrac{4}{5} + \dfrac{8}{25} + \dfrac{16}{125} + \cdots$

12. $3 + \dfrac{3}{4} + \dfrac{3}{16} + \dfrac{3}{64} + \cdots$

13. $3 + \dfrac{9}{4} + \dfrac{27}{16} + \dfrac{81}{64} + \cdots$

14. $3 + \dfrac{6}{5} + \dfrac{12}{25} + \dfrac{24}{125} + \cdots$

In Exercises 15–18, convert the decimal fractions to common fractions.

15. $0.1555\ldots$

16. $0.2777\ldots$

17. $0.383838\ldots$

18. $0.5383838\ldots$

In Exercises 19–26, expand the given expressions by means of the binomial theorem.

19. $(a - b)^3$

20. $(a - b)^4$

21. $(M - 2T)^5$

22. $(\pi + 2)^4$

23. $(e - 2)^4$

24. $(s + t^2)^4$

25. $(m - v^2)^5$

26. $(a + 2)^6$

In Exercises 27–31, write the first four terms of the binomial expansion.

27. $(x - 2)^{10}$
28. $(R - 1)^9$
29. $(1 - t)^{-2}$
30. $(1 + v)^{1/2}$
31. $(1 - c)^{-1/3}$
32. Show that the binomial expansion of $(1 - a)^{-1}$ agrees with the geometric series.
33. Approximate $\sqrt[3]{0.907}$ to four decimal places by means of the binomial series.
34. Approximate $1/\sqrt[3]{0.936}$ to four decimal places by means of the binomial series.
35. Determine the value of a $600 investment after 10 years if it earns 10% interest compounded semiannually.
36. Determine the value of a $1,000 investment after 8 years if it earns 12% interest compounded quarterly.
37. Two hundred dollars is invested yearly at 10% compounded quarterly. Determine the value of the annuity after 4 years (before the fifth payment is made).
38. Determine the value of the following annuity: $100 invested yearly for 5 years at 8% interest compounded quarterly.
39. The number of bacteria in a culture trebles every 2 hr. If there are N bacteria initially, how many will there be at the end of 16 hr?
40. A certain radioactive substance has a half-life of 2 years. If there are N grams of the substance initially, what fractional part will be left after 20 years?
41. A ball is dropped from a height of 10 ft. On each rebound, it reaches $\frac{1}{3}$ its former height. What is the total distance traveled when it hits the ground for the sixth time?
42. Each swing of a pendulum is 90% of the preceding swing. If the initial swing is 10 in. long, find the total distance that the tip of the pendulum has traveled at the end of 8 swings.
43. Each swing of a pendulum is 95% of the preceding swing. If the first swing is 5 in., how far will the tip travel before coming to rest?
44. A weight on a spring oscillates up and down, each oscillation being $\frac{4}{5}$ the preceding one. If the first oscillation is 6 cm, how far will the weight travel before coming to rest?
45. The *rigidity modulus* of a material is

$$M = \frac{E}{2(s + 1)}$$

Write M as a binomial series, assuming that $|s| < 1$.

Cumulative Review Exercises / Chapters 16–18

Verify the identities in Exercises 1–7.

1. $\cos^3 \theta + \sin^2 \theta \cos \theta = \cos \theta$

2. $\dfrac{1 + \cot^2 x}{1 + \tan^2 x} = \cot^2 x$

3. $\dfrac{1 + \cos \theta}{\sin \theta} + \dfrac{\sin \theta}{1 + \cos \theta} = 2 \csc \theta$

4. $\dfrac{\tan \alpha + \cot \alpha}{\csc \alpha} = \sec \alpha$

5. $\dfrac{1}{\sec x + \tan x} = \sec x - \tan x$

6. $\dfrac{2 - \sec^2 \theta}{\sec^2 \theta} = \cos 2\theta$

7. $\sin^2 \dfrac{\theta}{4} = \dfrac{1 - \cos(\theta/2)}{2}$

In Exercises 8 and 9, find the exact values without tables or calculators.

8. $\cos 60° \cos 15° + \sin 60° \sin 15°$

9. $\cos 15°$

10. Use a calculator to solve the following equation for x ($0° \leq \theta < 360°$):

   $\sec^2 x - 2 \tan x - 1 = 0$

11. Find the exact values of **a.** $\text{Arctan}(-\sqrt{3})$; **b.** $\text{Arccos}(-\tfrac{1}{2})$

In Exercises 12 and 13 solve the given inequalities for x.

12. $\dfrac{(x - 2)(x + 3)}{(x - 1)(x + 4)} \leq 0$

13. $|x^2 - 4x + 2| > 2$

14. Find the sum of the geometric series

   $\dfrac{1}{2} - \dfrac{1}{4} + \dfrac{1}{8} - \dfrac{1}{16} + \cdots$

15. Expand $(x - 2y)^5$ by the binomial theorem.

16. The range R of a projectile fired at angle θ with the horizontal at velocity v is given by

   $R = \dfrac{2v^2 \cos \theta \sin \theta}{g}$

   Write R as a single trigonometric function of θ.

17. Starting at $t = 0$ the current in a circuit is given by $i = 2.0 \sin 3.0t$. Find the smallest value of t (in seconds) for which $i = 0.50$ A.

18. The current in a coil rotating in a magnetic field is given by $I = I_{\max} \cos \theta$, where θ is the angle that the coil makes with the field. Solve the equation for θ, $0 \leq \theta \leq \pi$.

19. A ball is tossed upward from the ground to a height of 10 ft. On each rebound it reaches $\tfrac{1}{2}$ the height from which it fell. Determine the distance the ball will have traveled when it strikes the ground for the sixth time.

20. Each swing of a pendulum is 95% of its former swing. If the first swing is 5 in. long, find the distance that the tip of the pendulum will travel before coming to rest.

CHAPTER **19**

Introduction to Analytic Geometry

Objectives The primary objective of this chapter is to solve the following two fundamental problems of analytic geometry: **1.** Given an equation, determine what corresponding figure depicts it and its geometric properties, and **2.** given the geometric properties of a figure, find the corresponding equation. Upon completion of this chapter, you should be able to solve these problems for the following figures (appropriate geometric properties are indicated in parentheses):

1. Line (slope and intercepts).
2. Circle (center and radius).
3. Parabola (focus and directrix).
4. Ellipse (foci and vertices; major and minor axes).
5. Hyperbola (foci, vertices, and asymptotes; transverse and conjugate axes).

In addition, you should be able to:

6. Find the distance between two points.
7. Find the coordinates of the midpoint of a line segment.
8. Find the slope of a line determined by two points.
9. Determine if two lines are parallel or perpendicular.
10. Solve stated problems involving conics.
11. Simplify the equation of a conic by translating axes.
12. Convert a given equation from rectangular to polar coordinates and vice versa.
13. Sketch curves in polar coordinates.

19.1 Introduction

Analytic geometry may be defined as the study of classical geometry by means of algebra. Credit for the discovery of this method is usually given to René Descartes (1596–1650), the famous French mathematician and philosopher, but the same method was discovered independently and somewhat earlier by Pierre de Fermat (1601–1665), a French lawyer. The discovery of analytic geometry was of singular importance in advancing the study of mathematics, for it led the way to the discovery of the calculus a few decades later. Moreover, Fermat's own method for finding maximum and minimum points on a graph is essentially the one employed in differential calculus today.

René Descartes

19.2 The Distance Formula

For our first topic in analytic geometry, we shall learn how to find the distance between two points in a plane.

First recall from Chapter 2 that points are located in a plane by means of the **rectangular** or **Cartesian coordinate system.** (Please refer to Section 2.6 for the basic definitions.) In the rectangular coordinate system, fixed points are usually distinguished by subscripts, such as P_1 and P_2 or (x_1, y_1) and (x_2, y_2). Let $P_1(x_1, y_1)$ and $P_2(x_2, y_2)$ be two points with arbitrary coordinates, as shown in Figure 19.1. Join P_1 and P_2 by a straight line. To deter-

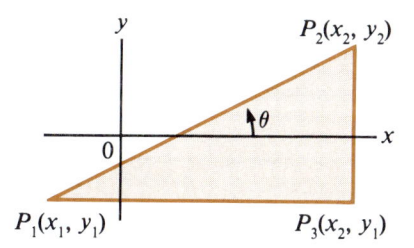

Figure 19.1

mine the distance between P_1 and P_2, a point P_3 is also needed, which is determined by a line through P_1 parallel to the x-axis and a line through P_2 parallel to the y-axis. Note that the coordinates of P_3 are (x_2, y_1) and that triangle $P_1P_2P_3$ is a right triangle. The length of the segment P_1P_3 is $x_2 - x_1$ and the length of P_2P_3 is $y_2 - y_1$. (See Exercise 12 at the end of this section.) Then by the Pythagorean theorem,

$$(P_1P_2)^2 = (P_1P_3)^2 + (P_2P_3)^2 \quad \text{or} \quad P_1P_2 = \sqrt{(x_2 - x_1)^2 + (y_2 - y_1)^2}$$

> **Distance formula:**
> $$P_1P_2 = \sqrt{(x_2 - x_1)^2 + (y_2 - y_1)^2} \tag{19.1}$$

Example 1 Find the distance between $(-2, -4)$ and $(5, -2)$.

Solution. $d = \sqrt{(-2 - 5)^2 + [-4 - (-2)]^2} = \sqrt{49 + 4} = \sqrt{53}$

Here we have chosen $(-2, -4)$ for (x_2, y_2), but we would have had the same result if we had chosen the other point for (x_2, y_2).

Exercises / Section 19.2

In Exercises 1–9, plot each pair of points and find the distance between them.

1. $(2, 4)$ and $(5, 2)$
2. $(-3, 2)$ and $(5, -4)$
3. $(-3, -6)$ and $(5, -2)$
4. $(0, 0)$ and $(-\sqrt{5}, 2)$
5. $(0, 2)$ and $(\sqrt{3}, 4)$
6. $(\sqrt{2}, \sqrt{5})$ and $(\sqrt{2}, 0)$
7. $(1, -\sqrt{2})$ and $(-1, 0)$
8. $(1, 6)$ and $(-3, 2)$
9. $(-10, -2)$ and $(-12, 0)$

10. Draw the triangle whose vertices are $(0, 0)$, $(4, 3)$, and $(6, 0)$.
11. If (x, y) is a point, determine in which quadrant x/y is **a.** positive; **b.** negative.
12. In the derivation of the distance formula, show that the length of P_1P_3 is $x_2 - x_1$ and the length of P_2P_3 is $y_2 - y_1$, no matter in which quadrants the points lie. The distance formula is therefore valid for all points. (See Figure 19.1.)
13. Find the set of all points whose **(a)** abscissas are zero; **(b)** ordinates are zero.
14. Find the perimeter of the triangle in Exercise 10.
15. Show that $(-2, 5)$, $(3, 5)$, and $(-2, 2)$ are vertices of a right triangle.
16. Repeat Exercise 15 for $(1, 7)$, $(-2, 1)$, and $(10, -5)$.
17. Show that $(12, 0)$, $(-4, 8)$, and $(-1, -13)$ lie on a circle whose center is $(1, -2)$.
18. Show that $(-2, 10)$, $(3, -2)$, and $(15, 3)$ are vertices of an isosceles triangle.
19. Show that $(-1, -1)$, $(2, 8)$, and $(5, 17)$ are collinear, that is, lie on the same line.
20. Find the relation between x and y so that (x, y) is 3 units from $(-1, 2)$.
21. Find the relation between x and y so that (x, y) is equidistant from the y-axis and the point $(2, 0)$.

19.3 The Slope

We saw in Section 19.2 that every point has a definite location with respect to the coordinate axes. A line also has a definite position. For example, the angle θ in Figure 19.2 is a measure of steepness, just as the steepness of a road is measured with respect to the horizontal. This angle θ ($0° \leq \theta < 180°$) is called the *angle of inclination*. It turns out to be more convenient, however, to define steepness by the formula $m = \tan \theta$, called the **slope** of the line.

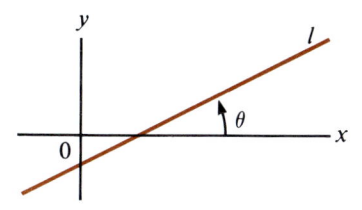

19.3 THE SLOPE

[Figure 19.2]

Figure 19.2

> **Definition of slope:**
> $$m = \tan \theta \tag{19.2}$$

By this definition, m can be viewed as the change in the y-direction with respect to the change in the x-direction.

We can obtain a formula for the slope of a line determined by points $P_1(x_1, y_1)$ and $P_2(x_2, y_2)$, as shown in Figure 19.3. From the previous section, $P_1P_3 = x_2 - x_1$ and $P_2P_3 = y_2 - y_1$. Since $\tan \alpha = \tan \theta$, we have established the formula below.

> **Slope of a line:**
> $$m = \frac{y_2 - y_1}{x_2 - x_1} = \frac{y_1 - y_2}{x_1 - x_2} \tag{19.3}$$

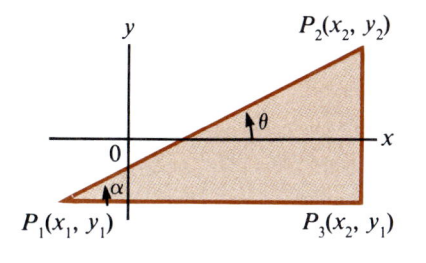

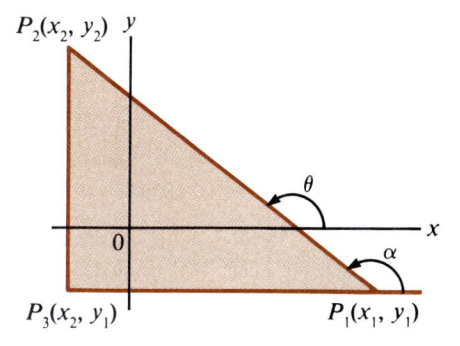

Figure 19.3

We also see that the slope is positive when θ is acute and negative when θ is obtuse. When $\theta = 0°$ (that is, when the line is horizontal), $m = 0$. If the line is parallel to the y-axis, then $x_1 = x_2$, so that m is undefined; this also follows from the fact that $\tan 90°$ is undefined.

Example 1 Find the slope of the line passing through $(-4, -5)$ and $(3, 6)$.

Solution. $m = \dfrac{6 - (-5)}{3 - (-4)} = \dfrac{11}{7}$

Care must be taken to subtract the coordinates in the correct order.

The slope formula can be used to determine whether two lines are parallel or perpendicular. If two lines are parallel, they have equal angles of inclination and equal slopes; the converse is also true. If two perpendicular lines have slopes $m_1 = \tan \theta_1$ and $m_2 = \tan \theta_2$, as in Figure 19.4, then θ_1 and θ_2 differ by 90°; that is, $\theta_2 = \theta_1 + 90°$. Consequently,

$$\tan \theta_2 = \tan(\theta_1 + 90°) = -\cot \theta_1 = -\dfrac{1}{\tan \theta_1}$$

and

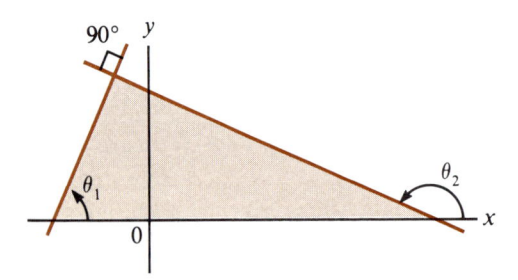

$$m_2 = -\dfrac{1}{m_1} \quad \text{or} \quad m_1 m_2 = -1 \tag{19.4}$$

Figure 19.4

These observations are summarized next.

Parallel and perpendicular lines: Two nonvertical lines are parallel if, and only if, they have equal slopes and are perpendicular if, and only if, their slopes are negative reciprocals.

Example 2 Show that the line through $(7, -1)$ and $(4, -6)$ is perpendicular to the line through $(2, 0)$ and $(-3, 3)$.

Solution. The respective slopes are

$$m_1 = \frac{-1 - (-6)}{7 - 4} = \frac{5}{3} \quad \text{and} \quad m_2 = \frac{0 - 3}{2 - (-3)} = -\frac{3}{5}$$

Since the slopes are negative reciprocals, the lines are perpendicular.

Consider the line segment with endpoints at $P_1(x_1, y_1)$ and $P_2(x_2, y_2)$, as in Figure 19.5. Let $P(x, y)$ be the midpoint of the line segment. Draw the triangles shown in Figure 19.5 by constructing lines perpendicular to the x-

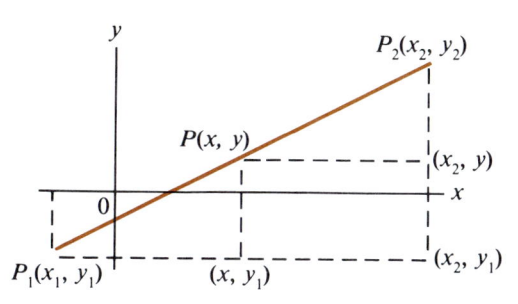

Figure 19.5

axis from $P_1(x_1, y_1)$, $P(x, y)$, and $P_2(x_2, y_2)$. It follows that $2(x - x_1) = x_2 - x_1$ and $2(y - y_1) = y_2 - y_1$. Solving, we get $x = \frac{1}{2}(x_1 + x_2)$ and $y = \frac{1}{2}(y_1 + y_2)$, which are the coordinates of the midpoint of the line segment.

Midpoint formula: The midpoint of the line segment joining (x_1, y_1) and (x_2, y_2) is

$$\left(\frac{x_1 + x_2}{2}, \frac{y_1 + y_2}{2} \right) \tag{19.5}$$

Exercises / Section 19.3

In Exercises 1–9, find the slope of the line through the given points.

1. (2, 6) and (1, 7)
2. (−3, −10) and (−5, 2)
3. (0, 2) and (−4, −4)
4. (6, −3) and (4, 0)
5. (7, 8) and (−3, −4)
6. (0, 0) and (8, −4)
7. (0, 0) and (−4, −8)
8. (1, 3) and (−2, 0)
9. (3, 4) and (3, −5)

10. Find the slope of the line whose angle of inclination is
    a. 0°
    b. 30°
    c. 150°
    d. 90°
    e. 45°
    f. 135°

11. Draw the line passing through (0, 2) and having a slope of
    a. 2
    b. $\frac{1}{2}$
    c. $-\frac{1}{2}$
    d. 0
    e. 3
    f. $-\frac{2}{3}$
12. The vertices of a triangle are $A(-2, 0)$, $B(-1, 2)$, and $C(2, -3)$. Find the slope of each of its sides.
13. Find the slope of a line perpendicular to the line through $(-7, 1)$ and $(6, -5)$.
14. Show that the points $(-2, 8)$, $(1, -1)$, and $(3, -7)$ are collinear.
15. Without using the distance formula, show that the points $(-4, 6)$, $(6, 10)$, and $(10, 0)$ are vertices of a right triangle.
16. Show that $(-4, 2)$, $(-1, 8)$, $(9, 4)$, and $(6, -2)$ are vertices of a parallelogram.
17. Show that $(0, -3)$, $(-2, 3)$, $(7, 6)$, and $(9, 0)$ are vertices of a rectangle.

In Exercises 18–20, find the midpoint of the line segment joining the given points.

18. $(-3, -5)$ and $(-1, 7)$
19. $(-2, 6)$ and $(2, -4)$
20. $(-3, 5)$ and $(-2, 9)$

21. Find the slope of the line through $(5, 6)$ and the midpoint of the line segment from $(-3, 0)$ to $(9, -2)$.
22. Show that $(0, 1)$, $(3, 3)$, $(8, 2)$, and $(-1, -4)$ are vertices of an isosceles trapezoid.
23. Find the slopes of the medians in the triangle in Exercise 12.
24. If the center of a circle is $(1, 2)$ and one end of the diameter is $(4, -3)$, find the other end.
25. Find the value of x so that the line through $(x, 2)$ and $(4, -6)$ is parallel to the line through $(-1, -1)$ and $(3, -5)$.
26. Find x so that the line through $(x, -2)$ and $(4, -7)$ is perpendicular to the line through $(2, -1)$ and $(-3, 2)$.

19.4 The Straight Line

The purpose of this section is twofold: to use the definition of slope to derive the equation of a straight line and to use the line to illustrate the two fundamental problems of analytic geometry, which are restated below.

> 1. Given the equation, locate the figure.
> 2. Given the geometric figure, find the corresponding equation.
>
> (19.6)

The two problems will be considered separately.

Moving from Figure to Equation

Suppose a line has slope m and passes through the point (x_1, y_1). To derive the equation, let (x, y) be any point on the line. Then the slope of the line segment joining (x_1, y_1) and (x, y) is the same, regardless of which point is chosen. So, if m is the given slope, then

$$m = \frac{y - y_1}{x - x_1} \quad \text{or} \quad y - y_1 = m(x - x_1)$$

The second equation is called the **point-slope form** of the line.

19.4 THE STRAIGHT LINE

> **Point-slope form of the line**: The equation of the line with slope m and passing through the point (x_1, y_1) is given by
>
> $$y - y_1 = m(x - x_1) \qquad (19.7)$$

Example 1 Find the equation of the line through $(-3, 5)$ and having a slope of 3.

Solution. From equation (19.7) we get

$$y - 5 = 3[x - (-3)] \quad \text{or} \quad 3x - y + 14 = 0$$

Example 2 Find the equation of the line determined by the points $(-5, -1)$ and $(5, 4)$.

Solution. The slope of the line is found to be $\frac{1}{2}$. The point-slope form can now be obtained by using either point; let us use **(5, 4)**. Thus

$$y - 4 = \frac{1}{2}(x - 5) \quad \text{or} \quad x - 2y + 3 = 0$$

Horizontal line

If the line is horizontal, then $m = 0$, and the equation reduces to

$$y = y_1 \qquad (19.8)$$

Vertical line

If the line is vertical, the slope is undefined. However, all points on a vertical line have a constant abscissa, here called x_1. So the equation is

$$x = x_1 \qquad (19.9)$$

Moving from Equation to Figure

To see how we can obtain the line from its equation, let us consider the following special case of the point-slope form: Suppose that a line has slope m and y-intercept b (that is, the line crosses the y-axis at $(0, b)$). Then, from (19.7),

$$y - b = m(x - 0) \quad \text{or} \quad y = mx + b$$

The latter equation is called the **slope-intercept form** of the line.

> **Slope-intercept form of the line**: The equation of the line with slope m and y-intercept $(0, b)$ is given by
>
> $$y = mx + b \qquad (19.10)$$

This form shows that any first-degree equation $Ax + By + C = 0$ ($B \neq 0$) represents a nonvertical straight line, since solving for y in terms of x reduces the equation to form (19.10). Consequently, *the slope-intercept form can be used to construct the line from the equation.* (If $B = 0$ and $A \neq 0$, then the line is vertical.)

Example 3 Show that the line $x + 2y + 1 = 0$ is perpendicular to the line $2x - y + 3 = 0$.

Solution. Solving each equation for y, we get

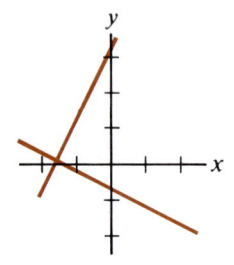

$$y = -\frac{1}{2}x - \frac{1}{2}$$

and

$$y = 2x + 3$$

respectively. From (19.10) we see that the slopes are negative reciprocals. The lines are therefore perpendicular to each other. (See Figure 19.6.)

Figure 19.6

Example 4 If R is the resistance in a direct current circuit and E the voltage, then by Ohm's law the current is $I = E/R$. If $R = 2\ \Omega$, then $E = 2I$ is the line in Figure 19.7.

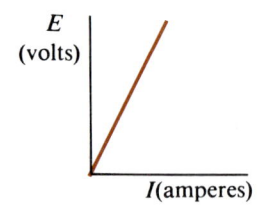

Figure 19.7

Remark. As in algebra, we normally use x and y for the variables. The purpose of this convention is to generalize: Since the variables can stand for any physical quantity, no restriction is placed on the problems to which the equations can be applied. As in this example, however, it is sometimes convenient to use letters that more directly suggest the physical quantities.

Example 5 The relationship between E and I in a direct current circuit is known to be linear (see Example 4). Find the relationship if $E = 6$ V when $I = 2$ A.

Solution. We already know that $E = 0$ V when $I = 0$ A. Hence the problem is equivalent to finding a line through $(0, 0)$ and $(2, 6)$. Thus

$$m = \frac{6 - 0}{2 - 0} = 3$$

so that $E = 3I$.

Returning to the general equation of first degree, we noted above that the equation $Ax + By + C = 0$ (A and B not both zero) represents a straight line. The converse is also true, since a line will either intersect the y-axis, be parallel to the y-axis, or be coincident with it.

Exercises / Section 19.4

In Exercises 1–14, write the equation of the line satisfying the given conditions.

1. Passing through $(-7, 2)$ and having a slope of $\frac{1}{2}$
2. Passing through $(0, 3)$ and having a slope of -4
3. Passing through $(3, -4)$ and having a slope of 3
4. Passing through $(1, 2)$ and having a slope of 0
5. Passing through $(0, 0)$ and having a slope of $-\frac{1}{3}$
6. Passing through $(-3, -10)$ and parallel to the y-axis
7. Passing through $(-4, 0)$ and parallel to the line $y = 1$
8. Passing through $(-3, 2)$ and $(7, 6)$
9. Passing through $(-3, 4)$ and $(3, -6)$
10. Passing through $(1, 2)$ and having an angle of inclination of $135°$
11. Passing through $(0, 10)$ and having an angle of inclination of $45°$
12. Slope of 3 and y-intercept 4
13. Slope of $-\frac{1}{3}$ and y-intercept -2
14. $b = 2$ and $m = 1$

In Exercises 15–20, change each of the straight-line equations to slope-intercept form and draw the lines.

15. $6x + 2y = 5$
16. $x - y = 1$
17. $2x = 3y$
18. $\frac{1}{3}x + \frac{1}{12}y = 1$
19. $2y - 7 = 0$
20. $\frac{1}{2}x - 2y = 3$

In Exercises 21–26, state whether the lines are parallel, perpendicular, or neither.

21. $2x - 3y = 1$ and $4x - 6y + 3 = 0$
22. $2x + 4y + 3 = 0$ and $y - 2x = 2$
23. $3x - 4y = 1$ and $3y - 4x = 3$
24. $7x - 10y = 6$ and $y - 4 = 0$
25. $x + 3y = 5$ and $y - 3x - 2 = 0$
26. $2x + 5y = 2$ and $6x + 15y = 1$
27. Find the equation of the line through $(-1, 1)$ and parallel to the line $3x - 4y = 7$.
28. Find the equation of the line through $(-1, 1)$ and perpendicular to the line $3x - 4y = 7$.
29. Find the equation of the line passing through the intersection of the lines $2x - 4y = 1$ and $3x + 4y = 4$ and parallel to $5x + 7y + 3 = 0$.
30. Show that the following lines form the sides of a right triangle: $x - 3y + 3 = 0$, $12x + 4y + 25 = 0$, and $2y + x + 8 = 0$.

31. Sketch the graph of $v = 1 + 10t$, where v is the velocity of an object and t is time, letting the t-axis be horizontal. (Note that the graph is a straight line and gives a pictorial representation of the velocity at any time.)

32. If a spring is stretched x units, it pulls back with a force $F = kx$ by Hooke's law; k is a constant called the *force constant* of the spring. If $k = 3$, draw the graph of the equation.

33. If a force of 200 dynes is required to stretch the spring in Exercise 32 a distance of 1 cm, find k and draw the graph of the resulting equation.

34. Deliveries of a certain item were made to a store on the first day of the month. The store manager noticed that the supply throughout the month was given by

$$y = 60 - 2t$$

Sketch the graph. What is the significance of the y-intercept? The t-intercept?

35. The relationship between temperature measured in degrees Celsius (C) and degrees Fahrenheit (F) is known to be linear; that is, it has the form $F = mC + b$. Water boils at 100°C or 212°F and freezes at 0°C or 32°F. Find the relationship.

19.5 The Conics

The purpose of the next several sections is twofold: to study a group of geometric figures called *conic sections* and to return to the two fundamental problems (19.6).

The term **conic**, or **conic section**, comes from the fact that the curves for which equations are sought correspond to sections cut by planes through a conic surface. A conic surface consists of two funnel-shaped sheets, called **nappes**, extending indefinitely in both directions from a point (vertex) where the nappes meet. If a plane cuts entirely across one nappe, the intersection of the plane and the nappe is called an *ellipse* (see Figure 19.8). A special case, the *circle,* is obtained if the plane is perpendicular to the axis of the cone. If the plane is parallel to one line on the surface of the cone, the intersection of the plane and the nappe is called a *parabola* (see Figure 19.9). If the plane intersects both nappes, the intersection is called a *hyperbola* (see Figure 19.10).

Ellipse

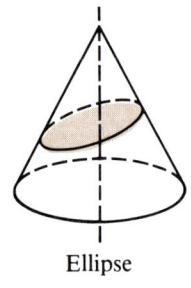

Figure 19.8

Parabola

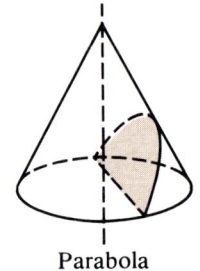

Figure 19.9

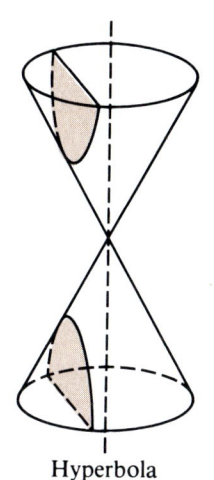

Certain special cases occur if the plane passes through the vertex (a point), through the axis (two intersecting lines), or through a single line on the surface.

The conics were well known to the ancient Greeks. Many of the geometric properties of conic sections, including the ones to be used in our later definitions, were studied in antiquity. Although Euclid's own contributions to that study have been lost, they appear to have been incorporated in the work of Apollonius of Perga (*circa* 230 B.C.), to whom the names of the curves have been attributed. Archimedes (287–212 B.C.) was the first to find the area under a parabolic arch.

Hyperbola

Figure 19.10

19.6 The Circle

Recall from geometry that a **circle** is the locus of all points equidistant from a given point, called the *center*. Let (h, k) in Figure 19.11 be the center and let

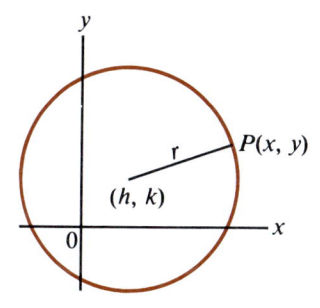

Figure 19.11

$P(x, y)$ be any point on the circle. If r is the radius, then from the distance formula (19.1) we get

$$\sqrt{(x - h)^2 + (y - k)^2} = r$$

Squaring both sides of this equation gives us the **standard form of the equation of the circle.**

Standard form of the equation of a circle:

$$(x - h)^2 + (y - k)^2 = r^2 \qquad (19.11)$$

Example 1 Find the equation of the circle with center at $(-2, 1)$ and radius 3.

Solution. Since $h = -2$, $k = 1$, and $r = 3$, we have

$$(x + 2)^2 + (y - 1)^2 = 9$$

For a circle of radius r centered at the origin, the equation becomes

$$x^2 + y^2 = r^2 \qquad (19.12)$$

If we multiply the terms of equation (19.11), we may conclude that any circle has the form shown below.

$$x^2 + y^2 + ax + by + c = 0 \qquad (19.13)$$

Example 2 Put $x^2 + y^2 + 6x - 4y + 9 = 0$ into standard form and find the center and radius.

Solution. To convert the equation to standard form, we complete the square on each quadratic expression. Transposing the 9 and rearranging terms, we get

$$(x^2 + 6x \quad) + (y^2 - 4y \quad) = -9$$

Since the coefficient of x is 6, we must add the square of one-half of 6 to both sides and, similarly, add to both sides the square of one-half of the coefficient -4. Thus

$$(x^2 + 6x + 9) + (y^2 - 4y + 4) = -9 + 9 + 4$$

or

$$(x + 3)^2 + (y - 2)^2 = 4 \qquad \text{factoring trinomials}$$

Hence the center is $(-3, 2)$ and the radius 2.

Example 3 Write $x^2 + y^2 - 2x + 4y + 5 = 0$ in standard form.

Solution. Before completing the square in each quadratic expression, we rearrange the terms thus:

$$(x^2 - 2x \quad) + (y^2 + 4y \quad) = -5$$

One-half of the coefficients of x and y are -1 and 2, respectively. We add the square of each to both sides; hence

$$(x^2 - 2x + 1) + (y^2 + 4y + 4) = -5 + 1 + 4$$

or

$$(x - 1)^2 + (y + 2)^2 = 0$$

Point circle   This equation is satisfied by the coordinates of only one real point $(1, -2)$. The locus is sometimes called a *point circle*.

The equation

$$(x - 1)^2 + (y + 2)^2 = -1$$

is not satisfied by any real point and, for lack of a better term, will be called an *imaginary circle*.

Exercises / Section 19.6

In Exercises 1–10, find the equation of each circle.

1. Center at the origin, radius 5
2. Center at the origin, radius 7
3. Center at the origin, passing through $(-6, 8)$
4. Center at the origin, passing through $(1, -4)$
5. Center at $(-2, 5)$, radius 1
6. Center at $(2, -3)$, radius $\sqrt{2}$
7. Center at $(-1, -4)$, passing through the origin
8. Center at $(3, 4)$, passing through $(5, 10)$
9. Ends of diameter at $(-2, -6)$ and $(1, 5)$
10. Center on the line $2y = 3x$, tangent to the y-axis, radius 2

In Exercises 11–16, write each equation in standard form and determine the center and radius.

11. $x^2 + y^2 - 2x - 2y - 2 = 0$
12. $4x^2 + 4y^2 - 4x - 8y - 11 = 0$
13. $x^2 + y^2 + 4x - 8y + 4 = 0$
14. $x^2 + y^2 + 2x + 6y + 3 = 0$
15. $4x^2 + 4y^2 + 12x + 16y + 5 = 0$
16. $36x^2 + 36y^2 - 144x - 120y + 219 = 0$

In Exercises 17–21, determine which equation describes a point circle or an imaginary circle.

17. $4x^2 + 4y^2 - 20x - 4y + 26 = 0$
18. $x^2 + y^2 + 4x - 2y + 7 = 0$
19. $x^2 + y^2 - 6x + 8y + 25 = 0$
20. $x^2 + y^2 - 2x + 4y + 5 = 0$
21. $x^2 + y^2 - 6x - 8y + 30 = 0$

19.7 The Parabola

Focus
Directrix
Vertex

Axis

Let F be a fixed point and l a fixed line. The locus of all points equidistant from F and l is called a **parabola** (Figure 19.12). F is called the **focus** of the parabola and l the **directrix**. If we draw a line from F perpendicular to l and intersecting l at Q, then the midpoint V of FQ is called the **vertex**. The vertex is the point on the parabola closest to the line l. The line through F and Q is called the **axis** of the parabola.

To get the equation of the parabola, we select a convenient coordinate system and apply the definition. If the vertex is at the origin, we can denote the focus by $(p, 0)$, as in Figure 19.13. If p is positive, then the focus lies to the right of the origin; if p is negative, it lies to the left. The directrix is the line $x = -p$ in each case. Now let $P(x, y)$ be any point on the parabola (Figure 19.13). It follows at once that the distance to the directrix is $x + p$.

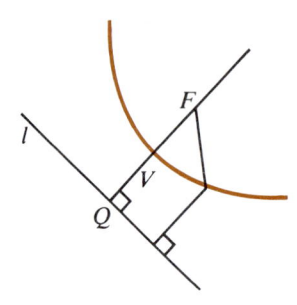

Figure 19.12

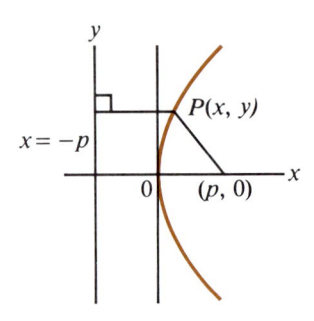

Figure 19.13

By the definition of the parabola,

$$\sqrt{(x - p)^2 + y^2} = x + p$$

Squaring both sides, we get

$$(x - p)^2 + y^2 = (x + p)^2 \qquad (\sqrt{a})^2 = a$$

or

$$x^2 - 2px + p^2 + y^2 = x^2 + 2px + p^2 \qquad (a \pm b)^2 = a^2 \pm 2ab + b^2$$

Combining like terms, the last expression reduces to

$$y^2 = 4px \qquad (19.14)$$

Conversely, any point whose coordinates satisfy this equation lies on the parabola. We owe the simplicity of this equation to the manner in which the coordinate system was chosen. It will be seen later that the equation can be modified to describe a parabola whose vertex is not at the origin.

If the focus is at $(0, p)$ and the directrix is the line $y = -p$, then we obtain the equation below.

$$x^2 = 4py \qquad (19.15)$$

These two cases are summarized next.

> **Equation of a parabola**: The equation of a parabola with the vertex at the origin is
>
> $\qquad y^2 = 4px \quad$ (axis horizontal)
>
> or
>
> $\qquad x^2 = 4py \quad$ (axis vertical)

Example 1 Find the focus and directrix of the parabola $y^2 = -9x$ and sketch.

Solution. Writing the equation in the form

$$y^2 = 4\left(-\frac{9}{4}\right)x \qquad y^2 = 4px$$

we see that the focus is at $\left(-\frac{9}{4}, 0\right)$ and the directrix is $x = \frac{9}{4}$ (Figure 19.14).

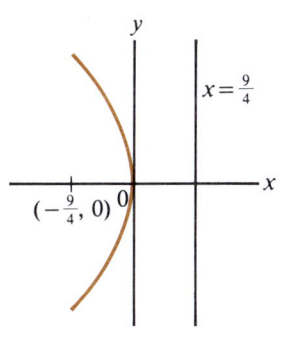

Figure 19.14

Example 2 Find the equation of the parabola with focus at $(0, 3)$ and directrix $y = -3$.

Solution. Since $p = 3$, $x^2 = 12y$ by (19.15).

Applications of the Parabola

1. The cable of a suspension bridge whose weight is uniformly distributed over the entire length of the bridge is in the form of a parabola (Figure 19.15).

Figure 19.15

2. The path of a projectile is a parabola if air resistance is neglected (Figure 19.16). (If the object is heavy and the initial velocity low, air resistance is negligible.)

Figure 19.16

3. If a parabola is rotated about its axis, it forms a parabolic surface. If a reflecting surface is parabolic with a light source placed at the focus, the rays are reflected parallel to the axis. Conversely, light rays coming in parallel to the axis are reflected so they pass through the focus. These principles are employed in the construction of searchlights and reflecting telescopes (Figure 19.17).

4. Another parabolic surface is obtained if a cylindrical vessel is partly filled with water and rotated about the axis of the cylinder: A plane through the axis will cut the surface of the water in a parabola.

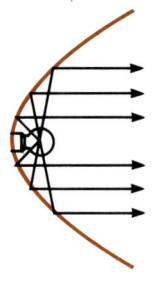

Figure 19.17

5. Steel bridges are frequently built with parabolic arches (Figure 19.18).
6. The relationship between the period of a pendulum and its length can be described in the form of a parabolic equation.

Figure 19.18

Example 3 The entrance to a formal garden is spanned by a parabolic arch 10 ft high and 10 ft across at its base. How wide should the walk through the center of the arch be to provide a minimum clearance of 7 ft above the walk?

Solution. Let the vertex of the arch be at the origin (Figure 19.19). The equation of the parabola has the form $x^2 = 4py$. Since the arch is 10 ft high,

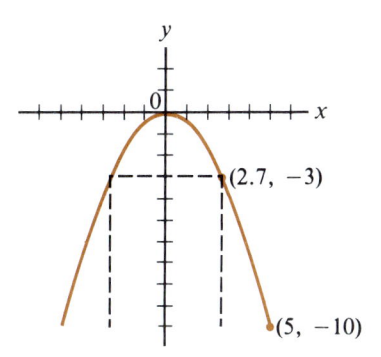

Figure 19.19

we know that the point $(5, -10)$ lies on the curve. Consequently, the coordinates of this point must satisfy the above equation. We begin by substituting:

$$5^2 = 4p(-10) \qquad \text{since } x = 5 \text{ and } y = -10$$

Solving for p, we get $p = -\frac{5}{8}$, so that the equation becomes

$$x^2 = -\frac{5}{2} y \qquad \text{since } x^2 = 4py$$

Now, a clearance of 7 ft corresponds to a point for which $y = -3$ (see Figure 19.19); its x-coordinate will tell us the width of the walk. Thus

$$x^2 = -\frac{5}{2}(-3) = \frac{15}{2} \quad \text{since } y = -3$$

and $x = 2.7$. So the walk must be $2(2.7) = 5.4$ ft wide.

Example 4 An amateur astronomer can construct a reflecting telescope with a parabolic mirror at one end of a hollow tube. The image is formed at the other end, where it may be viewed through an eyepiece. If the mirror has a radius of 20.0 cm and a depth of 3.33 mm in the center, find the position of the image.

Solution. We may assume that the parabolic cross section has the form

$$y^2 = 4px$$

We need to find the focus of this parabola. Since 3.33 mm = 0.333 cm, one point in the first quadrant is (0.333, 20.0). The coordinates of this point satisfy the equation; hence

$$(20.0)^2 = 4p \cdot 0.333 \quad y = 20.0, \, x = 0.333$$

and

$$p = \frac{(20.0)^2}{4 \cdot 0.333} = 300 \text{ cm}$$

Therefore, the image is 3 m from the center of the mirror.

Exercises / Section 19.7

In Exercises 1–10, write the equations of the parabolas.

1. Vertex at the origin, focus at (3, 0)
2. Vertex at the origin, focus at (−3, 0)
3. Vertex at the origin, focus at (0, −5)
4. Vertex at the origin, focus at (0, 4)
5. Directrix $x - 2 = 0$, vertex at the origin
6. Directrix $x + 2 = 0$, vertex at the origin
7. Directrix $y - 3 = 0$, focus at (0, −3)
8. Directrix $y = 4$, focus at (0, −4)
9. Directrix $y = -2$, focus at (0, 2)
10. Vertex at the origin, axis along y-axis and passing through (−1, 1)

In Exercises 11–26, find the coordinates of the focus and the equation of the directrix in each exercise and sketch.

11. $x^2 = 8y$
12. $x^2 = 16y$
13. $x^2 = -12y$
14. $x^2 = -16y$
15. $y^2 = 16x$
16. $y^2 = 8x$

17. $y^2 = -4x$ **18.** $y^2 = -12x$ **19.** $x^2 - 4y = 0$
20. $x^2 + 8y = 0$ **21.** $y^2 = 9x$ **22.** $y^2 = 10x$
23. $y^2 = -x$ **24.** $2x^2 - 3y = 0$ **25.** $3y^2 + 2x = 0$
26. $y^2 = 2px$

27. The chord of a parabola that passes through the focus and is parallel to the directrix is called the *focal chord*. Find the equation of the circle having for a diameter the focal chord of the parabola $x^2 = 12y$.

28. Derive equation (19.15).

29. Use the definition of the parabola to find the equation of the parabola whose focus is (4, 1) and whose directrix is the y-axis.

30. Repeat Exercise 29 for the parabola having a focus at (4, 7) and directrix $y + 1 = 0$.

31. Find the equations of the two parabolas that have vertices at the origin and that pass through $(-2, -4)$.

32. Repeat Exercise 31 for the point $(3, -5)$.

33. The support cables of a suspension bridge hang between towers in the form of a parabolic curve. The towers are 90.0 m high and 200 m apart. The roadway is suspended on other cables hung from the support cables. The shortest distance from the road surface to the supporting cable is 20.0 m. Determine the length of the vertical suspension cable that is 30.0 m from the center of the bridge. (Choose for the origin the lowest point on the supporting cable.)

34. The headlight on a car is 20.0 cm in diameter, has a parabolic surface, and is 12.0 cm deep. Where should the light bulb be placed so that the reflected rays are parallel?

35. Determine the distance across the base of a parabolic arch that measures 10.0 m at its highest point if the road through the center of the arch is 16.0 m wide and has a minimum clearance of 5.00 m.

36. A stream is 60.0 m wide and is spanned by a parabolic arch. Boats pass through the 40.0-m-wide channel in the center of the stream. How high must the arch be so that the minimum clearance is 10.0 m?

37. An entrance to a castle is in the form of a parabolic arch 6 m across at the base and 3 m high in the center. What is the length of a beam across the entrance, parallel to the base, and 2 m above the base?

38. The period T of a pendulum is directly proportional to the square root of its length L, provided that the arc of the swing is small (less than 15°). This can be described by $L = kT^2$. If $T = 6.0$ sec when $L = 9.0$ m, find the relationship.

19.8 The Ellipse

Foci

An **ellipse** is the locus of points such that the sum of the distances from two fixed points is constant. Suppose that the two fixed points, called the **foci** of the ellipse, are at $(c, 0)$ and $(-c, 0)$, respectively (Figure 19.20). The distance formula (19.1) can now be used to obtain the equation. A particularly simple form will result if we choose $2a$ for the constant, since the points $(a, 0)$ and $(-a, 0)$ will then lie on the ellipse. From the definition,

$$\sqrt{(x-c)^2 + y^2} + \sqrt{(x+c)^2 + y^2} = 2a$$

Conversely, any point whose coordinates satisfy this equation lies on the ellipse. The equation can be simplified by transposing one of the radicals and

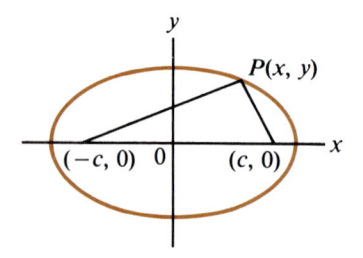

Figure 19.20

squaring both sides:

$$(\sqrt{(x-c)^2 + y^2})^2 = (2a - \sqrt{(x+c)^2 + y^2})^2$$

or

$$(x-c)^2 + y^2 = 4a^2 - 4a\sqrt{(x+c)^2 + y^2} + (x+c)^2 + y^2$$

Simplifying and isolating the radical, we have

$$a^2 + cx = a\sqrt{(x+c)^2 + y^2}$$

Squaring both sides again, we get

$$a^4 + 2a^2cx + c^2x^2 = a^2(x^2 + 2cx + c^2 + y^2)$$

or

$$a^4 + c^2x^2 = a^2x^2 + a^2c^2 + a^2y^2$$

We now collect all x and y terms on one side of the equation and factor the resulting expressions. Hence

$$a^2x^2 - c^2x^2 + a^2y^2 = a^4 - a^2c^2$$

or

$$(a^2 - c^2)x^2 + a^2y^2 = a^2(a^2 - c^2) \qquad \text{common factors}$$

After dividing both sides by $a^2(a^2 - c^2)$, the equation reduces to

$$\frac{x^2}{a^2} + \frac{y^2}{a^2 - c^2} = 1 \qquad \text{dividing by } a^2(a^2 - c^2)$$

A final simplification of the form can be obtained by letting

$$b^2 = a^2 - c^2 \qquad (19.16)$$

The equation now becomes

$$\frac{x^2}{a^2} + \frac{y^2}{b^2} = 1 \qquad (19.17)$$

As a result, the intercepts are simply $x = \pm a$ and $y = \pm b$, while the curve itself is symmetric with respect to both axes.

Example 1 Sketch the ellipse

$$\frac{x^2}{16} + \frac{y^2}{4} = 1$$

and find the foci.

Solution. From equation (19.17) we obtain $a = 4$ and $b = 2$, and from (19.16) we get

$$c^2 = a^2 - b^2 = 12 \quad \text{and} \quad c = \pm 2\sqrt{3}$$

Hence the foci are at $(\pm 2\sqrt{3}, 0)$. (See Figure 19.21.)

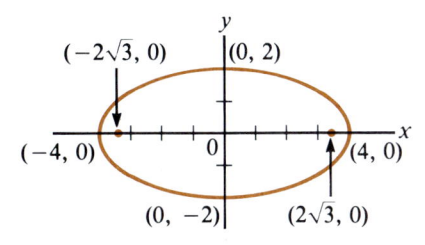

Figure 19.21

If the foci are at $(0, \pm c)$, we obtain the equation below.

$$\frac{x^2}{b^2} + \frac{y^2}{a^2} = 1 \tag{19.18}$$

(See Figure 19.22 and Exercise 30 at the end of the section.) Since $a > b$, one can easily distinguish between the forms in equations (19.17) and (19.18) by examining the intercepts. In the last equation, for example, we can tell at a glance that the numerically larger intercept lies on the y-axis.

The symmetric properties of the ellipse suggest some additional terminology: The line segment through the foci extending between $(a, 0)$ and $(-a, 0)$ or between $(0, a)$ and $(0, -a)$ is called the **major axis**, and the length a is called the **semimajor axis** (Figure 19.23). The segment of the other line of symmetry between $(b, 0)$ and $(-b, 0)$ or between $(0, b)$ and $(0, -b)$ is called the **minor axis** and the length b is called the **semiminor axis** (Figure 19.24). The intersection of the two axes is called the **center** of the ellipse. The ends of the major axis are called the **vertices** (Figure 19.23). These ideas are summarized next.

Major axis

Minor axis
Center
Vertices

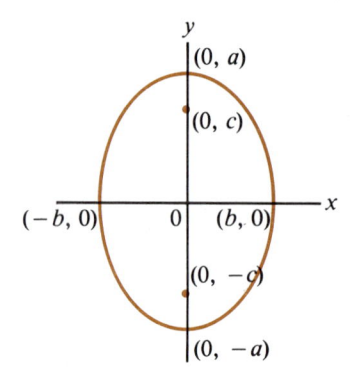

Figure 19.22

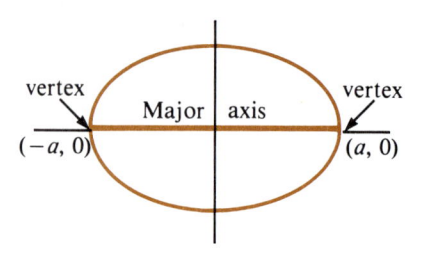

Figure 19.23

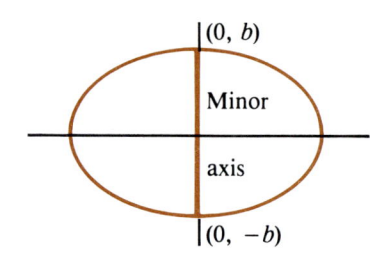

Figure 19.24

Equation of an ellipse: The equation of an ellipse with its center at the origin is

$$\frac{x^2}{a^2} + \frac{y^2}{b^2} = 1 \quad \text{(major axis horizontal)}$$

or

$$\frac{x^2}{b^2} + \frac{y^2}{a^2} = 1 \quad \text{(major axis vertical)}$$

and where

$$b^2 = a^2 - c^2$$

Example 2 Find the equation of the ellipse with center at the origin, foci at $(0, \pm 3)$, and vertices at $(0, \pm 4)$.

Solution. From equation (19.16),

$$b^2 = a^2 - c^2 = 4^2 - 3^2 = 7 \qquad a = 4, c = 3$$

19.8 THE ELLIPSE

so the equation is by (19.18)

$$\frac{x^2}{7} + \frac{y^2}{16} = 1 \qquad \text{since } b^2 = 7 \text{ and } a^2 = 16$$

Applications of the Ellipse

1. The paths of the planets and comets as they move about the sun are ellipses with the sun at one of the foci (Figure 19.25). Similarly, artificial satellites move about the earth in elliptical orbits with the center of the earth at one of the foci.
2. If an ellipse is rotated about its major axis, we obtain a solid called an *ellipsoid* or *prolate spheroid*. Within an ellipsoid enclosure or chamber, if a sound wave originates at one focus, then the reflected waves converge at the other focus. A room whose ceiling is a semiellipsoid is called a "whispering gallery." A person standing at one focus can hear even a slight noise made at the other focus, while a person standing in between does not hear anything.

    (A famous example of a whispering gallery is the Statuary Hall in the U.S. Capitol in Washington. Guides will position visitors in the proper places to hear faint sounds from another part of the hall. The hall was originally constructed for use as the House of Representatives. However, congressmen found the acoustics of the hall distressing, since confidential conferences in one place could easily be overheard at the opposite end of the hall.)
3. Arches of stone bridges are frequently semiellipses (Figure 19.26).
4. Elliptical gears, which revolve on a shaft turning a focus, are used in some power punches since they yield powerful movements with quick returns.

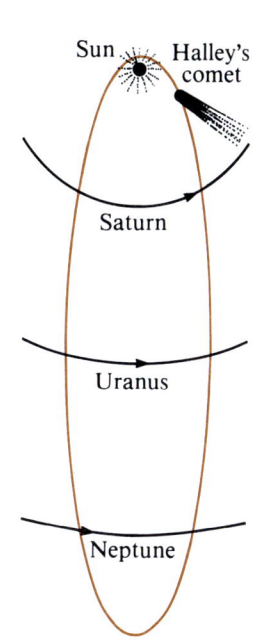

Figure 19.25

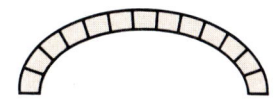

Figure 19.26

Example 3 A certain whispering gallery has a ceiling with elliptical cross sections. (See (2) above.) If the room is 30 ft long and 12 ft high in the center, where should two visitors place themselves to get the whispering effect?

Solution. If the center of the ellipse is placed at the origin, we get $a = 15$ and $b = 12$. Thus

$$c^2 = a^2 - b^2 = 225 - 144 = \mathbf{81}$$

and $c = \mathbf{9}$. So the visitors must place themselves on the two foci, each 9 ft from the center.

Exercises / Section 19.8

In Exercises 1–17, find the vertices, foci, and semiminor axis of each ellipse and sketch.

1. $\dfrac{x^2}{25} + \dfrac{y^2}{16} = 1$ **2.** $\dfrac{x^2}{16} + \dfrac{y^2}{9} = 1$ **3.** $\dfrac{x^2}{9} + \dfrac{y^2}{4} = 1$ **4.** $\dfrac{x^2}{4} + \dfrac{y^2}{9} = 1$

5. $\dfrac{x^2}{16} + y^2 = 1$ 6. $\dfrac{x^2}{2} + \dfrac{y^2}{4} = 1$ 7. $16x^2 + 9y^2 = 144$ 8. $x^2 + 2y^2 = 4$
9. $5x^2 + 2y^2 = 20$ 10. $5x^2 + 9y^2 = 45$ 11. $5x^2 + y^2 = 5$ 12. $x^2 + 4y^2 = 4$
13. $x^2 + 2y^2 = 6$ 14. $9x^2 + 2y^2 = 18$ 15. $15x^2 + 7y^2 = 105$ 16. $9x^2 + y^2 = 27$
17. $2x^2 + 5y^2 = 50$

In Exercises 18–28, find the equation of each ellipse.

18. Foci at $(\pm 1, 0)$, vertices at $(\pm 2, 0)$
19. Foci at $(\pm 3, 0)$, vertices at $(\pm 4, 0)$
20. Foci at $(0, \pm 2)$, major axis 6
21. Foci at $(0, \pm 2)$, major axis 8
22. Foci at $(0, \pm 2)$, minor axis 4
23. Foci at $(0, \pm 3)$, minor axis 6
24. Minor axis 5, vertices at $(\pm 7, 0)$
25. Center at the origin, one vertex at $(0, 8)$, one focus at $(0, -5)$
26. Center at the origin, one focus at $(-3, 0)$, semiminor axis 4
27. Foci at $(\pm 2\sqrt{3}, 0)$, minor axis 4
28. Foci at $(\pm \sqrt{5}, 0)$, vertices at $(\pm \sqrt{7}, 0)$
29. Find the locus of points such that the sum of the distances from $(\pm 6, 0)$ is 16.
30. Derive equation (19.18).
31. The shape of an ellipse depends on the values of c and a. The fraction $e = c/a$ is called the *eccentricity* of the ellipse. Find the equation of the ellipse with vertices at $(\pm 4, 0)$ and $e = \tfrac{1}{2}$.
32. Find the equation of the ellipse centered at the origin, having minor axis 6, and passing through the point $(1, 4)$.
33. Find the locus of points for which the distance from $(0, 0)$ is twice the distance from $(3, 0)$. What is the locus?
34. The area of an ellipse is given by $A = \pi ab$. Find the area of the floor of the whispering gallery in Example 3.
35. Find the equation of the ellipse to be inscribed in a rectangle 4 m long and 3 m wide if the center of the rectangle is placed at the origin with the long side horizontal.
36. A gateway is in the form of a semiellipse 10 ft across at the base and 15 ft high in the center. What is the length of the horizontal beam across the gateway 10 ft above the floor?
37. A road 8.0 m wide is spanned by a semielliptical arch 12 m across at its base. If the minimum clearance over the road is 4.0 m, what is the height of the arch?
38. Describe a mechanical method for drawing an ellipse inscribed in a rectangle. Refer to Exercise 35.
39. A satellite is in elliptical orbit about the earth. If the maximum altitude of the satellite is 120 mi and the minimum 80 mi, find the equation of the orbit. (The center of the earth is at a focus and the radius of the earth is approximately 4,000 mi.)
40. The earth moves about the sun in an elliptical orbit with the center of the sun at one of the foci. The minimum distance to the center of the sun is 9.14×10^7 mi and the maximum distance 9.46×10^7 mi. Show that the eccentricity is only about $\tfrac{1}{60}$, that is, the path is nearly circular. Refer to Exercise 31.

19.9 The Hyperbola

Starting with two points, $(c, 0)$ and $(-c, 0)$, let us consider the locus of points for which the difference of the distances from the two fixed points is numerically equal to $2a$. The locus is called a **hyperbola** (Figure 19.27).

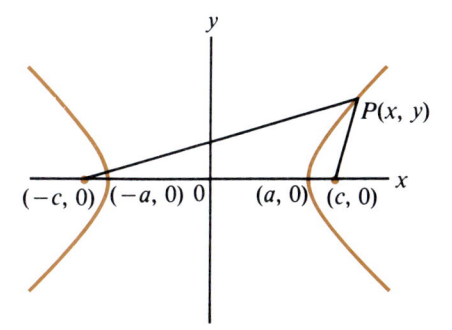

Figure 19.27

The derivation of the equation is similar to that of the ellipse. By the distance formula (19.1):

$$\sqrt{(x - c)^2 + y^2} - \sqrt{(x + c)^2 + y^2} = \pm 2a \tag{19.19}$$

which reduces to

$$\frac{x^2}{a^2} - \frac{y^2}{c^2 - a^2} = 1 \tag{19.20}$$

(The simplification will be left as an exercise.) This form suggests the substitution

$$b^2 = c^2 - a^2 \tag{19.21}$$

Hence

$$\frac{x^2}{a^2} - \frac{y^2}{b^2} = 1 \tag{19.22}$$

Conversely, any point whose coordinates satisfy this equation lies on the hyperbola. The x-intercepts of this curve are seen to be $(\pm a, 0)$, but the y-intercepts are lacking. Therefore, there appears to be no obvious interpretation of the constant b—yet a useful interpretation can be given. To see why, let us solve equation (19.22) for y in terms of x. Thus

$$\frac{y^2}{b^2} = \frac{x^2}{a^2} - 1$$

and

$$y = \pm b \sqrt{\frac{x^2}{a^2} - 1} \tag{19.23}$$

Now note that for very large values of x, the 1 in equation (19.23) becomes negligible. So y gets closer and closer to $y = \pm b\sqrt{x^2/a^2} = \pm b(x/a)$. In other words,

$$y = \pm \frac{b}{a} x \tag{19.24}$$

Asymptotes These two lines are the **asymptotes** of the hyperbola (Figure 19.28). (An *asymptote* is a line approached by the graph of an equation.)

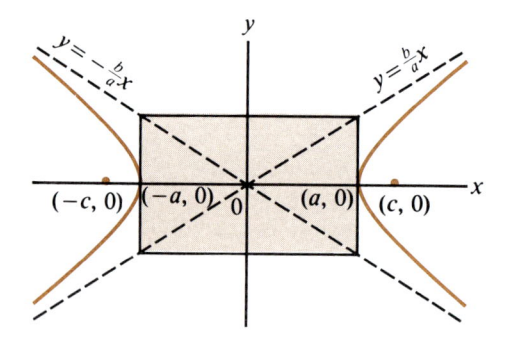

Figure 19.28

Vertices
Transverse and conjugate axes

As in the case of the ellipse, the points $(\pm c, 0)$ are called the **foci** and $(\pm a, 0)$ the **vertices** of the hyperbola. The line segment of length $2a$ joining the vertices is called the **transverse axis** of the hyperbola, and the line segment joining $(0, b)$ and $(0, -b)$ is called the **conjugate axis**. Finally, the point of intersection of the two axes is the **center**.

If the foci lie along the y-axis at $(0, \pm c)$, we obtain the form

$$\frac{y^2}{a^2} - \frac{x^2}{b^2} = 1 \tag{19.25}$$

Caution. Even though the equations of the hyperbola resemble those of the ellipse, some important differences should be noted. It is clear from the

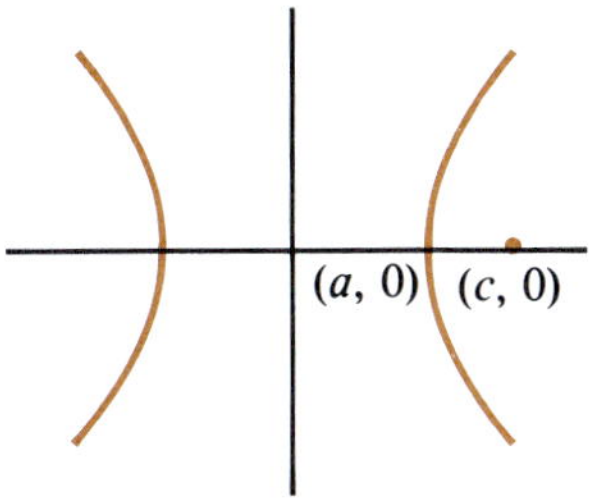

Hyperbola: $c > a$

Figure 19.29

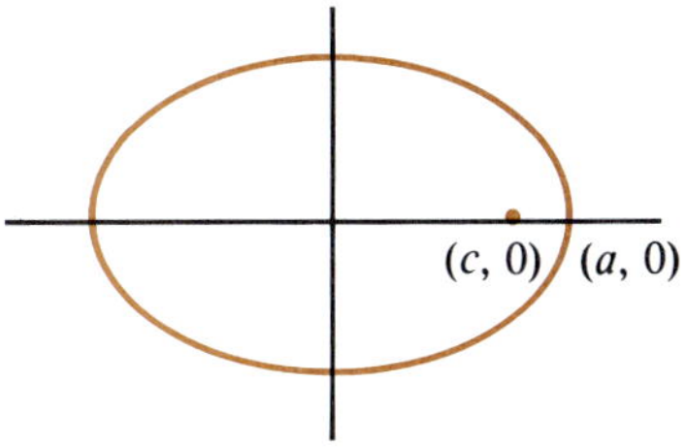

Ellipse: $c < a$

Figure 19.30

graph of the hyperbola, as well as from the formula $b^2 = c^2 - a^2$, that $c > a$ (Figure 19.29)—just the opposite of the case of the ellipse (Figure 19.30). Moreover, a is not necessarily larger than b in the hyperbola; rather, the constant a must be identified from the form of the equation. These ideas are summarized next.

Equation of a hyperbola: The equation of a hyperbola with its center at the origin is

$$\frac{x^2}{a^2} - \frac{y^2}{b^2} = 1 \quad \text{(transverse axis horizontal)}$$

or

$$\frac{y^2}{a^2} - \frac{x^2}{b^2} = 1 \quad \text{(transverse axis vertical)}$$

and where

$$b^2 = c^2 - a^2$$

Example 1 Sketch the hyperbola

$$\frac{x^2}{4} - \frac{y^2}{9} = 1$$

and find the foci, axes, and asymptotes.

Solution. From equation (19.22) we see that $a = 2$ and $b = 3$. (Note that $a < b$ in this case.) These numbers can be used to construct the asymptotes, which should always be drawn first to facilitate the sketching. Plot $(\pm 2, 0)$ and $(0, \pm 3)$ and draw the rectangle determined by these points. (See Figure 19.31.) A little reflection shows that the lines through the vertices of this rectangle are indeed the asymptotes. From $9 = c^2 - 4$, it now follows that the foci are at $(\pm\sqrt{13}, 0)$. The dotted rectangle in Figure 19.31 is called the **auxiliary rectangle**.

Auxiliary rectangle

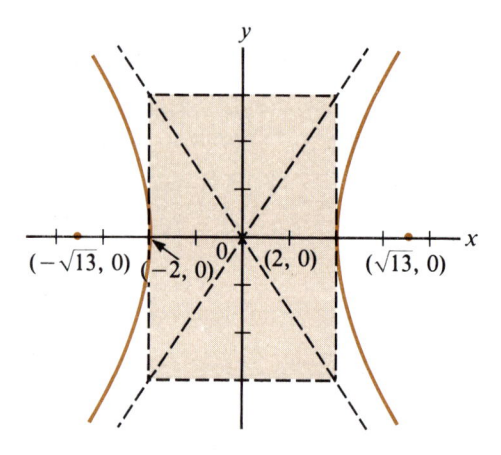

604 CHAPTER 19 INTRODUCTION TO ANALYTIC GEOMETRY

Figure 19.31

Example 2 Find the equation of the hyperbola with asymptotes $3y = \pm 2x$, conjugate axis 8, and foci on the x-axis.

Solution. Since $y = \pm \frac{2}{3}x$, $b/a = \frac{2}{3}$ by equation (19.24). Since the length of the conjugate axis is 8, $b = 4$, and we can solve the last equation for a to obtain $a = 6$. Thus by equation (19.22)

$$\frac{x^2}{36} - \frac{y^2}{16} = 1 \qquad \frac{x^2}{a^2} - \frac{y^2}{b^2} = 1$$

Example 3 Sketch the hyperbola $3y^2 - 2x^2 = 12$, including the asymptotes, and find the foci.

Solution. The equation can be written in the form

$$\frac{y^2}{4} - \frac{x^2}{6} = 1 \qquad \text{dividing by 12}$$

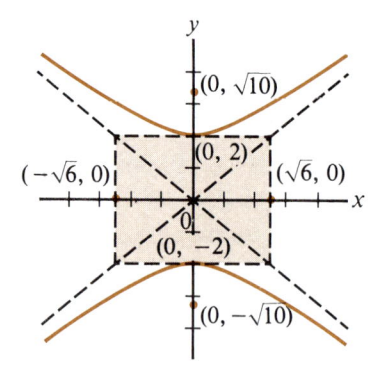

Figure 19.32

Therefore, $a = 2$ and $b = \sqrt{6}$ by equation (19.25), and these values may be used to draw the auxiliary rectangle in Figure 19.32. The foci are found to be $(0, \pm\sqrt{10})$.

Hyperbolas are used in long-range navigation as part of the LORAN system of navigation. A transmitter is located at each focus and radio signals are sent to the navigator simultaneously from each station. The difference in time at which the signals are received enables the navigator to determine his position. To be able to do so readily, however, the navigator uses charts on which hyperbolas are already plotted, as compiled from these differences in time. This way the navigator can find the hyperbola on which he is located. From a second pair of stations, another curve of position is then found, and the navigator then determines his location by noting the point of intersection of the two hyperbolas.

Exercises / Section 19.9

In Exercises 1–12, find the vertices and foci of each hyperbola, draw each auxiliary rectangle and asymptotes, and sketch the curves.

1. $\dfrac{x^2}{16} - \dfrac{y^2}{9} = 1$
2. $\dfrac{x^2}{9} - \dfrac{y^2}{4} = 1$
3. $\dfrac{x^2}{9} - \dfrac{y^2}{16} = 1$
4. $\dfrac{x^2}{16} - \dfrac{y^2}{4} = 1$
5. $\dfrac{y^2}{4} - \dfrac{x^2}{4} = 1$
6. $\dfrac{y^2}{4} - \dfrac{x^2}{8} = 1$
7. $x^2 - \dfrac{y^2}{5} = 1$
8. $9y^2 - 2x^2 = 18$
9. $2y^2 - 3x^2 = 24$
10. $x^2 - y^2 = 6$
11. $3y^2 - 2x^2 = 6$
12. $11x^2 - 7y^2 = 77$

In Exercises 13–20, determine the equation of each hyperbola.

13. Transverse axis 6, conjugate axis 4, foci on x-axis, center at the origin
14. Transverse axis 6, foci at $(\pm 4, 0)$
15. Conjugate axis 8, foci at $(0, \pm 5)$
16. Vertices at $(0, \pm 5)$, conjugate axis 10
17. Vertices at $(\pm 3, 0)$, foci at $(\pm 6, 0)$
18. Vertices at $(\pm 2, 0)$, foci at $(\pm 4, 0)$
19. Asymptotes $y = \pm 2x$, vertices at $(\pm 1, 0)$
20. Asymptotes $y = \pm\tfrac{1}{2}x$, conjugate axis 3, transverse axis horizontal
21. Find the equation of the locus of points such that the difference of the distances from $(0, \pm 5)$ is ± 6.
22. Reduce equation (19.19) to (19.20).
23. Find the equation of the locus of points such that the difference of the distances from $(1, 2)$ and $(-3, 2)$ is ± 2.
24. Derive the equation of the hyperbola whose asymptotes are $y = \pm\tfrac{4}{3}x$ and whose foci are at $(\pm 3, 0)$.
25. Write the equation of the hyperbola whose vertices are at $(0, \pm 12)$ and that passes through the point $(-1, 13)$.
26. Graph the equation $xy = 2$. It can be shown that the graph is a hyperbola.

27. Boyle's law states that, for an ideal gas, the product of the pressure and volume is always constant at a constant temperature. Mathematically, $pV = k$, where k is a constant. Suppose $V = 5L$ when $p = 2$ atmospheres for some ideal gas. Find k and graph the equation. Refer to Exercise 26.

19.10 Translation of Axes

In this section we shall use the **translation of axes** to obtain more general forms of the equations of the conics. A translation of axes gives a new set of axes parallel to, and oriented the same way as, the original axes.

If the origin, O, is translated to the point $O'(h, k)$, then each point has two sets of coordinates. Suppose (x, y) and (x', y') are the coordinates of a point P with respect to the old and new axes, respectively (Figure 19.33). We see from the diagram that

$$x = OB = OA + AB = h + x'$$

and

$$y = BP = BC + CP = k + y'$$

Thus $x' = x - h$ and $y' = y - k$.

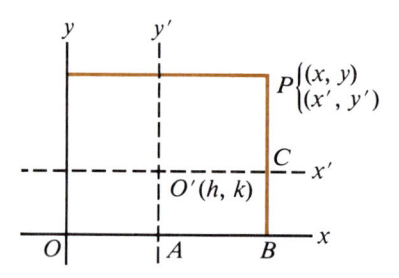

Figure 19.33

Translation formulas:

$$x' = x - h \quad \text{and} \quad y' = y - k$$

Example 1 Consider the following equation of a circle:

$$(x + 1)^2 + (y - 3)^2 = 4$$

If the origin is translated to $(-1, 3)$, we get

$$x' = x + 1 \quad \text{and} \quad y' = y - 3$$

and the equation becomes

$$x'^2 + y'^2 = 4$$

Both forms are familiar from Section 19.6.

Example 2 Consider the equation

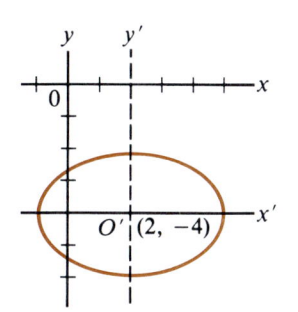

$$\frac{(x-2)^2}{9} + \frac{(y+4)^2}{4} = 1$$

Selecting the point $(2, -4)$ for O',

$$x' = x - 2 \quad \text{and} \quad y' = y + 4$$

and we get

$$\frac{x'^2}{9} + \frac{y'^2}{4} = 1$$

which we recognize to be the equation of an ellipse (Figure 19.34).

Figure 19.34

It follows from the above discussion that more general forms of the conics can be obtained. These are stated below.

Standard equations of the conics:

$$\left.\begin{array}{l}(y-k)^2 = 4p(x-h) \\ (x-h)^2 = 4p(y-k)\end{array}\right\} \quad \text{parabola; vertex at } (h, k)$$

$$\frac{(x-h)^2}{a^2} + \frac{(y-k)^2}{b^2} = 1 \quad \text{ellipse; center at } (h, k)$$

$$\frac{(x-h)^2}{a^2} - \frac{(y-k)^2}{b^2} = 1 \quad \text{hyperbola; center at } (h, k)$$

These forms are easily remembered, since we are merely replacing x by $x - h$ and y by $y - k$. Geometrically, we have translated the original figure in such a way that the new center or vertex is at the point (h, k). However, it is important to keep in mind that the translation formulas have nothing to do with conics and can be applied equally well to any graph: If h and k are positive, then *replacing x by $x - h$ in an equation will move the corresponding graph h units to the right and replacing y by $y - k$ will move the graph k units up.* Likewise, replacing x by $x + h$ will move the graph h units to the left, and similarly for y.

Example 3 Write the following equation in standard form and sketch the curve:

$$25x^2 - 9y^2 + 150x + 36y - 36 = 0$$

Solution. We proceed by rearranging the terms and completing the square as we did in Section 19.6; thus

$$25x^2 + 150x - 9y^2 + 36y = 36$$

At this point we need to factor out the coefficients of the squared terms, so that

$$25(x^2 + 6x \quad) - 9(y^2 - 4y \quad) = 36$$

We now complete the square inside the parentheses. Care must be taken when balancing the equation:

$$25(x^2 + 6x + \mathbf{9}) - 9(y^2 - 4y + \mathbf{4}) = 36 + 25 \cdot \mathbf{9} - 9 \cdot \mathbf{4}$$

or

$$25(x + 3)^2 - 9(y - 2)^2 = 225$$

This equation simplifies to

$$\frac{(x + 3)^2}{9} - \frac{(y - 2)^2}{25} = 1 \qquad \text{equation of a hyperbola}$$

which is a hyperbola centered at **(−3, 2)**. Since $a = 3$ and $b = 5$, the vertices are at (0, 2) and (−6, 2), while the ends of the conjugate axis are at (−3, 7) and (−3, −3). From the formula $b^2 = c^2 - a^2$, we get $c = \sqrt{34}$, so that the foci are at $(-3 \pm \sqrt{34}, 2)$. Using the values of a and b, we now draw the auxiliary rectangle centered at (−3, 2) and the two asymptotes (Figure 19.35).

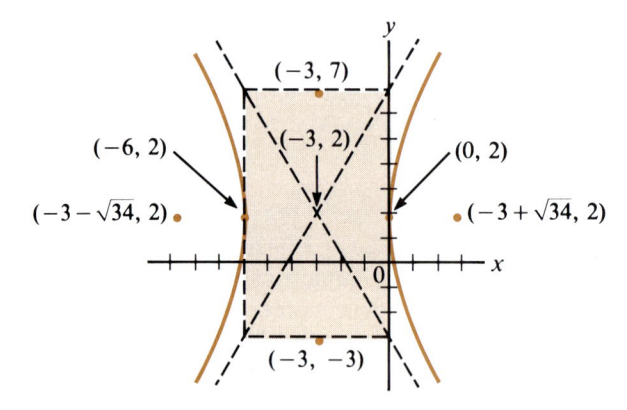

Figure 19.35

It follows from the above that any equation of the form

$$Ax^2 + By^2 + Cx + Dy + E = 0 \qquad (19.26)$$

represents a conic, since completing the square will convert the equation to one of the standard forms. The types of loci may be summarized as shown below.

1. If $A = B$, the locus is a circle.
2. If either $A = 0$ or $B = 0$, the locus is a parabola.
3. If $A \neq B$ and A and B have like signs, the locus is an ellipse.
4. If A and B have opposite signs, the locus is a hyperbola.

The following degenerate cases may also occur:

5. The locus may be a point. We encountered this case in Section 19.6, on circles, although the possibility exists for type 3 also. Imaginary loci may occur in types 1 and 3.
6. The equation may factor into two linear factors, in which case the locus is two straight lines. As an example of the last case, the equation $x^2 - 1 = 0$ can be written $x = \pm 1$, which represents two parallel lines. (See Exercises 27–29.)

Exercises / Section 19.10

In Exercises 1–16, identify the conic, find the center (or vertex), and sketch.

1. $(x - 1)^2 + (y - 2)^2 = 3$
2. $\dfrac{(x - 1)^2}{9} + \dfrac{(y - 2)^2}{5} = 1$
3. $(y + 3)^2 = 8(x - 2)$
4. $\dfrac{(x - 3)^2}{4} - \dfrac{y^2}{9} = 1$
5. $2x^2 - 3y^2 + 8x - 12y + 14 = 0$
6. $4x^2 - 4x - 48y + 193 = 0$
7. $16x^2 + 4y^2 + 64x - 12y + 57 = 0$
8. $y^2 - 12y - 5x + 41 = 0$
9. $x^2 + y^2 + 2x - 2y + 2 = 0$
10. $x^2 + 2y^2 - 6x + 4y + 1 = 0$
11. $2x^2 - 12y^2 + 60y - 63 = 0$
12. $2x^2 + 3y^2 - 8x - 18y + 35 = 0$
13. $64x^2 + 64y^2 - 16x - 96y - 27 = 0$
14. $4x^2 - 4x - 16y + 5 = 0$
15. $3x^2 + y^2 - 18x + 2y + 29 = 0$
16. $100x^2 - 180x - 100y + 81 = 0$

In Exercises 17–26, find the equations of the conics:

17. Parabola: vertex at $(-1, 2)$, focus at $(3, 2)$
18. Parabola: focus at $(2, 5)$, directrix $y + 1 = 0$

19. Ellipse: center at (−3, 0), one vertex at the origin, passing through (−3, −2)
20. Ellipse: vertices at (3, 1) and (3, −5), one focus at (3, 0)
21. Hyperbola: vertices at (1, 1) and (−7, 1), one focus at (3, 1)
22. Hyperbola: center at (−1, −2), one focus at (2, −2), conjugate axis 4
23. Ellipse: center at (2, 3), major axis 5 (horizontal), minor axis 2
24. Parabola: vertex at (5, −4), focus at (8, −4)
25. Hyperbola: center at (1, 0), a vertex at (3, 0), one asymptote $x - 2y = 1$
26. Parabola: vertex at (−1, 3), passing through the origin

In Exercises 27–29, factor the left side of each equation, identify the locus, and sketch.

27. $y^2 - 4 = 0$ 28. $x^2 - xy - 6y^2 = 0$ 29. $2x + xy - 3y - 6 = 0$

19.11 Polar Coordinates

So far we have consistently used the rectangular coordinate system. However, for certain types of curves, another coordinate system may be more convenient. Suppose we start with a point O, called the **pole,** and draw a ray with O as the endpoint. This ray is called the **polar axis.** Using this ray as the initial side, we may generate angles just as we do in trigonometry. Let θ be such an angle and P a point on the terminal side (Figure 19.36). If r is the distance from the pole, then we say that (r, θ) are the **polar coordinates** of P. Recall that in trigonometry we always assume that r is positive. In the polar coordinate system, however, we agree to plot a negative value for r by extending the terminal side of the angle in the opposite direction through the pole and marking off r units on the extended side.

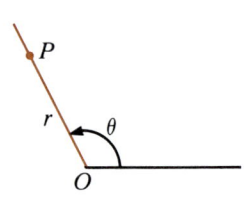

Figure 19.36

> **Polar coordinates:** Let P be a point in the plane, r the distance from P to a fixed point O (the **pole**), and θ the angle between the ray OP and a fixed ray (the **polar axis**). Then P is said to be represented by the **polar coordinates** (r, θ).

Example 1 Plot the points whose polar coordinates are (2, 30°), (3, −π/4), (4, 5π/6), (−3, 240°), (−1, 120°), (3, 180°), and (4, 0°).

Solution. To plot these points, we need only the polar axis. However, for convenience we keep the vertical axis as a line of reference and extend the polar axis to the left (Figure 19.37).

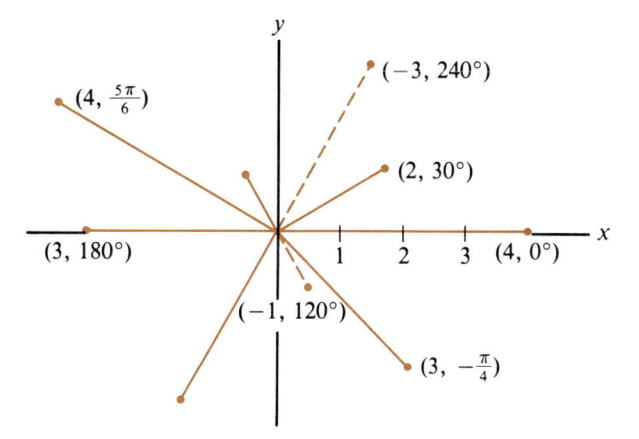

Figure 19.37

We can see from Example 1 that the pair of values (r, θ) determines a unique point. The converse is not true, however. The point $(2, 30°)$, for example, can also be represented by $(2, 390°)$, $(2, -330°)$, or $(-2, 210°)$; in fact, there are infinitely many possibilities.

The relationships between the rectangular and polar coordinate systems can readily be seen from the triangle in Figure 19.38.

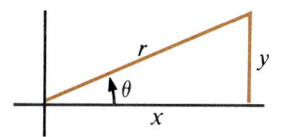

Figure 19.38

Rectangular to polar conversion:

$$x = r \cos \theta \qquad y = r \sin \theta \qquad (19.27)$$

Polar to rectangular conversion:

$$r^2 = x^2 + y^2 \qquad \tan \theta = \frac{y}{x} \qquad (19.28)$$

Example 2 Express $(-2, 2)$ in polar coordinates.

Solution. Since $r^2 = x^2 + y^2$, we obtain at once

$$r = \pm \sqrt{4 + 4} = \pm 2\sqrt{2}$$

and

$$\tan \theta = -1$$

Hence $\theta = 135° + n \cdot 360°$ or $315° + n \cdot 360°$. If we choose the positive root for r, then the coordinates are $(2\sqrt{2}, 135°)$ and, more generally, $(2\sqrt{2}, 135° + n \cdot 360°)$. For the negative root we have $(-2\sqrt{2}, 315° + n \cdot 360°)$.

Relationships (19.27) and (19.28) can also be used to convert equations from one system to the other.

Example 3 Write $y^2 = x$ in polar coordinates.

Solution. By direct substitution,

$$(r \sin \theta)^2 = r \cos \theta \qquad x = r \cos \theta, y = r \sin \theta$$
$$r^2 \sin^2 \theta - r \cos \theta = 0$$

and

$$r(r \sin^2 \theta - \cos \theta) = 0 \qquad \text{factoring } r$$

It then follows that $r = 0$ and

$$r \sin^2 \theta - \cos \theta = 0$$
$$r = \frac{\cos \theta}{\sin^2 \theta} = \frac{\cos \theta}{\sin \theta} \frac{1}{\sin \theta} \qquad \text{solving for } r$$

or

$$r = \cot \theta \csc \theta$$

Notice that $r = 0$ may be dropped, since it represents only the pole, which is included in the equation $r = \cot \theta \csc \theta$.

Example 4 Write the equation $r = 2 \cos 2\theta$ in rectangular form.

Solution. By the double-angle formula,

$$r = 2(\cos^2 \theta - \sin^2 \theta) \qquad \cos 2\theta = \cos^2 \theta - \sin^2 \theta$$

Since $\cos\theta = x/r$ and $\sin\theta = y/r$ by (19.27), we get

$$\pm\sqrt{x^2+y^2} = 2\left(\frac{x^2}{r^2} - \frac{y^2}{r^2}\right) \qquad r = \pm\sqrt{x^2+y^2}$$

$$\pm\sqrt{x^2+y^2} = 2\left(\frac{x^2}{x^2+y^2} - \frac{y^2}{x^2+y^2}\right) \qquad r^2 = x^2+y^2$$

Squaring both sides gives

$$x^2 + y^2 = 4\frac{(x^2-y^2)^2}{(x^2+y^2)^2}$$

or

$$(x^2+y^2)^3 = 4(x^2-y^2)^2 \qquad \text{multiplying by } (x^2+y^2)^2$$

A simple alternative is to multiply both sides of the equation $r = 2(\cos^2\theta - \sin^2\theta)$ by r^2 to get

$$r^3 = 2(r^2\cos^2\theta - r^2\sin^2\theta)$$

In this form, the equation converts directly to

$$\pm(x^2+y^2)^{3/2} = 2(x^2-y^2) \qquad r\cos\theta = x, r\sin\theta = y$$

or

$$(x^2+y^2)^3 = 4(x^2-y^2)^2 \qquad \text{squaring both sides}$$

Example 5 Convert the equation $r = 3\cot\theta$ to rectangular form.

Solution. In this example, inserting r is particularly convenient:

$$r = 3\cot\theta = 3\frac{\cos\theta}{\sin\theta}$$

or

$$r = 3\frac{r\cos\theta}{r\sin\theta}$$

This equation converts directly to

$$\pm\sqrt{x^2+y^2} = 3\frac{x}{y}$$

or

$$x^2 + y^2 = \frac{9x^2}{y^2}$$

Exercises / Section 19.11

1. Plot the following points: $(1, 60°)$, $(3, 3\pi/4)$, $(2, -50°)$, $(-2, 90°)$, $(-1, -\pi/4)$, $(-2, 270°)$, $(-4, 0°)$.

In Exercises 2–7, express each point in rectangular coordinates.

2. $\left(\sqrt{2}, \dfrac{\pi}{4}\right)$ **3.** $(2, 120°)$ **4.** $\left(4, -\dfrac{\pi}{6}\right)$

5. $\left(-6, \dfrac{\pi}{3}\right)$ **6.** $(1, 50°)$ **7.** $(-2, 170°)$

In Exercises 8–12, for each point in rectangular coordinates find a set of polar coordinates.

8. $(\sqrt{3}, 1)$ **9.** $(1, -1)$ **10.** $(-2, -2\sqrt{3})$
11. $(3, 4)$ **12.** $(-2, 5)$

In Exercises 13–20, express the given equations in polar coordinates.

13. $x = 2$ **14.** $y = 4$
15. $x^2 + y^2 = 2$ **16.** $x^2 + y^2 + 3x - 2y = 0$
17. $x^2 - y^2 = 1$ **18.** $y = 2x^2$
19. $2x^2 + 4y^2 = 1$ **20.** $y = x^3$

In Exercises 21–37, express the given equations in rectangular coordinates.

21. $r \sin \theta = 2$ **22.** $r = 3 \sec \theta$
23. $\theta = \dfrac{\pi}{4}$ **24.** $r = 5$
25. $r = 1 - \cos \theta$ **26.** $r = 3 \sin 2\theta$ (Recall that $\sin 2\theta = 2 \sin \theta \cos \theta$.)
27. $r^2 = \sin 2\theta$ **28.** $r = \tan \theta$
29. $r = \dfrac{2}{1 - \cos \theta}$ **30.** $r^2 = \cot \theta$
31. $r = \dfrac{4}{1 - \sin \theta}$ **32.** $r^2 \cos 2\theta = 4$
33. $r^2 = 3 \cos 2\theta$ **34.** $r = a \cos 3\theta$ (Use $\cos 3\theta = \cos (2\theta + \theta)$.)
35. $r = a \sin 3\theta$ **36.** $r = 4 \sec \theta \tan \theta$
37. $r = 1 - 2 \sin \theta$

19.12 Curves in Polar Coordinates

In this section we shall study curves in polar coordinates. Without prior experience we are essentially confined to the point-plotting method. We shall soon find out, however, that a number of polar curves can be classified according to type. Once the type is identified, drawing the curve becomes relatively easy.

19.12 CURVES IN POLAR COORDINATES

Example 1 Draw the graph of $r = 4(1 + \sin \theta)$.

Solution. First we make a table of values:

θ:	0°	30°	45°	60°	90°	120°	135°	150°	180°	225°	270°	315°
r:	4	6	6.8	7.5	8	7.5	6.8	6	4	1.2	0	1.2

Note that it is not necessary to use radian measure as long as we are just plotting points. As θ ranges from 0° to 90°, r ranges from 4 to 8; from 90° to 180° the values of r repeat in reverse order. Continuing from 180° to 270°, we see that $\sin \theta$ ranges from 0 to -1; hence r decreases from 4 to 0. Finally, as θ ranges from 270° to 360°, r increases from 0 to 4. Once we have a complete circuit, the values of r repeat; thus it is unnecessary to find additional points. The graph is shown in Figure 19.39.

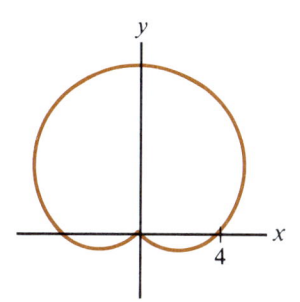

Figure 19.39

It is easy to check that the curve $r = 4(1 + \cos \theta)$ is the curve in Example 1 rotated by 90° in the clockwise direction.

Figure 19.39 is an example of a curve called a *cardioid* (meaning "heart-shaped").

Cardioid: A cardioid is a curve whose general equation is

$$r = a(1 \pm \sin \theta) \quad \text{or} \quad r = a(1 \pm \cos \theta) \qquad (19.29)$$

The coefficient a may be either positive or negative.

Example 2 Sketch the graph of $r = 2 + \sin \theta$.

Solution. Because of the similarity to Example 1, we locate only the four points analogous to the intercepts: $(2, 0°)$, $(3, 90°)$, $(2, 180°)$, and $(1, 270°)$. The graph is shown in Figure 19.40. Note that at $270°$, r is still positive, so that the curve does not pass through the pole.

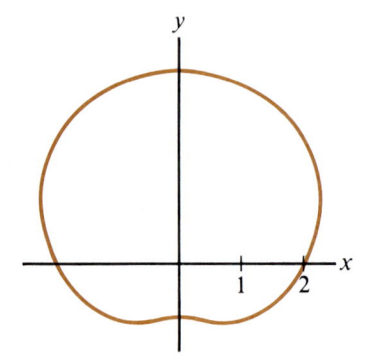

Figure 19.40

The curve in Example 2 is called a *limaçon of Pascal*.

Limaçon: A limaçon is a curve whose general equation is

$$r = a + b \sin \theta \quad \text{or} \quad r = a + b \cos \theta \tag{19.30}$$

If $|a| > |b|$ and $a > 0$, r is always positive (as in Example 2). If $|a| = |b|$, then $r = 0$ for one value of θ between $0°$ and $360°$, so that the curve is actually a cardioid. If $|a| < |b|$ and $a > 0$, r is negative for some values of θ, and we obtain a "limaçon with a loop." The three cases are shown in Figure 19.41.

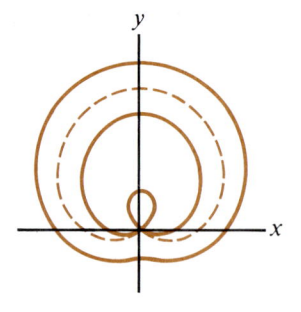

Figure 19.41

(If $a < 0$, the conclusion is the same in the sense that the limaçon has a loop if, and only if, $|a| < |b|$.)

Example 3 Sketch the limaçon $r = 1 - 2 \cos \theta$.

Solution. Since $|a| < |b|$, this limaçon has a loop. To find the values of θ tracing out the loop, we must first see where the loop starts and ends; that is, we need to find the values of θ for which the curve passes through the pole. To this end we let $r = 0$ to get

$$1 - 2 \cos \theta = 0 \quad \text{and} \quad \cos \theta = \frac{1}{2}$$

so that $\theta = \pm 60°$. If $-60° < \theta < 60°$, then $r < 0$. To sketch the rest of the graph, we find the four "intercepts": $(-1, 0°)$, $(1, 90°)$, $(3, 180°)$, and $(1, 270°)$. The graph is shown in Figure 19.42.

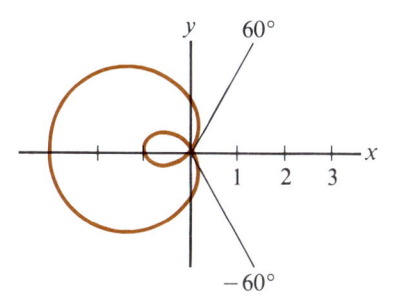

Figure 19.42

Example 4 Sketch the graph of $r = 2 \sin 3\theta$.

Solution. Instead of plotting points, we shall discuss the general behavior of this function: r is 0 whenever 3θ is coterminal with 0° or 180°. This will occur whenever $\theta = 0°$, 60°, 120°, 180°, 240°, and 300°. Numerically, r attains its maximum value of 2 whenever $\theta = 30°$, 90°, 150°, 210°, 270°, and 330°. We note especially that when θ ranges from 0° to 30°, 3θ ranges from 0° to 90°, so that $\sin 3\theta$ ranges from 0 to 1; at $\theta = 60°$, r is back to 0. As θ ranges from 60° to 120°, the values of r are negative. The graph is shown in Figure 19.43.

The curve in Figure 19.43 is an example of a *rose*.

Rose: A rose is a curve whose general equation is

$$r = a \sin n\theta \quad \text{or} \quad r = a \cos n\theta \tag{19.31}$$

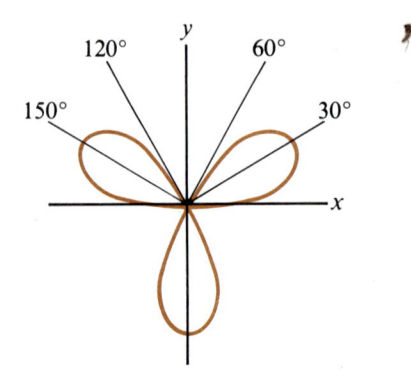

Figure 19.43

The coefficient *n* gives the number of "leaves": The rose has *n* leaves if *n* is odd and 2*n* leaves if *n* is even. For example, $r = 4 \cos 2\theta$ is a four-leaf rose (Figure 19.44).

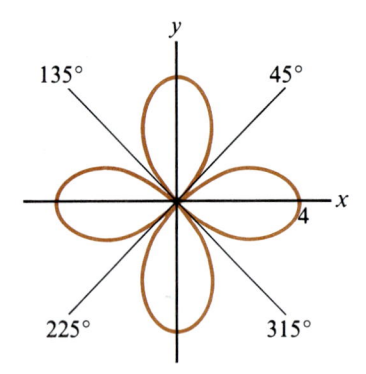

Figure 19.44

Another type of curve, called a *lemniscate of Bernoulli,* has the form given next.

Lemniscate: A lemniscate is a curve whose general equation is

$$r^2 = a \sin 2\theta \quad \text{or} \quad r^2 = a \cos 2\theta \tag{19.32}$$

Example 5 Sketch the lemniscate $r^2 = 4 \cos 2\theta$.

Solution. A brief table of values is

θ:	0°	10°	20°	30°	45°
r:	± 2	± 1.9	± 1.75	± 1.4	0

We get the same values for r if θ starts at 0° and goes to $-45°$. In the range $\theta = 45°$ to $\theta = 135°$, $\cos 2\theta < 0$, so that the values of r become imaginary. The graph is shown in Figure 19.45.

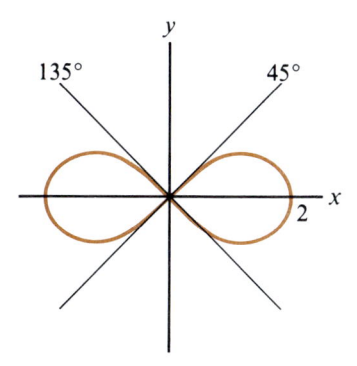

Figure 19.45

Certain special cases may also occur. The curve $r = a$ is a circle centered at the pole, while $\theta = b$ is a straight line through the pole. The curve $r = a\theta$ is called a *spiral of Archimedes*. Other spirals will be taken up in the exercises.

Example 6 Draw the spiral $r = 2\theta$, $\theta \geq 0$.

Solution. In this case the values of θ must be in radian measure. Since r is directly proportional to θ, the points recede farther and farther from the pole (Figure 19.46 on page 620).

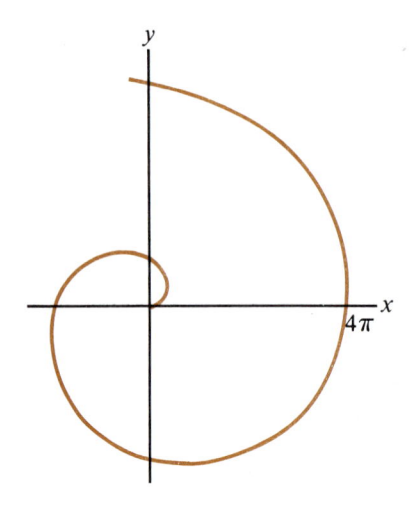

Figure 19.46

Exercises / Section 19.12

Sketch each of the following curves. Whenever possible, identify the locus before attempting the sketch.

1. $r = 2$
2. $\theta = \dfrac{\pi}{3}$
3. $r = 2 \cos \theta$
4. $r = 3 \sin \theta$
5. $r \cos \theta = 4$
6. $r = 2 \csc \theta$
7. $r = 2(1 + \cos \theta)$
8. $r = -4(1 + \sin \theta)$
9. $r = 2 + \cos \theta$
10. $r = 4 - \sin \theta$
11. $r = 3 - 2 \cos \theta$
12. $r = -1 + \cos \theta$
13. $r = 1 - 2 \sin \theta$
14. $r = 1 + 2 \cos \theta$
15. $r = -1 - 2 \sin \theta$
16. $r = 2 \cos 3\theta$
17. $r = 4 \sin 2\theta$
18. $r = 3 \sin 5\theta$
19. $r = \cos 3\theta$
20. $r^2 = 9 \cos 2\theta$
21. $r^2 = 4 \sin 2\theta$
22. $r = \sqrt{\sin \theta}$
23. $r = 3\sqrt{\sin 2\theta}$
24. $r = 1 + \cos 3\theta$
25. $r = \cos 3 \left(\theta - \dfrac{\pi}{6} \right)$
26. $r = 2\sqrt{\cos \theta}$
27. $r = 2 \tan \theta$
28. $r = \theta$ (spiral of Archimedes)
29. $r\theta = 2$ (hyperbolic spiral)
30. $r = e^{2\theta}$ (logarithmic spiral)
31. $r = \dfrac{3}{2 - \cos \theta}$ (ellipse)
32. $r = \dfrac{2}{1 - \cos \theta}$ (parabola)
33. $r = \dfrac{6}{2 + 3 \sin \theta}$ (hyperbola)

Review Exercises / Chapter 19

1. Show that (4, 2), (7, 4), and (3, 10) are the vertices of a right triangle.
2. Find the equation of the perpendicular bisector of the line segment joining $(-1, -2)$ and $(5, 4)$.

REVIEW EXERCISES / CHAPTER 19   **621**

3. Fahrenheit and Celsius temperatures are related by $C = \frac{5}{9}(F - 32)$. What temperature is the same in both Fahrenheit and Celsius?
4. What is the physical significance of the F-intercept in Exercise 3?
5. Show that $(-1, 5)$, $(3, 1)$, $(3, 9)$, and $(7, 5)$ are the vertices of a square.
6. Find the equation of the line through $(4, 1)$ and perpendicular to the line $x + 2y - 5 = 0$.
7. Find the equation of the line through $(-1, 5)$ and parallel to the line $3x + y = 3$.
8. Show that the following lines form the sides of a right triangle: $4x - y - 7 = 0$, $x + 4y - 1 = 0$, and $y + 2x = 0$.
9. Find the equation of the circle with center at $(1, -2)$ and passing through the origin.
10. Find the equation of the circle centered at $(2, 5)$ and tangent to the x-axis.
11. Find the center and radius of the circle $x^2 + y^2 + 2x + 2y = 0$.
12. Show that the circle $x^2 + y^2 - 10x - 8y + 16 = 0$ is tangent to the y-axis.

In Exercises 13–18, identify the conic, find the vertices and foci, and sketch.

13. $\dfrac{x^2}{9} + \dfrac{y^2}{16} = 1$ 14. $x^2 + 4y^2 = 1$ 15. $\dfrac{y^2}{4} - \dfrac{x^2}{7} = 1$

16. $\dfrac{x^2}{9} - \dfrac{y^2}{16} = 1$ 17. $y^2 = -3x$ 18. $x^2 = 9y$

In Exercises 19–23, identify the conic, find the center (or vertex), and sketch.

19. $y^2 + 6y + 4x + 1 = 0$ 20. $x^2 + y^2 - 8x + 10y - 4 = 0$
21. $16x^2 + 9y^2 - 64x + 18y = 71$ 22. $y + 12x^2 + 72x + 103 = 0$
23. $x^2 - y^2 - 4x + 8y - 21 = 0$

In Exercises 24–29, find the equations of the curves indicated.

24. Parabola: vertex at origin, directrix $x = -2$
25. Parabola: vertex at $(1, 3)$, directrix $y = 0$
26. Ellipse: center at $(-2, -4)$, one vertex at $(-2, 0)$, one focus at $(-2, -2)$
27. Ellipse: center at origin, one vertex at $(0, 4)$, one focus at $(0, 3)$
28. Hyperbola: center at origin, one vertex at $(3, 0)$, asymptotes $y = \pm \frac{3}{4}x$
29. Hyperbola: vertices at $(0, 5)$ and $(0, -1)$, one focus at $(0, -2)$
30. A comet moves in a parabolic path with the center of the sun as its focus. When the comet is 60 million kilometers from the sun, it makes an angle of 60° with the axis of the parabola. How close will the comet come to the sun?
31. A parabolic reflector has a diameter of 1.2 ft and a depth of 0.90 ft. Where should the light be placed to produce a beam of parallel rays?
32. A semielliptic arch measures 50.0 ft across the base and 15.0 ft high in the center. Find the heights of two vertical pillars each placed 15.0 ft from the center.
33. Express the equation $y^2 = 3x$ in polar coordinates.

In Exercises 34–36, express the given equations in rectangular coordinates.

34. $r = 4 \csc \theta$ **35.** $r^2 = 4 \tan \theta$ **36.** $r = 2 - 3 \cos \theta$

In Exercises 37–42, identify each curve and sketch.

37. $r = \sin \theta$ **38.** $r = 3(1 - \cos \theta)$ **39.** $r = 2 \cos 2\theta$
40. $r = 2 - 5 \sin \theta$ **41.** $r^2 = 9 \sin 2\theta$ **42.** $r = \sin 3\theta$

APPENDIX A

Units of Measurement and Conversions

Two basic systems of measurement are in use today, the **metric system** and the **British system.** In addition, there is a widely accepted system derived from metric units, the **International System of Units (SI).** A table of SI units is given on the inside front cover of this book.

Basic Units

The base unit of length is the *foot* in the British system and the *meter* in the metric system. (Converting from one system of measurement to the other will be discussed shortly.)

Two sets of very important units are those of mass and weight. The **weight** of an object is the measure of the gravitational force on the object, which varies with altitude. The **mass** of an object is the measure of its inertia, which does not vary with altitude. Although mass and weight are different quantities, they are closely related: The weight of an object is equal to its mass times the acceleration due to gravity, or $w = mg$. This is a special case of Newton's second law, $F = ma$, which was mentioned in Section 2.2.

In the SI system, the base unit of mass is the *kilogram*. In the metric system, mass can also be expressed in *grams*. In the British system, mass is expressed in *slugs* (since mass is a measure of "sluggishness"), where

$$1 \text{ slug} = 1 \frac{\text{lb} \cdot \text{sec}^2}{\text{ft}}$$

(The base unit of weight is the pound.)

To see why 1 slug = 1 (lb · sec^2)/ft, consider Newton's second law, $F = ma$. Given that the acceleration due to gravity g near the surface of the earth is

$$32 \frac{\text{ft/sec}}{\text{sec}} = 32 \frac{\text{ft}}{\text{sec}^2}$$

and that $mg = F$, then with $m = 1$ slug,

$$1 \left(\frac{\text{lb} \cdot \text{sec}^2}{\text{ft}}\right)\left(32 \frac{\text{ft}}{\text{sec}^2}\right) = 32 \text{ lb}$$

As already noted, the base unit of weight in the British system is the pound. In the SI system, the unit of weight is the newton (N), where

$$1 \text{ N} = 1 \frac{\text{kg} \cdot \text{m}}{\text{sec}^2}$$

As a check, given that the acceleration due to gravity is 10 m/sec^2 (a more precise value is 9.8 m/sec^2) and that $mg = F$, then with $m = 1$ kg,

$$(1 \text{ kg})\left(10 \frac{\text{m}}{\text{sec}^2}\right) = 10 \frac{\text{kg} \cdot \text{m}}{\text{sec}^2} = 10 \text{ N}$$

(For other units, see the table on the inside front cover of this book.)

A special feature of the metric system is the use of prefixes to denote different orders of magnitude. These are given in the following table:

Table A.1

Prefix	Factor	Symbol	Prefix	Factor	Symbol
exa-	10^{18}	E	deci-	10^{-1}	d
peta-	10^{15}	P	centi-	10^{-2}	c
tera-	10^{12}	T	milli-	10^{-3}	m
giga-	10^{9}	G	micro-	10^{-6}	μ
mega-	10^{6}	M	nano-	10^{-9}	n
kilo-	10^{3}	k	pico-	10^{-12}	p
hecto-	10^{2}	h	femto-	10^{-15}	f
deca-	10^{1}	da	atto-	10^{-18}	a

The following equivalencies illustrate the use of these prefixes:

$$1 \text{ millimeter (mm)} = 10^{-3} \text{ m}$$
$$1 \text{ kilometer (km)} = 10^{3} \text{ m}$$
$$1 \text{ milliohm (m}\Omega\text{)} = 10^{-3} \text{ }\Omega$$
$$1 \text{ microfarad } (\mu\text{F}) = 10^{-6} \text{ F}$$
$$1 \text{ hectoliter (hL)} = 10^{2} \text{ L}$$

Conversions

Converting between the British and SI or metric systems is fairly complicated. To illustrate the technique, we need the basic conversion units given in the box below.

1 in. = 2.54 cm (exact)	1 kg = 2.2 lb
1 km = 0.6214 mi	1 lb = 4.448 N
1 lb = 454 g	1 ft³ = 28.32 L

To change the units of measure of a given number, we make use of two basic principles.

1. *Perform algebraic operations with the units of measure as you would perform them with any algebraic symbols.*

For example, the operation

$$1 \frac{\text{ft}}{\text{sec}} = 1 \frac{\text{ft}}{\text{sec}} \times 60 \frac{\text{sec}}{\text{min}} = 60 \frac{\text{ft}}{\text{min}}$$

converts feet per second to feet per minute. Note that the units are canceled as if they were algebraic symbols.

2. *To convert units, multiply by quantities each of which has quotient 1.*

To see how this rule is used, note that the values in the box above are given as equal pairs. Thus

$$\frac{1 \text{ in.}}{2.54 \text{ cm}} = 1, \quad \frac{1 \text{ km}}{0.6214 \text{ mi}} = 1, \quad \frac{454 \text{ g}}{1 \text{ lb}} = 1$$

Multiplying a given quantity by any of these fractions does not change the value. For example,

$$2.1 \text{ mi} = 2.1 \text{ mi} \left(\frac{1 \text{ km}}{0.6214 \text{ mi}} \right) = \frac{2.1}{0.6214} \text{ km} = 3.4 \text{ km}$$

Example 1 Express 20.0 mi/hr in kilometers per minute.

Solution. Since 1 km = 0.6214 mi, we multiply the given quantity by

$$\frac{1 \text{ km}}{0.6214 \text{ mi}}$$

in order to cancel mi and introduce km. Similarly, we multiply by

$$\frac{1 \text{ hr}}{60 \text{ min}}$$

to cancel hr and introduce 1/min. Successive multiplication yields

$$20.0 \frac{\text{mi}}{\text{hr}} \left(\frac{1 \text{ km}}{0.6214 \text{ mi}} \right) \left(\frac{1 \text{ hr}}{60 \text{ min}} \right) = \frac{20.0}{(0.6214)(60)} \frac{\text{km}}{\text{min}} = 0.536 \frac{\text{km}}{\text{min}}$$

Example 2 Express 53.7 lb/in.² in newtons per square meter.

Solution. Since 1 in. = 2.54 cm, we multiply the given quantity by

$$\left(\frac{1 \text{ in.}}{2.54 \text{ cm}}\right)^2$$

to cancel in.². The multiplication introduces 1/cm². To cancel cm², we multiply by

$$\left(\frac{100 \text{ cm}}{1 \text{ m}}\right)^2$$

Thus

$$53.7 \frac{\text{lb}}{\text{in.}^2} = \left(53.7 \frac{\text{lb}}{\text{in.}^2}\right)\left(\frac{1 \text{ in.}}{2.54 \text{ cm}}\right)^2 \left(\frac{100 \text{ cm}}{1 \text{ m}}\right)^2 \left(\frac{4.448 \text{ N}}{1 \text{ lb}}\right)$$

$$= 53.7 \frac{\cancel{\text{lb}}}{\cancel{\text{in.}^2}} \times \frac{1^2 \cancel{\text{in.}^2}}{(2.54)^2 \cancel{\text{cm}^2}} \times \frac{(100)^2 \cancel{\text{cm}^2}}{1^2 \text{ m}^2} \times \frac{4.448 \text{ N}}{1 \cancel{\text{lb}}}$$

$$= \frac{(53.7)(100)^2(4.448)}{(2.54)^2} \frac{\text{N}}{\text{m}^2} = 3.70 \times 10^5 \frac{\text{N}}{\text{m}^2}$$

Example 3 Express 0.124 in.³ in cubic centiliters.

Solution.

$$0.124 \cancel{\text{in.}^3} \left(\frac{1^3 \cancel{\text{ft}^3}}{(12)^3 \cancel{\text{in.}^3}}\right) \cdot \frac{28.32 \cancel{\text{L}}}{1 \cancel{\text{ft}^3}} \cdot \frac{(100)^3 (\text{cL})^3}{1 \cancel{\text{L}}} = 2{,}030 \text{ (cL)}^3$$

Exercises / A.1

Carry out the indicated conversions.

1. 1.10 m to inches
2. 5.30 lb to newtons
3. 7.710 in.² to square centimeters
4. 10.5 cm² to square feet
5. 50.0 $\frac{\text{mi}}{\text{hr}}$ to kilometers per minute
6. 3.8 $\frac{\text{mi}}{\text{sec}}$ to meters per minute
7. 223 $\frac{\text{lb}}{\text{ft}}$ to newtons per centimeter
8. 50.3 $\frac{\text{km}}{\text{hr}}$ to miles per second
9. 251 $\frac{\text{ft}}{\text{min}}$ to meters per second
10. 101.1 $\frac{\text{km}}{\text{sec}}$ to miles per minute
11. 3.7 $\frac{\text{N}}{\text{cm}^2}$ to pounds per square inch
12. 10.3 $\frac{\text{lb}}{\text{ft}^2}$ to newtons per square meter

13. $1.50 \frac{\text{lb}}{\text{ft}^3}$ to newtons per cubic centimeter
14. $2.78 \frac{\text{N}}{\text{m}^3}$ to pounds per cubic inch
15. $3.200 \frac{\text{N}}{\text{cm}^3}$ to pounds per cubic foot
16. 2.30 ft^3 to liters
17. 50.1 L to cubic feet
18. 10.12 mL to cubic inches
19. 142.3 mL to cubic inches

APPENDIX B

Tables

Table 1 *Trigonometric functions*

Degrees	Sin θ	Cos θ	Tan θ	Cot θ	Sec θ	Csc θ	
0°00′	0.0000	1.0000	0.0000	—	1.000	—	90°00′
10	0.0029	1.0000	0.0029	343.8	1.000	343.8	50
20	0.0058	1.0000	0.0058	171.9	1.000	171.9	40
30	0.0087	1.0000	0.0087	114.6	1.000	114.6	30
40	0.0116	0.9999	0.0116	85.94	1.000	85.95	20
50	0.0145	0.9999	0.0145	68.75	1.000	68.76	10
1°00′	0.0175	0.9998	0.0175	57.29	1.000	57.30	89°00′
10	0.0204	0.9998	0.0204	49.10	1.000	49.11	50
20	0.0233	0.9997	0.0233	42.96	1.000	42.98	40
30	0.0262	0.9997	0.0262	38.19	1.000	38.20	30
40	0.0291	0.9996	0.0291	34.37	1.000	34.38	20
50	0.0320	0.9995	0.0320	31.24	1.001	31.26	10
2°00′	0.0349	0.9994	0.0349	28.64	1.001	28.65	88°00′
10	0.0378	0.9993	0.0378	26.43	1.001	26.45	50
20	0.0407	0.9992	0.0407	24.54	1.001	24.56	40
30	0.0436	0.9990	0.0437	22.90	1.001	22.93	30
40	0.0465	0.9989	0.0466	21.47	1.001	21.49	20
50	0.0494	0.9988	0.0495	20.21	1.001	20.23	10
3°00′	0.0523	0.9986	0.0524	19.08	1.001	19.11	87°00′
10	0.0552	0.9985	0.0553	18.07	1.002	18.10	50
20	0.0581	0.9983	0.0582	17.17	1.002	17.20	40
30	0.0610	0.9981	0.0612	16.35	1.002	16.38	30
40	0.0640	0.9980	0.0641	15.60	1.002	15.64	20
50	0.0669	0.9978	0.0670	14.92	1.002	14.96	10
4°00′	0.0698	0.9976	0.0699	14.30	1.002	14.34	86°00′
10	0.0727	0.9974	0.0729	13.73	1.003	13.76	50
20	0.0756	0.9971	0.0758	13.20	1.003	13.23	40
30	0.0785	0.9969	0.0787	12.71	1.003	12.75	30
40	0.0814	0.9967	0.0816	12.25	1.003	12.29	20
50	0.0843	0.9964	0.0846	11.83	1.004	11.87	10
5°00′	0.0872	0.9962	0.0875	11.43	1.004	11.47	85°00′
	Cos θ	Sin θ	Cot θ	Tan θ	Csc θ	Sec θ	Degrees

Table 1 *Trigonometric functions* (continued)

Degrees	Sin θ	Cos θ	Tan θ	Cot θ	Sec θ	Csc θ	
5°00′	0.0872	0.9962	0.0875	11.43	1.004	11.47	85°00′
10	0.0901	0.9959	0.0904	11.06	1.004	11.10	50
20	0.0929	0.9957	0.0934	10.71	1.004	10.76	40
30	0.0958	0.9954	0.0963	10.39	1.005	10.43	30
40	0.0987	0.9951	0.0992	10.08	1.005	10.13	20
50	0.1016	0.9948	0.1022	9.788	1.005	9.839	10
6°00′	0.1045	0.9945	0.1051	9.514	1.006	9.567	84°00′
10	0.1074	0.9942	0.1080	9.255	1.006	9.309	50
20	0.1103	0.9939	0.1110	9.010	1.006	9.065	40
30	0.1132	0.9936	0.1139	8.777	1.006	8.834	30
40	0.1161	0.9932	0.1169	8.556	1.007	8.614	20
50	0.1190	0.9929	0.1198	8.345	1.007	8.405	10
7°00′	0.1219	0.9925	0.1228	8.144	1.008	8.206	83°00′
10	0.1248	0.9922	0.1257	7.953	1.008	8.016	50
20	0.1276	0.9918	0.1287	7.770	1.008	7.834	40
30	0.1305	0.9914	0.1317	7.596	1.009	7.661	30
40	0.1334	0.9911	0.1346	7.429	1.009	7.496	20
50	0.1363	0.9907	0.1376	7.269	1.009	7.337	10
8°00′	0.1392	0.9903	0.1405	7.115	1.010	7.185	82°00′
10	0.1421	0.9899	0.1435	6.968	1.010	7.040	50
20	0.1449	0.9894	0.1465	6.827	1.011	6.900	40
30	0.1478	0.9890	0.1495	6.691	1.011	6.765	30
40	0.1507	0.9886	0.1524	6.561	1.012	6.636	20
50	0.1536	0.9881	0.1554	6.435	1.012	6.512	10
9°00′	0.1564	0.9877	0.1584	6.314	1.012	6.392	81°00′
10	0.1593	0.9872	0.1614	6.197	1.013	6.277	50
20	0.1622	0.9868	0.1644	6.084	1.013	6.166	40
30	0.1650	0.9863	0.1673	5.976	1.014	6.059	30
40	0.1679	0.9858	0.1703	5.871	1.014	5.955	20
50	0.1708	0.9853	0.1733	5.769	1.015	5.855	10
10°00′	0.1736	0.9848	0.1763	5.671	1.015	5.759	80°00′
10	0.1765	0.9843	0.1793	5.576	1.016	5.665	50
20	0.1794	0.9838	0.1823	5.485	1.016	5.575	40
30	0.1822	0.9833	0.1853	5.396	1.017	5.487	30
40	0.1851	0.9827	0.1883	5.309	1.018	5.403	20
50	0.1880	0.9822	0.1914	5.226	1.018	5.320	10
11°00′	0.1908	0.9816	0.1944	5.145	1.019	5.241	79°00′
10	0.1937	0.9811	0.1974	5.066	1.019	5.164	50
20	0.1965	0.9805	0.2004	4.989	1.020	5.089	40
30	0.1994	0.9799	0.2035	4.915	1.020	5.016	30
40	0.2022	0.9793	0.2065	4.843	1.021	4.945	20
50	0.2051	0.9787	0.2095	4.773	1.022	4.876	10
12°00′	0.2079	0.9781	0.2126	4.705	1.022	4.810	78°00′
10	0.2108	0.9775	0.2156	4.638	1.023	4.745	50
20	0.2136	0.9769	0.2186	4.574	1.024	4.682	40
30	0.2164	0.9763	0.2217	4.511	1.024	4.620	30
40	0.2193	0.9757	0.2247	4.449	1.025	4.560	20
50	0.2221	0.9750	0.2278	4.390	1.026	4.502	10
13°00′	0.2250	0.9744	0.2309	4.331	1.026	4.445	77°00′
	Cos θ	Sin θ	Cot θ	Tan θ	Csc θ	Sec θ	Degrees

Table 1 *Trigonometric functions* (continued)

Degrees	Sin θ	Cos θ	Tan θ	Cot θ	Sec θ	Csc θ	
13°00′	0.2250	0.9744	0.2309	4.331	1.026	4.445	77°00′
10	0.2278	0.9737	0.2339	4.275	1.027	4.390	50
20	0.2306	0.9730	0.2370	4.219	1.028	4.336	40
30	0.2334	0.9724	0.2401	4.165	1.028	4.284	30
40	0.2363	0.9717	0.2432	4.113	1.029	4.232	20
50	0.2391	0.9710	0.2462	4.061	1.030	4.182	10
14°00′	0.2419	0.9703	0.2493	4.011	1.031	4.134	76°00′
10	0.2447	0.9696	0.2524	3.962	1.031	4.086	50
20	0.2476	0.9689	0.2555	3.914	1.032	4.039	40
30	0.2504	0.9681	0.2586	3.867	1.033	3.994	30
40	0.2532	0.9674	0.2617	3.821	1.034	3.950	20
50	0.2560	0.9667	0.2648	3.776	1.034	3.906	10
15°00′	0.2588	0.9659	0.2679	3.732	1.035	3.864	75°00′
10	0.2616	0.9652	0.2711	3.689	1.036	3.822	50
20	0.2644	0.9644	0.2742	3.647	1.037	3.782	40
30	0.2672	0.9636	0.2773	3.606	1.038	3.742	30
40	0.2700	0.9628	0.2805	3.566	1.039	3.703	20
50	0.2728	0.9621	0.2836	3.526	1.039	3.665	10
16°00′	0.2756	0.9613	0.2867	3.487	1.040	3.628	74°00′
10	0.2784	0.9605	0.2899	3.450	1.041	3.592	50
20	0.2812	0.9596	0.2931	3.412	1.042	3.556	40
30	0.2840	0.9588	0.2962	3.376	1.043	3.521	30
40	0.2868	0.9580	0.2994	3.340	1.044	3.487	20
50	0.2896	0.9572	0.3026	3.305	1.045	3.453	10
17°00′	0.2924	0.9563	0.3057	3.271	1.046	3.420	73°00′
10	0.2952	0.9555	0.3089	3.237	1.047	3.388	50
20	0.2979	0.9546	0.3121	3.204	1.048	3.356	40
30	0.3007	0.9537	0.3153	3.172	1.049	3.326	30
40	0.3035	0.9528	0.3185	3.140	1.049	3.295	20
50	0.3062	0.9520	0.3217	3.108	1.050	3.265	10
18°00′	0.3090	0.9511	0.3249	3.078	1.051	3.236	72°00′
10	0.3118	0.9502	0.3281	3.047	1.052	3.207	50
20	0.3145	0.9492	0.3314	3.018	1.053	3.179	40
30	0.3173	0.9483	0.3346	2.989	1.054	3.152	30
40	0.3201	0.9474	0.3378	2.960	1.056	3.124	20
50	0.3228	0.9465	0.3411	2.932	1.057	3.098	10
19°00′	0.3256	0.9455	0.3443	2.904	1.058	3.072	71°00′
10	0.3283	0.9446	0.3476	2.877	1.059	3.046	50
20	0.3311	0.9436	0.3508	2.850	1.060	3.021	40
30	0.3338	0.9426	0.3541	2.824	1.061	2.996	30
40	0.3365	0.9417	0.3574	2.798	1.062	2.971	20
50	0.3393	0.9407	0.3607	2.773	1.063	2.947	10
20°00′	0.3420	0.9397	0.3640	2.747	1.064	2.924	70°00′
10	0.3448	0.9387	0.3673	2.723	1.065	2.901	50
20	0.3475	0.9377	0.3706	2.699	1.066	2.878	40
30	0.3502	0.9367	0.3739	2.675	1.068	2.855	30
40	0.3529	0.9356	0.3772	2.651	1.069	2.833	20
50	0.3557	0.9346	0.3805	2.628	1.070	2.812	10
21°00′	0.3584	0.9336	0.3839	2.605	1.071	2.790	69°00′
	Cos θ	Sin θ	Cot θ	Tan θ	Csc θ	Sec θ	Degrees

Table 1 *Trigonometric functions* (continued)

Degrees	Sin θ	Cos θ	Tan θ	Cot θ	Sec θ	Csc θ	
21°00′	0.3584	0.9336	0.3839	2.605	1.071	2.790	69°00′
10	0.3611	0.9325	0.3872	2.583	1.072	2.769	50
20	0.3638	0.9315	0.3906	2.560	1.074	2.749	40
30	0.3665	0.9304	0.3939	2.539	1.075	2.729	30
40	0.3692	0.9293	0.3973	2.517	1.076	2.709	20
50	0.3719	0.9283	0.4006	2.496	1.077	2.689	10
22°00′	0.3746	0.9272	0.4040	2.475	1.079	2.669	68°00′
10	0.3773	0.9261	0.4074	2.455	1.080	2.650	50
20	0.3800	0.9250	0.4108	2.434	1.081	2.632	40
30	0.3827	0.9239	0.4142	2.414	1.082	2.613	30
40	0.3854	0.9228	0.4176	2.394	1.084	2.595	20
50	0.3881	0.9216	0.4210	2.375	1.085	2.577	10
23°00′	0.3907	0.9205	0.4245	2.356	1.086	2.559	67°00′
10	0.3934	0.9194	0.4279	2.337	1.088	2.542	50
20	0.3961	0.9182	0.4314	2.318	1.089	2.525	40
30	0.3987	0.9171	0.4348	2.300	1.090	2.508	30
40	0.4014	0.9159	0.4383	2.282	1.092	2.491	20
50	0.4041	0.9147	0.4417	2.264	1.093	2.475	10
24°00′	0.4067	0.9135	0.4452	2.246	1.095	2.459	66°00′
10	0.4094	0.9124	0.4487	2.229	1.096	2.443	50
20	0.4120	0.9112	0.4522	2.211	1.097	2.427	40
30	0.4147	0.9100	0.4557	2.194	1.099	2.411	30
40	0.4173	0.9088	0.4592	2.177	1.100	2.396	20
50	0.4200	0.9075	0.4628	2.161	1.102	2.381	10
25°00′	0.4226	0.9063	0.4663	2.145	1.103	2.366	65°00′
10	0.4253	0.9051	0.4699	2.128	1.105	2.352	50
20	0.4279	0.9038	0.4734	2.112	1.106	2.337	40
30	0.4305	0.9026	0.4770	2.097	1.108	2.323	30
40	0.4331	0.9013	0.4806	2.081	1.109	2.309	20
50	0.4358	0.9001	0.4841	2.066	1.111	2.295	10
26°00′	0.4384	0.8988	0.4877	2.050	1.113	2.281	64°00′
10	0.4410	0.8975	0.4913	2.035	1.114	2.268	50
20	0.4436	0.8962	0.4950	2.020	1.116	2.254	40
30	0.4462	0.8949	0.4986	2.006	1.117	2.241	30
40	0.4488	0.8936	0.5022	1.991	1.119	2.228	20
50	0.4514	0.8923	0.5059	1.977	1.121	2.215	10
27°00′	0.4540	0.8910	0.5095	1.963	1.122	2.203	63°00′
10	0.4566	0.8897	0.5132	1.949	1.124	2.190	50
20	0.4592	0.8884	0.5169	1.935	1.126	2.178	40
30	0.4617	0.8870	0.5206	1.921	1.127	2.166	30
40	0.4643	0.8857	0.5243	1.907	1.129	2.154	20
50	0.4669	0.8843	0.5280	1.894	1.131	2.142	10
28°00′	0.4695	0.8829	0.5317	1.881	1.133	2.130	62°00′
10	0.4720	0.8816	0.5354	1.868	1.134	2.118	50
20	0.4746	0.8802	0.5392	1.855	1.136	2.107	40
30	0.4772	0.8788	0.5430	1.842	1.138	2.096	30
40	0.4797	0.8774	0.5467	1.829	1.140	2.085	20
50	0.4823	0.8760	0.5505	1.816	1.142	2.074	10
29°00′	0.4848	0.8746	0.5543	1.804	1.143	2.063	61°00′
	Cos θ	Sin θ	Cot θ	Tan θ	Csc θ	Sec θ	Degrees

Table 1 *Trigonometric functions* (continued)

Degrees	Sin θ	Cos θ	Tan θ	Cot θ	Sec θ	Csc θ	
29°00′	0.4848	0.8746	0.5543	1.804	1.143	2.063	61°00′
10	0.4874	0.8732	0.5581	1.792	1.145	2.052	50
20	0.4899	0.8718	0.5619	1.780	1.147	2.041	40
30	0.4924	0.8704	0.5658	1.767	1.149	2.031	30
40	0.4950	0.8689	0.5696	1.756	1.151	2.020	20
50	0.4975	0.8675	0.5735	1.744	1.153	2.010	10
30°00′	0.5000	0.8660	0.5774	1.732	1.155	2.000	60°00′
10	0.5025	0.8646	0.5812	1.720	1.157	1.990	50
20	0.5050	0.8631	0.5851	1.709	1.159	1.980	40
30	0.5075	0.8616	0.5890	1.698	1.161	1.970	30
40	0.5100	0.8601	0.5930	1.686	1.163	1.961	20
50	0.5125	0.8587	0.5969	1.675	1.165	1.951	10
31°00′	0.5150	0.8572	0.6009	1.664	1.167	1.942	59°00′
10	0.5175	0.8557	0.6048	1.653	1.169	1.932	50
20	0.5200	0.8542	0.6088	1.643	1.171	1.923	40
30	0.5225	0.8526	0.6128	1.632	1.173	1.914	30
40	0.5250	0.8511	0.6168	1.621	1.175	1.905	20
50	0.5275	0.8496	0.6208	1.611	1.177	1.896	10
32°00′	0.5299	0.8480	0.6249	1.600	1.179	1.887	58°00′
10	0.5324	0.8465	0.6289	1.590	1.181	1.878	50
20	0.5348	0.8450	0.6330	1.580	1.184	1.870	40
30	0.5373	0.8434	0.6371	1.570	1.186	1.861	30
40	0.5398	0.8418	0.6412	1.560	1.188	1.853	20
50	0.5422	0.8403	0.6453	1.550	1.190	1.844	10
33°00′	0.5446	0.8387	0.6494	1.540	1.192	1.836	57°00′
10	0.5471	0.8371	0.6536	1.530	1.195	1.828	50
20	0.5495	0.8355	0.6577	1.520	1.197	1.820	40
30	0.5519	0.8339	0.6619	1.511	1.199	1.812	30
40	0.5544	0.8323	0.6661	1.501	1.202	1.804	20
50	0.5568	0.8307	0.6703	1.492	1.204	1.796	10
34°00′	0.5592	0.8290	0.6745	1.483	1.206	1.788	56°00′
10	0.5616	0.8274	0.6787	1.473	1.209	1.781	50
20	0.5640	0.8258	0.6830	1.464	1.211	1.773	40
30	0.5664	0.8241	0.6873	1.455	1.213	1.766	30
40	0.5688	0.8225	0.6916	1.446	1.216	1.758	20
50	0.5712	0.8208	0.6959	1.437	1.218	1.751	10
35°00′	0.5736	0.8192	0.7002	1.428	1.221	1.743	55°00′
10	0.5760	0.8175	0.7046	1.419	1.223	1.736	50
20	0.5783	0.8158	0.7089	1.411	1.226	1.729	40
30	0.5807	0.8141	0.7133	1.402	1.228	1.722	30
40	0.5831	0.8124	0.7177	1.393	1.231	1.715	20
50	0.5854	0.8107	0.7221	1.385	1.233	1.708	10
36°00′	0.5878	0.8090	0.7265	1.376	1.236	1.701	54°00′
10	0.5901	0.8073	0.7310	1.368	1.239	1.695	50
20	0.5925	0.8056	0.7355	1.360	1.241	1.688	40
30	0.5948	0.8039	0.7400	1.351	1.244	1.681	30
40	0.5972	0.8021	0.7445	1.343	1.247	1.675	20
50	0.5995	0.8004	0.7490	1.335	1.249	1.668	10
37°00′	0.6018	0.7986	0.7536	1.327	1.252	1.662	53°00′
	Cos θ	Sin θ	Cot θ	Tan θ	Csc θ	Sec θ	Degrees

Table 1 *Trigonometric functions* (continued)

Degrees	Sin θ	Cos θ	Tan θ	Cot θ	Sec θ	Csc θ	
37°00′	0.6018	0.7986	0.7536	1.327	1.252	1.662	53°00′
10	0.6041	0.7969	0.7581	1.319	1.255	1.655	50
20	0.6065	0.7951	0.7627	1.311	1.258	1.649	40
30	0.6088	0.7934	0.7673	1.303	1.260	1.643	30
40	0.6111	0.7916	0.7720	1.295	1.263	1.636	20
50	0.6134	0.7898	0.7766	1.288	1.266	1.630	10
38°00′	0.6157	0.7880	0.7813	1.280	1.269	1.624	52°00′
10	0.6180	0.7862	0.7860	1.272	1.272	1.618	50
20	0.6202	0.7844	0.7907	1.265	1.275	1.612	40
30	0.6225	0.7826	0.7954	1.257	1.278	1.606	30
40	0.6248	0.7808	0.8002	1.250	1.281	1.601	20
50	0.6271	0.7790	0.8050	1.242	1.284	1.595	10
39°00′	0.6293	0.7771	0.8098	1.235	1.287	1.589	51°00′
10	0.6316	0.7753	0.8146	1.228	1.290	1.583	50
20	0.6338	0.7735	0.8195	1.220	1.293	1.578	40
30	0.6361	0.7716	0.8243	1.213	1.296	1.572	30
40	0.6383	0.7698	0.8292	1.206	1.299	1.567	20
50	0.6406	0.7679	0.8342	1.199	1.302	1.561	10
40°00′	0.6428	0.7660	0.8391	1.192	1.305	1.556	50°00′
10	0.6450	0.7642	0.8441	1.185	1.309	1.550	50
20	0.6472	0.7623	0.8491	1.178	1.312	1.545	40
30	0.6494	0.7604	0.8541	1.171	1.315	1.540	30
40	0.6517	0.7585	0.8591	1.164	1.318	1.535	20
50	0.6539	0.7566	0.8642	1.157	1.322	1.529	10
41°00′	0.6561	0.7547	0.8693	1.150	1.325	1.524	49°00′
10	0.6583	0.7528	0.8744	1.144	1.328	1.519	50
20	0.6604	0.7509	0.8796	1.137	1.332	1.514	40
30	0.6626	0.7490	0.8847	1.130	1.335	1.509	30
40	0.6648	0.7470	0.8899	1.124	1.339	1.504	20
50	0.6670	0.7451	0.8952	1.117	1.342	1.499	10
42°00′	0.6691	0.7431	0.9004	1.111	1.346	1.494	48°00′
10	0.6713	0.7412	0.9057	1.104	1.349	1.490	50
20	0.6734	0.7392	0.9110	1.098	1.353	1.485	40
30	0.6756	0.7373	0.9163	1.091	1.356	1.480	30
40	0.6777	0.7353	0.9217	1.085	1.360	1.476	20
50	0.6799	0.7333	0.9271	1.079	1.364	1.471	10
43°00′	0.6820	0.7314	0.9325	1.072	1.367	1.466	47°00′
10	0.6841	0.7294	0.9380	1.066	1.371	1.462	50
20	0.6862	0.7274	0.9435	1.060	1.375	1.457	40
30	0.6884	0.7254	0.9490	1.054	1.379	1.453	30
40	0.6905	0.7234	0.9545	1.048	1.382	1.448	20
50	0.6926	0.7214	0.9601	1.042	1.386	1.444	10
44°00′	0.6947	0.7193	0.9657	1.036	1.390	1.440	46°00′
10	0.6967	0.7173	0.9713	1.030	1.394	1.435	50
20	0.6988	0.7153	0.9770	1.024	1.398	1.431	40
30	0.7009	0.7133	0.9827	1.018	1.402	1.427	30
40	0.7030	0.7112	0.9884	1.012	1.406	1.423	20
50	0.7050	0.7092	0.9942	1.006	1.410	1.418	10
45°00′	0.7071	0.7071	1.000	1.000	1.414	1.414	45°00′
	Cos θ	Sin θ	Cot θ	Tan θ	Csc θ	Sec θ	Degrees

Table 2 *Common Logarithms*

n	0	1	2	3	4	5	6	7	8	9
1.0	.0000	.0043	.0086	.0128	.0170	.0212	.0253	.0294	.0334	.0374
1.1	.0414	.0453	.0492	.0531	.0569	.0607	.0645	.0682	.0719	.0755
1.2	.0792	.0828	.0864	.0899	.0934	.0969	.1004	.1038	.1072	.1106
1.3	.1139	.1173	.1206	.1239	.1271	.1303	.1335	.1367	.1399	.1430
1.4	.1461	.1492	.1523	.1553	.1584	.1614	.1644	.1673	.1703	.1732
1.5	.1761	.1790	.1818	.1847	.1875	.1903	.1931	.1959	.1987	.2014
1.6	.2041	.2068	.2095	.2122	.2148	.2175	.2201	.2227	.2253	.2279
1.7	.2304	.2330	.2355	.2380	.2405	.2430	.2455	.2480	.2504	.2529
1.8	.2553	.2577	.2601	.2625	.2648	.2672	.2695	.2718	.2742	.2765
1.9	.2788	.2810	.2833	.2856	.2878	.2900	.2923	.2945	.2967	.2989
2.0	.3010	.3032	.3054	.3075	.3096	.3118	.3139	.3160	.3181	.3201
2.1	.3222	.3243	.3263	.3284	.3304	.3324	.3345	.3365	.3385	.3404
2.2	.3424	.3444	.3464	.3483	.3502	.3522	.3541	.3560	.3579	.3598
2.3	.3617	.3636	.3655	.3674	.3692	.3711	.3729	.3747	.3766	.3784
2.4	.3802	.3820	.3838	.3856	.3874	.3892	.3909	.3927	.3945	.3962
2.5	.3979	.3997	.4014	.4031	.4048	.4065	.4082	.4099	.4116	.4133
2.6	.4150	.4166	.4183	.4200	.4216	.4232	.4249	.4265	.4281	.4298
2.7	.4314	.4330	.4346	.4362	.4378	.4393	.4409	.4425	.4440	.4456
2.8	.4472	.4487	.4502	.4518	.4533	.4548	.4564	.4579	.4594	.4609
2.9	.4624	.4639	.4654	.4669	.4683	.4698	.4713	.4728	.4742	.4757
3.0	.4771	.4786	.4800	.4814	.4829	.4843	.4857	.4871	.4886	.4900
3.1	.4914	.4928	.4942	.4955	.4969	.4983	.4997	.5011	.5024	.5038
3.2	.5051	.5065	.5079	.5092	.5105	.5119	.5132	.5145	.5159	.5172
3.3	.5185	.5198	.5211	.5224	.5237	.5250	.5263	.5276	.5289	.5302
3.4	.5315	.5328	.5340	.5353	.5366	.5378	.5391	.5403	.5416	.5428
3.5	.5441	.5453	.5465	.5478	.5490	.5502	.5514	.5527	.5539	.5551
3.6	.5563	.5575	.5587	.5599	.5611	.5623	.5635	.5647	.5658	.5670
3.7	.5682	.5694	.5705	.5717	.5729	.5740	.5752	.5763	.5775	.5786
3.8	.5798	.5809	.5821	.5832	.5843	.5855	.5866	.5877	.5888	.5899
3.9	.5911	.5922	.5933	.5944	.5955	.5966	.5977	.5988	.5999	.6010
4.0	.6021	.6031	.6042	.6053	.6064	.6075	.6085	.6096	.6107	.6117
4.1	.6128	.6138	.6149	.6160	.6170	.6180	.6191	.6201	.6212	.6222
4.2	.6232	.6243	.6253	.6263	.6274	.6284	.6294	.6304	.6314	.6325
4.3	.6335	.6345	.6355	.6365	.6375	.6385	.6395	.6405	.6415	.6425
4.4	.6435	.6444	.6454	.6464	.6474	.6484	.6493	.6503	.6513	.6522
4.5	.6532	.6542	.6551	.6561	.6571	.6580	.6590	.6599	.6609	.6618
4.6	.6628	.6637	.6646	.6656	.6665	.6675	.6684	.6693	.6702	.6712
4.7	.6721	.6730	.6739	.6749	.6758	.6767	.6776	.6785	.6794	.6803
4.8	.6812	.6821	.6830	.6839	.6848	.6857	.6866	.6875	.6884	.6893
4.9	.6902	.6911	.6920	.6928	.6937	.6946	.6955	.6964	.6972	.6981
5.0	.6990	.6998	.7007	.7016	.7024	.7033	.7042	.7050	.7059	.7067
5.1	.7076	.7084	.7093	.7101	.7110	.7118	.7126	.7135	.7143	.7152
5.2	.7160	.7168	.7177	.7185	.7193	.7202	.7210	.7218	.7226	.7235
5.3	.7243	.7251	.7259	.7267	.7275	.7284	.7292	.7300	.7308	.7316
5.4	.7324	.7332	.7340	.7348	.7356	.7364	.7372	.7380	.7388	.7396
5.5	.7404	.7412	.7419	.7427	.7435	.7443	.7451	.7459	.7466	.7474
5.6	.7482	.7490	.7497	.7505	.7513	.7520	.7528	.7536	.7543	.7551
5.7	.7559	.7566	.7574	.7582	.7589	.7597	.7604	.7612	.7619	.7627
5.8	.7634	.7642	.7649	.7657	.7664	.7672	.7679	.7686	.7694	.7701
5.9	.7709	.7716	.7723	.7731	.7738	.7745	.7752	.7760	.7767	.7774

Table 2 *Common Logarithms* (continued)

n	0	1	2	3	4	5	6	7	8	9
6.0	.7782	.7789	.7796	.7803	.7810	.7818	.7825	.7832	.7839	.7846
6.1	.7853	.7860	.7868	.7875	.7882	.7889	.7896	.7903	.7910	.7917
6.2	.7924	.7931	.7938	.7945	.7952	.7959	.7966	.7973	.7980	.7987
6.3	.7993	.8000	.8007	.8014	.8021	.8028	.8035	.8041	.8048	.8055
6.4	.8062	.8069	.8075	.8082	.8089	.8096	.8102	.8109	.8116	.8122
6.5	.8129	.8136	.8142	.8149	.8156	.8162	.8169	.8176	.8182	.8189
6.6	.8195	.8202	.8209	.8215	.8222	.8228	.8235	.8241	.8248	.8254
6.7	.8261	.8267	.8274	.8280	.8287	.8293	.8299	.8306	.8312	.8319
6.8	.8325	.8331	.8338	.8344	.8351	.8357	.8363	.8370	.8376	.8382
6.9	.8388	.8395	.8401	.8407	.8414	.8420	.8426	.8432	.8439	.8445
7.0	.8451	.8457	.8463	.8470	.8476	.8482	.8488	.8494	.8500	.8506
7.1	.8513	.8519	.8525	.8531	.8537	.8543	.8549	.8555	.8561	.8567
7.2	.8573	.8579	.8585	.8591	.8597	.8603	.8609	.8615	.8621	.8627
7.3	.8633	.8639	.8645	.8651	.8657	.8663	.8669	.8675	.8681	.8686
7.4	.8692	.8698	.8704	.8710	.8716	.8722	.8727	.8733	.8739	.8745
7.5	.8751	.8756	.8762	.8768	.8774	.8779	.8785	.8791	.8797	.8802
7.6	.8808	.8814	.8820	.8825	.8831	.8837	.8842	.8848	.8854	.8859
7.7	.8865	.8871	.8876	.8882	.8887	.8893	.8899	.8904	.8910	.8915
7.8	.8921	.8927	.8932	.8938	.8943	.8949	.8954	.8960	.8965	.8971
7.9	.8976	.8982	.8987	.8993	.8998	.9004	.9009	.9015	.9020	.9025
8.0	.9031	.9036	.9042	.9047	.9053	.9058	.9063	.9069	.9074	.9079
8.1	.9085	.9090	.9096	.9101	.9106	.9112	.9117	.9122	.9128	.9133
8.2	.9138	.9143	.9149	.9154	.9159	.9165	.9170	.9175	.9180	.9186
8.3	.9191	.9196	.9201	.9206	.9212	.9217	.9222	.9227	.9232	.9238
8.4	.9243	.9248	.9253	.9258	.9263	.9269	.9274	.9279	.9284	.9289
8.5	.9294	.9299	.9304	.9309	.9315	.9320	.9325	.9330	.9335	.9340
8.6	.9345	.9350	.9355	.9360	.9365	.9370	.9375	.9380	.9385	.9390
8.7	.9395	.9400	.9405	.9410	.9415	.9420	.9425	.9430	.9435	.9440
8.8	.9445	.9450	.9455	.9460	.9465	.9469	.9474	.9479	.9484	.9489
8.9	.9494	.9499	.9504	.9509	.9513	.9518	.9523	.9528	.9533	.9538
9.0	.9542	.9547	.9552	.9557	.9562	.9566	.9571	.9576	.9581	.9586
9.1	.9590	.9595	.9600	.9605	.9609	.9614	.9619	.9624	.9628	.9633
9.2	.9638	.9643	.9647	.9652	.9657	.9661	.9666	.9671	.9675	.9680
9.3	.9685	.9689	.9694	.9699	.9703	.9708	.9713	.9717	.9722	.9727
9.4	.9731	.9736	.9741	.9745	.9750	.9754	.9759	.9763	.9768	.9773
9.5	.9777	.9782	.9786	.9791	.9795	.9800	.9805	.9809	.9814	.9818
9.6	.9823	.9827	.9832	.9836	.9841	.9845	.9850	.9854	.9859	.9863
9.7	.9868	.9872	.9877	.9881	.9886	.9890	.9894	.9899	.9903	.9908
9.8	.9912	.9917	.9921	.9926	.9930	.9934	.9939	.9943	.9948	.9952
9.9	.9956	.9961	.9965	.9969	.9974	.9978	.9983	.9987	.9991	.9996

Table 3 *Natural Logarithms*

n	$\log_e n$	n	$\log_e n$	n	$\log_e n$	n	$\log_e n$
0.0	—	3.5	1.2528	7.0	1.9459	15	2.7081
0.1	−2.3026	3.6	1.2809	7.1	1.9601	16	2.7726
0.2	−1.6094	3.7	1.3083	7.2	1.9741	17	2.8332
0.3	−1.2040	3.8	1.3350	7.3	1.9879	18	2.8904
0.4	−0.9163	3.9	1.3610	7.4	2.0015	19	2.9444
0.5	−0.6931	4.0	1.3863	7.5	2.0149	20	2.9957
0.6	−0.5108	4.1	1.4110	7.6	2.0281	25	3.2189
0.7	−0.3567	4.2	1.4351	7.7	2.0412	30	3.4012
0.8	−0.2231	4.3	1.4586	7.8	2.0541	35	3.5553
0.9	−0.1054	4.4	1.4816	7.9	2.0669	40	3.6889
1.0	0.0000	4.5	1.5041	8.0	2.0794	45	3.8067
1.1	0.0953	4.6	1.5261	8.1	2.0919	50	3.9120
1.2	0.1823	4.7	1.5476	8.2	2.1041	55	4.0073
1.3	0.2624	4.8	1.5686	8.3	2.1163	60	4.0943
1.4	0.3365	4.9	1.5892	8.4	2.1282	65	4.1744
1.5	0.4055	5.0	1.6094	8.5	2.1401	70	4.2485
1.6	0.4700	5.1	1.6292	8.6	2.1518	75	4.3175
1.7	0.5306	5.2	1.6487	8.7	2.1633	80	4.3820
1.8	0.5878	5.3	1.6677	8.8	2.1748	85	4.4427
1.9	0.6419	5.4	1.6864	8.9	2.1861	90	4.4998
2.0	0.6931	5.5	1.7047	9.0	2.1972	95	4.5539
2.1	0.7419	5.6	1.7228	9.1	2.2083	100	4.6052
2.2	0.7885	5.7	1.7405	9.2	2.2192	200	5.2983
2.3	0.8329	5.8	1.7579	9.3	2.2300	300	5.7038
2.4	0.8755	5.9	1.7750	9.4	2.2407	400	5.9915
2.5	0.9163	6.0	1.7918	9.5	2.2513	500	6.2146
2.6	0.9555	6.1	1.8083	9.6	2.2618	600	6.3969
2.7	0.9933	6.2	1.8245	9.7	2.2721	700	6.5511
2.8	1.0296	6.3	1.8405	9.8	2.2824	800	6.6846
2.9	1.0647	6.4	1.8563	9.9	2.2925	900	6.8024
3.0	1.0986	6.5	1.8718	10	2.3026		
3.1	1.1314	6.6	1.8871	11	2.3979		
3.2	1.1632	6.7	1.9021	12	2.4849		
3.3	1.1939	6.8	1.9169	13	2.5649		
3.4	1.2238	6.9	1.9315	14	2.6391		

APPENDIX C

Answers to Odd-Numbered Exercises

Chapter 1

Section 1.3 (page 8)

1. 3 **3.** 7 **5.** π **7.** $\sqrt{3}$ **9.** 3 **11.** r **13.** r **15.** i **17.** r **19.** i
21. -3 **23.** 8 **25.** 11 **27.** -17 **29.** -5 **31.** -11 **33.** 16 **35.** 8.4
37. $-\dfrac{1}{2}$ **39.** -78 **41.** -96 **43.** 96 **45.** 0.91 **47.** 28.014 **49.** $-\dfrac{5}{3}$
51. undefined **53.** $-\dfrac{8}{225}$ **55.** 0 **57.** 1 **59.** 20,602 ft **61.** $-4°\text{F}$ **63.** 2.0 ft/sec

Section 1.5 (page 13)

1. $-6x - 5y$ **3.** $-2x - 4y$ **5.** $-6a^2 - 2a$ **7.** $xy^2 - 3x^2y$ **9.** $-x - 5y + z$
11. $4x^2 - 6y^2 - 3z^2$ **13.** a **15.** $-xy + 2y^2$ **17.** $-6x - 5y + z$ **19.** $-2a - 2b - 7c$
21. $11a - b + 7c$ **23.** $a + 6b + c$ **25.** $-7a^2 + 10c^2 - d^2 + e$ **27.** $-3a + 6b^2 + c$
29. $-b - 4c$ **31.** $41a^2 + 12b^2$ **33.** $2x + 2y$ **35.** total weight of contents **37.** $5I_1 - 4I_2$

Section 1.6 (page 17)

1. $-12 + 3x$ **3.** $-x + 10y$ **5.** $x - 2y$ **7.** $-2x^2 - 4xy$ **9.** $-3a + b$ **11.** $2xy + y - 3$
13. $-2x$ **15.** $8a - 60$ **17.** $a + b + (-c + 2d)$, $a + b - (c - 2d)$
19. $x - 2y + (-3x^2 - 2y^2)$, $x - 2y - (3x^2 + 2y^2)$ **21.** $7 + (2a - 5b)$, $7 - (-2a + 5b)$
23. $6w + 5z + (-3x + 5y)$, $6w + 5z - (3x - 5y)$
25. $7x^2y + 2xy^2 + (-6xz + 5yz)$, $7x^2y + 2xy^2 - (6xz - 5yz)$
27. $-3x^2 + 2y^2 + (-7z^2 - 8w^2)$, $-3x^2 + 2y^2 - (7z^2 + 8w^2)$
29. $-5x + (-7a - 7b)$, $-5x - (7a + 7b)$ **31.** $2R_1 + 4R_2$ **33.** $-(R_1I_1 - R_2I_2 + R_3I_3)$

Section 1.7 (page 23)

1. -8 **3.** 2 **5.** x^8 **7.** $-2x^3y^3$ **9.** x^5 **11.** $-\dfrac{z^4}{2}$ **13.** 1 **15.** $\dfrac{1}{2}$ **17.** $-\dfrac{y}{2x}$
19. $\dfrac{1}{x}$ **21.** $\dfrac{2y^2}{x^2}$ **23.** $\dfrac{2}{a^3b^2}$ **25.** $2\sqrt{2}$ **27.** $3\sqrt{6}$ **29.** $5\sqrt{3}$ **31.** $2\sqrt[3]{2}$ **33.** $2\sqrt[5]{2}$

637

35. $4\sqrt{2}$ **37.** $\dfrac{1}{5\alpha}$ **39.** $4A_2^2$ **41.** v_1 **43.** $8a^{12}b^{11}$ **45.** $2xy^3$ **47.** $-2v_1v_2$
49. $\dfrac{\sqrt{5}}{5}$ **51.** $\sqrt{3}$ **53.** $\dfrac{a\sqrt{b}}{b}$ **55.** $\dfrac{2\sqrt{5}}{15}$ **57.** $\dfrac{\sqrt{2}}{4}$ **59.** $\dfrac{\sqrt{3}}{2}$ **61.** 9 cm **63.** $\dfrac{2\sqrt{\pi ad}}{ad}$
65. $14\sqrt{5}$ cm² **67.** a^6 **69.** a^4 **71.** a **73.** x^6 **75.** $4x^3$ **77.** $6a^2$ **79.** $7p^4q^2$
81. $9V_2^3$

Section 1.8 (page 28)

1. $12a^3b^4$ **3.** $24a^4b^5x$ **5.** $3x^2y + 2x^2y^2$ **7.** $-30x^3y^4 + 20x^4y^3$ **9.** $2a^3b^2 + 3a^2b^2 - 4a^2b^3$
11. $-12a^4bx + 10a^5x^2 - 4a^3bx$ **13.** $16x^2y^2 - 60x^3y^3$ **15.** $7a^5b^2 - 5a^3b^3$ **17.** $x^2 - 3x + 2$
19. $2x^2 - 7x + 3$ **21.** $10x^2 - 4x - 14$ **23.** $x^2 - y^2$ **25.** $4x^2 - 1$ **27.** $9x^2 - 4y^2$
29. $9x^2 - 6x + 1$ **31.** $4x^2 - 8xy + 4y^2$ **33.** $4x^2 + 4xy + y^2$ **35.** $x^4 - y^4$
37. $5x^2 - 11bx + 2b^2$ **39.** $x^3 - 3x^2 - x + 3$ **41.** $a^3 - 4a^2b + 4ab^2 - 3b^3$
43. $25x^2 - 30xy + 9y^2$ **45.** $5x^2 - 7xy + 2xz + 2y^2 + yz - 3z^2$ **47.** $8x^8 - 8x^6 + 8x^4 - 3x^2 - 5$
49. $2x^5y^2 - 13x^3y^4 + 6x^2y^5 + 2xy^6$ **51.** $x^5 + x^2 - x - 1$ **53.** $x^2 + 2xy + 2xz + y^2 + 2yz + z^2$
55. $4a^2 - 12ab + 9b^2 + 8ac - 12bc + 4c^2$ **57.** $3l^2 + 3lw - 18w^2$ **59.** $6R_1^2 + 18R_1R_2 + 12R_2^2$
61. $65x^2 - 14x$ **63.** $0.04t^2 + 0.1t - 18$ **65.** $E = kT^4 - kT_0^4$ **67.** $\tfrac{1}{2}(y_2^2 - y_1^2)$

Section 1.9 (page 33)

1. a^2 **3.** $-2x^2y$ **5.** $-4x^4 - 2x^2$ **7.** $4c - 3ac^2$ **9.** $1 - 2xy^2 + 3x^3y^2$
11. $4x^5y^3a^2 + 2x^4y - 3x^3a^4$ **13.** $3x - 2$ **15.** $2x + 1$ **17.** $2x + 4$ **19.** $4x - 4$
21. $3x + 4$ **23.** $3x + 6y$ **25.** $x^2 - 2xy + y^2$ **27.** $a^2 + 2ab - 5b^2$ **29.** $2x^2 - xy + y^2$
31. $x^2 - xy + y^2$ **33.** $2x^2 - 3x + 1$ **35.** $2x - 3y$ **37.** $x^2 - 2xy + y^2 - \dfrac{y^3}{x + 2y}$
39. $x^3 + x^2 + x - 2 - \dfrac{5}{x - 4}$

Section 1.10 (page 37)

1. 3.142 **3.** 1.04 **5.** 0.00016 **7.** 986,000 **9.** 880,000 **11.** 0.020 **13.** 7.43
15. 11.096 **17.** 20.6 **19.** 13 **21.** 27 **23.** 32.3 **25.** 40 **27.** 160 **29.** 87.0
31. 0.650 **33.** 53.5 Ω **35.** 0.17 W

Section 1.11 (page 42)

1. 2.6×10^4 **3.** 3.792×10^8 **5.** 1.3×10^{-4} **7.** 8.927×10^{-5} **9.** 1,200,000
11. 0.006273 **13.** 0.00000027 **15.** 956,000 **17.** 2.4×10^5 mi
19. 0.00000000006670 N · m²/kg² **21.** 3,600,000 J **23.** 101,300 N/m² **25.** 3.92×10^{12}
27. 1.9×10^{-9} **29.** 2.34×10^{-8} **31.** 6.94×10^{10} **33.** 1.2×10^{-10} **35.** 3.0×10^{-12}
37. 2.0×10^{15} **39.** 8.26×10^{-6} 8.00×10^{-6}

Review Exercises for Chapter 1 (page 43)

1. 0 **3.** -16 **5.** -10 **7.** $\dfrac{6}{5}$ **9.** $7x - 8y$ **11.** $5x^2 - 8xy$ **13.** b
15. **a.** $2x + a + (-x^2 + y - 4w)$ **b.** $2x + a - (x^2 - y + 4w)$ **17.** $\dfrac{3}{5s^5}$ **19.** $5\sqrt{2}$ **21.** $3\sqrt[3]{3}$
23. $\dfrac{\sqrt{5}}{5}$ **25.** $x^3 - 2xy$ **27.** $-2x^2 + 2xy + 12y^2$ **29.** $4s^3 + 10s^2t - 7st^2 - 3t^3$ **31.** $x - 2y$
33. $5a - 3b + \dfrac{2b^2}{a + b}$ **35.** $a^2 + 2ab + 4b^2$ **37.** $32x^2$ **39.** $2L^2 - 2Lx - 2Lx^2 + 2x^3$
41. 1.55 **43.** 0.39 **45.** 0.00951 **47.** 1.0 **49.** 5,980,000,000,000,000,000,000,000 kg
51. 1.164×10^{-23} g **53.** 0.000000000000000000160 C **55.** 2.5×10^{-14}

Chapter 2

Section 2.1 (page 48)

1. $x = 7$ **3.** $x = -6$ **5.** $x = -2$ **7.** $x = -4$ **9.** $x = \frac{5}{4}$ **11.** $x = 1$ **13.** $x = 5$
15. $x = -\frac{7}{4}$

Section 2.2 (page 53)

1. $x = 9$ **3.** $x = 4$ **5.** $x = 4$ **7.** $x = 3$ **9.** $x = -4$ **11.** $x = -3$ **13.** $x = -\frac{4}{7}$
15. $x = \frac{1}{13}$ **17.** no solution **19.** no solution **21.** $x = -4$ **23.** $x = -3$ **25.** $x = -5$
27. no solution **29.** $x = \frac{b}{a}$ **31.** $x = -\frac{3}{a}$ **33.** $x = -\frac{b}{a}$ **35.** $x = 0$ **37.** $x = \frac{4}{7}$
39. $x = \frac{5}{b}$ **41.** $I = \frac{V}{R}$ **43.** $\pi = \frac{A}{r^2}$ **45.** $b_2 = \frac{2A - hb_1}{h}$ **47.** $r = \frac{A - p}{pt}$ **49.** $P_2 = \frac{P_1 V_1 T_2}{T_1 V_2}$
51. $\pi = \frac{3V}{4r^3}$ **53.** $x = -1$ **55.** no solution **57.** $x = -12$ **59.** $x = -2$

Section 2.3 (page 59)

1. $13\frac{1}{2}$, $16\frac{1}{2}$ **3.** 13 **5.** 60 lb of 10% alloy **7.** 7.5 gal **9.** $\frac{1}{2}$ L **11.** 5 mi/hr
13. 6 mi/hr **15.** 10 mi/hr **17.** $\frac{19}{3}$ in. × $\frac{38}{3}$ in. **19.** 100 hr **21.** 26.4 hr
23. 90 mi for slower car **25.** 1.2 A, 3.1 A **27.** 8 years

Section 2.4 (page 64)

1. 2.3 **3.** 624 lb/ft^2 **5.** 126 **7.** 3.94 in. **9.** 1.76 oz **11.** $84.80 **13.** $1,035
15. 0.17 Ω **17.** 36.7 ft/sec **19.** 70 Ω

Section 2.5 (page 68)

1. $y = kx$ **3.** $w = ki^2$ **5.** $y = \frac{k}{x^2}$ **7.** $y = kxw$ **9.** $A = k\left(\frac{a}{b}\right)$ **11.** $F = k\left(\frac{mM}{r^2}\right)$
13. $y = 4x$ **15.** $P = \frac{2}{V}$ **17.** 6.9 lb **19.** $S = kwd^2$ **21.** 100 units **23.** $f = k\frac{q_1 q_2}{r^2}$
25. 4 atmospheres **27.** $120 **29.** $R = 0.0000196 \, L/D^2$

Section 2.6 (page 74)

1. $V = \frac{4}{3}\pi r^3$ (V dependent) **3.** $V = 4\pi h$ (V dependent) **5.** all x **7.** $x \geq -2$ **9.** $x \neq 0$
11. **13.** **15.** **17.**

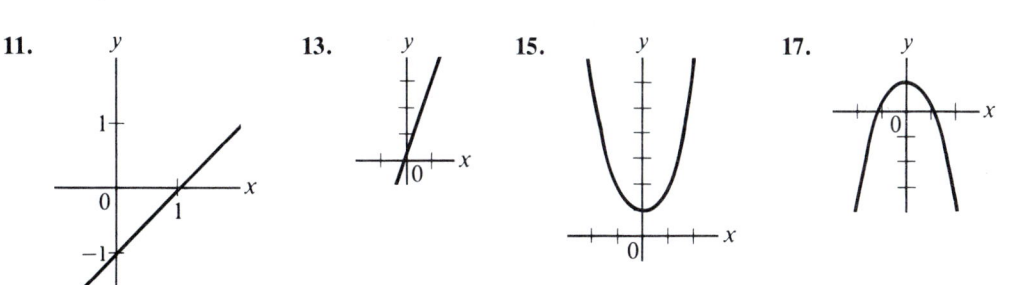

19.

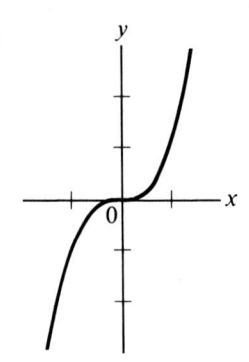

21.

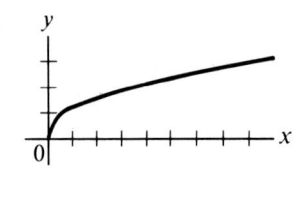

23.

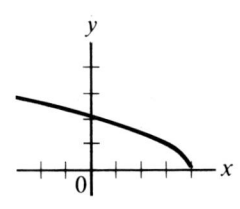

25.

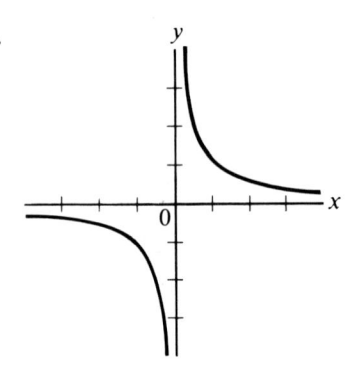

27.

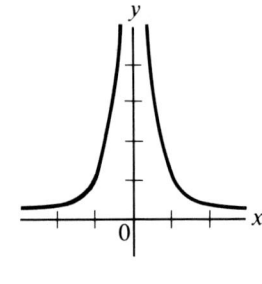

29.

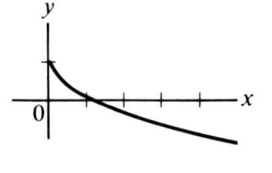

31.

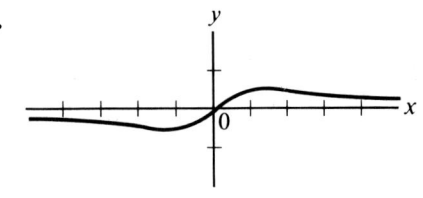

33. $R = 1\,\Omega$

Section 2.7 (page 76)

1. 1, 1, 1, 61   **3.** 0, 3   **5.** $\frac{1}{2}$, 1   **7.** 1, 3   **9.** 0, -28 **11.** $a^2 + 2$, $a^4 + 2$
13. 3.42 Ω, 25.2 Ω

Review Exercises for Chapter 2 (page 77)

1. $x = 9$ **3.** $x = 1$ **5.** $x = 3$ **7.** $x = -2$ **9.** $x = \frac{b}{a}$ **11.** $x = \frac{3b + 3}{2a}$ **13.** $R = \frac{V}{I}$
15. $b = \frac{3a}{Q - C}$ **17.** 2.5 A, 0.7 A   **19.** 40 lb   **21.** $22\frac{2}{9}$ min   **23.** 60%   **25.** 30 ft
27. $y = k\left(\frac{st}{r}\right)$ **29.** $w = \frac{k}{d^2}$ **31.** $R = k\left(\frac{L}{D^2}\right)$ **33.**   **35.** $x \geq 4$

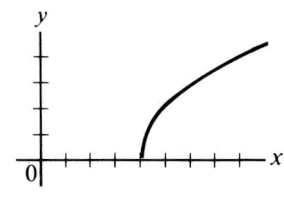

Chapter 3

Section 3.1 (page 84)
1. $(-1.0, -3.0)$ **3.** $(2.0, 3.0)$ **5.** $(5.8, -3.3)$ **7.** $(-0.2, 0.8)$ **9.** $(2.7, 2.3)$
11. $(2.9, -8.6)$

Section 3.2 (page 90)
1. $(3, 1)$ **3.** $(-5, -8)$ **5.** $(3, -6)$ **7.** $\left(-\frac{1}{2}, 3\right)$ **9.** $\left(\frac{7}{20}, \frac{3}{20}\right)$ **11.** $(1, -1)$ **13.** $(3, 3)$
15. $(-1, 3)$ **17.** $\left(-1, \frac{1}{2}\right)$ **19.** $\left(\frac{5}{3}, -\frac{8}{3}\right)$ **21.** $(0, -7)$ **23.** inconsistent **25.** $(1, 1)$
27. $\left(\frac{5}{4}, 5\right)$ **29.** $\left(-2, -\frac{1}{2}\right)$ **31.** $\left(-\frac{1}{3}, 3\right)$ **33.** $(0, 1)$ **35.** dependent **37.** inconsistent
39. $(1, 2)$

Section 3.3 (page 95)
1. 0 **3.** 0 **5.** 1 **7.** -10 **9.** 2 **11.** -69 **13.** 869 **15.** 450 **17.** $(-13, 10)$
19. $(22, 31)$ **21.** $(-83, -68)$ **23.** $\left(-3, \frac{17}{3}\right)$ **25.** $\left(\frac{13}{44}, -13\right)$ **27.** $\left(-\frac{19}{12}, -\frac{19}{68}\right)$
29. $(20, 10)$ **31.** $\left(\frac{7}{3}, \frac{4}{3}\right)$ **33.** $\left(2, \frac{7}{2}\right)$ **35.** inconsistent **37.** $(2.43, 1.07)$
39. $(-2.32, -0.974)$ **41.** $(0.669, -1.28)$

Section 3.4 (page 99)
1. $w_1 = 2$ lb, $w_2 = 1$ lb **3.** $w_1 = 11.00$ lb, $w_2 = 9.000$ lb **5.** $S = 566 + 0.00200T$ **7.** 70 Ω, 80 Ω
9. $4,110 at 12%, $4,390 at 11% **11.** 5 dozen at $35/dozen, 9 dozen at $50/dozen
13. 20 at $1.95, 60 at $2.50 **15.** 30.1 V, 25.0 V **17.** 190, 25 **19.** 8 L of 5% solution
21. 47, 15 **23.** 19 dimes, 28 nickels **25.** 7 days at $250, 6 days at $200

Section 3.5 (page 103)
1. $(1, 2, 3)$ **3.** $(1, 1, -2)$ **5.** $(1, -1, 2)$ **7.** $\left(2, -\frac{3}{2}, \frac{7}{2}\right)$ **9.** $\left(\frac{2}{3}, -\frac{5}{3}, 4\right)$
11. $\left(\frac{1}{6}, \frac{7}{20}, -\frac{7}{25}\right)$ **13.** $(1, 2, -2, 3)$ **15.** $\left(-\frac{1}{2}, 2, 3, -1\right)$

Section 3.6 (page 111)
1. 4 **3.** -55 **5.** 115 **7.** -234 **9.** 940 **11.** -129 **13.** $(1, 2, 3)$ **15.** $(1, 1, -2)$
17. $(1, -1, 2)$ **19.** $\left(2, -\frac{3}{2}, \frac{7}{2}\right)$ **21.** $\left(-\frac{8}{19}, -\frac{12}{19}, -\frac{1}{19}\right)$ **23.** $w_1 = 1$ lb, $w_2 = 1$ lb, $w_3 = 2$ lb
25. $1,200 at 8%, $1,550 at 10%, $3,200 at 12% **27.** $20, $6, $14
29. $I_1 = \frac{24}{11}$ A, $I_2 = -\frac{10}{11}$ A, $I_3 = \frac{14}{11}$ A **31.** $I_1 = 4$ A, $I_2 = 3$ A, $I_3 = -1$ A
33. $I_1 = \frac{98}{23}$ A, $I_2 = \frac{44}{23}$ A, $I_3 = \frac{54}{23}$ A **35.** $I_1 = \frac{13}{17}$ A, $I_2 = \frac{9}{17}$ A, $I_3 = -\frac{14}{17}$ A, $I_4 = \frac{8}{17}$ A

Review Exercises for Chapter 3 (page 114)
1. -7 **3.** 160 **5.** 3 **7.** 238 **9.** $(-1.5, 2.0)$ **11.** $(2.5, 2.0)$ **13.** $\left(\frac{14}{11}, \frac{1}{11}\right)$
15. $\left(-\frac{5}{2}, -\frac{3}{2}\right)$ **17.** $\left(\frac{9}{5}, -\frac{1}{5}\right)$ **19.** $\left(2, \frac{1}{3}\right)$ **21.** $\left(\frac{37}{38}, -\frac{17}{38}\right)$ **23.** $\left(-\frac{9}{10}, \frac{13}{5}\right)$ **25.** $(2, -1)$

642 APPENDIX C

27. $\left(-\frac{1}{4}, \frac{11}{8}\right)$ **29.** $\left(\frac{3}{35}, \frac{3}{37}\right)$ **31.** (2, 4) **33.** (2, 3) **35.** $\left(-\frac{1}{2}, 3\right)$ **37.** (1, −1, 3)
39. $\left(-\frac{1}{2}, 1, \frac{1}{3}\right)$ **41.** (−1, 3, −2) **43.** $\left(\frac{2}{3}, -2, -2\right)$ **45.** $a = -1$
47. \$172 at 10%, \$430 at 8%, \$412.80 at $12\frac{1}{2}$% **49.** 60 gal of 1% milk
51. $I_1 = 1$ A, $I_2 = 1$ A, $I_3 = 0$ A **53.** $I_1 = 0$ A, $I_2 = -1$ A, $I_3 = 0$ A, $I_4 = -1$ A

Cumulative Review Exercises for Chapters 1–3 (page 118)

1. $-7T_a - 10T_b + 2T_c$ **2.** $2L - 2C$ **3.** $5\sqrt{6}$ **4.** $L\sqrt{\pi}/(2\pi)$ **5.** $3p^4q^3$ **6.** $4A^2$
7. $a^3 - 5a^2b + 4ab^2 + 4b^3$ **8.** $x^2 + 2x + 3$ **9.** 6.4×10^{-14} **10.** $x = \frac{29}{11}$ **11.** $x = 18$
12. $t_1 = \dfrac{L_1 + L_1\beta t_2 - L}{L_1\beta}$ **13.** $x \leq 4$ **14.** $\sqrt{2}, 2$ **15.** (2, 0) **16.** (26, 15) **17.** $\left(\frac{7}{2}, 2, -\frac{3}{2}\right)$
18. $6.00I_1 - 5.00I_2$ **19.** $3F_1 + 2F_2$ **20.** 2.4×10^5 miles **21.** 1.2 mm by 3.6 mm **22.** 6.56 lb

Chapter 4

Section 4.2 (page 123)

3. 205° **5.** 287°20′ **7.** 180°3′ **9.** 213°55′ **11.** 178°6′ **13.** 34°19′ **15.** 355°

Section 4.3 (page 129)

For Exercises 1–19, the trigonometric functions are given in the following order: sin θ, cos θ, tan θ, csc θ, sec θ, and cot θ.

1. $\dfrac{2}{\sqrt{5}} = \dfrac{2\sqrt{5}}{5}, \dfrac{1}{\sqrt{5}} = \dfrac{\sqrt{5}}{5}, 2, \dfrac{\sqrt{5}}{2}, \sqrt{5}, \dfrac{1}{2}$ **3.** $\dfrac{2}{\sqrt{29}} = \dfrac{2\sqrt{29}}{29}, \dfrac{5}{\sqrt{29}} = \dfrac{5\sqrt{29}}{29}, \dfrac{2}{5}, \dfrac{\sqrt{29}}{2}, \dfrac{\sqrt{29}}{5}, \dfrac{5}{2}$
5. $\dfrac{1}{\sqrt{5}} = \dfrac{\sqrt{5}}{5}, \dfrac{2}{\sqrt{5}} = \dfrac{2\sqrt{5}}{5}, \dfrac{1}{2}, \sqrt{5}, \dfrac{\sqrt{5}}{2}, 2$ **7.** $\dfrac{4}{3\sqrt{2}} = \dfrac{2\sqrt{2}}{3}, \dfrac{1}{3}, \dfrac{4}{\sqrt{2}} = 2\sqrt{2}, \dfrac{3\sqrt{2}}{4}, 3, \dfrac{\sqrt{2}}{4}$
9. $\dfrac{\sqrt{6}}{\sqrt{42}} = \dfrac{1}{\sqrt{7}} = \dfrac{\sqrt{7}}{7}, \dfrac{6}{\sqrt{42}} = \dfrac{\sqrt{42}}{7}, \dfrac{\sqrt{6}}{6}, \sqrt{7}, \dfrac{\sqrt{42}}{6}, \sqrt{6}$
11. $\dfrac{\sqrt{10}}{\sqrt{59}} = \dfrac{\sqrt{590}}{59}, \dfrac{7\sqrt{59}}{59}, \dfrac{\sqrt{10}}{7}, \dfrac{\sqrt{590}}{10}, \dfrac{\sqrt{59}}{7}, \dfrac{7\sqrt{10}}{10}$ **13.** $\dfrac{\sqrt{3}}{6}, \dfrac{\sqrt{33}}{6}, \dfrac{\sqrt{11}}{11}, 2\sqrt{3}, \dfrac{2\sqrt{33}}{11}, \sqrt{11}$
15. $\dfrac{\sqrt{10}}{4}, \dfrac{\sqrt{6}}{4}, \dfrac{\sqrt{15}}{3}, \dfrac{2\sqrt{10}}{5}, \dfrac{2\sqrt{6}}{3}, \dfrac{\sqrt{15}}{5}$ **17.** $\dfrac{\sqrt{7}}{5}, \dfrac{3\sqrt{2}}{5}, \dfrac{\sqrt{14}}{6}, \dfrac{5\sqrt{7}}{7}, \dfrac{5\sqrt{2}}{6}, \dfrac{3\sqrt{14}}{7}$
19. $\dfrac{2\sqrt{142}}{71}, \dfrac{3\sqrt{497}}{71}, \dfrac{2\sqrt{14}}{21}, \dfrac{\sqrt{142}}{4}, \dfrac{\sqrt{497}}{21}, \dfrac{3\sqrt{14}}{4}$ **21.** $\dfrac{3}{4}$ **23.** $\dfrac{\sqrt{3}}{2}$ **25.** $\dfrac{\sqrt{34}}{5}$ **27.** $\dfrac{4}{9}$ **29.** $\dfrac{6}{7}$
31. $\dfrac{4\sqrt{19}}{19}$ **33.** $\dfrac{\sqrt{6}}{3}$ **35.** $\dfrac{\sqrt{33}}{6}$ **37.** $\sqrt{10}$ **39.** $\sqrt{10}$

Section 4.4 (page 137)

1. $\dfrac{\sqrt{3}}{3}$ **3.** $\dfrac{\sqrt{3}}{2}$ **5.** 1 **7.** 2 **9.** 1 **11.** $\dfrac{2\sqrt{3}}{3}$ **13.** 1 **15.** $\sqrt{2}$ **17.** $\dfrac{1}{2}$
19. $\dfrac{\sqrt{2}}{2}$ **21.** 45° **23.** 45° **25.** 30° **27.** 45° **29.** 0° **31.** 30° **33.** 60° **35.** 90°
37. 45° **39.** 30° **41.** 0.1851 **43.** 1.319 **45.** 11.10 **47.** 1.217 **49.** 1.047
51. 3.030 **53.** 0.3169 **55.** 0.9960 **57.** 0.7950 **59.** 4.843 **61.** 20°20′ **63.** 51°10′
65. 85°38′ **67.** 81°25′ **69.** 6°5′ **71.** 38°9′ **73.** 18°38′ **75.** 46°15′ **77.** 45°38′
79. 82°28′ **81. a.** $W = 16.1$ g **b.** $W = 17.1$ g **c.** $W = 19.3$ g **d.** $W = 23.4$ g **e.** $W = 25.1$ g
83. 46.1 V

Section 4.5 (page 143)

1. 1.67 **3.** 73.5 **5.** $B = 59°29'$, $a = 2.075$, $c = 4.086$ **7.** $A = 33°50'$, $B = 56°10'$, $b = 782$
9. $B = 67°20'$, $a = 0.771$, $b = 1.85$ **11.** $B = 84°47'$, $b = 0.08981$, $c = 0.09019$ **13.** 65.0 ft
15. 7.0 ft **17.** 3.72 ft **19.** 0.160 in. **21.** 1,250 ft **23.** 134 ft **25.** 345.5 ft **27.** 471 ft
29. 27.67 m **31.** 932 ft^2 **33.** 8.21 m **35.** 31.3°

Review Exercises for Chapter 4 (page 146)

In Exercises 1 and 3, the trigonometric functions are given in the following order: sin θ, cos θ, tan θ, csc θ, sec θ, and cot θ.

1. $\dfrac{\sqrt{10}}{10}, \dfrac{3\sqrt{10}}{10}, \dfrac{1}{3}, \sqrt{10}, \dfrac{\sqrt{10}}{3}, 3$ **3.** $\dfrac{2}{3}, \dfrac{\sqrt{5}}{3}, \dfrac{2\sqrt{5}}{5}, \dfrac{3}{2}, \dfrac{3\sqrt{5}}{5}, \dfrac{\sqrt{5}}{2}$ **5.** $\dfrac{2\sqrt{6}}{5}$ **7.** $\dfrac{5\sqrt{7}}{14}$ **9.** $\sqrt{3}$
11. $\dfrac{\sqrt{38}}{6}$ **13.** $\dfrac{1}{2}\sqrt{4 - a^2}$ **17.** $\dfrac{\sqrt{3}}{2}$ **19.** 1 **21.** 0 **23.** 0 **25.** 60° **27.** 45°
29. 30° **31.** 30° **33.** 0.5336 **35.** 1.757 **37.** 7.068 **39.** 10°45' **41.** 10°59'
43. 30°25' **45.** 69.5° **47.** 11.3 in.^{2} **49.** 1.28 in. **51.** 65.4 ft **53.** 1,100 ft

Chapter 5

Section 5.1 (page 150)

1. $2x - 6y$ **3.** $2a^4 - 2a^2b^2 - 6a^2c$ **5.** $VR_1^2 - VR_2^2$ **7.** $2PQ^2 - 2P^3 + 2P$
9. $2ab^2c - 4b^3c - 2b^2c$ **11.** $x^2 - 4y^2$ **13.** $36A^2 - B^2$ **15.** $a^4 - b^2$ **17.** $4x^4 - y^4$
19. $S^2 - 4T^4$ **21.** $x^2y^2 - 1$ **23.** $9R^2S^4 - 1$

Section 5.2 (page 156)

1. $2(a + b)$ **3.** $5(x - 1)$ **5.** $3x(x - 2)$ **7.** $2xy(2xy - 3x + 6y)$ **9.** $14c^2d^2(c + 2cd + d)$
11. $11R_1R_2(R_1 - 2R_2 + 1)$ **13.** $(x - y)(x + y)$ **15.** $4(x - y)(x + y)$ **17.** $16(2m - n)(2m + n)$
19. $2(2f_1 - f_2)(2f_1 + f_2)$ **21.** $2(5x - 4z)(5x + 4z)$ **23.** $\left(\dfrac{1}{2}x - y\right)\left(\dfrac{1}{2}x + y\right)$
25. $\left(\dfrac{1}{3}x - \dfrac{1}{4}y\right)\left(\dfrac{1}{3}x + \dfrac{1}{4}y\right)$ **27.** $\left(\dfrac{x}{a} - B\right)\left(\dfrac{x}{a} + B\right)$ **29.** $\left(1 - \dfrac{b}{a}\right)\left(1 + \dfrac{b}{a}\right)$
31. $(t_1^2 + t_2^2)(t_1 - t_2)(t_1 + t_2)$ **33.** $3(x - 3a)(x + 3a)$ **35.** $(8a^2 - 1)(8a^2 + 1)$
37. $(x - y)(x^2 + xy + y^2)$ **39.** $2(x + a)(x^2 - ax + a^2)$
41. $(3a - y)(9a^2 + 3ay + y^2)$ **43.** $(5m - 3a)(25m^2 + 15am + 9a^2)$ **45.** $(ab - 1)(a^2b^2 + ab + 1)$
47. $(1 + xy^2)(1 - xy^2 + x^2y^4)$ **49.** $(a + b - 1)(a + b + 1)$ **51.** $(1 - x - y)(1 + x + y)$
53. $(x - y + z)(x^2 - 2xy + y^2 - xz + yz + z^2)$
55. $(v_1 - v_2 - v_3)(v_1^2 - 2v_1v_2 + v_2^2 + v_1v_3 - v_2v_3 + v_3^2)$
57. $(F_1 + F_2 - 2)(F_1^2 + 2F_1F_2 + F_2^2 + 2F_1 + 2F_2 + 4)$
59. $(2 + 3r + 2s)(4 - 6r - 4s + 9r^2 + 12rs + 4s^2)$ **61.** $16(x - 2a)(x^2 + 2ax + 4a^2)$
63. $E_d = a(T_1^2 + T_2^2)(T_1 + T_2)(T_1 - T_2)$
65. $\dfrac{1}{2}m(v_2 - v_1)(v_2 + v_1)$ **67.** $\dfrac{4}{3}\pi(r_2 - r_1)(r_2^2 + r_2r_1 + r_1^2)$

Section 5.3 (page 160)

1. $x^2 + 4x + 4$ **3.** $C_1^2 + 2C_1C_2 + C_2^2$ **5.** $x^2 - x - 2$ **7.** $15x^2 - 7x - 2$
9. $12x^2 - 11xy + 2y^2$ **11.** $s^2 - 5st + 6t^2$ **13.** $49i_1^2 + 28i_1i_2 + 4i_2^2$ **15.** $9x^2 - 12xy + 4y^2$
17. $2i^2 + 11ik + 15k^2$ **19.** $2x^2 - 8xy - 10y^2$ **21.** $6x^2 - 31xy + 40y^2$ **23.** $3v_1^2 - 12v_2^2$
25. $81x^2 - 36y^2$ **27.** $7x^2 - 28y^2$ **29.** $9x^3 - xy^2$ **31.** $2x^2 - 10xy + 12y^2$
33. $12ax^2 + 17axy - 5ay^2$ **35.** $x^2 + 2xy + y^2 + 4xz + 4yz + 4z^2$
37. $a^2 + 4ab + 4b^2 + 2ac + 4bc + c^2$ **39.** $x^2 + 2xy + y^2 + xz + yz - 6z^2$
41. $p^2 + 2pq + q^2 - 7pr - 7qr + 10r^2$ **43.** $2x^2 + 3ax + 3bx + a^2 + 2ab + b^2$

Section 5.4 (page 166)

1. $(x + 2y)^2$ 3. $(3x - 4y)^2$ 5. $(x - 1)(x - 3)$ 7. $(x - 4)(x + 3)$ 9. $2(a - 2b)^2$
11. $2(x + 6)(x + 1)$ 13. $(x - 6y)(x + y)$ 15. $(D + 7)(D - 2)$ 17. $(2x - y)(x - y)$
19. $(5x - y)(x - 2y)$ 21. $(4x + y)(x + 3y)$ 23. $(2x - 3y)(3x + 4y)$ 25. $(5w_1 - 2w_2)(w_1 - 4w_2)$
27. $8(L - 3C)(L + 2C)$ 29. $2(3f - 4g)^2$ 31. $x^2(x - 2)^2$ 33. not factorable
35. $(a + b - 3)(a + b + 2)$ 37. $(n + m - 2)(n + m - 1)$ 39. $(2a + 2b - 1)(a + b - 4)$
41. $(f_1 + 2f_2)^2(f_1 + 2f_2 - 1)$ 43. $(1 - x + y)(1 + x - y + x^2 - 2xy + y^2)$ 45. $(7a - 2b)(4a + b)$
47. $(5x - y)(8x + 3y)$ 49. $(3\alpha - 2\beta)(4\alpha - 5\beta)$ 51. $t = 3$ sec 53. $t = 1.33$ sec

Section 5.5 (page 169)

1. $(x - y)(a + b)$ 3. $(x + 3y)(2x + 1)$ 5. $(a - b)(4c + 1)$ 7. $(x - y)(5b - 1)$
9. $2(x + y)(a - c)$ 11. $3(R - r)(a - 2b)$ 13. $(x + y)(x - y - z)$ 15. $(x + y)(a - x + y)$
17. $(x - y)(x + y - 2z)$ 19. $(x - y)(x + 4y + 1)$ 21. $(2x - y - z)(2x - y + z)$
23. $(x + 2y - z)(x + 2y + z)$ 25. $(3a - 2b - c)(3a + 2b + c)$ 27. $(x - y)(3 - x - 4y)$
29. $(x + 2y)(a - x + 3y)$ 31. $a(A - 1)(Aa - 4)$

Section 5.6 (page 174)

1. $\dfrac{x}{2}$ 3. $\dfrac{a}{2x}$ 5. x 7. $\dfrac{2}{x}$ 9. $x + 4$ 11. $x^2 - xy + y^2$ 13. $\dfrac{x - y}{3}$
15. $R^2 + 2R + 4$ 17. $x - 4$ 19. $i + 5$ 21. $\dfrac{3x - 4}{x + 3}$ 23. $\dfrac{7x + 2}{2x + 1}$ 25. $\dfrac{m_1 + m_2}{m_1 - 2m_2}$
27. -1 29. $\dfrac{2L + C}{3L - 4C}$ 31. $\dfrac{x - 4}{x - 6}$ 33. $\dfrac{a - 4b}{2a - b}$ 35. $\dfrac{1}{2x + 1}$ 37. $\dfrac{1}{2l + 1}$ 39. $\dfrac{1}{3t + 1}$
41. $\dfrac{1}{3E - 2}$ 43. $P - Q$ 45. $\dfrac{M^2 + m^2}{(M + m)^2}$ 47. $2t - 6$ (amperes)

Section 5.7 (page 177)

1. $\dfrac{3xy}{2zw^2}$ 3. $\dfrac{8x^2y}{3a^2b}$ 5. $\dfrac{6}{5}bcxy$ 7. $\dfrac{dx}{ac^2}$ 9. $2x(x - y)$ 11. $\dfrac{x + y}{x - y}$ 13. $-\dfrac{a + 2b}{3}$
15. $-(1 + 3a)(x^2 - 3xy + 9y^2)$ 17. $\dfrac{4(2a + b)}{(3x - 1)(a + b)}$ 19. $\dfrac{(3v_0 - 4)(v_1 + 6)}{(2v_0 - 3)(2v_1 + 1)}$
21. $\dfrac{(3T + J)(5K - 6)}{(4T + J)(K - 7)}$ 23. $\dfrac{x - 3y}{(x + y)(x^2 + 2xy + 4y^2)}$ 25. $\dfrac{x - y}{x + 2y}$ 27. $\dfrac{1}{(R + 2r)(R + r - 1)}$
29. $\dfrac{2(2L - 7)}{(L + 4)(C - 5)}$ 31. $\dfrac{2}{c - d}$ 33. 2 35. $\dfrac{2w_1w_2}{w_1 - w_2}$ 37. $\dfrac{1}{8}w(s - 2H)$

Section 5.8 (page 183)

1. 1 3. $\dfrac{5x + 1}{9}$ 5. $\dfrac{3x - 3}{4x}$ 7. $\dfrac{4a - 1}{12b}$ 9. $\dfrac{8xy}{x^2 - y^2}$ 11. $\dfrac{1}{y - x}$ 13. $\dfrac{2x^2 + 2xy - y^2}{(x + 2y)(x - y)}$
15. $\dfrac{1}{x - 3}$ 17. $\dfrac{1}{x - y}$ 19. $\dfrac{4x^2 - 2y^2}{(3x - y)(2x - 3y)}$ 21. $-\dfrac{x}{y(x + y)}$ 23. $\dfrac{4a + 2b}{a + 2b}$ 25. $\dfrac{A + B}{A}$
27. 1 29. $\dfrac{1}{x^2 - y^2}$ 31. $\dfrac{4}{x^2 - 4}$ 33. $\dfrac{c^2 - cd}{(c + 2d)(2c + d)}$ 35. $\dfrac{3y}{(x + y)(2x + y)(x + 3y)}$
37. $\dfrac{k(2np - p^2)}{n^2(n - p)^2}$ 39. $\dfrac{k^2}{k^2 + L^2}$

ANSWERS TO ODD-NUMBERED EXERCISES   645

Section 5.9 (page 188)

1. $\dfrac{1}{2}$ 3. $\dfrac{8}{11}$ 5. $\dfrac{x}{3x+1}$ 7. $\dfrac{x-4}{x}$ 9. $C_1 + C_2$ 11. $\dfrac{x+5}{x-2}$ 13. $\dfrac{h-5}{h+4}$ 15. $\dfrac{w}{w+1}$

17. $\dfrac{1}{\beta+1}$ 19. $\dfrac{2E-3}{E^2-2E+1}$ 21. $\dfrac{k+3}{k+4}$ 23. $\dfrac{x+1}{x-4}$ 25. $\dfrac{(t+1)(t-4)}{(t-3)(2t+1)}$ 27. $\dfrac{R_1 R_2}{R_1 + R_2}$

29. $\dfrac{pf}{p-f}$ 31. $\dfrac{R_1(R_2 + R_3)}{R_1 + R_2 + R_3}$

Section 5.10 (page 193)

1. 3   3. -2 5. 10   7. 1   9. -8 11. no solution   13. -2 15. 12

17. no solution   19. $\dfrac{3}{2}$ 21. 0   23. 10   25. -5 27. 9   29. $d = \dfrac{c}{3c-1}$

31. $R_1 = \dfrac{3R}{3-R}$ 33. $f = \dfrac{pq}{p+q}$ 35. $R = \dfrac{R_1 R_2}{R_1 + R_2}$ 37. $10\,\Omega$

Review Exercises for Chapter 5 (page 195)

1. $8s^3 t^3 - 12s^3 t^5 + 20s^4 t^4$ 3. $4i_1^2 - 12i_1 i_2 + 9i_2^2$ 5. $4s^2 - 25t^2$ 7. $2c^2 + cd - 15d^2$
9. $x^3 y - 9xy^3$ 11. $v^2 + 4vw + 4w^2 + 2v + 4w + 1$ 13. $4x(a-b)$ 15. $pq(3p^2 q^3 + 1)$
17. $(2\beta - \gamma)(2\beta + \gamma)$ 19. $16(L - 2C)(L + 2C)$ 21. $(3h + g)(9h^2 - 3hg + g^2)$
23. $(a + 2b - 1)(a^2 + 4ab + 4b^2 + a + 2b + 1)$
25. $(a-1)(a+1)(a^4 + a^2 + 1)$ or $(a-1)(a+1)(a^2 - a + 1)(a^2 + a + 1)$
27. $(v_0 - 4)(v_0 + 3)$ 29. $(3v_1 - v_2)(v_1 + v_2)$ 31. $(C_1 - 5C_2)(C_1 + 3C_2)$ 33. not factorable
35. $(x - y - 1)(x - y + 1)$ 37. $(s + t)(s - 2t + 1)$ 39. $(a + b)(x - y)$
41. $(2x + y - a)(2x + y + a)$ 43. $x + 3y$ 45. $\dfrac{q + 7r}{q + 9r}$ 47. $\dfrac{2(a+d)}{a^2 d}$ 49. $x + y$

51. $\dfrac{x + 4y}{x + 2y}$ 53. $\dfrac{w^2}{y^2 - w^2}$ 55. 1   57. $\dfrac{a}{(a-b)(a+b)(a+2b)}$ 59. $\dfrac{2vw - w^2}{(v - 2w)(v + w)}$

61. $\dfrac{1}{\omega - 2}$ 63. $\dfrac{i}{i-2}$ 65. $\dfrac{1}{r_1}$ 67. 1   69. 3   71. $-\dfrac{12}{5}$ 73. no solution   75. 6

77. 1   79. $a = \dfrac{S(1-r)}{1 - r^n}$ 81. $t = 4.5$ sec, $t = 5$ sec   83. $q = 12$ cm   85. $\dfrac{m_2}{m_1 + m_2}$

Chapter 6

Section 6.1 (page 203)

1. $1, -1$ 3. $6, -6$ 5. $3, -3$ 7. $\sqrt{10}, -\sqrt{10}$ 9. $4, -4$ 11. $\dfrac{5}{6}, -\dfrac{5}{6}$ 13. $1, -2$

15. $4, -6$ 17. $3, -\dfrac{1}{2}$ 19. $-2, -\dfrac{1}{3}$ 21. $\dfrac{3}{4}, -2$ 23. $-3, \dfrac{7}{5}$ 25. $-\dfrac{5}{2}, \dfrac{3}{2}$ 27. $-\dfrac{3}{5}, \dfrac{5}{6}$

29. $-\dfrac{9}{4}, \dfrac{5}{2}$ 31. $-\dfrac{11}{7}, 1$ 33. $-\dfrac{3}{2}, \dfrac{5}{9}$ 35. $-\dfrac{1}{11}, 7$ 37. $-\dfrac{5}{9}, \dfrac{3}{8}$ 39. $\dfrac{11}{6}, \dfrac{10}{3}$

41. $-\dfrac{4}{3}, -\dfrac{4}{3}$ 43. $\dfrac{1}{4}, \dfrac{1}{4}$ 45. $\dfrac{v_0^2}{32}$ ft   47. $x = L, \dfrac{1}{3}L, 0$

Section 6.2 (page 208)

1. $2, 4$ 3. $-6, 2$ 5. $-4, 3$ 7. $-5, -2$ 9. $\dfrac{1}{2}(-5 \pm \sqrt{17})$ 11. $-3 \pm \sqrt{3}$

13. $\dfrac{1}{2}(3 \pm \sqrt{7})$ 15. $\dfrac{1}{4}(-3 \pm \sqrt{33})$ 17. $-1, \dfrac{1}{3}$ 19. $\dfrac{1}{3}(2 \pm \sqrt{19})$ 21. $-\dfrac{3}{4}, 1$

23. $\frac{1}{12}(-1 \pm \sqrt{47}j)$ 25. $2 \pm j$ 27. $\frac{5}{8} \pm \frac{\sqrt{23}}{8}j$ 29. $\frac{1}{7}(-1 \pm 2\sqrt{2})$ 31. $\frac{1}{12}(5 \pm \sqrt{73})$
33. $-\frac{10}{3}, \frac{5}{2}$ 35. $\frac{1}{10} \pm \frac{\sqrt{19}}{10}j$ 37. $\frac{1}{2}(b \pm \sqrt{b^2 - 8})$ 39. $\frac{1}{2a}(-5 \pm \sqrt{4a + 25})$

Section 6.3 (page 213)

1. $-3, 2$ 3. $4, 5$ 5. $\frac{1}{2}, 2$ 7. $-\frac{1}{2}, \frac{2}{3}$ 9. $\frac{3 \pm \sqrt{17}}{4}$ 11. $\frac{-1 \pm \sqrt{7}}{3}$ 13. $1 \pm j$
15. $-1 \pm \sqrt{3}j$ 17. $\frac{-3 \pm \sqrt{3}j}{6}$ 19. $\frac{-1 \pm \sqrt{3}j}{4}$ 21. $\frac{-5 \pm \sqrt{73}}{8}$ 23. $\pm \frac{\sqrt{5}}{5}j$ 25. $-\frac{3}{2}, 0$
27. $\frac{3c \pm \sqrt{9c^2 - 8}}{4}$ 29. $\frac{-3 \pm \sqrt{9 - 4b}}{2b}$ 31. $\frac{3}{2}, \frac{3}{2}$ 33. $\frac{5}{2}, \frac{5}{2}$ 35. $0.70, -2.26$
37. $5.77, -0.18$ 39. $0.66, -0.49$ 41. $x = 1 + y, x = 1 - y$ 43. $x = 2y - 1, x = 2y + 1$
45. $x = y - 1, x = y - 2$ 47. $x = 1 - y, x = 2 + y$ 49. $-5, 4$ 51. $-\frac{6}{5}, 4$ 53. $\frac{4}{3}, 8$
55. $\frac{7 \pm \sqrt{33}}{2}$

Section 6.4 (page 217)

1. $t = 0$ sec, $t = 0.49$ sec   3. $8.22 \, \Omega, 10.22 \, \Omega$ 5. 6.0 cm   7. 18.19 in. for base
9. 5.00 in. $\times 17.0$ in.   11. 8 in. $\times 8$ in.   13. $2\frac{2}{3}$ hr, 8 hr   15. 60 min, 40 min
17. 25 mi/hr, 10 mi/hr   19. 30 ft $\times 40$ ft or $\frac{80}{3}$ ft $\times 45$ ft   21. 10 ft $\times 15$ ft   23. 8 shares

Review Exercises for Chapter 6 (page 219)

1. $-2, 5$ 3. $-\frac{3}{5}, 2$ 5. $-\frac{2}{3}, -\frac{1}{2}$ 7. $\frac{5}{6}, 1$ 9. $-2, 4$ 11. $\frac{-5 \pm \sqrt{13}}{2}$ 13. $\frac{5}{4} \pm \frac{\sqrt{7}}{4}j$
15. $\frac{1}{8} \pm \frac{\sqrt{47}}{8}j$ 17. $2, 4$ 19. $\frac{1}{2} \pm \frac{\sqrt{3}}{6}j$ 21. $\frac{2 \pm \sqrt{10}}{2}$ 23. $\frac{4 \pm \sqrt{34}}{6}$ 25. $0.81, -1.90$
27. $x = 2 + y, x = 2 - y$ 29. $-\frac{5}{9}, 4$ 31. 16.0 A, 18.0 A   33. 6 hr, 12 hr
35. 20 mi/hr, 35 mi/hr

Cumulative Review Exercises for Chapters 4–6 (page 220)

1. $122°19'$ 2. $\sin \theta = \frac{\sqrt{3}}{4}$, $\cos \theta = \frac{\sqrt{13}}{4}$, $\tan \theta = \frac{\sqrt{39}}{13}$, $\csc \theta = \frac{4\sqrt{3}}{3}$, $\sec \theta = \frac{4\sqrt{13}}{13}$, $\cot \theta = \frac{\sqrt{39}}{3}$
3. $\frac{3\sqrt{7}}{7}$ 4. $\frac{\sqrt{3}}{3}$ 5. 1.688 6. $56°59'$ 7. $20°31'$ 8. $(V_a - V_b)(s - 1)$
9. $\frac{1}{2}(L^2 + LC + C^2)$ 10. $\frac{x + y}{2(x + 3y)}$ 11. $\frac{2a - 1}{a(2x + y)}$ 12. $\frac{3st - t^2 + 1}{s^2 - t^2}$ 13. $\frac{L}{L + 4}$
14. $x = 4, -2$ 15. $x = \frac{1}{4}, -3$ 16. $x = 1, 3$ 17. $x = 1 \pm j$ 18. $-0.975, 0.427$
19. 0.433 A   20. 1.55 in.   21. $\frac{R_1 R_2 R_3}{R_2 R_3 + R_1 R_3 + R_1 R_2}$ 22. 3.4 in. by 5.4 in.

Chapter 7

Section 7.1 (page 228)

In Exercises 1–11, the trigonometric functions are given in the following order: sin θ, cos θ, tan θ, csc θ, sec θ, and cot θ.

1. $\dfrac{3}{5}, -\dfrac{4}{5}, -\dfrac{3}{4}, \dfrac{5}{3}, -\dfrac{5}{4}, -\dfrac{4}{3}$ 3. $-\dfrac{12}{13}, \dfrac{5}{13}, -\dfrac{12}{5}, -\dfrac{13}{12}, \dfrac{13}{5}, -\dfrac{5}{12}$ 5. $\dfrac{2\sqrt{2}}{3}, \dfrac{1}{3}, 2\sqrt{2}, \dfrac{3\sqrt{2}}{4}, 3, \dfrac{\sqrt{2}}{4}$

7. $-\dfrac{3\sqrt{10}}{10}, -\dfrac{\sqrt{10}}{10}, 3, -\dfrac{\sqrt{10}}{3}, -\sqrt{10}, \dfrac{1}{3}$ 9. $\dfrac{\sqrt{15}}{4}, -\dfrac{1}{4}, -\sqrt{15}, \dfrac{4\sqrt{15}}{15}, -4, -\dfrac{\sqrt{15}}{15}$

11. $\dfrac{2\sqrt{5}}{5}, -\dfrac{\sqrt{5}}{5}, -2, \dfrac{\sqrt{5}}{2}, -\sqrt{5}, -\dfrac{1}{2}$ 13. $\dfrac{5\sqrt{29}}{29}$ 15. $-\dfrac{\sqrt{30}}{5}$ 17. $-\dfrac{\sqrt{7}}{3}$ 19. $\dfrac{\sqrt{6}}{2}$

21. $-\dfrac{\sqrt{13}}{4}$ 23. $-\dfrac{\sqrt{10}}{10}$ 25. $\sqrt{3}$ 27. $\dfrac{\sqrt{6}}{2}$

Section 7.2 (page 234)

1. $\dfrac{1}{2}$ 3. $-\dfrac{\sqrt{3}}{3}$ 5. $-\dfrac{\sqrt{2}}{2}$ 7. 2   9. $-\dfrac{1}{2}$ 11. 1   13. 2   15. 0   17. -1

19. $-\dfrac{\sqrt{3}}{3}$ 21. undefined   23. $-\dfrac{\sqrt{3}}{2}$ 25. $-\dfrac{2\sqrt{3}}{3}$ 27. $-\dfrac{\sqrt{3}}{3}$ 29. $-\dfrac{2\sqrt{3}}{3}$ 31. 0

33. $-\dfrac{\sqrt{2}}{2}$ 35. $-\dfrac{\sqrt{3}}{3}$ 37. $\dfrac{2\sqrt{3}}{3}$ 39. $\dfrac{1}{2}$ 45. 30°, 150°   47. 30°, 150°   49. 240°, 300°

51. 180°   53. 60°, 240°   55. 60°, 300°   57. 225°, 315°   59. 90°, 270°   61. 135°, 315°
63. 135°, 225°   65. 60°, 120°   67. 120°, 300°   69. 90°, 270°   71. 120°, 300°   73. 45°, 135°
75. 30°, 330°

Section 7.3 (page 238)

1. 0.8675   3. -5.575 5. -1.502 7. 1.381   9. 0.6356   11. 1.180   13. 2.801
15. 0.8993   17. -1.053 19. 1.294   21. -0.2568 23. 1.166   25. 0.8637   27. 2.723
29. -0.1794 31. 17°, 163°   33. 79°46′, 280°14′   35. 191°9′, 348°51′   37. 242°15′, 297°45′
39. 165°28′, 345°28′   41. 21°17′, 158°43′   43. 137°2′, 222°58′   45. 65°5′, 294°55′
47. 49°42′, 229°42′   49. 130°5′, 229°55′   51. 53.8 m   53. 91.6 lb   55. 90°

Section 7.4 (page 243)

1. $\dfrac{\pi}{6}$ 3. $\dfrac{\pi}{3}$ 5. $\dfrac{\pi}{9}$ 7. $\dfrac{2\pi}{5}$ 9. $-\dfrac{\pi}{3}$ 11. $\dfrac{7\pi}{6}$ 13. $\dfrac{11\pi}{20}$ 15. $\dfrac{11\pi}{9}$ 17. $\dfrac{3\pi}{5}$

19. $\dfrac{19\pi}{90}$ 21. $-\dfrac{28\pi}{15}$ 23. $\dfrac{13\pi}{20}$ 25. 45°   27. $-210°$ 29. 25°   31. 198°   33. 320°

35. 378°   37. 3°   39. 140°   41. 0.3069   43. 2.3178   45. 5.5920   47. -1.7482
49. 23°50′   51. 153°37′   53. 186°37′   55. $-51°43′$ 57. 0.7606   59. 1.6450
61. 5.8819   63. -0.4321

Section 7.5 (page 247)

1. 2.62 in.   3. 6.55 in.²   5. 9.63 cm   7. 13.6 ft   9. 62°   11. 2,100 mi   13. 31′
15. 39.5 cm²   17. 5.0 cm/sec   19. 10.4 ft/sec   21. 1,050 mi/hr   23. 6.0 rev/sec
25. $\dfrac{4\pi}{15}$ in./min   27. 18.5 mi/sec

Review Exercises for Chapter 7 (page 248)

In Exercises 1 and 3, the trigonometric functions are given in the following order: sin θ, cos θ, tan θ, csc θ, sec θ, and cot θ.

1. $\frac{\sqrt{5}}{3}, -\frac{2}{3}, -\frac{\sqrt{5}}{2}, \frac{3\sqrt{5}}{5}, -\frac{3}{2}, -\frac{2\sqrt{5}}{5}$ **3.** $-\frac{2\sqrt{5}}{5}, \frac{\sqrt{5}}{5}, -2, -\frac{\sqrt{5}}{2}, \sqrt{5}, -\frac{1}{2}$ **5.** $-\frac{2\sqrt{2}}{3}$

7. $-\frac{\sqrt{13}}{4}$ **9.** $\frac{\sqrt{2}}{2}$ **11.** $-\frac{1}{2}$ **13.** 1 **15.** $-\frac{2\sqrt{3}}{3}$ **17.** 1 **19.** $-\frac{\sqrt{3}}{2}$ **21.** 60°, 300°

23. 120°, 300° **25.** 90°, 270° **27.** 150°, 210° **29.** −1.525 **31.** 0.7528

33. 144°10′, 215°50′ **35.** 157°22′, 337°22′ **37.** 29°15′, 209°15′ **39.** $\frac{2\pi}{9}$ **41.** $-\frac{5\pi}{9}$ **43.** $\frac{3\pi}{10}$

45. 150° **47.** 234° **49.** 340° **51.** 0.4416 **53.** 2.5150 **55.** −4.4884 **57.** 6.8068
59. 41°2′ **61.** 149°15′ **63.** −1.493 **65.** −0.2106 **67.** 1.002 **69.** 3.35 in. **71.** 48 ft
73. 12.6 mi/hr **75.** 6,900 mi/hr

Chapter 8

Section 8.1 (page 257)

1. $A = 2, P = 2\pi$ **3.** $A = 1, P = 2\pi$

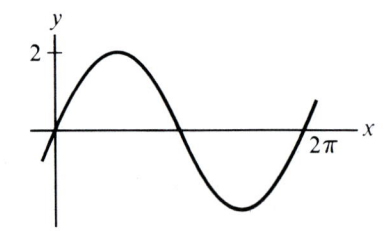

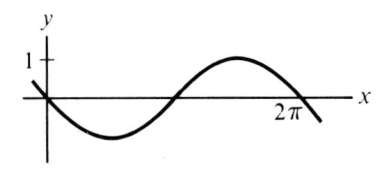

5. $A = \frac{1}{2}, P = 2\pi$ 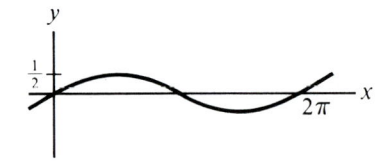   **7.** $A = 2, P = \frac{2\pi}{3}$ 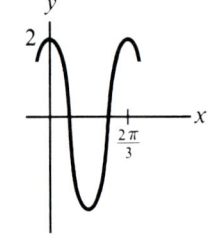   **9.** $A = 5, P = \pi$

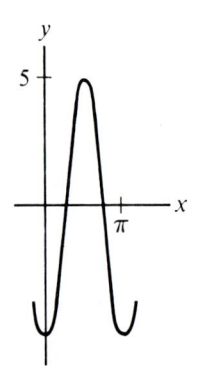

11. $A = \dfrac{1}{2}, P = \dfrac{2\pi}{3}$

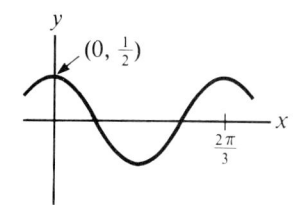

13. $A = 4, P = 3\pi$

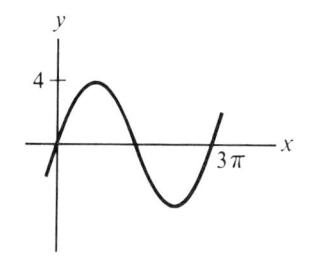

15. $A = 10, P = 8\pi$

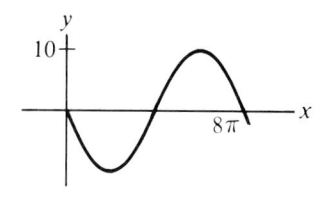

17. $A = 5, P = 10\pi$

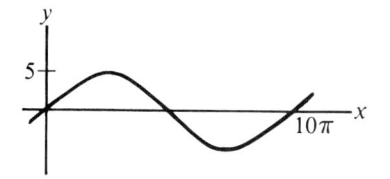

19. $A = 4, P = \dfrac{6\pi}{7}$

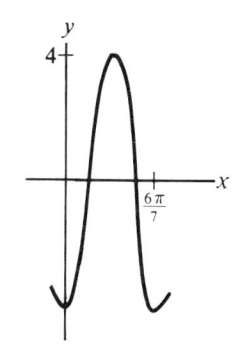

21. $A = 1, P = 2$

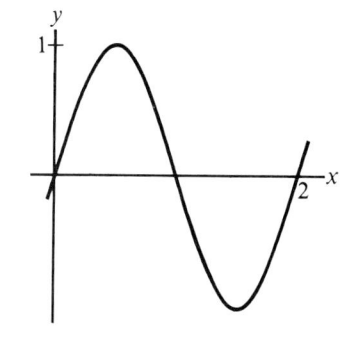

23. $A = \dfrac{1}{3}, P = \dfrac{2}{3}$

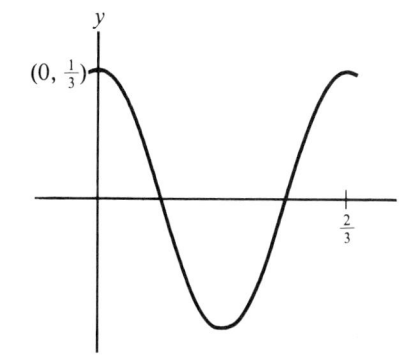

Section 8.2 (page 260)

1. $A = 2$, $P = 2\pi$, shift: $\dfrac{\pi}{4}$ to the right

3. $A = \dfrac{1}{2}$, $P = 2\pi$, shift: $\dfrac{\pi}{8}$ to the left

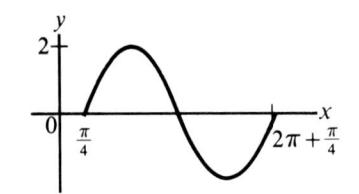

5. $A = \dfrac{1}{2}$, $P = 4\pi$, shift: $\dfrac{\pi}{4}$ to the right

7. $A = 3$, $P = 2\pi$, shift: $\dfrac{3\pi}{2}$ to the left

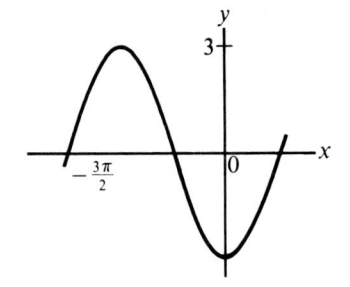

9. $A = 2$, $P = 2\pi$, shift: $\dfrac{\pi}{3}$ to the right

11. $A = 3$, $P = 2\pi$, shift: 2 to the right

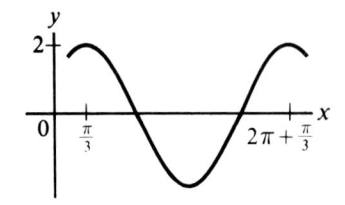

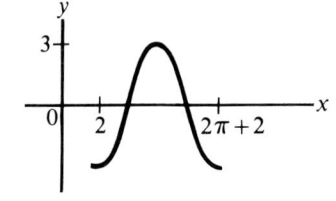

13. $A = 3$, $P = \pi$, shift: $\dfrac{\pi}{4}$ to the left   **15.** $A = 10$, $P = 6\pi$, shift: $\dfrac{12\pi}{5}$ to the right

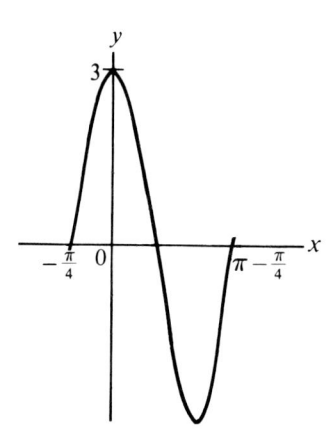

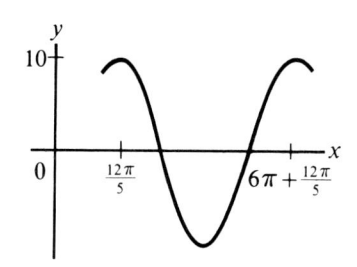

17. $A = 2$, $P = 2$, shift: 1 to the left   **19.** $A = 3$, $P = 4$, shift: $\dfrac{2}{\pi}$ to the right

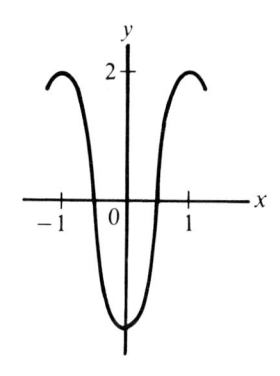

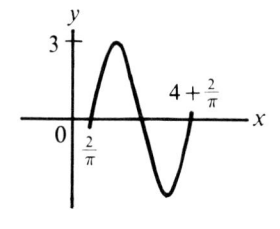

Section 8.3 (page 265)

1. $y = 3.0 \sin 2t$, $A = 3.0$ in., $P = \pi$ sec   **3.** $y = 5 \cos 8\pi t$, $A = 5$ cm, $P = \dfrac{1}{4}$ min

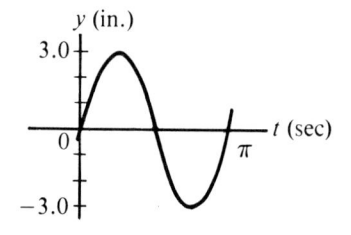

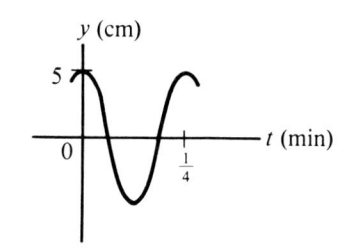

5. $A = 6.00$ V, $P = \dfrac{1}{40}$ sec, shift: $\dfrac{1}{320}$ sec to the left

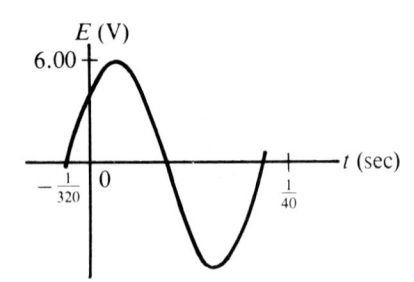

7. $A = 6.0$ A, $P = \dfrac{1}{60}$ sec, shift: $\dfrac{1}{480}$ sec to the left

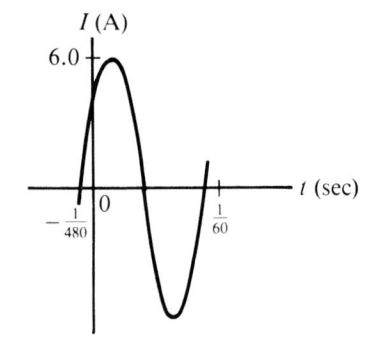

9. $A = 5$, $P = \pi$

11. $A = 0.002$ in., $P = \dfrac{1}{224}$ sec $= 0.0045$ sec

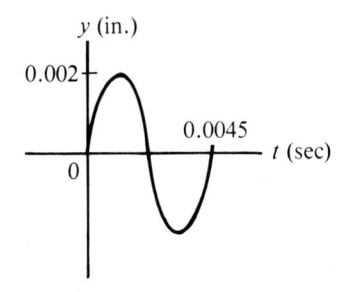

13. $y = 2.1 \sin \dfrac{\pi}{3.0} (t - 0.65)$, $A = 2.1$ ft, $P = 6.0$ sec, shift: 0.65 sec to the right

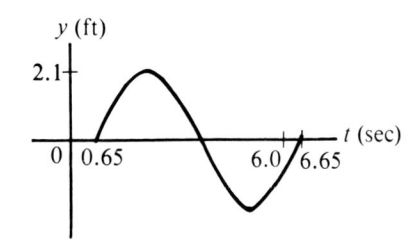

Section 8.4 (page 269)

1. $P = \pi$

3. $P = \pi$

5. $P = \pi$

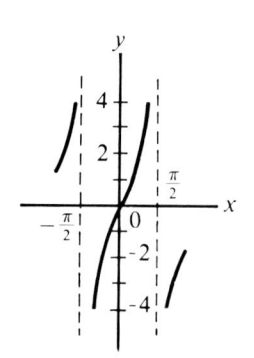

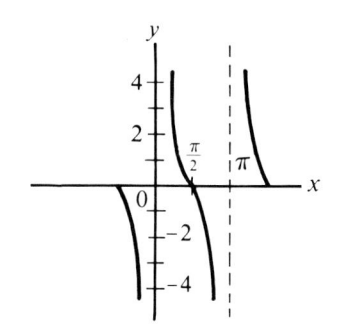

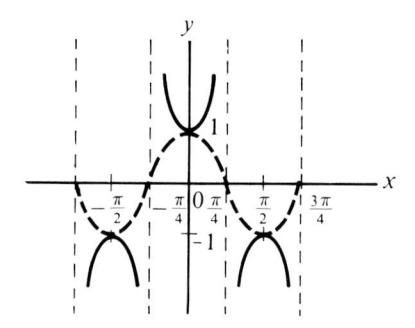

7. $P = \pi$

9. $P = \dfrac{\pi}{3}$

11. $P = \pi$

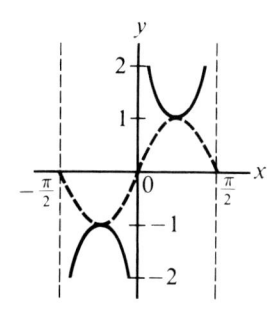

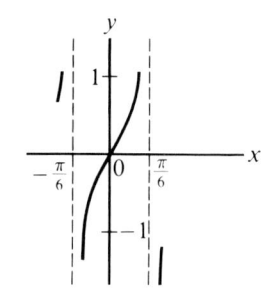

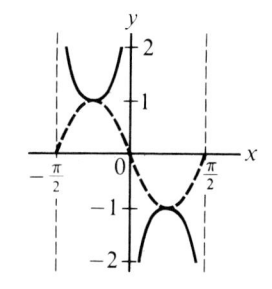

13. $P = 4\pi$

15. $P = 3\pi$

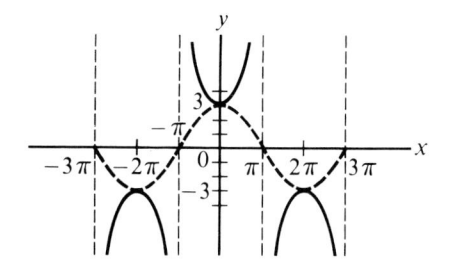

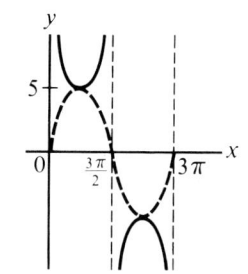

17. $P = \pi$, shift: $\dfrac{\pi}{6}$ to the left

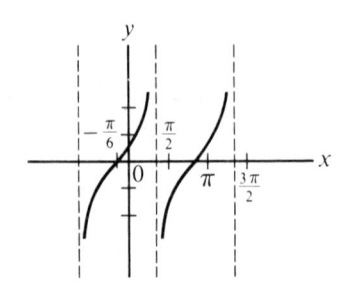

19. $P = \dfrac{1}{3}$

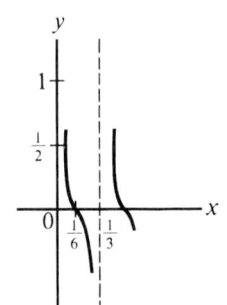

21. $P = 2\pi$

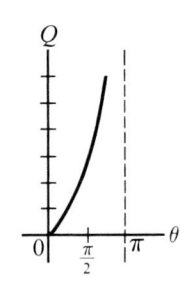

Section 8.5 (page 272)

1.

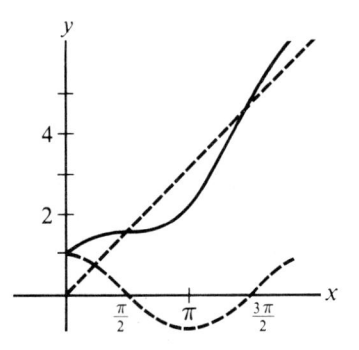

3.

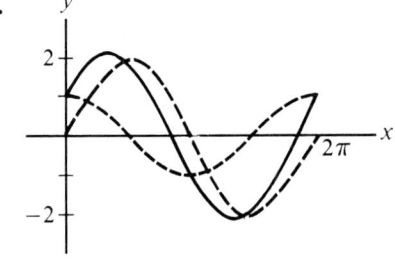

5.

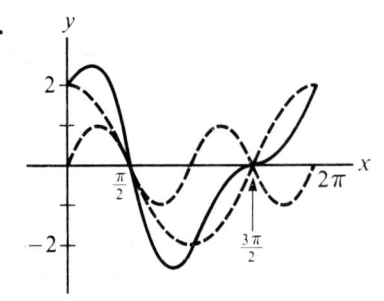

7.

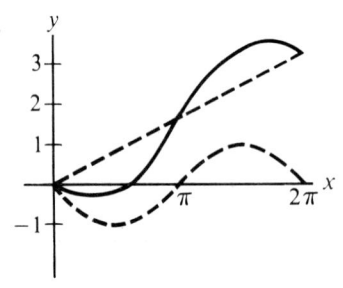

9.

11.

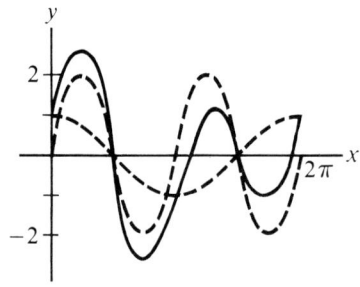

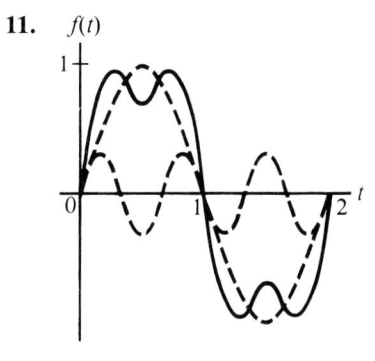

13. *i* (A)

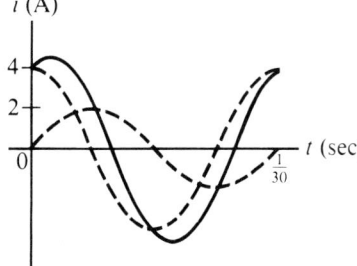

Review Exercises for Chapter 8 (page 272)

1. $A = 2, P = 2\pi$

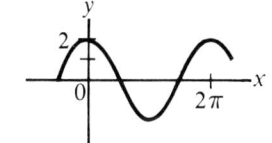

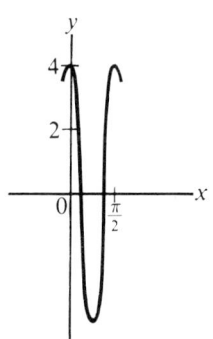

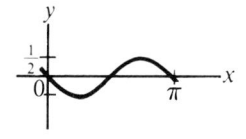

3. $A = 4, P = \dfrac{\pi}{2}$

5. $A = \dfrac{1}{2}, P = \pi$

7. $A = 4, P = 4\pi$

9. $A = 2, P = 2\pi$, shift: $\dfrac{\pi}{4}$ to the left

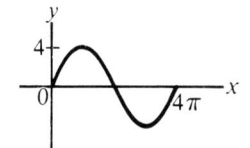

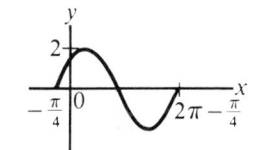

11. $A = \frac{1}{2}$, $P = 2\pi$, shift: $\frac{\pi}{8}$ to the left

13. $A = 4$, $P = 4\pi$, shift: $\frac{\pi}{4}$ to the right

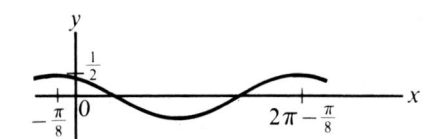

15. $A = \frac{1}{2}$, $P = \pi$, shift: $\frac{\pi}{8}$ to the right

17. $P = \pi$

19. $P = 2\pi$

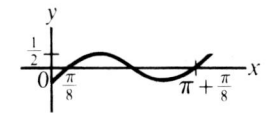

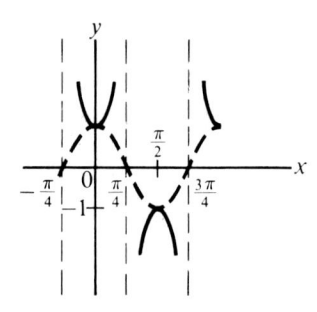

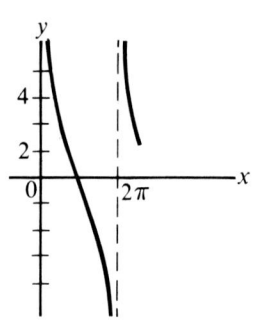

21. $y = 4 \sin 2t$, $A = 4$ cm, $P = \pi$ sec

23. $y = 6 \cos 10\pi t$
$A = 6$ cm, $P = \frac{1}{5}$ min

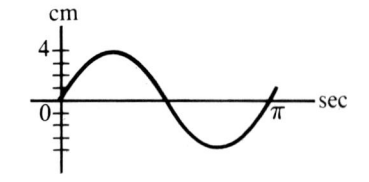

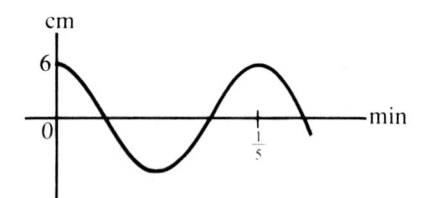

25. $A = 5.00$ V, $P = \frac{1}{30}$ sec, shift: $\frac{1}{180}$ sec to the left

27. $y = 0.003 \sin 2\pi(236)t$
$A = 0.003$ in., $P = \frac{1}{236}$ sec

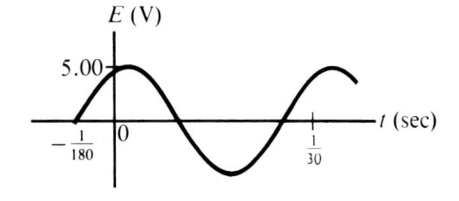

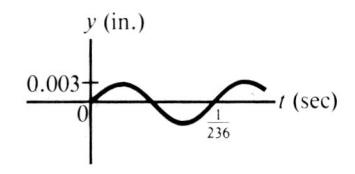

29.

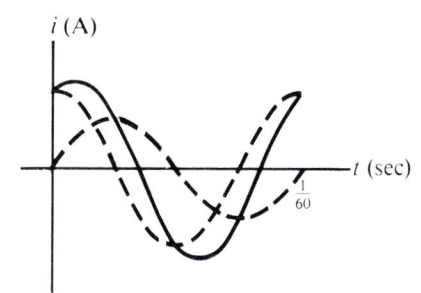

31.

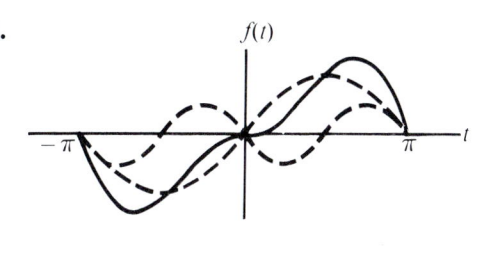

Chapter 9

Section 9.1 (page 277)

1. **3.** 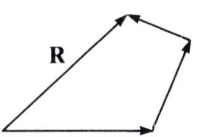 **5.** 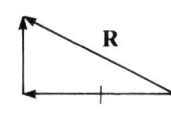 **7.**

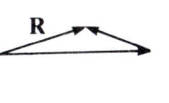

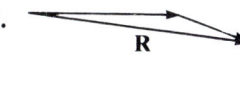

9. 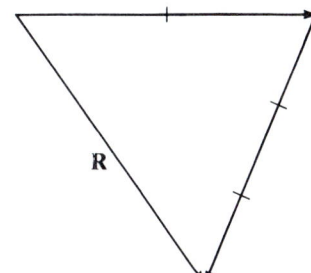 **11.** 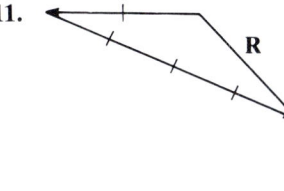 **13.**

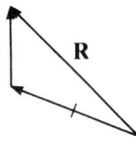

15. 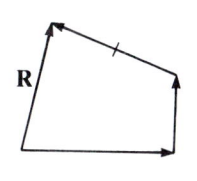 **17.**

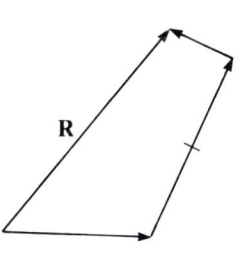

Section 9.2 (page 280)
1. $\sqrt{2}$, 45°   **3.** $3\sqrt{2}$, 225°   **5.** $\sqrt{13}$, 56°19′   **7.** 3, 234°44′   **9.** 4, 228°35′   **11.** 4, 154°20′
13. $\sqrt{5}$, 296°34′   **15.** $\sqrt{17}$, 255°58′   **17.** $0.52\mathbf{i} + 2.95\mathbf{j}$ **19.** $-2.01\mathbf{i} + 1.40\mathbf{j}$
21. $-0.63\mathbf{i} - 3.95\mathbf{j}$ **23.** $2.95\mathbf{i} - 0.52\mathbf{j}$ **25.** $-0.53\mathbf{i} + 1.31\mathbf{j}$

Section 9.3 (page 282)
1. 11.7 lb, 29°10′ from F_1 **3.** 20.15 N, 73°16′ from F_1 **5.** 16.2 mi/hr, 14°40′ with line across
7. 55 lb, 35° with larger force   **9.** 83.2 mi/hr   **11.** 9.0 mi/hr, 25.5° upstream with line across

13. 401 km/hr, 4° east of north   **15.** 223 km/hr, 7°40′ east of south   **17.** 623.9 lb   **19.** 1,450 lb
21. 101 lb, 14°   **23.** 188.6 lb

Section 9.4 (page 289)

1. $B = 109°10′, a = 1.46, c = 3.28$ **3.** $A = 23°50′, a = 8.00, b = 8.63$
5. $B = 8°50′, C = 151°20′, c = 144$ **7.** no solution   **9.** $C = 68.4°, B = 82.3°, b = 145$
11. $A = 80.0°, b = 172, c = 194$ **13.** $A = 73.2°, B = 43.2°, b = 8.87$
15. $A = 13°7′, C = 133°8′, c = 3.684$ **17.** $A = 39°33′, C = 106°42′, c = 1.724$
19. $B = 43.10°, b = 115.9, c = 30.63$ **21.** 510 mi/hr   **23.** 66.2 ft   **25.** 953 ft

Section 9.5 (page 295)

1. $a = 1.76$; by the cosine law: $C = 109.8°, B = 23.9°$ (by the sine law: $B = 24.3°$)
3. $c = 280, A = 31°20′, B = 23°30′$
5. by the cosine law: $B = 81.9°, A = 41.0°, C = 57.1°$ (by the sine law: $A = 41.1°$)
7. $a = 4.60, B = 139°50′, C = 25°30′$
9. $c = 81.8, A = 102.2°$ (by the cosine law), $B = 38.4°$
11. $B = 153°31′, A = 15°29′, C = 11°0′$ **13.** 349 ft   **15.** 27.1 km
17. 17.9° north of east, 158 mi/hr   **19.** 22.0 in.

Review Exercises for Chapter 9 (page 296)

1. $\sqrt{2}, 135°$ **3.** $\sqrt{5}, 63°26′$ **5.** 4, 14°29′   **7.** $\sqrt{6}, 234°44′$ **9.** $1.29i + 4.83j$
11. $-2.03i - 1.70j$ **13.** $-4.85i - 3.53j$ **15.** $B = 80.0°, a = 3.86, c = 4.30$
17. $A = 39°0′, a = 195, b = 57.3$ **19.** $B = 14°, C = 140°, b = 14$
21. $a = 19.80, C = 149°46′, B = 10°51′$ **23.** $B = 76.1°, A = 38.3°, C = 65.6°$ **25.** 1.2 tons
27. 66.24 ft   **29.** 12.4 lb   **31.** 44.7°   **33.** 11.0 mi/hr

Cumulative Review Exercises for Chapters 7–9 (page 298)

1. $\dfrac{\sqrt{5}}{2}$ **2. a.** $-\dfrac{2\sqrt{3}}{3}$; **b.** $-\dfrac{\sqrt{3}}{3}$ **3. a.** 45°, 315°; **b.** 150°, 330°   **4.** 1.388
5. 228°13′, 311°47′   **6.** $\dfrac{28\pi}{45}$ **7.** 500°   **8.** 1.11 cm   **9.** 98.0 in.²   **10.** 85.0 m/sec
11. 19 m/sec   **12.**

14. 4, 150°   **15.** 22.5 lb   **16.** 146 mi/hr, 8.0° north of west   **17.** 549 lb
18. 24 knots, 18° east of south

Chapter 10

Section 10.1 (page 301)

1. 72   **3.** $\dfrac{27}{64}$ **5.** $6x^5y^9$ **7.** $2a^4b^3$ **9.** R^4S^8 **11.** $-512C_1^9C_2^{12}$ **13.** $\dfrac{3sv}{7t}$ **15.** $\dfrac{1}{3}ns$
17. $3V^3$ **19.** $\dfrac{x^2}{2y^3z^6}$ **21.** $3abc$ **23.** $\dfrac{7v_0}{s_2^5}$ **25.** $\dfrac{8x^3}{27y^3}$ **27.** $\dfrac{16q^{12}}{81p^8r^4}$ **29.** $5{,}625x^8y^{16}z^{10}$
31. $\dfrac{3bxy^4}{2a^4}$ **33.** $\dfrac{R^2V^2}{32C}$ **35.** x **37.** y **39.** xy^a **41.** a **43.** x^{5b} **45.** x^{6b}

Section 10.2 (page 307)

1. $\dfrac{1}{16}$ 3. 16 5. 36 7. 1 9. x^5 11. $\dfrac{b^6}{a^3}$ 13. $\dfrac{1}{4z}$ 15. $\dfrac{4f_1f_2^5}{f_3^5}$ 17. $\dfrac{t^2}{su^7}$
19. $\dfrac{16q^7}{3r^2s^7}$ 21. $\dfrac{b}{a}$ 23. $\dfrac{x^6}{8y^6}$ 25. $\dfrac{b^8}{256x^4}$ 27. 1 29. $\dfrac{64\omega^6}{\pi^6}$ 31. $\dfrac{1}{16a^8b^8}$ 33. $\dfrac{8L^{12}C^{15}}{R^{12}}$
35. $\dfrac{216a^{12}c_1^3}{c_2^3}$ 37. $s^4 r^{22} t^6$ 39. $\dfrac{1}{81a^8b^2y^2}$ 41. π^{2e} 43. $\dfrac{b^y}{3a^x}$ 45. $a^{b^2}x^{bc}$ 47. $2x^2$
49. $\dfrac{x+y}{xy}$ 51. $1 + ab$ 53. $\dfrac{b}{b+1}$ 55. $\dfrac{1}{b-a}$ 57. $\dfrac{1}{b^2-a^2}$ 61. $\dfrac{n}{n-1}$ 63. \$4,038.83

Section 10.3 (page 311)

1. 9 3. 9 5. $\dfrac{1}{7}$ 7. 0.3 9. 3 11. -4 13. -8 15. $\dfrac{1}{343}$ 17. $\dfrac{1}{5}$ 19. $\dfrac{1}{25}$
21. 1 23. $m^{3/7}$ 25. $x^{25/12}$ 27. $\omega^{1/2}$ 29. $\dfrac{10}{x^{1/4}}$ 31. $\dfrac{a^{1/2}}{2b^{1/20}}$ 33. $\dfrac{3}{5R_1^{7/12}R_2^{1/12}}$
35. $\dfrac{2a^{7/30}}{w^{7/8}}$ 37. $2x^3y^{16}$ 39. $\dfrac{y^3}{3a^2}$ 41. $\dfrac{x}{2y^2z^3}$ 43. $\dfrac{c^{11/4}}{2a^{5/3}}$ 45. $\dfrac{2}{m^{17/10}d^2}$ 47. $\dfrac{1}{6}m^3v^{6/5}$
49. $\dfrac{8L^3}{T^5}$ 51. $\dfrac{2x+2}{(x+2)^{1/2}}$ 53. $\dfrac{2x-1}{(x-3)^{1/2}}$ 55. $\dfrac{2x+1}{(x+1)^{2/3}}$ 57. $\dfrac{7}{(x-1)^{2/3}}$ 59. $\dfrac{-x-1}{(x-1)^{1/3}}$
61. $\dfrac{-2x-6}{(x+2)^{4/5}}$ 63. $\dfrac{-3}{(x+2)^{6/7}}$ 65. x^{a+1} 67. $\dfrac{1}{x}$ 69. $p = \dfrac{k}{(\sqrt[5]{v})^7}$ 71. $\dfrac{32}{7}$ cm
75. $v = \dfrac{3t+2}{(1+t)^{1/2}}$

Section 10.4 (page 319)

1. $\sqrt[4]{x}$ 3. $\sqrt[12]{b}$ 5. $\sqrt{abc^3}$ 7. $\sqrt[4]{u^2v^3w^4}$ 9. $\sqrt[3]{v^2m^2g^3}$ 11. $\sqrt[4]{5p^2q^4r^5}$ 13. $2x\sqrt{x}$
15. $3ac^2\sqrt[3]{2ac^2}$ 17. $2u^3v^4\sqrt[4]{2uv^2}$ 19. $3mns^3t^5\sqrt{2nst}$ 21. $\dfrac{\sqrt{5}}{5}$ 23. $\dfrac{\sqrt{3x}}{x}$ 25. $\dfrac{\sqrt{abc}}{bc}$
27. $\dfrac{\sqrt{3pqv}}{3pq}$ 29. $\dfrac{\sqrt[3]{18}}{3}$ 31. $\dfrac{\sqrt[3]{2\pi^2 V}}{2\pi}$ 33. $\dfrac{\sqrt[4]{6L^3V^2}}{2L^2V}$ 35. $\dfrac{\sqrt{21xy}}{3x^2y^3}$ 37. $\dfrac{\sqrt[3]{3aRC^2}}{3C}$
39. $\dfrac{\sqrt{2pq}}{2pqr}$ 41. $5\sqrt{6}$ 43. $3\sqrt{5}$ 45. $-\sqrt[3]{3}$ 47. $7\sqrt{5} - 5\sqrt[3]{2}$ 49. $-ab\sqrt{ab}$
51. $(5mn - m^2n + 2m^2n^2)\sqrt{mn}$ 53. $\dfrac{2a^2\sqrt{3b}}{3b^2}$ 55. $\dfrac{3\sqrt{wz}}{w^3z}$ 57. $\dfrac{2x\sqrt{2z}}{z^2}$ 59. $\dfrac{v\sqrt{3uvw}}{3u^2w^4}$
61. $\dfrac{\sqrt[3]{4c^2}}{2ac}$ 63. $\dfrac{\sqrt[3]{6uv^2}}{2uv}$ 65. $\dfrac{\sqrt[4]{2xyz}}{2xy^2}$ 67. $\dfrac{\sqrt{7stu}}{49s^2t^3u^3}$ 69. $\dfrac{2\sqrt{2\pi kmT}}{\pi m}$
71. $1.215\sqrt{273(T+273)}$ 73. $x = \pm\dfrac{\sqrt{16\pi^2-k^2}}{4\pi}$ 75. $t = C\sqrt[3]{Lm^2i/mi}$ 77. 2.45

Section 10.5 (page 323)

1. $2y\sqrt{3}$ 3. $ab\sqrt{6}$ 5. $2x\sqrt[3]{y^2}$ 7. $2mc\sqrt[5]{2c}$ 9. $2st\sqrt{5}$ 11. $3\pi\omega\sqrt[4]{\pi^3\omega}$ 13. $\sqrt[6]{243}$
15. $\sqrt[4]{x^2y}$ 17. $\sqrt[10]{c^5d^2}$ 19. $\sqrt[12]{\pi^3e^2}$ 21. -4 23. 5 25. $5 + 2\sqrt{6}$ 27. $17 + 4\sqrt{13}$
29. -68 31. 44 33. $x^2 - y$ 35. $1 - k$ 37. $4 + 4\sqrt{q} + q$ 39. $2a + \sqrt{ab} - 6b$
41. $5\sqrt{35} - 74$ 43. $a\sqrt{a} + b\sqrt{b}$ 45. $a\sqrt{a} - 2b\sqrt{a} + b\sqrt{b}$

Section 10.6 (page 326)

1. $\dfrac{\sqrt{2ab}}{b}$ 3. $\dfrac{2\sqrt{cd}}{d}$ 5. $\dfrac{\sqrt{6y}}{3}$ 7. $\dfrac{\sqrt[3]{6v_2v_3}}{3v_3}$ 9. $\dfrac{\sqrt[4]{6\pi r}}{2}$ 11. $\sqrt[6]{a}$ 13. $-1-\sqrt{2}$

15. $4(\sqrt{5}-2)$ 17. $\sqrt{2}$ 19. $\dfrac{1}{7}(5+4\sqrt{2})$ 21. $5-2\sqrt{6}$ 23. $\sqrt{2}$ 25. $3\sqrt{3}+2\sqrt{6}$

27. $\dfrac{2(1+\sqrt{a})}{1-a}$ 29. $\dfrac{3(\sqrt{b}+2)}{b-4}$ 31. $\dfrac{a+2\sqrt{ab}+b}{a-b}$ 33. $\dfrac{\sqrt{a}+\sqrt{a-b}}{b}$

35. $\dfrac{a^{2/3}+a^{1/3}b^{1/3}+b^{2/3}}{a-b}$ 39. $\dfrac{\mu_1-2\sqrt{\mu_1\mu_2}+\mu_2}{\mu_1-\mu_2}$ 41. $\dfrac{a^{1/3}(a^{2/3}-a^{1/3}b^{1/3}+b^{2/3})}{a+b}$

Review Exercises for Chapter 10 (page 328)

1. $\dfrac{1}{25}$ 3. 64 5. 0.2 7. $\dfrac{2V^3}{9W^{10}}$ 9. $\dfrac{C_2}{2C_1}$ 11. $\dfrac{w^8}{256v^{12}z^4}$ 13. $\dfrac{r^{18}}{9R^{12}}$ 15. x^a

17. $\dfrac{C}{C-1}$ 19. $\dfrac{2v}{w^{1/2}}$ 21. $10x^{3/4}$ 23. $\dfrac{3^{1/3}F_2^{1/12}}{F_1^{5/7}}$ 25. $\dfrac{2n^{1/6}}{m^{17/8}}$ 27. $\dfrac{2x}{(x-1)^{2/3}}$ 29. $\dfrac{9-3x}{(x-2)^{4/5}}$

31. $\sqrt[6]{x}$ 33. $\sqrt{4a^2b^3c^5}$ 35. $2Rr^3\sqrt{3Rr}$ 37. $2ab^2\sqrt[4]{ab^3}$ 39. 0 41. $8\sqrt{5}-5\sqrt{3}$

43. $\dfrac{2\sqrt{5}}{5ab^2}$ 45. $\dfrac{\sqrt{6st}}{3st}$ 47. $\dfrac{\sqrt[3]{9s^2}}{3Rs}$ 49. $\dfrac{\sqrt[3]{3a^2b}}{3a^2b}$ 51. $\dfrac{\sqrt[5]{3\pi^2\omega^3}}{3\pi\omega}$ 53. $3x\sqrt[3]{y^2}$ 55. $\sqrt[12]{a^7}$

57. 4 59. $1-9x$ 61. $\dfrac{4}{3}(3+\sqrt{6})$ 63. $\dfrac{1}{3}(4\sqrt{2}+\sqrt{5})$ 65. $\sqrt{\pi}-\sqrt{\pi-1}$

67. $\$6{,}805.83$ 69. $\dfrac{2\pi}{5gL}\sqrt{5gL(5L^2+2r^2)}$

Chapter 11

Section 11.1 (page 337)

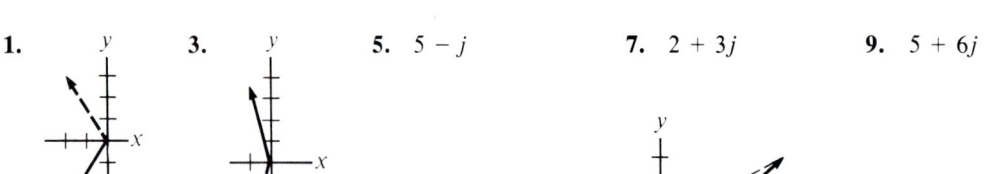

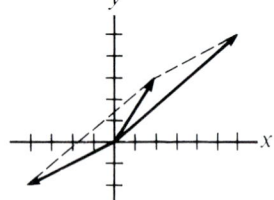

1.   3.   5. $5-j$ 7. $2+3j$ 9. $5+6j$

11. $-4 + 6j$ **13.** $7j$ **15.** 10 **17.** $-7 - j$

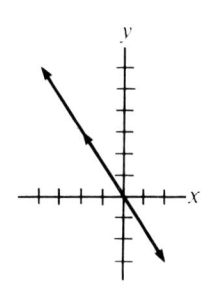

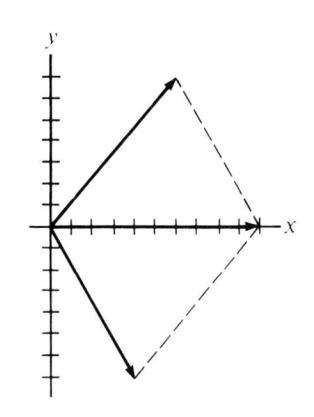

19. $-2 - 14j$ **21.** $x = 2, y = 3$ **23.** $x = 2, y = -3$ **25.** $x = 3, y = -1$

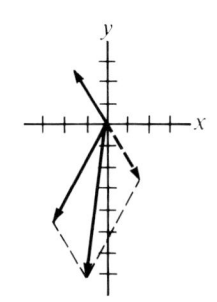

27. $x = 6, y = 1$ **29.** $x = \frac{5}{2}, y = \frac{1}{2}$

Section 11.2 (page 340)

1. $9 + 10j$ **3.** $3\sqrt{5} - 4j$ **5.** $6 - 6j$ **7.** $4 - 6j$ **9.** $-9 + 13j$ **11.** $-12 + j$
13. $7 + j$ **15.** $4 - 7j$ **17.** 25 **19.** $-11 - 17j$ **21.** $2j$ **23.** $-1 - 4\sqrt{3}j$
25. $1 - 2\sqrt{2}j$ **27.** $38 + 6j$ **29.** $\frac{1}{10} + \frac{3}{10}j$ **31.** $\frac{4}{5} - \frac{2}{5}j$ **33.** $-\frac{3}{5} - \frac{4}{5}j$
35. $-\frac{7}{25} - \frac{24}{25}j$ **37.** $-\frac{10}{17} + \frac{11}{17}j$ **39.** $-\frac{1}{68} - \frac{21}{68}j$ **41.** $\frac{7}{5} - \frac{1}{5}j$ **43.** $-2 - \frac{7}{2}j$
45. $\frac{7}{10} + \frac{1}{10}j$ **47.** $\frac{62}{85} - \frac{41}{85}j$ **49.** $-j$ **51.** j **53.** -1 **55.** $-j$ **57.** -1 **59.** 1

Section 11.3 (page 345)

1. cis 0° **3.** 4 cis 180° **5.** 2 cis 90° **7.** 5 cis 270° **9.** $\sqrt{2}$ cis 45° **11.** 2 cis 300°
13. $3\sqrt{2}$ cis 225° **15.** 6 cis 150° **17.** $2e^{(\pi/2)j}$ **19.** $\sqrt{2}e^{(7\pi/4)j}$ **21.** $2e^{(11\pi/6)j}$ **23.** $2\sqrt{2}e^{(3\pi/4)j}$
25. $\sqrt{2} + \sqrt{2}j$ **27.** $-1 + \sqrt{3}j$ **29.** $-\frac{3}{2} - \frac{3\sqrt{3}}{2}j$ **31.** $-\frac{5\sqrt{3}}{2} + \frac{5}{2}j$ **33.** 3 **35.** $5j$
37. $\sqrt{29}$ cis 68.2° **39.** $2\sqrt{5}$ cis 243.4° **41.** $3\sqrt{2}$ cis 331.9° **43.** $1.60 + 1.21j$
45. $-0.271 - 0.238j$ **47.** $0.508 - 6.32j$ **49.** $2.73 + 1.31j$

Section 11.4 (page 348)

1. 6 cis 93° **3.** 18 cis 350° **5.** 30 cis 170° **7.** 54 cis 185° **9.** 4 cis 60° **11.** 3 cis 260°
13. $\frac{3}{2}$ cis 211° **15.** $\frac{5}{3}$ cis 153° **17.** 2 cis 226° **19.** $\frac{1}{14}$ cis 181°

Section 11.5 (page 352)

1. $-4 - 4j$ **3.** 4,096 **5.** $-16 - 16\sqrt{3}j$ **7.** -64 **9.** $-128 + 128\sqrt{3}j$ **11.** 1
13. $-7.01 - 24.0j$ **15.** $-1,120 + 404j$ **17.** 3^{10} cis 33° **19.** 3^{15} cis 261°
21. $1, -\frac{1}{2} + \frac{\sqrt{3}}{2}j, -\frac{1}{2} - \frac{\sqrt{3}}{2}j$ **23.** $-\frac{\sqrt{2}}{2} + \frac{\sqrt{2}}{2}j, \frac{\sqrt{2}}{2} - \frac{\sqrt{2}}{2}j$
25. 2 cis $(18° + k \cdot 72°)$, $k = 0, 1, 2, 3, 4$ **27.** $2^{1/6}$ cis $(10° + k \cdot 60°)$, $k = 0, 1, \ldots, 5$
29. 2 cis $(54° + k \cdot 72°)$, $k = 0, 1, \ldots, 4$ **31.** $2^{1/16}$ cis $\frac{225° + k \cdot 360°}{8}$, $k = 0, 1, \ldots, 7$
33. $2^{1/3}$ cis $(35° + k \cdot 60°)$, $k = 0, 1, \ldots, 5$ **35.** $6^{1/8}$ cis $(37.5° + k \cdot 45°)$, $k = 0, 1, \ldots, 7$
37. $3^{1/9}$ cis $(30° + k \cdot 40°)$, $k = 0, 1, \ldots, 8$ **39.** $2^{1/10}$ cis $(24° + k \cdot 36°)$, $k = 0, 1, \ldots, 9$

Section 11.6 (page 355)

1. $Z = 3.0 + 4.0j$, $|Z| = 5.0 \, \Omega$, $\theta = 53.1°$ **3.** $Z = 12.0 + 9.0j$, $|Z| = 15.0 \, \Omega$, $\theta = 36.9°$
5. $Z = 20.0 - 17.4j$, $|Z| = 26.5 \, \Omega$, $\theta = -41.0°$

Review Exercises for Chapter 11 (page 355)

1. $3 + j$ **3.** $1 - 3j$ **5.** $x = 4, y = 2$ **7.** $x = 3, y = -1$ **9.** $3\sqrt{2}$ **11.** 25
13. $-8 - 6j$ **15.** $\frac{12}{25} + \frac{9}{25}j$ **17.** $-j$ **19.** 2 cis 0° **21.** 3 cis 180° **23.** $2\sqrt{2}$ cis 45°
25. 2 cis 120° **27.** $3e^{(\pi/2)j}$ **29.** $3\sqrt{2}e^{(3\pi/4)j}$ **31.** $6e^{(5\pi/3)j}$ **33.** $-\sqrt{2} + \sqrt{2}j$
35. $-2\sqrt{2} - 2\sqrt{2}j$ **37.** $-\frac{3\sqrt{3}}{2} - \frac{3}{2}j$ **39.** $-2.30 + 2.22j$ **41.** $5.46 + 2.56j$
43. $\sqrt{34}$ cis 301.0° **45.** $\sqrt{10}$ cis 161.6° **47.** 6 cis 142° **49.** $2\sqrt{5}$ cis 12° **51.** 7 cis 236°
53. $\frac{1}{2}$ cis 211° **55.** $8 - 8j$ **57.** -64 **59.** $-38.0 + 41.0j$ **61.** $28.0 + 96.0j$
63. $\sqrt{3} + j, -\sqrt{3} + j, -2j$ **65.** cis $k \cdot 72°$, $k = 0, 1, 2, 3, 4$
67. $2^{1/6}$ cis $(5° + k \cdot 60°)$, $k = 0, 1, \ldots, 5$

Chapter 12

Section 12.1 (page 360)

1. $\log_3 81 = 4$ **3.** $\log_{10} 1{,}000 = 3$ **5.** $\log_4 256 = 4$ **7.** $\log_2 \frac{1}{16} = -4$ **9.** $\log_3 1 = 0$
11. $\log_{3/4} \frac{9}{16} = 2$ **13.** $\log_{3/2} \frac{8}{27} = -3$ **15.** $\log_3 \frac{1}{81} = -4$ **17.** $5^3 = 125$ **19.** $5^0 = 1$
21. $2^1 = 2$ **23.** $3^{-3} = \frac{1}{27}$ **25.** $\left(\frac{1}{2}\right)^2 = \frac{1}{4}$ **27.** $10^{-3} = \frac{1}{1{,}000}$ **29.** $6^{-2} = \frac{1}{36}$ **31.** $25^{-1} = \frac{1}{25}$
33. $x = 81$ **35.** $a = 3$ **37.** $x = 10{,}000$ **39.** $b = 2$ **41.** $x = 5$ **43.** $x = \frac{1}{64}$
45. $x = \frac{16}{9}$ **47.** $b = \frac{2}{3}$ **49.** $a = -\frac{1}{2}$

ANSWERS TO ODD-NUMBERED EXERCISES   **663**

Section 12.2 (page 365)

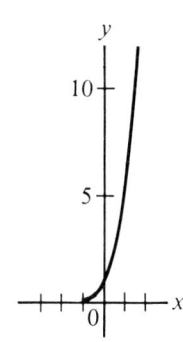

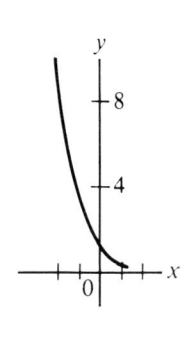

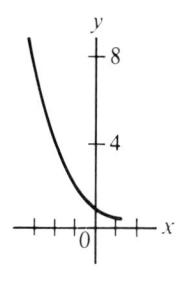

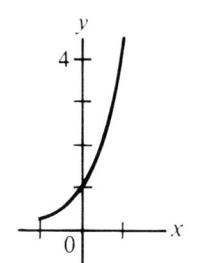

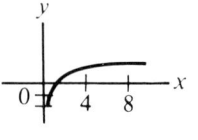

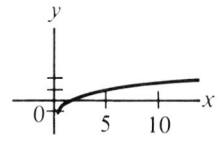

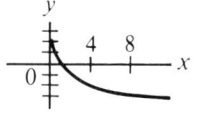

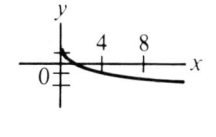

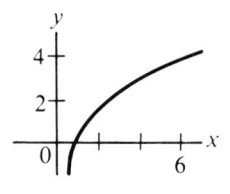

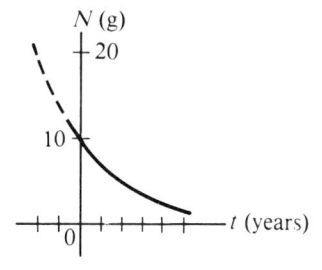

1. [graph]
3. [graph]
5. [graph]
7. [graph]
9. [graph]
11. [graph]
13. [graph]
15. [graph]
17. [graph]
19. [graph]
21. [graph]
23. [graph]
25. $T = 10e^{0.5\theta}$ [graph]

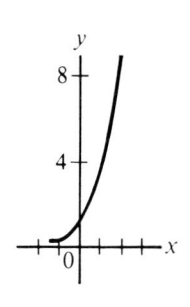

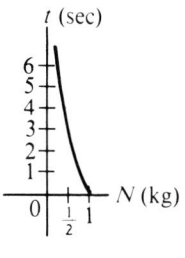

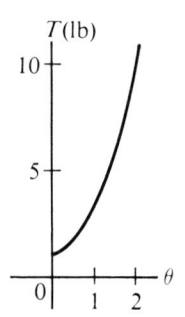

Section 12.3 (page 369)

1. $4 \log_{10} 3$ 3. 3 5. $2 \log_7 6$ 7. 2 9. 4 11. $\frac{3}{2}$ 13. $2 + 2 \log_5 x$
15. $1 + \log_3 5 + 2 \log_3 x$ 17. $1 + \log_4 5 + \frac{1}{2} \log_4 a$ 19. $-2 - 2 \log_3 a$ 21. $-\frac{1}{2}(1 + \log_3 C)$
23. $-\left(1 + \log_5 6 + \frac{1}{2} \log_5 u\right)$ 25. -2 27. $-\frac{1}{3}$ 29. $2 \log_e 5 - 3$ 31. $-\frac{1}{2}(\log_e \pi + 1)$
33. $\log_e 2 + \frac{1}{3}$ 35. $-2(\log_{10} 2 + 1)$ 37. $2 \log_{10} 3 - 3$ 39. $\log_{10} x - \log_{10} 3 - 2$ 41. $\log_2 15$
43. $\log_4 10$ 45. $\log_2 \frac{35}{3}$ 47. $\log_2 \frac{25}{3}$ 49. $\log_{10} 4$ 51. $\log_2 \frac{\sqrt{x}}{3}$ 53. $\log_e \sqrt{mv}$
55. $\log_7 \frac{\sqrt{st}}{2}$ 57. 0.7781 59. 0.6020 61. 1.0791 63. 1.3010 65. $p = p_0 10^{-0.0149k/T}$
67. $(5.97 \times 10^7)p = 10^{-1706.4/T}$

Section 12.4 (page 374)

1. 0.7474 3. 4.9212 5. $7.7193 - 10$ 7. 3.3941 9. 5.9310 (table); 5.9311 (calculator)
11. 73.1 13. $5,370$ 15. 0.856 17. $4,438$ 19. 0.008528 (table); 0.008527 (calculator)
21. 7 23. 0.380 W

Section 12.5 (page 376)

1. 81.4 3. 3.192 5. 20.52 7. 0.211 9. 2.721 11. 219.0 13. 37.56 15. 61.73

Section 12.6 (page 379)

1. 1.9906 3. -0.6591 5. 0.8399 7. -3.5659 9. 2.7026 11. 20.0855 13. 0.1353
15. 0.3985 17. 1.0131 19. 0.9704 21. 16 years   23. $\$5,422.65$ 25. 0.16 atmosphere
27. 214 ft   29. 17.8 W

Section 12.7 (page 383)

1. 2.3219 3. 2.1610 5. -0.6309 7. 1.4307 9. 1.3123 11. -1.1338 13. 9
15. 1.5647 17. 4 19. $\frac{e^2}{4}$ 21. 2 23. no solution   25. $t = -RC \ln \frac{Ri}{E}$
27. $V_2 = V_1[10^{Q/(P_1 V_1)}]$ 29. (a) 3.25 (b) 5 31. 10 decibels   33. 0.0032 W/m^2
35. $10^{5.25} \approx 180{,}000$ times   37. (a) $11{,}000$ (b) 4.3 hr   39. $11{,}200$ years   41. 6.5 million years
43. $1{,}750$ m

Section 12.8 (page 390)

1.

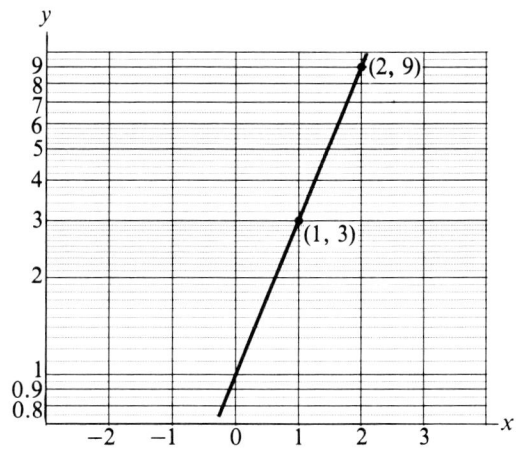

3.

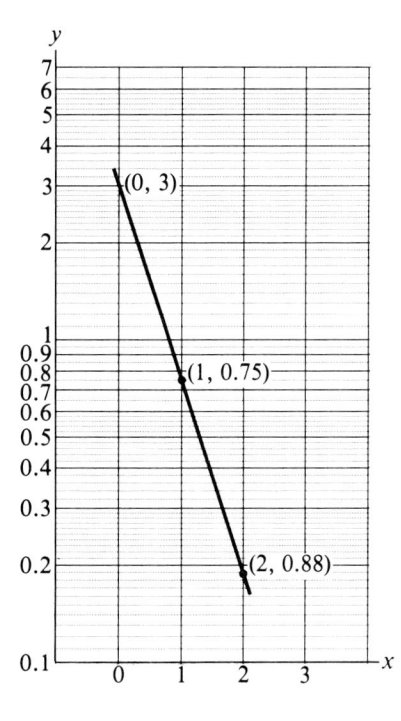

5.

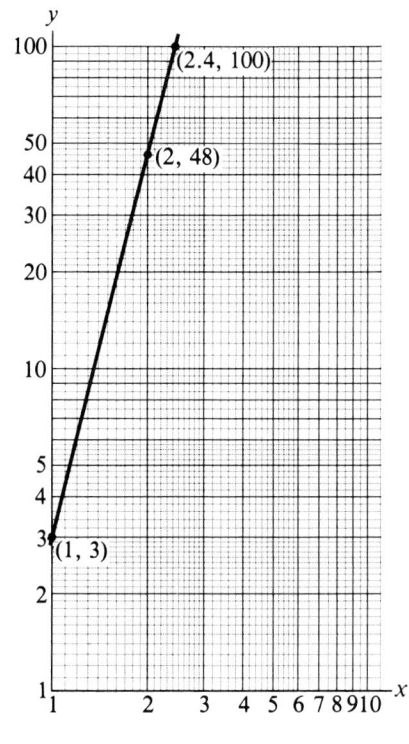

7.

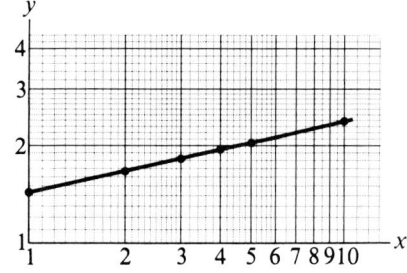

9.

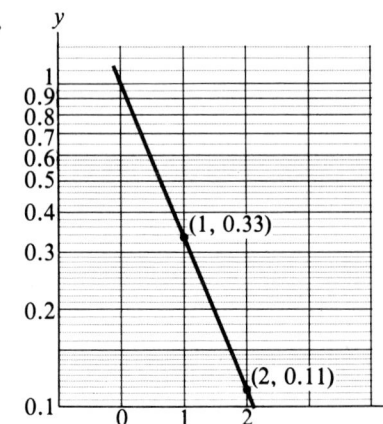

11.

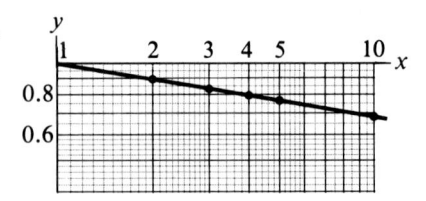

13.

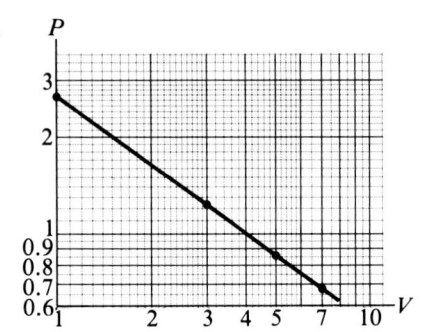

Review Exercises for Chapter 12 (page 391)

1. $\log_3 \frac{1}{27} = -3$ **3.** $\log_7 \frac{1}{49} = -2$ **5.** $4^1 = 4$ **7.** $\left(\frac{1}{4}\right)^2 = \frac{1}{16}$ **9.** $x = \frac{1}{25}$ **11.** $a = -4$

13. $x = \frac{27}{8}$ **15.** $b = 2$ **17.** **19.** **21.**

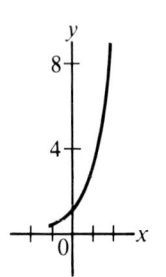

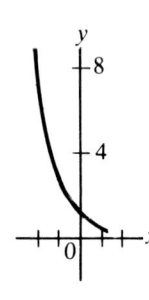

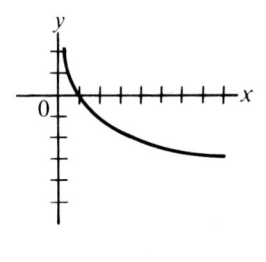

23. 5 **25.** 3 **27.** $3 + 3 \log_5 x$ **29.** $\frac{1}{2}(1 + \log_6 x)$ **31.** $-4 - 2 \log_2 a$ **33.** $4 \ln 2 - 2$
35. $-\frac{1}{3}$ **37.** $\log_5 2$ **39.** $\log_6 \frac{\sqrt{x}}{b^2}$ **41.** 0.4116 **43.** 3.9610 **45.** $7.7608 - 10 = -2.2392$
47. 448.4 **49.** 0.3410 **51.** 0.4216 **53.** 3.6222 **55.** 2.3740 **57.** 0.3012 **59.** 14.8797

61. 0.2019 **63.** 0.8488 **65.** −0.1631 **67.** 2.5850 **69.** −0.8340 **71.** $x = \frac{1}{2}(-1 + \sqrt{13})$

73. 100 times **75.** 31°C **77.** $t = -\frac{1}{2} \ln \frac{i}{3.0}$ **79.** 7.0 years **81.** 2.9 hr

83. 7.1×10^{-4} mol/L

Cumulative Review Exercises for Chapters 10–12 (page 394)

1. $\frac{x^4}{y^2}$ **2.** $\frac{g_2^5}{18g_1^6 g_3^2}$ **3.** $\frac{b-a}{b}$ **4.** $\frac{R^3}{4L^4}$ **5.** $\frac{3y^{4/15}}{x^{5/6}}$ **6.** $-\frac{5}{(a+3)^{1/2}}$ **7.** $\frac{\sqrt[4]{a^3 b}}{ab}$

8. $\frac{\sqrt[3]{2G^2 H}}{2G^2 H^2}$ **9.** $\frac{1}{3}(2 + \sqrt{10} - \sqrt{5} - \sqrt{2})$ **10.** $2 + 2j$ **11.** $1 + 5j$ **12.** $\frac{1}{13}(-1 - 5j)$

13. a. 4 cis 210°; b. $4e^{(7\pi/6)j}$ **14.** $\frac{1}{2}$ cis 220°   **15.** $\sqrt[4]{2}$ cis (30° + k • 90°), k = 0, 1, 2, 3

16. $\log_6 \frac{Q\sqrt{P}}{R^2}$ **17.** $x = 1.771$ **18.** $t = \frac{1}{k} \ln \frac{S_2}{S_1 - S}$ **19.** $p = \frac{fq}{q-f}$ **20.** $1838.84

21. $N = 10^{-0.4t}$ **22.** 203 days

Chapter 13

Section 13.1 (page 403)

1. (2.0, 2.0), (−2.0, −2.0)   **3.** (0.7, 1.0)   **5.** (2.8, 2.4), (−1.4, 0.3)

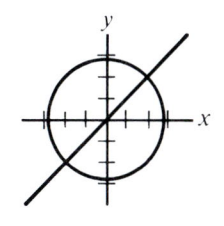

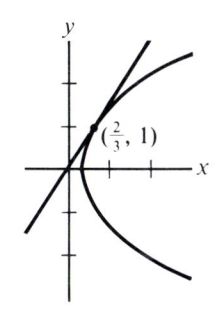

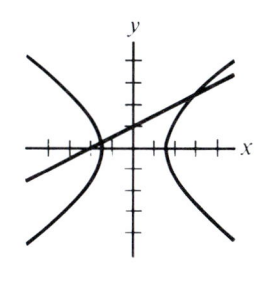

7. (3.3, 1.8)   **9.** (1.6, 0.6), (0.4, 2.4)

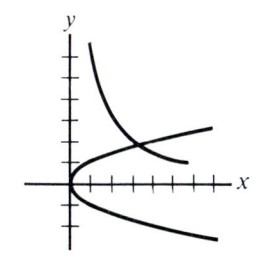

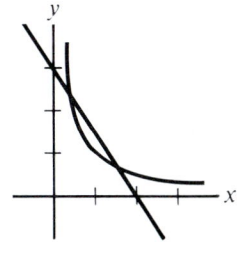

11. (2.6, 1.5), (2.6, −1.5), (−2.6, 1.5), (−2.6, −1.5)   **13.** (2.9, −4.3), (−2.9, −4.3)

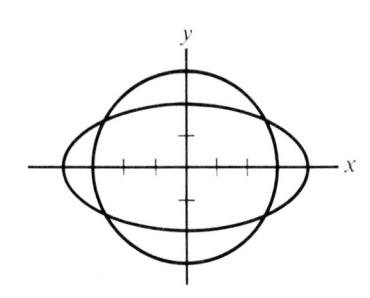

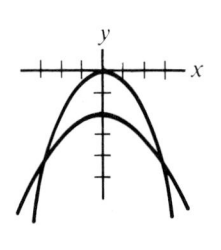

15. (0.4, 0.7)   **17.** (1.5, 0.4)

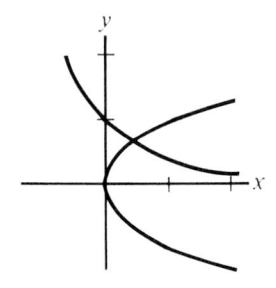

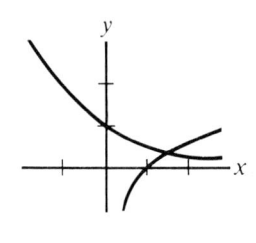

Section 13.2 (page 406)

1. (2, 4), (−1, 1)   **3.** (2, 5), (−1, 2)   **5.** (−2, 0), (1, 3)   **7.** (0, 0), (2, −2)   **9.** (2, 1), (−1, −2)
11. (−2, 0), $\left(\frac{5}{2}, \frac{9}{4}\right)$ **13.** (1, 3), (1, 1)   **15.** $\left(1 - \frac{\sqrt{6}}{3}j, 1 + \frac{\sqrt{6}}{6}j\right), \left(1 + \frac{\sqrt{6}}{3}j, 1 - \frac{\sqrt{6}}{6}j\right)$
17. (−1, −1)   **19.** $\left(\frac{11}{4}, -\frac{9}{8}\right)$, (1, −2)   **21.** (1, 2), (1, −2), (−1, 2), (−1, −2)
23. $\left(\sqrt{2}, \frac{2\sqrt{3}}{3}j\right), \left(\sqrt{2}, -\frac{2\sqrt{3}}{3}j\right), \left(-\sqrt{2}, \frac{2\sqrt{3}}{3}j\right), \left(-\sqrt{2}, -\frac{2\sqrt{3}}{3}j\right)$
25. $(\sqrt{5}, \sqrt{11}), (\sqrt{5}, -\sqrt{11}), (-\sqrt{5}, \sqrt{11}), (-\sqrt{5}, -\sqrt{11})$
27. $(\sqrt{2}, \sqrt{14}), (\sqrt{2}, -\sqrt{14}), (-\sqrt{2}, \sqrt{14}), (-\sqrt{2}, -\sqrt{14})$
29. 2.5 cm × 10 cm   **31.** 3.8 sec   **33.** 6.47 Ω, 2.53 Ω   **35.** 20 cm/sec, 30 cm/sec

Section 13.3 (page 411)

1. ±1, ±2   **3.** ±2, ±j   **5.** 2, −1 ± $\sqrt{3}j$, −1, $\frac{1}{2} \pm \frac{1}{2}\sqrt{3}j$ **7.** $\frac{1}{3}, -\frac{1}{2}$ **9.** $\pm\frac{1}{2}, \pm\frac{\sqrt{6}}{6}j$
11. ±2, ±$\frac{\sqrt{2}}{2}j$ **13.** 1, 25   **15.** $\frac{16}{9}$ **17.** 5, 26   **19.** 8, −27   **21.** 2, 2, 2 ± $\sqrt{6}$
23. ±$\frac{\sqrt{3}}{2}$, ±$\frac{\sqrt{42}}{6}j$ **25.** 16   **27.** $6\sqrt{10}$ in. × $2\sqrt{10}$ in.   **29.** 0°, 90°, 180°   **31.** 0°, 120°, 240°

Section 13.4 (page 415)

1. 8 **3.** no solution **5.** 0 **7.** 2 **9.** 9 **11.** 3 **13.** 6 **15.** 1 **17.** 5
19. 5, 13 **21.** 6 **23.** -3 **25.** $v = \left(\dfrac{L^2 + 1}{2L}\right)^2$ **27.** 10 in., 24 in.
29. $x = 3$ cm, $x = \dfrac{153}{26}$ cm **31.** $C = \dfrac{Z}{2\pi f \sqrt{1 - RZ^2}}$

Review Exercises for Chapter 13 (page 416)

1. (1.5, 0.5) **3.** (1, 3), $\left(-\dfrac{1}{2}, \dfrac{3}{2}\right)$ **5.** (2, 0), (1, -3)
7. $\left(\dfrac{\sqrt{2}}{2}j, \dfrac{\sqrt{6}}{2}\right), \left(\dfrac{\sqrt{2}}{2}j, -\dfrac{\sqrt{6}}{2}\right), \left(-\dfrac{\sqrt{2}}{2}j, \dfrac{\sqrt{6}}{2}\right), \left(-\dfrac{\sqrt{2}}{2}j, -\dfrac{\sqrt{6}}{2}\right)$ **9.** $(\sqrt{2}, \sqrt{2}), (-\sqrt{2}, -\sqrt{2})$
11. $\pm 2, \pm 4$ **13.** $\pm \dfrac{\sqrt{3}}{2}, \pm 2j$ **15.** $\dfrac{2}{5}, -\dfrac{1}{3}$ **17.** $\pm 2, \pm \dfrac{1}{2}$ **19.** $\dfrac{16}{9}$ **21.** 3, 15
23. $-1, -1, -1 \pm j$ **25.** -2 **27.** -5 **29.** 3 **31.** 3.19 Ω, 6.81 Ω **33.** 36 in. $\times$ 15 in.
35. $v = \dfrac{c}{m}\sqrt{m^2 - m_0^2}$

Chapter 14

Section 14.2 (page 422)

1. 3 **3.** -36 **5.** -10 **7.** 40 **9.** 3 **11.** 36 **13.** 40 **15.** 60 **17.** 12 **19.** 0
21. yes **23.** no **25.** yes **27.** yes **29.** no

Section 14.3 (page 427)

1. $x^2 + 3x + 2, R = 3$ **3.** $x^2 - x - 4, R = -10$ **5.** $x^3 - x^2 + 2x - 7, R = 13$
7. $4x^2 - 18, R = 60$ **9.** $4x^4 + 8x - 5, R = 0$ **11.** $2x^3 + 3x + 3, R = -4$
13. $4x^3 + 4x^2 - 2x, R = 2$ **15.** $x^4 + 2x^3 + 4x^2 - 4, R = -2$ **17.** $9x^4 + 3x^3 + 3x^2 - 3x - 6, R = 0$
19. $x^3 + 2x^2 + 4x + 3, R = -4$ **21.** yes **23.** no **25.** yes **27.** yes **29.** yes
31. yes **33.** yes **35.** no **37.** yes **39.** 19 **41.** $\dfrac{5}{2}$ **43.** 430 **45.** -7

Section 14.4 (page 431)

1. degree 5, roots: -2 (multiplicity 2), 3 (multiplicity 2), -1 (single)
3. degree 7, roots: 0 (multiplicity 2), -4 (multiplicity 3), 5 (multiplicity 2)
5. degree 9, roots: $-\dfrac{1}{2}$ (multiplicity 4), 10 (multiplicity 5)
7. degree 9, roots: 0 (multiplicity 2), $\dfrac{3}{4}$ (multiplicity 3), -7 (multiplicity 4) **9.** 1, -3 **11.** 4, -1
13. $\dfrac{1}{2}(-1 \pm \sqrt{13})$ **15.** $-1 \pm \dfrac{1}{2}\sqrt{2}j$ **17.** 2, 4 **19.** $-\dfrac{1}{2} \pm \dfrac{1}{2}\sqrt{3}j$ **21.** $-j$, 2, -1
23. $2j, -3, 2$ **25.** $-j, -1 \pm \sqrt{3}j$ **27.** $1 - j, 1, 2$ **29.** 1, 2 **31.** $\dfrac{1}{2}(1 \pm \sqrt{5})$

Section 14.5 (page 438)

1. 1, -1, 2 **3.** $-3, -2, 2$ **5.** $-1, -3, 1, 2$ **7.** $-1, -3, -3, 2$ **9.** $3, 3, \dfrac{1}{2}(-3 \pm \sqrt{5})$
11. $4, 4, 1 \pm \sqrt{2}j$ **13.** $\dfrac{3}{2}, \dfrac{1}{2} \pm \dfrac{\sqrt{11}}{2}j$ **15.** $\dfrac{1}{2}, \dfrac{1}{2}, -1, -2$ **17.** $-2, -\dfrac{1}{2}, -\dfrac{1}{2}, 1$ **19.** $1, 1, \dfrac{4}{3}, -\dfrac{3}{2}$

670 APPENDIX C

21. $-2, -1, \frac{1}{2}, \frac{5}{3}$ **23.** $\frac{3}{2}, \frac{3}{2}, 1, -2$ **25.** $\frac{1}{3}, \frac{1}{2}, 2, -4$ **27.** 3 in.   **29.** 4 ft × 4 ft × 2 ft
31. 3 μF, 6 μF, 12 μF

Section 14.6 (page 445)

1. 2.21   **3.** 1.10   **5.** 0.75   **7.** 3.27   **9.** -1.87 **11.** 0.90   **13.** 2.672   **15.** -0.414
17. 3.24   **19.** 1.44 in.

Review Exercises for Chapter 14 (page 445)

1. yes   **3.** no   **5.** $x^2 - x + 2, R = 3$ **7.** $2x^4 + 2x^3 - 10x^2 + 20x - 20, R = 20$
9. $4x^4 + 4x^3 + 2x^2 + 3, R = -1$ **11.** yes   **13.** no   **15.** no   **17.** -2 **19.** 4   **21.** 2, 2
23. $1 \pm \sqrt{2}$ **25.** $-j, 3, -5$ **27.** $1, -2, -6$ **29.** $-2, -2, 1, 3$ **31.** $\frac{1}{2}, \frac{1}{2}, -3$
33. $\frac{1}{3}, \frac{1}{3}, \frac{1}{2}(1 \pm \sqrt{13})$ **35.** 0.59   **37.** -0.62 **39.** 2.196

Chapter 15

Section 15.2 (page 449)

1. 12   **3.** 20   **5.** 5   **7.** 21   **9.** 189   **11.** 440   **13.** 366

Section 15.3 (page 455)

1. 0   **3.** -11 **5.** 85   **7.** -935 **9.** 14   **11.** -104 **13.** 394   **15.** -50 **17.** 450
19. 406   **21.** 124   **23.** -27

Section 15.4 (page 459)

1. $(1, 2, 3)$ **3.** $(1, 1, -2)$ **5.** $(1, -1, 2)$ **7.** $\left(2, -\frac{3}{2}, \frac{7}{2}\right)$ **9.** $\left(\frac{2}{3}, -\frac{5}{3}, 4\right)$
11. $\left(6, \frac{20}{7}, -\frac{25}{7}\right)$ **13.** $(1, 2, -2, 3)$ **15.** $\left(-\frac{1}{2}, 2, 3, -1\right)$ **17.** $\left(\frac{3}{2}, \frac{1}{3}, -\frac{7}{6}, \frac{2}{3}\right)$
19. $(7, -2, -4, -6, 2)$ **21.** $I_1 = -\frac{3}{171}$ A, $I_2 = \frac{39}{171}$ A, $I_3 = \frac{109}{171}$ A, $I_4 = \frac{145}{171}$ A
23. $25, $15, $10, $5

Section 15.5 (page 468)

1. $x = -3, y = 2$ **3.** $a = 2, b = -7, x = 6, y = 3$ **5.** $\begin{bmatrix} 3 & -9 & 3 \\ 9 & 3 & 6 \end{bmatrix}$ **7.** $\begin{bmatrix} -8 & -1 & -5 \\ -2 & -1 & 1 \end{bmatrix}$
9. $\begin{bmatrix} -3 & 9 & 6 \\ -6 & 0 & -18 \end{bmatrix}$ **11.** $\begin{bmatrix} -2 & -19 & 10 \\ 19 & 8 & 1 \end{bmatrix}$ **13.** $\begin{bmatrix} 5 \\ 4 \end{bmatrix}$ **15.** $[-15 \quad 4]$ **17.** $\begin{bmatrix} 62 & 80 \\ -5 & 0 \end{bmatrix}$
19. $\begin{bmatrix} -2 & -4 & -3 \\ -13 & -47 & -24 \\ 8 & 58 & 21 \end{bmatrix}$ **21.** $\begin{bmatrix} -94 & 13 \\ -60 & -22 \end{bmatrix}$ **23.** $\begin{bmatrix} 87 & 46 \\ 27 & 47 \end{bmatrix}$
25. $AB = \begin{bmatrix} -10 & 3 \\ 0 & 6 \end{bmatrix}, BA = \begin{bmatrix} 4 & 7 \\ 4 & -8 \end{bmatrix}$ **27.** $AB = \begin{bmatrix} -40 \\ -20 \end{bmatrix}$, BA undefined

Section 15.6 (page 476)

1. $\dfrac{1}{2}\begin{bmatrix} -4 & 3 \\ 2 & -1 \end{bmatrix}$ **3.** $\dfrac{1}{8}\begin{bmatrix} -2 & 4 \\ 3 & -2 \end{bmatrix}$ **5.** $\begin{bmatrix} 1 & 0 & 0 \\ 0 & 1 & 0 \\ -1 & 0 & 1 \end{bmatrix}$ **7.** $\dfrac{1}{3}\begin{bmatrix} 1 & -4 & 2 \\ 1 & -1 & -1 \\ 1 & 2 & -1 \end{bmatrix}$

9. $\dfrac{1}{6}\begin{bmatrix} -4 & 3 & -4 \\ 2 & 0 & -4 \\ 2 & 0 & 2 \end{bmatrix}$ **11.** $\begin{bmatrix} -3 & 5 & -1 \\ 2 & -3 & 1 \\ 2 & -3 & 0 \end{bmatrix}$ **13.** $\dfrac{1}{20}\begin{bmatrix} 5 & 0 & 5 \\ 15 & -4 & 3 \\ -10 & 8 & -6 \end{bmatrix}$ **15.** $\dfrac{1}{16}\begin{bmatrix} -5 & -4 & 7 \\ 6 & 8 & -2 \\ 3 & -4 & -1 \end{bmatrix}$

17. $\begin{bmatrix} -3 & -2 & 0 & 2 \\ 2 & 1 & 0 & -1 \\ -4 & -2 & 2 & 3 \\ 2 & 1 & -1 & -1 \end{bmatrix}$ **19.** $\dfrac{1}{24}\begin{bmatrix} 22 & -3 & -16 & 1 \\ -14 & 3 & 8 & 7 \\ 6 & 9 & 0 & -3 \\ -2 & -3 & 8 & 1 \end{bmatrix}$ **21.** $\dfrac{1}{6}\begin{bmatrix} 17 & -4 & -5 & 24 \\ -10 & 2 & 4 & -12 \\ -8 & 4 & 2 & -12 \\ -10 & 2 & 4 & -18 \end{bmatrix}$

23. $\dfrac{1}{4}\begin{bmatrix} 14 & -4 & 2 & -6 \\ 10 & -4 & 2 & -4 \\ -4 & 2 & 0 & 1 \\ -14 & 4 & -2 & 8 \end{bmatrix}$

Section 15.7 (page 481)

1. $X = \begin{bmatrix} -\dfrac{5}{2} \\ \dfrac{3}{2} \end{bmatrix}$ **3.** $X = \begin{bmatrix} -2 \\ \dfrac{5}{2} \end{bmatrix}$ **5.** $X = \begin{bmatrix} 2 \\ 3 \\ 2 \end{bmatrix}$ **7.** $X = \begin{bmatrix} 7 \\ -1 \\ -2 \end{bmatrix}$ **9.** $X = \begin{bmatrix} 3 \\ -1 \\ 0 \end{bmatrix}$

11. $X = \begin{bmatrix} 4 \\ -1 \\ -4 \end{bmatrix}$ **13.** $X = \begin{bmatrix} 1 \\ 2 \\ -2 \end{bmatrix}$ **15.** $X = \begin{bmatrix} -1 \\ 1 \\ 0 \end{bmatrix}$ **17.** $X = \begin{bmatrix} -9 \\ 5 \\ -9 \\ 3 \end{bmatrix}$ **19.** $X = \begin{bmatrix} 1 \\ -1 \\ 1 \\ 0 \end{bmatrix}$

21. $X = \begin{bmatrix} 2 \\ -1 \\ -1 \\ -1 \end{bmatrix}$ **23.** $X = \begin{bmatrix} \dfrac{5}{2} \\ 2 \\ -\dfrac{5}{4} \\ -1 \end{bmatrix}$ **25.** $A^{-1} = \begin{bmatrix} 4 & -2 & -1 \\ -3 & 2 & 1 \\ 2 & -1 & -1 \end{bmatrix}$, $X = \begin{bmatrix} -1 \\ 2 \\ -2 \end{bmatrix}$

27. $A^{-1} = \dfrac{1}{4}\begin{bmatrix} -4 & 1 & 2 \\ -4 & 2 & 0 \\ 8 & -1 & -2 \end{bmatrix}$, $X = \begin{bmatrix} \dfrac{3}{4} \\ \dfrac{1}{2} \\ -\dfrac{3}{4} \end{bmatrix}$ **29.** $A^{-1} = \dfrac{1}{3}\begin{bmatrix} -6 & 3 & 6 & 0 \\ -2 & -1 & 2 & 1 \\ -8 & 2 & 5 & 1 \\ 1 & -1 & -1 & 1 \end{bmatrix}$, $X = \begin{bmatrix} 2 \\ 1 \\ 3 \\ 0 \end{bmatrix}$

31. $A^{-1} = \dfrac{1}{5}\begin{bmatrix} -3 & 14 & -7 & -10 \\ -6 & 3 & 1 & 0 \\ 2 & 4 & -2 & -5 \\ 4 & -2 & 1 & 0 \end{bmatrix}$, $X = \begin{bmatrix} -5 \\ 2 \\ -3 \\ -1 \end{bmatrix}$ **33.** $\dfrac{13}{17}, \dfrac{9}{17}, -\dfrac{14}{17}, \dfrac{8}{17}$ (in amps) **35.** $2\,\Omega, 2\,\Omega, 4\,\Omega$

Review Exercises for Chapter 15 (page 483)

1. -4 **3.** 0 **5.** -6 **7.** $\left(\dfrac{1}{3}, -\dfrac{11}{15}, -\dfrac{1}{15}\right)$ **9.** $(-3, 2, 0, 1)$ **11.** $x = 2, y = 1, z = -3$

13. $\begin{bmatrix} -1 & 4 \\ -2 & 3 \\ 3 & 7 \end{bmatrix}$ **15.** $\begin{bmatrix} -2 & 3 \\ 2 & -8 \\ -5 & -5 \end{bmatrix}$ **17.** $\begin{bmatrix} -9 \\ -20 \end{bmatrix}$ **19.** $\begin{bmatrix} 1 & -20 & 13 \\ 0 & -48 & 30 \\ -4 & 24 & -17 \end{bmatrix}$ **21.** $\dfrac{1}{4}\begin{bmatrix} 11 & -6 \\ -3 & 2 \end{bmatrix}$

23. $\dfrac{1}{3}\begin{bmatrix} -11 & -15 & 5 \\ 6 & 9 & -3 \\ -14 & -21 & 8 \end{bmatrix}$ **25.** $\dfrac{1}{25}\begin{bmatrix} 10 & 5 & -15 \\ 1 & -2 & -14 \\ 2 & -4 & -3 \end{bmatrix}$ **27.** $\dfrac{1}{3}\begin{bmatrix} 5 & -6 & 11 & -3 \\ 4 & -6 & 13 & -3 \\ -7 & 9 & -19 & 6 \\ -4 & 6 & -10 & 3 \end{bmatrix}$ **29.** $\begin{bmatrix} -\frac{2}{3} \\ 0 \\ \frac{1}{3} \end{bmatrix}$

31. $\begin{bmatrix} \frac{7}{5} \\ \frac{11}{25} \\ -\frac{3}{25} \end{bmatrix}$ **33.** $\begin{bmatrix} \frac{1}{3} \\ -\frac{1}{3} \\ -\frac{2}{3} \\ \frac{1}{3} \end{bmatrix}$ **35.** $3, -\dfrac{4}{3}, -\dfrac{7}{3}, 4$ (in amps)

Cumulative Review Exercises for Chapters 13–15 (page 486)

1. $(0.6, 0.4), (-1.6, 2.6)$ **2.** $(2, 4), (4, 2)$ **3.** $(\sqrt{14}, \sqrt{2}), (\sqrt{14}, -\sqrt{2}), (-\sqrt{14}, \sqrt{2}), (-\sqrt{14}, -\sqrt{2})$
4. $\pm\dfrac{1}{2}, \pm j$ **5.** $x = 6$ **6.** no **7.** yes **8.** 5 **9.** 1, 1, 2, 4 **10.** $1, 1, \dfrac{4}{3}, -\dfrac{3}{2}$ **11.** 0.75
12. 394 **13.** $(1, 2, -2, 3)$ **14.** $\begin{bmatrix} 2 & -3 & -2 \\ -1 & 7 & 5 \\ -6 & 6 & 11 \end{bmatrix}$ **15.** $\begin{bmatrix} 0 & 8 \\ 10 & -17 \end{bmatrix}$ **16.** $\begin{bmatrix} -3 & 2 & 2 \\ 5 & -3 & -3 \\ -1 & 1 & 0 \end{bmatrix}$
17. $(-1, 1, 0)$ **18.** $3.10\ \Omega, 6.90\ \Omega$

Chapter 16

Section 16.1 (page 493)

1. $\dfrac{\cos\beta}{\sin\beta}$ **3.** $\sin\theta$ **5.** 1 **7.** $\dfrac{\cos x + 1}{\sin x}$ **9.** $\cos\theta$ **11.** $\dfrac{1}{\sin^2 s}$ **13.** $-\dfrac{\sin^2\theta}{\cos^2\theta}$
15. $\csc x$ **17.** $\csc^2\theta$ **19.** $\csc\theta$ **21.** 1 **23.** $\cot t$ **25.** $\cos\theta$ **27.** 1 **29.** $\sec^3 x$
31. $\cot\theta$

Section 16.2 (page 497)

41. $a\omega$ **43.** $y = \dfrac{2v_0^2 x \sin\alpha \cos\alpha - gx^2}{2v_0^2 \cos^2\alpha}$

Section 16.3 (page 503)

1. $\dfrac{\sqrt{6}+\sqrt{2}}{4}$ **3.** $\dfrac{\sqrt{2}-\sqrt{6}}{4}$ **5.** $\dfrac{\sqrt{2}}{2}$ **7.** $\dfrac{\sqrt{2}}{2}$ **9.** $\dfrac{\sqrt{2}}{2}$ **11.** $\sin 2x$ **13.** $\sin 3x$
15. $\cos 2x$ **17.** $\cos 9x$ **19.** $\sin x$ **21.** $\dfrac{1}{2}(\sqrt{3}\cos x - \sin x)$ **23.** $-\sin 2x$ **25.** $-\cos x$
27. $\sin 2x$ **29.** $\cos 2x$ **31.** $\dfrac{1}{2}(\sin x - \sqrt{3}\cos x)$ **33.** $\dfrac{1}{2}(\sqrt{3}\cos x - \sin x)$
35. $\dfrac{\sqrt{2}}{2}(\sin x + \cos x)$ **37.** $\dfrac{1+\tan x}{1-\tan x}$ **57.** $y = 2A\cos\left(\dfrac{2\pi t}{T}\right)\cos\left(\dfrac{2\pi x}{\lambda}\right)$

Section 16.4 (page 509)

1. $\dfrac{24}{25}$ **3.** $\dfrac{7}{25}$ **5.** $-\dfrac{120}{169}$ **7.** $-\dfrac{\sqrt{3}}{2}$ **9.** $\dfrac{17}{25}$ **11.** $-\dfrac{31}{49}$ **13.** $\cos 6y$ **15.** $\sin 6\theta$
17. $\cos 4\beta$ **19.** $-\cos 8y$ **21.** $\dfrac{1}{2}\sin 8\omega$ **23.** $2\sin 4x$ **37.** $v = 4\sin 2t$

Section 16.5 (page 514)

1. $\dfrac{\sqrt{2-\sqrt{3}}}{2} = 0.2588$ 3. $\dfrac{\sqrt{2+\sqrt{2}}}{2} = 0.9239$ 5. $\dfrac{\sqrt{2+\sqrt{2}}}{2} = 0.9239$ 7. $\dfrac{7\sqrt{2}}{10}$ 9. $\dfrac{2\sqrt{13}}{13}$

11. $\sin 2\theta$ 13. $\sqrt{2}\cos 3\theta$ 15. $\sqrt{10}\sin 2\theta$ 17. $\dfrac{1}{2}(1-\cos 8x)$ 19. $\dfrac{1}{2}(1+\cos 4x)$

21. $1-\cos 6x$ 23. $6(1-\cos 2x)$ 29. $\dfrac{\sqrt{2}}{2}\csc\dfrac{x}{2}$ 31. $2\sin\dfrac{\theta}{2}$

Section 16.6 (page 519)

1. $\dfrac{\pi}{6}, \dfrac{5\pi}{6}$ 3. $\dfrac{\pi}{4}, \dfrac{5\pi}{4}$ 5. $0, \pi$ 7. $0, \dfrac{\pi}{2}, \pi$ 9. $\dfrac{\pi}{6}, \dfrac{5\pi}{6}, \dfrac{7\pi}{6}, \dfrac{11\pi}{6}$ 11. $\dfrac{\pi}{4}, \pi, \dfrac{5\pi}{4}$

13. $\dfrac{\pi}{2}, \dfrac{7\pi}{6}, \dfrac{11\pi}{6}$ 15. 0 17. $\dfrac{\pi}{4}, \dfrac{5\pi}{4}$ 19. π 21. $\dfrac{\pi}{4}, \dfrac{5\pi}{4}$ 23. $\dfrac{\pi}{4}, \dfrac{3\pi}{4}, \dfrac{5\pi}{4}, \dfrac{7\pi}{4}$ 25. $\dfrac{\pi}{2}, \dfrac{3\pi}{2}$

27. $0, \dfrac{2\pi}{3}, \dfrac{4\pi}{3}$ 29. $\dfrac{\pi}{2}$ 31. $0, \dfrac{\pi}{2}, \pi, \dfrac{3\pi}{2}$ 33. $54.7°, 125.3°, 234.7°, 305.3°$

35. $90°, 194.5°, 270°, 345.5°$ 37. $0°, 41.4°, 180°, 318.6°$ 39. $54.7°, 125.3°, 234.7°, 305.3°$

41. $22.5°, 112.5°$ 43. $t = \dfrac{\pi}{24} = 0.13$ sec   45. $t = \dfrac{\pi}{6\omega}$ sec

Section 16.7 (page 522)

1. $\dfrac{\pi}{3}, \dfrac{2\pi}{3}$ 3. $\dfrac{\pi}{4}, \dfrac{5\pi}{4}$ 5. $\dfrac{\pi}{2}$ 7. $0, \pi$ 9. $\dfrac{\pi}{3}, \dfrac{5\pi}{3}$ 11. $\dfrac{\pi}{2}, \dfrac{3\pi}{2}$ 13. $\dfrac{7\pi}{6}, \dfrac{11\pi}{6}$ 15. $\dfrac{\pi}{3}, \dfrac{4\pi}{3}$

17. $x = \tan y$ 19. $x = \sin(1-y)$ 21. $x = \dfrac{1}{2}\sin(y+1)$ 23. $x = \csc y - 1$

25. $x = \dfrac{1}{3}\sec(y-1)$ 27. $x = \dfrac{1}{3}\cot\dfrac{y}{3}$ 29. $x = \sin\dfrac{1}{2}(y-3) - 1$

Section 16.8 (page 528)

1. $\dfrac{\pi}{3}$ 3. $\dfrac{\pi}{4}$ 5. 0 7. $\dfrac{\pi}{2}$ 9. 0 11. $-\dfrac{\pi}{4}$ 13. $-\dfrac{\pi}{3}$ 15. $\dfrac{\pi}{3}$ 17. $\dfrac{5\pi}{6}$

19. $-2\sqrt{2}$ 21. $-\dfrac{4}{3}$ 23. $\dfrac{\sqrt{5}}{5}$ 25. $\dfrac{\sqrt{5}}{3}$ 27. $-\dfrac{5}{3}$ 29. $-\dfrac{5}{12}$ 31. 4 33. $-\sqrt{15}$

35. $\dfrac{\sqrt{21}}{5}$ 37. $\dfrac{1}{5}$ 39. $\dfrac{1}{4}$ 41. $\dfrac{x}{\sqrt{1-x^2}}$ 43. $\sqrt{1+x^2}$ 45. $\dfrac{\sqrt{1-4x^2}}{2x}$ 47. $\dfrac{\sqrt{9x^2+1}}{3x}$

49. $\sqrt{1-4x^2}$ 51. 1.1071 53. -0.3398 55. 0.9203 57. 2.0846 59. $x = \sin\dfrac{y}{2}$

61. $x = \text{Arcsin}\dfrac{y}{2}$ 63. $x = \tan(y-3)$ 65. $x = \dfrac{1}{3}\text{Arcsin}\dfrac{y}{2}$ 67. $x = \text{Arctan}\dfrac{y}{4} + 2$

71. $R = X\cot\theta$ 73. $t = \sqrt{\dfrac{m}{k}}\text{Arccos}\dfrac{x}{A}$

Review Exercises for Chapter 16 (page 530)

1. $\sqrt{\dfrac{2-\sqrt{2}}{2}}$ 3. $\dfrac{\sqrt{3}}{2}$ 5. $\sin 6x$ 7. $\sin 3x$ 9. $-\cos 2x$ 11. $-\sin 2x$

13. $\dfrac{1}{2}(\sqrt{3}\sin x - \cos x)$ 15. $-\dfrac{24}{25}$ 17. $\dfrac{119}{169}$ 19. $-\dfrac{3}{5}$ 21. $\cos 6x$ 23. $\cos 8x$

25. $\sin 6x$ 27. $\sin 2\theta$ 29. $\sqrt{2}\cos 2\theta$ 31. $\dfrac{1}{2}(1-\cos 6x)$ 33. $1+\cos 6x$ 57. $0, \pi$

59. $\dfrac{\pi}{3}, \pi, \dfrac{5\pi}{3}$ **61.** $\pi, \dfrac{4\pi}{3}$ **63.** $\theta = \dfrac{\pi}{4} + \dfrac{\alpha}{2}$ **65.** $\dfrac{\pi}{3}, \dfrac{5\pi}{3}$ **67.** $\dfrac{5\pi}{6}, \dfrac{11\pi}{6}$ **69.** $\dfrac{2\pi}{3}$
71. $-\sqrt{35}$ **73.** $\sqrt{1 - 4x^2}$ **75.** $x = \dfrac{1}{4} \operatorname{Arcsin} \dfrac{y}{2}$

Chapter 17

Section 17.1 (page 538)

1. $x < 3$ **3.** $x \le 2$ **5.** $x < -6$ **7.** $x \ge -5$ **9.** $x \le \dfrac{4}{3}$ **11.** $x > -\dfrac{3}{2}$ **13.** $x \ge \dfrac{7}{3}$

15. $x > \dfrac{5}{2}$ **17.**   **19.** 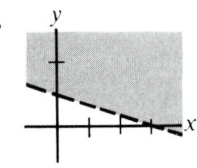   **21.**

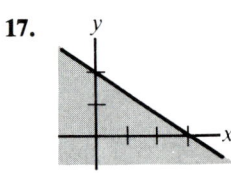

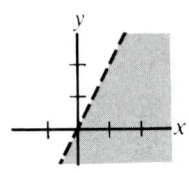

23. 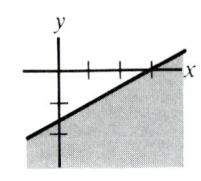   **25.** 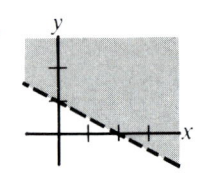   **27.** 7.6 hr   **29.** $0.0 \le F \le 80$ (pounds)

Section 17.2 (page 543)

1. $x > -3$ **3.** $-2 < x < 2$ **5.** $x \le -3, x \ge -2$ **7.** $x < -2, 0 < x < 3$
9. $x \le -5, -2 \le x \le 3$ **11.** $-8 < x < 6, x > 7$ **13.** $1 < x < 2$ **15.** $x \le 6, x > 7$
17. $3 < x < 4$ **19.** $x \le -4, 1 < x \le 3$ **21.** $-3 < x < 3, x > 6$ **23.** $x \le -6, 1 \le x < 2, x > 3$
25. $x > 4$ **27.** all x **29.** $x < -1$ **31.** $x < -4, 1 < x \le 4$ **33.** $0 \le t \le \dfrac{4}{3}$ (seconds)
35. $x \le 5, x \ge 15$ (feet)   **37.** $1 \le x < 20$

Section 17.3 (page 547)

1. $-1 < x < 3$ **3.** $-4 \le x \le -2$ **5.** $-1 < x < 0$ **7.** $x \le \dfrac{1}{2}, x \ge \dfrac{7}{2}$ **9.** $-\dfrac{13}{2} < x < -\dfrac{1}{2}$
11. $x < -1, x > 5$ **13.** $x \le -3, -2 < x < 3, x > 4$ **15.** $x < 0, x > 4$
17. $-2 < x < 1 - \sqrt{7}, 1 + \sqrt{7} < x < 4$ **19.** $-3 - \sqrt{10} < x < -5, -1 < x < -3 + \sqrt{10}$
21. $|d - 2.5550| \le 0.0001$

Section 17.4 (page 552)

1.   **3.** 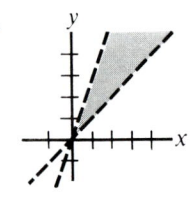   **5.** 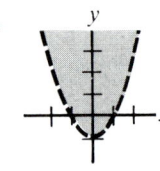   **7.** (1, 1)   (2, 0)

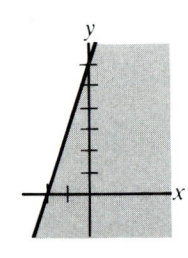

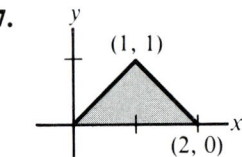

9.

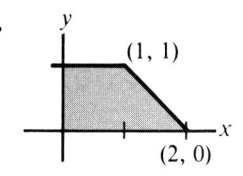

11.

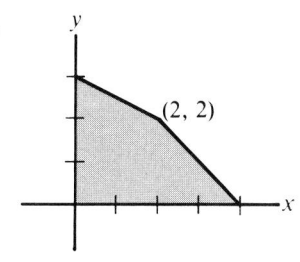

13.

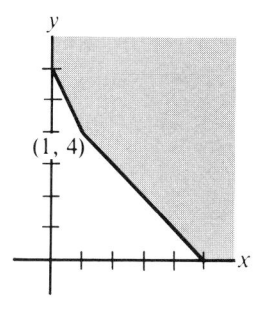

15.

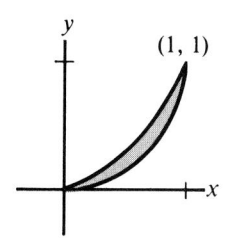

17.

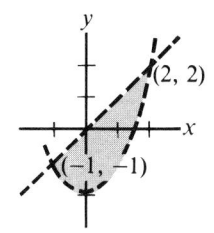

Section 17.5 (page 558)

1. $P = 4$ at $(2, 0)$ **3.** $P = 3$ at $(1, 0)$ **5.** $P = 13$ at $(3, 1)$ **7.** $P = 17$ at $(3, 2)$
9. $P = 25$ at $(3, 8)$ **11.** 90 acres of soybeans, no corn; $8,100
13. 20 smaller trees, 10 larger trees; $9,600 **15.** 4 of Type I, 11 of Type II; $1,720/month
17. 2 lb of Product I, 4 lb of Product II; $2.10 per bag

Review Exercises for Chapter 17 (page 559)

1. $x < -5$ **3.** $x < 3$ **5.** $-3 \leq x \leq 3$ **7.** $-4 < x < -1, x > 3$ **9.** $-2 < x \leq 1$
11. $x < 1, x > 6$ **13.** $-4 < x \leq 1, x \geq 2$ **15.** $x \leq -2$ **17.** $2 \leq x \leq 6$ **19.** $-\frac{5}{2} < x < \frac{1}{2}$
21. $-1 < x < 2 - \sqrt{7}; 2 + \sqrt{7} < x < 5$ **23.** **25.**

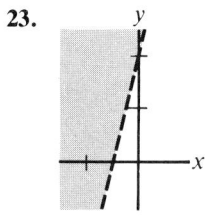

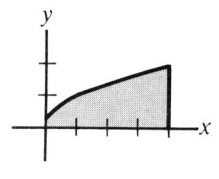

27. 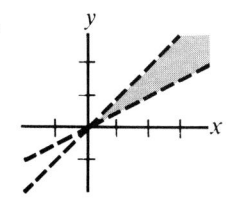 **29.** 6.5 hr **31.** $t \geq 6$ sec **33.** $P = 48$ at $(8, 6)$

35. 60 small appliances, 20 large appliances; $8,600

Chapter 18

Section 18.1 (page 564)

1. $\dfrac{128}{243}$ 3. $-\dfrac{5}{512}$ 5. 162 7. 63 9. 242 11. $\dfrac{127}{64}$ 13. $\dfrac{20}{27}$ 15. $\dfrac{422}{81}$
17. $\dfrac{129}{64}$ 19. (a) $540 (b) $540.80 (c) $541.22 (d) $541.50 21. $977.34 23. $671.56
25. (a) $676.45 (b) $679.02 27. $2,484.04 29. $1,024N$ 31. $\dfrac{N}{4,096}$ 33. $\dfrac{155}{4}$ ft
35. $\dfrac{36,447}{512}$ ft

Section 18.2 (page 567)

1. $\dfrac{4}{9}$ 3. $\dfrac{13}{99}$ 5. $\dfrac{104}{333}$ 7. $\dfrac{133}{990}$ 9. 1 11. $\dfrac{3}{4}$ 13. $\dfrac{4}{5}$ 15. $\dfrac{27}{4}$ 17. $\dfrac{3}{8}$ 19. $\dfrac{16}{7}$
21. 100 in. 23. 48 ft 25. 8 mm

Section 18.3 (page 572)

1. $x^4 - 4x^3y + 6x^2y^2 - 4xy^3 + y^4$ 3. $s^4 + 8s^3v + 24s^2v^2 + 32sv^3 + 16v^4$ 5. $x^3 - 3x^2y + 3xy^2 - y^3$
7. $s^4 + 8s^3t + 24s^2t^2 + 32st^3 + 16t^4$ 9. $a^5 - 5a^4b^2 + 10a^3b^4 - 10a^2b^6 + 5ab^8 - b^{10}$
11. $S^6 - 12S^5W + 60S^4W^2 - 160S^3W^3 + 240S^2W^4 - 192SW^5 + 64W^6$
13. $R^6 + 12R^5 + 60R^4 + 160R^3 + 240R^2 + 192R + 64$
15. $R^6 - 18R^5 + 135R^4 - 540R^3 + 1215R^2 - 1458R + 729$ 17. $x^{10} + 20x^9 + 180x^8 + 960x^7 + \cdots$
19. $k^{12} - 12k^{11}m^2 + 66k^{10}m^4 - 220k^9m^6 + \cdots$ 21. $r^8 + 16\pi r^7 + 122\pi^2 r^6 + 448\pi^3 r^5 + \cdots$
23. $1 - 2x + 3x^2 - 4x^3 + \cdots$ 25. $1 + \dfrac{1}{2}s - \dfrac{1}{8}s^2 + \dfrac{1}{16}s^3 - \cdots$ 27. $1 + x + x^2 + x^3 + \cdots$

29. 0.9698   31. 1.096   35. $1 + \dfrac{1}{2}x^2 - \dfrac{1}{8}x^4 + \dfrac{1}{16}x^6 - \cdots$

Review Exercises for Chapter 18 (page 573)

1. $\dfrac{2}{243}$ 3. $\dfrac{43}{64}$ 5. $\dfrac{1,055}{81}$ 7. $\dfrac{266}{243}$ 9. $\dfrac{3}{4}$ 11. $\dfrac{10}{3}$ 13. 12 15. $\dfrac{7}{45}$ 17. $\dfrac{38}{99}$
19. $a^3 - 3a^2b + 3ab^2 - b^3$ 21. $M^5 - 10M^4T + 40M^3T^2 - 80M^2T^3 + 80MT^4 - 32T^5$
23. $e^4 - 8e^3 + 24e^2 - 32e + 16$ 25. $m^5 - 5m^4v^2 + 10m^3v^4 - 10m^2v^6 + 5mv^8 - v^{10}$
27. $x^{10} - 20x^9 + 180x^8 - 960x^7 + \cdots$ 29. $1 + 2t + 3t^2 + 4t^3 + \cdots$ 31. $1 + \dfrac{1}{3}c + \dfrac{2}{9}c^2 + \dfrac{14}{81}c^3$
33. 0.9680   35. $1,591.98   37. $1,030.32   39. $6,561N$ 41. $\dfrac{4,850}{243}$ ft   43. 100 in.
45. $M = \dfrac{E}{2}(1 - s + s^2 - s^3 + \cdots)$

Cumulative Review Exercises for Chapters 16–18 (page 575)

8. $\dfrac{\sqrt{2}}{2}$ 9. $\dfrac{\sqrt{2} + \sqrt{3}}{2}$ 10. $0°, 180°, 63.4°, 243.4°$ 11. a. $-\dfrac{\pi}{3}$; b. $\dfrac{2\pi}{3}$
12. $-4 < x \leq -3,\ 1 < x \leq 2$ 13. $x < 0, x > 4$ 14. $\dfrac{1}{3}$
15. $x^5 - 10x^4y + 40x^3y^2 - 80x^2y^3 + 80xy^4 - 32y^5$ 16. $R = \dfrac{v^2}{g}\sin 2\theta$ 17. 0.084 sec
18. $\theta = \operatorname{Arccos}\left(\dfrac{I}{I_{max}}\right)$ 19. $\dfrac{315}{8}$ ft   20. 100 ft

… **ANSWERS TO ODD-NUMBERED EXERCISES** … **677**

Chapter 19

Section 19.2 (page 578)
1. $\sqrt{13}$ **3.** $4\sqrt{5}$ **5.** $\sqrt{7}$ **7.** $\sqrt{6}$ **9.** $2\sqrt{2}$ **11.** positive in quadrants I and III
13. (a) y-axis **(b)** x-axis **21.** $y^2 - 4x + 4 = 0$

Section 19.3 (page 581)
1. -1 **3.** $\frac{3}{2}$ **5.** $\frac{6}{5}$ **7.** 2 **9.** undefined **13.** $\frac{13}{6}$ **19.** $(0, 1)$ **21.** $\frac{7}{2}$
23. $-\frac{1}{5}, -\frac{8}{7}, -\frac{7}{2}$ **25.** $x = -4$

Section 19.4 (page 585)
1. $x - 2y + 11 = 0$ **3.** $3x - y - 13 = 0$ **5.** $x + 3y = 0$ **7.** $y = 0$ (x-axis)
9. $5x + 3y + 3 = 0$ **11.** $x - y + 10 = 0$ **13.** $x + 3y + 6 = 0$ **15.** $y = -3x + \frac{5}{2}$
17. $y = \frac{2}{3}x + 0$ **19.** $y = 0x + \frac{7}{2}$ **21.** parallel **23.** neither **25.** perpendicular
27. $3x - 4y + 7 = 0$ **29.** $20x + 28y - 27 = 0$ **33.** $F = 200x$ **35.** $F = \frac{9}{5}C + 32$

Section 19.6 (page 589)
1. $x^2 + y^2 = 25$ **3.** $x^2 + y^2 = 100$ **5.** $x^2 + y^2 + 4x - 10y + 28 = 0$
7. $x^2 + y^2 + 2x + 8y = 0$ **9.** $x^2 + y^2 + x + y - 32 = 0$ **11.** $(x - 1)^2 + (y - 1)^2 = 4$; $(1, 1)$; $r = 2$
13. $(x + 2)^2 + (y - 4)^2 = 16$; $(-2, 4)$; $r = 4$ **15.** $\left(x + \frac{3}{2}\right)^2 + (y + 2)^2 = 5$; $\left(-\frac{3}{2}, -2\right)$; $r = \sqrt{5}$
17. point circle, $\left(\frac{5}{2}, \frac{1}{2}\right)$ **19.** point circle, $(3, -4)$ **21.** imaginary circle

Section 19.7 (page 594)
1. $y^2 = 12x$ **3.** $x^2 = -20y$ **5.** $y^2 = -8x$ **7.** $x^2 = -12y$ **9.** $x^2 = 8y$
11. $(0, 2), y + 2 = 0$ **13.** $(0, -3), y - 3 = 0$ **15.** $(4, 0), x + 4 = 0$ **17.** $(-1, 0), x - 1 = 0$
19. $(0, 1), y + 1 = 0$ **21.** $\left(\frac{9}{4}, 0\right), x + \frac{9}{4} = 0$ **23.** $\left(-\frac{1}{4}, 0\right), x - \frac{1}{4} = 0$ **25.** $\left(-\frac{1}{6}, 0\right), x - \frac{1}{6} = 0$
27. $x^2 + y^2 - 6y - 27 = 0$ **29.** $y^2 - 2y - 8x + 17 = 0$ **31.** $x^2 = -y, y^2 = -8x$ **33.** 26.3 m
35. $16\sqrt{2} = 22.6$ m **37.** $2\sqrt{3}$ m

Section 19.8 (page 599)
1. $(\pm 5, 0), (\pm 3, 0), 4$ **3.** $(\pm 3, 0), (\pm\sqrt{5}, 0), 2$ **5.** $(\pm 4, 0), (\pm\sqrt{15}, 0), 1$
7. $(0, \pm 4), (0, \pm\sqrt{7}), 3$ **9.** $(0, \pm\sqrt{10}), (0, \pm\sqrt{6}), 2$ **11.** $(0, \pm\sqrt{5}), (0, \pm 2), 1$
13. $(\pm\sqrt{6}, 0), (\pm\sqrt{3}, 0), \sqrt{3}$ **15.** $(0, \pm\sqrt{15}), (0, \pm 2\sqrt{2}), \sqrt{7}$ **17.** $(\pm 5, 0), (\pm\sqrt{15}, 0), \sqrt{10}$
19. $\frac{x^2}{16} + \frac{y^2}{7} = 1$ **21.** $4x^2 + 3y^2 = 48$ **23.** $2x^2 + y^2 = 18$ **25.** $\frac{x^2}{39} + \frac{y^2}{64} = 1$
27. $x^2 + 4y^2 = 16$ **29.** $\frac{x^2}{64} + \frac{y^2}{28} = 1$ **31.** $3x^2 + 4y^2 = 48$ **33.** circle; $x^2 + y^2 - 8x + 12 = 0$
35. $9x^2 + 16y^2 = 36$ **37.** $\frac{12\sqrt{5}}{5} = 5.4$ m **39.** $\frac{x^2}{16,810,000} + \frac{y^2}{16,809,600} = 1$

Section 19.9 (page 605)

1. $(\pm 4, 0), (\pm 5, 0)$ **3.** $(\pm 3, 0), (\pm 5, 0)$ **5.** $(0, \pm 2), (0, \pm 2\sqrt{2})$

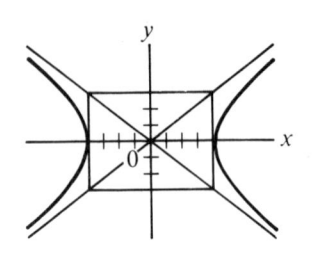

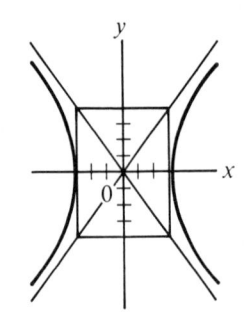

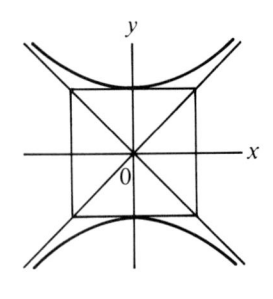

7. $(\pm 1, 0), (\pm\sqrt{6}, 0)$ **9.** $(0, \pm 2\sqrt{3}), (0, \pm 2\sqrt{5})$ **11.** $(0, \pm\sqrt{2}), (0, \pm\sqrt{5})$ **13.** $4x^2 - 9y^2 = 36$
15. $16y^2 - 9x^2 = 144$ **17.** $3x^2 - y^2 = 27$ **19.** $4x^2 - y^2 = 4$ **21.** $16y^2 - 9x^2 = 144$
23. $3x^2 - y^2 + 6x + 4y - 4 = 0$ **25.** $y^2 - 25x^2 = 144$ **27.** $pV = 10$

Section 19.10 (page 609)

5. $\dfrac{(y+2)^2}{6} - \dfrac{(x+2)^2}{9} = 1$; hyperbola, center at $(-2, -2)$

7. $(x+2)^2 + \dfrac{(y-3/2)^2}{4} = 1$; ellipse, center at $\left(-2, \dfrac{3}{2}\right)$ **9.** $(x+1)^2 + (y-1)^2 = 0$; point

11. $\left(y - \dfrac{5}{2}\right)^2 - \dfrac{x^2}{6} = 1$; hyperbola; center at $\left(0, \dfrac{5}{2}\right)$

13. $\left(x - \dfrac{1}{8}\right)^2 + \left(y - \dfrac{3}{4}\right)^2 = 1$; circle; center at $\left(\dfrac{1}{8}, \dfrac{3}{4}\right)$ **15.** $3(x-3)^2 + (y+1)^2 = -1$; imaginary locus
17. $y^2 - 4y - 16x - 12 = 0$ **19.** $4x^2 + 9y^2 + 24x = 0$ **21.** $5x^2 - 4y^2 + 30x + 8y - 39 = 0$
23. $4x^2 + 25y^2 - 16x - 150y + 216 = 0$ **25.** $x^2 - 4y^2 - 2x - 3 = 0$
27. $y = \pm 2$; two parallel lines   **29.** $(x-3)(y+2) = 0$; lines $x = 3$ and $y = -2$

Section 19.11 (page 614)

3. $(-1, \sqrt{3})$ **5.** $(-3, -3\sqrt{3})$ **7.** $(1.97, -0.35)$ **9.** $\left(\sqrt{2}, \dfrac{7\pi}{4}\right)$ **11.** $(5, 0.93)$

13. $r = 2\sec\theta$ **15.** $r = \sqrt{2}$ **17.** $r^2 = \sec 2\theta$ **19.** $r^2 = \dfrac{1}{2(1 + \sin^2\theta)}$ **21.** $y = 2$
23. $y = x$ **25.** $(x^2 + y^2 + x)^2 = x^2 + y^2$ **27.** $(x^2 + y^2)^2 = 2xy$ **29.** $y^2 = 4(x+1)$
31. $x^2 = 8(y+2)$ **33.** $(x^2 + y^2)^2 = 3(x^2 - y^2)$ **35.** $(x^2 + y^2)^2 = a(3x^2y - y^3)$
37. $(x^2 + y^2 + 2y)^2 = x^2 + y^2$

Section 19.12 (page 620)

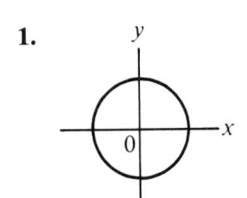

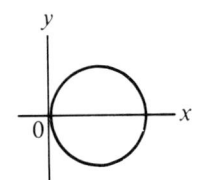

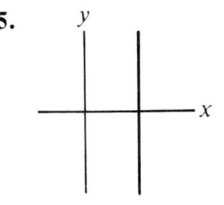

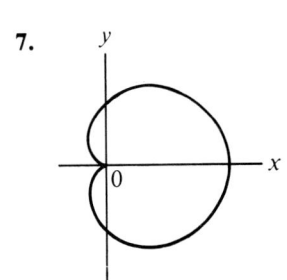

9.

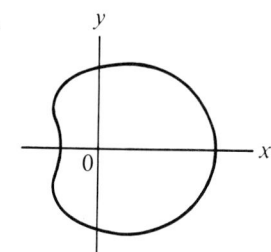

11.

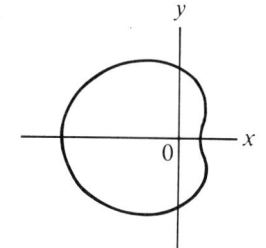

**13.
15.**

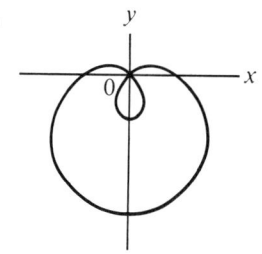

17.

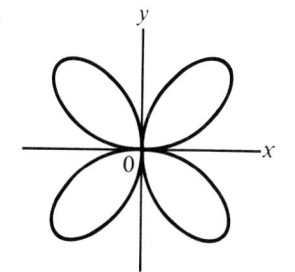

19.

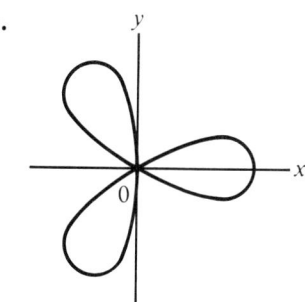

21.

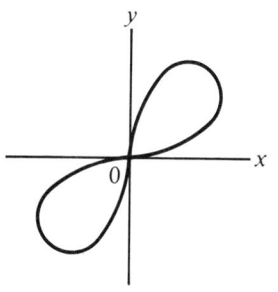

23.

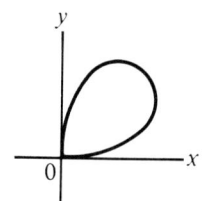

25.

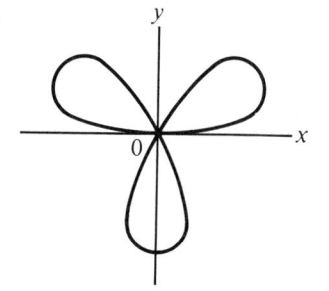

27.

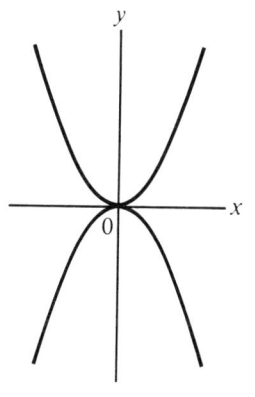

29.

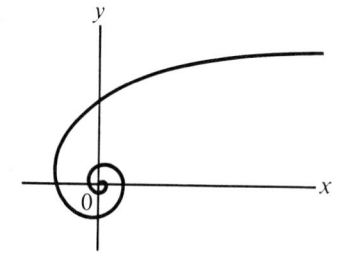

Review Exercises for Chapter 19 (page 620)

3. -40 **7.** $3x + y - 2 = 0$ **9.** $x^2 + y^2 - 2x + 4y = 0$ **11.** $(-1, -1)$, $\sqrt{2}$

13. ellipse, $(0, \pm 4)$, $(0, \pm\sqrt{7})$ **15.** hyperbola, $(0, \pm 2)$, $(0, \pm\sqrt{11})$ **17.** parabola, $(0, 0)$, $\left(-\frac{3}{4}, 0\right)$

19. $(y + 3)^2 = -4(x - 2)$; parabola, $(2, -3)$ **21.** $\dfrac{(x - 2)^2}{9} + \dfrac{(y + 1)^2}{16} = 1$; ellipse, $(2, -1)$
23. $\dfrac{(x - 2)^2}{9} - \dfrac{(y - 4)^2}{9} = 1$; hyperbola, $(2, 4)$ **25.** $x^2 - 2x - 12y + 37 = 0$ **27.** $\dfrac{x^2}{7} + \dfrac{y^2}{16} = 1$
29. $\dfrac{(y - 2)^2}{9} - \dfrac{x^2}{7} = 1$ **31.** 0.10 ft from vertex
33. $r = 3 \cot \theta \csc \theta$ **35.** $x(x^2 + y^2) = 4y$ **37.** circle

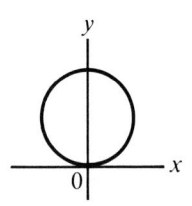

39. four-leaf rose   **41.** lemniscate

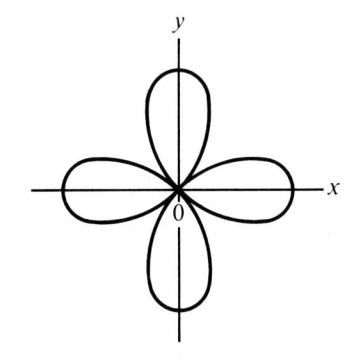

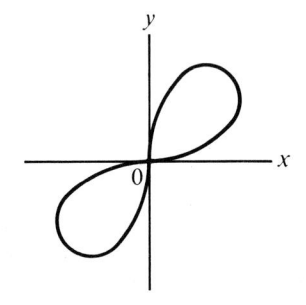

Section A.1 (page 626)

1. 43.3 in.   **3.** 49.74 cm^2 **5.** 1.34 $\dfrac{\text{km}}{\text{min}}$ **7.** 32.5 $\dfrac{\text{N}}{\text{cm}}$ **9.** 1.28 $\dfrac{\text{m}}{\text{sec}}$ **11.** 5.4 $\dfrac{\text{lb}}{\text{in}^2}$
13. $2.36 \times 10^{-4} \dfrac{\text{N}}{\text{cm}^3}$ **15.** 20,370 $\dfrac{\text{lb}}{\text{ft}^3}$ **17.** 1.77 ft^3 **19.** 8.683 in.^{3}

Index

Abel, Niels, 419
Abscissa, 70
Absolute inequality, 544
Absolute value, 5
  of complex number, 333
  inequality, 544
Acceleration, 623
Accuracy of numbers, 35
Addition:
  of algebraic expressions, 11
  of complex numbers, 334, 339
  of coordinates, 270
  of fractions, 179
  of matrices, 462
  of radicals, 317
  of signed numbers, 5
  of vectors, 275
Algebraic expressions, 9
al-Khowarizmi, 10, 202
Alternating current, 262
Ambiguous case, sine law, 288
Ampere, 56
Amplitude, 254
Analytic geometry, 576
Angles, 121
  acute, 124, 288
  central, 239
  complementary, 128
  coterminal, 122
  of depression, 141
  of elevation, 141
  of inclination, 578
  negative, 122
  obtuse, 288
  quadrantal, 122, 231
  reference, 225

Angles (*cont'd*)
  special, 130, 132, 229
  standard-position, 122
Angular velocity, 246
Annuity, 563
Antilogarithm, 372
Approximate numbers, 34
  roots, 439
Archimedes, 145
Arc length, 244
Area:
  of circular sector, 245
  of geometric figures, *see* Inside cover
Argument of complex number, 333
Associative law, 4
Asymptotes, 266, 602
Axes, coordinate, 70
Axis:
  conjugate, 602
  major, 597
  minor, 597
  of a parabola, 590
  polar, 610
  transverse, 602

Base:
  change of, 377
  of exponent, 9
  of logarithm, 359
Basic imaginary unit, 332
Binomial, 10
  formula, 570
  series, 571
  theorem, 570

INDEX

Braces, 15
Brackets, 15

Calculator operations:
 arithmetic, 35
 determinants, 95
 exponential functions, 378
 inverse trigonometric functions, 526
 logarithmic functions, 372, 373, 378
 powers and radicals, 318
 radian mode, 242, 526
 roots of quadratic equations, 211
 scientific notation, 41
 trigonometric functions, 135, 236, 237, 288, 293
Cancellation, 171, 176
Capacitor, 194, 353
Carbon dating, 385
Cardano, Geronimo, 419
Cardioid, 615
Cartesian coordinate system, 70
Cayley, Arthur, 91, 461
Characteristic of a logarithm, 371
Chester, Robert of, 129
Chord, 129
Circle:
 arc length, 244
 area of sector, 245
 standard equation of, 587
Clearing fractions, 51, 186, 190
Coefficient, 9
 matrix 477
Cofunctions, 128
Column, 92
Common factors, 151
Common logarithm, 370
Common ratio, 561
Commutative law, 4
 vectors, 276
Complementary angles, 128
Completing the square, 205
Complex fractions, 185
Complex numbers, 3, 22, 207, 331–357
 basic concepts, 331–337
 basic operations with, 337–341
 De Moivre's theorem, 348, 349
 exponential form, 344
 polar form, 341
 powers, 348
 products and quotients, 346
 roots, 349
 as vectors, 333
Complex plane, 333
Complex roots, 206, 430
Components of a vector, 277
Compound interest, 379, 565
Computations using logarithms, 374
Conditional equations, 46, 515
Conic sections, 586

Conjugate:
 axis, 602
 of binomial, 325
 of complex number, 333
Constant of proportionality, 65
Conversion of units, 624
Coordinate axes, 70
Coordinates, polar, 610
 rectangular, 70
Cosecant, of angle, 125, 126
 graph of, 267
Cosine(s):
 of angle, 125, 126
 of double angle, 507
 graph of, 252, 254, 258
 of half angle, 511
 law of, 292
 of sum of two angles, 500
Cotangent, of angle, 125, 126
 graph of, 267
Coterminal angles, 122
Coulomb, 56
Cramer, Gabriel, 91
Cramer's rule, 93, 107, 456, 457
Critical value, 539
Current, 110, 262, 353

Decibel, 369, 384
Decimal, 2
Degree measure of angle, 121
Degree of polynomial, 418
De Moivre, Abraham, 348
De Moivre's theorem, 348, 349
Denominator:
 common, 51, 179, 190
 rationalizing, 21, 314, 325
Dependent:
 system of equations, 88
 variable, 70
Descartes, René, 577
Descartes' rule of signs, 433
Determinants:
 of higher order, 103, 449
 properties of, 450
 of second order, 92
 of third order, 103
Difference:
 of two cubes, 153
 of two squares, 153
Directrix of parabola, 590
Direct variation, 65
Discriminant, 211
Distance formula, 578
Distributive law, 4, 15, 150
Division:
 of algebraic expressions, 18, 30
 of complex numbers, 339, 346
 of fractions, 175, 176
 of radicals, 324
 of signed numbers, 5
 synthetic, 425
Domain, 73

Double-angle formulas, 507, 510
Double root, 202, 436

e (number), 344, 376
Eccentricity, of ellipse, 600
Electrical circuit, 28, 67, 110
Element:
 of determinant, 92
 of matrix, 461
Ellipse, 398, 595
Equality:
 of complex numbers, 336
 of matrices, 462
Equations:
 conditional, 46, 515
 exponential, 380
 first-degree, 50
 formulas, 52
 fractional, 51, 190
 graphical solution of, 81, 396
 higher-order, 418
 linear, 45, 79
 literal, 52
 logarithmic, 380
 polar, 614
 polynomial, 418
 quadratic, 198–219
 in quadratic form, 407, 517
 radical, 411
 root of, 46
 systems of, see Systems of equations
 transposition, 49
 trigonometric, 515
Equivalent matrices, 472
Eratosthenes, 247
Euler's identity, 344
Exponential equations, 380
Exponential form, complex number, 344
Exponential function, 361
 graph of, 362
Exponents, 9
 fractional, 308
 laws of, 18, 299
 negative, 22, 303
 zero, 22, 303
Expression, algebraic, 9
Extraneous roots, 192, 411

Factor, prime, 151, 165
 theorem, 421
Factorial, 569
Factoring:
 common factors, 151
 difference of two cubes, 153
 difference of two squares, 153
 FOIL method, 163
 by grouping, 166
 perfect squares, 161
 trinomials, 161
Farad, 56

INDEX

Feasible region, 552
First-degree equations, 50
Focus:
  of ellipse, 595
  of hyperbola, 602
  of parabola, 590
FOIL method, 163
Formulas, 52
Fourier series, 271
Fractional equations, 51, 190
Fractional exponents, 308
Fractions, 169
  cancellation, 171
  complex, 185
  equivalent, 169
  lowest common denominator, 179
  operations with, 18, 169–185
  reducing, 170, 185
  simplest form, 170
Frequency, 265
Function:
  definition of, 70
  exponential, 361
  inverse trigonometric, 524
  logarithmic, 362
  notation for, 75
  periodic, 254
  polynomial, 419
  trigonometric, 125, 126, 223
  zeros of, 426
Fundamental operations, 3
Fundamental theorem of algebra, 428

Gauss, Karl F., 419
Geometric formulas, *see* Inside cover
Geometric progression, 562
  series, 566
Graphical:
  addition, 270
  representation of complex numbers, 332, 333
  solution of equations, 81, 396
  solution of inequalities, 536, 547
Graphs:
  of exponential functions, 362
  of inverse trigonometric functions, 523, 524, 525
  of logarithmic functions, 363
  on logarithmic paper, 385
  in polar coordinates, 614
  of trigonometric functions, 250–273
Greater than, 3, 534
Grouping, symbols of, 15

Half-angle formulas, 511
Half-life, 364, 378
Hertz, 265
Hipparchus, 120
Hyperbola, 400, 601

Identities, trigonometric:
  basic, 488, 492
  double-angle, 507, 510
  half-angle, 511
  product-to-sum, 504
  sum and difference, 500, 502
  sum-to-product, 504
Identity matrix, 467
Imaginary numbers, 22, 331
Impedance, 354
Inconsistent system, 88
Independent variable, 70
Index of radical, 20, 313
Inductor, 353
Inequalities, 534–560
  with absolute values, 544
  algebraic solution of, 536, 539, 541
  graphical solution of, 536, 547
  properties of, 535
  systems of, 548
  with two variables, 547
Initial point of a vector, 274
Initial side of an angle, 121
Intercepts, 81
International System of Units (SI), 56, 623
Interpolation, 134, 373
  irrational roots by, 439
Inverse:
  functions, 523
  matrix, 470
  trigonometric functions, 524
  trigonometric relations, 520
  variation, 66
Irrational numbers, 2
  roots, 439

j, definition of, 206
Joint variation, 67

Khayyám, Omar, 419

Law of cosines, 292
Law of sines, 285
Lemniscate, 618
Less than, 3, 534
Limaçon, 616
Line, equation of, 79, 583
Linear:
  approximation of roots, 439
  equations, 79
  interpolation, 134
  programming, 552
  velocity, 246
Literal symbol, 46
Logarithmic equations, 380
Logarithmic function, 362
  graph of, 363
  properties of, 363

Logarithmic paper, 385
Logarithms, 358–393
  to base e, 376
  to base 10, 370
  change of base, 377
  common, 370
  computations with, 374
  definition of, 359
  natural, 376
  properties of, 367
Lowest common denominator, 51, 179, 190

Major axis, 597
Mantissa, 371
Mass, 623
Matrix, 461
  addition, 462
  augmented, 472, 478
  element of, 461
  identity, 467
  inverse, 470
  multiplication, 465
  square, 466
  zero, 463
Maximum value, 553
Metric system, 624
Midpoint formula, 581
Minimum value, 553
Minor, 104
Minor axis, 597
Minute (as measurement of angle), 121
Modulus of complex number, 333
Moment, 98
Monomial, 10
Multinomial, 10
Multiple root, 428
Multiplication:
  of algebraic expressions, 18, 25
  of complex numbers, 339, 346
  of fractions, 175
  of matrices, 465
  of radicals, 322
  of signed numbers, 5

Napier, John, 358
Natural logarithm, 376
Negative:
  angle, 122
  exponent, 22, 303
  number, 2
Numbers:
  approximate, 34
  complex, 3, 22
  imaginary, 22
  irrational, 2
  negative, 2
  rational, 1
  real, 3
Numerical coefficient, 9

684 INDEX

Oblique triangles, 287, 291
Ohm, 56
Order of operations, 6
Order of radical, 313
Ordinate, 70
Ordinates, addition of, 270
Origin, 2

Parabola, 399, 590
Parabolic reflector, 592
Parentheses, 15
Pascal, Blaise, 569
Pascal's triangle, 569
Period, 254, 268
pH, 374, 385
Phase, 353
Phase shift, 258
Phasor, 353
Point-slope form, 583
Polar axis, 610
Polar coordinates, 610
   curves in, 614
   relation to rectangular coordinates, 611
Polar form of complex number, 341, 342, 344
Pole, 610
Polynomial, 10
   equation, 418
   function, 419
Power, 9
Precision, 36
Prime factors, 151, 165, 180
Principal root, 20
Product:
   of complex numbers, 339, 346
   of matrices, 465
   scalar, 276, 463
   special, 150, 157, 160
Product-to-sum formulas, 504
Progression:
   geometric, 562
   infinite, 566
Proportion, 61
Pure quadratic equations, 201
Pythagorean theorem, 126

Quadrant, 70
Quadrantal angles, 122, 231
Quadratic equations, solution by:
   completing the square, 205
   factoring, 200
   quadratic formula, 209
Quadratic forms, 407, 517
Quadratic formula, 209
Quotient:
   of complex numbers, 339, 346
   of radicals, 324

Radian measure, 239
   conversion to, 241

Radical equations, 411
Radicals, 20, 313–328
   addition of, 317
   division of, 324
   equations with, 411
   laws of, 313
   multiplication of, 322
   simplest form, 21, 314, 317
   subtraction of, 317
Radicand, 313
Radioactive decay, 364, 378
Radius vector, 126, 223
Range, 74
Ratio, 60
   common, 561
Rational numbers, 1
   roots, 432
Rationalizing the denominator, 21, 314, 325
Reactance, 353
Real number, 3
Reciprocal, 176
Rectangular coordinate system, 70
Rectangular form of complex number, 342, 344
Reference angle, 225
   triangle, 225
Remainder, 32, 421
   theorem, 421
Repeating decimal, 2
   root, 202, 428
Resistance, 55, 63, 64, 65, 192, 353, 405
Resolution of vectors, 279, 282
Resonance, 22
Resultant, 275
Right triangles, trigonometry of, 123–148
Romanus, 503
Roots, 46
   complex, 206, 430
   of complex numbers, 349
   double, 202, 436
   extraneous, 192, 411
   irrational, 439
   of linear equations, 45
   multiple, 202, 428
   of quadratic equations, 198–219
   rational, 432
   repeating, 202
Rose, 617
Rounding off, 35
Row:
   of determinant, 92
   of matrix, 461
   operations, 471, 479

Scalar, 274
   product, 276, 463
Scientific notation, 38

Secant, of angle, 125, 126
   graph of, 267
Second (as measurement of angle), 121
Sense of inequality, 534
Series:
   binomial, 571
   geometric, 566
SI (International Systems of Units), 56, 623
Signed numbers, 5
Significant figures, 34, 39, 140, 280
Similar triangles, 77
Simple harmonic motion, 260
Simultaneous equations, see Systems of equations
Sine(s):
   of angle, 125, 126
   of double angle, 507
   graph of, 251, 254, 258
   of half angle, 511
   inverse, 524
   law of, 285
   of sum of two angles, 500
Sinusoidal curves, 254
Slope, 579
Slope-intercept form, 583
Solution:
   definition of, 46
   by determinants, 107, 456
   of exponential equations, 380
   graphical method of, 81, 396
   of higher-order equations, 418
   of inequalities, 534
   of linear equations, 45
   of logarithmic equations, 380
   of quadratic equations, 198–219
   of systems of equations, see Systems of equations
   of triangles, 138, 284–296
   of trigonometric equations, 515
Special angles, 130, 132, 229
   in radian measure, 240
Special products, 150, 157, 160
Specific gravity, 61
Spiral, 619, 620
Square, completing the, 203
Square root, 20
Standard equations of conics, 607
Standard position of angle, 122
Straight line, 583
Subtraction:
   of algebraic expressions, 13
   of complex numbers, 335, 339
   of fractions, 179
   of matrices, 462
   of radicals, 317
   of signed numbers, 5
   of vectors, 276
Sum and difference formulas, 500, 502

INDEX

Sum-to-product formulas, 504
Sum of two cubes, 153
Symbols of grouping, 15
Synthetic division, 425
Systems of equations, 79, 396, 456, 477
  by addition or subtraction, 84, 101, 405
  consistent, 459
  by Cramer's rule, 93, 107, 456, 457
  dependent, 88
  by determinants 93, 107, 456, 457
  graphical method, 81, 396
  inconsistent, 88
  matrix solution, 477
  by substitution, 88, 403
Systems of inequalities, 548

Tabular difference, 134
Tangent of angle, 125, 126
  graph of, 266
Terminal:
  point of vector, 274
  side of angle, 121
Terms:
  algebraic, 10
  like, 11
  of a progression, 562
  similar, 11
Transformer, 65

Translation of axes, 606
Transposition, 49
Transverse axis, 602
Triangles:
  oblique, 284, 287, 291
  reference, 225
  right, 138
  solution of, 138, 284–296
Trigonometric equations, 515
Trigonometric functions:
  of acute angles, 125, 126
  of arbitrary angles, 133, 223, 235
  with calculator, 135, 236
  graphs of, 250–273
  inverse, 524
Trigonometric identities, *see* Identities, trigonometric
Trigonometric ratios, 123–125
Trinomials, 10
  factoring, 161

Units, *see* Inside cover
Unit vector, 278

Variable, 46
  dependent, 70
  independent, 70
Variation:
  constant of, 65
  direct, 65

Variation (*cont'd*)
  inverse, 66
  joint, 67
Vectors, 274
  direction, 278
  magnitude, 278
Velocity:
  angular, 246
  linear, 246
Vertex:
  of angle, 121
  of ellipse, 597
  of hyperbola, 602
  of parabola, 590
  of polygonal region, 553
Vieta, Franciscus, 202
Volt, 56
Voltage, 65, 262, 353
Volumes, *see* Inside cover

Weight, 623
Wheatstone bridge, 64

x-intercept, 81

y-intercept, 81

Zero:
  exponent, 22, 303
  of a function, 426
  operations with, 7

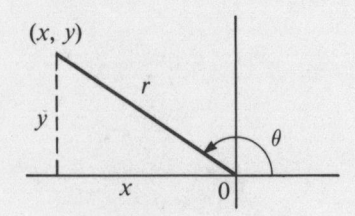

$$\sin \theta = \frac{y}{r} \qquad \csc \theta = \frac{r}{y}$$

$$\cos \theta = \frac{x}{r} \qquad \sec \theta = \frac{r}{x}$$

$$\tan \theta = \frac{y}{x} \qquad \cot \theta = \frac{x}{y}$$

Basic Identities

$$\sec \theta = \frac{1}{\cos \theta} \qquad \cot \theta = \frac{\cos \theta}{\sin \theta}$$

$$\csc \theta = \frac{1}{\sin \theta} \qquad \sin^2 \theta + \cos^2 \theta = 1$$

$$\cot \theta = \frac{1}{\tan \theta} \qquad 1 + \tan^2 \theta = \sec^2 \theta$$

$$\tan \theta = \frac{\sin \theta}{\cos \theta} \qquad 1 + \cot^2 \theta = \csc^2 \theta$$

Sum and Difference Formulas

$$\sin(A \pm B) = \sin A \cos B \pm \cos A \sin B$$
$$\cos(A \pm B) = \cos A \cos B \mp \sin A \sin B$$

Double-Angle Formulas

$$\sin 2A = 2 \sin A \cos A$$
$$\cos 2A = \cos^2 A - \sin^2 A$$
$$= 1 - 2 \sin^2 A$$
$$= 2 \cos^2 A - 1$$

Half-Angle Formulas

$$\sin \frac{A}{2} = \pm \sqrt{\frac{1 - \cos A}{2}}$$

$$\cos \frac{A}{2} = \pm \sqrt{\frac{1 + \cos A}{2}}$$

Alternate Half-Angle Formulas

$$\sin^2 A = \frac{1}{2}(1 - \cos 2A)$$

$$\cos^2 A = \frac{1}{2}(1 + \cos 2A)$$